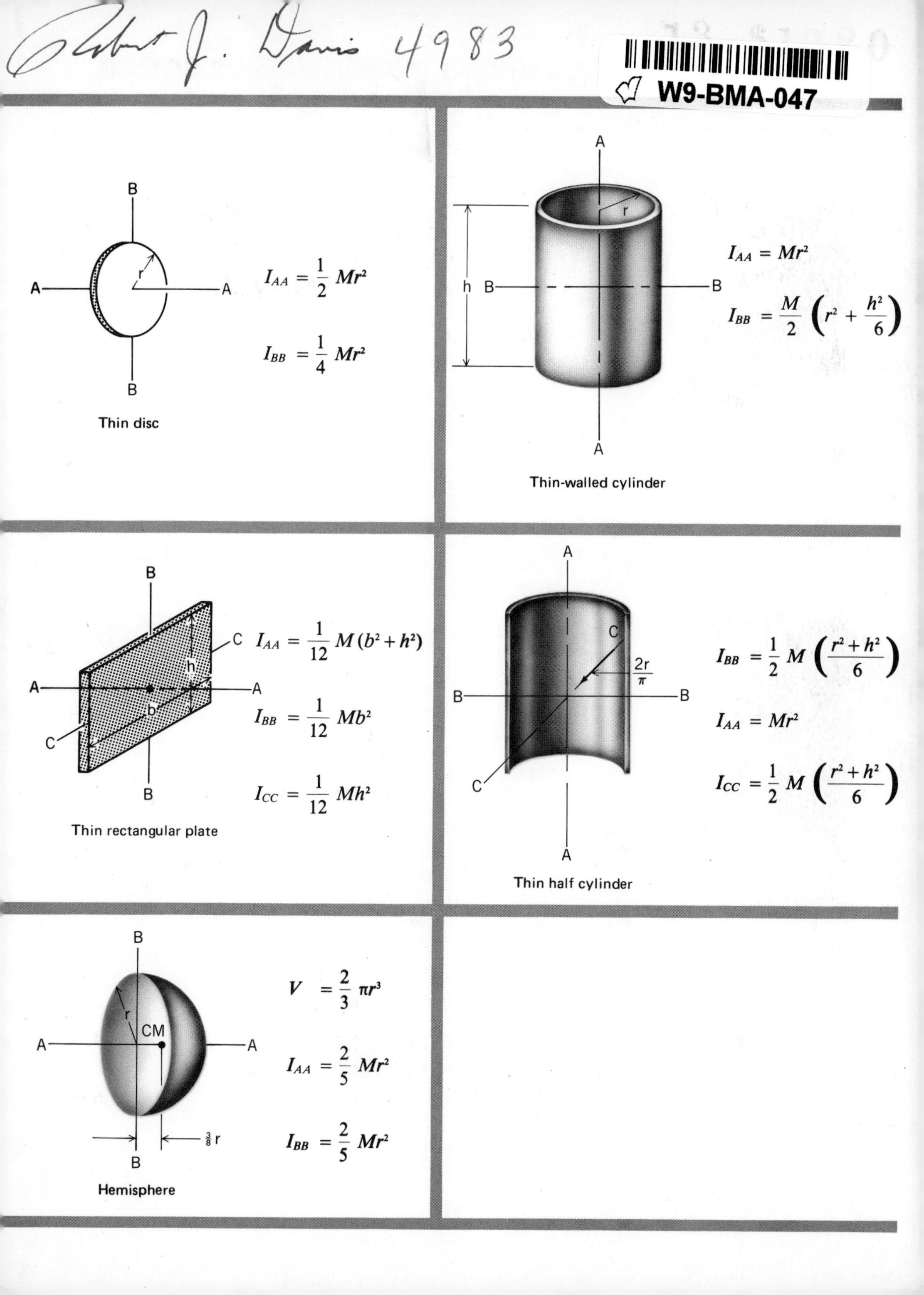
Robert J. Davis 4983
W9-BMA-047
Thin disc
$I_{AA} = \frac{1}{2} Mr^2$
$I_{BB} = \frac{1}{4} Mr^2$
Thin-walled cylinder
$I_{AA} = Mr^2$
$I_{BB} = \frac{M}{2}\left(r^2 + \frac{h^2}{6}\right)$
Thin rectangular plate
$I_{AA} = \frac{1}{12} M(b^2 + h^2)$
$I_{BB} = \frac{1}{12} Mb^2$
$I_{CC} = \frac{1}{12} Mh^2$
Thin half cylinder
$\frac{2r}{\pi}$
$I_{BB} = \frac{1}{2} M \left(\frac{r^2 + h^2}{6}\right)$
$I_{AA} = Mr^2$
$I_{CC} = \frac{1}{2} M \left(\frac{r^2 + h^2}{6}\right)$
Hemisphere
CM
$\frac{3}{8}$r
$V = \frac{2}{3} \pi r^3$
$I_{AA} = \frac{2}{5} Mr^2$
$I_{BB} = \frac{2}{5} Mr^2$

## SELECTED DIMENSIONAL EQUIVALENTS

| | |
|---|---|
| LENGTH | $1\ \text{m} \equiv 3.281\ \text{ft} \equiv 39.37\ \text{in.}$<br>$1\ \text{mi} \equiv 5280\ \text{ft} \equiv 1.609\ \text{km}$<br>$1\ \text{km} \equiv .6214\ \text{mi}$ |
| TIME | $1\ \text{hr} \equiv 60\ \text{min} \equiv 3600\ \text{sec}$ |
| MASS | $1\ \text{kg} \equiv 2.2046\ \text{lbm} \equiv .068521\ \text{slug}$ |
| FORCE | $1\ \text{N} \equiv .2248\ \text{lbf}$<br>$1\ \text{dyne} \equiv 1\ \mu\text{N}$ |
| SPEED | $1\ \text{mi/hr} \equiv 1.609\ \text{km/hr} \equiv 1.467\ \text{ft/sec}$<br>$1\ \text{km/hr} = .6214\ \text{mi/hr}$<br>$1\ \text{knot} = 1.152\ \text{mi/hr} \equiv 1.853\ \text{km/hr} \equiv 1.689\ \text{ft/sec}$ |
| ENERGY | $1\ \text{J} \equiv 1\ \text{N-m}$<br>$1\ \text{Btu} \equiv 778.16\ \text{ft-lbf} \equiv 1.055\ \text{kJ}$<br>$1\ \text{watt-hour} \equiv 2.778 \times 10^{-4}\ \text{J}$ |
| VOLUME | $1\ \text{gal} \equiv .16054\ \text{ft}^3 \equiv .0045461\ \text{m}^3$<br>$1\ \text{liter} \equiv .03531\ \text{ft}^3 = .2642\ \text{gal}$ |
| POWER | $1\ \text{w} \equiv 1\ \text{J/S}$<br>$1\ \text{hp} \equiv 550\ \text{ft-lb/sec} \equiv .7068\ \text{Btu/sec} \equiv 746\ \text{w}$ |

IRVING H. SHAMES
Faculty Professor
Engineering and Applied Science
State University of New York at Buffalo
third edition
ENGINEERING
MECHANICS
VOLUME I
STATICS
PRENTICE-HALL, INC., Englewood Cliffs, New Jersey 07632

*Library of Congress Cataloging in Publication Data*

SHAMES, IRVING HERMAN, (date)
Engineering mechanics.

Includes indexes.
CONTENTS: v. 1. Statics.—v. 2. Dynamics
1. Mechanics, Applied. I. Title.
[TA350.S493 1980b] 620 80-11903
ISBN 0-13-279141-2 (v. 1)
ISBN 0-13-279158-7 (v. 2)

Editorial/production supervision and interior design by Leon J. Liguori
and Suzanne Behnke with the assistance of Karen Mulé
Manufacturing buyer: Anthony Caruso



Printed in the United States of America

10 9 8 7 6 5 4 3 2 1

Prentice-Hall International, Inc. *London*
Prentice-Hall of Australia Pty. Limited, *Sydney*
Prentice-Hall of Canada, Ltd., *Toronto*
Prentice-Hall of India Private Limited, *New Delhi*
Prentice-Hall of Japan, Inc., *Tokyo*
Prentice-Hall of Southeast Asia Pte. Ltd., *Singapore*
Whitehall Books Limited, Wellington, *New Zealand*

I am happy and grateful to acknowledge the collaboration
on the Statics Volume of Dr. Robert M. Jones
from Southern Methodist University.
Dr. Jones went over with me the entire Statics Volume
with the aim of making the treatment
as clear and as simple as possible.
In addition, Dr. Jones made available to me 200 fine statics problems.
Finally, Bob gave me the benefit of a careful review
of the Dynamics Volume.

# Contents

# 2

## ELEMENTS OF VECTOR ALGEBRA 22



# 3

## IMPORTANT VECTOR QUANTITIES 56



# 4

## EQUIVALENT FORCE SYSTEMS 85

# 5

## EQUATIONS OF EQUILIBRIUM 134

## INTRODUCTION TO STRUCTURAL MECHANICS 195

# 7

## FRICTION FORCES 250

## PROPERTIES OF SURFACES 291

## MOMENTS AND PRODUCTS OF INERTIA 335

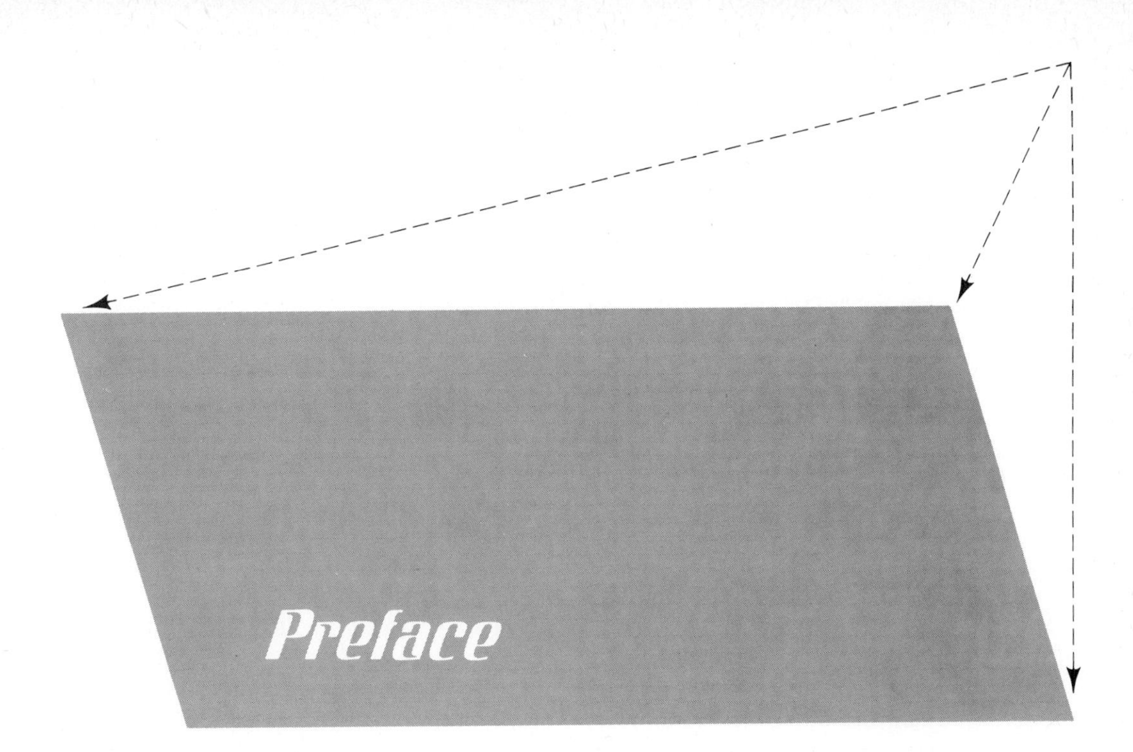


My main effort in writing the third edition of Statics has been to simplify and streamline the treatment and to make the text more practical and down to earth. In addition, I have sought to make the book more flexible to allow a range of coverage and degree of depth that will suit the wide range of present engineering programs.* These changes have been developed over a period of six years as a result of teaching mechanics to large classes from 100 to 300 students from all fields.

Specifically, the primary changes are listed as follows:

1. Over 40 percent of the problems are new. In this regard, I have endeavored to present interesting, practical, topical problems. The others have been overhauled from the best problems of the second edition. Furthermore, the problems have been moved in from the ends of the chapters to appropriate places within each chapter. At the end of each chapter, I have selected about 10 problems that represent the key concepts and techniques of that chapter. I have called this group of problems "Review Problems" and I urge prospective instructors not to use these problems as assignments but to leave them for students for review purposes in preparation for tests. The answers to all these review problems are given at the end of the book. For the other problems of the chapter, the even numbered problems have answers. I have used SI units at the rate of 50 percent of all the problems at the beginning of Statics and at the beginning of Dynamics. As one

*In the instructor's manual, I have listed a number of coverages having varying degrees of pace and depth starting with the case of a minimal number of topics and a maximum concentration on problem solving.

progresses in each area, there is a shift towards SI units to comprise 65 to 70 percent of the problems at the ends of Statics and Dynamics.

2. It has been brought to my attention from various users around the country that students are not well versed in multiple integration and that one cannot assume that they have fully mastered this technique by the early part of the sophomore fall semester. Accordingly, I have set forth procedures in Example problems for carrying out multiple integrations first with constant limits and then later with variable limits. The arguments are based primarily on physical reasoning; more rigorous discussions are left for math classes.
3. I have moved the chapter on the inertia tensor from Dynamics volume to the Statics volume to follow Chapter 8 on properties of surfaces including second moments and products of area. In this way, the inertia tensor can be more easily related to second moments and products of area. Additionally, by discussing the case of the thin plate, we can extend the concept of principal axes, first brought up in Chapter 8, to the inertia tensor of Chapter 9. Also, I have made clear how for a body having two orthogonal planes of symmetry, the principal axes for points in the intersection of the planes of symmetry can be determined by inspection. This suffices for most of the problems in Dynamics and so the instructor, if he is short of time, can delete the subsequent discussion on the transformation equations for inertia components, as well as the discussion of elipsoid of inertia.
4. In the first two chapters I have made use of the dagger (†) for those sections which I believe have been covered at various times at various levels in courses such as chemistry, physics and mathematics. For such sections, I have included at the ends of these chapters a series of simple questions that the instructor may wish to assign to assure himself that these sections, after being assigned to be read by the students during the first week of the course, are well understood. By using this procedure of assigning readings at the early stages so as to bring together in a unified way the various and sundry approaches that the student has had before, it is possible (I and my colleagues do this) to begin Chapter 3 by the second week of the course and still have a rational, sound foundation to develop the subject. For those who wish to spend more time on vector algebra in Chapter 2, I have presented an array of problems that will help the students quickly master these operations.
5. A great effort has been made to make the book more flexible as to possible usage by instructors having varied interests and concerns. For this reason I have starred much more material than the earlier edition indicating that such material can be bypassed with no loss in continuity. Any essential feature of material from starred sections needed later in an unstarred section has been deliberately repeated when needed. Furthermore, I have made greater use of fine print for material which is less likely to be covered in class but which may be of interest and value to the reader.
6. In general, I have overhauled the entire second edition and have rewritten much of it to attain greater clarity and simplicity. I have had the good fortune of having

my friend, Dr. Robert M. Jones, of Southern Methodist University, go over the entire text, including problems, line by line with the mission of testing each phrase and each expression for maximum clarity and continuity.

The following is a more detailed description of the contents of the text and will further illustrate some of the changes made in the third edition.

In Chapter 1, we begin by introducing certain fundamental ideas underlying mechanics, such as the concepts of dimensions and units, permitting us then to discuss certain common idealizations employed in mechanics. This sets the stage for a brief discussion of the basic laws of mechanics. A self-contained treatment of vector algebra is then presented in Chapter 2 for students who have as yet not been introduced to this subject matter. This chapter may also be used for purposes of review for those students who have had this material in their mathematics and physics courses. In Chapter 3 we then examine carefully the position vector, the moment of a force, and the couple. This permits us in Chapter 4 to present a thorough discussion of the equivalence of force systems for rigid-body mechanics. In particular, the simplest resultant force system is set forth for general, coplanar, parallel, and concurrent force systems. In Chapter 5, we then can establish the necessary equations of equilibrium for each of the aforementioned force systems by seeing what is necessary in each case to render the simplest resultant equal to zero. At the end of the chapter, we explain why the rigid body equation of equilibrium must still be satisfied when considering statically indeterminate problems where deformation must be taken into account.

Application of these equations is continued in Chapter 6 to simple trusses, beams, chains and cables. In Chapter 7 there are still further applications of these laws coupled now with the laws of Coulomb friction. In Chapter 8 we next present the concepts of the first moment of areas, masses, and volumes, and then go on to the second moment of areas, wherein we carefully present transformation properties with respect to a rotation of axes. The concept of principal axes is then presented. In Chapter 8, we consider first the definition of moments and products of inertia. Considering the thin plate, we are able to extend the principal axis concept for areas to that of masses. Those programs with sufficient time to spend can then study the transformation equations for the general inertia components as well as the ellipsoid of inertia. In Chapter 10, there is a rather careful development of virtual work. I feel this should be done carefully or not at all; ''quickee'' treatments do more harm than good. This conclusion stems from teaching follow on courses in variational mechanics.*

I am indebted to many people for valuable assistance in writing the third edition. First, I wish to thank Dr. Robert M. Jones of Southern Methodist University and his lovely wife, Donna. Bob and I went over the statics volume line by line together alternating at my home in Buffalo and his home in Dallas. The main effort was to achieve maximum simplicity and clarity. In addition, Bob made available to me 200 of his best statics problems. I was fortunate also in having careful reviews from Profs. David McGill and Wilton King of Georgia Institute of Technology. As a result of their keen perception, I was able to make improvements in the text. My sincere thanks goes

*See the author's text with C. Dym, ''Solid Mechanics—A Variational Approach''. McGraw-Hill Book Co., 1973.

to these gentlemen. Professor William Lee of the Naval Academy gave me a detailed, line by line review of the entire text including problems. I found his suggestions to be extremely helpful and I wish to extend my sincere appreciation to him for such a useful and valuable input. Profs. I. McIvor of U. of Michigan, J.S. Chen of U. of Pittsburgh and W.E. Clausen of Ohio State University also were helpful viewers of the manuscript to whom I wish to extend my thanks for their efforts. At home in Buffalo, I wish to thank my colleagues, Profs. P. Culkowski, C. Fogel, R. Mates, S. Prawel, T. Ranov and H. Reismann. They have been a constant source of encouragement and of valuable assistance as we taught different classes of the mechanics course. I wish to thank my son Bruce for the photographs used in Chapter 6. Finally, I wish to thank Mrs. K. Ward and Mrs. G. Huck for their excellent typing efforts.

Irving H. Shames

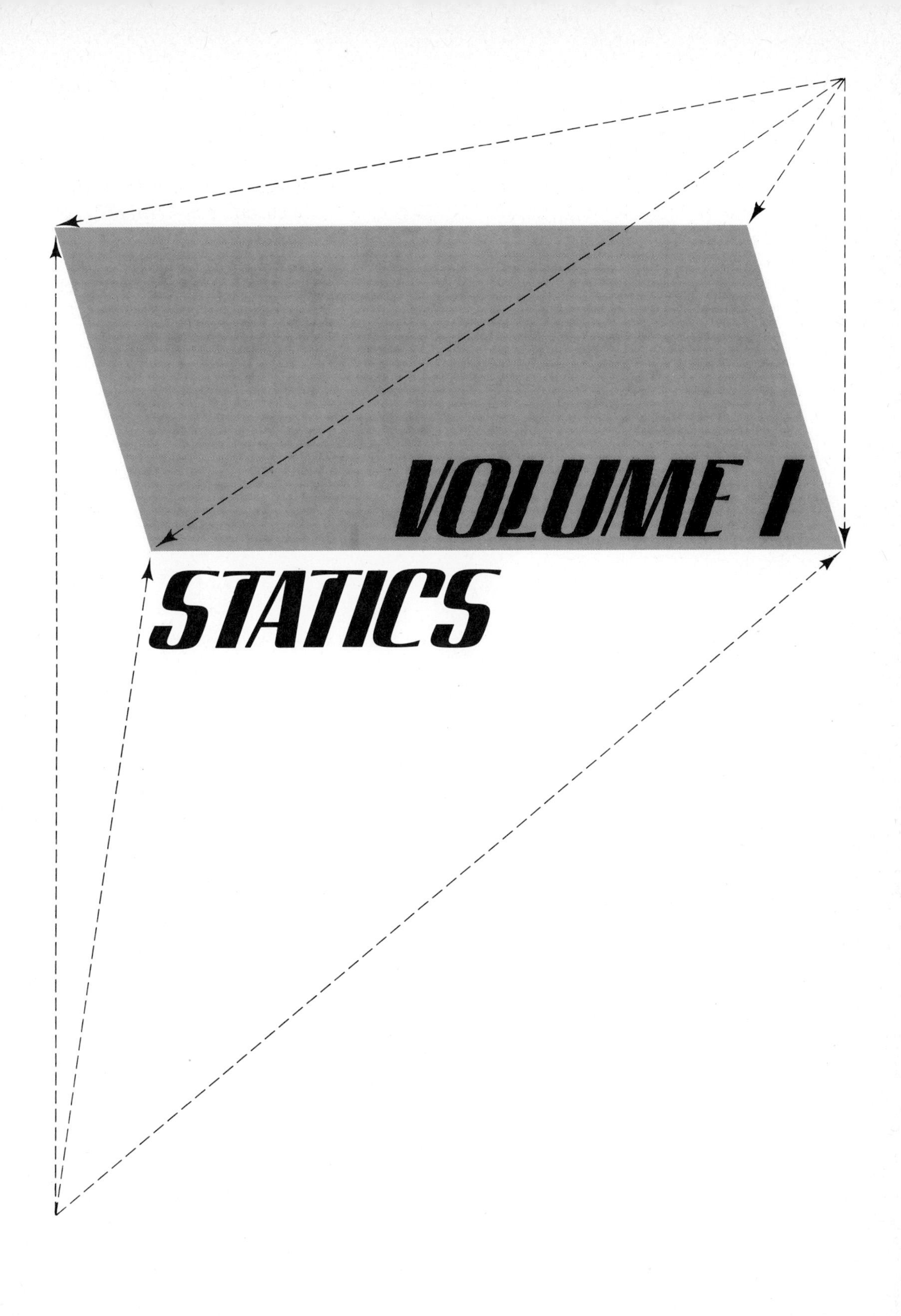
VOLUME I
STATICS

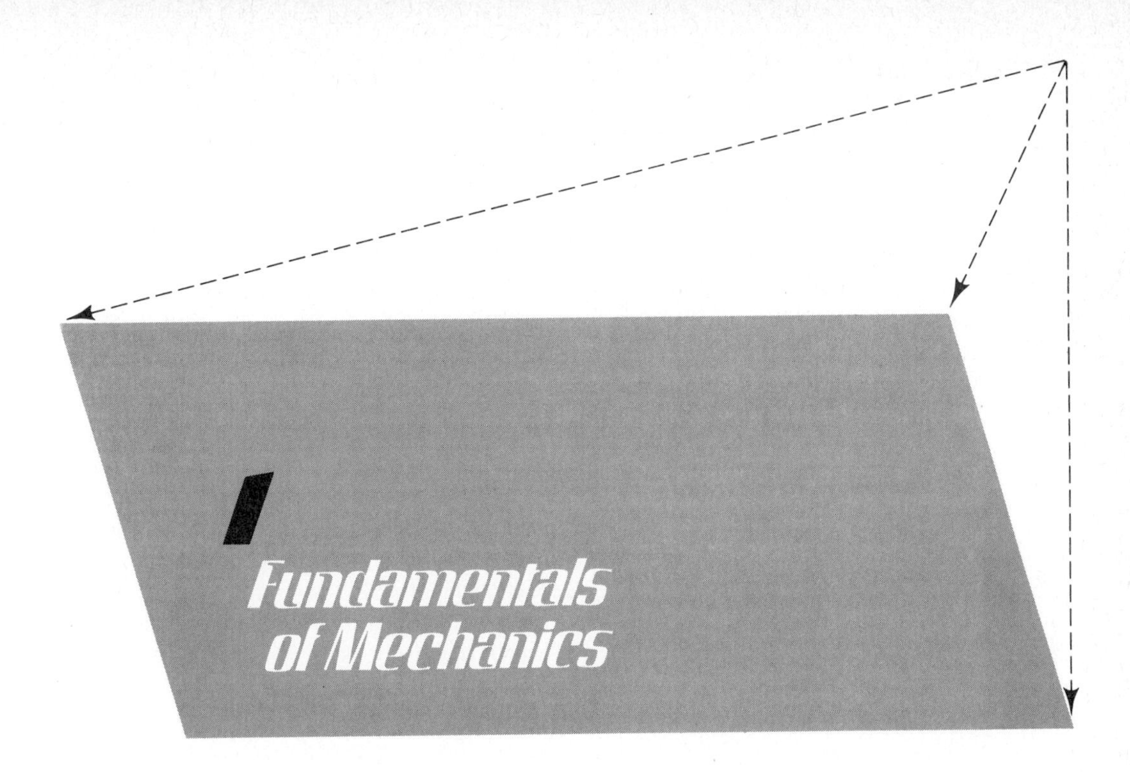


## †1.1 Introduction

*Mechanics* is the physical science concerned with the dynamical behavior (as opposed to chemical and thermal behavior) of bodies that are acted on by mechanical disturbances. Since such behavior is involved in virtually all the situations that confront an engineer, mechanics lies at the core of much engineering analysis. In fact, no physical science plays a greater role in engineering then does mechanics, and it is the oldest of all the physical sciences. The writings of Archimedes covering buoyancy and the lever were recorded before 200 B.C. Our modern knowledge of gravity and motion was established by Isaac Newton (1642–1727), whose laws founded Newtonian mechanics, the subject matter of this text.

In 1905, Einstein placed limitations on Newton's formulations with his theory of relativity and thus set the stage for the development of relativistic mechanics. The newer theories, however, give results that depart from those of Newton's formulations only when the speed of a body approaches the speed of light (186,000 miles/sec). These speeds are encountered in the large-scale phenomena of dynamical astronomy and the small-scale phenomena involving subatomic particles. Despite these limitations, it remains nevertheless true that, in the great bulk of engineering problems, Newtonian mechanics still applies.

## †1.2 Basic Dimensions and Units of Mechanics

To study mechanics, we must establish abstractions to describe those characteristics of a body that interest us. These abstractions are called *dimensions*. The dimensions that we pick, which are independent of all dimensions, are termed *primary* or *basic dimensions*, and the ones that are then developed in terms of the basic dimensions we call *secondary dimensions*. Of the many possible sets of basic dimensions that we could use, we will confine ourselves at present to the set that includes the dimensions of length, time, and mass. Another convenient set will be examined later.

**Length—A Concept for Describing Size Quantitatively.** In order to determine the size of an object, we must place a second object of known size next to it. Thus, in pictures of machinery, a man often appears standing disinterestedly beside the apparatus. Without him, it would be difficult to gage the size of the unfamiliar machine. Although the man has served as some sort of standard measure, we can, of course, only get an approximate idea of the machine's size. Men's heights vary, and, what is even worse, the shape of a man is too complicated to be of much help in acquiring a precise measurement of the machine's size. What we need, obviously, is an object that is constant in shape and, moreover, simple in concept. Thus, instead of a three-dimensional object, we choose a one-dimensional object.[1] Then, we can use the known mathematical concepts of geometry to extend the measure of size in one dimension to the three dimensions necessary to characterize a general body. A straight line scratched on a metal bar that is kept at uniform thermal and physical conditions (as, e.g., the meter bar kept at Sèvres, France) serves as this simple invariant standard in one dimension. We can now readily calculate and communicate the distance along a certain direction of an object by counting the number of standards and fractions thereof that can be marked off along this direction. We commonly refer to this distance as length, although the term "length" could also apply to the more general concept of size. Other aspects of size, such as volume and area, can then be formulated in terms of the standard by the methods of plane, spherical, and solid geometry.

A *unit* is the name we give an accepted measure of a dimension. Many systems of units are actually employed around the world, but we shall only use the two major systems, the American system and the SI system. The unit of length in the American system is the foot, whereas the unit of length in the SI system is the meter.

**Time—A Concept for Ordering the Flow of Events.** In observing the picture of the machine with the man standing close by, we can sometimes tell approximately when the picture was taken by the style of clothes the man is wearing. But how do we determine this? We may say to ourselves: "During the thirties, people wore the type of straw hat that the fellow in the picture is wearing." In other words, the "when" is tied to certain events that are experienced by, or otherwise known to, the observer. For a more accurate description of "when," we must find an action that appears to be completely repeatable. Then, we can order the events under study by counting the

[1]We are using the word "dimensional" here in its everyday sense and not as defined above.

number of these repeatable actions and fractions thereof that occur while the events transpire. The rotation of the earth gives rise to an event that serves as a good measure of time—the day. But we need smaller units in most of our work in engineering, and thus, generally, we tie events to the second, which is an action repeatable 86,400 times a day.

**Mass—A Property of Matter.** The student ordinarily has no trouble understanding the concepts of length and time because he is constantly aware of the size of things through his senses of sight and touch, and is always conscious of time by observing the flow of events in his daily life. The concept of mass, however, is not as easily grasped since it does not impinge as directly on our daily experience.

Mass is a property of matter that can be determined from *two* different actions of bodies. To study the first action, suppose that we consider two hard bodies of entirely different composition, size, shape, color, and so on. If we attach the bodies to identical springs, as shown in Fig. 1.1, each spring will extend some distance as a result of the attraction of gravity for the bodies. By grinding off some of the material on the body that causes the greater extension, we can make the deflections that are induced on both springs equal. Even if we raise the springs to a new height above the earth's surface, thus lessening the deformation of the springs, the extensions induced by the pull of gravity will be the same for both bodies. And since they are, we can conclude that the bodies have an equivalent innate property. This property of each body that manifests itself in the *amount of gravitational attraction* we call *mass.*

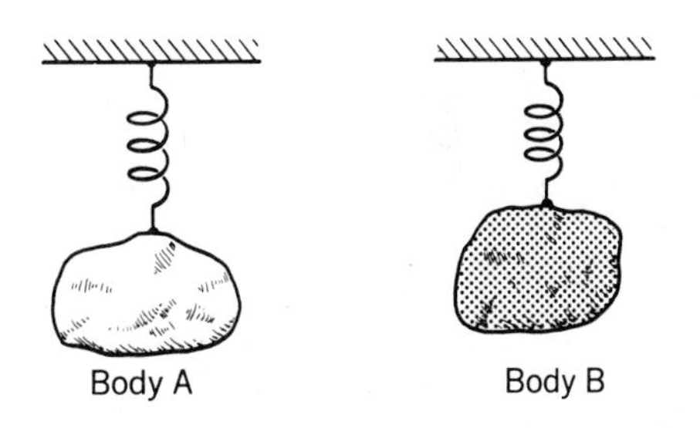


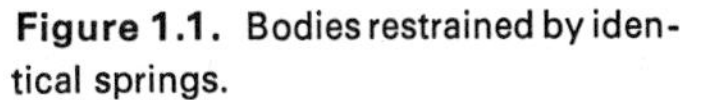
**Figure 1.1.** Bodies restrained by identical springs.

The equivalence of these bodies can be indicated in yet a second action. If we move both bodies an equal distance downward, by stretching each spring, and then release them at the same time, they will begin to move in an identical manner (except for small variations due to differences in wind friction and local deformations of the bodies). We have imposed, in effect, the same mechanical disturbance on each body and we have elicited the same dynamical response. Hence, despite many obvious differences, the two bodies again show an equivalence.

> *The property of mass, then, characterizes a body both in the action of gravitational attraction and in the response to a mechanical disturbance.*

To communicate this property quantitatively, we may choose some convenient body and compare other bodies to it in either of the two abovementioned actions. The two units commonly used in much American engineering practice to measure mass are the *pound mass,* which is defined in terms of the attraction of gravity for a standard body at a standard location, and the *slug,* which is defined in terms of the dynamical response of a standard body to a standard mechanical disturbance. A similar duality of mass units does not exist in the SI system. There only the *kilogram* is used as a measure of mass. The kilogram is measured in terms of response of a body to a mechanical disturbance. Both systems of units will be discussed further in a subsequent section.

We have now established three basic independent dimensions to describe certain physical phenomena. It is convenient to identify these dimensions in the following manner:

$$\begin{array}{ll} \text{length} & [L] \\ \text{time} & [t] \\ \text{mass} & [M] \end{array}$$

These formal expressions of identification for basic dimensions and the more complicated groupings to be presented in Section 1.3 for secondary dimensions are called "dimensional representations."

Often, there are occasions when we want to change units during computations. For instance, we may wish to change feet into inches or millimeters. In such a case, we must replace the unit in question by a *physically equivalent* number of new units. Thus, a foot is replaced by 12 inches or 305 millimeters. A listing of common systems of units is given in Table 1.1, and a table of equivalences between these and other units is given

**TABLE 1.1 Common Systems of Units**

| cgs | | SI | |
|---|---|---|---|
| Mass | Gram | Mass | Kilogram |
| Length | Centimeter | Length | Meter |
| Time | Second | Time | Second |
| Force | Dyne | Force | Newton |
| **English** | | **American Practice** | |
| Mass | Pound mass | Mass | Slug or pound mass |
| Length | Foot | Length | Foot |
| Time | Second | Time | Second |
| Force | Poundal | Force | Pound force |

on the inside covers. Such relations between units will be expressed in this way:

$$1 \text{ ft} \equiv 12 \text{ in.} \equiv 305 \text{ mm}$$

The three horizontal bars are not used to denote *algebraic* equivalence; instead, they are used to indicate physical equivalence. Here is another way of expressing the relations above:

$$\begin{aligned} \left(\frac{1 \text{ ft}}{12 \text{ in.}}\right) \equiv 1, \qquad \left(\frac{1 \text{ ft}}{305 \text{ mm}}\right) \equiv 1 \\ \left(\frac{12 \text{ in.}}{1 \text{ ft}}\right) \equiv 1, \qquad \left(\frac{305 \text{ mm}}{1 \text{ ft}}\right) \equiv 1 \end{aligned} \tag{1.1}$$

The unity on the right side of these relations indicates that the numerator and denominator on the left side are physically equivalent, and thus have a 1:1 relation. This notation will prove convenient when we consider the change of units for secondary dimensions in the next section.

## †1.3 Secondary Dimensional Quantities

When physical characteristics are described in terms of basic dimensions by the use of suitable definitions (e.g., velocity is defined[2] as a distance divided by a time interval), such quantities are called *secondary dimensional quantities.* In Section 1.4, we will see that these quantities may also be established as a consequence of natural laws. The dimensional representation of secondary quantities is given in terms of the basic dimensions that enter into the formulation of the concept. For example, the dimensional representation of velocity is

$$[\text{velocity}] \equiv \frac{[L]}{[t]}$$

That is, the dimensional representation of velocity is the dimension length divided by the dimension time. The units for a secondary quantity are then given in terms of the units of the constituent basic dimensions. Thus,

$$\text{velocity units} \equiv \frac{[\text{ft}]}{[\text{sec}]}$$

A *change* of units from one system into another usually involves a change in the scale of measure of the secondary quantities involved in the problem. Thus, one scale unit of velocity in the American system is 1 foot per second, while in the SI system it is 1 meter per second. How may these scale units be correctly related for complicated secondary quantities? That is, for our simple case, how many meters per second are equivalent to 1 foot per second? The formal expressions of dimensional representation may be put to good use for such an evaluation. The procedure is as follows. Express the dependent quantity dimensionally; substitute existing units for the basic dimensions; and finally, change these units to the equivalent numbers of units in the new system. The result gives the number of scale units of the quantity in the new system of units that is equivalent to 1 scale unit of the quantity in the old system. Performing these operations for velocity, we would thus have

$$1\left(\frac{\text{ft}}{\text{sec}}\right) \equiv 1\left(\frac{.305\ \text{m}}{\text{sec}}\right) \equiv .305\left(\frac{\text{m}}{\text{sec}}\right)$$

which means that .305 scale unit of velocity in the SI system is equivalent to 1 scale unit in the American system.

Another way of changing units when secondary dimensions are present is to make use of the formalism illustrated in relations 1.1. To change a unit in an expression,

[2] A more precise definition will be given in the chapters on dynamics.

multiply this unit by a ratio physically equivalent to unity, as we discussed earlier, so that the old unit is canceled out, leaving the desired unit with the proper numerical coefficient. In the example of velocity used above, we may replace ft/sec by m/sec in the following manner:

$$1\left(\frac{\text{ft}}{\text{sec}}\right) \equiv \left(\frac{1\,\cancel{\text{ft}}}{\text{sec}}\right) \cdot \left(\frac{.305\text{ m}}{1\,\cancel{\text{ft}}}\right) \equiv .305\left(\frac{\text{m}}{\text{sec}}\right)$$

It should be clear that, when we multiply by such ratios to accomplish a change of units as shown above, we do not alter the magnitude of the *actual physical quantity* represented by the expression. Students are strongly urged to employ the above technique in their work, for the use of less formal methods is generally an invitation to error.

## †1.4 Law of Dimensional Homogeneity

Now that we can describe certain aspects of nature in a quantitative manner through basic and secondary dimensions, we can by careful observation and experimentation learn to relate certain of the quantities in the form of equations. In this regard, there is an important law, the law of *dimensional homogeneity*, which imposes a restriction on the formulation of such equations. This law states that, because natural phenomena proceed with no regard for man-made units, *basic equations representing physical phenomena must be valid for all systems of units.* Thus, the equation for the period of a pendulum, $t = 2\pi\sqrt{L/g}$, must be valid for all systems of units, and is accordingly said to be *dimensionally homogeneous.* It then follows that the fundamental equations of physics are dimensionally homogeneous; and all equations derived analytically from these fundamental laws must also be dimensionally homogeneous.

What restriction does this condition place on an equation? To answer this, let us examine the following arbitrary equation:

$$x = ygd + k$$

For this equation to be dimensionally homogeneous, the numerical equality between both sides of the equation must be maintained for all systems of units. To accomplish this, the change in the scale of measure of each group of terms must be the same when there is a change of units. That is, if the numerical measure of one group such as $ygd$ is doubled for a new system of units, so must that of the quantities $x$ and $k$. *For this to occur under all systems of units, it is necessary that every grouping in the equation have the same dimensional representation.*

In this regard, consider the dimensional representation of the above equation expressed in the following manner:

$$[x] = [ygd] + [k]$$

From the previous conclusion for dimensional homogeneity, we require that

$$[x] \equiv [ygd] \equiv [k]$$

As a further illustration, consider the dimensional representation of an equation that is *not* dimensionally homogeneous:

$$[L] = [t]^2 + [t]$$

When we change units from the American to the SI system, the units of feet give way to units of meters, but there is no change in the unit of time, and it becomes clear that the numerical value of the left side of the equation changes while that of the right side does not. The equation, then, becomes invalid in the new system of units and hence is not derived from the basic laws of physics. Throughout this book, we shall invariably be concerned with dimensionally homogeneous equations. Therefore, we should dimensionally analyze our equations to help spot errors.

## †1.5 Dimensional Relation Between Force and Mass

We shall now employ the law of dimensional homogeneity to establish a new secondary dimension—namely *force*. A superficial use of Newton's law will be employed for this purpose. In a later section, this law will be presented in greater detail, but it will suffice at this time to state that the acceleration of a particle[3] is inversely proportional to its mass for a given disturbance. Mathematically, this becomes

$$a \propto \frac{1}{m} \tag{1.2}$$

where $\propto$ is the proportionality symbol. Inserting the constant of proportionality, $F$, we have, on rearranging the equation,

$$F = ma \tag{1.3}$$

The mechanical disturbance, represented by $F$ and called *force*, must have the following dimensional representation, according to the law of dimensional homogeneity:

$$[F] \equiv [M]\frac{[L]}{[t]^2} \tag{1.4}$$

The type of disturbance for which relation 1.2 is valid is usually the action of one body on another by direct contact. However, other actions, such as magnetic, electrostatic, and gravitational actions of one body on another, also create mechanical effects that are valid in Newton's equation.

We could have initiated the study of mechanics by considering *force* as a basic dimension, the manifestation of which can be measured by the elongation of a standard spring at a prescribed temperature. Experiment would then indicate that for a given body the acceleration is directly proportional to the applied force. Mathematically,

$$F \propto a; \quad \text{therefore, } F = ma$$

[3]We shall define particles in Section 1.7.

from which we see that the proportionality constant now represents the property of mass. Here, mass is now a secondary quantity whose dimensional representation is determined from Newton's law:

$$[M] \equiv [F]\frac{[t]^2}{[L]} \tag{1.5}$$

As was mentioned earlier, we now have a choice between two systems of basic dimensions—the *MLt* or the *FLt* system of basic dimensions. Physicists prefer the former, whereas engineers usually prefer the latter.

## 1.6 Units of Mass

As we have already seen, the concept of mass arose from two types of actions—those of motion and gravitational attraction. In American engineering practice, units of mass are based on both actions, and this sometimes leads to confusion. Let us consider the *FLt* system of basic dimensions for the following discussion. The unit of force may be taken to be the pound, which is defined as a force that extends a standard spring a certain distance. Using Newton's law, we then define the *slug* as the amount of mass that a 1-pound force will cause to accelerate at the rate of 1 foot per second per second.

On the other hand, another unit of mass can be stipulated if we use the gravitational effect as a criterion. Here, the *pound mass* (lbm) is defined as the amount of matter that is drawn by gravity toward the earth by a force of 1 pound (lbf) at a specified position on the earth's surface.

We have formulated two units of mass by two different actions, and to relate these units we must subject them to the *same* action. Thus, we can take 1 pound mass and see what fraction or multiple of it will be accelerated 1 ft/sec$^2$ under the action of 1 pound of force. This fraction or multiple will then represent the number of units of pound mass that are equivalent to 1 slug. It turns out that this coefficient is $g_0$, where $g_0$ has the value corresponding to the acceleration of gravity at a position on the earth's surface where the pound mass was standardized. To three significant figures, the value of $g_0$ is 32.2. We may then make the statement of equivalence that

$$1 \text{ slug} \equiv 32.2 \text{ pounds mass}$$

To use the pound-mass unit in Newton's law, it is necessary to divide by $g_0$ to form units of mass that have been derived from Newton's law. Thus,

$$F = \frac{m}{g_0} a \tag{1.6}$$

where $m$ has the units of pound mass. Having properly introduced into Newton's law the pound-mass unit from the viewpoint of physical equivalence, let us now consider the dimensional homogeneity of the resulting equation. The right side of Eq. 1.6 must have the dimensional representation of $F$ and, since the unit here for $F$ is the pound force, the right side must then have this unit. Examination of the units on the right side of the equation then indicates that the units of $g_0$ must be

$$[g_0] \equiv \frac{[\text{lbm}][\text{ft}]}{[\text{lbf}][\text{sec}]^2} \tag{1.7}$$

How does *weight* fit into this picture? Weight is defined as *the force of gravity on a body*. Its value will depend on the position of the body relative to the earth's surface. At a location on the earth's surface where the pound mass is standardized, a mass of 1 pound (lbm) has the weight of 1 pound (lbf), but with increasing altitude the weight will become smaller than 1 pound (lbf). The mass, however, remains at all times a 1-pound mass (lbm). If the altitude is not exceedingly large, the measure of weight, in lbf, will practically equal the measure of mass, in lbm. Therefore, it is unfortunately the practice in engineering erroneously to think of weight at positions other than on the earth's surface as the measure of mass, and consequently to use the symbol $W$ to represent either lbm or lbf. In this age of rockets and missiles, it behooves us to be careful about the proper usage of units of mass and weight throughout the entire text.

If we know the weight of a body at some point, we can determine its mass in slugs very easily, provided that we know the acceleration of gravity, $g$, at that point. Thus, according to Newton's law,

$$W\,(\text{lbf}) = m\,(\text{slugs}) \times g\,(\text{ft/sec}^2)$$

Therefore,

$$m\,(\text{slugs}) = \frac{W\,(\text{lbf})}{g\,(\text{ft/sec}^2)} \tag{1.8}$$

Up to this point, we have only considered the American system of units. In the SI system of units, a *kilogram* is the amount of mass that will accelerate 1 m/sec$^2$ under the action of a force of 1 newton. Here we do not have the problem of 2 units of mass; the kilogram is the basic unit of mass. However, we do have another kind of problem —that the kilogram is unfortunately also used as a measure of force, as is the newton. One kilogram of force is the weight of 1 kilogram of mass at the earth's surface, where the acceleration of gravity (i.e., the acceleration due to the force of gravity) is 9.81 m/sec$^2$. A newton, on the other hand, is the force that acclerates 1 kilogram of mass only 1 m/sec$^2$. Hence, 9.81 newtons are equivalent to 1 kilogram of force. That is,

$$9.81\text{ newtons} \equiv 1\text{ kilogram (force)} \equiv 2.205\text{ lbf}$$

Note from the above that the newton is a comparatively small force, equaling approximately one-fifth of a pound. A kilonewton (1000 newtons), which will be used often, is about 200 lb. In this text, we shall not use the kilogram as a unit of force. However, you should be aware that many people do.

Note that at the earth's surface the weight $W$ of a mass $M$ is:

$$W\,(\text{newtons}) = [M\,(\text{kilograms})]\,(9.81) \tag{1.9}$$

Hence:

$$M\,(\text{kilograms}) = \frac{W\,(\text{newtons})}{9.81} \tag{1.10}$$

Away from the earth's surface, use the acceleration of gravity $g$ rather than 9.81 in the above equations.

## 1.7 Idealizations of Mechanics

As we have pointed out, basic and secondary dimensions may sometimes be related in equations to represent a physical action that we are interested in. We want to represent an action using the known laws of physics, and also to be able to form equations simple enough to be susceptible to mathematical computational techniques. Invariably in our deliberations, we must replace the actual physical action and the participating bodies with hypothetical, highly simplified substitutes. We must be sure, of course, that the results of our substitutions have some reasonable correlation with reality. All analytical physical sciences must resort to this technique, and, consequently, their computations are not cut and dried but involve a considerable amount of imagination, ingenuity, and insight into physical behavior. We shall, at this time, set forth the most fundamental idealizations of mechanics and a bit of the philosophy involved in scientific analysis.

**Continuum.** Even the simplification of matter into molecules, atoms, electrons, and so on, is too complex a picture for many problems of engineering mechanics. In most problems, we are interested only in the average measurable manifestations of these elementary bodies. Pressure, density, and temperature are actually the gross effects of the actions of the many molecules and atoms, and they can be conveniently assumed to arise from a hypothetically continuous distribution of matter, which we shall call the *continuum*, instead of from a conglomeration of discrete, tiny bodies. Without such an artifice, we would have to consider the action of each of these elementary bodies—a virtual impossibility for most problems.

**Rigid Body.** In many cases involving the action on a body by a force, we simplify the continuum concept even further. The most elemental case is that of a rigid body, which is a continuum that undergoes theoretically no deformation whatever. Actually, every body must deform to a certain degree under the actions of forces, but in many cases the deformation is too small to affect the desired analysis. It is then preferable to consider the body as rigid, and proceed with the simplified computations. For example, assume that we are to determine the forces transmitted by a beam to the earth as the result of a load $P$ (Fig. 1.2). If $P$ is small enough, the beam will undergo little deflection, and we can carry out a straightforward simple analysis using the undeformed geometry as if the body were indeed rigid. If we were to attempt a more accurate analysis—even

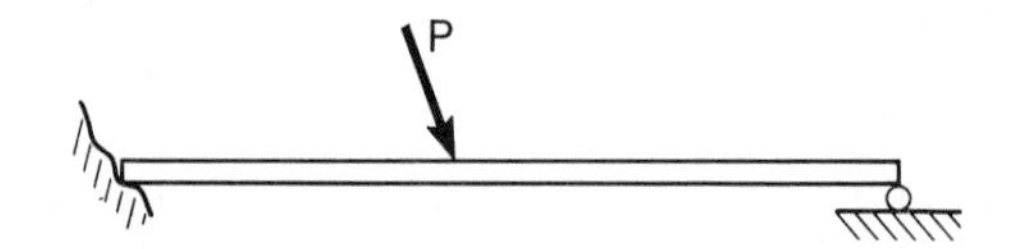


**Figure 1.2.** Rigid-body assumption—use original geometry.

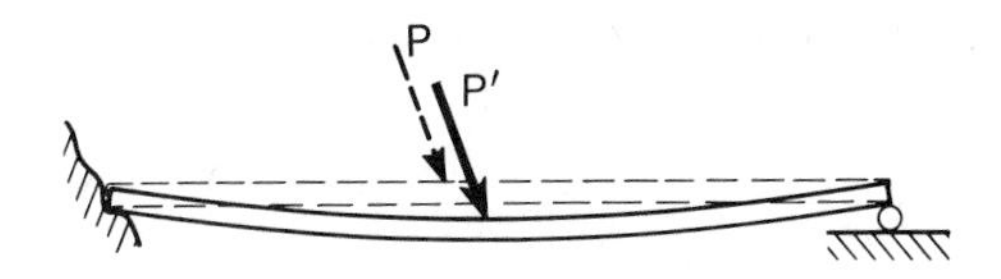


**Figure 1.3.** Deformable body.

though a slight increase in accuracy is not required—we would then need to know the exact position that the load assumes relative to the earth *after* the beam has ceased to deform, as shown in an exaggerated manner in Fig. 1.3. To do this accurately is a hopelessly difficult task, especially when we consider that the support must also "give" in a certain way. Although the alternative to a rigid-body analysis here leads us to a virtually impossible calculation, situations do arise in which more realistic models must be employed to yield the required accuracy. For example, when determining the internal force distribution in a body, we must often take the deformation into account, however small it might be. *The guiding principle is to make such simplifications as are consistent with the required accuracy of the results.*

**Point Force.** A finite force exerted on one body by another must cause a finite amount of local deformation, and always creates a finite area of contact between the bodies through which the force is transmitted. However, since we have formulated the concept of the rigid body, we should also be able to imagine a finite force to be transmitted through an infinitesimal area or point. This simplification of a force distribution is called a *point force*. In many cases where the actual area of contact in a problem is very small but is not known exactly, the use of the concept of the point force results in little sacrifice in accuracy. In Figs. 1.2 and 1.3, we actually employed the graphical representation of the point force.

**Particle.** The *particle* is defined as an object that has no size but that has a mass. Perhaps this does not sound like a very helpful definition for engineers to employ, but it is actually one of the most useful in mechanics. For the trajectory of a planet, for example, it is the mass of the planet and not its size that is significant. Hence, we can consider planets as particles for such computations. On the other hand, take a figure skater spinning on the ice. Her revolutions are controlled beautifully by the orientation of the body. In this motion, the size and distribution of the body are significant, and since a particle, by definition, can have no distribution, it is patently clear that a particle cannot represent the skater in this case. If, however, the skater should be billed as the "human cannonball on skates" and be shot out of a large gun, it would be possible to consider her as a single particle in ascertaining her trajectory, since arm and leg movements that were significant while she was spinning on the ice would have little effect on the arc traversed by the main portion of her body.

Many other simplifications pervade mechanics. The perfectly elastic body, the frictionless fluid, and so on, will become quite familiar as you study various phases of mechanics.

## †1.8 Vector and Scalar Quantities

We have now proposed sets of basic dimensions and secondary dimensions to describe certain aspects of nature. However, more than just the dimensional identification and the number of units are often needed to convey adequately the desired information. For instance, to specify fully the motion of a car, which we may represent as a particle at this time, we must answer the following questions:

1. How fast?
2. Which way?

The concept of velocity entails the information desired in questions 1 and 2. The first question, "How fast?", is answered by the speedometer reading, which gives the value of the velocity in miles per hour or kilometers per hour. The second question, "Which way?", is more complicated, because two separate factors are involved. First, we must specify the angular orientation of the velocity relative to a reference frame. Second, we must specify the sense of the velocity, which tells us whether we are moving *toward or away from* a given point. The concepts of angular orientation of the velocity and sense of the velocity are often collectively denoted as the *direction* of the velocity. Graphically, we may use a *directed line segment* (an arrow) to describe the velocity of the car. The *length* of the directed line segment gives information as to "how fast" and is the *magnitude* of the velocity. The angular orientation of the directed line segment and the position of the arrowhead give information as to "which way"—that is, as to the *direction* of the velocity. The directed line segment itself is called the *velocity*, whereas the length of the directed line segment—that is, the magnitude—is called the *speed.*

There are many physical quantities that are represented by a directed line segment and thus are describable by specifying a magnitude and a direction. The most common example is force, where the magnitude is a measure of the intensity of the force and the direction is evident from how the force is applied. Another example is the *displacement vector* between two points on the path of a particle. The magnitude of the dis-

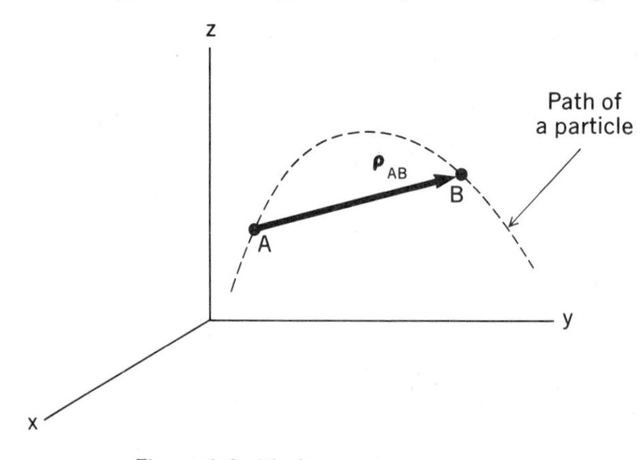


**Figure 1.4.** Displacement vector $\boldsymbol{\rho}_{AB}$.

placement vector corresponds to the distance moved along a *straight line* between two points, and the direction is defined by the orientation of this line relative to a reference, with the sense corresponding to which point is being approached. Thus, $\boldsymbol{\rho}_{AB}$ (see Fig. 1.4) is the displacement vector from $A$ to $B$ (while $\boldsymbol{\rho}_{BA}$ goes from $B$ to $A$).

Certain quantities having magnitude and direction combine their effects in a special way. Thus, the combined effect of two forces acting on a particle, as shown in Fig. 1.5, corresponds to a single force that may be shown by experiment to be equal to the diagonal of a parallelogram formed by the graphical representation of the forces. That is, the quantities add according to the *parallelogram law*. All quantities that have magnitude and direction and that add according to the parallelogram law are called *vector quantities*. Other quantities that have only magnitude, such as temperature and work, are called *scalar quantities*. A vector quantity will be denoted with a boldface italic letter, which in the case of force becomes $\boldsymbol{F}$.[4]

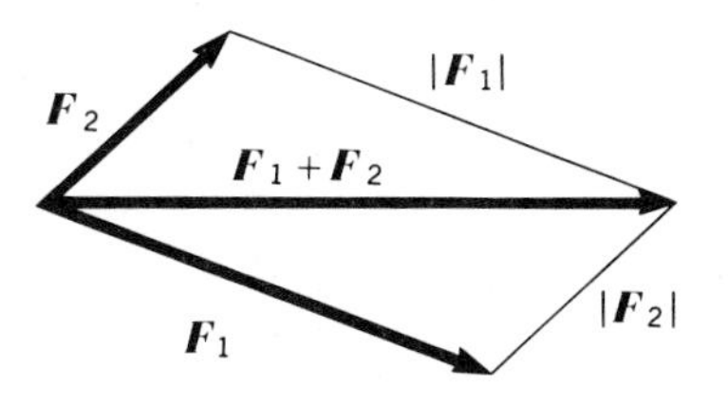

**Figure 1.5.** Parallelogram law.

The reader may ask: Don't all quantities having magnitude and direction combine according to the parallelogram law and, therefore, become vector quantities? No, not all of them do. One very important example will be pointed out after we reconsider Fig. 1.5. In the construction of the parallelogram it matters not which force is laid out first. In other words, "$\boldsymbol{F}_1$ combined with $\boldsymbol{F}_2$" gives the same result as "$\boldsymbol{F}_2$ combined with $\boldsymbol{F}_1$." In short, the combination is *commutative*. If a combination is not commutative, it cannot in general be represented by a parallelogram operation and is thus not a vector. With this in mind, consider the finite angle of rotation of a body about an axis. We can associate a magnitude (degrees or radians) and a direction (the axis and a stipulation of clockwise or counterclockwise) with this quantity. However, the finite angle of rotation cannot be considered a vector because in general two finite rotations about different axes cannot be replaced by a single finite rotation consistent with the parallelogram law. The easiest way to show this is to demonstrate that the combination of such rotations is not commutative. In Fig. 1.6(a) a book is to be given two rotations —a 90° counterclockwise rotation about the $x$ axis and a 90° clockwise rotation about the $z$ axis, both looking in toward the origin. This is carried out in Figs. 1.6(b) and (c). In Fig. 1.6(c), the sequence of combination is reversed from that in Fig. 1.6(b), and you can see how it alters the final orientation of the book. Finite angular rotation, therefore, is not a vector quantity, since the parallelogram law is not valid for such a combination.[5]

---

[4]Your instructor on the blackboard and you in your homework will not be able to use boldface notation for vectors. Accordingly, you may choose to use a superscript arrow or bar, e.g., $\vec{\mathrm{F}}$ or $\bar{\mathrm{F}}$ ($\underline{\mathrm{F}}$ or $\underset{\sim}{\mathrm{F}}$ are other possibilities).

[5]However, *vanishingly small* rotations can be considered as vectors since the commutative law applies for the combination of such rotations. This will be an important consideration when we discuss the angular velocity vector in Chapter 15.

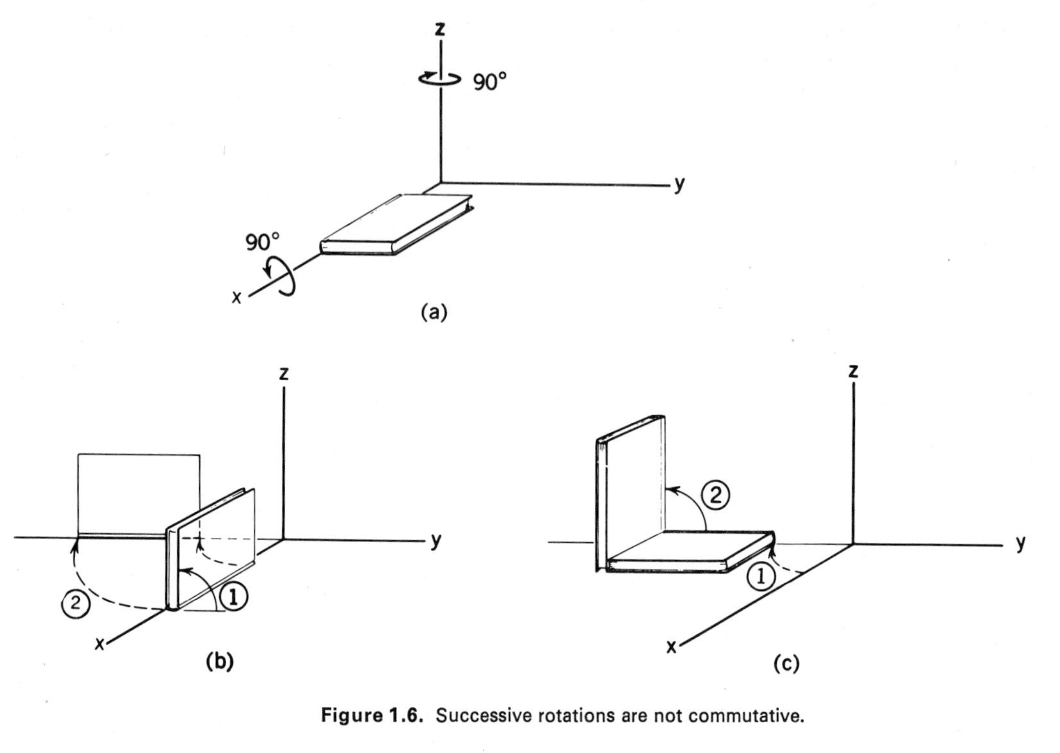


**Figure 1.6.** Successive rotations are not commutative.

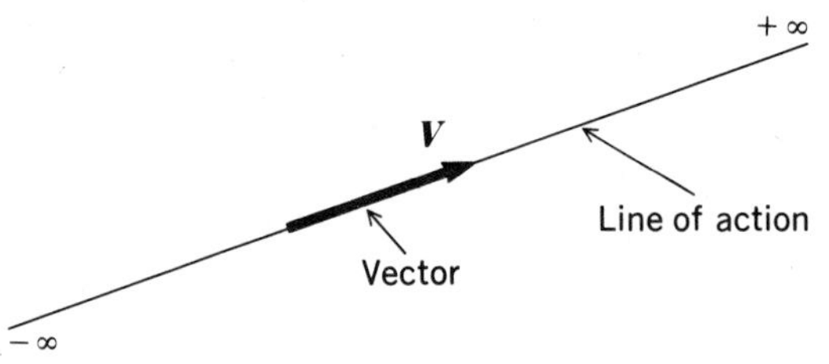


**Figure 1.7.** Line of action of a vector.

Before closing the section, we will set forth one more definition. The *line of action* of a vector is a hypothetical infinite straight line collinear with the vector (see Fig. 1.7). Thus, the velocities of two cars moving on different lanes of a straight highway have different lines of action. Keep in mind that the line of action involves no connotation as to sense. Thus, a vector $\boldsymbol{V}'$ collinear with $\boldsymbol{V}$ in Fig. 1.7 and with opposite sense would nevertheless have the same line of action.

## 1.9 Equality and Equivalence of Vectors

We shall avoid many pitfalls in the study of mechanics if we clearly make a distinction between the equality and the equivalence of vectors.

*Two vectors are equal if they have the same dimensions, magnitude, and direction.*

In Fig. 1.8, the velocity vectors of three particles have equal length, are identically inclined toward the reference *xyz*, and have the same sense. Although they have different lines of action, they are nevertheless equal according to the definition.

*Two vectors are equivalent in a certain capacity if each produces the very same effect in this capacity.* If the criterion in Fig. 1.8 is change of elevation of the particles or total distance traveled by the particles, all three vectors give the same result. They are, in addition to being equal, also equivalent for these capacities. If the absolute

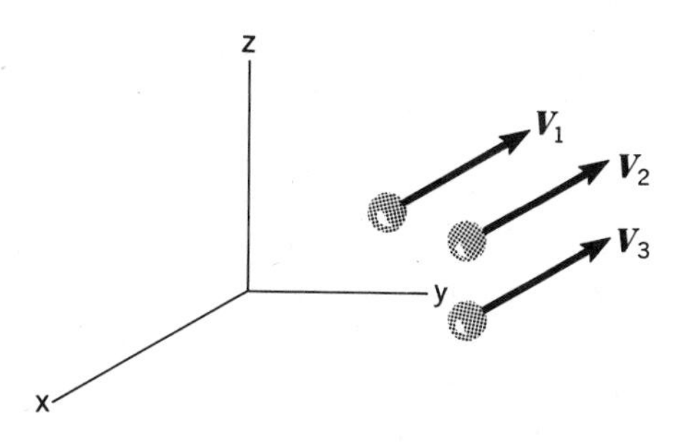


**Figure 1.8.** Equal-velocity vectors.

height of the particles above the *xy* plane is the question in point, these vectors will not be equivalent despite their equality. Thus, it must be emphasized that *equal vectors need not always be equivalent; it depends entirely on the situation at hand.* Furthermore, vectors that are not equal may still be equivalent in some capacity. Thus, in the beam in Fig. 1.9, forces $F_1$ and $F_2$ are unequal, since their magnitudes are 10 lb and 20 lb, respectively. However, it is clear from elementary physics that their moments about the base of the beam are equal, and so the forces have the same "turning" action at the fixed end of the beam. In that capacity, the forces are equivalent. If. however, we

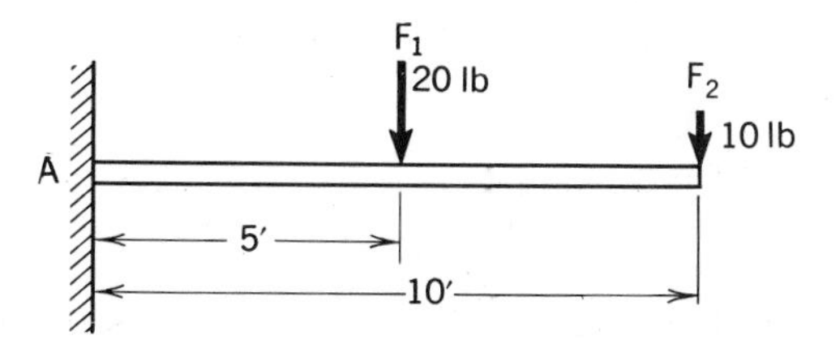


**Figure 1.9.** $F_1$ and $F_2$ equivalent for moment about $A$.

are interested in the deflection of the free end of the beam resulting from each force, there is no longer an equivalence between the forces, since each will give a different deflection.

To sum up, the *equality* of two vectors is determined by the vectors themselves, and the *equivalence* between two vectors is determined by the task involving the vectors.

In problems of mechanics, we can profitably delineate three classes of situations concerning equivalence of vectors:

1. *Situations in which vectors may be positioned anywhere in space without loss or change of meaning provided that magnitude and direction are kept intact.* Under such circumstances the vectors are called *free vectors.* For example, the velocity vectors in Fig. 1.8 are free vectors as far as total distance traveled is concerned.

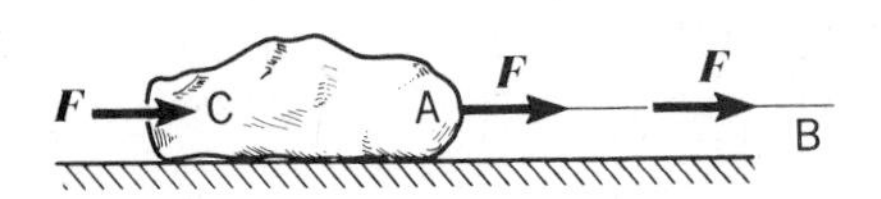


Figure 1.10. *F* transmissible for towing.

2. *Situations in which vectors may be moved along their lines of action without change of meaning.* Under such circumstances the vectors are called *transmissible vectors.* For example, in towing the object in Fig. 1.10, we may apply the force anywhere along the rope *AB* or may push at point *C*. The resulting motion is the same in all cases, so the force is a transmissible vector for this purpose.
3. *Situations in which the vectors must be applied at definite points.* The point may be represented as the tail or head of the arrow in the graphical representation. For this case, no other position of application leads to equivalence. Under such circumstances, the vector is called a *bound vector.* For example, if we are interested in the deformation induced by forces in the body in Fig. 1.10, we must be more selective in our actions than we were when all we wanted to know was the motion of the body. Clearly, force $\boldsymbol{F}$ will cause a different deformation when applied at point *C* than it will when applied at point *A*. The force is thus a bound vector for this problem.

We shall be concerned throughout this text with considerations of equivalence.

## †1.10 Laws of Mechanics

The entire structure of mechanics rests on relatively few basic laws. Nevertheless, for the student to comprehend these laws sufficiently to undertake novel and varied problems, much study will be required.

We shall now discuss briefly the following laws, which are considered to be the foundation of mechanics:

1. Newton's first and second laws of motion.
2. Newton's third law.
3. The gravitational law of attraction.
4. The parallelogram law.

**Newton's First and Second Laws of Motion.** These laws were first stated by Newton as

> *Every particle continues in a state of rest or uniform motion in a straight line unless it is compelled to change that state by forces imposed on it.*

> *The change of motion is proportional to the natural force impressed and is made in a direction of the straight line in which the force is impressed.*

Notice that the words "rest," "uniform motion," and "change of motion" appear in the statements above. For such information to be meaningful, we must have some frame of reference relative to which these states of motion can be described. We may then ask: relative to what reference in space does every particle remain at "rest" or "move uniformly along a straight line" in the absence of any forces? Or, in the case of a force acting on the particle, relative to what reference in space is the "change in motion proportional to the force"? Experiment indicates that the "fixed" stars act as a reference for which the first and second laws of Newton are highly accurate. Later, we will see that any other system that moves uniformly and without rotation relative to the fixed stars may be used as a reference with equal accuracy. All such references are called *inertial references*. The earth's surface is usually employed as a reference in engineering work. Because of the rotation of the earth and the variations in its motion around the sun, it is not, strictly speaking, an inertial reference. However, the departure is so small for most situations (exceptions are the motion of guided missiles and spacecraft) that the error incurred is very slight. We shall, therefore, usually consider the earth's surface as an inertial reference, but will keep in mind the somewhat approximate nature of this step.

As a result of the preceding discussion, we may define *equilibrium* as *that state of a body in which all its constituent particles are at rest or moving uniformly along a straight line relative to an inertial reference*. The converse of Newton's first law, then, stipulates for the equilibrium state that there must be no force (or equivalent action of no force) acting on the body. Many situations fall into this category. The study of bodies in equilibrium is called *statics*, and it will be an important consideration in this text.

In addition to the reference limitations explained above, a serious limitation was brought to light at the turn of this century. As pointed out earlier, the pioneering work of Einstein revealed that the laws of Newton become increasingly more approximate as the speed of a body increases. Near the speed of light, they are untenable. In the vast majority of engineering computations, the speed of a body is so small compared to the speed of light that these departures from Newtonian mechanics, called *relativistic effects*, may be entirely disregarded with little sacrifice in accuracy. In considering the motion of high-energy elementary particles occurring in nuclear phenomena, however, we cannot ignore relativistic effects. Finally, when we get down to very small distances, such as those between the protons and neutrons in the nucleus of an atom, we find that Newtonian mechanics cannot explain many observed phenomena. In this case, we must resort to quantum mechanics, and then Newton's laws give way to the Schrödinger equation as the key equation.

**Newton's Third Law.** Newton stated in his third law:

> *To every action there is always opposed an equal reaction, or the mutual actions of two bodies upon each other are always equal and directed to contrary points.*

This is illustrated graphically in Fig. 1.11, where the action and reaction between two bodies arise from direct contact. Other important actions in which Newton's third law holds are gravitational attractions (to be discussed next) and electrostatic forces between charged particles. It should be pointed out that there are actions that do not follow this law, notably the electromagnetic forces between charged moving bodies.[6]

**Law of Gravitational Attraction.** It has already been pointed out that there is an attraction between the earth and the bodies at its surface, such as $A$ and $B$ in Fig. 1.11. This attraction is mutual and Newton's third law applies. There is also an attraction between the two bodies $A$ and $B$ themselves, but this force because of the small size of

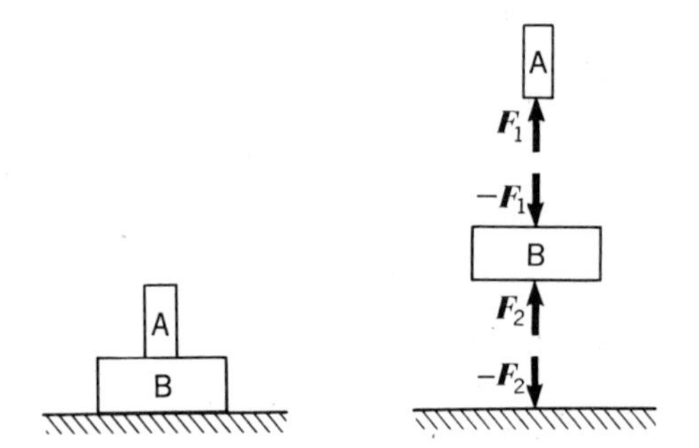


**Figure 1.11.** Newton's third law.

both bodies is extremely weak. However, the mechanism for the mutual attraction between the earth and each body is the same as that for the mutual attraction between the bodies. These forces of attraction may be given by the *law of gravitational attraction*:

> Two particles will be attracted toward each other along their connecting line with a force whose magnitude is directly proportional to the product of the masses and inversely proportional to the distance squared between the particles.

Avoiding vector notation for now, we may thus say that

$$F = G\frac{m_1 m_2}{r^2} \tag{1.11}$$

where $G$ is called the *universal gravitational constant.* In the actions involving the earth and the bodies discussed above, we may consider each body as a particle, with its entire mass concentrated at its center of gravity.[7] Hence, if we know the various constants in formula 1.11, we can compute the weight of a given mass at different altitudes above the earth.

**Parallelogram Law.** Stevinius (1548–1620) was the first to demonstrate that forces could be combined by representing them by arrows to some suitable scale, and then

---

[6]Electromagnetic forces between charged moving particles are equal and opposite but are not collinear and hence are not "directed to contrary points."

[7]To be studied in detail in Chapter 4.

forming a parallelogram in which the diagonal represents the sum of the two forces. As we pointed out, all vectors must combine in this manner.

## 1.11 Closure

In this chapter, we have introduced the basic dimensions by which we can describe in a quantitative manner certain aspects of nature. These basic, and from them secondary, dimensions may be related by dimensionally homogeneous equations which, with suitable idealizations, can represent certain actions in nature. The basic laws of mechanics were thus introduced. Since the equations of these laws relate vector quantities, we shall introduce a useful and highly descriptive set of vector operations in Chapter 2 in order to learn to handle these laws effectively and to gain more insight into mechanics in general. These operations are generally called *vector algebra.*

*Check-Out for Sections with†*

1.1 What are two kinds of limitations on Newtonian mechanics?
1.2 What are the two phenomena wherein mass plays a key role?
1.3 If a pound force is defined by the extension of a standard spring, define the pound mass and the slug.
1.4 Express mass density dimensionally. How many scale units of mass density in the SI units are equivalent to 1 scale unit in the American system using (a) slugs, ft, sec and (b) lbm, ft, sec.
1.5 (a) What is a necessary condition for *dimensional homogeneity* in an equation?
(b) In the Newtonian viscosity law, the frictional resistance $\tau$ (force per unit area) in a fluid is proportional to the distance rate of change of velocity $dV/dy$. The proportionality constant $\mu$ is called the *coefficient of viscosity*. What is its dimensional representation?
1.6 Define a vector and a scalar.
1.7 What is meant by *line of action* of a vector?
1.8 What is a *displacement* vector?
1.9 What is an *inertial reference*?

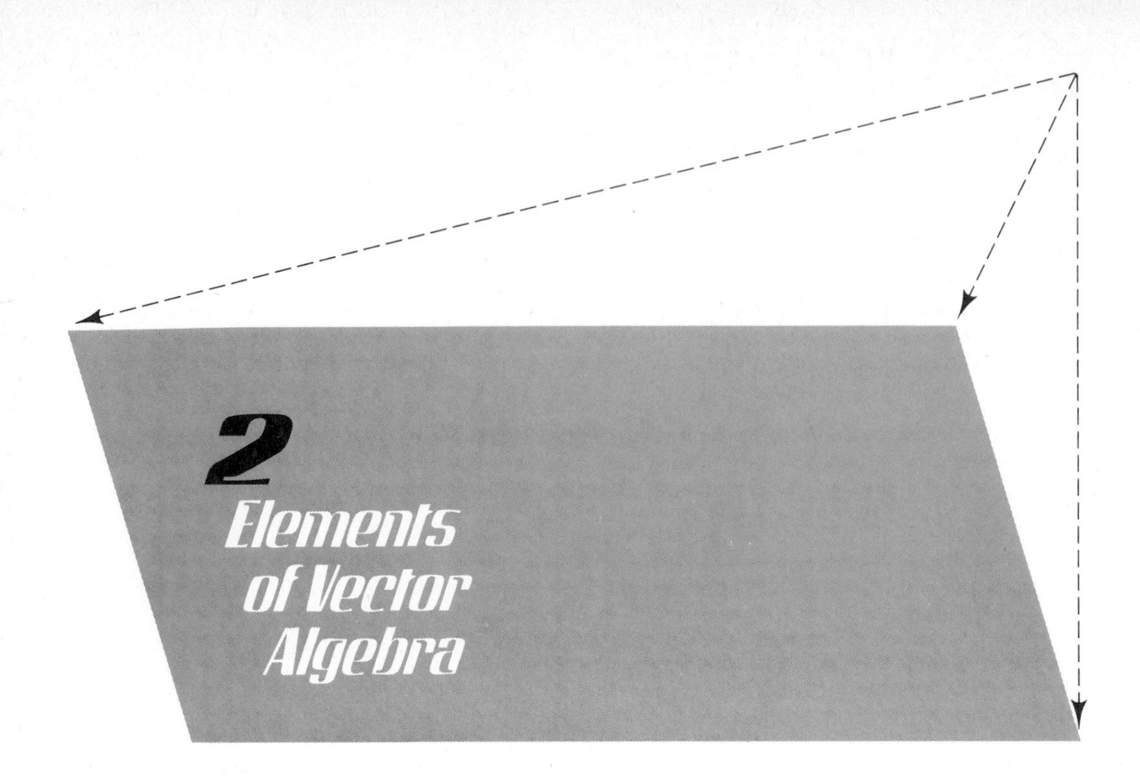


## †2.1 Introduction

In Chapter 1, we saw that a scalar quantity is adequately given by a magnitude, while a vector quantity requires the additional specification of a direction. The basic algebraic operations for the handling of scalar quantities are those familiar ones studied in grade school, so familiar that you now wonder even that you had to be "introduced" to them. For vector quantities, these methods may be cumbersome since the directional aspects must be taken into account. Therefore, an algebra has evolved that clearly and concisely allows for certain very useful manipulations of vectors. It is not merely for elegance or sophistication that we employ vector algebra. Indeed, we can achieve greater insight into the subject matter—particularly into dynamics—by employing the more powerful and descriptive methods introduced in this chapter. These methods will at first appear rather arbitrary and artificial. However, you must remember that many years ago, perhaps when you were six or seven years of age, addition and subtraction were rather puzzling and later, when you were eight or nine, multiplication and division were by no means perfectly "natural." Now, of course, you perform those operations almost without thinking about them. Similarly, constant use of vector algebra in mechanics and in other disciplines will bring a comfortable familiarity in a surprisingly short time.

## †2.2 Magnitude and Multiplication of a Vector by a Scalar

The magnitude of a quantity, in strict mathematical parlance, is always a *positive* number of units whose value corresponds to the numerical measure of the quantity. Thus, the magnitude of a quantity of measure −50 units is +50 units. Note that the magnitude of a quantity is its absolute value. The mathematical symbol for indicating the magnitude of a quantity is a set of vertical lines enclosing the quantity. That is,

$$|-50 \text{ units}| = \text{absolute value } (-50 \text{ units}) = +50 \text{ units}$$

Similarly, the magnitude of a vector quantity is a positive number of units corresponding to the length of the vector in those units. Using our vector symbols, we can say that

$$\text{magnitude of vector } \boldsymbol{A} = |\boldsymbol{A}|$$

Thus, $|\boldsymbol{A}|$ is a positive scalar quantity. We may now discuss the multiplication of a vector by a scalar.

The definition of the product of vector $\boldsymbol{A}$ by scalar $m$, written simply as $m\boldsymbol{A}$, is given in the following manner:

> $m\boldsymbol{A}$ is a vector having the same direction as $\boldsymbol{A}$ and a magnitude equal to the ordinary scalar product between the magnitudes of $m$ and $\boldsymbol{A}$. If $m$ is negative, it means simply that the vector $m\boldsymbol{A}$ has a direction directly opposite to that of $\boldsymbol{A}$.

The vector $-\boldsymbol{A}$ may be considered as the product of the scalar −1 and the vector $\boldsymbol{A}$. Thus, from the statement above we see that $-\boldsymbol{A}$ differs from $\boldsymbol{A}$ in that it has an opposite sense. Furthermore, these operations have nothing to do with the line of action of a vector, so $\boldsymbol{A}$ and $-\boldsymbol{A}$ may have different lines of action. This will be the case of the couple to be studied in Chapter 3.

## †2.3 Addition and Subtraction of Vectors

In adding a number of vectors, we may repeatedly employ the parallelogram construction. We can do this graphically by scaling the lengths of the arrows according to the magnitudes of the vector quantities they represent. The magnitude of the final arrow can then be interpreted in terms of its length by employing the chosen scale factor. As an example, consider the coplanar[1] vectors $\boldsymbol{A}$, $\boldsymbol{B}$, and $\boldsymbol{C}$ shown in Fig. 2.1(a). The addition of the vectors $\boldsymbol{A}$, $\boldsymbol{B}$, and $\boldsymbol{C}$ has been accomplished in two ways. In Fig. 2.1(b) we first add $\boldsymbol{B}$ and $\boldsymbol{C}$ and then add the resulting vector (shown dashed) to $\boldsymbol{A}$. This combination can be represented by the notation $\boldsymbol{A} + (\boldsymbol{B} + \boldsymbol{C})$. In Fig.

[1]Coplanar, meaning "same plane," is a word used often in mechanics.

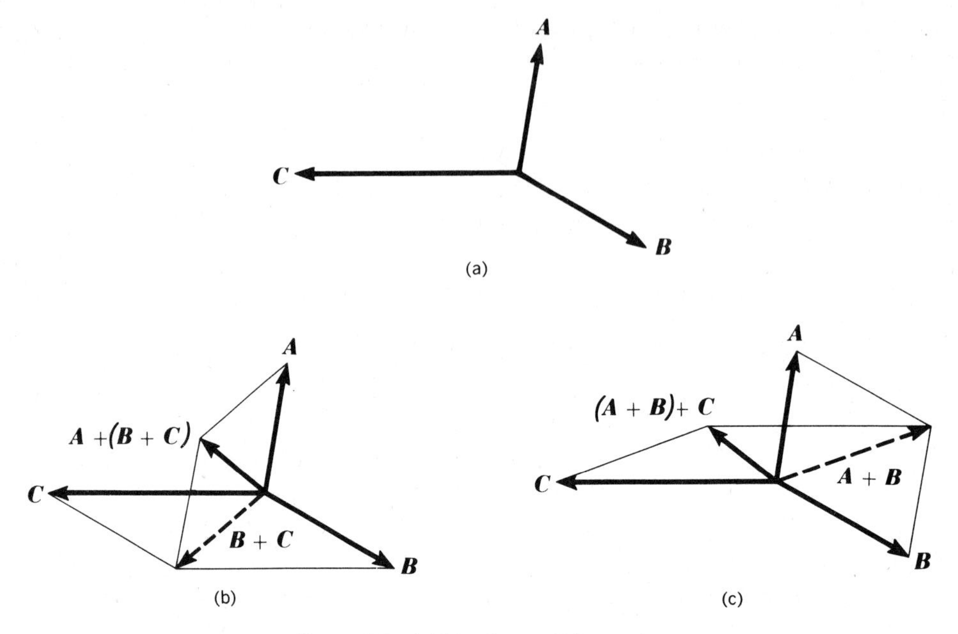


**Figure 2.1.** Addition by parallelogram law.

2.1(c), we add $\boldsymbol{A}$ and $\boldsymbol{B}$, and then add the resulting vector (shown dashed) to $\boldsymbol{C}$. The representation of this combination is given as $(\boldsymbol{A} + \boldsymbol{B}) + \boldsymbol{C}$. Note that the final vector is identical for both procedures. Thus,

$$\boldsymbol{A} + (\boldsymbol{B} + \boldsymbol{C}) = (\boldsymbol{A} + \boldsymbol{B}) + \boldsymbol{C} \tag{2.1}$$

When the quantities involved in an algebraic operation can be grouped without restriction, the operation is said to be *associative*. Thus, the addition of vectors is both commutative, as explained earlier, and associative.

To determine a summation of, let us say, two vectors without recourse to graphics, we need only make a simple sketch of the vectors approximately to scale. By using familiar trigonometric relations, we can get a direct evaluation of the result. This is illustrated in the following examples.

**EXAMPLE 2.1**

Add the forces acting on a particle situated at the origin of a two-dimensional reference frame (Fig. 2.2). One force has a magnitude of 10 lb acting in the positive $x$ direction, whereas the other has a magnitude of 5 lb acting at an angle of 135° with a sense directed away from the origin.

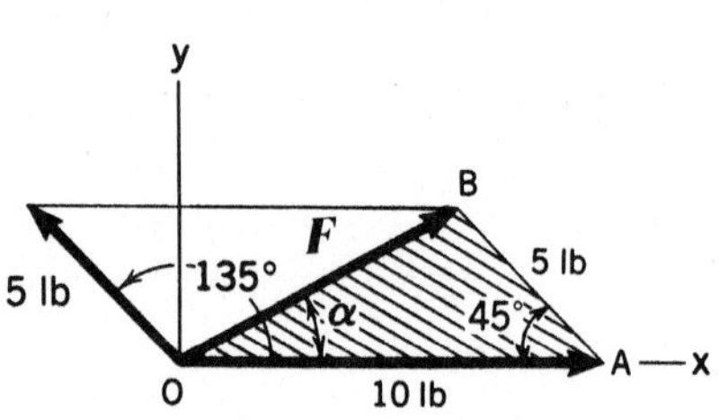


**Figure 2.2.** Find $F$ and $\alpha$ using trigonometry.

To get the sum (shown as $\boldsymbol{F}$), we may use the law of cosines[2] for one of the triangular portions of the sketched parallelogram. Thus, using triangle $OBA$,

$$|\boldsymbol{F}| = [10^2 + 5^2 - (2)(10)(5)\cos 45°]^{1/2}$$
$$= (100 + 25 - 70.7)^{1/2} = \sqrt{54.3} = 7.37 \text{ lb}$$

The direction of the vector may be described by giving the angle and the sense. This is done by employing the law of sines for triangle $OBA$.[3]

$$\frac{5}{\sin\alpha} = \frac{7.37}{\sin 45°}$$

$$\sin\alpha = \frac{(5)(0.707)}{7.37} = 0.480$$

Therefore,

$$\alpha = 28.6°$$

**EXAMPLE 2.2**

A light cable from a Jeep is tied to the peak of an A-frame and exerts a force of 450 N along the cable (see Fig. 2.3). A 1000-kg log is suspended from a second cable, which is fastened to the peak. What is the total force from the cables on the A-frame?

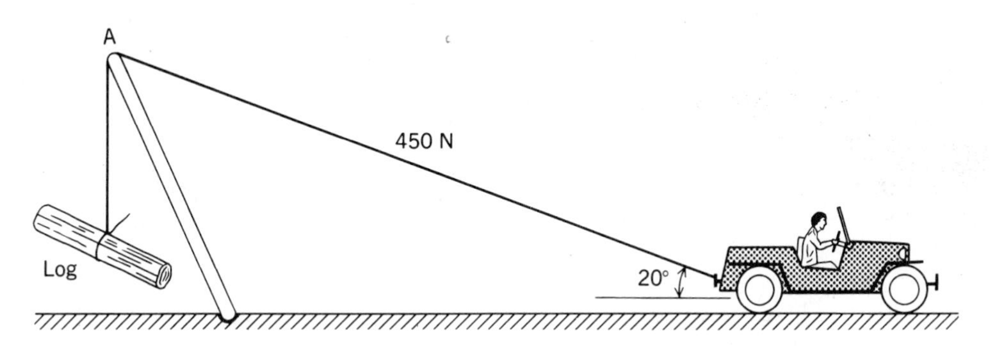


**Figure 2.3.** An A-frame supports a log.

---

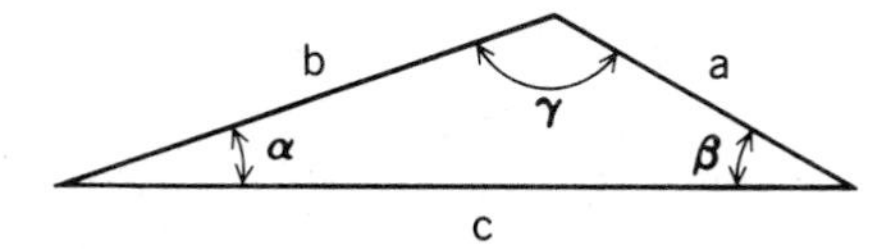


[2]You will recall from trigonometry that the *law of cosines* for side $b$ of a triangle is given as

$$b^2 = a^2 + c^2 - 2ac\cos\beta$$

[3]The *law of sines* is given as follows for a triangle:

$$\frac{a}{\sin\alpha} = \frac{b}{\sin\beta} = \frac{c}{\sin\gamma}$$

We first note that the force $F$ on the log from gravity (i.e., the weight) is

$$F = Mg = (1000)(9.81) = 9810 \text{ N} \tag{a}$$

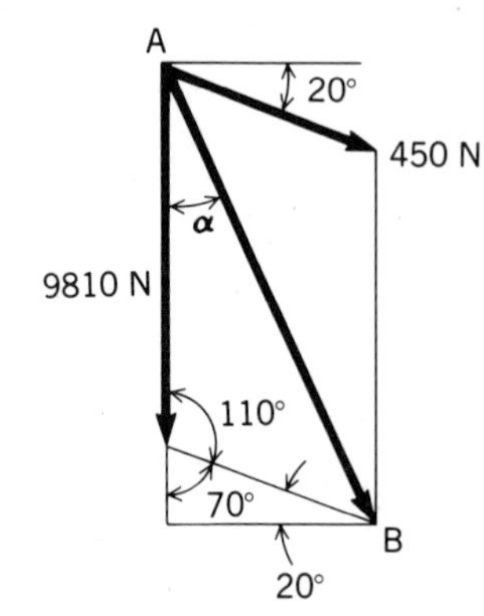


Figure 2.4. Parallelogram law.

The force parallelogram is now drawn (see Fig. 2.4). From the law of cosines, we can say that

$$(AB)^2 = 450^2 + 9810^2 - (2)(450)(9810)\cos 110°$$

$$AB = 9973 \text{ N}$$

The angle $\alpha$ is next determined using the law of sines. Thus,

$$\frac{450}{\sin\alpha} = \frac{9973}{\sin 110°}$$

$$\alpha = 2.43°$$

We may also add the vectors by moving them successively to parallel positions so that the head of one vector connects to the tail of the next vector, and so on. The sum of the vectors will then be a vector whose tail connects to the tail of the first vector and whose head connects to the head of the last vector. This last step will form a polygon from the vectors, and we say that the vector sum then "closes the polygon." Thus, adding the 10-lb vector to the 5-lb vector in Fig. 2.2, we would form the sides $OA$ and $AB$ of a triangle. The sum $\boldsymbol{F}$ then closes the triangle and is $OB$. Also, in Fig. 2.5(a), we have shown three coplanar vectors $\boldsymbol{F}_1$, $\boldsymbol{F}_2$, and $\boldsymbol{F}_3$. The vectors are connected in Fig. 2.5(b)

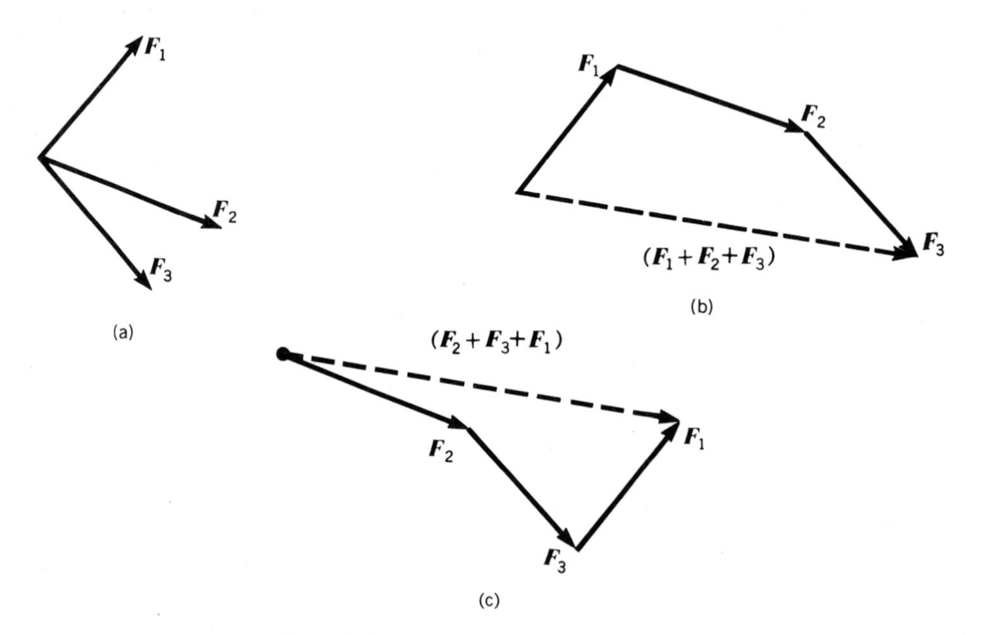


Figure 2.5. Addition by "closing the polygon."

as described. The sum of the vectors then is the dashed vector that closes the polygon. In Fig. 2.5(c), we have laid off the vectors $\boldsymbol{F}_1$, $\boldsymbol{F}_2$, and $\boldsymbol{F}_3$ in a different sequence. Nevertheless, it is seen that the sum is the same vector as in Fig. 2.5(b). Clearly, the *order* of laying off the vectors is not significant.

A simple physical interpretation of the above vector sum can be formed for vectors each of which represents a movement of a certain distance and direction (i.e., a *displacement* vector). Then, traveling along the system of given vectors you start from one point (the tail of the first vector) and end at another point (the head of the last vector). The vector *sum* that closes the polygon is equivalent to the system of given vectors, in that it takes you from the same initial to the same final point.

The polygon summation process, like the parallelogram of addition, can be used as a graphical process, or, still better, can be used to generate analytical computations with the aid of trigonometry. The extension of this procedure to any number of vectors is obvious.

The process of *subtraction* of vectors is defined in the following manner: to subtract vector $\boldsymbol{B}$ from vector $\boldsymbol{A}$, we reverse the direction of $\boldsymbol{B}$ (i.e., multiply by $-1$) and then add this new vector to $\boldsymbol{A}$ (Fig. 2.6).

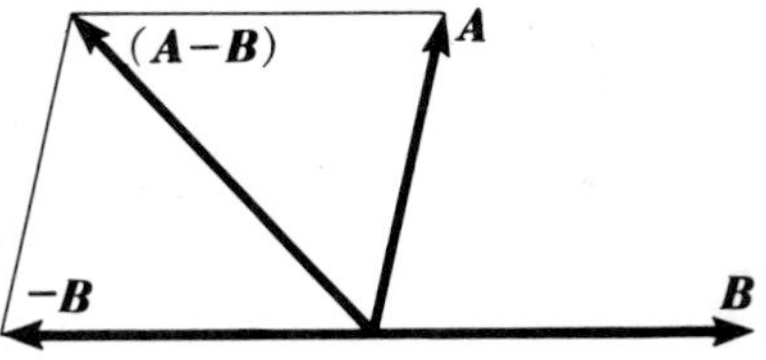

Figure 2.6. Subtraction of vectors.

This process may also be used in the polygon construction. Thus, consider coplanar vectors $\boldsymbol{A}$, $\boldsymbol{B}$, $\boldsymbol{C}$, and $\boldsymbol{D}$ in Fig. 2.7(a). To form $\boldsymbol{A} + \boldsymbol{B} - \boldsymbol{C} - \boldsymbol{D}$, we proceed as shown in Fig. 2.7(b). Again, the order of the process is not significant, as can be seen in Fig. 2.7(c).

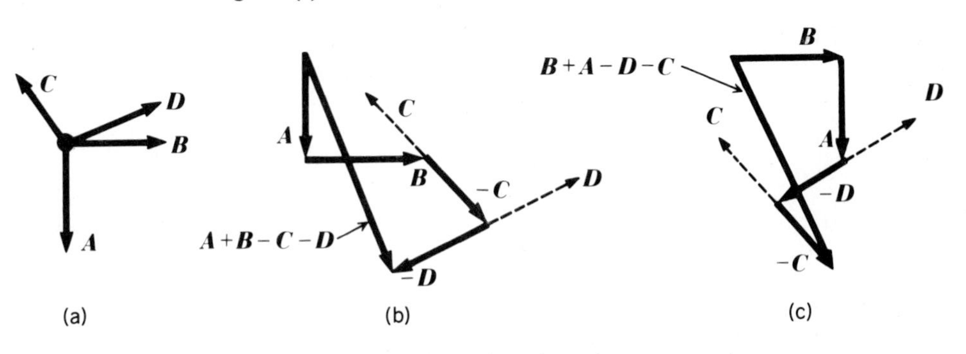

Figure 2.7. Addition and subtraction using polygon construction.

## Problems

**2.1.** Add a 20-N force pointing in the positive $x$ direction to a 50-N force at an angle 45° to the $x$ axis in the first quadrant and directed away from the origin.

**2.2.** Subtract the 20-N force in Problem 2.1 from the 50-N force.

**2.3.** Add the vectors in the $xy$ plane. Do this first graphically, using the force polygon and then do it analytically.

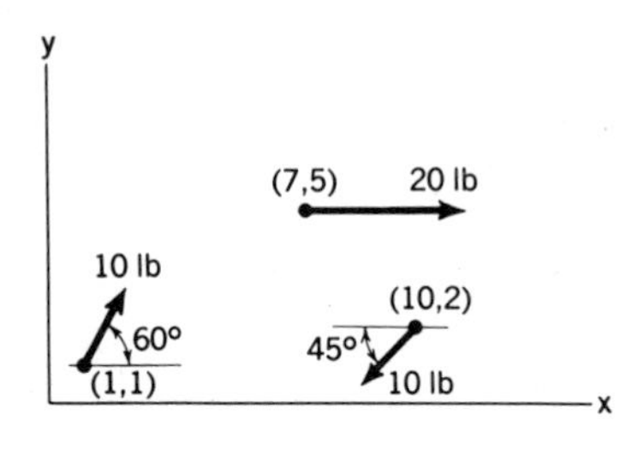


Figure P.2.3

**2.4.** A homing pigeon is released at point $A$ and is observed. It flies 10 km due south, then goes due east for 15 km. Next it goes southeast for 10 km and finally goes due south 5 km to reach its destination $B$. Graphically determine the shortest distance between $A$ and $B$. Neglect the earth's curvature.

**2.5.** Force $\boldsymbol{A}$ (given as a horizontal 10-N force) and $\boldsymbol{B}$ (vertical) add up to a force $\boldsymbol{C}$ that has a magnitude of 20 N. What is the magnitude of force $\boldsymbol{B}$ and the direction of force $\boldsymbol{C}$? (For the simplest results, use the force polygon, which for this case is a right triangle, and perform analytical computations.)

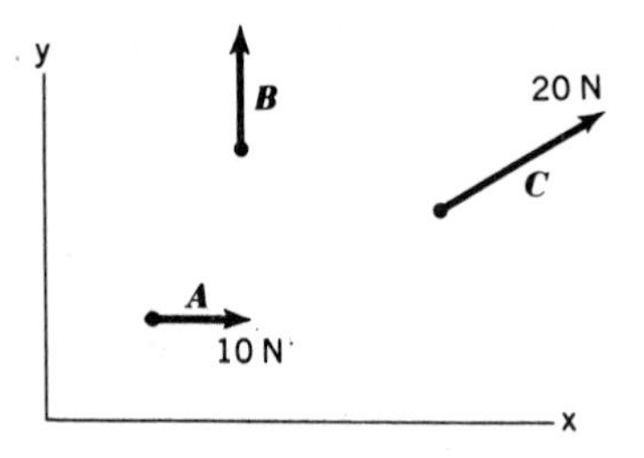


Figure P.2.5

**2.6.** If the difference between forces $\boldsymbol{B}$ and $\boldsymbol{A}$ in Fig. P.2.5 is a force $\boldsymbol{D}$ having a magnitude of 25 N, what is the magnitude of $\boldsymbol{B}$ and the direction of $\boldsymbol{D}$?

**2.7.** What is the sum of the forces transmitted by the structural members to the pin at $A$?

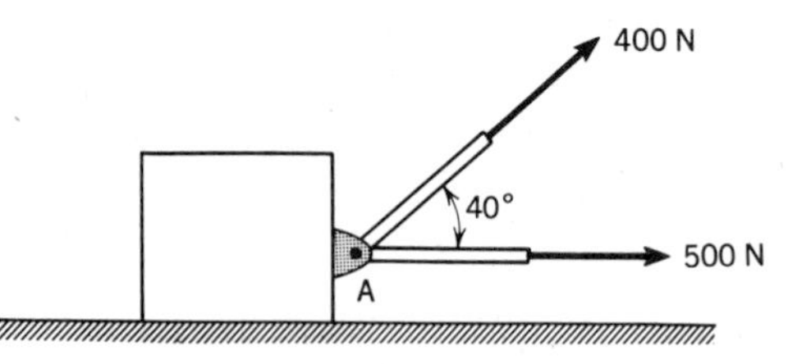

Figure P.2.7

**2.8.** Suppose in Problem 2.7 we require that the total force transmitted by the members to pin $A$ be inclined 12° to the horizontal. If we do not change the force transmitted by the horizontal member, what must be the new force for the other member whose direction remains at 40°? What is the total force?

**2.9.** A man pulls with force $W$ on a rope through a simple frictionless pulley to raise a weight $W$. What total force is exerted on the pulley?

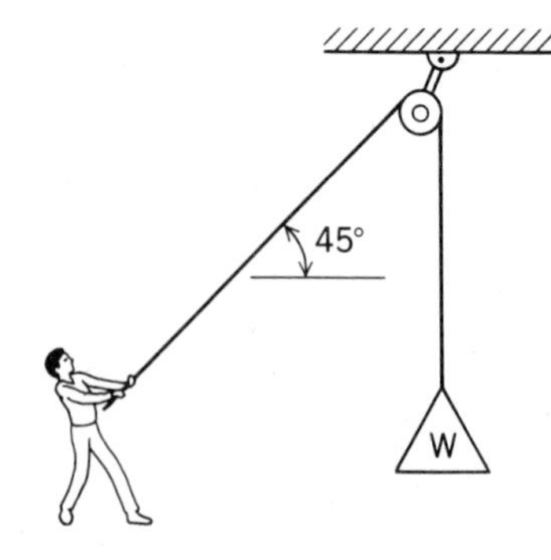


Figure P.2.9

**2.10.** A mass $M$ is supported by cables (1) and (2). The tension in cable (1) is 200 N, whereas the tension in (2) is such as to maintain the configuration shown. What is the mass of

$M$ in kilograms? (You will learn very shortly that the weight of $M$ must be equal and opposite to the vector sum of the supporting forces for equilibrium.)

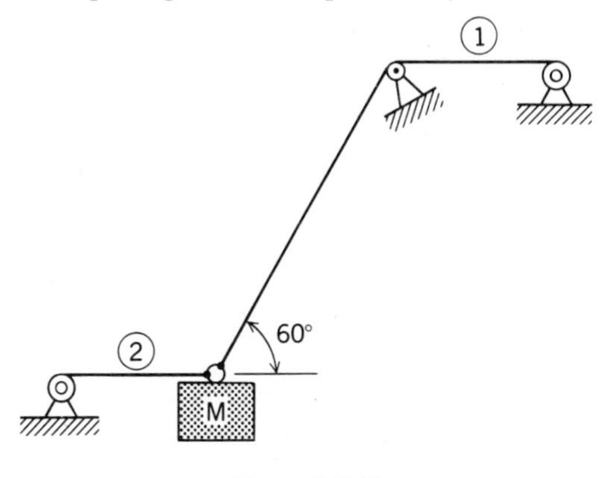


Figure P.2.10

**2.11.** Two football players are pushing a blocking dummy. Player $A$ pushes with 100-lb force while player $B$ pushes with 150-lb force toward bow $C$ of the dummy. What is the total force exerted on the dummy by the players?

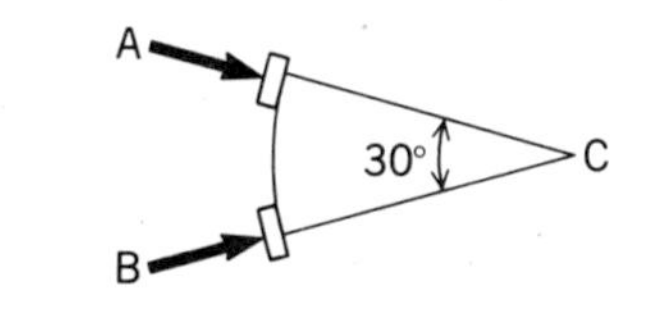

Figure P.2.11

**2.12.** A simple slingshot is about to be "fired." If the entire rubber band has a stretch of 1 in./3 lb, what force does the band exert on the right hand? The total unstretched length of the rubber band is 5 in. Note, if you use half the rubber band, you double the resistance to stretching.

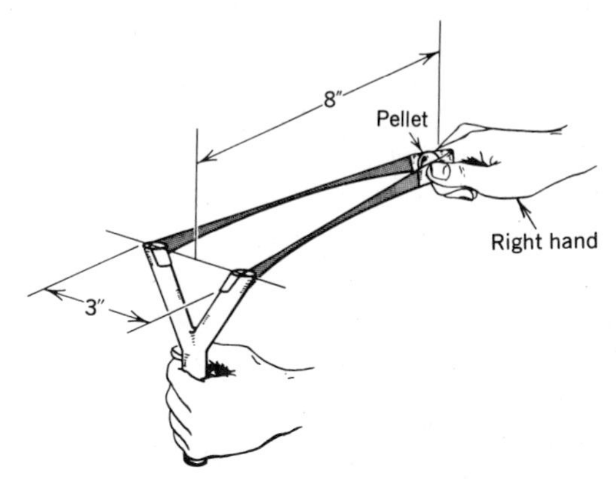


Figure P.2.12

**2.13.** Two soccer players approach a stationary ball 10 ft away from the goal. Simultaneously, a player on team $O$ (offense) kicks the ball with force 100 lb for a split second while a player on team $D$ (defense) kicks with force 70 lb during the same time interval. Does the offense score (assuming that the goalie is asleep)?

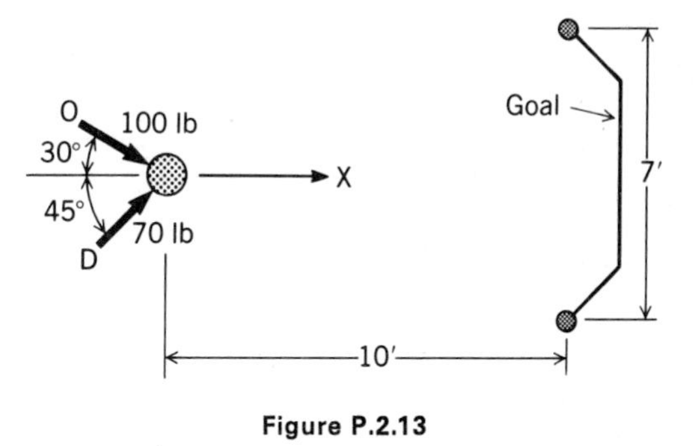


Figure P.2.13

## 2.4 Resolution of Vectors; Scalar Components

The opposite action of addition of vectors is called *resolution*. Thus, for a given vector $\boldsymbol{C}$, we may find a pair of vectors in any two stipulated directions coplanar with $\boldsymbol{C}$ such that the two vectors, called *components*, sum to the original vector. This is a *two-dimensional* resolution involving two component vectors *coplanar* with the original vector. We shall discuss three-dimensional resolution involving three noncoplanar component vectors later in the section. The two-dimensional resolution can be accomplished by graphical construction, or by using simple helpful sketches and then employing trigonometric relations. An example of two-dimensional resolution is shown in Fig. 2.8. The two vectors $\boldsymbol{C}_1$ and $\boldsymbol{C}_2$ formed in this way are the *component* vectors. We often replace a vector by its components since the components are always equivalent in rigid-body mechanics to the original vector. When this is done, it is often helpful to indicate that the original vector is no longer operative by drawing a wavy line through the original vector as shown in Fig. 2.9.

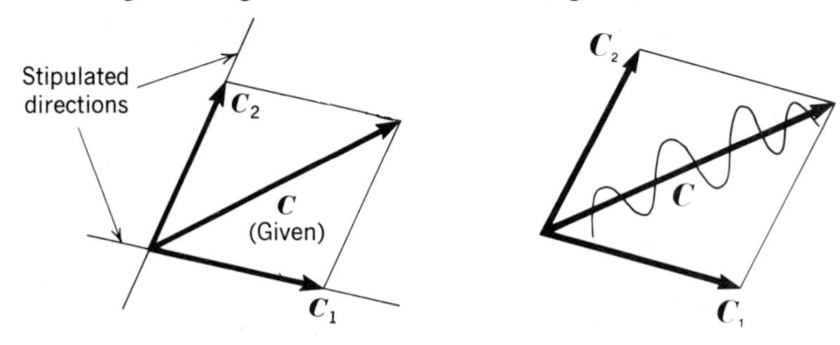


**Figure 2.8.** Two-dimensional resolution of vector $\boldsymbol{C}$.

**Figure 2.9.** Vector $\boldsymbol{C}$ is replaced by its components and is no longer operative.

**EXAMPLE 2.3**

Flight 304 from Dallas is flying NE to Chicago 900 miles away (see Fig. 2.10). To avoid a massive storm front, the pilot decides instead to fly due north to Topeka, Kansas, and then ENE (see Fig. 2.11 for compass settings) to Chicago. What are the distances that he must travel from Dallas to Topeka and from Topeka to Chicago?

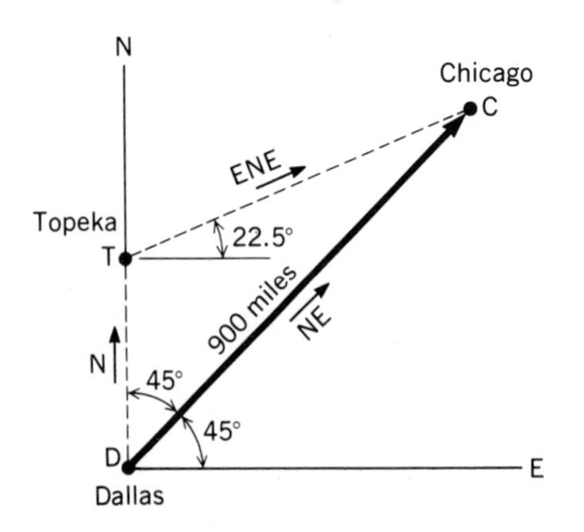


**Figure 2.10.** Compass headings.

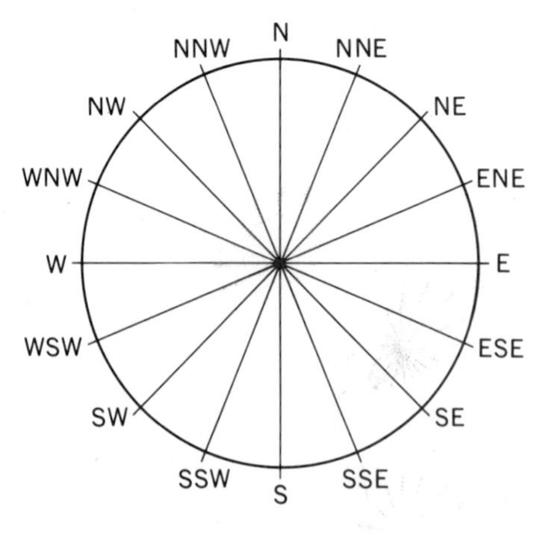


**Figure 2.11.** Compass settings.

We know the vector from Dallas to Chicago, and we want to resolve this vector into two components whose directions are known (see Fig. 2.10). Note from Fig. 2.11 that NE corresponds to a 45° angle to the north axis, and ENE corresponds to a 22.5° angle to the east axis. Now, form a parallelogram from $DC$ and the two directions (see Fig. 2.12). We can consider triangle $DTC$ of the paralleologram.[4] Note angle $\measuredangle DTC = 180° - 45° - 22.5° = 112.5°$, so that using the law of sines, we can say

$$\frac{900}{\sin 112.5°} = \frac{DT}{\sin 22.5°} = \frac{TC}{\sin 45°}$$

Therefore,

$$DT = 373 \text{ miles}$$

$$TC = 689 \text{ miles}$$

The alternate route is thus 162 miles longer.

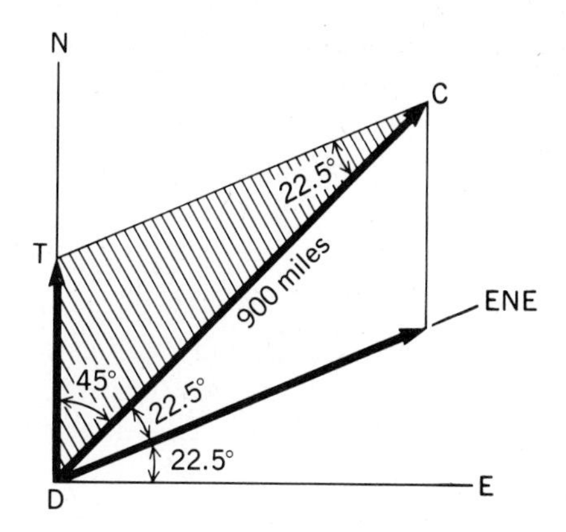


**Figure 2.12.** Form parallelogram.

It is also readily possible to find *three* components *not in the same plane as* **C**. This is the aforementioned three-dimensional resolution. Consider the specification of three *orthogonal* directions[5] for the resolution of **C**, as is shown in Fig. 2.13. The

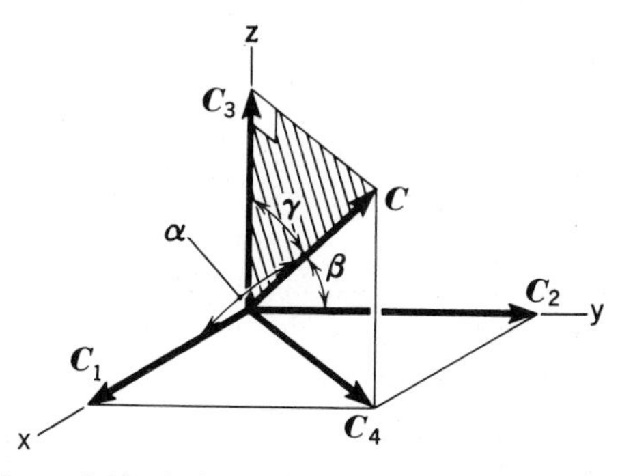


**Figure 2.13.** Orthogonal or rectangular components.

resolution may be accomplished in two steps. Resolve **C** along the $z$ direction, and along the intersection of the $xy$ plane and the plane formed by **C** and the $z$ axis. This gives vectors $\mathbf{C}_3$ and $\mathbf{C}_4$. Since $\mathbf{C}_3$ is perpendicular to the $xy$ plane, it must be perpendicular to $\mathbf{C}_4$. Thus, $\mathbf{C}_3$ and $\mathbf{C}_4$ form the sides of a *rectangle.* Now resolve $\mathbf{C}_4$ along the $x$ and $y$ directions, forming the other two component vectors $\mathbf{C}_1$ and $\mathbf{C}_2$. It is clear that the vectors $\mathbf{C}_1$, $\mathbf{C}_2$, and $\mathbf{C}_3$ add up to the vector **C** and are all mutually perpendicular to one another. Hence, $\mathbf{C}_1$, $\mathbf{C}_2$, and $\mathbf{C}_3$ are called the *orthogonal* (or *rectangular*) *component vectors* of **C**.

---

[4]This triangle could have been reached directly using the concept of the vector polygon of Section 2.3.

[5]Although the vector can be resolved along three *skew* directions (hence nonorthogonal), the orthogonal directions are used most often in engineering practice.

The direction of a vector $\boldsymbol{C}$ relative to an orthogonal reference is given by the cosines of the angles formed by the vector and the respective coordinate axes. These are called *direction cosines* and are denoted as

$$\begin{aligned}\cos(\boldsymbol{C}, x) &= \cos\alpha \equiv l \\ \cos(\boldsymbol{C}, y) &= \cos\beta \equiv m \\ \cos(\boldsymbol{C}, z) &= \cos\gamma \equiv n\end{aligned} \tag{2.2}$$

where $\alpha$, $\beta$, and $\gamma$ are associated with the $x$, $y$, and $z$ axes, respectively. Now let us consider the right triangle, whose sides are $\boldsymbol{C}$ and the component vector, $\boldsymbol{C}_3$, shown shaded in Fig. 2.13. It then becomes clear, from trigonometric considerations of the right triangle, that

$$|\boldsymbol{C}_3| = |\boldsymbol{C}|\cos\gamma = |\boldsymbol{C}|n \tag{2.3}$$

If we had decided to resolve $\boldsymbol{C}$ first in the $y$ direction instead of the $z$ direction, we would have produced a geometry from which we could conclude that $|\boldsymbol{C}_2| = |\boldsymbol{C}|m$. Similarly, we can say that $|\boldsymbol{C}_1| = |\boldsymbol{C}|l$. We can then express $|\boldsymbol{C}|$ in terms of its orthogonal components in the following manner, using the Pythagorean theorem[6]:

$$|\boldsymbol{C}| = [(|\boldsymbol{C}|l)^2 + (|\boldsymbol{C}|m)^2 + (|\boldsymbol{C}|n)^2]^{1/2} \tag{2.4}$$

From this equation we can define the *orthogonal* or *rectangular scalar components* of the vector $\boldsymbol{C}$ as

$$C_x = |\boldsymbol{C}|l, \qquad C_y = |\boldsymbol{C}|m, \qquad C_z = |\boldsymbol{C}|n \tag{2.5}$$

Note that $C_x$, $C_y$, and $C_z$ may be negative, depending on the sign of the direction cosines. Finally, it must be pointed out that *although $C_x$, $C_y$, and $C_z$ are associated with certain axes and hence certain directions, they have been developed as scalars and must be handled as scalars.* Thus, an equation such as $10\boldsymbol{V} = V_x \cos\beta$ is not correct, because the left side is a vector and the right side is a scalar. This should spur you to observe care in your notation.

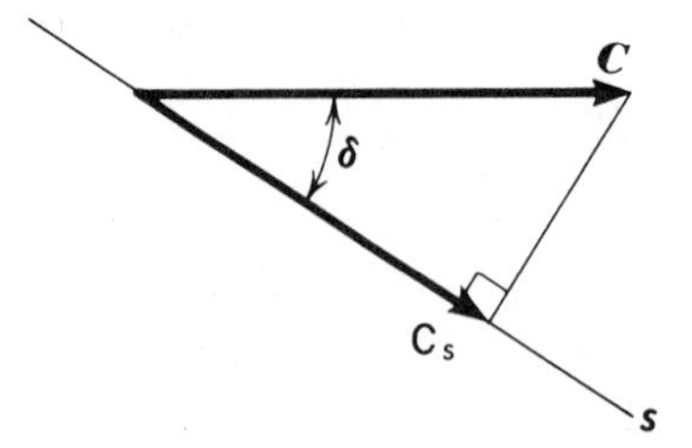


**Figure 2.14.** Rectangular component of $\boldsymbol{C}$.

Sometimes only *one* of the scalar orthogonal components of a vector (often called a rectangular component) is desired. Then, just one direction is prescribed, as shown in Fig. 2.14. Thus, the scalar rectangular component $C_s$ is $|\boldsymbol{C}|\cos\delta$. It is always the case that the triangle formed by the vector and its scalar rectangular component is a right triangle. In establishing $C_s$ we speak, therefore, of "dropping a perpendicular from $\boldsymbol{C}$ to $s$" or of "projecting $\boldsymbol{C}$ along $s$."

As a final consideration, let us examine vectors $\boldsymbol{A}$ and $\boldsymbol{B}$, which, along with direction $s$, form a plane as is shown in Fig. 2.15. The sum of the vectors $\boldsymbol{A}$ and $\boldsymbol{B}$ is found by the parallelogram

[6]From Eq. 2.4 one readily can conclude that $l^2 + m^2 + n^2 = 1$, which is a well-known geometric relation.

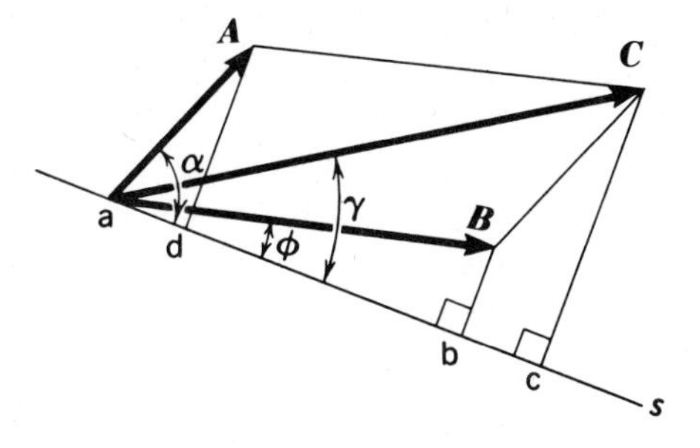


**Figure 2.15.** $C_s = A_s + B_s$.

law to be $C$. We shall now show that the projection of $C$ along $s$ is the same as the sum of the projections of its components, $A$ and $B$, along $s$. That is,

$$C_s = A_s + B_s$$

On the diagram, then, the following relation must be verified:

$$ac = ad + ab \tag{a}$$

But

$$ac = ab + bc \tag{b}$$

Also, it is clear that

$$ad = bc \tag{c}$$

By substituting from Eqs. (b) and (c) into Eq. (a), we reduce Eq. (a) to an identity which shows that the projection of the sum of two vectors is the same as the sum of the projections of the two vectors.

## †2.5 Unit Vectors

It is sometimes convenient to express a vector $C$ as the product of its magnitude and a vector $a$ of unit magnitude and having direction corresponding to the vector $C$. The vector $a$ is called a *unit vector*. (You will write it as $\hat{a}$.) It has no dimensions. We formulate this vector as follows:

$$a \text{ (unit vector in direction } C) = \frac{C}{|C|} \tag{2.6}$$

Clearly, this development fulfills the requirements that have been set forth for this vector. We can then express the vector $C$ in the form

$$C = |C|\,a \tag{2.7}$$

The unit vector, once established, does not have, per se, an inherent line of action. This will be determined entirely by its use. In the preceding equation, the unit vector $a$ is collinear with the vector $C$. However, we can represent the vector $D$, shown in Fig. 2.16 parallel to $C$, by using the unit vector $a$ as follows:

$$D = |D|\,a \tag{2.7a}$$

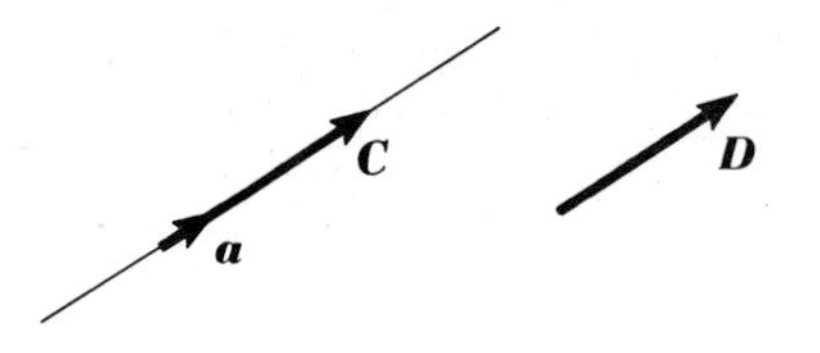


Figure 2.16. Unit vector $\boldsymbol{a}$.

In this operation, the unit vector $\boldsymbol{a}$ has been moved to have a line of action collinear with the vector $\boldsymbol{D}$. Occasionally, it is useful to label a unit vector that has the line of action of a certain vector with the lowercase letter of the capital letter associated with the vector. Thus, in Eqs. 2.7 and 2.7(a) we might have employed in the place of $\boldsymbol{a}$ the letters $\boldsymbol{c}$ and $\boldsymbol{d}$, (in your case $\hat{c}$ and $\hat{d}$) respectively. Next, if a given vector is represented using a lowercase letter, such as the vector $\boldsymbol{r}$, then we often make use of the circumflex mark to indicate the associated unit vector. Thus,

$$\boldsymbol{r} = |\boldsymbol{r}|\hat{\boldsymbol{r}} \tag{2.7b}$$

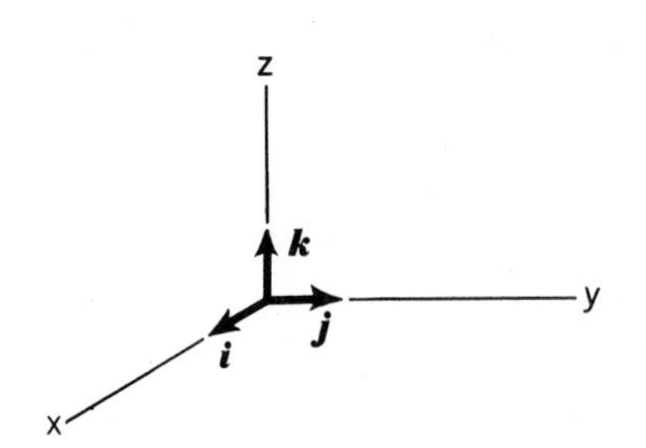


Figure 2.17. Unit vectors for *xyz* axes.

Unit vectors that are of particular use are those directed along the coordinate axes of a rectangular reference, where $\boldsymbol{i}$, $\boldsymbol{j}$, and $\boldsymbol{k}$ (your instructor will probably use the notation $\hat{i}$, $\hat{j}$, and $\hat{k}$) correspond to the $x$, $y$, and $z$ directions, as shown in Fig. 2.17.[7]

Since the sum of a set of concurrent vectors is equivalent in all situations to the original vector, we can always replace the vector $\boldsymbol{C}$ by its rectangular scalar components in the following manner:

$$\boldsymbol{C} = C_x\boldsymbol{i} + C_y\boldsymbol{j} + C_z\boldsymbol{k} \tag{2.8}$$

In Chapter 1, we saw that vectors that are equal have the same magnitude and direction. Hence, if $\boldsymbol{A} = \boldsymbol{B}$, we can say that

$$A_x\boldsymbol{i} + A_y\boldsymbol{j} + A_z\boldsymbol{k} = B_x\boldsymbol{i} + B_y\boldsymbol{j} + B_z\boldsymbol{k} \tag{2.9}$$

Then, since the unit vectors have mutually different directions, we conclude that

$$A_x\boldsymbol{i} = B_x\boldsymbol{i}$$
$$A_y\boldsymbol{j} = B_y\boldsymbol{j}$$
$$A_z\boldsymbol{k} = B_z\boldsymbol{k}$$

It then follows that

$$A_x = B_x$$
$$A_y = B_y$$
$$A_z = B_z$$

[7]Curvilinear coordinate systems have associated sets of unit vectors just as do the rectangular coordinate systems. As will be seen later, however, these unit vectors do not all have fixed directions in space for a given reference as do the vectors $\boldsymbol{i}, \boldsymbol{j}$, and $\boldsymbol{k}$.

Hence, the vector equation, $\boldsymbol{A} = \boldsymbol{B}$, has resulted in three scalar equations that in totality are equivalent in every way to the vector statement of equality. Thus, in Newton's law we would have

$$\boldsymbol{F} = m\boldsymbol{a} \tag{2.10a}$$

as the vector equation, and

$$F_x = ma_x, \qquad F_y = ma_y, \qquad F_z = ma_z \tag{2.10b}$$

as the corresponding scalar equations.

We now present a series of homework problems. In a few of these problems, we have inserted a rectangular parallelepiped along a vector (see Fig. P.2.24 for Prob. 2.24). The purpose of this rectangular parallelepiped clearly is to convey sufficient information concerning the *direction* of the vector. In this regard, note first in Fig. P.2.24 that the vector $\overrightarrow{DE}$ between points $D$ and $E$ can be given in terms of rectangular components as follows

$$\begin{aligned}\overrightarrow{DE} &= 4\boldsymbol{i} - 3\boldsymbol{k} + 5\boldsymbol{j} \\ &= 4\boldsymbol{i} + 5\boldsymbol{j} - 3\boldsymbol{k}\end{aligned}$$

by simply moving from $D$ to $E$ *along the coordinate directions*. Since from Eq. 2.5

$$4 = (DE)l$$

then

$$l = \frac{4}{DE} = \frac{4}{\sqrt{4^2 + 5^2 + 3^2}}$$

This value of $l$ must then be the direction cosine for the 100-lb force collinear with $DE$. We get $m$ and $n$ the same way. The 100-N force in Fig. P.2.24 can now be given vectorially as

$$\boldsymbol{F} = 100(l\boldsymbol{i} + m\boldsymbol{j} + n\boldsymbol{k}) = 100\frac{(4\boldsymbol{i} + 5\boldsymbol{j} - 3\boldsymbol{k})}{\sqrt{4^2 + 5^2 + 3^2}} = 100\frac{\overrightarrow{DE}}{|\overrightarrow{DE}|}$$

## Problems

**2.14.** Resolve the 100-lb force into a set of components along the slot shown and in the vertical direction.

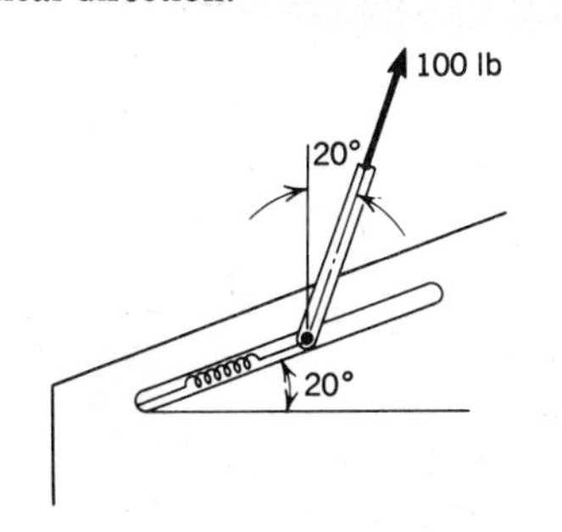


**Figure P.2.14**

**2.15.** A farmer needs to build a fence from the corner of his barn to the corner of his chicken house 30 m NE away. However, he wants to enclose as much of the barnyard as possible. Thus, he runs the fence east, from the corner of his barn to the property line and then NNE to the corner of his chicken house. How long is the fence?

**2.16.** Resolve the force $F$ into a component perpendicular to $AB$ and a component parallel to $BC$.

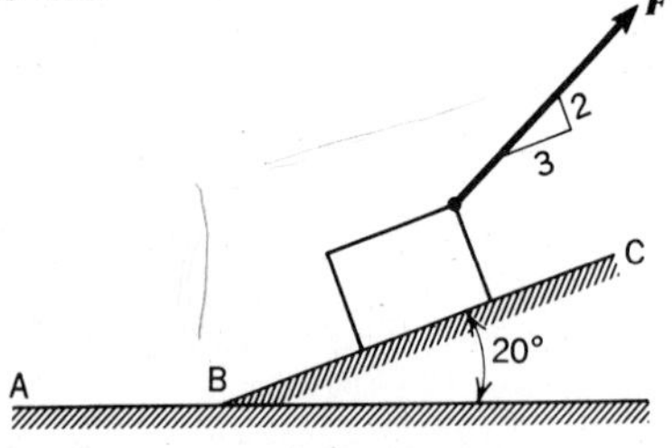


Figure P.2.16

**2.17.** A 1000-N force is resolved into components along $AB$ and $AC$. If the component along $AB$ is 700 N, determine the angle $\alpha$ and the value of the component along $AC$.

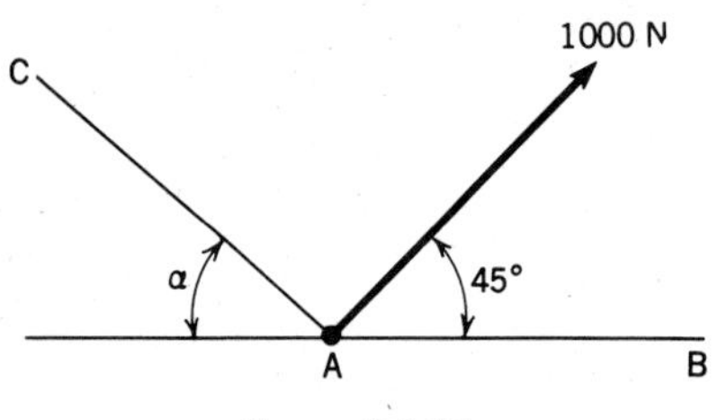


Figure P.2.17

**2.18.** Two men are trying to pull a crate which will not move until a 150-lb total force is applied in any one direction. Man $A$ can pull only at 45° to the desired direction of crate motion, whereas man $B$ can pull only at 60° to the desired motion. What force must each man exert to start the box moving as shown?

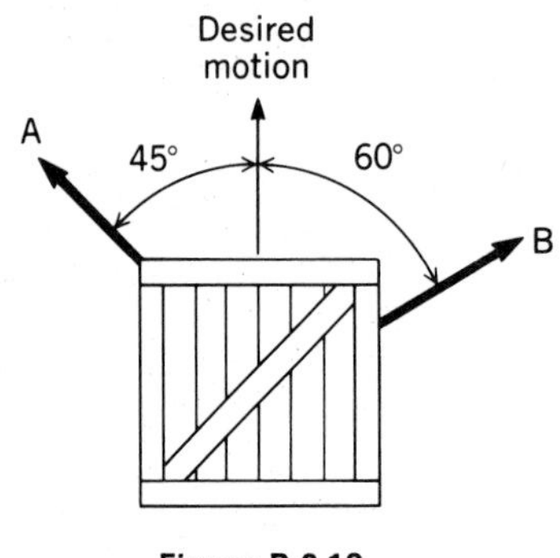


Figure P.2.18

**2.19.** The 500-N force is to be resolved into components along the $AB$ and $AC$ directions measured by the angles $\alpha$ and $\beta$. If the component along $AC$ is to be 1000 N and the component along $AB$ is to be 800 N, compute $\alpha$ and $\beta$.

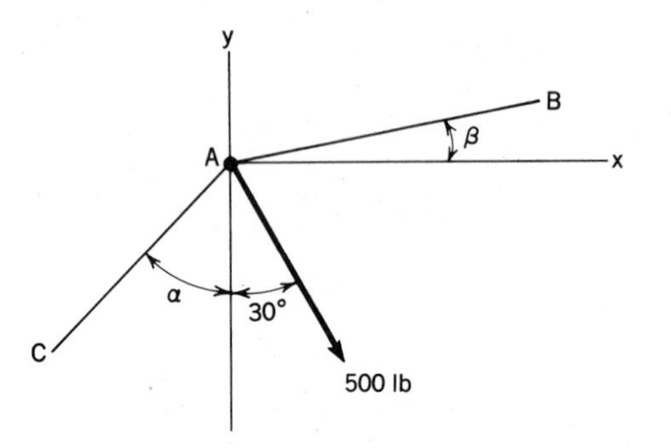


Figure P.2.19

**2.20.** The orthogonal components of a force are:
$x$ component 10 lb in positive $x$ direction
$y$ component 20 lb in positive $y$ direction
$z$ component 30 lb in negative $z$ direction
(a) What is the magnitude of the force itself?
(b) What are the direction cosines of the force?

**2.21.** What are the rectangular components of the 100-lb force? What are the direction cosines for this force?

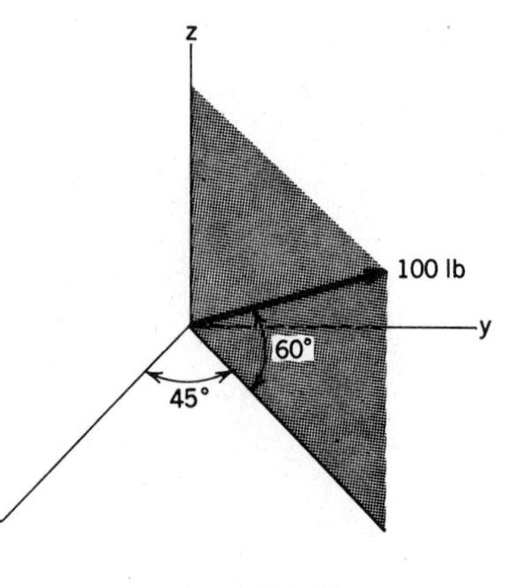


Figure P.2.21

**2.22.** A 50-m-long diagonal member $OE$ in a space frame is inclined at $\alpha = 70°$ and $\beta = 30°$ to the $x$ and $y$ axes, respectively. What is $\gamma$? How long must members $OA$, $AC$, $OB$, $BC$, and $CE$ be to support end $E$ of $OE$?

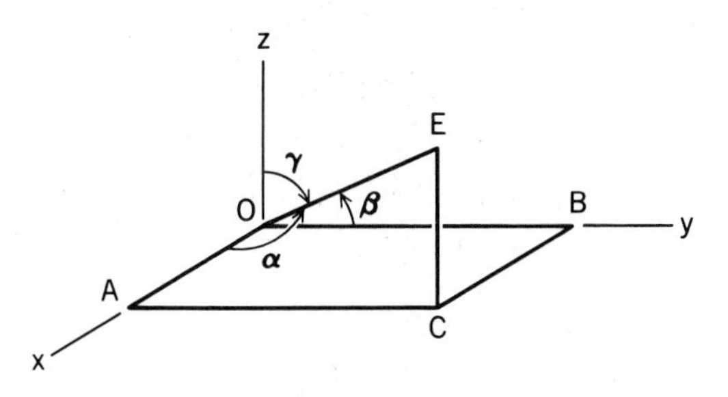


Figure P.2.22

**2.23.** What is the orthogonal total force component in the $x$ direction of the force transmitted to pin $A$ of a roof truss by the four members? What is the total component in the $y$ direction?

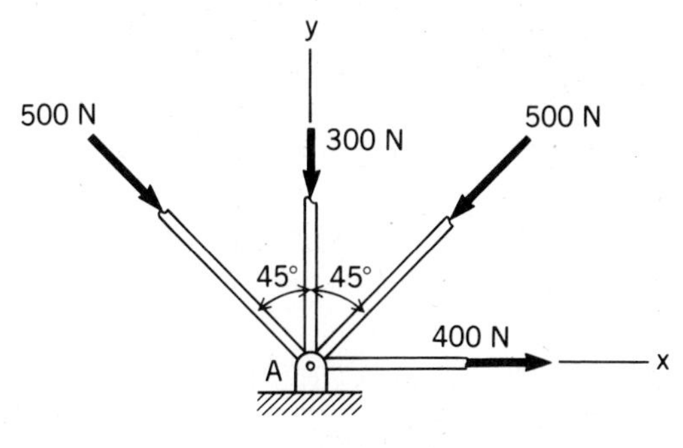


Figure P.2.23

**2.24.** A man pulls with a force of 100 N on a rope attached to a ring. The ring is supported by three linkages, $A$, $B$, and $C$, that can take forces only in the $x, y$, and $z$ directions, respectively. What are the forces in the direction of the linkages stemming from the 100-N force?

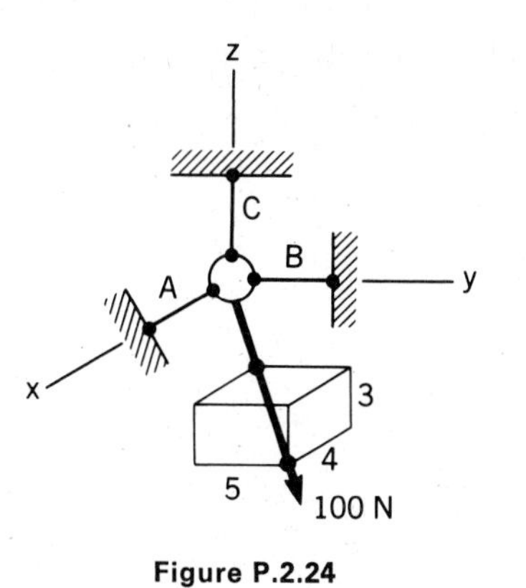


Figure P.2.24

**2.25.** Given the following force expressed as a function of position:

$$\boldsymbol{F} = (10x - 6)\boldsymbol{i} + x^2z\boldsymbol{j} + xy\boldsymbol{k}$$

What are the direction cosines of the force at position (1, 2, 2)? What is the position along the $x$ coordinate where $F_x = 0$? Plot $F_y$ versus the $x$ coordinate for an elevation $z = 1$.

**2.26.** What is the sum of the following set of three vectors?

$\boldsymbol{A} = 6\boldsymbol{i} + 10\boldsymbol{j} + 16\boldsymbol{k}$ lb

$\boldsymbol{B} = 2\boldsymbol{i} - 3\boldsymbol{j}$ lb

$\boldsymbol{C}$ is a vector in the $xy$ plane at an inclination of 45° to the positive $x$ axis and directed away from the origin; it has a magnitude of 25 lb

**2.27.** What is the unit vector for the displacement vector from point (2, 1, 9) to point (7, 4, 2)? Express a 10-m displacement vector in the same direction in terms of $\boldsymbol{i}, \boldsymbol{j}$, and $\boldsymbol{k}$.

**2.28.** A vector $\boldsymbol{A}$ has a line of action that goes through the coordinates (0, 2, 3) and (−1, 2, 4). If the magnitude of this vector is 10 units, express the vector in terms of the unit vectors $\boldsymbol{i}, \boldsymbol{j}$, and $\boldsymbol{k}$.

**2.29.** Express the force $\boldsymbol{F}$ in terms of the unit vectors $\boldsymbol{i}, \boldsymbol{j}$, and $\boldsymbol{k}$.

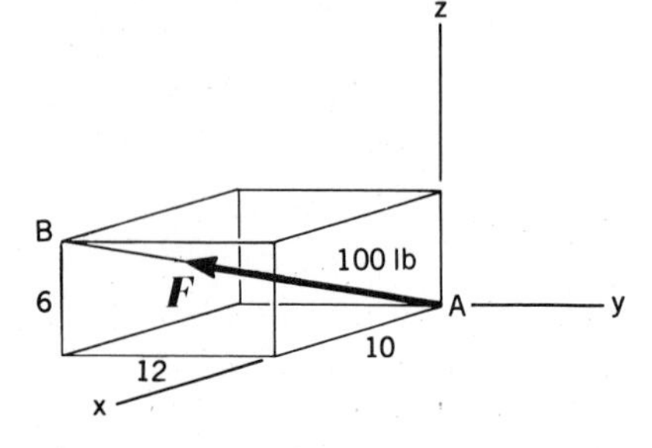


Figure P.2.29

**2.30.** Express the 100-N force in terms of the unit vectors $\boldsymbol{i}, \boldsymbol{j}$, and $\boldsymbol{k}$. What is the unit vector in the direction of the 100-N force? The force lies along diagonal $AB$.

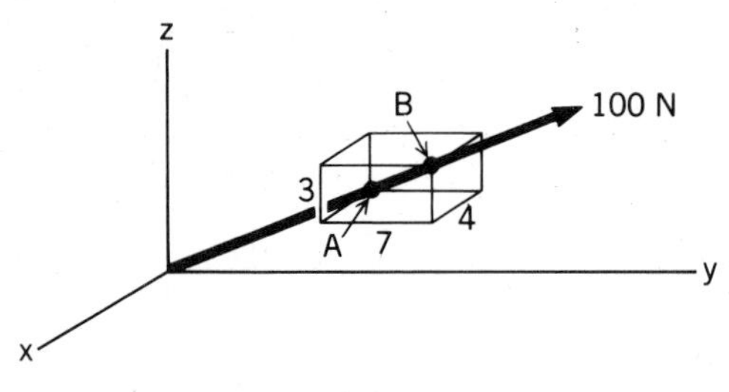

Figure P.2.30

**2.31.** Express the unit vectors $\boldsymbol{i}, \boldsymbol{j}$, and $\boldsymbol{k}$ in terms of unit vectors $\boldsymbol{\epsilon}_{\bar{r}}$, $\boldsymbol{\epsilon}_\theta$, and $\boldsymbol{\epsilon}_z$. (These are unit vectors for *cylindrical coordinates*.) Express the 1000-lb force going through the origin and through point (2, 4, 4) in terms of the unit vectors $\boldsymbol{i}, \boldsymbol{j}, \boldsymbol{k}$ and $\boldsymbol{\epsilon}_{\bar{r}}, \boldsymbol{\epsilon}_\theta, \boldsymbol{\epsilon}_z$ with $\theta = 60°$. (See the footnote on p. 34.)

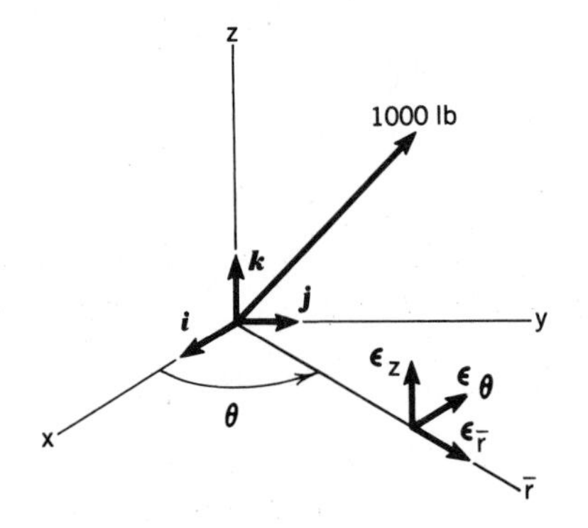


Figure P.2.31

## 2.6 Scalar or Dot Product of Two Vectors

In elementary physics, work was defined as the product of the force component, in the direction of a displacement, times the displacement. In effect, two vectors, force and displacement, are employed to give a scalar, work. In other physical problems, vectors are associated in this same manner so as to result in a scalar quantity. A vector operation that represents such operations concisely is the scalar product (or dot product), which, for the vector $\boldsymbol{A}$ and $\boldsymbol{B}$ in Fig. 2.18, is defined as[8]

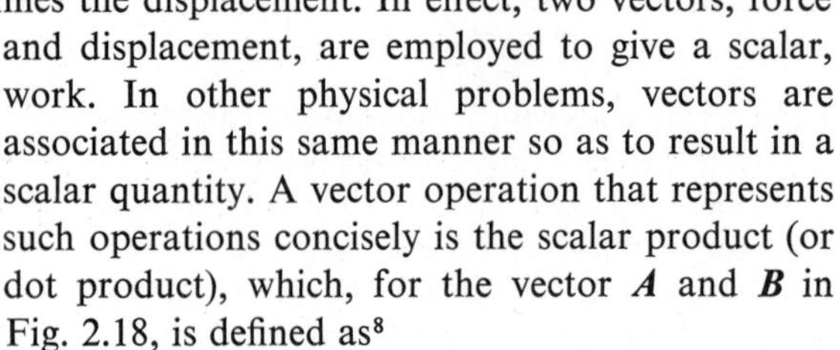

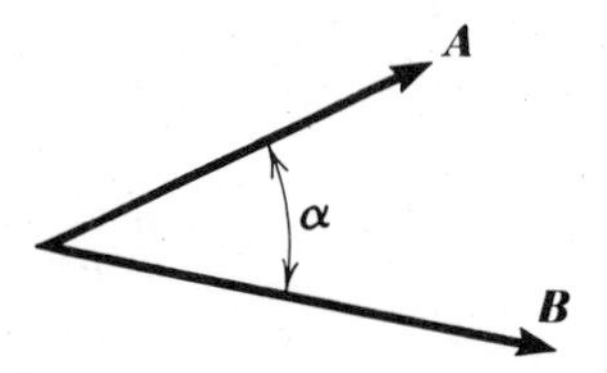


**Figure 2.18.** $\alpha$ is smallest angle between $\boldsymbol{A}$ and $\boldsymbol{B}$.

$$\boldsymbol{A} \cdot \boldsymbol{B} = |\boldsymbol{A}||\boldsymbol{B}| \cos \alpha \tag{2.11}$$

[8]To ensure that there is no confusion between the dot product of two vectors and the ordinary product of two scalars that you have used up to now, we urge you to read $\boldsymbol{A} \cdot \boldsymbol{B} = C$ as "$\boldsymbol{A}$ dotted into $\boldsymbol{B}$ yields $C$." Also, note that we can move the vectors so they intersect as in Fig. 2.18.

where $\alpha$ is the smaller angle between the two vectors. Note that the dot product may involve vectors of different dimensional representation, and may be positive or negative, depending on whether the smaller included angle $\alpha$ is less than or greater than 90°. Note also that $\boldsymbol{A} \cdot \boldsymbol{B}$ is equivalent to first projecting vector $\boldsymbol{A}$ onto the line of action of vector $\boldsymbol{B}$ (this gives us $|\boldsymbol{A}| \cos \alpha$), and than multiplying by the magnitude of vector $\boldsymbol{B}$ (or vice versa). The appropriate sign must, of course, be assigned positive if the projected component of $\boldsymbol{A}$ and the vector $\boldsymbol{B}$ point in the same direction; negative, if not.

The work concept for a force $\boldsymbol{F}$ acting on a particle moving along a path described by $s$ can now be given as

$$W = \int \boldsymbol{F} \cdot d\boldsymbol{s} \tag{2.12}$$

where $d\boldsymbol{s}$ is a displacement on the path along which the particle is moved.

Let us next consider the scalar product of $m\boldsymbol{A}$ and $n\boldsymbol{B}$. If we carry it out according to our definitions:

$$\begin{aligned}(m\boldsymbol{A}) \cdot (n\boldsymbol{B}) &= |m\boldsymbol{A}||n\boldsymbol{B}| \cos (m\boldsymbol{A}, n\boldsymbol{B}) \\ &= (mn)|\boldsymbol{A}||\boldsymbol{B}| \cos (\boldsymbol{A}, \boldsymbol{B}) = (mn)(\boldsymbol{A} \cdot \boldsymbol{B})\end{aligned} \tag{2.13}$$

Hence, the scalar coefficients in the dot product of two vectors multiply in the ordinary way, while only the vectors themselves undergo the vectorial operation as we have defined it.

From the definition, clearly the dot product is *commutative*, since the number $|\boldsymbol{A}||\boldsymbol{B}| \cos (\boldsymbol{A}, \boldsymbol{B})$ is independent of the order of multiplication of its terms. Thus,

$$\boldsymbol{A} \cdot \boldsymbol{B} = \boldsymbol{B} \cdot \boldsymbol{A} \tag{2.14}$$

Let us now consider $\boldsymbol{A} \cdot (\boldsymbol{B} + \boldsymbol{C})$. By definition, we may project the vector $(\boldsymbol{B} + \boldsymbol{C})$ onto the line of action of $\boldsymbol{A}$ and then, assigning the appropriate sign, multiply the magnitude of $\boldsymbol{A}$ times the projection of $\boldsymbol{B} + \boldsymbol{C}$. However, we have shown that the projection of the sum of two vectors is the same as the sum of the projections of the vectors, which means that

$$\boldsymbol{A} \cdot (\boldsymbol{B} + \boldsymbol{C}) = \boldsymbol{A} \cdot \boldsymbol{B} + \boldsymbol{A} \cdot \boldsymbol{C} \tag{2.15}$$

An operation on a sum of quantities that is the same as the sum of the operations on the quantities is called a *distributive operation*. Thus, the dot product is distributive.

The scalar product between unit vectors will now be carried out. The product $\boldsymbol{i} \cdot \boldsymbol{j}$ is 0, since the angle $\alpha$ in Eq. 2.11 is 90°, which makes $\cos \alpha = 0$. On the other hand, $\boldsymbol{i} \cdot \boldsymbol{i} = 1$. We can thus conclude that the dot product of equal orthogonal unit vectors for a given reference is unity and that of unequal orthogonal unit vectors is zero.

If we express the vectors $\boldsymbol{A}$ and $\boldsymbol{B}$ in Cartesian components when taking the dot product, we get

$$\begin{aligned}\boldsymbol{A} \cdot \boldsymbol{B} &= (A_x\boldsymbol{i} + A_y\boldsymbol{j} + A_z\boldsymbol{k}) \cdot (B_x\boldsymbol{i} + B_y\boldsymbol{j} + B_z\boldsymbol{k}) \\ &= A_xB_x + A_yB_y + A_zB_z\end{aligned} \tag{2.16}$$

Thus, we see that a scalar product of two vectors is the sum of the ordinary products of the respective components.

If a vector is multiplied by itself as a dot product, the result is the square of the magnitude of the vector. That is,

$$\boldsymbol{A} \cdot \boldsymbol{A} = |\boldsymbol{A}||\boldsymbol{A}| = A^2 \tag{2.17}$$

Conversely, the square of a number may be considered to be the dot product of two equal vectors having a magnitude equal to the number. Note also that

$$\boldsymbol{A} \cdot \boldsymbol{A} = A_x^2 + A_y^2 + A_z^2 \tag{2.18}$$

We can conclude from Eqs. 2.17 and 2.18 that

$$A = \sqrt{A_x^2 + A_y^2 + A_z^2}$$

which checks with the Pythagorean theorem.

The dot product may be of immediate use in expressing the scalar rectangular component of a vector along a given direction as discussed in Section 2.4. If you refer back to Fig. 2.14, you will recall that the component of $\boldsymbol{C}$ along the direction $\boldsymbol{s}$ is given as

$$C_s = |\boldsymbol{C}| \cos \delta$$

Now let us consider a unit vector $\boldsymbol{s}$ along the direction of the line $s$. If we carry out the dot product of $\boldsymbol{C}$ and $\boldsymbol{s}$ according to our fundamental definition, the result is

$$\boldsymbol{C} \cdot \boldsymbol{s} = |\boldsymbol{C}||\boldsymbol{s}| \cos \delta$$

But since $|\boldsymbol{s}|$ is unity, when we compare the preceding two equations, it is apparent that

$$C_s = \boldsymbol{C} \cdot \boldsymbol{s}$$

Similarly, the following useful relations are valid:

$$C_x = \boldsymbol{C} \cdot \boldsymbol{i}, \qquad C_y = \boldsymbol{C} \cdot \boldsymbol{j}, \qquad C_z = \boldsymbol{C} \cdot \boldsymbol{k}$$

Finally, express the unit vector $\hat{\boldsymbol{r}}$ directed out from the origin (see Fig. 2.19) in terms of the orthogonal scalar components:

$$\hat{\boldsymbol{r}} = (\hat{\boldsymbol{r}} \cdot \boldsymbol{i})\boldsymbol{i} + (\hat{\boldsymbol{r}} \cdot \boldsymbol{j})\boldsymbol{j} + (\hat{\boldsymbol{r}} \cdot \boldsymbol{k})\boldsymbol{k}$$

But

$$\hat{\boldsymbol{r}} \cdot \boldsymbol{i} = |\hat{\boldsymbol{r}}||\boldsymbol{i}| \cos(\hat{\boldsymbol{r}}, x) = l$$

Similarly, $\hat{\boldsymbol{r}} \cdot \boldsymbol{j} = m$ and $\hat{\boldsymbol{r}} \cdot \boldsymbol{k} = n$. Hence, we can say that

$$\hat{\boldsymbol{r}} = l\boldsymbol{i} + m\boldsymbol{j} + n\boldsymbol{k} \tag{2.19}$$

Thus, *the orthogonal scalar components of a unit vector are the direction cosines of the direction of the unit vector.*

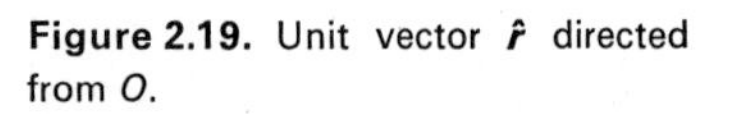


**Figure 2.19.** Unit vector $\hat{\boldsymbol{r}}$ directed from $O$.

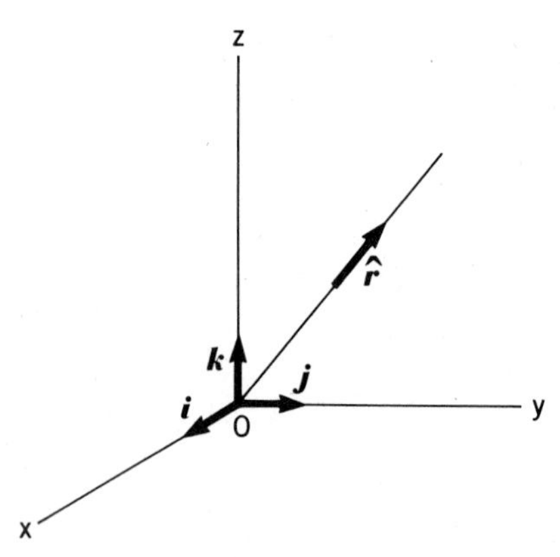


But

Now, computing the square of the magnitude of $\hat{\mathbf{r}}$, we have

$$|\hat{\mathbf{r}}|^2 = 1 = l^2 + m^2 + n^2 \tag{2.20}$$

We thus arrive at the familiar geometrical relation that the sum of the squares of the direction cosines of a vector is unity.

**EXAMPLE 2.4**

Cables $GA$ and $GB$ (see Fig. 2.20) are part of a guy-wire system supporting two radio transmission towers. What are the lengths of $GA$ and $GB$ and the angle $\alpha$ between them?

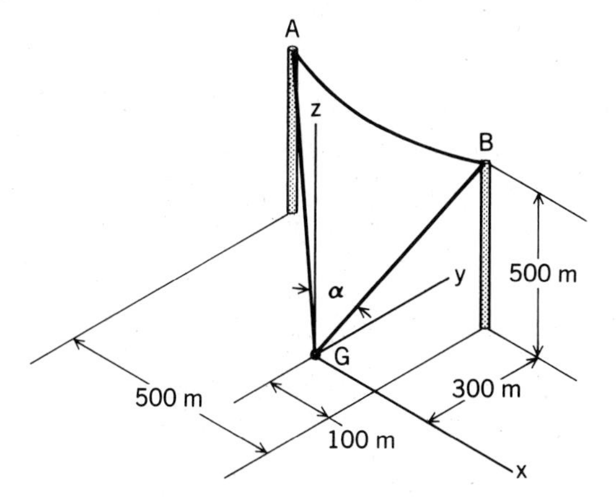


**Figure 2.20.** Radio transmission towers.

We may directly set up the vectors $\overrightarrow{GA}$ and $\overrightarrow{GB}$ by inspecting the diagram. Thus, it is easy to see that

$$\overrightarrow{GA} = 300\mathbf{j} - 400\mathbf{i} + 500\mathbf{k} \text{ m}$$
$$\overrightarrow{GB} = 300\mathbf{j} + 100\mathbf{i} + 500\mathbf{k} \text{ m}$$

Using the Pythagorean theorem, we can say for the lengths of $\overrightarrow{GA}$ and $\overrightarrow{GB}$:

$$GA = (300^2 + 400^2 + 500^2)^{1/2} = 707 \text{ m}$$
$$GB = (300^2 + 100^2 + 500^2)^{1/2} = 592 \text{ m}$$

Now we use the dot product definition to find the angle.

$$\overrightarrow{GA} \cdot \overrightarrow{GB} = (GA)(GB) \cos \alpha$$

Therefore,

$$\cos \alpha = \frac{\overrightarrow{GA} \cdot \overrightarrow{GB}}{(GA)(GB)} = \frac{90{,}000 - 40{,}000 + 250{,}000}{(707)(592)}$$
$$= .717$$

Hence,

$$\alpha = 44.18°$$

## Problems

**2.32.** Given the vectors

$$A = 10i + 20j + 3k$$
$$B = -10j + 12k$$

what is $A \cdot B$? What is $\cos(A, B)$? What is the projection of $A$ along $B$?

**2.33.** Given the vectors

$$A = 16i + 3j, \quad B = 10k - 6i, \quad C = 4j$$

compute
(a) $C(A \cdot C) + B$
(b) $-C + [B \cdot (-A)]C$

**2.34.** Given the vectors

$$A = 6i + 3j + 10k$$
$$B = 2i - 5j + 5k$$
$$C = 5i - 2j + 7k$$

what vector $D$ gives the following results?

$$D \cdot A = 20$$
$$D \cdot B = 5$$
$$D \cdot i = 10$$

**2.35.** Show that

$$\cos(A, B) = ll' + mm' + nn'$$

where $l, m, n$ and $l', m', n'$ are direction cosines of $A$ and $B$, respectively, with respect to the given $xyz$ reference.

**2.36.** Explain why the following operations are meaningless:
(a) $(A \cdot B) \cdot C$
(b) $(A \cdot B) + C$

**2.37.** A block $A$ is constrained to move along a 20° incline in the $yz$ plane. How far does the block have to move if the force $F$ is to do 10 ft-lb of work?

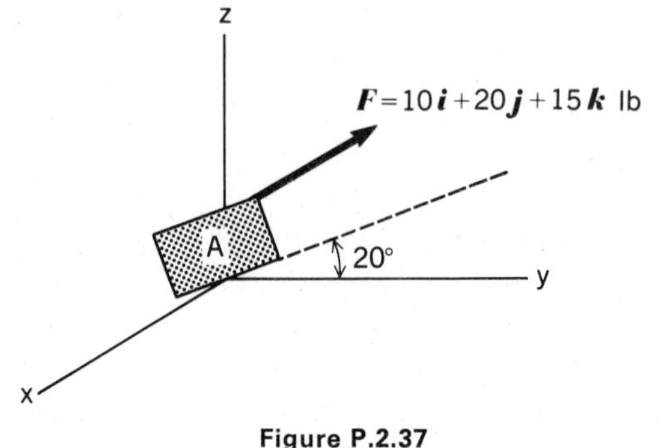


Figure P.2.37

**2.38.** An electrostatic field $E$ exerts a force on a charged particle of $qE$, where $q$ is the charge of the particle. If we have for $E$:

$$E = 6i + 3j + 2k \text{ dynes/coulomb}$$

what work is done by the field on a particle with a unit charge moving along a straight line from the origin to position $x = 20$ mm, $y = 40$ mm, $z = -40$ mm?

**2.39.** A force vector of magnitude 100 N has a line of action with direction cosines $l = .7$, $m = .2$, $n = .59$ relative to a reference $xyz$. The vector points away from the origin. What is the component of the force vector along a direction $a$ having direction cosines $l = -.3$, $m = .1$, and $n = .95$ for the $xyz$ reference? (*Hint:* Whenever simply a component is asked for, it is virtually always the *rectangular* component that is desired.)

**2.40.** Given a force $F = 10i + 5j + Ak$ N. If this force is to have a rectangular component of 8 N along a line having a unit vector $\hat{r} = .6i + .8k$, what should $A$ be? What is the angle between $F$ and $\hat{r}$?

**2.41.** Given a force $Ai + Bj + 20k$ N, what must $A$ and $B$ be to give a rectangular component of 10 N in the direction

$$\hat{r}_1 = .3i + .6j + .742k$$

as well as a component of 18 N in the direction

$$\hat{r}_2 = .4i + .9j + .1732k?$$

**2.42.** Find the dot product of the vectors represented by the diagonals from $A$ to $F$ and from $D$ to $G$. What is the angle between them?

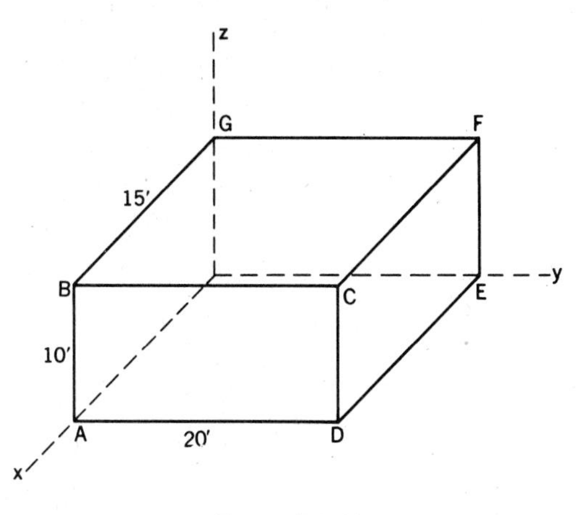


Figure P.2.42

**2.43.** What is the rectangular component of the 500-N force along the diagonal from $B$ to $A$?

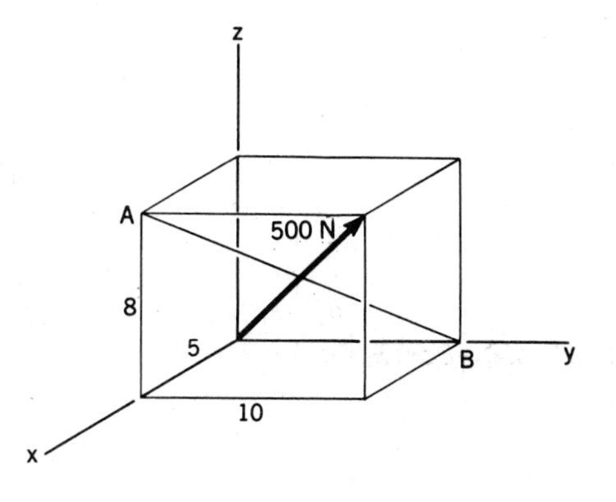


Figure P.2.43

**2.44.** A radio tower is held by guy wires. If $AB$ were to be moved to intersect $CD$ while remaining parallel to its original position, what is the angle between $AB$ and $CD$?

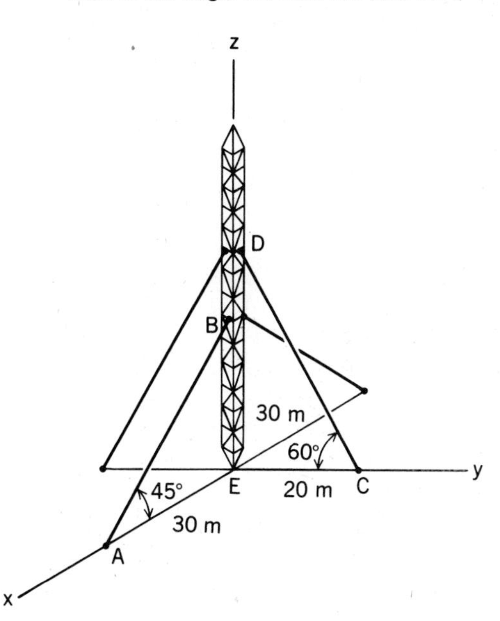


Figure P.2.44

## 2.7 Cross Product of Two Vectors

There are interactions between the vector quantities that result in vector quantities. One such interaction is the moment of a force, which involves a special product of the force and a position vector (to be studied in Chapter 3). To set up a convenient operation for these situations, the *vector cross product* has been established. For the

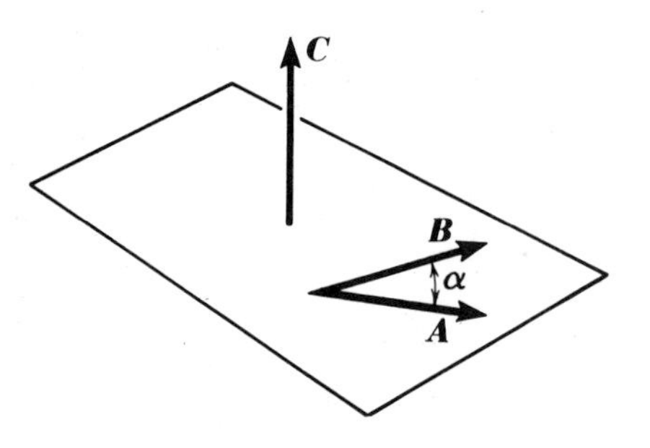


Figure 2.21. $A \times B = C$.

two vectors[9] (having possibly different dimensions) shown in Fig. 2.21 as $A$ and $B$, the operation[10] is defined as

$$A \times B = C \tag{2.21}$$

where $C$ has a magnitude that is given as

$$|C| = |A||B| \sin \alpha \tag{2.22}$$

The angle $\alpha$ is the smaller of the two angles between the vectors, thus making $\sin \alpha$ always positive. The vector $C$ has an orientation normal to the plane of the vectors $A$ and $B$. The sense, furthermore, corresponds to the advance of a right-hand screw rotated about $C$ as an axis while turning from $A$ to $B$ through $\alpha$—that is, from the first stated vector to the second stated vector through the smaller angle between them. In Fig. 2.21, the screw would advance upward in rotating from $A$ to $B$, whether the procedure is viewed from above or below the plane formed by $A$ and $B$. The reader can easily verify this for himself. The description of vector $C$ is now complete, since the magnitude and direction are fully established. The line of action of $C$ is not determined by the cross product; it depends on the use of the vector $C$.

As in the previous case, the coefficients of the vectors will multiply as ordinary scalars. This may be deduced from the nature of the definition. However, the *commutative* law breaks down for this product. We can verify, by carefully considering the definition of the cross product, that

$$(A \times B) = -(B \times A) \tag{2.23}$$

We can readily show that the cross product, like the dot product, is a distributive operation. To do this, consider in Fig. 2.22 a prism *mnopqr* with edges coinciding with the vectors $A$, $B$, $C$, and $(A + B)$. We can represent the area of each face of the prism as a vector whose magnitude equals the area of the face and whose direction is normal to the face with a sense pointing out (by convention) from the body. It will be left to the student to justify the given formulation for each of the vectors in Fig. 2.23. Since the prism is a closed surface, the net projected area in any direction must be zero, and this, in turn, means that the total area vector must be zero. We then get

$$(A + B) \times C + \tfrac{1}{2}A \times B + \tfrac{1}{2}B \times A + C \times A + C \times B = 0$$

Noting that the second and third expressions cancel each other, we get, on rearranging the terms,

$$C \times (A + B) = C \times A + C \times B \tag{2.24}$$

We have thus demonstrated the *distributive* property of the cross product.

---

[9]In carrying out the cross product between any two vectors $A$ and $B$, we may move the vectors without changing their directions so that the vectors come together at a point, such as shown in Fig. 2.21.

[10]Again, we urge you to read $A \times B = C$ as "$A$ crossed into $B$ yields $C$."

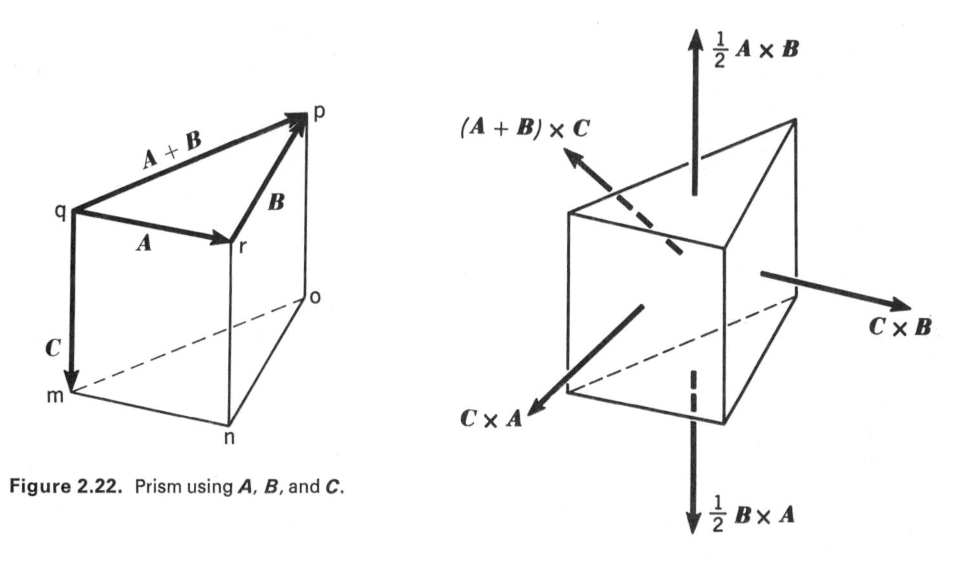


**Figure 2.22.** Prism using ***A***, ***B***, and ***C***.

**Figure 2.23.** Area vectors for prism faces.

Next, consider the cross product of rectangular unit vectors. Here, the product of equal vectors is zero because $\alpha$ and, consequently, $\sin \alpha$ are zero. The product $\boldsymbol{i} \times \boldsymbol{j}$ is unity in magnitude, and because of the right-hand-screw rule must be parallel to the $z$ axis. If the $z$ axis has been erected in a sense consistent with the right-hand-screw rule when rotating from the $x$ to the $y$ direction, the reference is called a *right-hand triad* [see Fig. 2.24(a)] and we can write

$$\boldsymbol{i} \times \boldsymbol{j} = \boldsymbol{k}$$

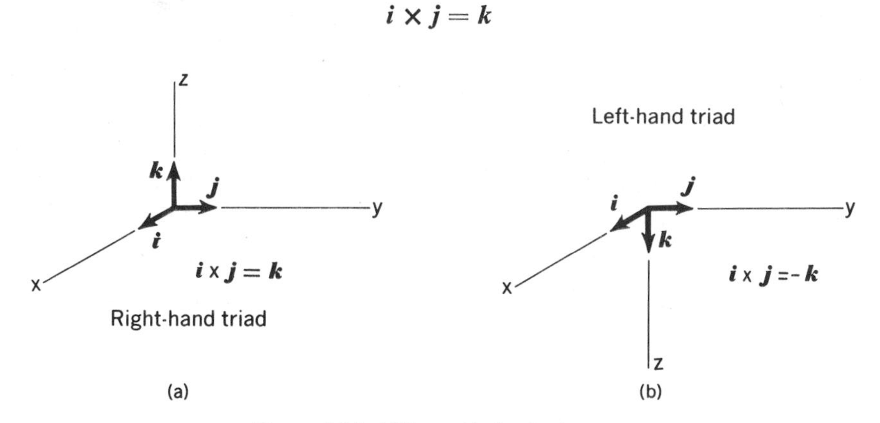

**Figure 2.24.** Different kinds of references.

If a left-hand triad is used, the result is a $-\boldsymbol{k}$ for the cross product above [see Fig. 2.24(b)]. In this text, we will use a right-hand triad as a reference. For ease in evaluation of unit cross products for such references, a simple permutation scheme is helpful.

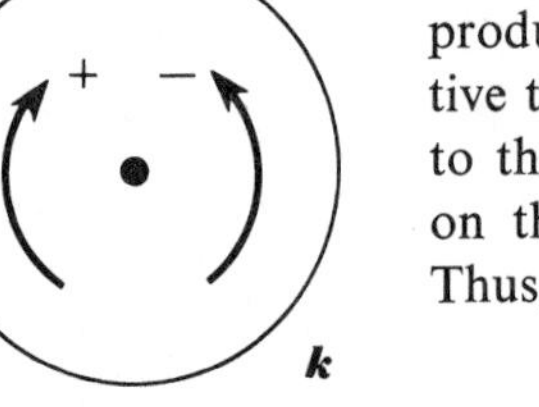


**Figure 2.25.** Permutation scheme.

In Fig. 2.25, the unit vectors $\boldsymbol{i}, \boldsymbol{j}$, and $\boldsymbol{k}$ are indicated on a circle in a clockwise sequence. Any cross product of a pair of unit vectors results in a positive third unit vector if going from the first vector to the second vector involves a clockwise motion on this circle. Otherwise, the vector is negative. Thus,

$$\boldsymbol{k} \times \boldsymbol{j} = -\boldsymbol{i}, \qquad \boldsymbol{k} \times \boldsymbol{i} = \boldsymbol{j}, \qquad \text{etc.}$$

Next, the cross product of two vectors in terms of their rectangular components is

$$\begin{aligned} \boldsymbol{A} \times \boldsymbol{B} &= (A_x\boldsymbol{i} + A_y\boldsymbol{j} + A_z\boldsymbol{k}) \times (B_x\boldsymbol{i} + B_y\boldsymbol{j} + B_z\boldsymbol{k}) \\ &= (A_yB_z - A_zB_y)\boldsymbol{i} + (A_zB_x - A_xB_z)\boldsymbol{j} + (A_xB_y - A_yB_x)\boldsymbol{k} \end{aligned} \tag{2.25}$$

Another method of carrying out this long computation is to evaluate the following determinant:

$$\begin{vmatrix} A_x & A_y & A_z \\ B_x & B_y & B_z \\ \boldsymbol{i} & \boldsymbol{j} & \boldsymbol{k} \end{vmatrix} \tag{2.26}$$

The determinant may easily be evaluated in the following manner. Repeat the first two rows below the determinant, and then form products along diagonals.

$$\begin{array}{ccc} A_x & A_y & A_z \\ B_x & B_y & B_z \\ \boldsymbol{i} & \boldsymbol{j} & \boldsymbol{k} \\ A_x & A_y & A_z \\ B_x & B_y & B_z \end{array} \tag{2.27}$$

For the products along the dashed diagonals, we must remember in this method to multiply by $-1$. We then add all six products as follows:

$$\begin{aligned} &A_xB_y\boldsymbol{k} + B_xA_z\boldsymbol{j} + A_yB_z\boldsymbol{i} - A_zB_y\boldsymbol{i} - B_zA_x\boldsymbol{j} - A_yB_x\boldsymbol{k} \\ &\qquad = (A_yB_z - A_zB_y)\boldsymbol{i} + (A_zB_x - A_xB_z)\boldsymbol{j} + (A_xB_y - A_yB_x)\boldsymbol{k} \end{aligned}$$

Clearly, this is the same result as in Eq. 2.25. It must be cautioned that this method of evaluating a determinant is correct only for $3 \times 3$ determinants. If the cross product of two vectors involves less than six nonzero components, such as in the cross product

$$(6\boldsymbol{i} + 10\boldsymbol{j}) \times (5\boldsymbol{j} - 3\boldsymbol{k})$$

then it is advisable to multiply the components directly and collect terms, as in Eq. 2.25.

## 2.8 Scalar Triple Product

A very useful quantity is the *scalar triple product*, which for a set of vectors $\boldsymbol{A}$, $\boldsymbol{B}$, and $\boldsymbol{C}$ is defined as

$$(\boldsymbol{A} \times \boldsymbol{B}) \cdot \boldsymbol{C} \tag{2.28}$$

This clearly is a scalar quantity.

A simple geometric meaning can be associated with this operation. In Fig. 2.26, we have shown $\boldsymbol{A}$, $\boldsymbol{B}$, and $\boldsymbol{C}$ as an arbitrary set of concurrent vectors. We have set up

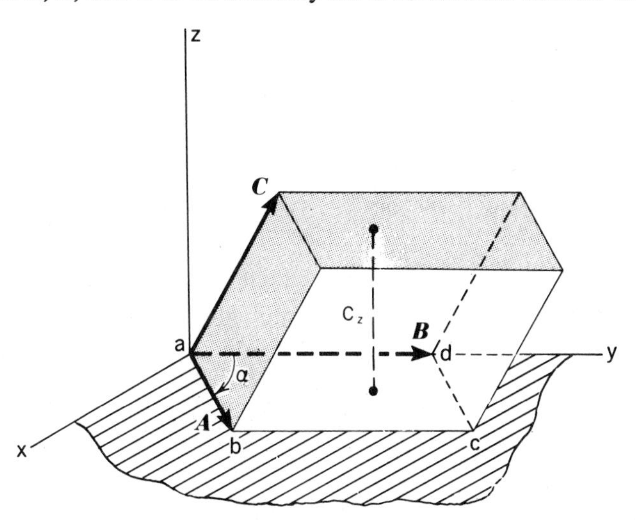


**Figure 2.26.** $\boldsymbol{A}$ and $\boldsymbol{B}$ in $xy$ plane.

an $xyz$ reference such that the $\boldsymbol{A}$ and $\boldsymbol{B}$ vectors are in the $xy$ plane. We have formed a parallelogram *abcd* in the $xy$ plane as shown in the diagram. We can say that

$$|\boldsymbol{A} \times \boldsymbol{B}| = |\boldsymbol{A}||\boldsymbol{B}| \sin \alpha = \text{area of } abcd$$

Furthermore, the direction of $\boldsymbol{A} \times \boldsymbol{B}$ is in the $z$ direction. Clearly, when we carry out Eq. 2.28, we are thus multiplying the component of $\boldsymbol{C}$ in the $z$ direction by the area of the aforementioned parallelogram. Thus, we have, for Eq. 2.28:

$$(\boldsymbol{A} \times \boldsymbol{B}) \cdot \boldsymbol{C} = (\text{area of } abcd)(C_z)$$

But $C_z$ is the *slant height* of the parallelepiped formed by vectors $\boldsymbol{A}$, $\boldsymbol{B}$, and $\boldsymbol{C}$. We then conclude from solid geometry that *the scalar triple product is the volume of the parallelepiped formed by the concurrent vectors of the scalar triple product.*

Using this geometrical interpretation of the scalar triple product, the reader can easily conclude that

$$(\boldsymbol{A} \times \boldsymbol{B}) \cdot \boldsymbol{C} = -(\boldsymbol{A} \times \boldsymbol{C}) \cdot \boldsymbol{B} = -(\boldsymbol{C} \times \boldsymbol{B}) \cdot \boldsymbol{A} \tag{2.29}$$

The computation of the scalar triple product is a very straightforward process. It will be left as an exercise (Problem 2.54) for you to demonstrate that

$$(\boldsymbol{A} \times \boldsymbol{B}) \cdot \boldsymbol{C} = \begin{vmatrix} A_x & A_y & A_z \\ B_x & B_y & B_z \\ C_x & C_y & C_z \end{vmatrix} \tag{2.30}$$

In later chapters, we shall employ the scalar triple product, although we shall not always want to associate the preceding geometric interpretation of this product.

Another operation involving three vectors is the *vector triple product* defined for vectors $\boldsymbol{A}$, $\boldsymbol{B}$, and $\boldsymbol{C}$ as $\boldsymbol{A} \times (\boldsymbol{B} \times \boldsymbol{C})$. The vector triple product is a vector quantity and will appear quite often in studies of dynamics. It will be left for you to demonstrate that

$$\boldsymbol{A} \times (\boldsymbol{B} \times \boldsymbol{C}) = \boldsymbol{B}(\boldsymbol{A} \cdot \boldsymbol{C}) - \boldsymbol{C}(\boldsymbol{A} \cdot \boldsymbol{B}) \tag{2.31}$$

Notice here that the vector triple product can be carried out by using only dot products.

**EXAMPLE 2.5**

A pyramid is shown in Fig. 2.27. If the height of the pyramid is 300 ft, find the angle between planes $ADB$ and $BDC$ (i.e., find the angle between the normals to these planes).

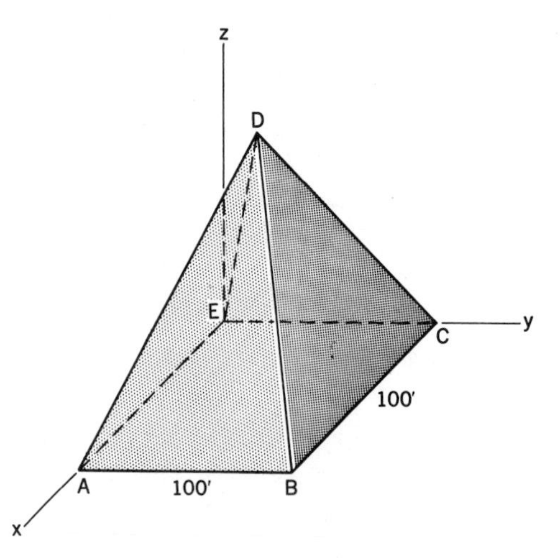


**Figure 2.27.** Pyramid.

We shall first find the unit normals to the aforestated planes. Then, using the dot product between these normals, we can easily find the desired angle.

To get the unit normal $\boldsymbol{n}_1$ to plane $ABD$, we first compute the area vector $\boldsymbol{A}_1$ for this plane. Thus, from simple trigonometry and the definition of the cross product,

$$\boldsymbol{A}_1 = \tfrac{1}{2}\overrightarrow{AB} \times \overrightarrow{AD}$$

Next, note that

$$\overrightarrow{AB} = 100\boldsymbol{j} \text{ ft}$$

$$\overrightarrow{AD} = \overrightarrow{ED} - \overrightarrow{EA} = (50\boldsymbol{i} + 50\boldsymbol{j} + 300\boldsymbol{k}) - 100\boldsymbol{i}$$

$$= -50\boldsymbol{i} + 50\boldsymbol{j} + 300\boldsymbol{k} \text{ ft}$$

Hence,

$$\boldsymbol{A}_1 = \tfrac{1}{2}(100\boldsymbol{j}) \times (-50\boldsymbol{i} + 50\boldsymbol{j} + 300\boldsymbol{k})$$

$$= 15{,}000\boldsymbol{i} + 2500\boldsymbol{k} \text{ ft}^2$$

Accordingly,

$$\boldsymbol{n}_1 = \frac{\boldsymbol{A}_1}{|\boldsymbol{A}_1|} = \frac{15{,}000\boldsymbol{i} + 2500\boldsymbol{k}}{\sqrt{15{,}000^2 + 2500^2}}$$

$$= .986\boldsymbol{i} + .1644\boldsymbol{k} \qquad \text{(a)}$$

As for unit normal $\boldsymbol{n}_2$ corresponding to plane $BDC$, whose area vector we denote as $\boldsymbol{A}_2$, we have

$$\boldsymbol{A}_2 = \tfrac{1}{2}\overrightarrow{BC} \times \overrightarrow{BD}$$

Note that

$$\overrightarrow{BC} = -100\boldsymbol{i} \text{ ft}$$

$$\overrightarrow{BD} = \overrightarrow{ED} - \overrightarrow{EB} = (50\boldsymbol{i} + 50\boldsymbol{j} + 300\boldsymbol{k}) - (100\boldsymbol{i} + 100\boldsymbol{j})$$

$$= -50\boldsymbol{i} - 50\boldsymbol{j} + 300\boldsymbol{k} \text{ ft}$$

Hence,

$$\boldsymbol{A}_2 = \tfrac{1}{2}(-100\boldsymbol{i}) \times (-50\boldsymbol{i} - 50\boldsymbol{j} + 300\boldsymbol{k})$$

$$= 15{,}000\boldsymbol{j} + 2500\boldsymbol{k} \text{ ft}^2$$

Accordingly,

$$\boldsymbol{n}_2 = \frac{\boldsymbol{A}_2}{|\boldsymbol{A}_2|} = \frac{15{,}000\boldsymbol{j} + 2500\boldsymbol{k}}{\sqrt{15{,}000^2 + 2500^2}}$$

$$= .986\boldsymbol{j} + .1644\boldsymbol{k} \qquad \text{(b)}$$

Now, we use the dot product of $\boldsymbol{n}_1$ and $\boldsymbol{n}_2$. Thus,

$$\boldsymbol{n}_1 \cdot \boldsymbol{n}_2 = \cos\beta \qquad \text{(c)}$$

where $\beta$ is the angle between the normals to the planes. Substituting from Eqs. (a) and (b) into (c), we get

$$\cos\beta = .0270$$

Therefore,

$$\beta = 88.5^\circ$$

**EXAMPLE 2.6**

In Example 2.5, what is the area projected by plane $ADE$ onto an infinite plane that is inclined equally to the $x$, $y$, and $z$ axes?

The normal $\boldsymbol{n}$ to the infinite plane must have three equal direction cosines. Hence, noting Eq. 2.20 for the sum of the squares of a set of direction cosines, we can say that

$$l^2 = m^2 = n^2 = \frac{1}{3}$$

Therefore,

$$l = m = n = \frac{1}{\sqrt{3}}$$

Hence,

$$\boldsymbol{n} = \frac{1}{\sqrt{3}}\boldsymbol{i} + \frac{1}{\sqrt{3}}\boldsymbol{j} + \frac{1}{\sqrt{3}}\boldsymbol{k}$$

The projected area then is given as

$$A_n = \left(\frac{1}{2}\overrightarrow{AD} \times \overrightarrow{AE}\right) \cdot \boldsymbol{n}$$

$$= \left[\frac{1}{2}(-50\boldsymbol{i} + 50\boldsymbol{j} + 300\boldsymbol{k}) \times (-100\boldsymbol{i})\right] \cdot \frac{1}{\sqrt{3}}(\boldsymbol{i} + \boldsymbol{j} + \boldsymbol{k})$$

The preceding result is a scalar triple product which can readily be solved as follows (disregarding the final sign):

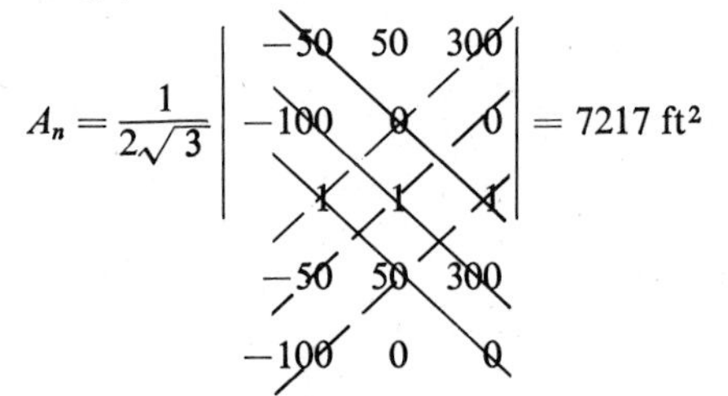

$$A_n = \frac{1}{2\sqrt{3}}\begin{vmatrix} -50 & 50 & 300 \\ -100 & 0 & 0 \\ 1 & 1 & 1 \end{vmatrix} = 7217 \text{ ft}^2$$

## 2.9 A Note on Vector Notation

When expressing *equations*, we must at all times clearly denote scalar and vector quantities and handle them accordingly. When we are simply identifying quantities in a *discussion* or in a *diagram*, however, instead of using the vector representation, $\boldsymbol{F}$, we can use just $F$. On the other hand, $F$ will be understood to represent in an equation the magnitude of the vector $\boldsymbol{F}$. Thus, using $\boldsymbol{f}$ as the unit vector in the direction of $\boldsymbol{F}$, we can say:

$$\boldsymbol{F} = F\boldsymbol{f}$$

$$= F[\cos(\boldsymbol{F}, x)\boldsymbol{i} + \cos(\boldsymbol{F}, y)\boldsymbol{j} + \cos(\boldsymbol{F}, z)\boldsymbol{k}]$$

As another example, we might want to employ the force $F$, which is shown in the coplanar diagram of Fig. 2.28(a) at a known inclination and acting at a point $a$. A correct representation of this force in a vector equation would be $F(-\cos\alpha\boldsymbol{i} + \sin\alpha\boldsymbol{j})$.

As for scalar components of any vector $\boldsymbol{F}$, we shall adopt the following understanding. The notation $F_x$, $F_y$, or $F_z$ labeling some vector component in a *diagram* will be understood to represent the *magnitude* of that particular component. Thus, in Fig. 2.28(b) the two components shown are equal in magnitude but opposite in sense. Nevertheless, they are both labeled $F_x$. However, in an equation involving these quantities, the sense must properly be accounted for by the appropriate use of signs.

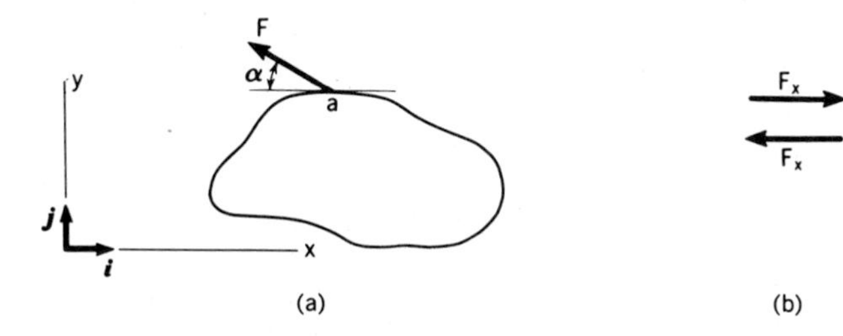


**Figure 2.28.** Notation in diagrams.

## Problems

**2.45.** If $\boldsymbol{A} = 10\boldsymbol{i} + 6\boldsymbol{j} - 3\boldsymbol{k}$ and $\boldsymbol{B} = 6\boldsymbol{i}$, find $\boldsymbol{A} \times \boldsymbol{B}$ and $\boldsymbol{B} \times \boldsymbol{A}$. What is the magnitude of the resulting vector? What are its direction cosines relative to the *xyz* reference in which $\boldsymbol{A}$ and $\boldsymbol{B}$ are expressed?

**2.46.** What are the cross and dot products for the vectors $\boldsymbol{A}$ and $\boldsymbol{B}$ given as:

$$\boldsymbol{A} = 6\boldsymbol{i} + 3\boldsymbol{j} + 4\boldsymbol{k}$$

$$\boldsymbol{B} = 8\boldsymbol{i} - 3\boldsymbol{j} + 2\boldsymbol{k}?$$

**2.47.** If vectors $\boldsymbol{A}$ and $\boldsymbol{B}$ in the *xy* plane have a dot product of 50 units, and if the magnitudes of these vectors are 10 units and 8 units, respectively, what is $\boldsymbol{A} \times \boldsymbol{B}$?

**2.48.** (a) If $\boldsymbol{A} \cdot \boldsymbol{B} = \boldsymbol{A} \cdot \boldsymbol{B}'$, does $\boldsymbol{B}$ necessarily equal $\boldsymbol{B}'$? Explain.

(b) If $\boldsymbol{A} \times \boldsymbol{B} = \boldsymbol{A} \times \boldsymbol{B}'$, does $\boldsymbol{B}$ necessarily equal $\boldsymbol{B}'$? Explain.

**2.49.** What is the cross product of the displacement vector from $A$ to $B$ times the displacement vector from $C$ to $D$?

**2.50.** Making use of the cross product, give the unit vector $\boldsymbol{n}$ normal to the inclined surface $ABC$.

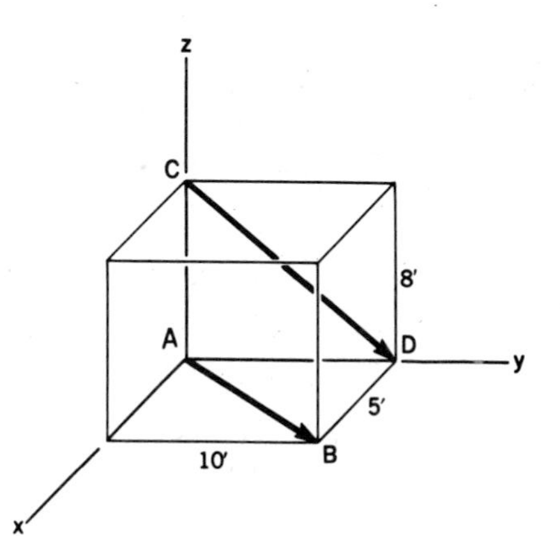

**Figure P.2.49**

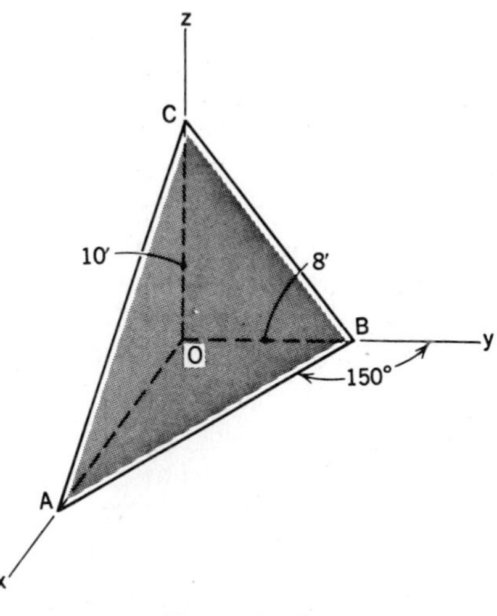


**Figure P.2.50**

**2.51.** If coordinates of vertex $E$ of the inclined pyramid are (5, 50, 80) m, what is the angle between faces $ADE$ and $BCE$?

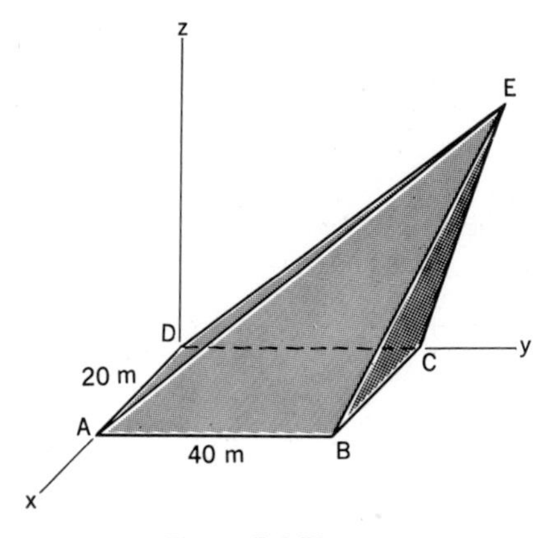


Figure P.2.51

**2.52.** In Problem 2.51, what is the area of face $ADE$ of the pyramid? What is the projection of the area of face $ADE$ onto a plane whose normal is along the direction $\boldsymbol{\epsilon}$ where:

$$\boldsymbol{\epsilon} = 0.6\boldsymbol{i} - 0.8\boldsymbol{j}$$

**2.53.** (a) Compute the product

$$(\boldsymbol{A} \times \boldsymbol{B}) \cdot \boldsymbol{C}$$

in terms of orthogonal components.

(b) Compute $(\boldsymbol{C} \times \boldsymbol{A}) \cdot \boldsymbol{B}$ and compare with the result in part (a).

**2.54.** Compute the determinant

$$\begin{vmatrix} A_x & A_y & A_z \\ B_x & B_y & B_z \\ C_x & C_y & C_z \end{vmatrix}$$

where each row represents the scalar components of $\boldsymbol{A}$, $\boldsymbol{B}$, and $\boldsymbol{C}$. Compare the result with the computation of $(\boldsymbol{A} \times \boldsymbol{B}) \cdot \boldsymbol{C}$ by using the dot-product and cross-product operations.

**2.55.** What is the component of the cross product $\boldsymbol{A} \times \boldsymbol{B}$ along the direction $\boldsymbol{n}$, where

$$\boldsymbol{A} = 10\boldsymbol{i} + 16\boldsymbol{j} + 3\boldsymbol{k}$$

$$\boldsymbol{B} = 5\boldsymbol{i} - 2\boldsymbol{j} + 2\boldsymbol{k}$$

$$\boldsymbol{n} = 0.8\boldsymbol{i} + 0.6\boldsymbol{k}$$

**2.56.** The surface $abcd$ of the parallelepiped is in the $xz$ plane. Compute the volume using vector analysis.

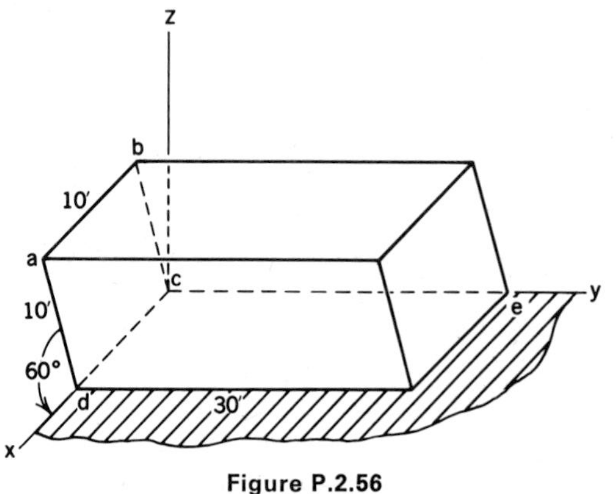


Figure P.2.56

**2.57.** Given the vectors

$$\boldsymbol{A} = 10\boldsymbol{i} + 6\boldsymbol{j}$$

$$\boldsymbol{B} = 3\boldsymbol{i} + 5\boldsymbol{j} + 10\boldsymbol{k}$$

$$\boldsymbol{C} = \boldsymbol{i} + \boldsymbol{j} - 3\boldsymbol{k}$$

find

(a) $(\boldsymbol{A} + \boldsymbol{B}) \times \boldsymbol{C}$

(b) $(\boldsymbol{A} \times \boldsymbol{B}) \cdot \boldsymbol{C}$

(c) $\boldsymbol{A} \cdot (\boldsymbol{B} \times \boldsymbol{C})$

***2.58.** A mirror system is used to relay a laser-beam signal from mountain $M$ to hill $H$. The mountain is 5000 m high and 20,000 m NW from the mirror site $S$, while the hill is 200 m high and 1500 m ENE from the mirror site $S$. Set up, but do not necessarily solve, the equations to find the direction of the mirror to properly relay the signal. Recall that the angle of reflection for a mirror equals the angle of incidence and that the incident ray, reflecting ray, and normal to the mirror are coplanar.

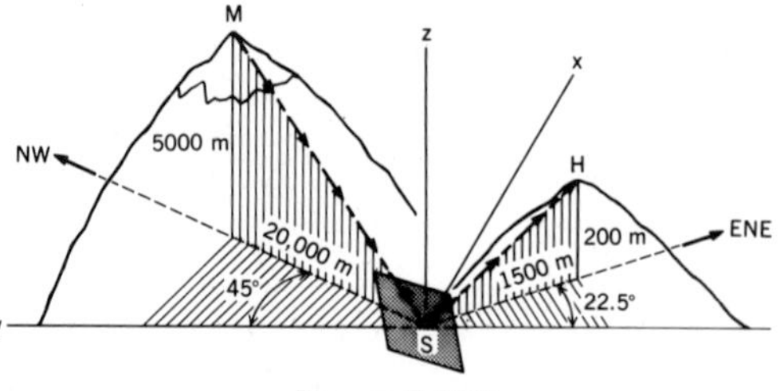


Figure P.2.58

## 2.10 Closure

In this chapter, we have presented symbols and notations that are associated with vectors. Also, various vector operations have been set forth that enable us to represent certain actions in nature mathematically. With this background, we shall now be able to study certain vector quantities that are of essential importance in mechanics. Some of these vectors will be formulated in terms of the operations contained in this chapter.

*Check-Out for Sections with †*

2.1 What is meant by the *magnitude* of a vector? What sign must it have?
2.2 Can you multiply a vector $\boldsymbol{C}$ by a scalar $s$? If so, describe the result.
2.3 What are the *law of cosines* and the *law of sines*?
2.4 What is meant by the *associative* law of addition?
2.5 Describe two ways to add any three vectors graphically.
2.6 How do you subtract vector $\boldsymbol{D}$ from vector $\boldsymbol{F}$?
2.7 Given a vector $\boldsymbol{D}$, how would you form a *unit* vector collinear with $\boldsymbol{D}$?
2.8 What are the scalar equations of the following vector equation?

$$D\boldsymbol{i} + E\boldsymbol{j} - 16\boldsymbol{k} = 20\boldsymbol{i} + (15 + G)\boldsymbol{k}$$

## Review Problems

**2.59.** A bridge truss has bar forces as shown in the cutaway sketch. What is the total force on the supporting pin at point $A$ from the members?

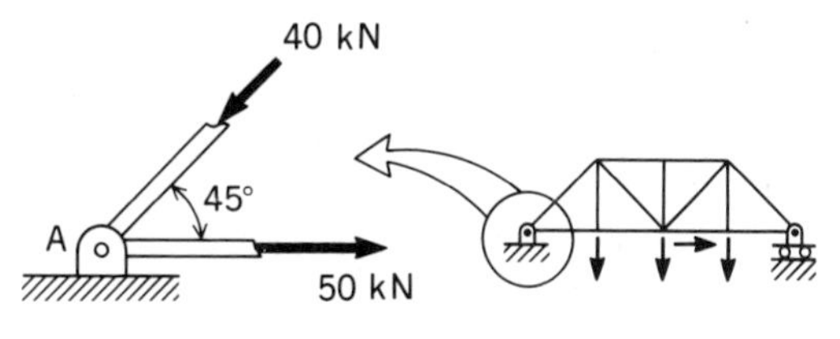


Figure P.2.59

**2.60.** Forces are transmitted by two members to pin $A$. If the sum of these forces is 700 lb directed vertically, what are the angles $\alpha$ and $\beta$?

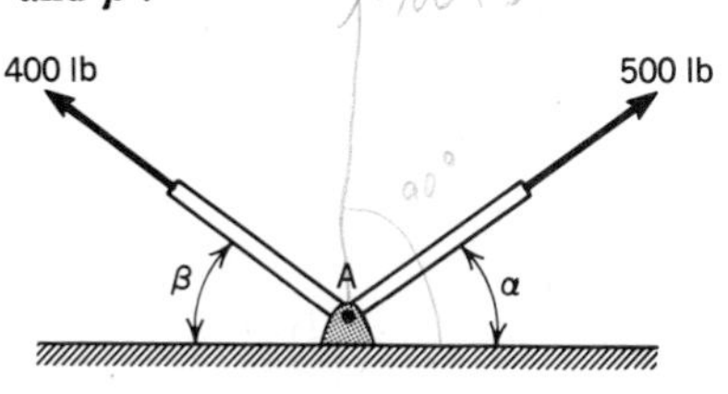


Figure P.2.60

**2.61.** Contractors encountered an impassable swamp while building a road from town $T$ to city $C$ 50 km SE. To avoid the swamp, they built the road SSW from $T$ and then ENE to $C$. How long is the road? (*Hint:* See the compass-settings diagram, Fig. 2.11.)

**2.62.** Four members of a space frame are loaded as shown. What are the orthogonal scalar components of the forces on the ball joint at $O$? The 1000-N force goes through points $D$ and $E$ of the rectangular parallelepiped.

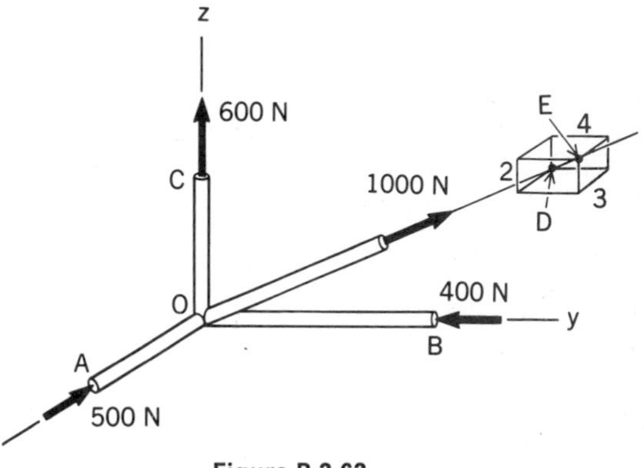


Figure P.2.62

**2.63.** The $x$ and $z$ components of the force $F$ are known to be 100 lb and $-30$ lb, respectively. What is the force $F$ and what are its direction cosines?

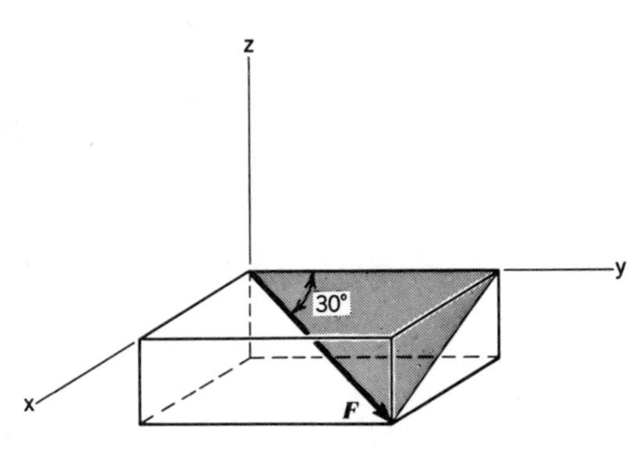


Figure P.2.63

**2.64.** A constant force given as $2\boldsymbol{i} + 3\boldsymbol{k}$ N moves a particle along a straight line from position $x = 10, y = 20, z = 0$ to position $x = 3$, $y = 0, z = -10$. If the coordinates of the $xyz$ reference are given in meter units, how much work does the force do in ft-lb?

**2.65.** The force on a charge moving through a magnetic field $\boldsymbol{B}$ is given as

$$\boldsymbol{F} = q\boldsymbol{V} \times \boldsymbol{B}$$

where $q$ = magnitude of the charge, coulombs

$\boldsymbol{F}$ = force on the body, newtons

$\boldsymbol{V}$ = velocity vector of the particle, meters per second

$\boldsymbol{B}$ = magnetic flux density, webers per meter$^2$

Suppose that an electron moves through a uniform magnetic field of $10^6$ Wb/m$^2$ in a direction inclined 30° to the field, as shown, with a speed of 100 m/sec. What are the force components on the electron? The charge of the electron is $1.6018 \times 10^{-19}$ coulomb.

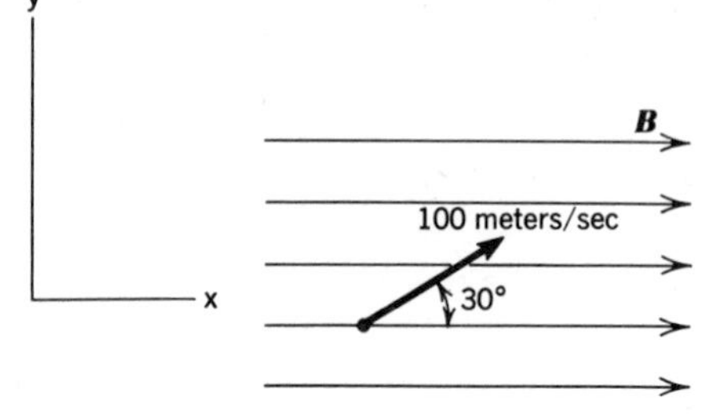


Figure P.2.65

**2.66.** The velocity of a particle of flow is given as

$$\boldsymbol{V} = 10\boldsymbol{i} + 16\boldsymbol{j} + 2\boldsymbol{k} \text{ m/sec}$$

What is the cross product $\boldsymbol{r} \times \boldsymbol{V}$, where $\boldsymbol{r}$ is given as

$$\boldsymbol{r} = 3\boldsymbol{i} + 2\boldsymbol{j} + 10\boldsymbol{k} \text{ m}$$

Give the proper units.

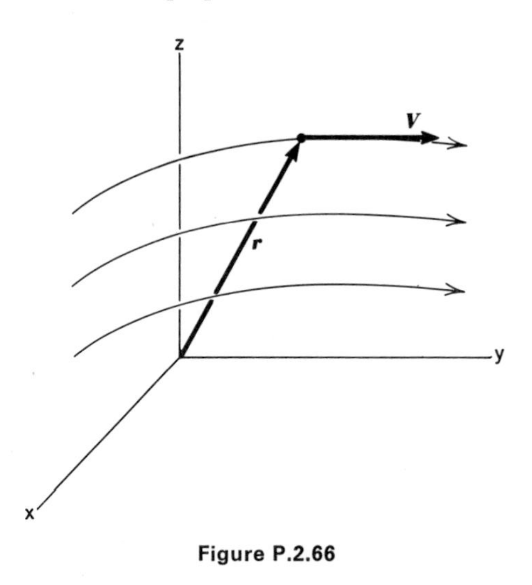


Figure P.2.66

**2.67.** Using the scalar triple product, find the area projected onto the plane $N$ from the surface $ABC$. Plane $N$ is infinite and is normal to the vector

$$\boldsymbol{r} = 50\boldsymbol{i} + 40\boldsymbol{j} + 30\boldsymbol{k} \text{ ft}$$

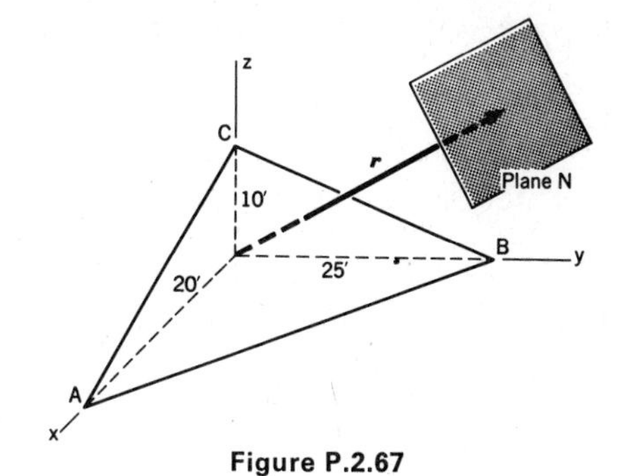


Figure P.2.67

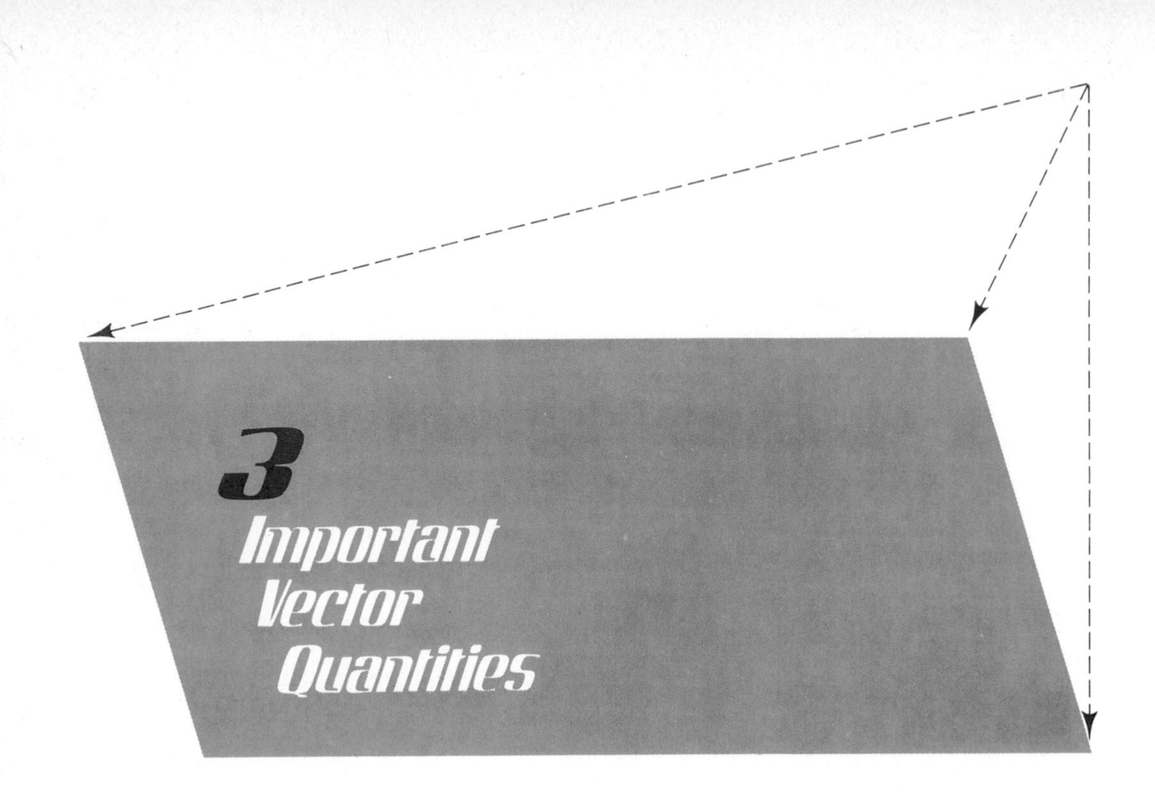


## 3.1 Position Vector

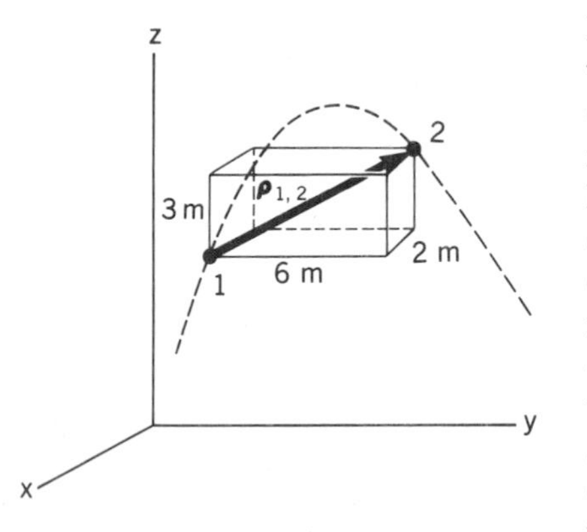


**Figure 3.1.** Displacement vector $\boldsymbol{\rho}$ between points 1 and 2.

In this chapter, we shall discuss a number of useful vector quantities. Consider first the path of motion of a particle shown dashed in Fig. 3.1. As indicated in Chapter 1, the *displacement vector* $\boldsymbol{\rho}$ is a directed line segment connecting any two points on the path of motion, such as points 1 and 2 in Fig. 3.1. The displacement vector thus represents the shortest movement of the particle to get from one position on the path of motion to another. The purpose of the rectangular parallelepiped shown in the diagram is to convey the magnitude and direction of $\boldsymbol{\rho}$. We can readily express $\boldsymbol{\rho}$ between points 1 and 2 in terms of rectangular components by noting the distance in the coordinate directions needed to go from 1 to 2. Thus, in Fig. 3.1, $\boldsymbol{\rho}_{1,2} = -2\boldsymbol{i} + 6\boldsymbol{j} + 3\boldsymbol{k}$ m.

The directed line segment $\boldsymbol{r}$ from the origin of a coordinate system to a point $P$ in space (Fig. 3.2) is called the *position vector*. The notations $\boldsymbol{R}$ and $\boldsymbol{\rho}$ are also used for position vectors. You can conclude from Chapter 2 that the magnitude of the position vector is the distance between the origin $O$ and point $P$. The scalar components of a position vector are simply the coordinates of the point $P$. To express $\boldsymbol{r}$ in Cartesian

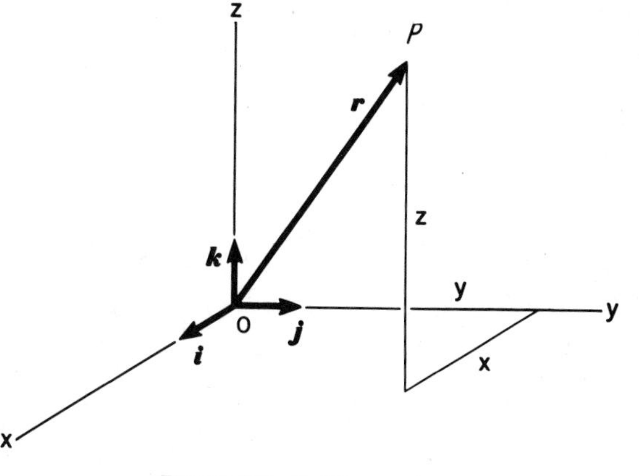


Figure 3.2. Position vector.

components, we then have

$$r = xi + yj + zk \tag{3.1}$$

We can obviously express a displacement vector $\boldsymbol{\rho}$ between points 1 and 2 (see Fig. 3.3) in terms of position vectors for points 1 and 2 (i.e., $r_1$ and $r_2$) as follows:

$$\boldsymbol{\rho} = r_2 - r_1 = (x_2 - x_1)i + (y_2 - y_1)j + (z_2 - z_1)k \tag{3.2}$$

**EXAMPLE 3.1**

Two sets of references, $xyz$ and $XYZ$, are shown in Fig. 3.4. The position vector of the origin $O$ of $xyz$ relative to $XYZ$ is given as

$$R = 10i + 6j + 5k \text{ m} \tag{a}$$

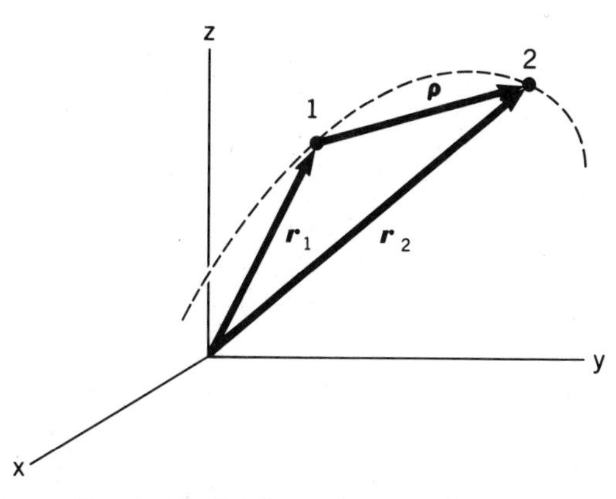


Figure 3.3. Relation between a displacement vector and position vectors.

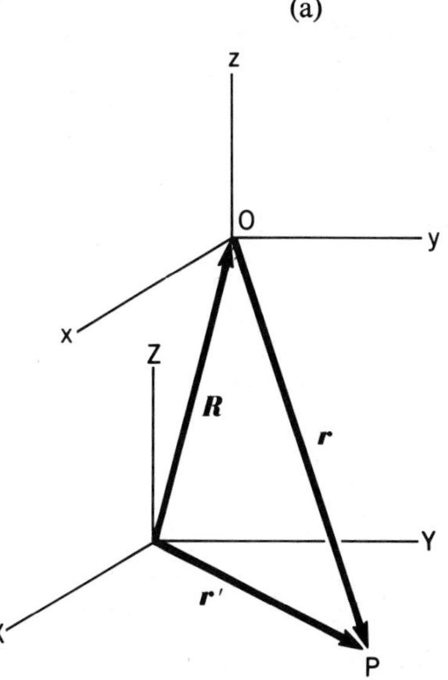


Figure 3.4. References $xyz$ and $XYZ$ separated by position vector $R$.

The position vector, $\boldsymbol{r}'$, of a point $P$ relative to $XYZ$ is

$$\boldsymbol{r}' = 3\boldsymbol{i} + 2\boldsymbol{j} - 6\boldsymbol{k} \text{ m} \tag{b}$$

What is the position vector $\boldsymbol{r}$ of point $P$ relative to $xyz$? What are the coordinates $x$, $y$, and $z$ of $P$?

From Fig. 3.4, it is clear that

$$\boldsymbol{r}' = \boldsymbol{R} + \boldsymbol{r} \tag{c}$$

Therefore,

$$\begin{aligned} \boldsymbol{r} = \boldsymbol{r}' - \boldsymbol{R} &= (3\boldsymbol{i} + 2\boldsymbol{j} - 6\boldsymbol{k}) - (10\boldsymbol{i} + 6\boldsymbol{j} + 5\boldsymbol{k}) \\ &= -7\boldsymbol{i} - 4\boldsymbol{j} - 11\boldsymbol{k} \text{ m} \end{aligned} \tag{d}$$

We can then conclude that

$$\begin{aligned} x &= -7 \text{ m} \\ y &= -4 \text{ m} \\ z &= -11 \text{ m} \end{aligned} \tag{e}$$

## 3.2 Moment of a Force About a Point

The moment of a force about a point $O$ (see Fig. 3.5), you will recall from physics, is a vector $\boldsymbol{M}$ whose magnitude equals the product of the force magnitude times the perpendicular distance $d$ from $O$ to the line of action of the force. And the direction of this vector is perpendicular to the plane of the point and the force, with a sense determined from the familiar right-hand-screw rule.[1] The line of action of $\boldsymbol{M}$ is

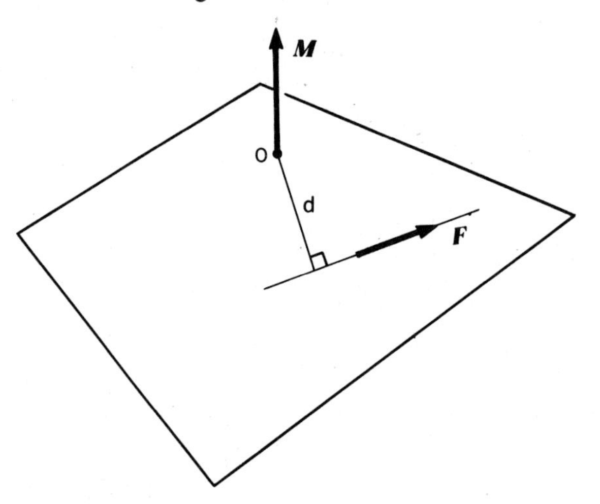


**Figure 3.5.** Moment of force $\boldsymbol{F}$ about $O$ is $Fd$.

[1]The sense of $\boldsymbol{M}$ would be that of the direction of advance of a screw at $O$, oriented normal to the plane of $O$ and $\boldsymbol{F}$, when this screw is turned with a sense of rotation corresponding to that of $\boldsymbol{F}$ around $O$.

determined by the problem at hand. In Fig. 3.5, the line of action of $\boldsymbol{M}$ is taken for simplicity through point $O$.

Another approach is to employ a position vector $\boldsymbol{r}$ from point $O$ to *any point P* along the line of action of force $\boldsymbol{F}$ as shown in Fig. 3.6. The moment of $\boldsymbol{F}$ about point $O$ is then defined as

$$\boxed{\boldsymbol{M} = \boldsymbol{r} \times \boldsymbol{F}} \tag{3.3}$$

For the purpose of forming the cross product, the vectors in Fig. 3.6 can be moved to the configuration shown in Fig. 3.7. Then the cross product between $\boldsymbol{r}$ and $\boldsymbol{F}$ obviously has the magnitude

$$|\boldsymbol{r} \times \boldsymbol{F}| = |\boldsymbol{r}||\boldsymbol{F}| \sin \alpha = |\boldsymbol{F}||\boldsymbol{r}| \sin \beta = Fd \tag{3.4}$$

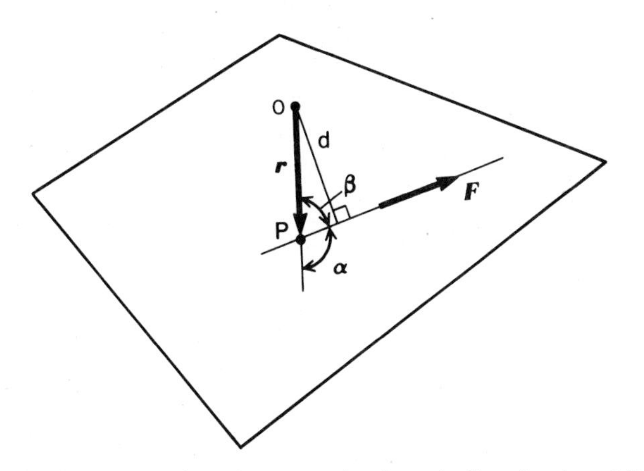


**Figure 3.6.** Put $\boldsymbol{r}$ from $O$ to any point along the line of action of $\boldsymbol{F}$.

where $|\boldsymbol{r}| \sin \beta = d$, the perpendicular distance from $O$ to the line of action of $\boldsymbol{F}$, as can readily be seen in Fig. 3.7. Thus, we get the same *magnitude* of $\boldsymbol{M}$ as with the elementary definition. Also, note that the vector definition of the *direction* of $\boldsymbol{M}$ is identical to that of the elementary definition. Thus, we have the same result for the vector definition as for the elementary definition in all pertinent respects. We shall use both definitions. The first one will be used generally for cases where the force and point are in a convenient plane, and where the perpendicular distance between the point and the line of action of the force is easily measured. As an example, we have shown in Fig. 3.8 a system of coplanar forces acting on a beam. The moment of the forces about point $A$ is then[2]

$$\begin{aligned} \boldsymbol{M}_A &= -(5)(1000)\boldsymbol{k} - (4)(600)\boldsymbol{k} + (11)R_B\boldsymbol{k} \text{ ft-lb} \\ &= (11R_B - 7400)\boldsymbol{k} \text{ ft-lb} \end{aligned}$$

[2]Please note that we still use the right-hand-screw rule in determining the signs of the respective moments.

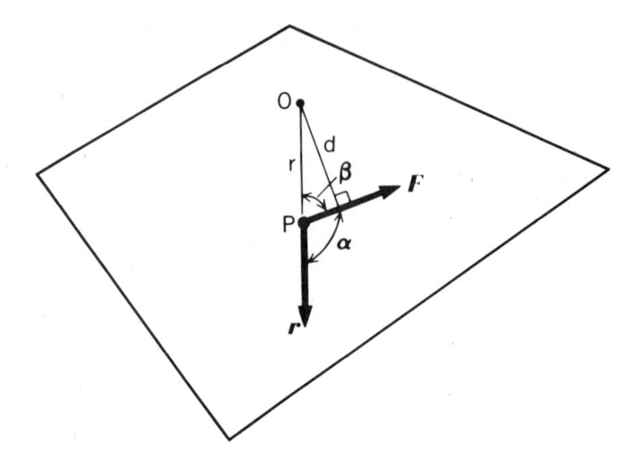


**Figure 3.7.** Move vector $\boldsymbol{r}$.

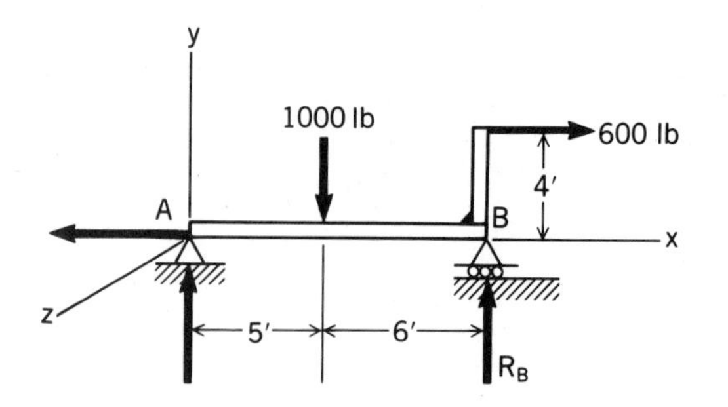


**Figure 3.8.** Coplanar forces on a beam.

For a coplanar force system such as this, we may simply give the scalar form of the equation above, as follows:

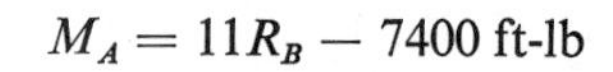

$$M_A = 11R_B - 7400 \text{ ft-lb}$$

The second definition of the moment about a point, namely $\boldsymbol{r} \times \boldsymbol{F}$, is used for complicated coplanar cases and for three-dimensional cases. We shall illustrate such a case in Example 3.2.

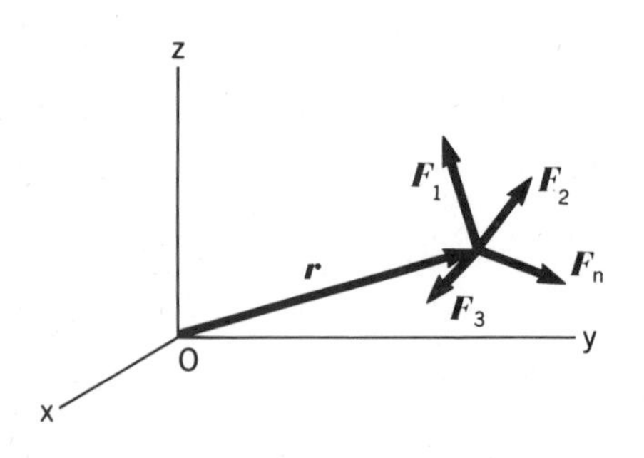


**Figure 3.9.** Concurrent forces.

Consider next a system of $n$ concurrent forces in Fig. 3.9 whose total moment about point $O$ (where we have established reference $xyz$) is desired. We can say that

$$\begin{aligned} \boldsymbol{M} &= \boldsymbol{M}_1 + \boldsymbol{M}_2 + \boldsymbol{M}_3 + \ldots + \boldsymbol{M}_n \\ &= \boldsymbol{r} \times \boldsymbol{F}_1 + \boldsymbol{r} \times \boldsymbol{F}_2 + \boldsymbol{r} \times \boldsymbol{F}_3 \\ &\quad + \ldots + \boldsymbol{r} \times \boldsymbol{F}_n \end{aligned} \tag{3.5}$$

Now, because of the distributive property of the cross product, Eq. 3.5 can be written

$$\boldsymbol{M} = \boldsymbol{r} \times (\boldsymbol{F}_1 + \boldsymbol{F}_2 + \boldsymbol{F}_3 + \ldots + \boldsymbol{F}_n) \tag{3.6}$$

We can conclude from the preceding equations that the sum of the moments about a point of a system of concurrent forces is the same as the moment about the point of the sum of the forces. This result is known as *Varignon's theorem*, which you may well recall from physics.

As a special case of Varignon's theorem, we may find it convenient to decompose a force $\boldsymbol{F}$ into its rectangular components (Fig. 3.10), and then to use these components for taking moments about a point. We can then say that

$$\boldsymbol{M} = \boldsymbol{r} \times \boldsymbol{F} = \boldsymbol{r} \times (F_x\boldsymbol{i} + F_y\boldsymbol{j} + F_z\boldsymbol{k}) \tag{3.7}$$

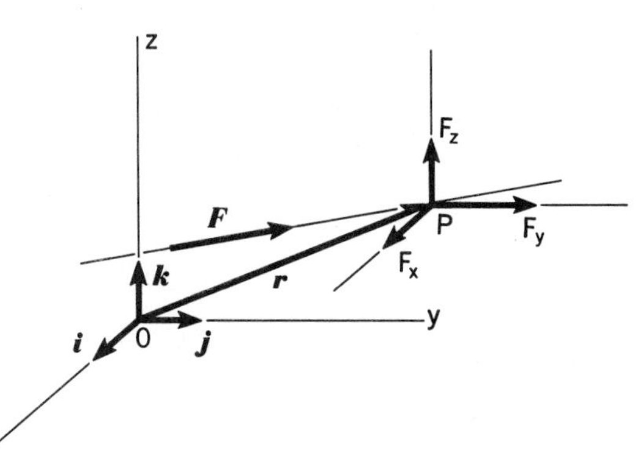


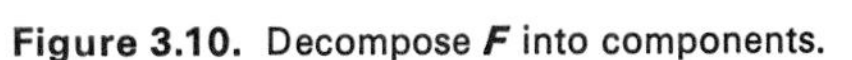
**Figure 3.10.** Decompose $\boldsymbol{F}$ into components.

Now replacing $\boldsymbol{r}$ by its components, we get

$$\begin{aligned} \boldsymbol{M} &= (x\boldsymbol{i} + y\boldsymbol{j} + z\boldsymbol{k}) \times (F_x\boldsymbol{i} + F_y\boldsymbol{j} + F_z\boldsymbol{k}) \\ &= (yF_z - zF_y)\boldsymbol{i} + (zF_x - xF_z)\boldsymbol{j} + (xF_y - yF_x)\boldsymbol{k} \end{aligned} \tag{3.8}$$

The scalar rectangular components of $\boldsymbol{M}$ are then

$$M_x = yF_z - zF_y \tag{3.9a}$$

$$M_y = zF_x - xF_z \tag{3.9b}$$

$$M_z = xF_y - yF_x \tag{3.9c}$$

As a final note, it should be apparent that, because we can choose $\boldsymbol{r}$ so as to terminate anywhere along the line of action of $\boldsymbol{F}$ in computing $\boldsymbol{M}$, we are, in effect, stipulating that $\boldsymbol{F}$ is a *transmissible* vector (defined in Chapter 1) in the computation of $\boldsymbol{M}$.

#### EXAMPLE 3.2

Determine the moment of the 100-lb force *F*, shown in Fig. 3.11, about points *A* and *B*, respectively.

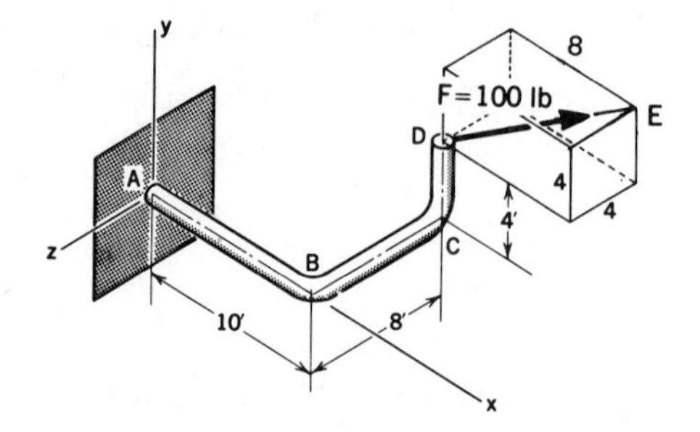


**Figure 3.11.** Find moments at *A* and *B*.

As a first step, let us express force $F$ vectorially. Note that the force is collinear with the vector $\boldsymbol{\rho}_{DE}$ from $D$ to $E$, where

$$\boldsymbol{\rho}_{DE} = 8\boldsymbol{i} + 4\boldsymbol{j} - 4\boldsymbol{k} \tag{a}$$

To get a unit vector $\hat{\boldsymbol{\rho}}$ in the direction of $\boldsymbol{\rho}$, we proceed as follows:

$$\begin{aligned}\hat{\boldsymbol{\rho}}_{DE} = \frac{\boldsymbol{\rho}_{DE}}{|\boldsymbol{\rho}_{DE}|} &= \frac{8\boldsymbol{i} + 4\boldsymbol{j} + 4\boldsymbol{k}}{\sqrt{8^2 + 4^2 - 4^2}} \\ &= .816\boldsymbol{i} + .408\boldsymbol{j} - .408\boldsymbol{k}\end{aligned} \tag{b}$$

We can then express the force $\boldsymbol{F}$ in the following manner:

$$\begin{aligned}\boldsymbol{F} = F\hat{\boldsymbol{\rho}}_{DE} &= (100)(.816\boldsymbol{i} + .408\boldsymbol{j} - .408\boldsymbol{k}) \\ &= 81.6\boldsymbol{i} + 40.8\boldsymbol{j} - 40.8\boldsymbol{k}\end{aligned} \tag{c}$$

To get the moment $\boldsymbol{M}_A$ about point $A$, we choose a position vector from point $A$ to point $D$ which is on the line of action of force $F$. Thus, we have, for $\boldsymbol{r}_{AD}$,

$$\boldsymbol{r}_{AD} = 10\boldsymbol{i} + 4\boldsymbol{j} - 8\boldsymbol{k} \text{ ft} \tag{d}$$

and for $\boldsymbol{M}_A$, we then get

$$\boldsymbol{M}_A = \boldsymbol{r}_{AD} \times \boldsymbol{F} = (10\boldsymbol{i} + 4\boldsymbol{j} - 8\boldsymbol{k}) \times (81.6\boldsymbol{i} + 40.8\boldsymbol{j} - 40.8\boldsymbol{k})$$

$$= \left|\begin{array}{ccc} 10 & 4 & -8 \\ 81.6 & 40.8 & -40.8 \\ \boldsymbol{i} & \boldsymbol{j} & \boldsymbol{k} \\ 10 & 4 & -8 \\ 81.6 & 40.8 & -40.8 \end{array}\right|$$

$$\begin{aligned}&= (10)(40.8)\boldsymbol{k} + (81.6)(-8)\boldsymbol{j} + (4)(-40.8)\boldsymbol{i} \\ &\quad -(-8)(40.8)\boldsymbol{i} - (-40.8)(10)\boldsymbol{j} - (4)(81.6)\boldsymbol{k}\end{aligned}$$

Therefore,

$$\boldsymbol{M}_A = 163.2\boldsymbol{i} - 245\boldsymbol{j} + 81.6\boldsymbol{k} \text{ ft-lb} \tag{e}$$

As for the moment about reference point $B$, we employ the position vector $\boldsymbol{r}_{BD}$ from $B$ to position $D$, again on the line of action of force $\boldsymbol{F}$. Thus, we have

$$\boldsymbol{r}_{BD} = 4\boldsymbol{j} - 8\boldsymbol{k} \text{ ft}$$

Accordingly,

$$\begin{aligned} \boldsymbol{M}_B = \boldsymbol{r}_{BD} \times \boldsymbol{F} &= (4\boldsymbol{j} - 8\boldsymbol{k}) \times (81.6\boldsymbol{i} + 40.8\boldsymbol{j} - 40.8\boldsymbol{k}) \\ &= (4)(81.6)(-\boldsymbol{k}) + (4)(-40.8)(\boldsymbol{i}) + (-8)(81.6)(\boldsymbol{j}) + (-8)(40.8)(-\boldsymbol{i}) \qquad \text{(f)} \\ &= 163.2\boldsymbol{i} - 653\boldsymbol{j} - 326\boldsymbol{k} \text{ ft-lb} \end{aligned}$$

## Problems

**3.1.** What is the position vector $\boldsymbol{r}$ from the origin (0, 0, 0) to the point (3, 4, 5) ft? What are its magnitude and direction cosines?

**3.2.** What is the displacement vector from position (6, 13, 7) ft to position (10, −3, 4) ft?

**3.3.** A surveyor determines that the top of a radio transmission tower is at position $\boldsymbol{r}_1 = (1000\boldsymbol{i} + 1000\boldsymbol{j} + 1000\boldsymbol{k})$ m relative to her position. Similarly, the top of a second tower is located by $\boldsymbol{r}_2 = (2000\boldsymbol{i} + 500\boldsymbol{j} + 700\boldsymbol{k})$ m. What is the distance between the two tower tops?

**3.4.** Reference $xyz$ is rotated 30° about its $x$ axis relative to reference $XYZ$. What is the position vector $\boldsymbol{r}$ for reference $xyz$ of a point having a position vector $\boldsymbol{r}'$ for reference $XYZ$ given as

$$\boldsymbol{r}' = 6\boldsymbol{i}' + 10\boldsymbol{j}' + 3\boldsymbol{k}' \text{ m?}$$

Use $\boldsymbol{i}, \boldsymbol{j}$, and $\boldsymbol{k}$ (no primes) for unit vectors associated with reference $xyz$.

**3.5.** A particle moves along a circular path in the $xy$ plane. What is the position vector $\boldsymbol{r}$ of this particle as a function of the coordinate $x$?

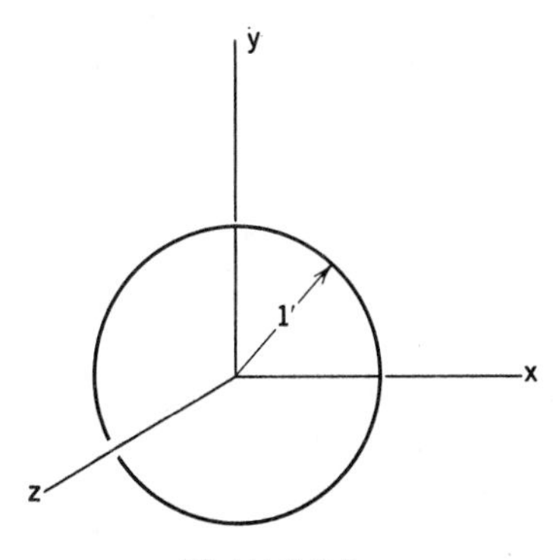


Figure P.3.5

**3.6.** A particle moves along a parabolic path in the $yz$ plane. If the particle has at one point a position vector $\boldsymbol{r} = 4\boldsymbol{j} + 2\boldsymbol{k}$, give the position vector at any point on the path as a function of the $z$ coordinate.

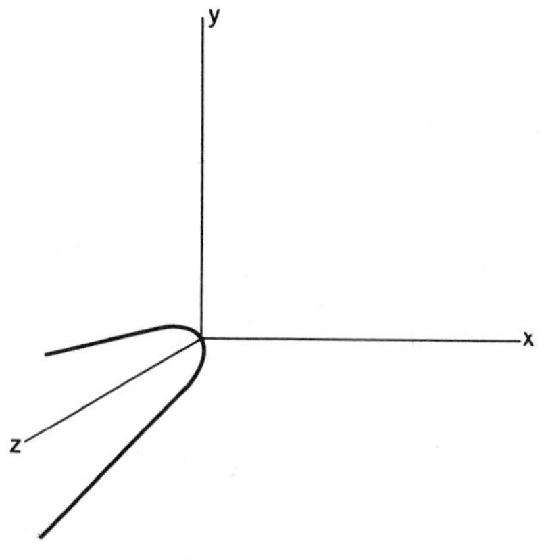


Figure P.3.6

**3.7.** An artillery spotter on Hill 350 (350 m high) estimates the position of an enemy tank as 3000 m NE of him at an elevation 200 m below his position. A 105-mm howitzer unit with a range of 11,000 m is 10,000 m due south of the spotter, and a 155-mm howitzer unit with a range of 15,000 m is 13,000 m SSE of the spotter. Both gun units are located at an elevation of 150 m. Can either or both gun units hit the tank, or must an air strike be called in?

**3.8.** The crew of a submarine patrol plane, with three-dimensional radar, sights a surfaced submarine 10,000 yards north and 5000 yards east while flying at an elevation of 3000 ft above sea level. Where should the pilot instruct a second patrol plane flying at an elevation of 4000 ft at a position 40,000 yards east of the first plane to look for confirmation of the sighting?

**3.9.** A power company lineman can comfortably trim branches 1 m from his waist at an angle of 45° above the horizontal. His waist coincides with the pivot of the work capsule. How high a branch can he trim if the maximum elevation angle of the arm is 75° and the maximum extended length is 12 m?

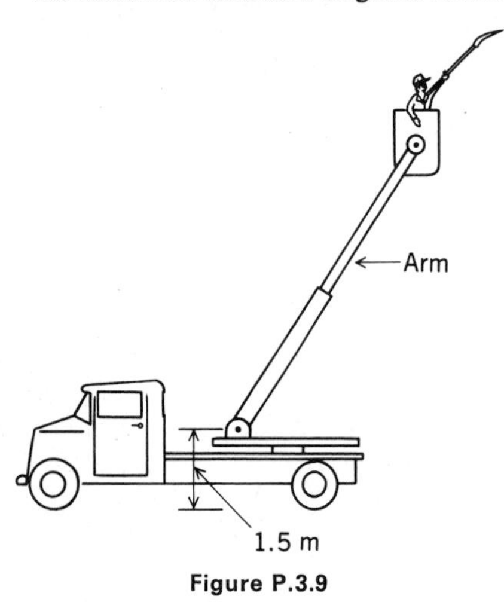


**Figure P.3.9**

**3.10.** The total equivalent forces from water and gravity are shown on the dam. (We will soon be able to compute such equivalents.) Compute the moment of these forces about the toe of the dam in the right-hand corner.

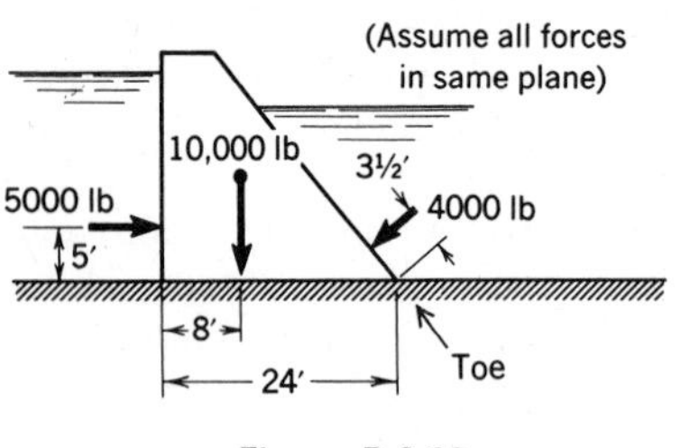


**Figure P.3.10**

**3.11.** Three transmission lines are placed unsymmetrically on a power-line pole. For each pole, the weight of a single line when covered with ice is 2000 N. What is the moment at the base of a pole?

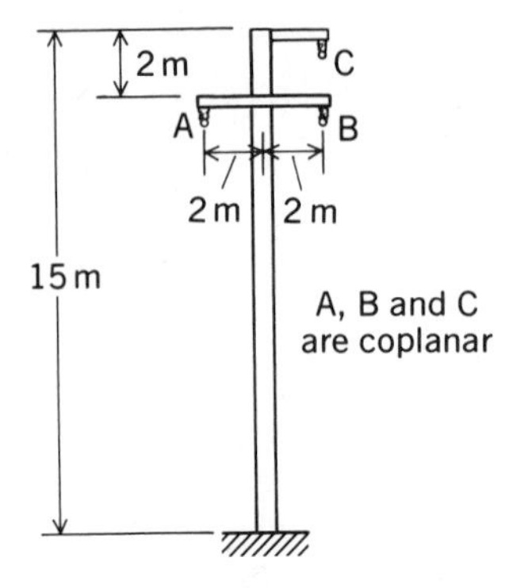


**Figure P.3.11**

**3.12.** Find the moment of the 50-lb force about the support at $A$ and about support $B$ of the simply supported beam.

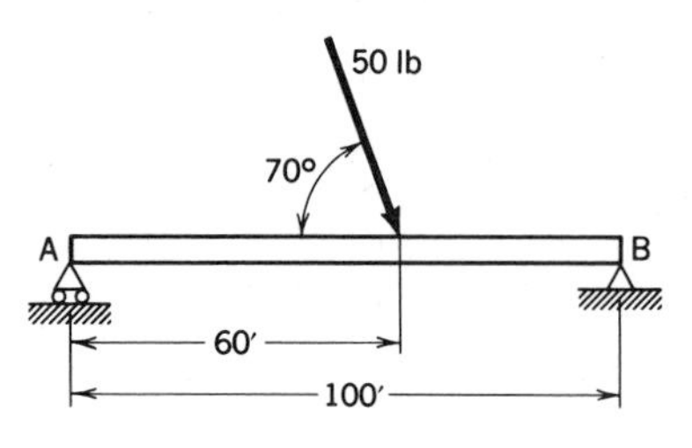


**Figure P.3.12**

**3.13.** Compute the moment of the 1000-lb force about points $A$, $B$, and $C$. Use the transmissibility property of force and rectangular components to make the computations simplest.

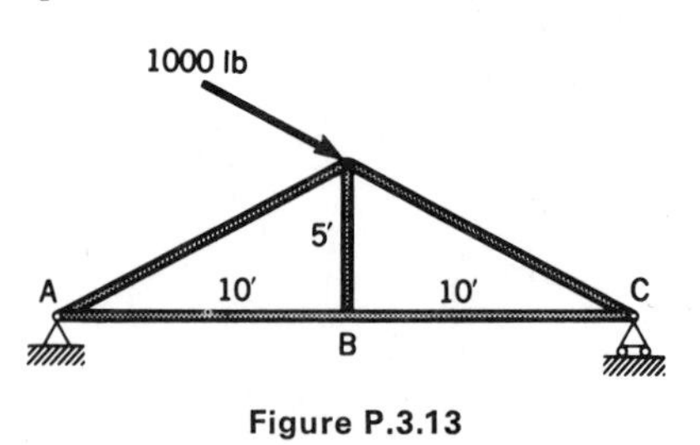


**Figure P.3.13**

**3.14.** A truck-mounted crane has a 20-m boom inclined at 60° to the horizontal. What is the moment about the boom pivot due to a lifted weight of 30 kN? Do by vector and by scalar methods.

Figure P.3.14

**3.15.** A force $F = 10i + 6j - 6k$ N acts at position (10, 3, 4) m relative to a coordinate system. What is the moment of the force about the origin?

**3.16.** What is the moment of the force in Problem 3.15 about the point (6, −4, −3) m?

**3.17.** Two forces $F_1$ and $F_2$ have magnitudes of 10 lb and 20 lb, respectively. $F_1$ has a set of direction cosines $l = 0.5$, $m = 0.707$, $n = -0.5$. $F_2$ l as a set of direction cosines $l = 0$, $m = 0.6$, $n = 0.8$. If $F_1$ acts at point (3, 2, 2) and $F_2$ acts at (1, 0, −3), what is the sum of these moments about the origin?

**3.18.** What is the moment of a 10-lb force $F$ directed along the diagonal of a cube about the corners of the cube? The side of the cube is $a$ ft.

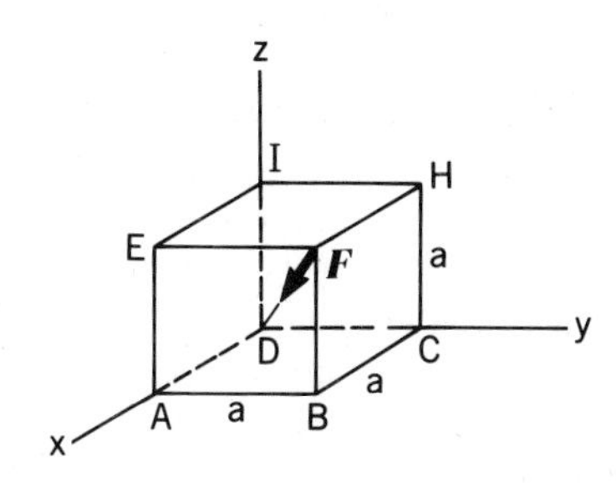


Figure P.3.18

**3.19.** Compute the moment of the 300-lb force about points $P_1$ and $P_2$.

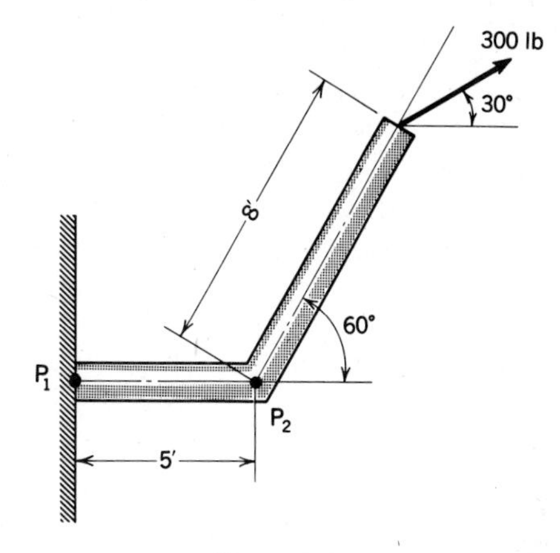


Figure P.3.19

**3.20.** Three guy wires are used in the support system for a television transmission tower that is 600 m tall. Wires $A$ and $B$ are tightened to a tension of 60 kN, whereas wire $C$ has only 30 kN of tension. What is the moment of the wire forces about the base of the tower? The $y$ axis is collinear with $AO$.

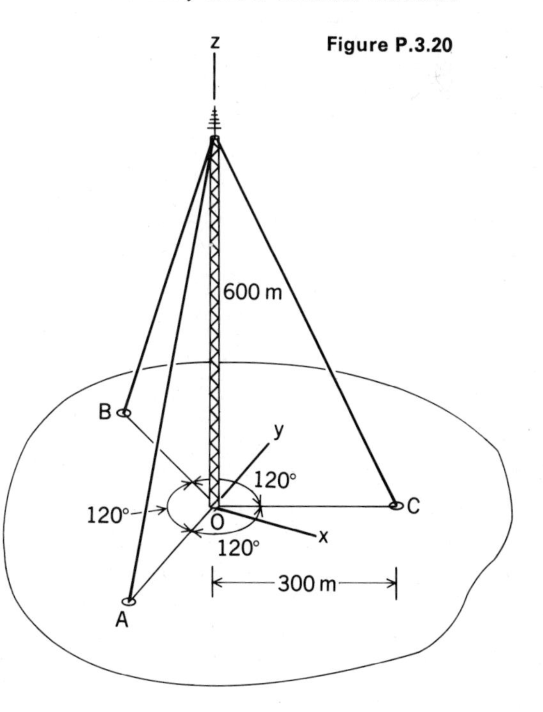


Figure P.3.20

**3.21.** Cables $CD$ and $AB$ help support the member $ED$ and the 1000-lb load at $D$. At $E$ there is a ball-and-socket joint which also supports the member. Denoting the forces from the cables as $F_{CD}$ and $F_{AB}$, respectively, compute moments of the three forces about the point $E$.

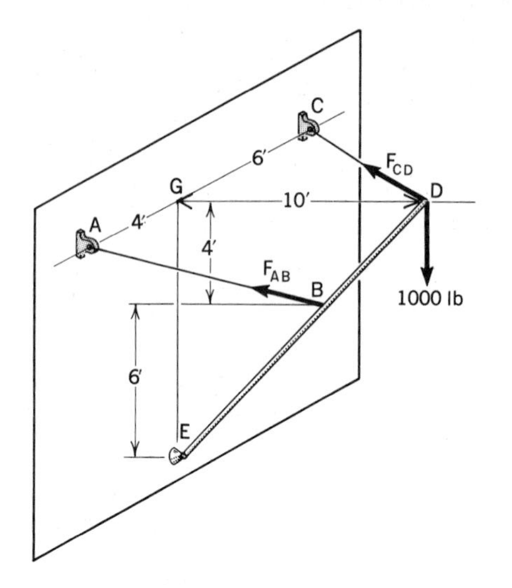

**Figure P.3.21**

## 3.3 Moment of a Force About an Axis

To compute the moment of (or torque) a force $\boldsymbol{F}$ about an axis $B$–$B$ [Fig. 3.12(a)], we pass any plane $A$ perpendicular to the axis. This plane cuts $B$–$B$ at $a$ and the line of action of force $\boldsymbol{F}$ at some point $P$. The force $\boldsymbol{F}$ is then projected to form a rectangular component $F_B$ along a line at $P$ normal to plane $A$ and thus parallel to $B$–$B$, as shown in the diagram. The intersection of plane $A$ with the plane of forces $F_B$ and $\boldsymbol{F}$ (the latter plane is shown shaded and is a plane through $\boldsymbol{F}$ and perpendicular to plane $A$) gives a direction $C$–$C$ along which the other rectangular component of $\boldsymbol{F}$, denoted as $F_A$, can be projected.[3] The moment of $\boldsymbol{F}$ about the line $B$–$B$ is then defined as the moment of the component $F_A$ about point $a$—a coplanar problem discussed at the beginning of the previous section. Thus, in accordance with the definition, the component $F_B$, which is parallel to the axis $B$–$B$, contributes no moment about the axis, and we may say:

$$\text{moment about the axis } B\text{–}B = (F_A)(d) = |\boldsymbol{F}|(\cos\alpha)(d)$$

The moment about an axis is a scalar, even though this moment is associated with a particular axis that has a distinct direction. The situation is the same as it is with the scalar components $V_x$, $V_y$, $V_z$, etc., which are associated with certain directions but which are, nevertheless, scalars. Before continuing, we wish to point out that $F_A$ in Fig. 3.12(a) can be decomposed into pairs of components in plane $A$. From Varignon's theorem we can employ these components instead of $F_A$ in computing the moment

[3]Notice that we are decomposing $\boldsymbol{F}$ into only *two* rectangular components, which, to replace $\boldsymbol{F}$, must be coplanar with $\boldsymbol{F}$.

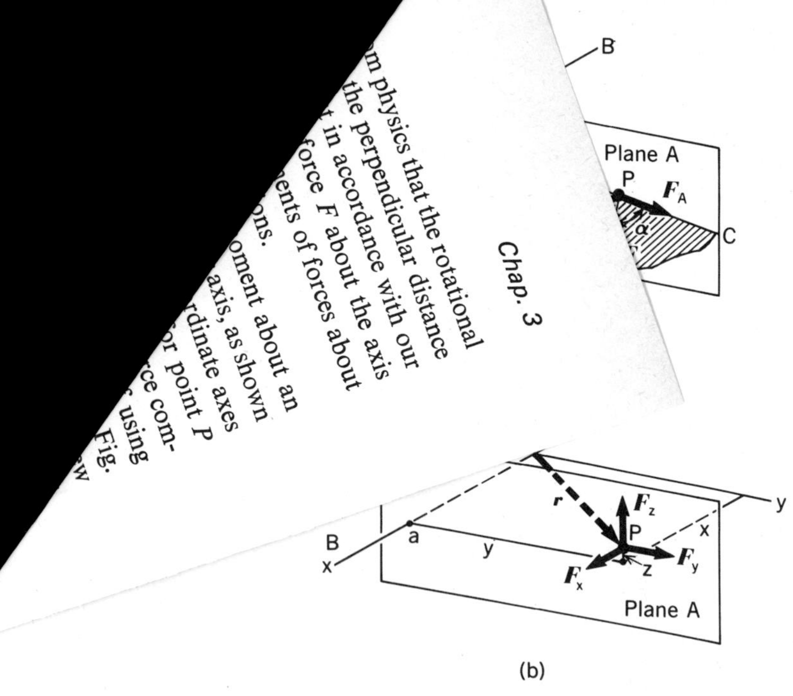


(b)

**Figure 3.12.** Moment about an axis.

about the *B–B* axis. For each force component, we multiply the force times the perpendicular distance from *a* to the line of action of the force component using the right-hand-screw rule to determine the sense and thus the sign.

By means of a simple situation, we can easily show why we set forth the definition above for moment of a force about an axis. Suppose that a disc is mounted on a shaft that is free to rotate in a set of bearings, as shown in Fig. 3.13. A force *F*, inclined to the plane *A* of the disc, acts on the disc. We decompose the force into two rectangular components, one normal to the plane *A* of the disc and the other tangent to the plane *A* of the disc, that is, into forces $F_B$ and $F_A$, respectively. We know from experience

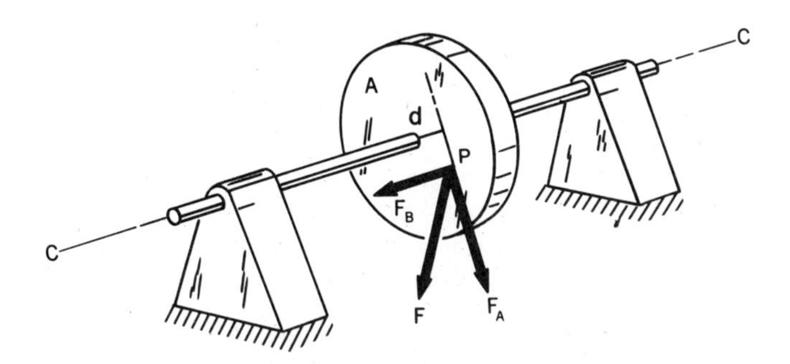


**Figure 3.13.** $F_A$ turns disc.

that $F_B$ does not cause the disc to rotate. And we know fr
motion of the disc is determined by the product of $F_A$ and
$d$ from the centerline of the shaft to the line of action $F_A$. Bu
definition, this product is nothing more than the moment of
of the shaft. Later, in more general dynamics problems, the mom
a point as well as about an axis will enter into the dynamics relati

What is the relation between a moment about a point and a m
axis? To answer this most simply, consider $B$–$B$ of Fig. 3.12(a) to be an $x$
in Fig. 3.12(b). Now choose *any* point $O$ along this axis, and set up coo
$y$ and $z$, as shown in the diagram. The coordinate distances $x$, $y$, and $z$ f
are shown for this reference. The position vector $\boldsymbol{r}$ to $P$ is also shown. The fo
ponent $F_B$ of Fig. 3.12(a) now becomes force component $F_x$. And, instead of
$F_A$, we shall decompose it into components $F_y$ and $F_z$ in plane $A$ as shown in
3.12(b). We now compute the moment about the $x$ axis for force $\boldsymbol{F}$ using this ne
arrangement. Clearly, $F_x$ contributes no moment, as before. The force components
$F_y$ and $F_z$ are in plane $A$ that is perpendicular to the axis of interest and so, as before in the case of $F_A$, we multiply each of these forces by the perpendicular distance of point $a$ to the respective lines of action of these forces. For force $F_z$, this perpendicular distance is clearly $y$, as can readily be seen from the diagram, and, for force $F_y$, this perpendicular distance is $z$. Using the right-hand-screw rule for ascertaining the sense of each of the moments, we can say:

$$\text{moment about } x \text{ axis} = (yF_z - zF_y) \tag{3.10}$$

Were we to take moments of $\boldsymbol{F}$ about the origin $O$, we would get (see Eq. 3.8)

$$\begin{aligned}\boldsymbol{M} &= M_x\boldsymbol{i} + M_y\boldsymbol{j} + M_z\boldsymbol{k} = \boldsymbol{r} \times \boldsymbol{F} \\ &= (yF_z - zF_y)\boldsymbol{i} + (zF_x - xF_z)\boldsymbol{j} + (xF_y - yF_x)\boldsymbol{k}\end{aligned} \tag{3.11}$$

Comparing Eqs. 3.10 and 3.11, we can conclude that the moment about the $x$ axis is simply $M_x$, the $x$ component of $\boldsymbol{M}$ about $O$. We can thus conclude that the moment about the $x$ axis of the force $\boldsymbol{F}$ is the component in the $x$ direction of the moment of $\boldsymbol{F}$ about a point $O$ positioned *anywhere* along the $x$ axis. That is,

$$\text{moment about } x \text{ axis} = M_x = \boldsymbol{M}_o \cdot \boldsymbol{i} = (\boldsymbol{r} \times \boldsymbol{F}) \cdot \boldsymbol{i} \tag{3.12}$$

We may generalize the preceding discussion as follows. Consider an arbitrary axis $n$–$n$ to which we have assigned a unit vector $\boldsymbol{n}$ (Fig. 3.14). An arbitrary force $\boldsymbol{F}$ is also shown. To get the moment $M_n$ of force $\boldsymbol{F}$ about axis $n$–$n$, we choose any point $O$ along $n$–$n$. Then draw a position vector $\boldsymbol{r}$ from point $O$ to any point along the line of action of $\boldsymbol{F}$. This has been shown in the diagram. We can then say, from our previous discussion,

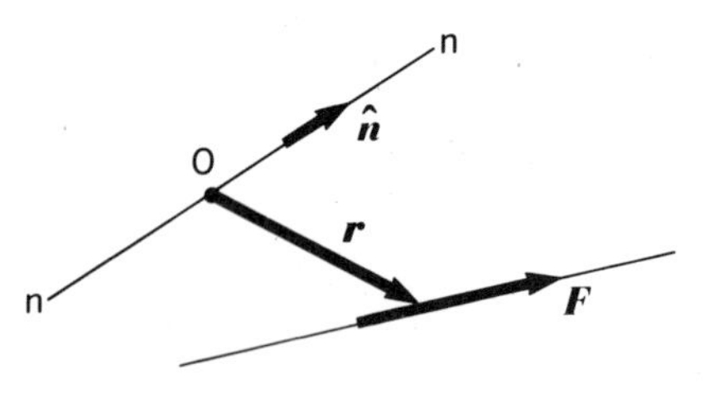


Figure 3.14. $M_n = (\boldsymbol{r} \times \boldsymbol{F}) \cdot \boldsymbol{n}$.

$$\boxed{M_n = (\boldsymbol{r} \times \boldsymbol{F}) \cdot \boldsymbol{n}} \tag{3.13}$$

(Notice from Eqs. 3.12 and 3.13 that the moment of a force about an axis involves a *scalar triple product*.) Equation 3.13 stipulates in words that:

> *The moment of a force about an axis equals the scalar component in the direction of the axis of the moment vector taken about any point along the axis.*[4]

Note that the unit vector $\boldsymbol{n}$ can have two opposite senses along the axis $n$, in contrast to the usual unit vectors $\boldsymbol{i}, \boldsymbol{j}$, and $\boldsymbol{k}$ associated with the coordinate axes. A moment $M_n$ about the $n$ axis determined from $\boldsymbol{M} \cdot \boldsymbol{n}$ has a sense consistent with the sense chosen for $\boldsymbol{n}$. That is, a positive moment $M_n$ has a sense corresponding to that of $\boldsymbol{n}$, and a negative moment $M_n$ has a sense opposite to that of $\boldsymbol{n}$. If the opposite sense had been chosen for $\boldsymbol{n}$, the sign of $\boldsymbol{M} \cdot \boldsymbol{n}$ would be opposite to that found in the first case. However, the same physical moment is obtained in both cases.

If we specify the moments of a force about *three* orthogonal concurrent axes, we then single out *one* possible point in space for $O$ along the axes. Point $O$, of course, is the origin of the axes. These three moments then become the orthogonal scalar components of the moment of $\boldsymbol{F}$ about point $O$, and we can say:

$$\begin{aligned} \boldsymbol{M} = {} & (\text{moment about the } x \text{ axis})\boldsymbol{i} \\ & + (\text{moment about the } y \text{ axis})\boldsymbol{j} \\ & + (\text{moment about the } z \text{ axis})\boldsymbol{k} = M_x\boldsymbol{i} + M_y\boldsymbol{j} + M_z\boldsymbol{k} \end{aligned} \tag{3.14}$$

From this relation, we can conclude that:

> *The three orthogonal components of the moment of a force about a point are the moments of this force about the three orthogonal axes that have the point as an origin.*

You may now ask what the physical differences are in applications of moments about an axis and moments about a point. The simplest example is in the dynamics of rigid bodies. If a body of revolution is constrained so it can only spin about its axis, as in Fig. 3.13, the rotary motion will depend on the moment of the forces about the axis of rotation, as related by a scalar equation. The less familiar concept of moment about a point is illustrated in the motion of bodies that have no constraints, such as missiles and rockets. In these cases, the motion of the body is related by a vector equation to the moment of forces acting on the body about a point called the *center of mass*. (The center of mass will be defined completely later.)

### EXAMPLE 3.3

Compute the moment of a force $\boldsymbol{F} = 10\boldsymbol{i} + 6\boldsymbol{j}$ N, which goes through position $\boldsymbol{r}_a = 2\boldsymbol{i} + 6\boldsymbol{j}$ m (see Fig. 3.15), about a line going through points 1 and 2 having the

---

[4]If the force is in a plane *perpendicular* to the axis about which we are taking the moment like force $F_A$ in Fig. 3.13, where plane $A$ is perpendicular to axis $C$–$C$ about which we desire the moment (or torque) of $F_A$, remember we can use the elementary definition of moment of a force about an axis that was presented at the outset of this section. Hence for $F_A$ we then have $M_{CC} = F_A d$.

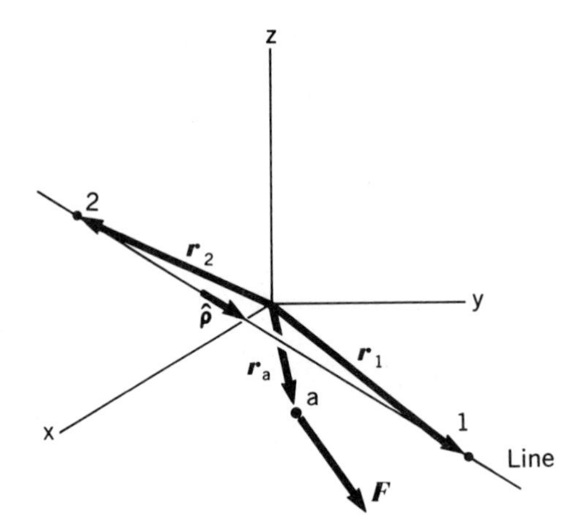


**Figure 3.15.** Find moment of $F$ about line.

respective position vectors

$$r_1 = 6i + 10j - 3k \text{ m}$$
$$r_2 = -3i - 12j + 6k \text{ m}$$

To compute this moment, we can take the moment of $F$ about either point 1 or point 2, and then find the component of this vector along the direction of the displacement vector between 1 and 2 or between 2 and 1. Mathematically, we have, using a position vector from point 1 to point $a$, namely $(r_a - r_1)$,

$$M_\rho = [(r_a - r_1) \times F] \cdot \hat{\rho} \tag{a}$$

where $\hat{\rho}$ is the unit vector along the line chosen to have a sense going from point 2 to point 1. The formulation above is the scalar triple product examined in Chapter 2 and we can use the determinant approach for the calculation once the components of the vectors $(r_a - r_1)$, $F$, and $\hat{\rho}$ have been determined. Thus, we have

$$\begin{aligned} r_a - r_1 &= (2i + 6j) - (6i + 10j - 3k) \\ &= -4i - 4j + 3k \text{ m} \\ F &= 10i + 6j \text{ N} \\ \hat{\rho} &= \frac{r_1 - r_2}{|r_1 - r_2|} = \frac{9i + 22j - 9k}{\sqrt{81 + 484 + 81}} \\ &= .354i + .866j - .354k \end{aligned}$$

We then have, for $M_\rho$:

$$M_\rho = \begin{vmatrix} -4 & -4 & 3 \\ 10 & 6 & 0 \\ .354 & .866 & -.354 \end{vmatrix} = 13.94 \text{ N-m} \tag{b}$$

Because $M_\rho$ is positive, we have a clockwise moment about the line as we look from point 2 to point 1. If we had chosen $\hat{\rho}$ to have an opposite sense, then $M_\rho$ would have been computed as $-13.94$ N-m. Then, we would conclude that $M_\rho$ is a counterclockwise moment about the line as one looks from point 1 to point 2. Note that the same physical moment is determined in both cases.

## Problems

**3.22.** Disc $A$ has a radius of 600 mm. What is the moment of the forces about the center of the disc? What is the torque of these forces about the axis of the shaft?

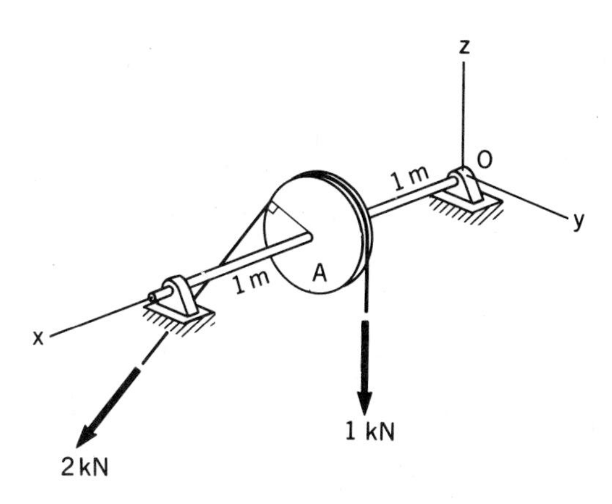


Figure P.3.22

**3.23.** A force $F$ acts at position (3, 2, 0) ft. It is in the $xy$ plane and is inclined at 30° from the $x$ axis with a sense directed away from the origin. What is the moment of this force about an axis going through the points (6, 2, 5) ft and (0, −2, −3) ft?

**3.24.** A force $\boldsymbol{F} = 10\boldsymbol{i} + 16\boldsymbol{j}$ N goes through the origin of the coordinate system. What is the moment of this force $\boldsymbol{F}$ about an axis going through points 1 and 2 with position vectors

$$\boldsymbol{r}_1 = 6\boldsymbol{i} + 3\boldsymbol{k} \text{ m}$$

$$\boldsymbol{r}_2 = 16\boldsymbol{j} - 4\boldsymbol{k} \text{ m}$$

**3.25.** A blimp is moored to a tower at $A$. A force on $A$ from this blimp is

$$\boldsymbol{F} = 5\boldsymbol{i} + 3\boldsymbol{j} + 1.8\boldsymbol{k} \text{ kN}$$

What is the moment about axis $C$ on the ground? Knowledge of this moment and other moments at the base is needed to properly design the foundation of the tower.

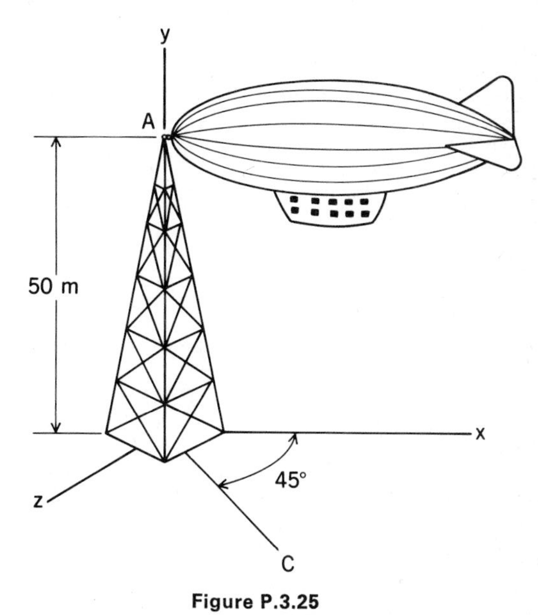


Figure P.3.25

**3.26.** Compute the thrust of the applied forces shown along the axis of the shaft and the torque of the forces about the axis of the shaft.

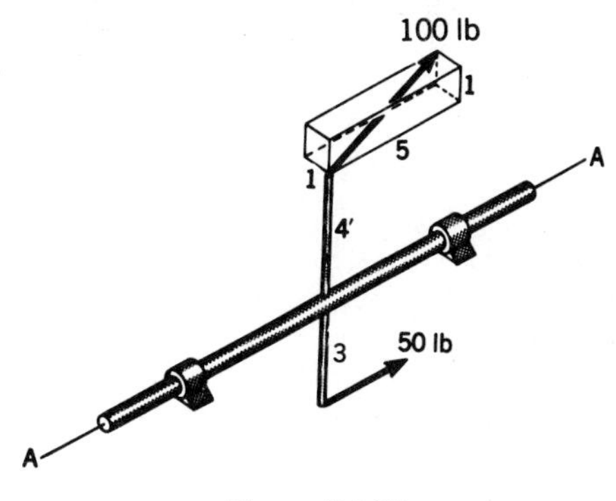


Figure P.3.26

**3.27.** A deep-submergence vessel is connected to its mother ship by a cable. The vessel becomes snagged on some rocks, and the mother ship steams away in an attempt to free the submerged vessel. The connecting cable is suspended from a crane sticking out over the water 20 m above the mother ship's center of mass and 15 m out from the longitudinal axis of the mother ship. The cable develops a force of 200 kN. It is inclined at 50° from the vertical in a vertical plane oriented 20° from the longitudinal axis of the ship. What is the moment tending to cause the mother ship to roll over?

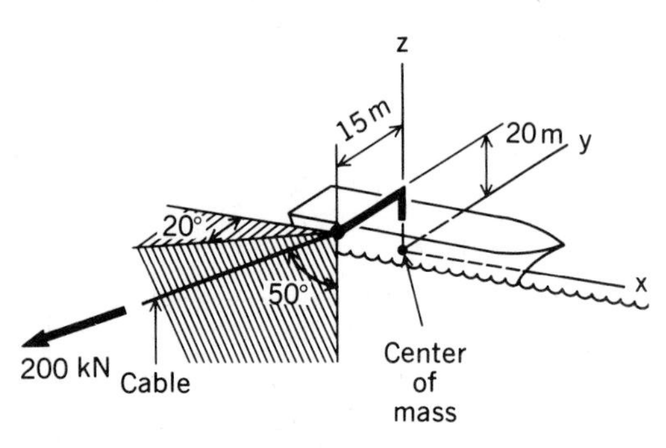

Figure P.3.27

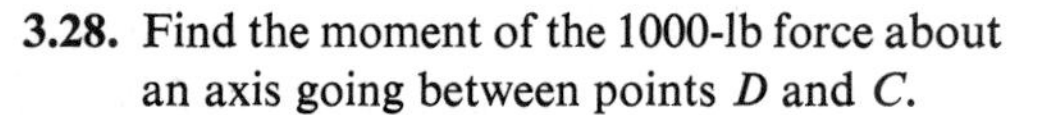
**3.28.** Find the moment of the 1000-lb force about an axis going between points $D$ and $C$.

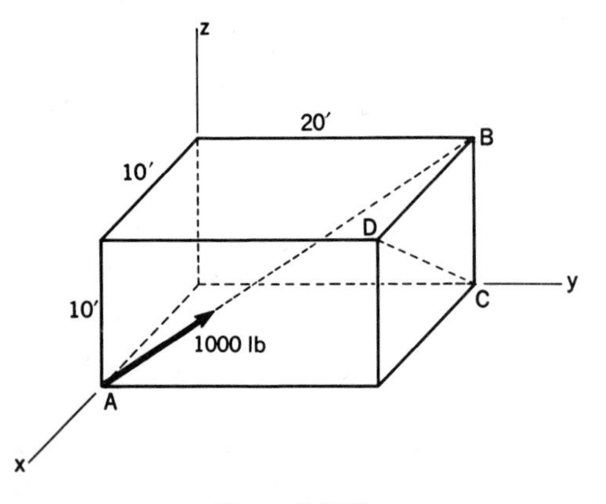


Figure P.3.28

**3.29.** In Problem 3.21, what is the moment of the three indicated forces about axis $GD$?

**3.30.** A tow truck is pointed at 45° to the edge $A$–$A$ of a ravine with sides sloping at 45° to the vertical. The operator attaches a cable to a wrecked car in the ravine and starts the winch. The cable is oriented normal to $A$–$A$ and develops a force of 15 kN. What are the moments tending to tip over the tow truck about rear wheels (rocking backwards) and also the moment for rolling sideways. (*Hint:* Use the position vector from $C$ to $B$.)

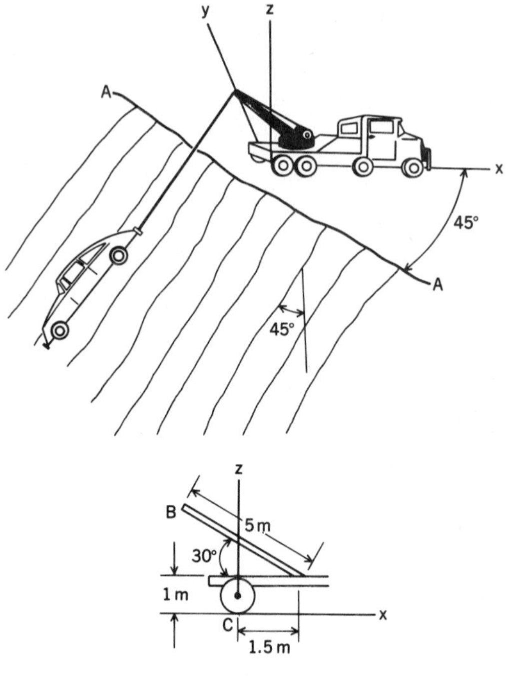

Figure P.3.30

**3.31.** The base of a fire truck extension ladder is rotated 75° from the front of the truck. The 25-m ladder is elevated 60° from the horizontal. The ladder weight is 20 kN and is regarded as concentrated at a point 10 m up from the base (the lower part of the ladder weighs much more than the upper part). A 900-N fireman and the 500-N young lady he is rescuing are at the top of the ladder. (a) What is the moment at the base of the ladder tending to tip over the fire truck? (b) What is the moment about the horizontal axis $\hat{\mathbf{p}}$ about which the ladder rotates?

## Problems

**3.22.** Disc $A$ has a radius of 600 mm. What is the moment of the forces about the center of the disc? What is the torque of these forces about the axis of the shaft?

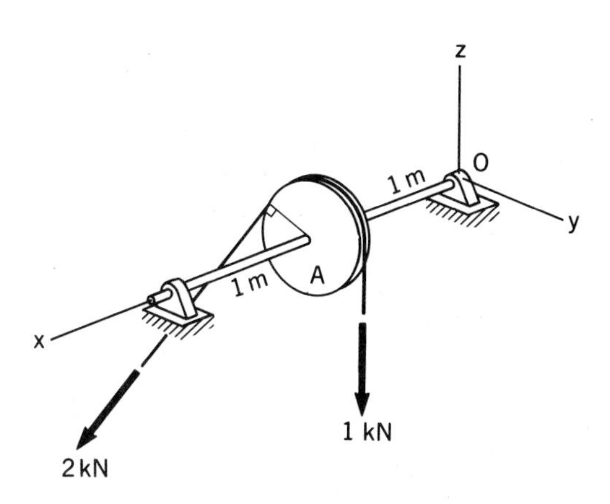

Figure P.3.22

**3.23.** A force $F$ acts at position (3, 2, 0) ft. It is in the $xy$ plane and is inclined at 30° from the $x$ axis with a sense directed away from the origin. What is the moment of this force about an axis going through the points (6, 2, 5) ft and (0, −2, −3) ft?

**3.24.** A force $\boldsymbol{F} = 10\boldsymbol{i} + 16\boldsymbol{j}$ N goes through the origin of the coordinate system. What is the moment of this force $\boldsymbol{F}$ about an axis going through points 1 and 2 with position vectors

$$\boldsymbol{r}_1 = 6\boldsymbol{i} + 3\boldsymbol{k} \text{ m}$$

$$\boldsymbol{r}_2 = 16\boldsymbol{j} - 4\boldsymbol{k} \text{ m}$$

**3.25.** A blimp is moored to a tower at $A$. A force on $A$ from this blimp is

$$\boldsymbol{F} = 5\boldsymbol{i} + 3\boldsymbol{j} + 1.8\boldsymbol{k} \text{ kN}$$

What is the moment about axis $C$ on the ground? Knowledge of this moment and other moments at the base is needed to properly design the foundation of the tower.

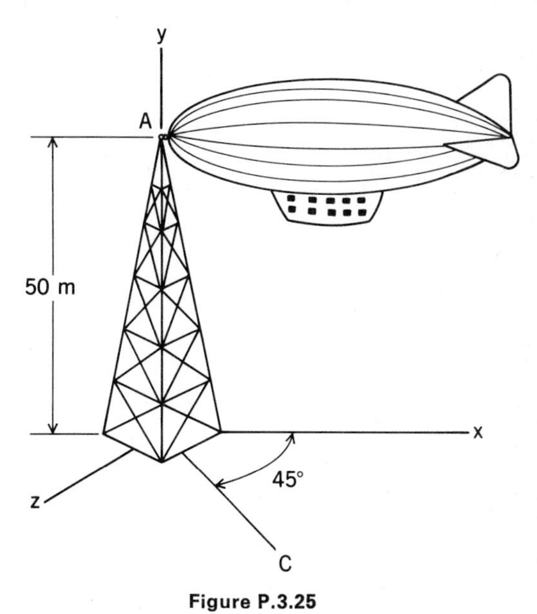


Figure P.3.25

**3.26.** Compute the thrust of the applied forces shown along the axis of the shaft and the torque of the forces about the axis of the shaft.

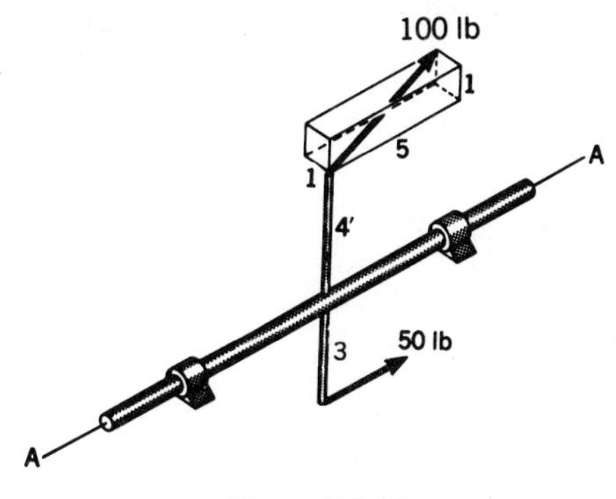


Figure P.3.26

**3.27.** A deep-submergence vessel is connected to its mother ship by a cable. The vessel becomes snagged on some rocks, and the mother ship steams away in an attempt to free the submerged vessel. The connecting cable is suspended from a crane sticking out over the water 20 m above the mother ship's center of mass and 15 m out from the longitudinal axis of the mother ship. The cable develops a force of 200 kN. It is inclined at 50° from the vertical in a vertical plane oriented 20° from the longitudinal axis of the ship. What is the moment tending to cause the mother ship to roll over?

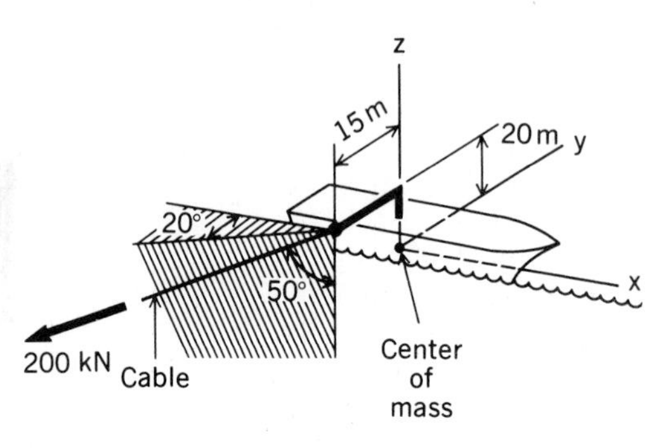

Figure P.3.27

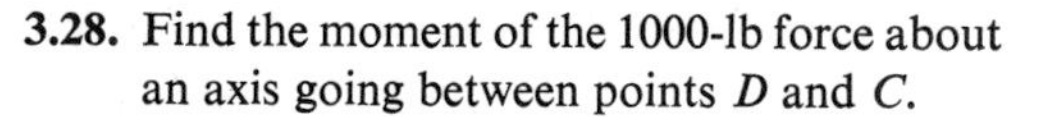
**3.28.** Find the moment of the 1000-lb force about an axis going between points $D$ and $C$.

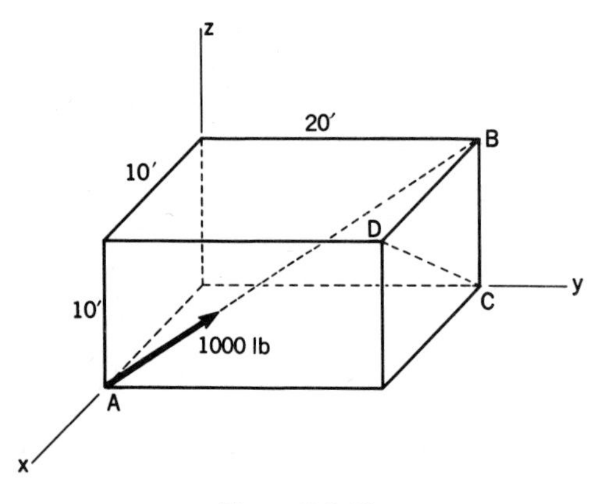


Figure P.3.28

**3.29.** In Problem 3.21, what is the moment of the three indicated forces about axis $GD$?

**3.30.** A tow truck is pointed at 45° to the edge $A$–$A$ of a ravine with sides sloping at 45° to the vertical. The operator attaches a cable to

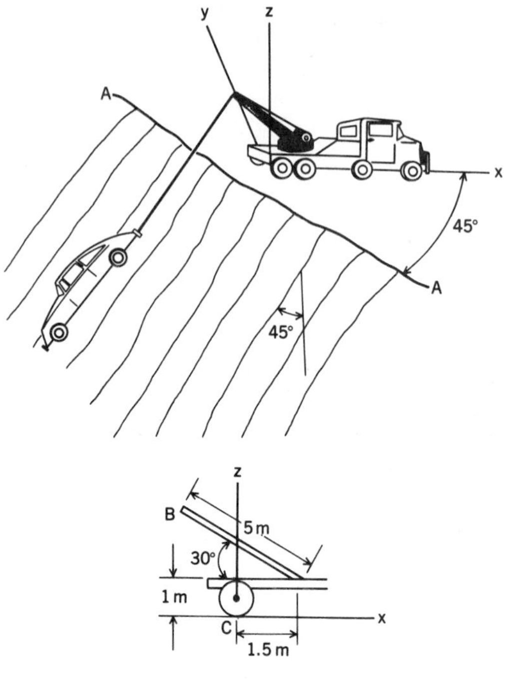

Figure P.3.30

a wrecked car in the ravine and starts the winch. The cable is oriented normal to $A$–$A$ and develops a force of 15 kN. What are the moments tending to tip over the tow truck about rear wheels (rocking backwards) and also the moment for rolling sideways. (*Hint:* Use the position vector from $C$ to $B$.)

**3.31.** The base of a fire truck extension ladder is rotated 75° from the front of the truck. The 25-m ladder is elevated 60° from the horizontal. The ladder weight is 20 kN and is regarded as concentrated at a point 10 m up from the base (the lower part of the ladder weighs much more than the upper part). A 900-N fireman and the 500-N young lady he is rescuing are at the top of the ladder. (a) What is the moment at the base of the ladder tending to tip over the fire truck? (b) What is the moment about the horizontal axis $\hat{\mathbf{p}}$ about which the ladder rotates?

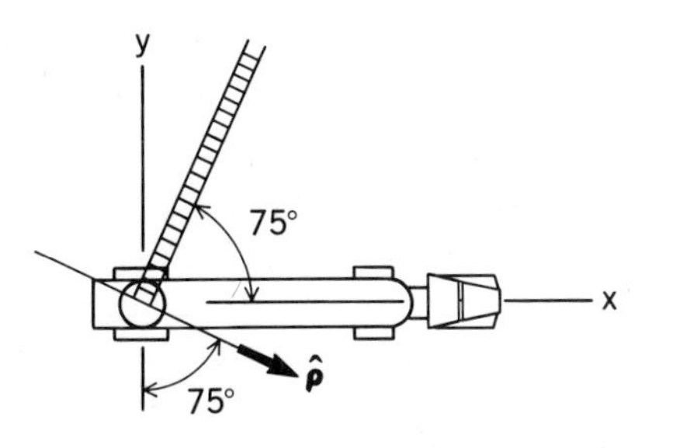


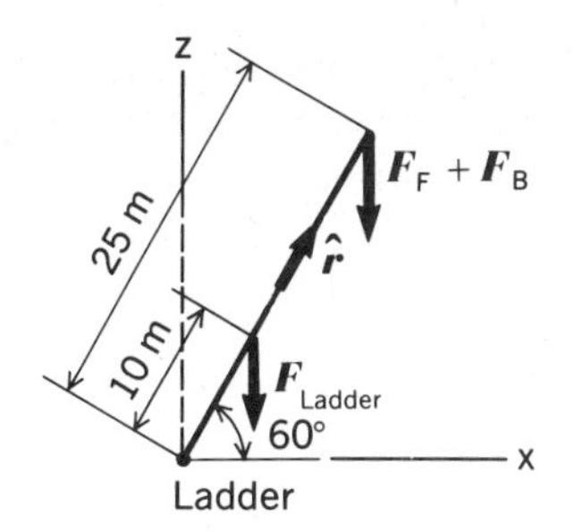


Figure P.3.31

## 3.4 The Couple and Couple Moment

A special arrangement of forces that is of great importance is the *couple*. *The couple is formed by any two equal parallel forces that have opposite senses* (Fig. 3.16). On a rigid body, a couple has only *one* effect, a "turning" action. Individual forces or combinations of forces that do not constitute couples may "push" or "pull" as well as "turn" a body. The turning action is given quantitatively by the moment of forces about a point or an axis. We shall, accordingly, be most concerned with the moment of a couple, or what we shall call the *couple moment.*

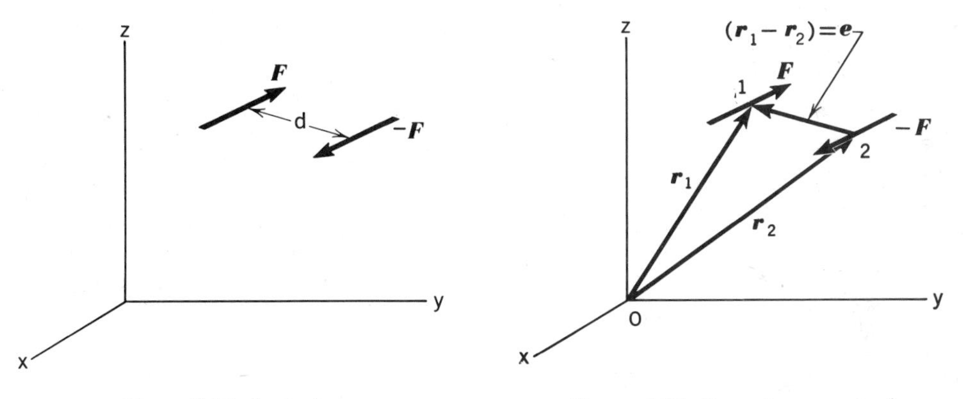


**Figure 3.16.** A couple.

**Figure 3.17.** Compute moment of couple about $O$.

Let us now evaluate the moment of the couple about the origin. Position vectors have been drawn in Fig. 3.17 to points 1 and 2 anywhere along the respective line of action of each force. Adding the moment of each force about $O$, we have for the couple moment $\boldsymbol{M}$

$$\begin{aligned} \boldsymbol{M} &= \boldsymbol{r}_1 \times \boldsymbol{F} + \boldsymbol{r}_2 \times (-\boldsymbol{F}) \\ &= (\boldsymbol{r}_1 - \boldsymbol{r}_2) \times \boldsymbol{F} \end{aligned} \tag{3.15}$$

We can see that $(\boldsymbol{r}_1 - \boldsymbol{r}_2)$ is a position vector between points 2 and 1, and if we call this vector $\boldsymbol{e}$, the formulation above becomes

$$\boldsymbol{M} = \boldsymbol{e} \times \boldsymbol{F} \tag{3.16}$$

Since $\boldsymbol{e}$ is in the plane of the couple, it is clear from the definition of a cross product that $\boldsymbol{M}$ is in an orientation normal to the plane of the couple. The sense in this case may be seen in Fig. 3.18 to be directed downward, in accordance with the right-hand-screw rule. Note the use of the double arrow to represent the couple moment. Note also that the rotation of $\boldsymbol{e}$ to $\boldsymbol{F}$, as stipulated in the cross-product formulation, is in the same direction as the "turning" action of the two force vectors, and from now on we shall use the latter criterion for determining the sense of rotation to be used with the right-hand-screw rule.

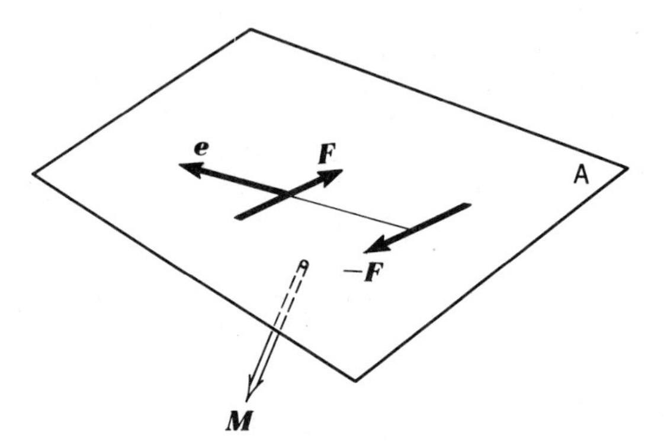


**Figure 3.18.** The couple moment $\boldsymbol{M}$.

Now that the direction of couple moment $\boldsymbol{M}$ has been established for the couple, we need only compute the magnitude for a complete description. Points 1 and 2 may be chosen anywhere along the lines of action of the forces without changing the resulting moment, since the forces are transmissible for this computation. Therefore, to compute the magnitude of the couple moment vector it will be simplest to choose positions 1 and 2 so that $\boldsymbol{e}$ is *perpendicular* to the lines of action of the forces ($\boldsymbol{e}$ is then denoted as $\boldsymbol{e}_\perp$). From the definition of the cross product, we can then say:

$$|\boldsymbol{M}| = |\boldsymbol{e}_\perp||\boldsymbol{F}| \sin 90° = |\boldsymbol{e}_\perp||\boldsymbol{F}| = |\boldsymbol{F}|d \tag{3.17}$$

where the more familiar notation, $d$, has been used in place of $|\boldsymbol{e}_\perp|$ as the perpendicular distance between the lines of action of the forces.

> To summarize the preceding discussions, we may say that: The moment of a couple is a vector whose orientation is normal to the plane of the couple and whose sense is determined in accordance with the right-hand-screw rule, using the "turning" action of the forces to give the proper rotation. The magnitude of the couple moment equals the product of either force magnitude comprising the couple times the perpendicular distance between the forces.

### 3.5 The Couple Moment as a Free Vector

Had we chosen any other position in space as the origin, and had we computed the moment of the couple about it, we would have formed the same moment vector. To understand this, note that although the position vectors to points 1 and 2 will change for a new origin, the *difference* between these vectors (which has been termed $\boldsymbol{e}$) does *not* change, as can readily be observed in Fig. 3.19. Since $\boldsymbol{M} = \boldsymbol{e} \times \boldsymbol{F}$, we

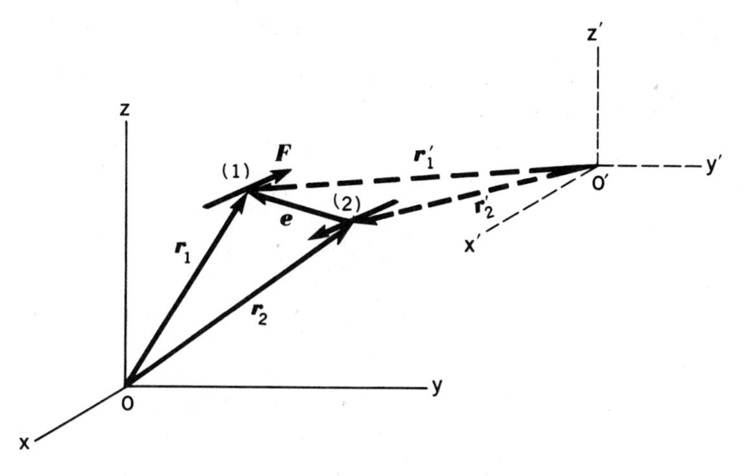


**Figure 3.19.** Vector $\boldsymbol{e}$ is the same for both references.

can conclude that *the couple has the same moment about every point in space.* The particular line of action of the vector representation of the couple moment that is illustrated in Fig. 3.18 is then of little significance and can be moved anywhere. In short, the *couple moment is a free vector*. That is, we may move this vector anywhere in space without changing its meaning, provided that we keep the direction and magnitude intact. Consequently, *for the purpose of taking moments*, we may move the couple itself anywhere in its own or a parallel plane, provided that the direction of turning is not altered—i.e., we cannot "flip" the couple over. In any of these possible planes, we can also change the magnitude of the forces of the couple to other equal values, provided that the distance $d$ is simultaneously changed so that the product $|\boldsymbol{F}|d$ remains the same. Since none of these steps changes the direction or magnitude of the couple moment, all of them are permissible.

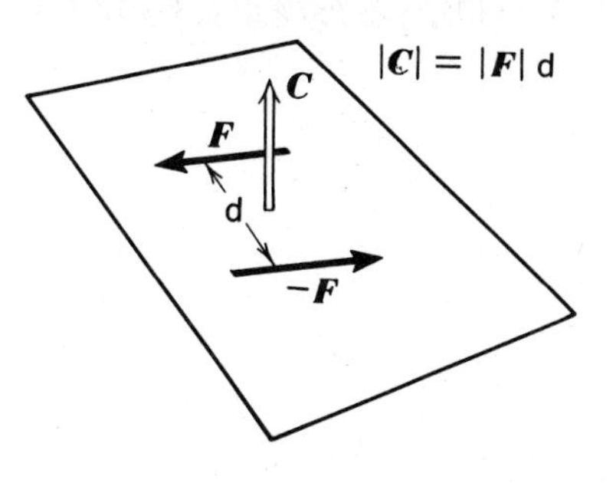


**Figure 3.20.** $\boldsymbol{C}$ represents couple.

As we pointed out earlier, the only effect of a couple on a rigid body is its turning action, which is represented quantitatively by the moment of the couple—i.e., the couple moment. Since this is so often its sole effect, it is only natural to represent the couple by specifications of its moment; its magnitude, then, becomes $|F|d$ and its direction that of its moment. This is the same as identifying a person by her/his job (i.e., as a teacher, plumber, etc.). Thus, in Fig. 3.20, the couple moment $C$ may be used to represent the indicated couple.

## 3.6 Addition and Subtraction of Couples

Since couples themselves have zero net forces, addition per se of couples always yields zero force. For this reason, the addition and subtraction of couples is interpreted to mean addition and subtraction of the *moments* of the couples. Since couple moments are free vectors, we can always arrange to have a concurrent system of vectors. We shall now take the opportunity to illustrate many of the earlier remarks about couples by adding the two couples shown on the face of the cube in Fig. 3.21. Notice that the couple moment vectors of the couples have been drawn. Since these vectors are free, they may be moved to a convenient position and then added. The total couple moment then becomes 103.2 lb-ft at an angle of 76° with the horizontal, as shown in Fig. 3.22. The couple that creates this turning action is in a plane at right angles to this orientation with a clockwise sense as observed from below.

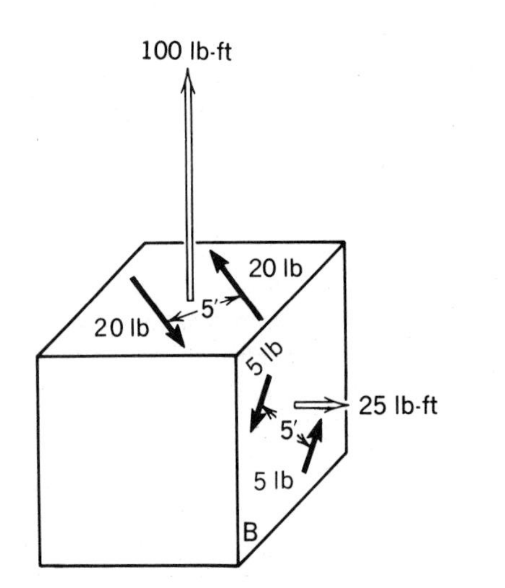


**Figure 3.21.** Add couples.

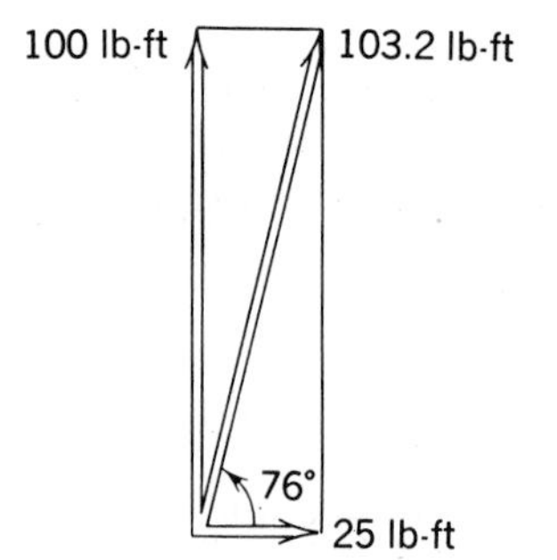


**Figure 3.22.** Add couple moments.

This addition may be shown to be valid by the following more elementary procedure. The couples of the cube are moved in their respective planes to the positions shown in Fig. 3.23, which does not alter the moment of the couples, as pointed out in Section 3.5. If the couple on plane $B$ is adjusted to have a force magnitude of 20 lb and if the separating distance is decreased to $\frac{5}{4}$ ft, the couple moment is not changed (Fig. 3.24). We thus form a system of forces in which two of the forces are equal, opposite, and collinear and, since these two forces cannot contribute moment, they

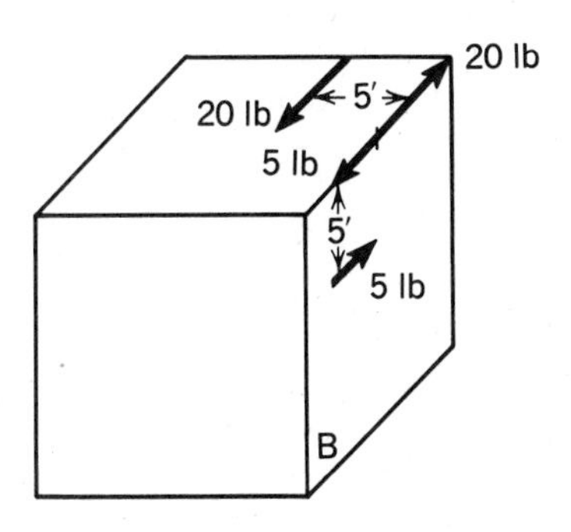


Figure 3.23. Move couples.

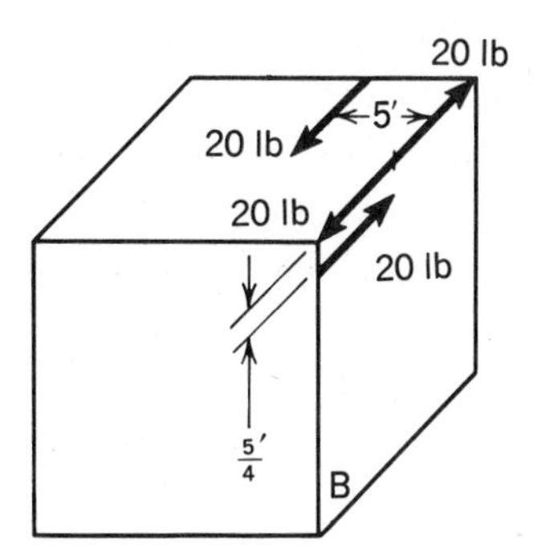


Figure 3.24. Change values of two forces.

may be deleted, leaving a single couple on a plane inclined to the original planes (Fig. 3.25). The distance between the remaining forces is

$$\sqrt{25 + \tfrac{25}{16}}\ \text{ft} = 5.16\ \text{ft}$$

and so the magnitude of the couple moment may then be computed to be 103.2 lb-ft. The orientation of the normal to the plane of the couple is readily evaluated as 76° with the horizontal, making the total couple moment identical to our preceding result.

A common notation for couples is shown in Fig. 3.26. The values given will be that of the couple moments.

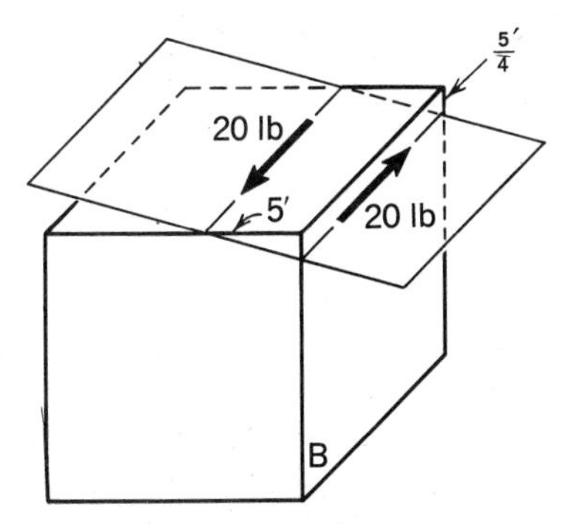


Figure 3.25. Eliminate collinear 20-lb forces.

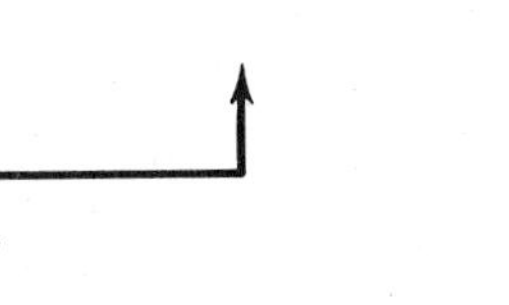

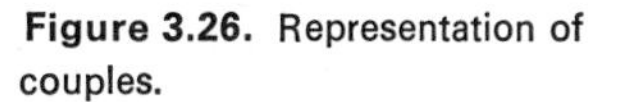

Figure 3.26. Representation of couples.

## 3.7 Moment of a Couple About a Line

In a previous section, we pointed out that the moment of a force $\boldsymbol{F}$ about a line $A$–$A$ (see Fig. 3.27) is found by first taking the moment of $\boldsymbol{F}$ about *any* point $P$ on $A$–$A$ and then dotting this vector into $\boldsymbol{a}$, the unit vector along the line. That is,

$$M_{AA} = (\boldsymbol{r} \times \boldsymbol{F}) \cdot \boldsymbol{a} \tag{3.18}$$

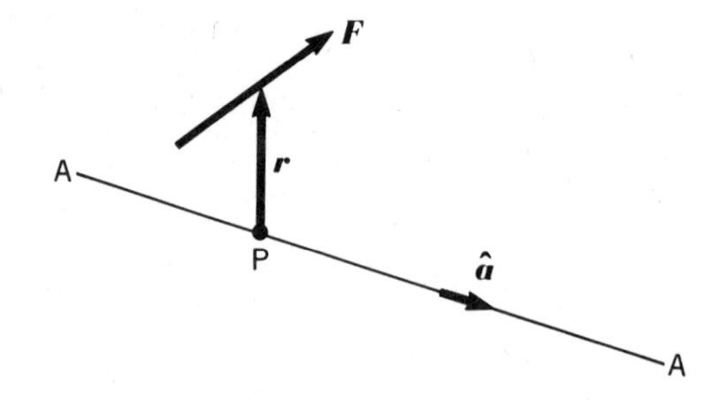


**Figure 3.27.** To find moment of ***F*** about *A–A*.

Consider now the moment of a couple about a line. For this purpose, we show a couple moment ***C*** and line *A–A* in Fig. 3.28. As before, we want first the moment of the couple about any point *P* along *A–A*. But the moment of ***C*** about *every* point in space is simply ***C*** itself. Therefore, to get the moment about the line *A–A* all we need do is dot ***C*** into ***a***. Thus,

$$M_{AA} = \mathbf{C} \cdot \mathbf{a} \tag{3.19}$$

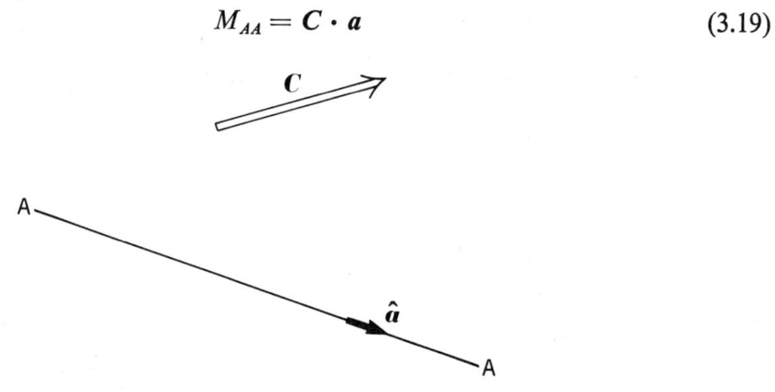


**Figure 3.28.** To find moment of couple about *A–A*.

Since ***C*** is a free vector, the moments of ***C*** about all lines parallel to *A–A* must have the same value.

## Problems

**3.32.** A truck driver, while changing a tire, must tighten the nuts holding on the wheel using a torque of 80 lb-ft. If his tire tool has a length such that the forces from his hand are 22 in. apart, how much force must he exert with each hand? To remove the nuts, he exerted 70 lb with each hand. What torque did he apply?

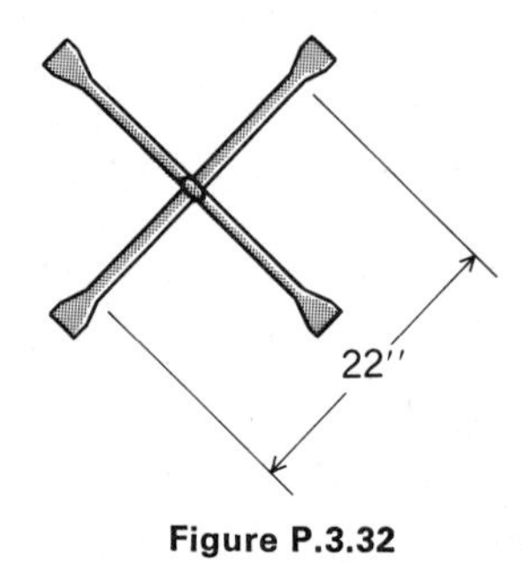


**Figure P.3.32**

**3.33.** Equal couples in the plane of a wheel are shown in Fig. P.3.33. Explain why they are equivalent for the purpose of turning the wheel. Are they equivalent from the viewpoint of the deformation of the wheel? Explain.

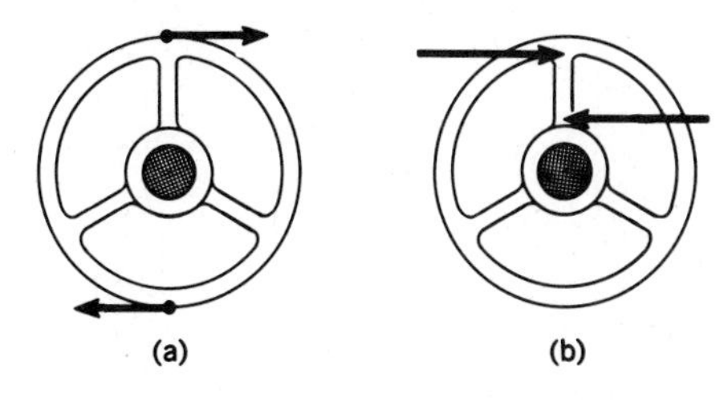


Figure P.3.33

**3.34.** Oil-field workers can exert between 50 lb and 125 lb with each hand on a valve wheel (one hand on each side). If a couple moment of 100 lb-ft is required to close the valve, what diameter $d$ must the wheel have?

Figure P.3.34

**3.35.** Two children push with 30 lb of force each on the rail of a 10-ft-diameter merry-go-round. What couple moment do they produce? They fasten a 20-ft-long 2-in. by 4-in. board to the merry-go-round so that the middle of the board is at the middle of the merry-go-round. What is the resulting couple moment if they push on the ends of the board? What moment about the merry-go-round axis would they generate by fastening one end of the board to the middle of the merry-go-round and both pushing in the same direction on the other end?

**3.36.** A posthole digger has a 2-ft-long handle on a 5-ft-long shaft fastened to the digging (scraping) base. From tests, we know that a couple moment of 100 lb-ft is required to dig a posthole in clay, but only 65 lb-ft is needed in sandy soil. What force $F$ must be applied in each case to dig a hole if the distance between the forces from a person's hands is 20 in.?

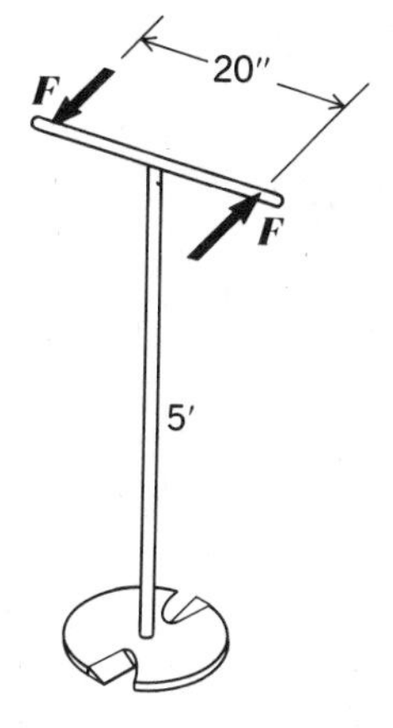


Figure P.3.36

**3.37.** While stopping, a truck develops a 350 N-m of torque at the rear axle due to the action of the brake drum on the axle. What forces are generated at the front and rear supports of the springs to which the axle is attached?

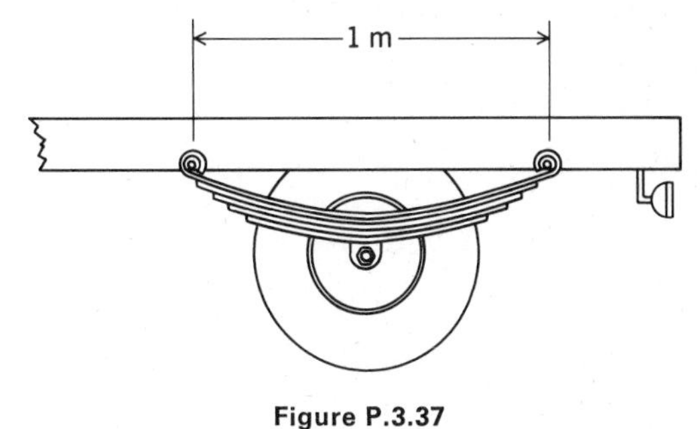


Figure P.3.37

**3.38.** A couple is shown in the $yz$ plane. What is the moment of this couple about the origin?

About point (6, 3, 4) m? What is the moment of the couple about a line through the origin with direction cosines $l = 0$, $m = .8$, $n = -.6$? If this line is shifted to a parallel position so that it goes through point (6, 3, 4) m, what is the moment of the couple about this line?

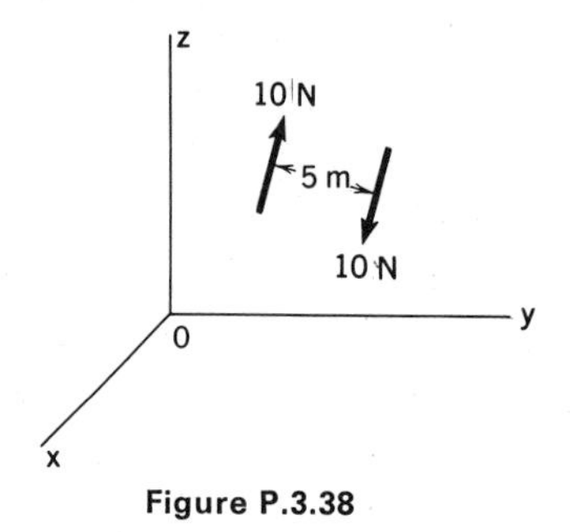


Figure P.3.38

**3.39.** Given the indicated forces, what is the moment of these forces about points $A$ and $B$?

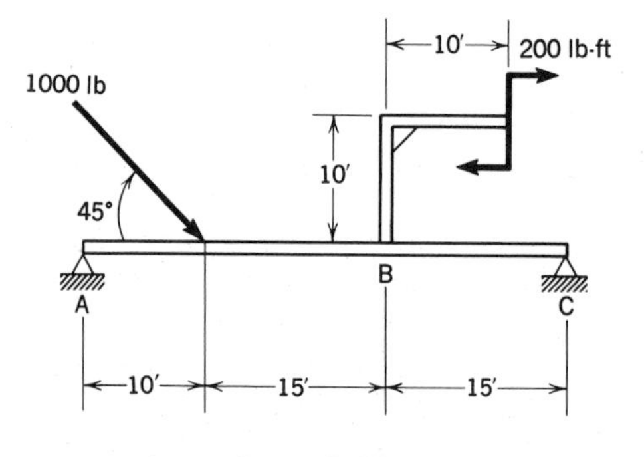

Figure P.3.39

**3.40.** Consider the steering mechanism for a go-cart. The linkages are all in a plane at 45° to the horizontal, which is perpendicular to the steering shaft. In a hard turn, the driver exerts 30 lb with each hand on the 12-in.-diameter steering wheel. What forces develop at the joints B and E of the steering mechanism? What couple moment is applied to each wheel normal to the ground? Assume half the transmitted torque goes to each wheel.

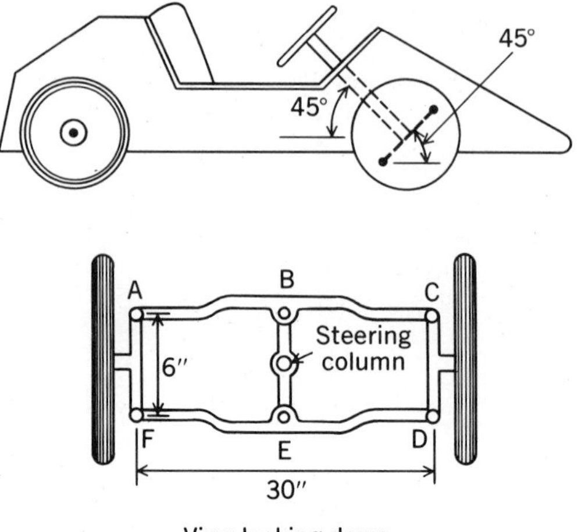


Figure P.3.40

**3.41.** An eight-bladed windmill used for power generation and pumping water stops turning because a bearing on the blade shaft has "frozen up." However, the wind still blows, so each blade is subjected to a 25-lb force perpendicular to the (flat) blade surface. The force effectively acts at 2 ft from the shaft to which the blades are attached. The blades are inclined at 30° to the axis of rotation. What is the total thrust of all the blade forces on the windmill shaft? What is the moment on the stalled shaft?

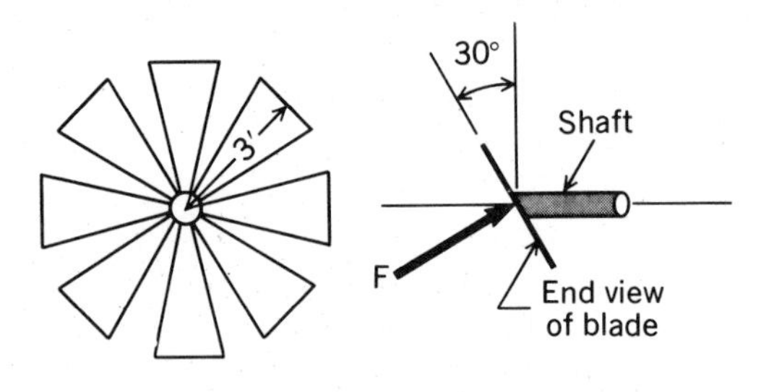


Figure P.3.41

**3.42.** What is the moment of the forces shown about point $A$ and about a point $P$ having a position vector

$$\boldsymbol{r}_p = 10\boldsymbol{i} + 7\boldsymbol{j} + 15\boldsymbol{k} \text{ m?}$$

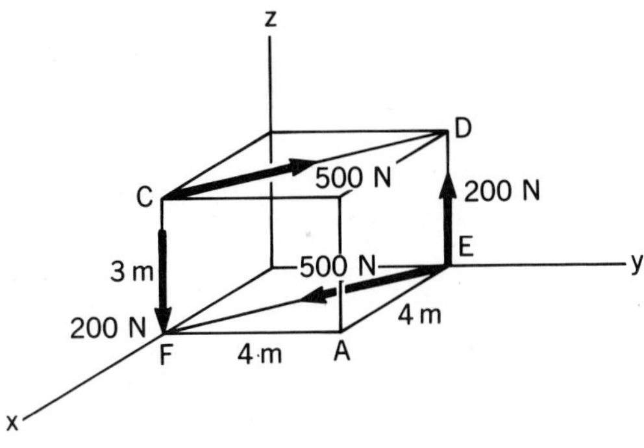

Figure P.3.42

**3.43.** A force $F_1 = 10i + 6j + 3k$ N acts at position (3, 0, 2) m. At point (0, 2, −3) m, an equal but opposite force $-F_1$ acts. What is the couple moment? What are the direction cosines of the normal to the plane of the couple?

**3.44.** Force $F_1 = 16i - 10j + 5k$ N acts at the origin while $F_2 = -F_1$ acts at the end of a rod of length 12 m protruding from the origin with direction cosines $l = .6$, $m = .8$. What is the moment about point $P$ at

$$r_p = 3i + 10j + 15k \text{ m?}$$

What is the twist about an axis going through $P$ having the unit vector

$$\epsilon = .2i + .8j + .566k?$$

**3.45.** Equal and opposite forces are directed along diagonals on the faces of a cube. What is the couple moment if $a = 3$ m and $F = 10$ N? What is the moment of this couple about a diagonal from $A$ to $D$?

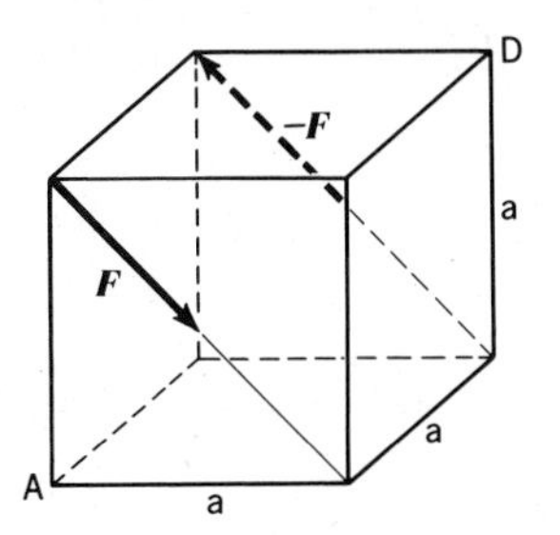


Figure P.3.45

**3.46.** What is the turning action of the forces shown about the diagonal $A$–$D$?

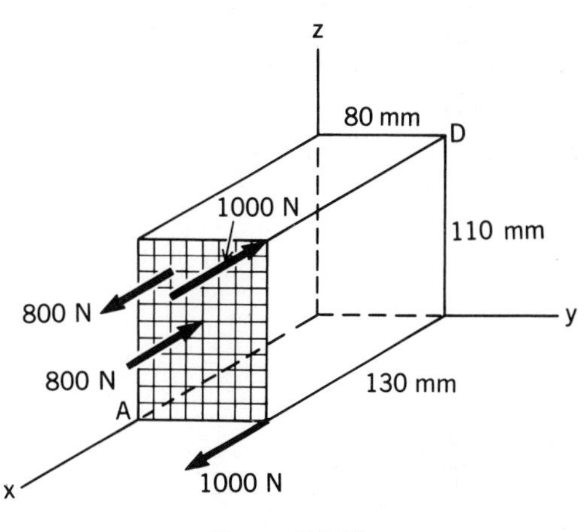


Figure P.3.46

**3.47.** What is the total couple moment of the three couples shown? What is the moment of this force system about point (3, 4, 2) ft? What is the moment of this force system about the position vector $r = 3i + 4j + 2k$ ft? What is the total force of this system?

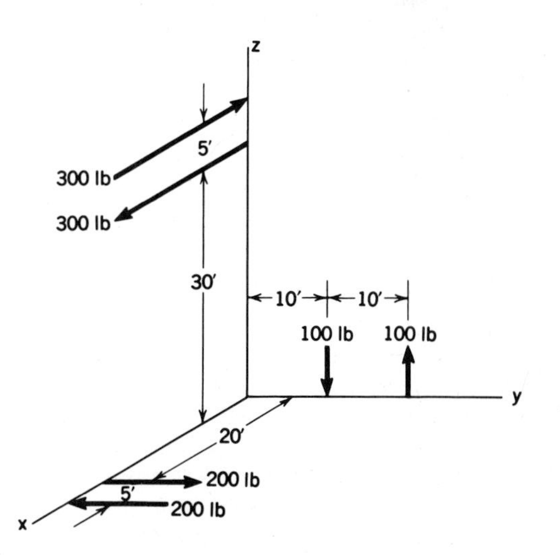


Figure P.3.47

**3.48.** An oil-field pump has two valves, one on top and one on the side, that must be closed simultaneously. The valve wheels are each 27 in. in diameter and are turned with both hands by workers who can exert between 50 lb and 125 lb with each hand. If a weak worker turns the side wheel and a strong worker turns the top wheel, what is the total twisting moment (couple moment) on the pump?

**3.49.** What is the total moment of the force system shown about the origin?

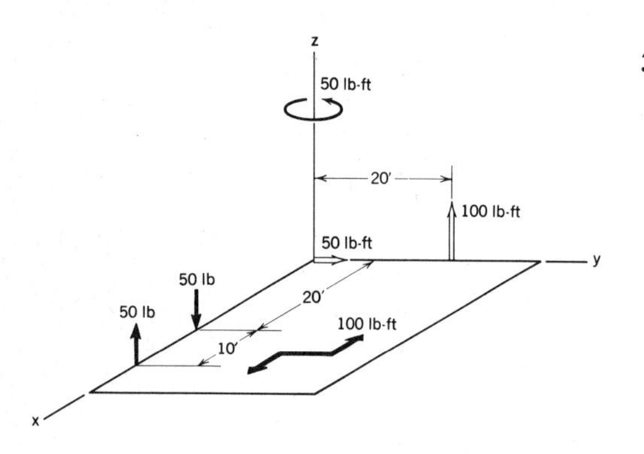

**Figure P.3.49**

**3.50.** Add the couples whose forces act along diagonals of the sides of the rectangular parallelepiped.

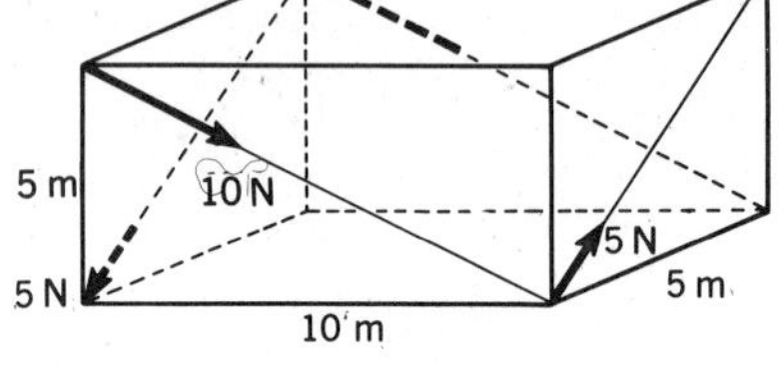


**Figure P.3.50**

**3.51.** Given the couple moments

$$C_1 = 100\boldsymbol{i} + 30\boldsymbol{j} + 82\boldsymbol{k} \text{ lb-ft}$$
$$C_2 = -16\boldsymbol{i} + 42\boldsymbol{j} \text{ lb-ft}$$
$$C_3 = 15\boldsymbol{k} \text{ lb-ft}$$

what couple will restrain the twisting action of this system about an axis going from

$$\boldsymbol{r}_1 = 6\boldsymbol{i} + 3\boldsymbol{j} + 2\boldsymbol{k} \text{ ft}$$

to

$$\boldsymbol{r}_2 = 10\boldsymbol{i} - 2\boldsymbol{j} + 3\boldsymbol{k} \text{ ft}$$

while giving a moment of 100 lb-ft about the $x$ axis and 50 lb-ft about the $y$ axis?

## 3.8 Closure

In this chapter, we have considered several important vector quantities and their properties. In particular, for rigid bodies we could take certain liberties with a couple without invalidating the results. We are now ready to pursue in greater detail the important subject of equivalence of force systems for rigid-body considerations.

## Review Problems

**3.52.** A surveyor on a 100-m-high hill determines that the corner of a building at the base of the hill is 600 m east and 1500 m north of her position. What is the position of the building corner relative to another surveyor on top of a 5000-m-high mountain that is 10,000 m west and 3000 m south of the hill? What is the distance from the second surveyor to the building corner?

**3.53.** Compute the moment of the 1000-lb force about supporting points $A$ and $B$.

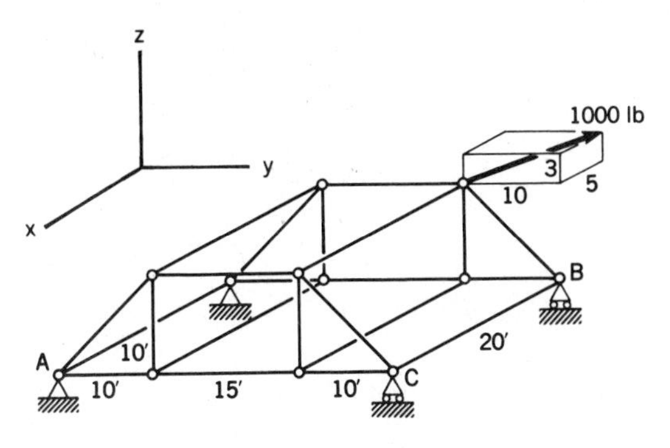


Figure P.3.53

**3.54.** An A-frame for hoisting and dragging equipment is held in the position shown by a cable $C$. To determine the cable force, the moment of the applied force about axis $B$–$B$ must be known. What is that moment when a 1000-N force is applied as shown?

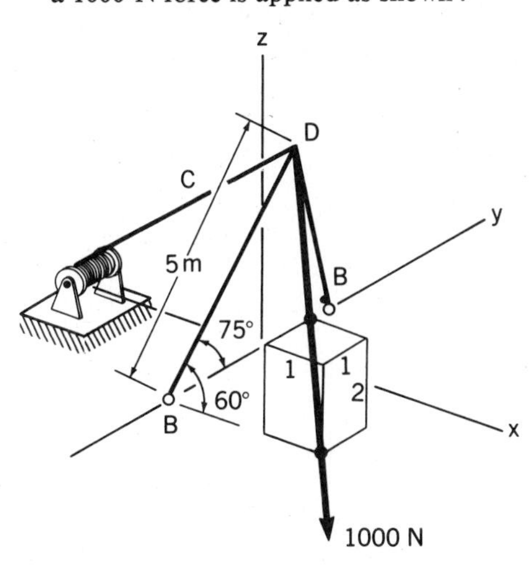

Figure P.3.54

**3.55.** A plumber places his hands 18 in. apart on a pipe threader and can push (and pull) with 80 lb of force. What couple moment does he exert? How much could he exert if he moved his hands to the ends so that his hands are 24 in. apart? What force must he apply at the ends to achieve the same couple moment as when he held his hands 18 in. apart?

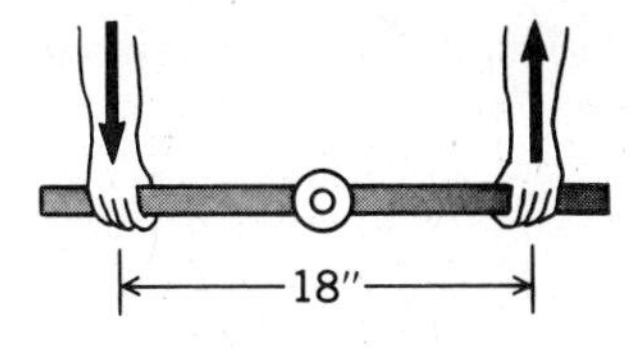


Figure P.3.55

**3.56.** What is the moment about $A$ of the 500-N force and the 3000-N-m couple acting on the cantilever beam?

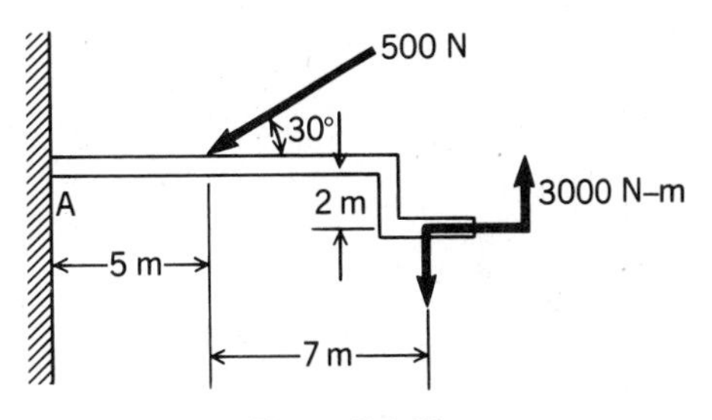


Figure P.3.56

**3.57.** A force $\boldsymbol{F} = 16\boldsymbol{i} + 10\boldsymbol{j} - 3\boldsymbol{k}$ lb goes through point $a$ having a position vector $\boldsymbol{r}_a = 16\boldsymbol{i} - 3\boldsymbol{j} + 12\boldsymbol{k}$ ft. What is the moment about an axis going through points 1 and 2 having respective position vectors given as

$$\boldsymbol{r}_1 = 6\boldsymbol{i} + 3\boldsymbol{j} - 2\boldsymbol{k} \text{ ft}$$

$$\boldsymbol{r}_2 = 3\boldsymbol{i} - 4\boldsymbol{j} + 12\boldsymbol{k} \text{ ft?}$$

**3.58.** Replace the system of forces and couples by a single couple. Note that the 1000-N-m couple is in the diagonal plane $ABCD$.

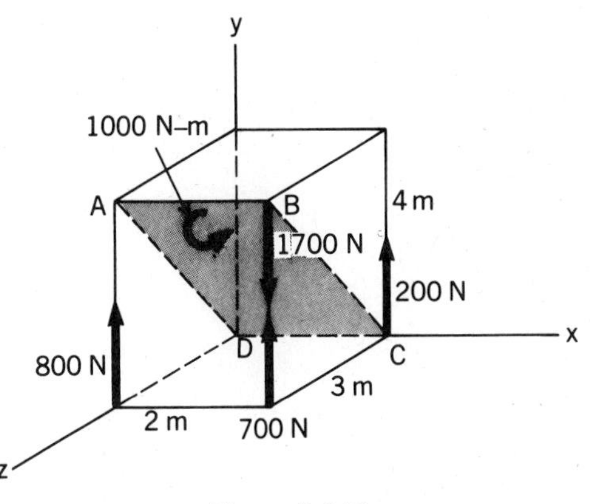


Figure P.3.58

**3.59.** Find the torque of the force system about axis $AB$.

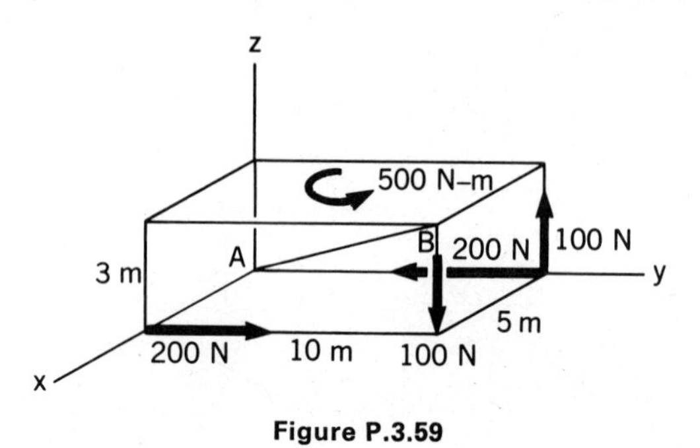


**Figure P.3.59**

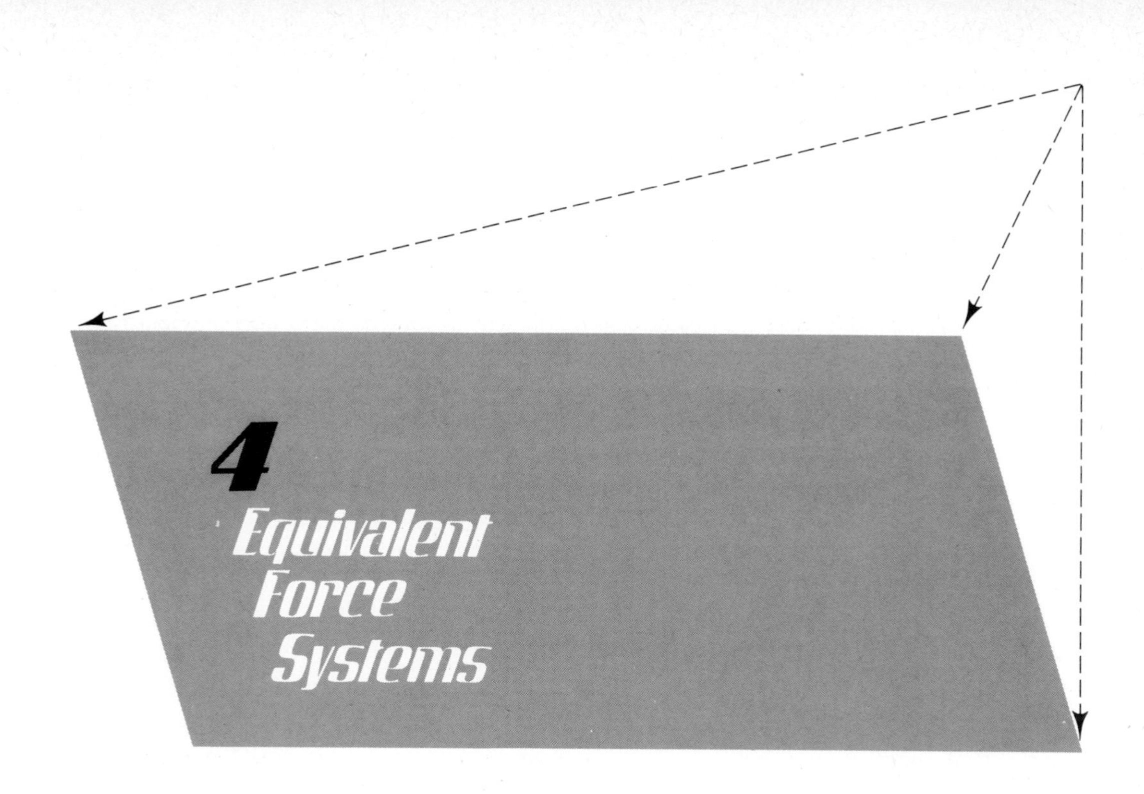


## 4.1 Introduction

In Chapter 1, we defined equivalent vectors as those that have the same capacity in some given situation. We shall now investigate an important class of situations, those in which a rigid-body model can be employed. We will be concerned with the equivalence requirements for force systems acting on a rigid body.

The effect that forces have on a rigid body is only manifested in the motion (or lack of motion) of the body induced by the forces. Two force systems, then, are equivalent if they are capable of *initiating* the same motion of the rigid body. The conditions required to give two force systems this equal capacity are:

1. Each force system must exert an equal "push" or "pull" on the body in any direction. For two concurrent force systems, this requirement is satisfied if the add tion of the forces in each system results in equal force vectors.
2. Each force system must exert an equal "turning" action about any point in space. This means that the moment vectors of the force systems for any chosen point must be equal.

Although these conditions will most likely be intuitively acceptable to the reader, we shall later prove them to be necessary and, for certain situations, sufficient for equivalence when we study dynamics.

As a beginning here, we shall reiterate several basic force equivalences for rigid bodies that will serve as a foundation for more complex cases. You should subject them to the tests listed above.

1. The sum of a set of concurrent forces is a single force that is equivalent to the original system. Conversely, a single force is equivalent to any set of its components.
2. A force may be moved along its line of action (i.e., forces are transmissible vectors).
3. The only effect that a couple develops on a rigid body is embodied in the couple moment. Since the couple moment is always a free vector, for our purposes at present the couple may be altered in any way as long as the couple moment is not changed.

In succeeding sections, we shall present other equivalence relations for rigid bodies and then examine perfectly general force systems with a view to replacing them with more convenient and simpler equivalent force systems. These simpler replacements are often called *resultants* of the more general systems.

## 4.2 Translation of a Force to a Parallel Position

In Fig. 4.1, let us consider the possibility of moving a force $\boldsymbol{F}$ (solid arrow) acting on a rigid body to a parallel position at point $a$ while maintaining rigid-body equivalence. If at position $a$ we apply equal and opposite forces, one of which is $\boldsymbol{F}$ and the other $-\boldsymbol{F}$, a system of three forces is formed that is clearly equivalent to the single force $\boldsymbol{F}$. Note that the original force $\boldsymbol{F}$ and the new force in the opposite sense form a couple

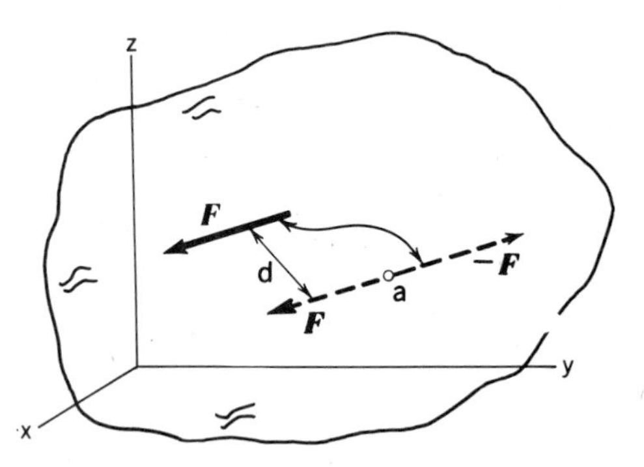


**Figure 4.1.** Insert equal and opposite forces at *a*.

(the pair is identified by a wavy connecting line). As usual, we represent the couple by its moment $\boldsymbol{C}$, as shown in Fig. 4.2, normal to the plane $A$ of point $a$ and the original force $\boldsymbol{F}$. The magnitude of the couple moment $\boldsymbol{C}$ is $|\boldsymbol{F}|d$, where $d$ is the perpendicular distance between point $a$ and the original line of action of the force. The couple moment may be moved to any parallel position, including the origin, as indicated in Fig. 4.2.

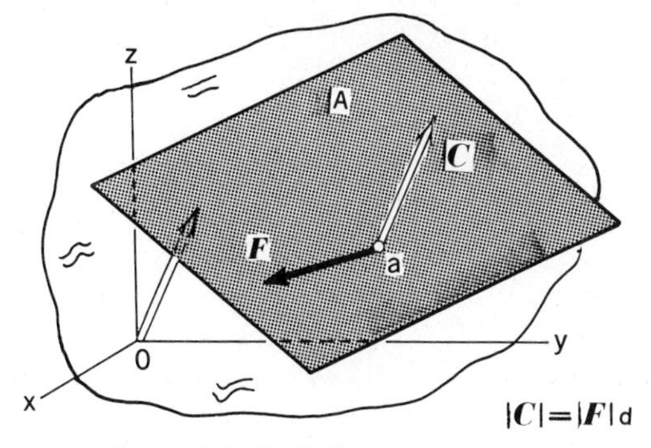


**Figure 4.2.** Equivalent system at *a*.

Thus, we see that *a force may be moved to any parallel position, provided that a couple moment of the correct orientation and size is simultaneously provided.* There are, then, an infinite number of arrangements possible to get the equivalent effects of a single force on a rigid body.

The reverse procedure may also be instituted in reducing a force and a couple *in the same plane* to a *single* equivalent force. This is illustrated in Fig. 4.3, where a couple composed of forces $\boldsymbol{B}$ and $-\boldsymbol{B}$ a distance $d_1$ apart and a force $\boldsymbol{A}$ are shown in plane $N$. The moment representation of the couple is shown with force $\boldsymbol{A}$ in Fig. 4.4.

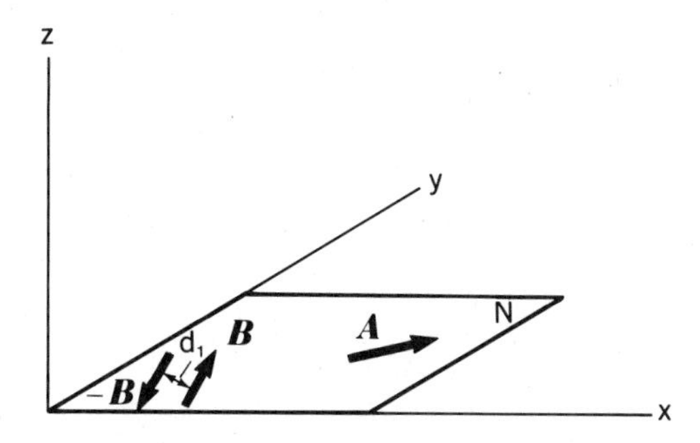


**Figure 4.3.** Force *A* and couple in same plane.

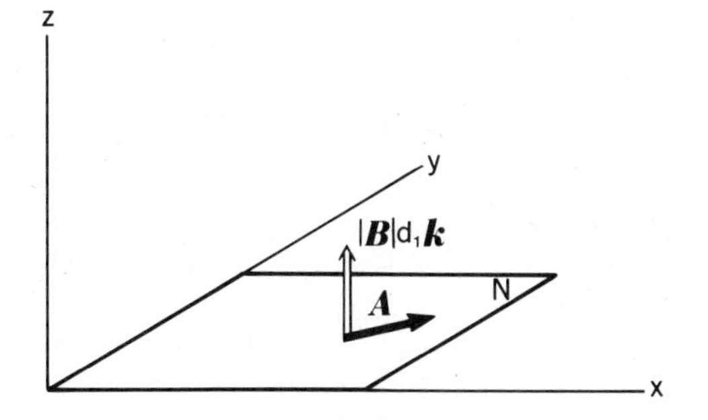


**Figure 4.4.** Force *A* and couple moment.

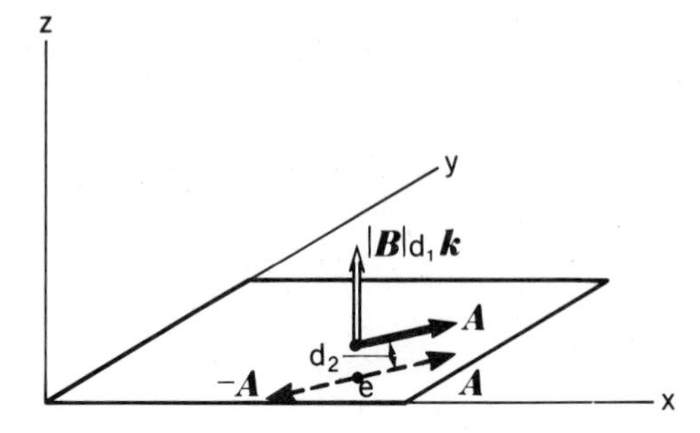


**Figure 4.5.** Equal and opposite forces placed at $e$.

Equal and opposite forces $\boldsymbol{A}$ and $-\boldsymbol{A}$ may next be added to the system at a position $e$ (see Fig. 4.5). The purpose of this step is to form another couple moment with a magnitude $|\boldsymbol{A}|d_2$ equal to $|\boldsymbol{B}|d_1$ and with a direction of turning opposite to the original couple moment (see Fig. 4.6). The couple moments then cancel each other out, and we have, in effect, only the single force $\boldsymbol{A}$ going through point $e$.

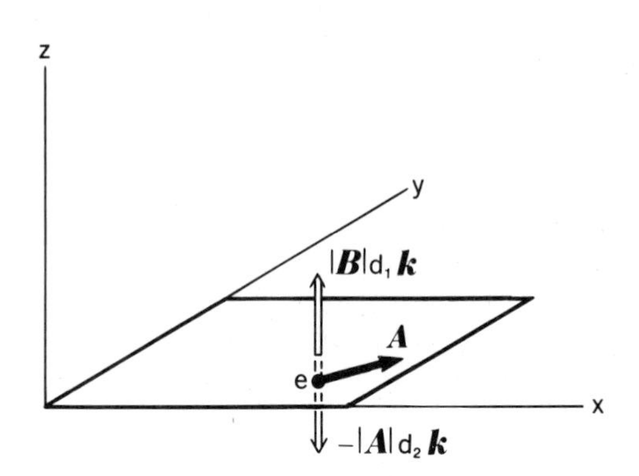


**Figure 4.6.** Adjust $d_2$ so that couple moments cancel.

We now present a simple method for computing the couple moment developed on moving a force to a parallel position. Return to Fig. 4.1 and compute the moment $\boldsymbol{M}$ of the force $\boldsymbol{F}$ about point $a$. We can express this as [see Fig. 4.7(a)]

$$\boldsymbol{M} = \boldsymbol{\rho} \times \boldsymbol{F} \tag{4.1}$$

where $\boldsymbol{\rho}$ is a position vector from $a$ to any point along the line of action of $\boldsymbol{F}$. Now the equivalent force system, shown in Fig. 4.7(b), must have the *same moment*, $\boldsymbol{M}$, about point $a$ as the original system. Clearly, the moment about point $a$ is due only to the couple moment $\boldsymbol{C}$. That is,

$$\boldsymbol{M} = \boldsymbol{C} \tag{4.2}$$

Accordingly, we conclude, on comparing the previous two equations, that

$$\boldsymbol{C} = \boldsymbol{\rho} \times \boldsymbol{F}$$

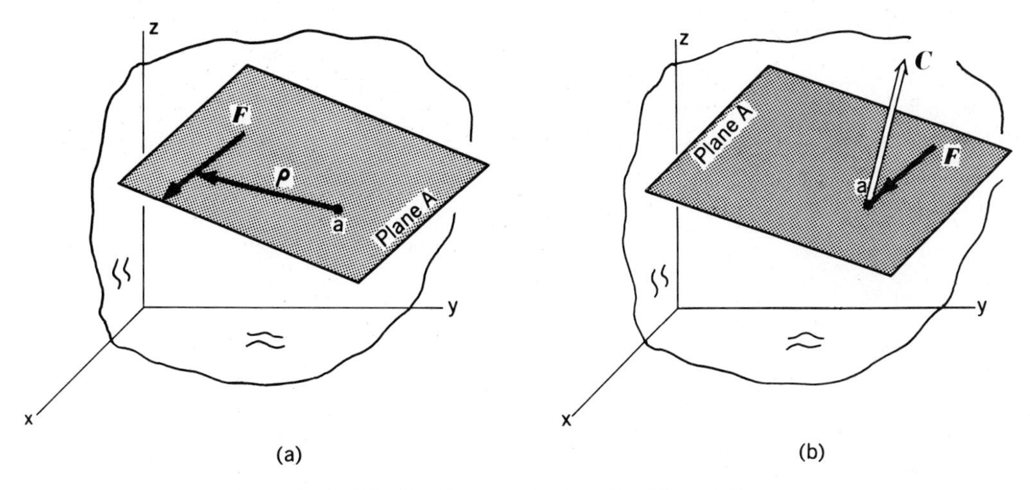


**Figure 4.7.** Couple moment on moving $\boldsymbol{F}$ is $\boldsymbol{\rho} \times \boldsymbol{F}$.

Thus, in shifting a force to pass through some new point, *we introduce a couple whose couple moment equals the moment of the force about this new point.*

We illustrate this in the following example.

**EXAMPLE 4.1**

A force $\boldsymbol{F} = 6\boldsymbol{i} + 3\boldsymbol{j} + 6\boldsymbol{k}$ lb goes through a point whose position vector is $\boldsymbol{r}_1 = 2\boldsymbol{i} + \boldsymbol{j} + 10\boldsymbol{k}$ ft (see Fig. 4.8). Replace this force by an equivalent force system, for purposes of rigid-body mechanics, going through position $P$, whose position vector is $\boldsymbol{r}_2 = 6\boldsymbol{i} + 10\boldsymbol{j} + 12\boldsymbol{k}$ ft.

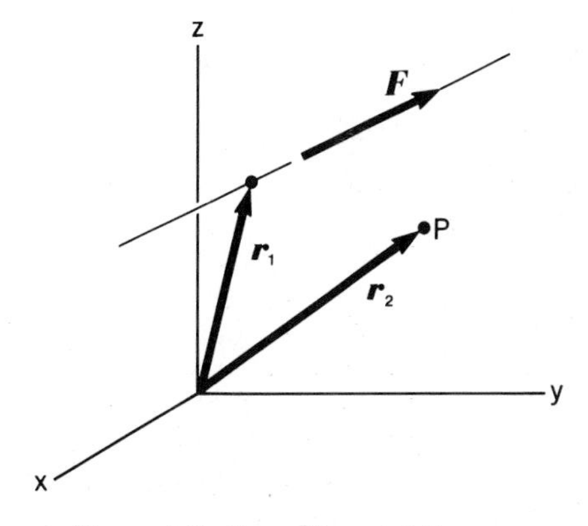


**Figure 4.8.** Move $\boldsymbol{F}$ to point $P$.

The new system will consist of this force $\boldsymbol{F}$ going through the position $\boldsymbol{r}_2$ and, in addition, there will be a couple moment $\boldsymbol{C}$ given as

$$\boldsymbol{C} = \boldsymbol{\rho} \times \boldsymbol{F} = (\boldsymbol{r}_1 - \boldsymbol{r}_2) \times \boldsymbol{F}$$

Inserting values, we have

$$\begin{aligned}\boldsymbol{C} &= [(2\boldsymbol{i} + \boldsymbol{j} + 10\boldsymbol{k}) - (6\boldsymbol{i} + 10\boldsymbol{j} + 12\boldsymbol{k})] \\ &\quad \times (6\boldsymbol{i} + 3\boldsymbol{j} + 6\boldsymbol{k}) \\ &= (-4\boldsymbol{i} - 9\boldsymbol{j} - 2\boldsymbol{k}) \times (6\boldsymbol{i} + 3\boldsymbol{j} + 6\boldsymbol{k}) \\ &= \begin{vmatrix} -4 & -9 & -2 \\ 6 & 3 & 6 \\ \boldsymbol{i} & \boldsymbol{j} & \boldsymbol{k} \end{vmatrix} \\ &\quad\;\; \begin{matrix} -4 & -9 & -2 \\ 6 & 3 & 6 \end{matrix}\end{aligned}$$

Therefore,

$$C = -12\mathbf{k} - 12\mathbf{j} - 54\mathbf{i} + 6\mathbf{i} + 24\mathbf{j} + 54\mathbf{k}$$
$$= -48\mathbf{i} + 12\mathbf{j} + 42\mathbf{k} \text{ ft-lb}$$

**EXAMPLE 4.2**

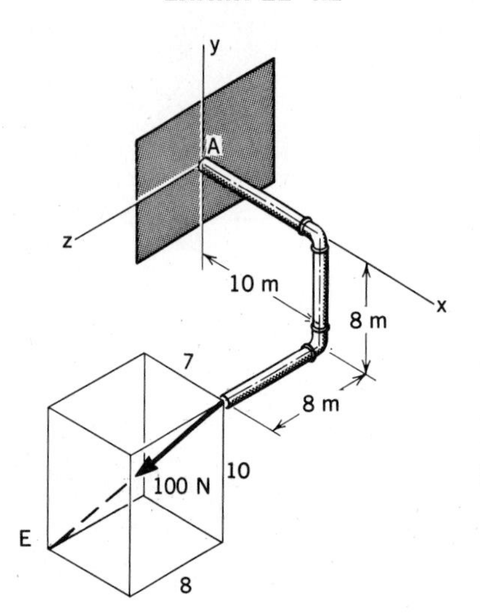


**Figure 4.9.** Find equivalent force system at $A$.

What is the equivalent force system at position $A$ for the 100-N force shown in Fig. 4.9?

The 100-N force can be expressed vectorially as follows:

$$\mathbf{F} = F\frac{\overrightarrow{BE}}{|\overrightarrow{BE}|} = 100\left(\frac{-7\mathbf{i} - 10\mathbf{j} + 8\mathbf{k}}{\sqrt{7^2 + 10^2 + 8^2}}\right) \tag{a}$$
$$= -48.0\mathbf{i} - 68.5\mathbf{j} + 54.8\mathbf{k} \text{ N}$$

We then have the force given above at $A$. And in addition, we have a couple moment, $\mathbf{C}$, found using a position vector, $\mathbf{r}$, from $A$ to any point along the line of action of the 100-N force. Thus, choosing point $B$ for $\mathbf{r}$, we have

$$\mathbf{C} = (10\mathbf{i} - 8\mathbf{j} + 8\mathbf{k}) \times (-48.0\mathbf{i} - 68.5\mathbf{j} + 54.8\mathbf{k})$$

$$= \left|\begin{array}{ccc} 10 & -8 & 8 \\ -48.0 & -68.5 & 54.8 \\ \mathbf{i} & \mathbf{j} & \mathbf{k} \end{array}\right| \begin{array}{ccc} 10 & -8 & 8 \\ -48.0 & -68.5 & 54.8 \end{array}$$

$$= (10)(-68.5)\mathbf{k} + (-48)(8)\mathbf{j} + (-8)(54.8)\mathbf{i}$$
$$- (8)(-68.5)\mathbf{i} - (54.8)(10)\mathbf{j} - (-8)(-48.0)\mathbf{k}$$

Therefore,

$$\mathbf{C} = 109.6\mathbf{i} - 932\mathbf{j} - 1069\mathbf{k} \text{ N-m} \tag{b}$$

## Problems

*In several of the problems of this set we shall concentrate the weight of a body at its center of gravity. Most likely you are used to doing this from an earlier physics course. In Section 4.5 we shall justify this procedure.*

**4.1.** Replace the 100-lb force by an equivalent system, from the rigid-body point of view, at $A$. Do the same for point $B$. Do this problem by the technique of adding equal and opposite collinear forces and also by using the cross product.

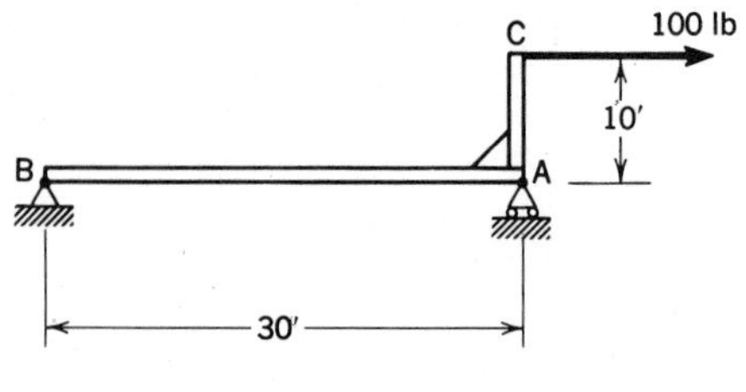


**Figure P.4.1**

**4.2.** To back an airplane away from the boarding gate, a tractor pushes with a force of 15 kN on the nose wheels. What is the equivalent force system on the landing-gear pivot point which is 2 m above the point where the tractor pushes?

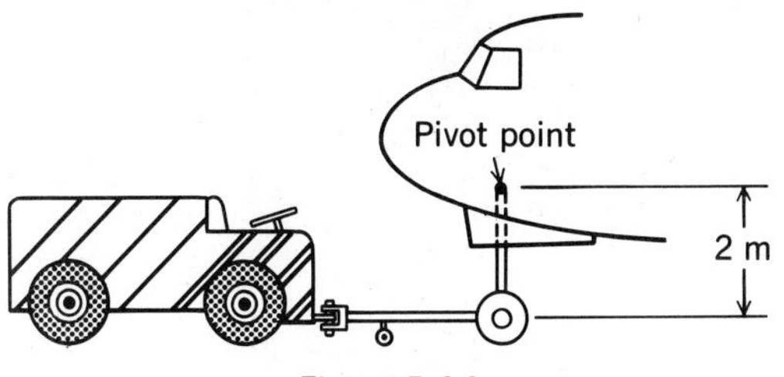


Figure P.4.2

**4.3.** Replace the 1000-lb force by equivalent systems at points $A$ and $B$. Do so by using the addition of equal and opposite collinear force components and by using the cross product.

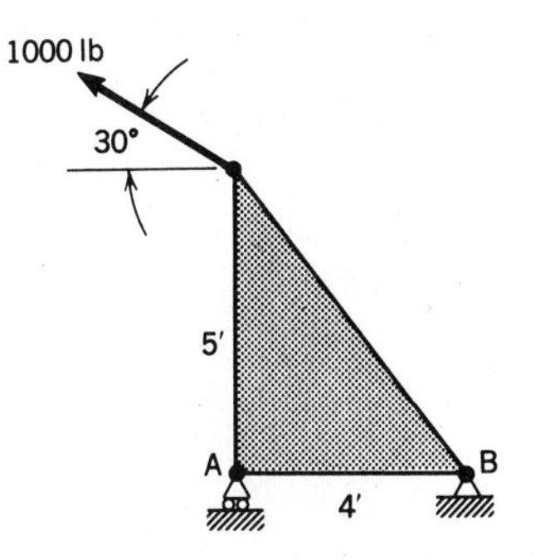


Figure P.4.3

**4.4.** A parking-lot gate arm weighs 150 N. Because of the taper, the weight is concentrated at a point $1\frac{1}{4}$ m from the pivot point. What is the equivalent force system at the gate-arm pivot point?

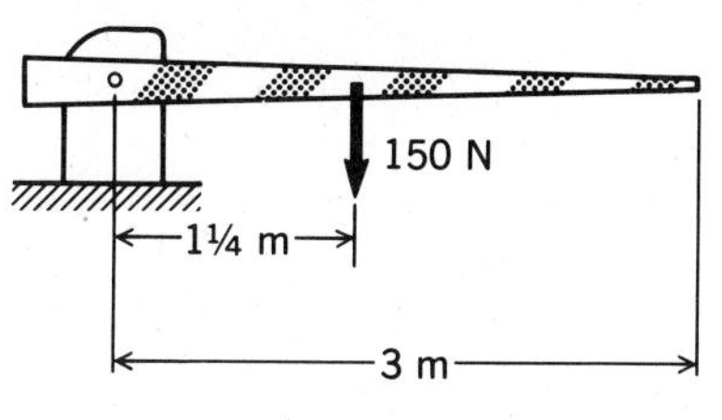


Figure P.4.4

**4.5.** A plumber exerts a vertical 60-lb force on a pipe wrench inclined at 30° to the horizontal. What force and couple moment on the pipe are equivalent to the plumber's action?

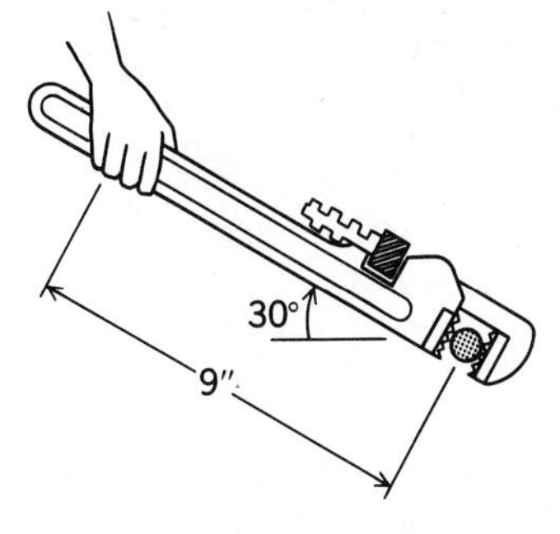

Figure P.4.5

**4.6.** A tractor operator is attempting to lift a 10-kN boulder. What are the equivalent force systems at $A$ and at $B$ from the boulder?

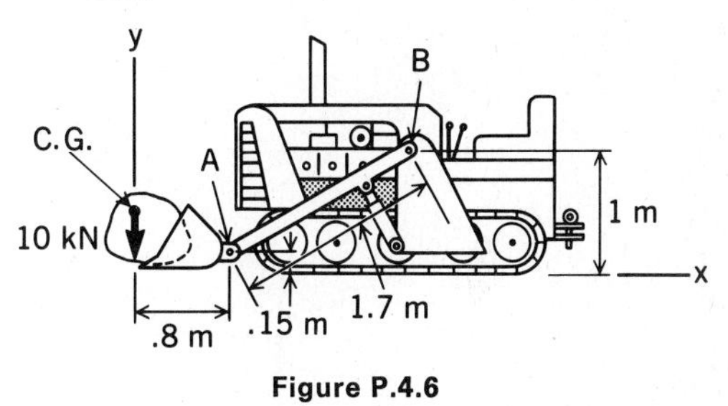


Figure P.4.6

**4.7.** A small hoist has a lifting capacity of 20 kN. What are the largest and smallest equivalent force systems at $A$ for the rated maximum capacity?

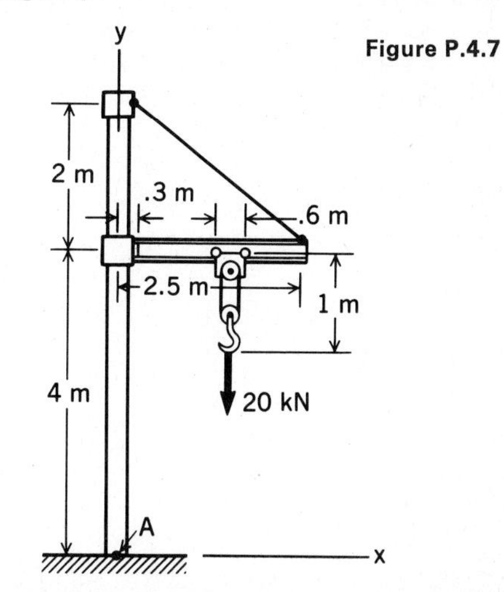


Figure P.4.7

**4.8.** Replace the forces by a single equivalent force.

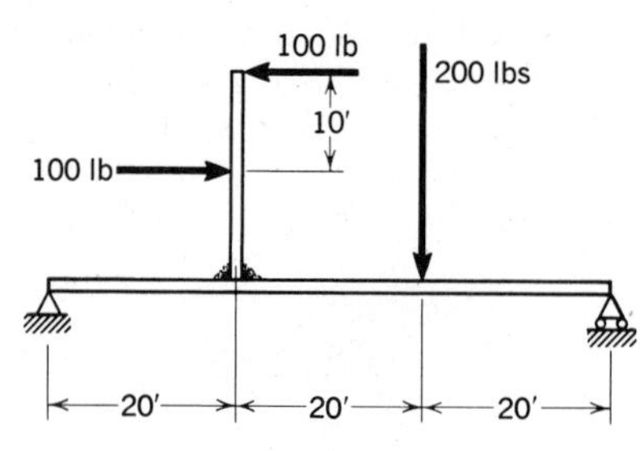


**Figure P.4.8**

**4.9.** Replace the forces and torques shown acting on the apparatus by a single force. Carefully give the line of action of this force.

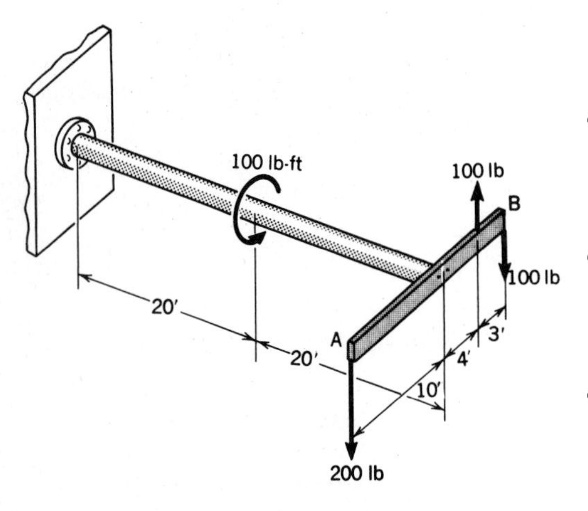


**Figure P.4.9**

**4.10.** A carpenter presses down on a brace-and-bit with a 150-N force while turning the brace with a 200-N force oriented for maximum twist. What is the equivalent force system on the end of the bit at $A$?

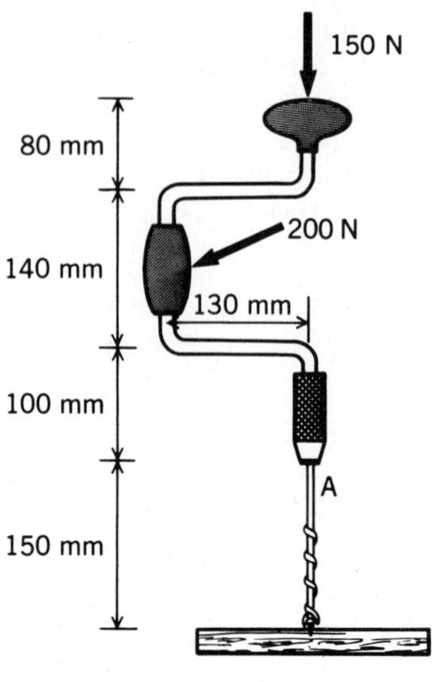


**Figure P.4.10**

**4.11.** A force $\boldsymbol{F} = 3\boldsymbol{i} - 6\boldsymbol{j} + 4\boldsymbol{k}$ lb goes through point (6, 3, 2) ft. Replace this force by an equivalent system where the force goes through point (2, −5, 10) ft.

**4.12.** A force $\boldsymbol{F} = 20\boldsymbol{i} - 60\boldsymbol{j} + 30\boldsymbol{k}$ N goes through a point (10, −5, 4) m. What is the equivalent system at point $A$ having position vector $\boldsymbol{r}_A = 20\boldsymbol{i} + 3\boldsymbol{j} - 15\boldsymbol{k}$ m?

**4.13.** Find the equivalent force system at the base of the cantilever pipe system stemming from force $\boldsymbol{F} = 1000$ lb.

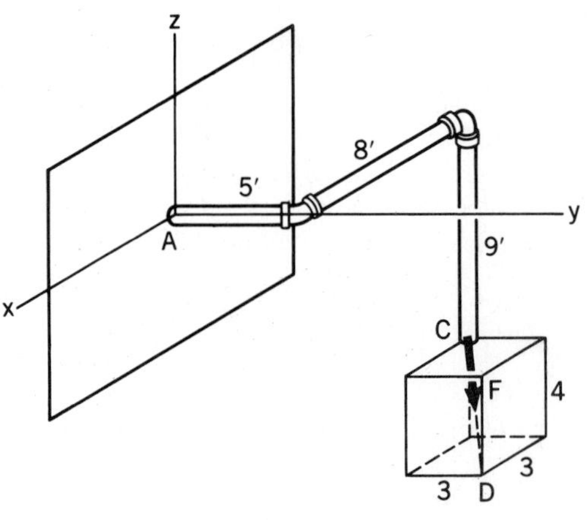


**Figure P.4.13**

**4.14.** In Problem 4.13, the pipe weighs 20 lb/ft. What is the equivalent force system at $A$ from the weight of the pipe? [*Hint:* Concentrate the weights of the pipe sections at the respective centers of gravity (geometric centers in this case).]

**4.15.** The operator of a small boom-type crane is trying to drag a chunk of concrete. The boom is 10° above the horizontal and rotated 30° from the front of the crane. The cable is directed as shown in diagram and has 60 kN of tension. What is the equivalent force system at the boom pivot point?

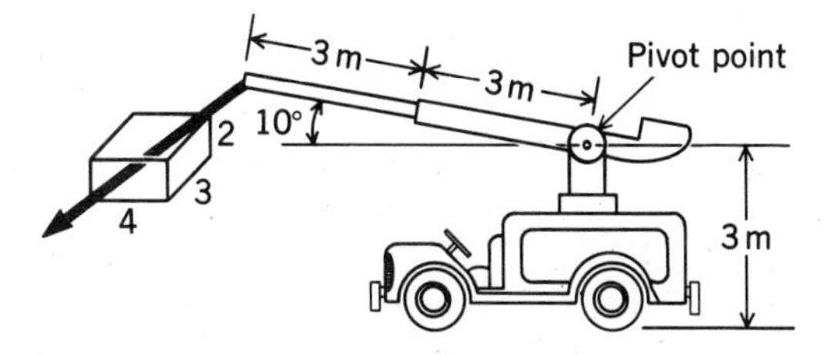


Figure P.4.15

**4.16.** A supplementary supporting guy-wire system for a 200-m-tall tower is tightened. The cables are fastened to the ground at points 120° apart and 100 m from the tower base. What is the equivalent force system acting on the tower base when the tension is 50 kN in cable $AT$, 75 kN in $BT$, and 25 kN in $CT$?

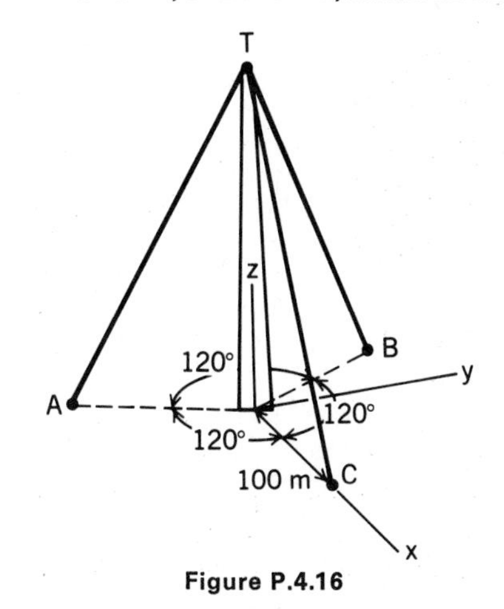


Figure P.4.16

## 4.3 Resultant of a Force System

As defined at the beginning of the chapter, a *resultant of a force system* is a simpler equivalent force system. In many computations it is desirable first to establish the resultant before entering into other computations.

For a general arrangement of forces, no matter how complex, we can always move all forces and couple moments, the latter including both those given and those formed from the movement of forces, to proceed through any single point. The result is then a system of concurrent forces at the point and a system of concurrent couple moments. These systems may then be combined into a single force and a single couple moment. Thus, in Fig. 4.10 we have shown some arbitrary system of forces

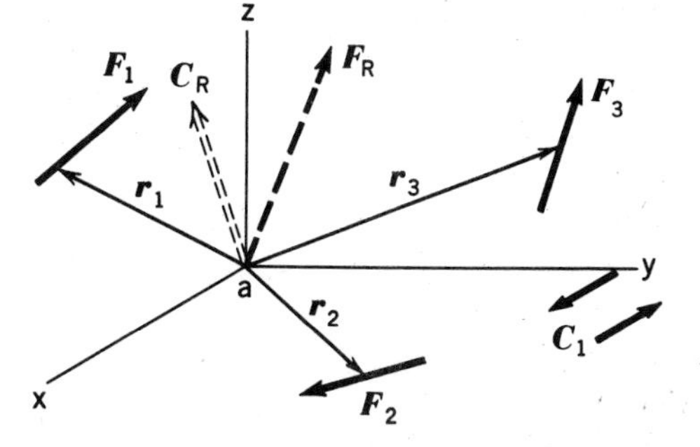


**Figure 4.10.** Resultant of general force system.

and couples using full lines. The resultant force and couple-moment combination at the origin of a rectangular reference is shown as dashed lines.

Thus, *any force system can be replaced at any point by equivalents no more complex than a single force and a single couple moment.* In special cases, which we shall examine shortly, we may have simpler equivalents such as a single force or a single couple moment. Finally, for *equilibrium of a body*, it is necessary that at any point the resultant system of forces and couple moments acting on the body be zero vectors—a fact that will be discussed in dynamics.

The methods of finding a resultant of forces involve nothing new. In moving to any new point, you will recall, there is no change in the force itself other than a shift of line of action; thus, any component of the *resultant* force, such as the $x$ component, can simply be taken as the sum of the respective $x$ components of all the forces in the system. We may then say for the resultant force

$$\boldsymbol{F}_R = [\sum_i (F_i)_x]\boldsymbol{i} + [\sum_i (F_i)_y]\boldsymbol{j} + [\sum_i (F_i)_z]\boldsymbol{k} \tag{4.3}$$

The couple moment accompanying $\boldsymbol{F}_R$ for a chosen point $a$ may then be given as

$$\boldsymbol{C}_R = [\boldsymbol{r}_1 \times \boldsymbol{F}_1 + \boldsymbol{r}_2 \times \boldsymbol{F}_2 + \dots] + [\boldsymbol{C}_1 + \boldsymbol{C}_2 + \dots] \tag{4.4}$$

where the first bracketed quantities result from moving the noncouple forces to $a$, and the second are simply the sum of the given couple moments. The vectors $\boldsymbol{r}$ are from $a$ to arbitrary points along the lines of action of the forces. In more compact form, the equation above becomes

$$\boldsymbol{C}_R = \sum_i \boldsymbol{r}_i \times \boldsymbol{F}_i + \sum_i \boldsymbol{C}_i \tag{4.5}$$

The following example is an illustration of the procedure.

**EXAMPLE 4.3**

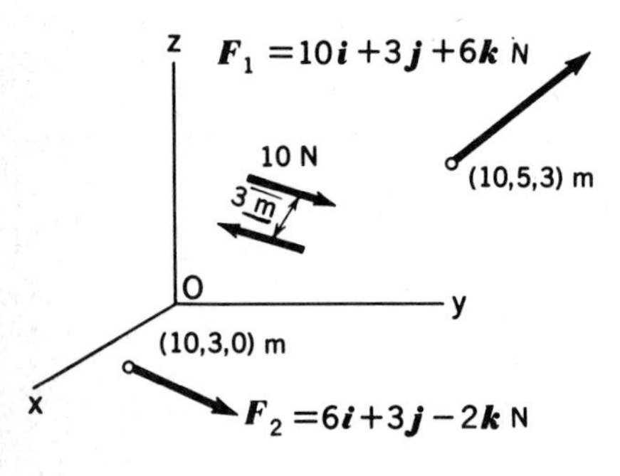

Figure 4.11. Find resultant at $O$.

Two forces and a couple are shown in Fig. 4.11, the couple being positioned in plane $zy$. We shall find the resultant of the system at the origin $O$.

At $O$ we will have a set of two concurrent forces, which may be added to give $\boldsymbol{F}_R$:

$$\boldsymbol{F}_R = (10 + 6)\boldsymbol{i} + (3 + 3)\boldsymbol{j} + (6 - 2)\boldsymbol{k}$$
$$= 16\boldsymbol{i} + 6\boldsymbol{j} + 4\boldsymbol{k} \text{ N}$$

The resultant couple moment at point $O$ is the vector sum of the couple-moment vectors developed by moving the two forces, plus the couple moment of the couple in the $zy$ plane. Thus,

$$\boldsymbol{C}_R = \boldsymbol{r}_1 \times \boldsymbol{F}_1 + \boldsymbol{r}_2 \times \boldsymbol{F}_2 - 30\boldsymbol{i} \text{ N-m}$$

Now

$$\begin{aligned} \boldsymbol{r}_1 \times \boldsymbol{F}_1 &= (10\boldsymbol{i} + 5\boldsymbol{j} + 3\boldsymbol{k}) \times (10\boldsymbol{i} + 3\boldsymbol{j} + 6\boldsymbol{k}) \\ &= 21\boldsymbol{i} - 30\boldsymbol{j} - 20\boldsymbol{k} \text{ N-m} \\ \boldsymbol{r}_2 \times \boldsymbol{F}_2 &= (10\boldsymbol{i} + 3\boldsymbol{j}) \times (6\boldsymbol{i} + 3\boldsymbol{j} - 2\boldsymbol{k}) \\ &= -6\boldsymbol{i} + 20\boldsymbol{j} + 12\boldsymbol{k} \text{ N-m} \end{aligned}$$

Hence,

$$\boldsymbol{C}_R = -15\boldsymbol{i} - 10\boldsymbol{j} - 8\boldsymbol{k} \text{ N-m}$$

The resultant is shown in Fig. 4.12.

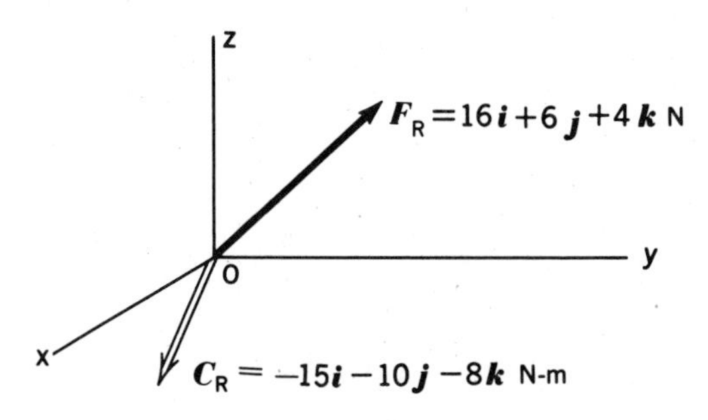

Figure 4.12. Resultant at *O*.

**EXAMPLE 4.4**

What is the resultant at *A* of the applied loads acting in Fig. 4.13? We first express the loads vectorially. Thus,

$$\begin{aligned} \boldsymbol{F}_1 &= F_1\hat{\boldsymbol{d}}_1 = 150\left(\frac{-10\boldsymbol{k} - 3\boldsymbol{i} + 4\boldsymbol{j}}{\sqrt{10^2 + 3^2 + 4^2}}\right) \\ &= -40.2\boldsymbol{i} + 53.7\boldsymbol{j} - 134.1\boldsymbol{k} \text{ lb} \\ \boldsymbol{F}_2 &= F_2\hat{\boldsymbol{d}}_2 = 200\left(\frac{-13\boldsymbol{k} + 7\boldsymbol{i}}{\sqrt{13^2 + 7^2}}\right) \\ &= 94.8\boldsymbol{i} - 176\boldsymbol{k} \text{ lb} \\ \boldsymbol{F}_3 &= -100\boldsymbol{j} \text{ lb} \\ \boldsymbol{C} &= -50\boldsymbol{k} \text{ ft-lb} \end{aligned}$$

We can now readily find the resultant force system at *A*. Thus,

$$\begin{aligned} \boldsymbol{F}_R &= (-40.2 + 94.8)\boldsymbol{i} + (53.7 - 100)\boldsymbol{j} + (-134.1 - 176.0)\boldsymbol{k} \\ &= 54.6\boldsymbol{i} - 46.3\boldsymbol{j} - 310\boldsymbol{k} \text{ lb} \\ \boldsymbol{C}_R &= (-11\boldsymbol{k}) \times \boldsymbol{F}_1 + (-8\boldsymbol{k}) \times (\boldsymbol{F}_2 + \boldsymbol{F}_3) + (-50\boldsymbol{k}) \\ &= -11\boldsymbol{k} \times (-40.2\boldsymbol{i} + 53.7\boldsymbol{j} - 134.1\boldsymbol{k}) + (-8\boldsymbol{k}) \\ &\quad \times (94.8\boldsymbol{i} - 176.0\boldsymbol{k} - 100\boldsymbol{j}) - 50\boldsymbol{k} \\ &= -209\boldsymbol{i} - 316\boldsymbol{j} - 50\boldsymbol{k} \text{ ft-lb} \end{aligned}$$

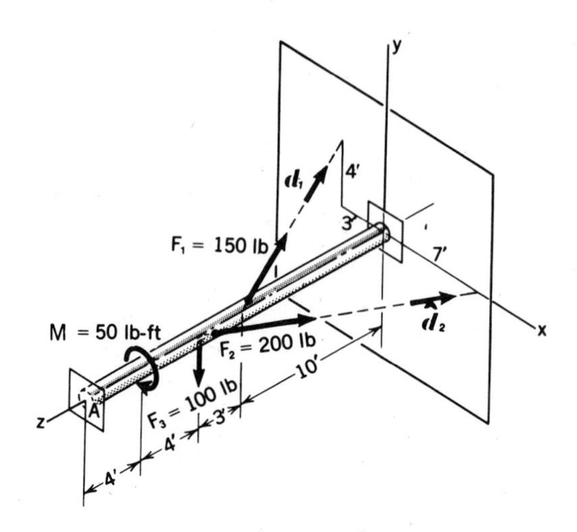


Figure 4.13. Find resultant at $A$; $F_2$ and $F_3$ are concurrent.

## 4.4 Simplest Resultants of Special Force Systems

We shall now consider special but important force systems in order to establish the *simplest* resultants possible. Examples will serve to illustrate the method of procedure.

**Case A. Coplanar Force Systems.** In Fig. 4.14 is shown a system of forces and couples in plane $M$. By moving the forces to a common point $a$ in plane $M$, we will form only couples in the plane. The force portion of the equivalent system at such point will be given as

$$\boldsymbol{F}_R = [\sum_i (F_i)_x]\boldsymbol{i} + [\sum_i (F_i)_y]\boldsymbol{j} \tag{4.6}$$

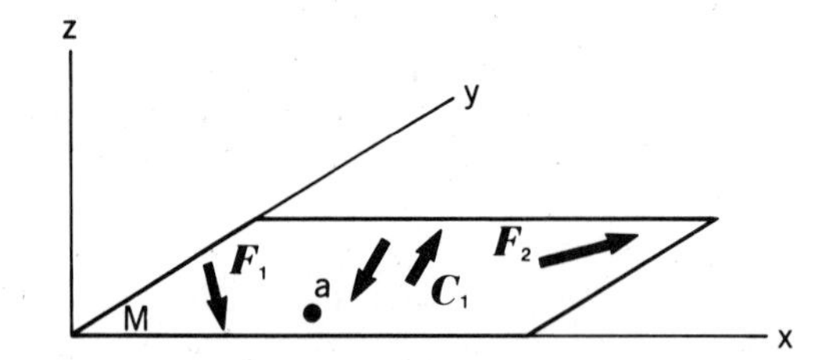


Figure 4.14. Coplanar force system.

The couple-moment portion of the equivalent system can be given as:

$$\boldsymbol{C}_R = (F_1 d_1 + F_2 d_2 + \ldots)\boldsymbol{k} + (C_1 + C_2 + \ldots)\boldsymbol{k} \tag{4.7}$$

where $d_1$, $d_2$, etc., are perpendicular distances from point $a$ to the lines of action of the noncouple forces, and $C_1$, $C_2$, etc., are the values of the given couple moments. The resultant at $a$ is shown in Fig. 4.15.

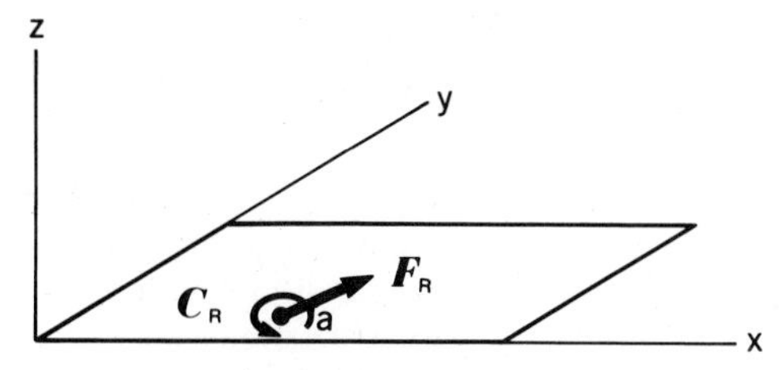


**Figure 4.15.** Resultant at point *a*.

If $\boldsymbol{F}_R \neq \boldsymbol{0}$—that is, if $\sum_i F_x \neq 0$ and/or $\sum_i F_y \neq 0$—we can move the force from *a* to yet a new position so as to introduce a second couple moment to cancel $\boldsymbol{C}_R$ of Fig. 4.15 in the manner described earlier in Section 4.2. Since the $x$ and $y$ directions used are arbitrary, except for the condition that they be in the plane of the forces, we can make the following conclusion. *If the force components in any direction in the plane add to other than zero, we may replace the entire coplanar system by a single force with a specific line of action.*

What happens if $\sum_i F_x = 0$ and $\sum_i F_y = 0$? Without a force at point *a*, we can no longer eliminate a couple in plane *M*. Thus, our second conclusion is that *if* $\sum_i F_x$ *and* $\sum_i F_y$ *are zero, the resultant must be a couple moment or be zero.*

In the coplanar case, therefore, the simplest equivalent force system must be a single force along a specific line of action, or a single couple moment, or the zero vector. The following example is used to illustrate the method of determining such a resultant directly without the intermediate steps followed in this discussion.

**EXAMPLE 4.5**

Consider a coplanar force system shown in Fig. 4.16. The *simplest* resultant is to be found. Since $\sum_i F_x$ and $\sum_i F_y$ are not zero, we know that we can replace the

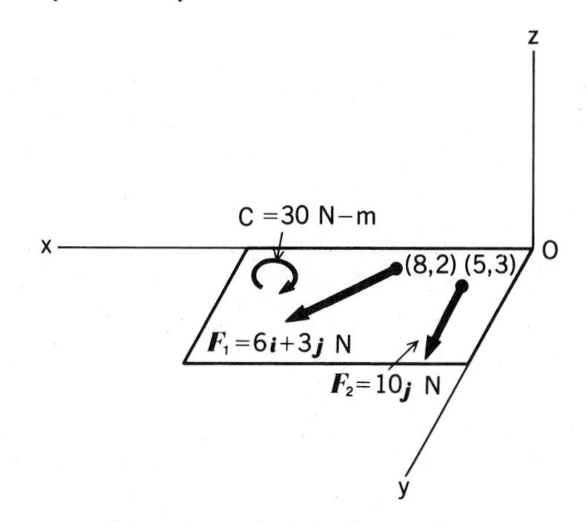


**Figure 4.16.** Find simplest resultant.

system by a single force, which is

$$\boldsymbol{F}_R = 6\boldsymbol{i} + 13\boldsymbol{j}\text{ N} \tag{a}$$

We now need to find the line of action in the plane that will make this single force equivalent to the given system. To be equivalent for rigid-body mechanics, this force without a couple moment must have the same turning action about any point or axis in space as that of the given system. Now the simplest resultant force must intercept the $x$ axis at some point $\bar{x}$.[1] We can determine $\bar{x}$ by equating the moment of the resultant force without a couple moment about the origin with that of the original system of forces and couples. Using the vector $\bar{x}\boldsymbol{i}$ as a position vector from the origin to the line of action of $\boldsymbol{F}_R$ (see Fig. 4.17), we accordingly have

$$\bar{x}\boldsymbol{i} \times (6\boldsymbol{i} + 13\boldsymbol{j}) = (8\boldsymbol{i} + 2\boldsymbol{j}) \times (6\boldsymbol{i} + 3\boldsymbol{j}) + (5\boldsymbol{i} + 3\boldsymbol{j}) \times (10\boldsymbol{j}) - 30\boldsymbol{k} \tag{b}$$

Carrying out the cross products,

$$24\boldsymbol{k} - 12\boldsymbol{k} + 50\boldsymbol{k} - 30\boldsymbol{k} = 13\bar{x}\boldsymbol{k} \tag{c}$$

Hence,

$$\bar{x} = 2.46\text{ ft}$$

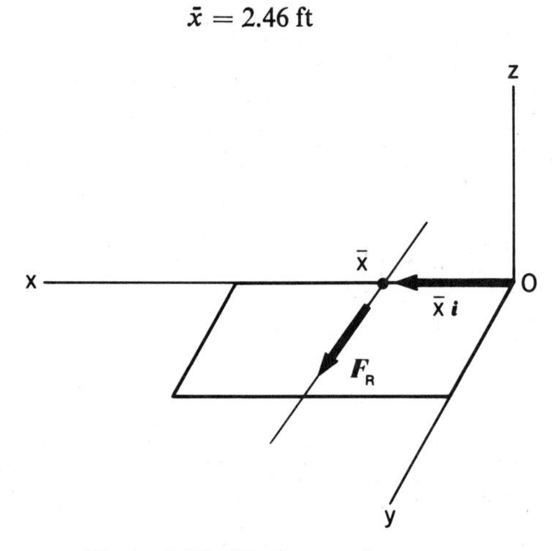


**Figure 4.17.** Simplest resultant.

By specifying the $x$ intercept, $\bar{x}$, we fully determine the line of action of the simplest resultant force. We could have also used the intercept with the $y$ axis, $\bar{y}$, for this purpose. In that case, the position vector from the origin out to the line of action is $\bar{y}\boldsymbol{j}$, and we have, on equating moments about $O$ of the resultant without a couple moment with that of the original system:

$$\bar{y}\boldsymbol{j} \times (6\boldsymbol{i} + 13\boldsymbol{j}) = (8\boldsymbol{i} + 2\boldsymbol{j}) \times (6\boldsymbol{i} + 3\boldsymbol{j}) + (5\boldsymbol{i} + 3\boldsymbol{j}) \times (10\boldsymbol{j}) - 30\boldsymbol{k}$$

$$y = -5.35\text{ ft}$$

**EXAMPLE 4.6**

Compute the *simplest* resultant for the loads shown acting on the beam in Fig. 4.18(a). Give the intercept with the $x$ axis.

---

[1] If the resultant force is parallel to the $x$ axis, the intercept will be at infinity.

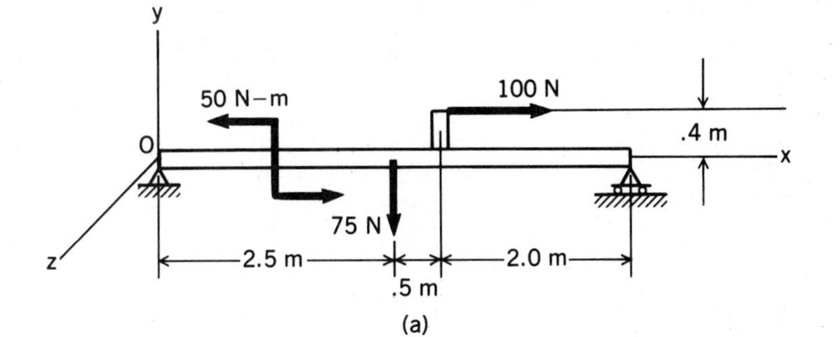


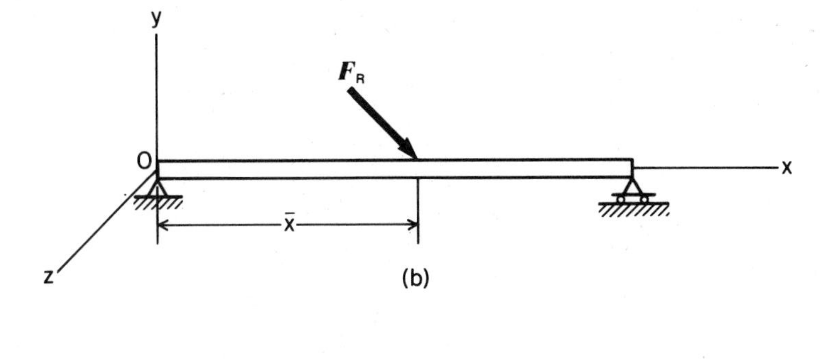


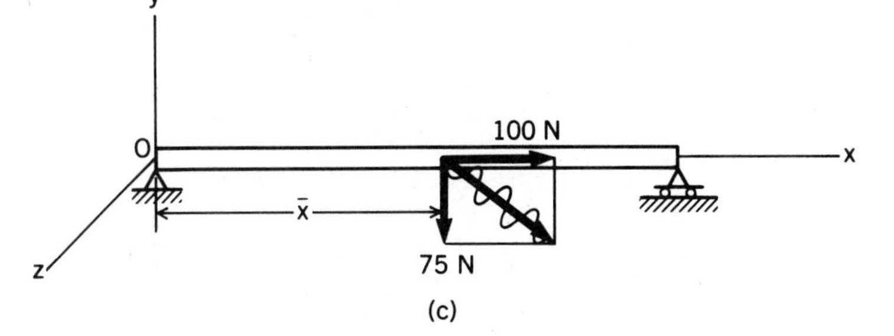


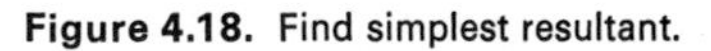
**Figure 4.18.** Find simplest resultant.

It is immediately apparent on inspection of the diagram that

$$\boldsymbol{F}_R = 100\boldsymbol{i} - 75\boldsymbol{j}\text{ N} \tag{a}$$

Let $\bar{x}$ be the intercept with the $x$ axis of the line of action of $\boldsymbol{F}_R$ when this line of action corresponds to zero couple moment $\boldsymbol{C}_R$ [see Fig. 4.18(b)]. In Fig. 4.18(c), we have decomposed $\boldsymbol{F}_R$ along this line of action into rectangular components so as to permit simple calculations of moments about the origin $O$ (i.e., about the $z$ axis). Accordingly, equating moments about the $z$ axis of $\boldsymbol{F}_R$, without a couple moment, with that of the original system of loads, we get, using scalar components:

$$-(75)(\bar{x}) = 50 - (2.5)(75) - (.4)(100)$$

$$\bar{x} = 2.37 \text{ m}$$

Thus, the simplest resultant is a force $100\boldsymbol{i} - 75\boldsymbol{j}$ N intercepting the beam axis at a position $\bar{x} = 2.37$ m.

As pointed out earlier, in the instance wherein $\boldsymbol{F}_R = \boldsymbol{0}$, we then possibly have as the simplest resultant a couple moment normal to the plane of the coplanar force system. There is the possibility that there is also zero couple moment, in which case the members of the coplanar force system *completely cancel* each other's effects on a rigid body. To find the couple moment for the case where $\boldsymbol{F}_R = \boldsymbol{0}$, we simply take moments of the coplanar force system about *any point* in space. This moment, if it be not zero, is clearly the couple-moment vector sought. We leave this straightforward kind of a problem to the exercises.

**Case B. Parallel Force Systems in Space.** Now, consider the system of $n$ parallel forces in Fig. 4.19, where the $z$ direction has been selected parallel to the forces. We also include $m$ couples whose planes are parallel to the $z$ direction because such couples

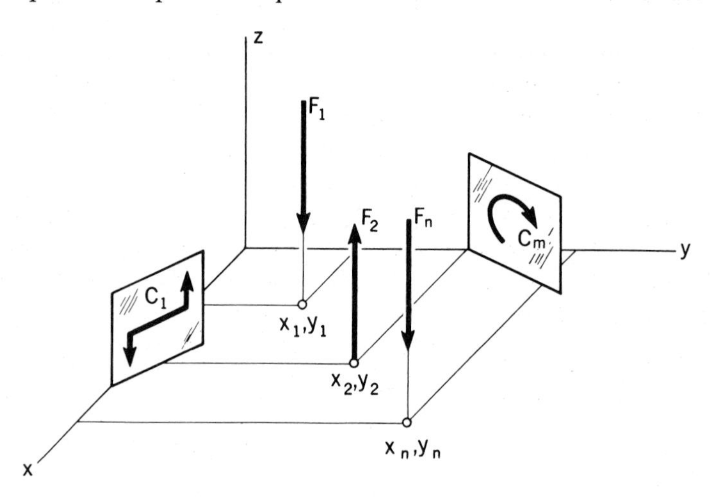


**Figure 4.19.** Parallel system of forces.

can be considered to be composed of equal and opposite forces parallel to the $z$ direction. We can move the forces so that they all pass through the origin of the $xyz$ axes; the force portion of the equivalent system is then

$$\boldsymbol{F}_R = \left(\sum_{i=1}^{n} F_i\right)\boldsymbol{k} \tag{4.8}$$

The couple-moment portion of the equivalent system is found by applying Eq. 4.5 to this case:

$$\boldsymbol{C}_R = \sum_{i=1}^{n} [(x_i\boldsymbol{i} + y_i\boldsymbol{j}) \times F_i\boldsymbol{k}] + \sum_{i=1}^{m} [(C_i)_x\boldsymbol{i} + (C_i)_y\boldsymbol{j}] \tag{4.9}$$

where $F_i$ represents the noncouple force magnitudes. Carrying out the cross product, we get

$$\boldsymbol{C}_R = \sum_{i=1}^{n} [(F_i y_i)\boldsymbol{i} - (F_i x_i)\boldsymbol{j}] + \sum_{i=1}^{m} [(C_i)_x\boldsymbol{i} + (C_i)_y\boldsymbol{j}] \tag{4.10}$$

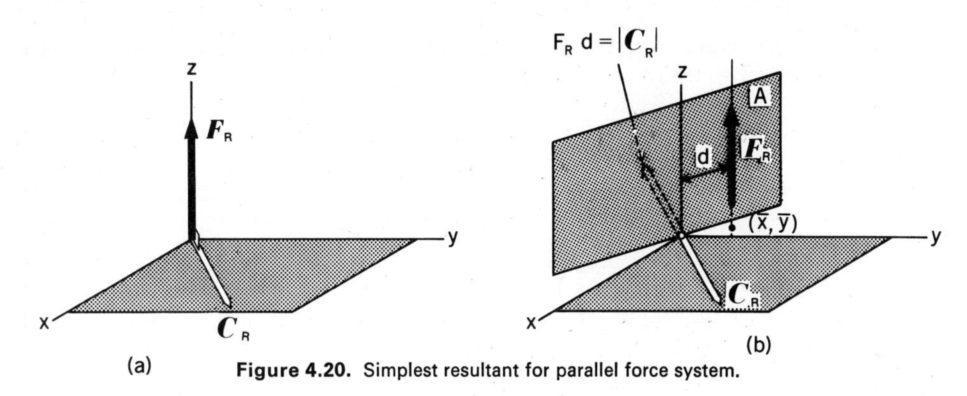


**Figure 4.20.** Simplest resultant for parallel force system.

From this, we see that the couple moment must always be parallel to the $xy$ plane (i.e., perpendicular to the direction of the forces). We then have at the origin a single force and a single couple moment at right angles to each other [see Fig. 4.20(a)]. If $\boldsymbol{F}_R \neq \mathbf{0}$, we can move $\boldsymbol{F}_R$ again to another line of action in a plane $A$ perpendicular to $\boldsymbol{C}_R$ and, choosing the proper value of $d$, such that $F_R d = |\boldsymbol{C}_R|$, we can eliminate the couple moment [see Fig. 4.20(b)]. We thus end up with a *single* force having a particular line of action specified by the intercept $\bar{x}\bar{y}$ of the line of action of the force with the $xy$ plane. If the summation of forces should happen to be zero, the equivalent system must then be a couple moment or be zero.

Thus, *the simplest resultant system of a parallel force system is either a force or a couple moment.* The following example will illustrate how we can directly determine the simplest resultant.

### EXAMPLE 4.7

Find the simplest resultant of the parallel force system in Fig. 4.21(a).

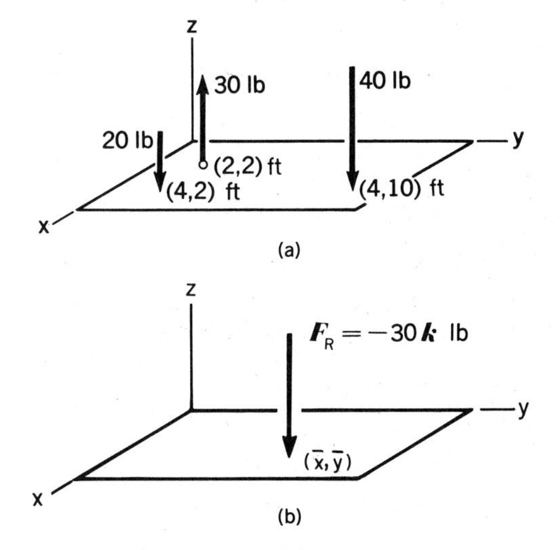


**Figure 4.21.** Find simplest resultant.

The sum of the forces is 30 lb in the negative $z$ direction. Hence, a position can be found in which a single force is equivalent to the original system. Assume that this resultant force without a couple moment proceeds through the point $\bar{x}$, $\bar{y}$ [Fig. 4.21(b)]. We can equate the moment of this resultant force about the $x$ and $y$ axes with the corresponding moments of the original system and thus form the scalar equations that yield the proper value of $\bar{x}$ and $\bar{y}$. Equating moments about the $x$ axis,[2] we get

$$(30)(2) - (20)(2) - (40)(10) = -30\bar{y}$$

Therefore,

$$\bar{y} = 12.7 \text{ ft}$$

Equating moments about the $y$ axis, we have

$$-(30)(2) + (20)(4) + (40)(4) = 30\bar{x}$$

Therefore,

$$\bar{x} = 6 \text{ ft}$$

You can also show, as an exercise, that the same result can be reached for $\bar{x}$, $\bar{y}$ by equating moments of the resultant force without a couple moment about the origin with that of the original system about the origin.

**EXAMPLE 4.8**

Consider the parallel force system in Fig. 4.22. What is the simplest resultant?

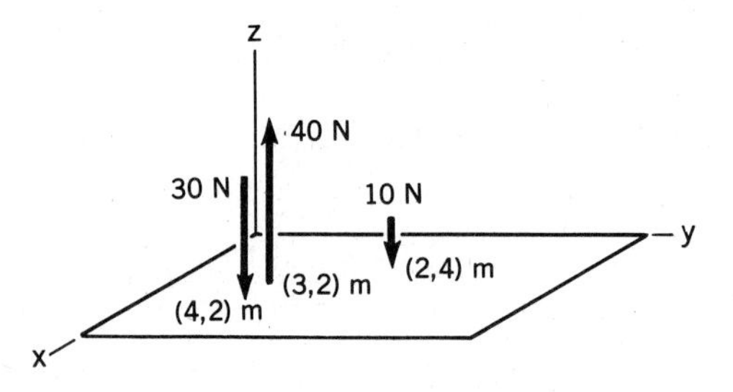


**Figure 4.22.** Parallel force system.

Here we have a case where the sum of the forces is zero and so $\boldsymbol{F}_R = \boldsymbol{0}$. Therefore, the simplest resultant must be a couple moment or be zero. To get this couple moment, $\boldsymbol{C}_R$, we can take moments of the forces about *any point* in space. This moment vector then equals the desired couple moment $\boldsymbol{C}_R$. The simplest procedure is to use the origin of the reference as the point about which to take moments. Then we

[2]To get the moment of a force, such as the 40-lb force, about the $x$ axis we use the elementary definition of moment about an axis given at the outset of Section 3.3. Thus, consider a plane $A$ perpendicular to the $x$ axis and containing the 40-lb force. The perpendicular distance from the $x$ axis to this force clearly is the $y$ coordinate of the intercept (4, 10) of the line of action of the force and the $xy$ plane. Hence, $M_x$ for the 40-lb force must be $-(40)(10) = -400$ ft-lb.

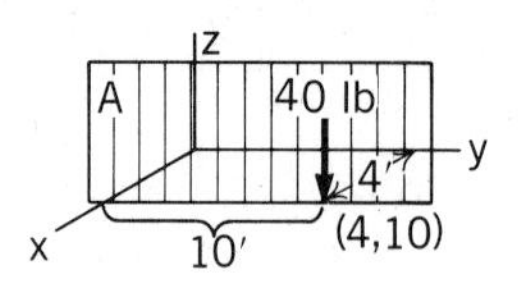

can say that

$$C_R = (4i + 2j) \times (-30k) + (3i + 2j) \times (40k) + (2i + 4j) \times (-10k)$$
$$= -20i + 20j \text{ N-m} \quad \text{(a)}$$

The rectangular components of $C_R$ along the $x$ and $y$ axes are the moments of the force system about these axes. Thus,

$$(C_R)_x = -20 \text{ N-m}$$
$$(C_R)_y = 20 \text{ N-m} \quad \text{(b)}$$

We can get the moments of the forces about the $x$ and $y$ axes directly and thus generate the components of the desired couple moment $C_R$. Accordingly, using the elementary definition of the moment of a force about a line as presented earlier, we have

$$(C_R)_x = -(10)(4) + (40)(2) - (30)(2) = -20 \text{ N-m}$$
$$(C_R)_y = (10)(2) - (40)(3) + (30)(4) = 20 \text{ N-m}$$

Thus, the moment of the force system about the origin, and hence about any point, is then the desired couple moment (Fig. 4.23).

$$C_R = -20i + 20j \text{ N-m}$$

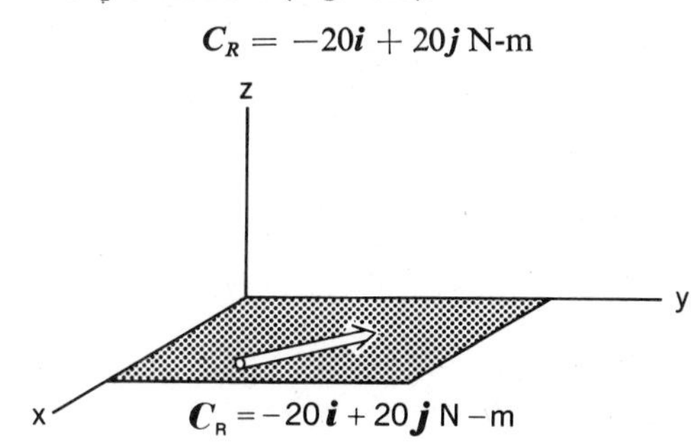


**Figure 4.23.** Simplest resultant is a couple moment.

Now that we have considered the concept of the simplest resultant for coplanar and parallel force systems, we wish to go back to the *general force* systems for a moment. We learned earlier that we can always replace such a system in rigid body mechanics by a single force $F_R$ and a single couple moment $C_R$ at any chosen point. Is this always the very simplest system for rigid body mechanics? No, it is not. To show this, decompose the couple moment $C_R$ into two rectangular components $C_\perp$ and $C_{||}$, perpendicular to the force and collinear with the force respectively. We can now move the force to a chosen parallel position and can eliminate $C_\perp$, the component of couple moment normal to the force. However, there is nothing that we can do about the $C_{||}$ component of couple moment collinear (or parallel) to the force. The reason for this is that any movement of the force to a parallel position *always* introduces a couple moment *perpendicular* to the force. Thus the component $C_{||}$ cannot be affected. By eliminating $C_\perp$ we end up with the force $F_R$ and $C_{||}$ collinear with $F_R$. This system is the simplest in the general case and it is called a *wrench*. However, we shall not use the wrench concept in this text and will work instead with the resultant force $F_R$ and the couple moment $C_R$ at any chosen point.

## Problems

*In several of the problems of this set we shall concentrate the weight of a body at its center of gravity. Most likely you are used to doing this from your previous physics course. In Section 4.5 we shall justify this procedure.*

**4.17.** Compute the resultant force system of the applied loads at positions $A$ and $B$.

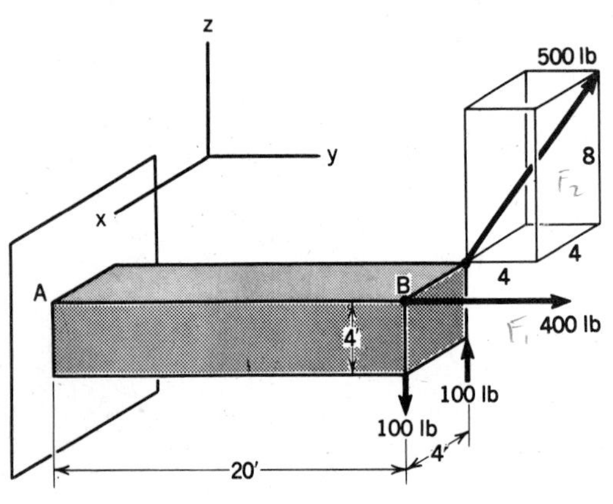


**Figure P.4.17**

**4.18.** Compute the resultant force system at $A$ stemming from the indicated 50-lb force. What is the twist developed about the axis of the shaft at $A$?

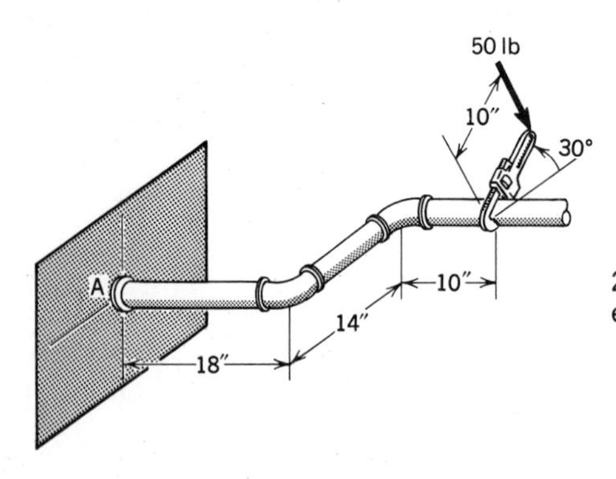


**Figure P.4.18**

**4.19.** Find the resultant of the force system at point $A$. The 300-N, 200-N, and 900-N loads are at the centers of the pipe sections.

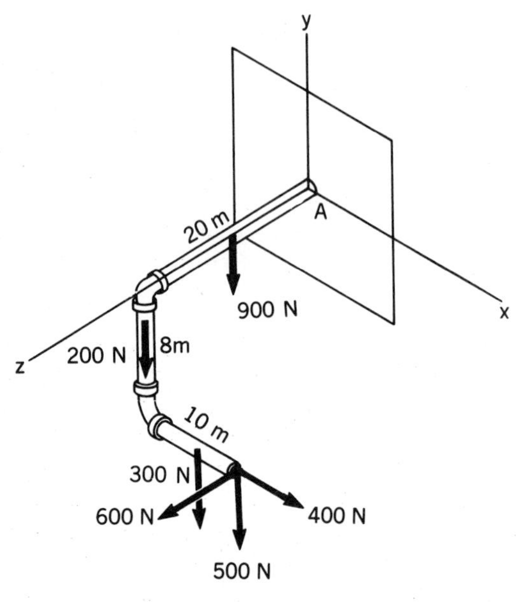

**Figure P.4.19**

**4.20.** A 20-kN car and an 80-kN truck are stopped on a bridge. What is the resultant force system of these vehicles at the center of the bridge? At the center of the left end of the bridge? The distances given to truck and car are to respective centers of gravity where we can concentrate the weights.

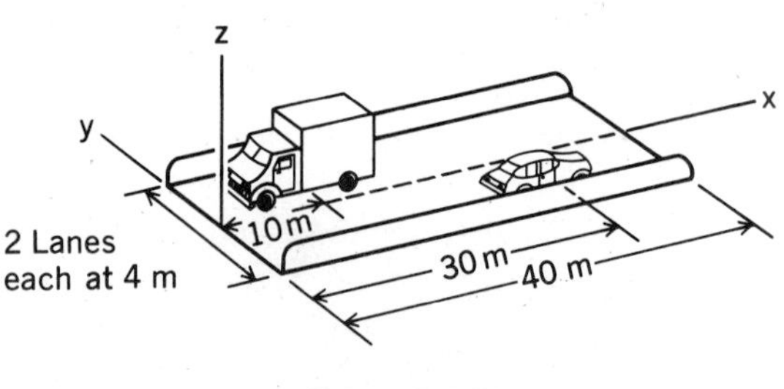


**Figure P.4.20**

**4.21.** Two heavy machinery crates ($A$ weighs 20 kN and $B$ weighs 30 kN) are placed on a truck. What is the resultant force system at the center of the rear axle? The centers of gravity of the crates, where we can concen-

trate the weights, are at the geometric centers.

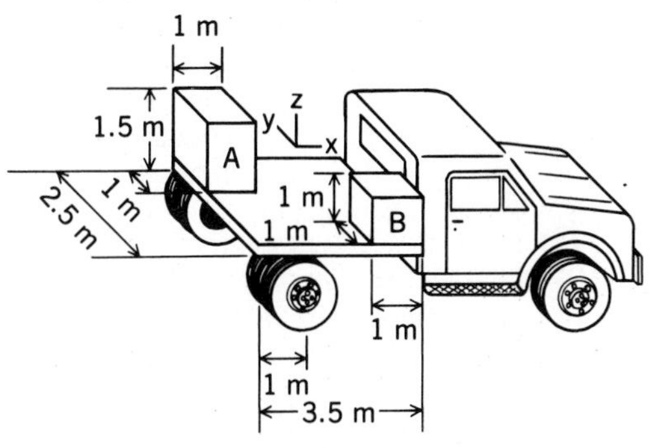


Figure P.4.21

**4.22.** Replace the system of forces by a resultant at $A$.

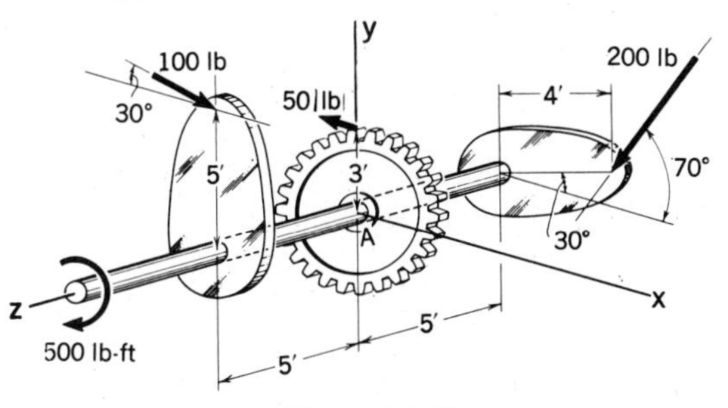


Figure P.4.22

**4.23.** Evaluate forces $F_1$, $F_2$, and $F_3$ so that the resultant of the forces and torque acting on the plate is zero in both force and couple moment. (*Hint:* If the resultant is zero for one point, will it not be zero for any point? Explain why.)

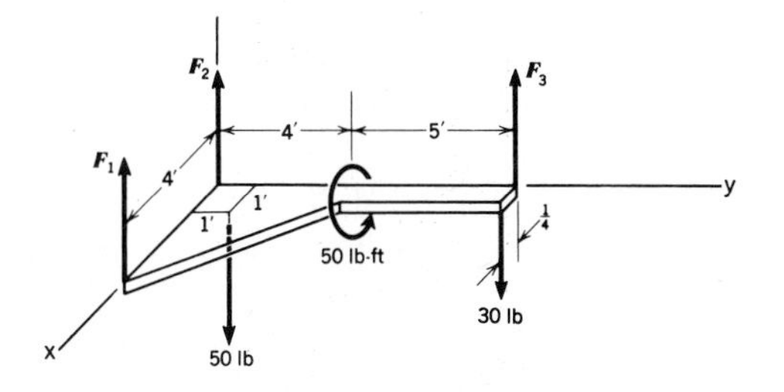


Figure P.4.23

**4.24.** Find the *simplest* resultant of the forces shown acting on the beam. Give the intercept with the axis of the beam.

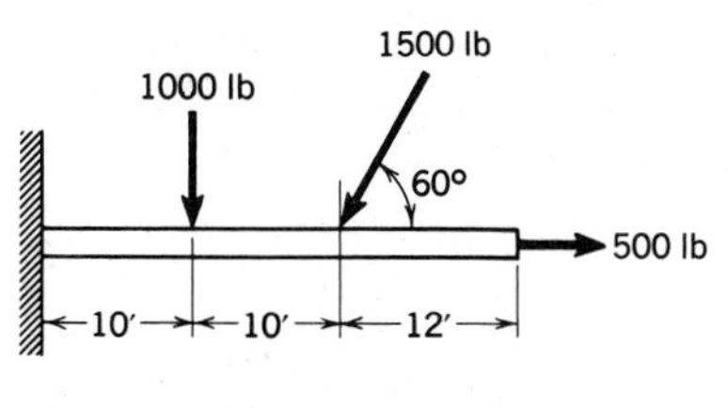


Figure P.4.24

**4.25.** Find the *simplest* resultant of the forces shown acting on the pulley. Give the intercept with the $x$ axis.

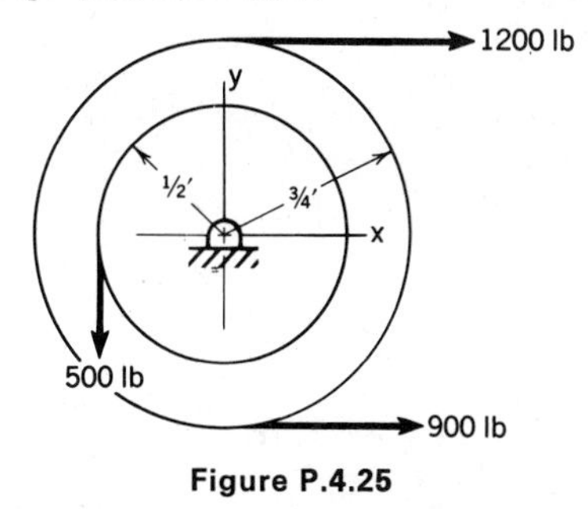


Figure P.4.25

**4.26.** A man raises a 50-lb bucket of water to the top of a bricklayer's scaffold. Also, a Jeep winch is used to raise a 200-lb load of bricks. What is the *simplest* resultant force system on the scaffold? Give the $x$ intercept. Consider the pulleys to be frictionless so that the 50-lb force and the 200-lb force are transmitted respectively to the man and to the Jeep.

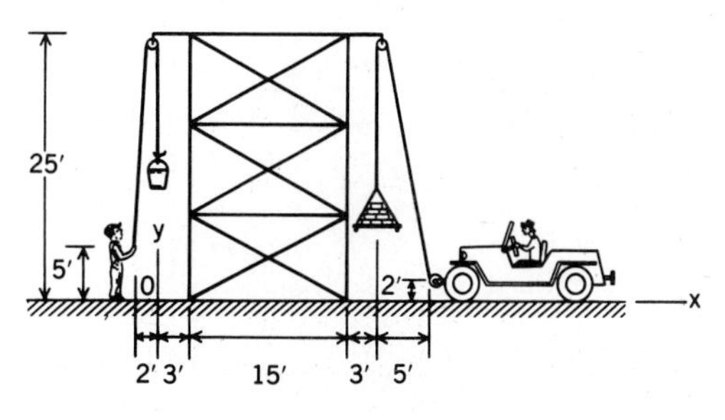


Figure P.4.26

**4.27.** Compute the *simplest* resultant for the loads acting on the beam. Give the intercept with the axis of the beam.

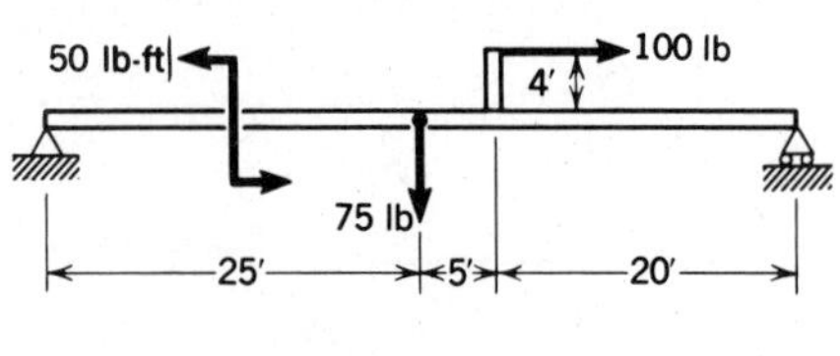


Figure P.4.27

**4.28.** Find the *simplest* resultant for the forces. Give the location of this resultant clearly.

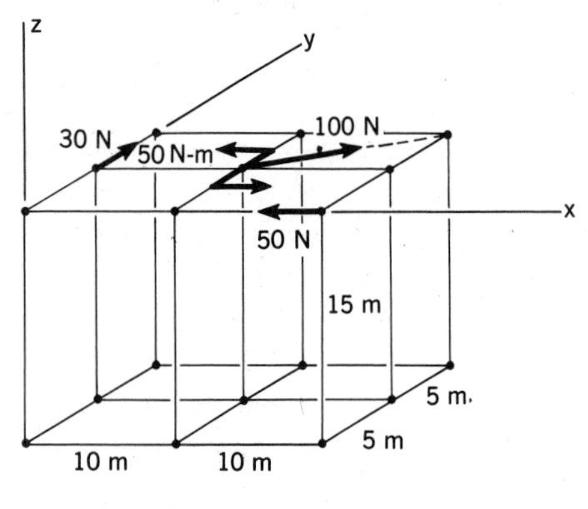


Figure P.4.28

**4.29.** Replace the system of forces acting on the rivets of the plate by the *simplest* resultant. Give the intercept of this resultant with the $x$ axis.

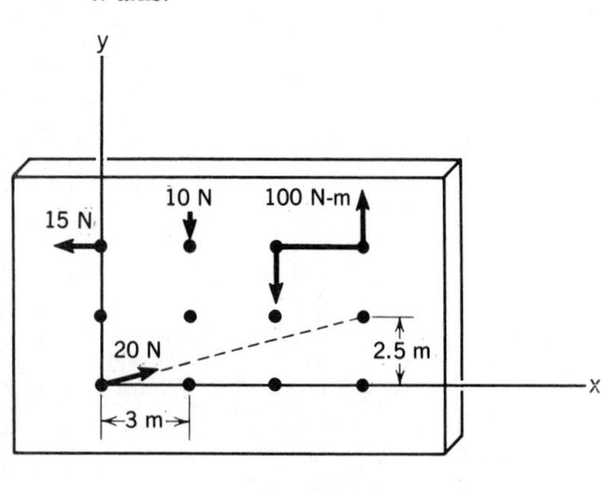


Figure P.4.29

**4.30.** A parallel system of forces is such that: a 20-N force acts at position $x = 10, y = 3$ m; a 30-N force acts at position $x = 5, y = -3$ m; a 50 N force acts at position $x = -2, y = 5$ m.

(a) If all forces point in the negative $z$ direction, give the *simplest* resultant force and its line of action.

(b) If the 50-N force points in the plus $z$ direction and the others in the negative $z$ direction, what is the *simplest* resultant?

**4.31.** What is the *simplest* resultant of the three forces and couple shown acting on the shaft and disc? The disc radius is 5 ft.

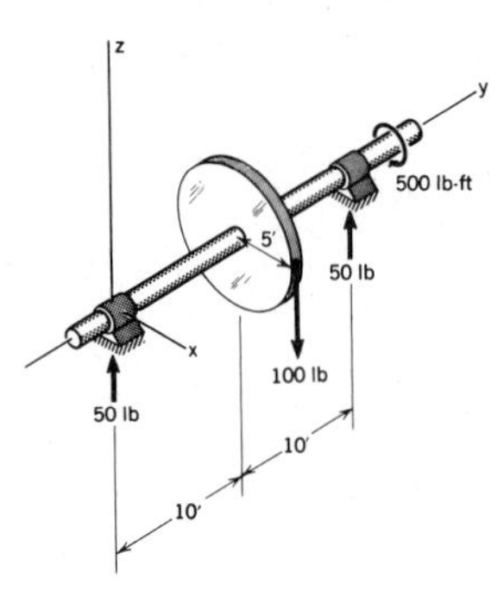


Figure P.4.31

**4.32.** What is the *simplest* resultant for the system of forces? Each square is 10 mm on edge.

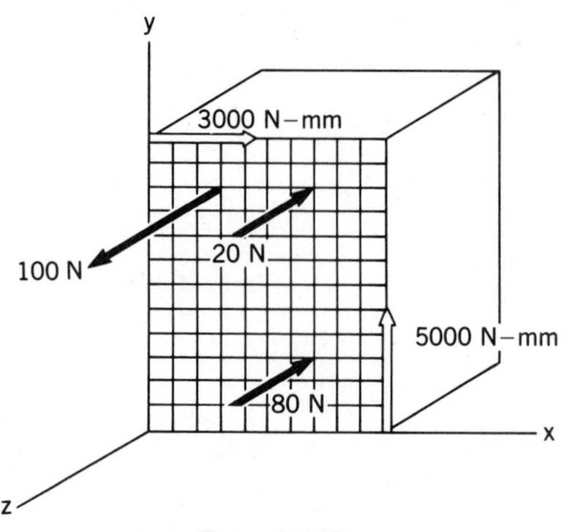


Figure P.4.32

**4.33.** Two hoists are operated on the same overhead track. Hoist $A$ has a 3000-kN load, and hoist $B$ has a 4000-kN load. What is the resultant force system at the left end $O$ of the track? Where does the *simplest* resultant force act?

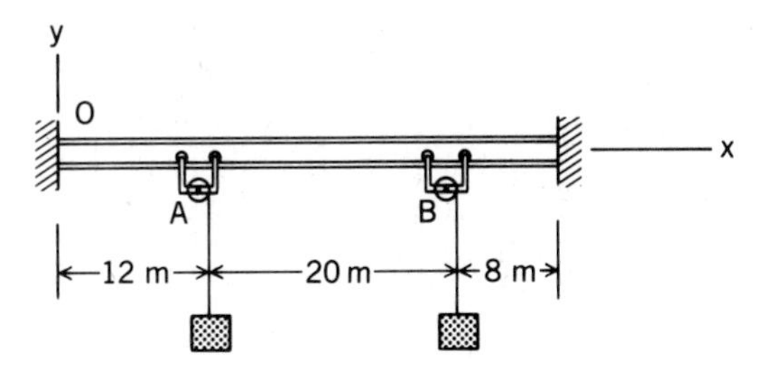


Figure P.4.33

**4.34.** A lo-boy trailer weighs 16,000 lb and is loaded with a 15,000-lb tractor and a 12,000-lb front-end loader. What is the simplest resultant force and where does it act? The weights of the machines and trailer act at their respective centers of gravity (C.G.).

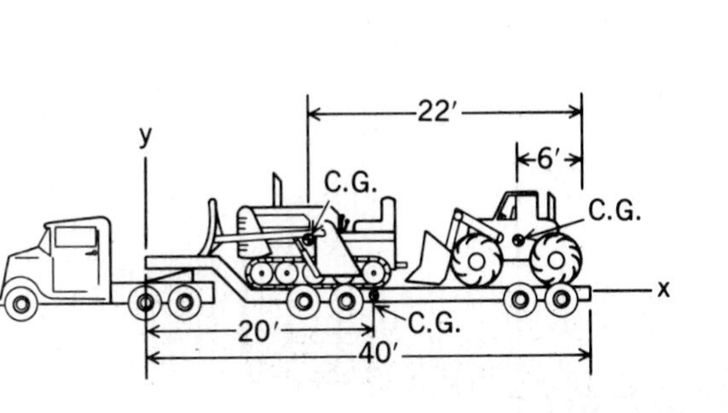


Figure P.4.34

**4.35.** Where should a 100-N force in a downward direction be placed for the *simplest* resultant of all shown forces to be at position (5, 5) m?

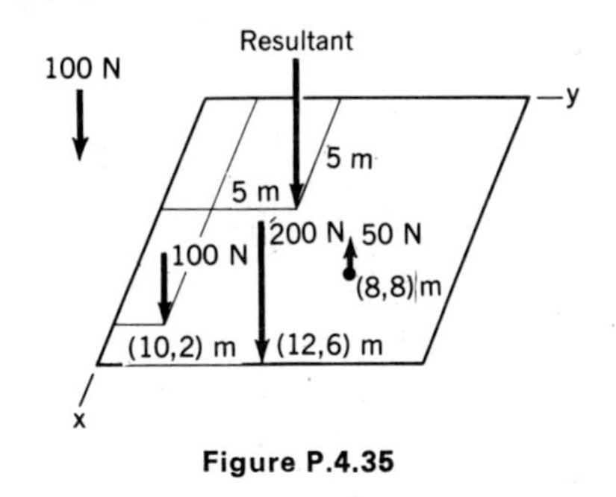


Figure P.4.35

**4.36.** A barge must be evenly loaded so it does not list in any direction. Where can the three large machinery crates be placed (without either hanging over the edge, stacking, or standing on end)? Each crate is as tall as it is wide. Is there only one solution to this problem? Centers of gravity of crates correspond to geometric centers.

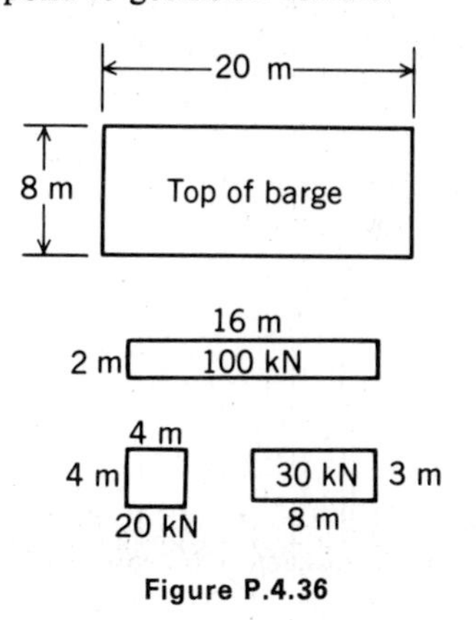


Figure P.4.36

## 4.5 Distributed Force Systems

Our discussions up to now have been restricted to discrete vectors—in particular, to point forces. Scalars and vectors may also be continuously distributed throughout a finite volume so that at each position in space in the volume there is a definite scalar or vector quantity. Such distributions are called *scalar* and *vector fields*, respectively. A simple example of a scalar field is the temperature distribution, expressed as $T(x, y, z, t)$, where the variable $t$ indicates that the field may be changing with time. Thus, if a position $x_0, y_0, z_0$ and a time $t_0$ are specified, we can determine the temperature at this position and time provided that we know the temperature distribution function (i.e., how $T$ depends on the independent variables $x$, $y$, $z$, and $t$). A vector field is sometimes expressed in the form $\boldsymbol{F}(x, y, z, t)$. A common example of a vector field is the gravitational force field of the earth—a field that is known to vary with elevation above sea level, among other factors. Note however that the gravitational field is virtually constant with time.

In place of the vector field, it is more convenient at times to employ three scalar fields that represent the orthogonal scalar components of a vector field at all points. Thus, for a force field we can say:

$$\text{force component in } x \text{ direction} = g(x, y, z, t)$$
$$\text{force component in } y \text{ direction} = h(x, y, z, t)$$
$$\text{force component in } z \text{ direction} = k(x, y, z, t)$$

where $g$, $h$, and $k$ represent functions of the coordinates and time. If we substitute coordinates of a special position and the time into these functions, we get the force components $F_x$, $F_y$, and $F_z$ for that position and time. The force field and its component scalar fields are then related in this way:

$$\boldsymbol{F}(x, y, z, t) = g(x, y, z, t)\boldsymbol{i} + h(x, y, z, t)\boldsymbol{j} + k(x, y, z, t)\boldsymbol{k}$$

More often, the notation for the equation above is written

$$\boldsymbol{F}(x, y, z, t) = F_x(x, y, z, t)\boldsymbol{i} + F_y(x, y, z, t)\boldsymbol{j} + F_z(x, y, z, t)\boldsymbol{k} \tag{4.11}$$

Vector fields are not restricted to forces but include such other quantities as velocity fields and heat-flow fields.

Force distributions, such as gravitational force, that exert influence directly on the elements of mass distributed throughout the body are termed *body force distributions* and are usually given per unit of mass that they directly influence. Thus, if $\boldsymbol{B}(x, y, z, t)$ is such a body force distribution, the force on an element $dm$ would be $\boldsymbol{B}(x, y, z, t)dm$.

Force distributions over a *surface* are called *surface force distributions*[3] and are given per unit area of the surface directly influenced. A simple example is the force distribution on the surface of a body submerged in a fluid. In the case of a static fluid or of a frictionless fluid, the force from the fluid on an area element is always normal to the area element and directed in toward the body. The force per unit area stemming from such fluid action is called *pressure* and is denoted as $p$. Pressure is a scalar quantity. The direction of the force resulting from a pressure on a surface is given by the orientation of the surface. [You will recall from Chapter 2 that an area element can be considered as a vector which is normal to the area element and directed outward from the enclosed body (Fig. 4.24).] The infinitesimal force on the area element is then given as

$$d\boldsymbol{f} = -p\, d\boldsymbol{A}$$

[3]Surface forces are often called *surface tractions* in solid mechanics.

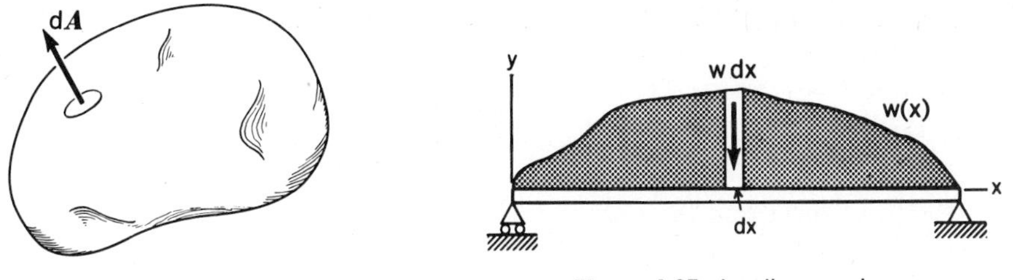


**Figure 4.24.** Area vector.

**Figure 4.25.** Loading on a beam.

A more specialized, but nevertheless common, force distribution is that of a continuous load on a beam. This is often a parallel loading distribution that is symmetrical about the center plane $xy$ of a beam, as illustrated in Fig. 4.25. Various heights of bricks stacked on a beam would be an example of this kind of loading. We can replace such a loading by an equivalent coplanar distribution that acts at the center plane. The loading is given per unit length and is denoted as $w$, the *intensity of loading*. The force on an element $dx$ of the beam, then, is $w\,dx$.

We have thus presented force systems distributed throughout volumes (body forces), over surfaces (surface forces), and over lines. The conclusions about resultants that were reached earlier for general, parallel, and coplanar point-force systems are also valid for these distributed-force systems. These conclusions are true because each distributed force system can be considered as an infinite number of infinitesimal point forces of the type used heretofore. We shall illustrate the handling of force distributions in the following examples.

**Case A. Parallel Body Force System—Center of Gravity.** Consider a rigid body (Fig. 4.26) whose density (mass/unit volume) is given as $\rho(x, y, z)$. It is acted on by gravity, which, for a small body, may be considered to result in a distributed parallel force field.

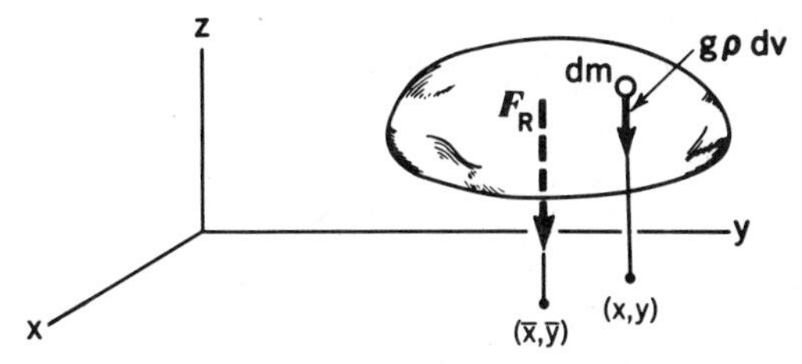


**Figure 4.26.** Gravity body force distribution.

Since we have here a parallel system of forces in space with the same sense, we know that a single force without a couple moment along a certain line of action will be equivalent to the distribution. The gravity body force $\boldsymbol{B}(x, y, z)$ given per unit mass is $-g\boldsymbol{k}$. The infinitesimal force on a differential mass element $dm$, then, is $-g(\rho\,dv)\boldsymbol{k}$, where $dv$ is the volume of the element.[4] We find the resultant force on the system by

[4]Note that $g\rho$ is the weight per unit volume which is often given as $\gamma$, the so-called *specific weight.*

replacing the summation in Eq. 4.8 with an integration. Thus,

$$\boldsymbol{F}_R = -\int_V g(\rho\, dv)\boldsymbol{k} = -g\boldsymbol{k}\int_V \rho\, dv = -gM\boldsymbol{k}$$

where, with $g$ as a constant, the second integral becomes simply the entire mass of the body $M$.

Next, we must find the line of action of this single equivalent force without a couple moment. Let us denote the intercept of this line of action with the $xy$ plane as $\bar{x}, \bar{y}$ (see Fig. 4.26). The resultant at this position must have the same moments as the distribution about both the $x$ and $y$ axes:

$$-F_R\bar{x} = -g\int_V x\rho\, dv, \qquad F_R\bar{y} = g\int_V y\rho\, dv$$

Hence, we have

$$\bar{x} = \frac{\int x\rho\, dv}{M}, \qquad \bar{y} = \frac{\int y\rho\, dv}{M}$$

Thus, we have fully established the simplest resultant. Now, the body is reoriented in space, keeping with it the line of action of the resultant as shown in Fig. 4.27. A new computation of the line of action of the simplest resultant for the second orientation yields a line that intersects the original line at a point $C$. It can be shown that lines of action for simplest resultants for all other orientations of the body must intersect at the same point. We call this point the *center of gravity*. Effectively, we can say for rigid-body considerations that all the weight of the body can be assumed to be concentrated at the center of gravity.

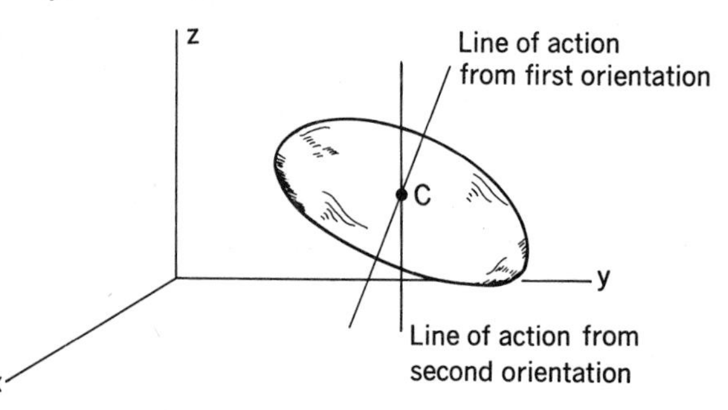


**Figure 4.27.** Location of center of gravity.

**EXAMPLE 4.9**

Find the center of gravity of the triangular block having a uniform density $\rho$ shown in Fig. 4.28.

The total weight of the body is easily evaluated as

$$F_R = g\rho\frac{abc}{2} \tag{a}$$

To find $\bar{y}$, we will equate the moment of $F_R$ about the $x$ axis with that of the weight distribution of the block. To facilitate the latter, we shall choose within the block *infinitesimal* elements whose weights are easily computed. Also, the moment

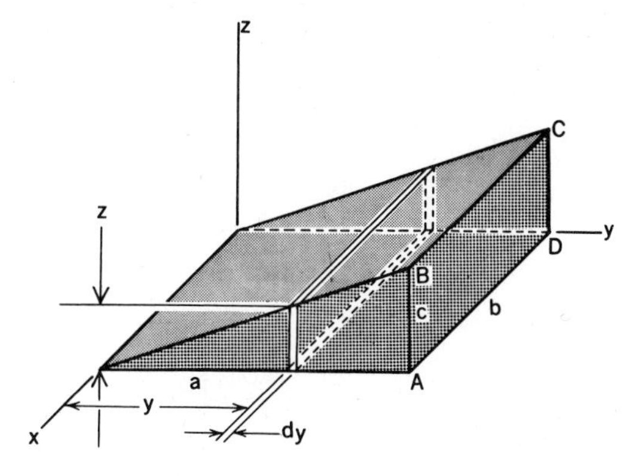


**Figure 4.28.** Find center of gravity.

of the weight of each element about the $x$ axis is to be likewise easily computed. Infinitesimal slices of thickness $dy$ parallel to the $xz$ plane fulfill our requirements nicely. The weight of such a slice is simply $(zb\,dy)\rho g$, where $z$ is the height of the slice (see Fig. 4.28). Because all points of the slice are a distance $y$ from the $x$ axis, clearly the moment of the weight of the slice is easily computed as $-y(zb\,dy)\rho g$. By letting $y$ run from 0 to $a$ during an integration, we can account for all the slices in the body. Thus, we have

$$-F_R\bar{y} = -\int_0^a y(zb\,dy)\rho g \qquad \text{(b)}$$

The term $z$ can be expressed with the aid of similar triangles in terms of the integration variable $y$ as follows:

$$\frac{z}{c} = \frac{y}{a}$$

$$z = \left(\frac{y}{a}\right)c \qquad \text{(c)}$$

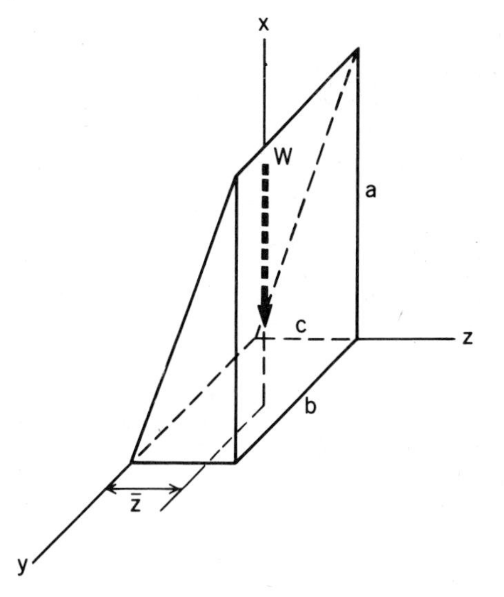


**Figure 4.29.** Reorientation of block.

We then have for Eq. (b), on replacing $F_R$ using Eq. (a),

$$\bar{y} = \frac{1}{g\rho(abc/2)}g\int_0^a \rho y^2 \frac{bc}{a}dy = \frac{2}{3}a \qquad \text{(d)}$$

To find the coordinate in the $z$ direction to the center of gravity, we could reorient the body as shown in Fig. 4.29. A computation simliar to the preceding one would give the result that $\bar{z} = \frac{2}{3}c$. You are urged to verify this yourself.

Finally, it should be clear by inspection of Fig. 4.28 that $\bar{x} = \frac{1}{2}b$.

**EXAMPLE 4.10**

Find the center of gravity for the body of revolution shown in Fig. 4.30. The radial distance of the surface from the $y$ axis is given as

$$r = \tfrac{1}{20}y^2 \text{ ft} \qquad \text{(a)}$$

The body has constant density $\rho$, is 10 ft long, and has a cylindrical hole at the right end of length 2 ft and diameter of 1 ft.

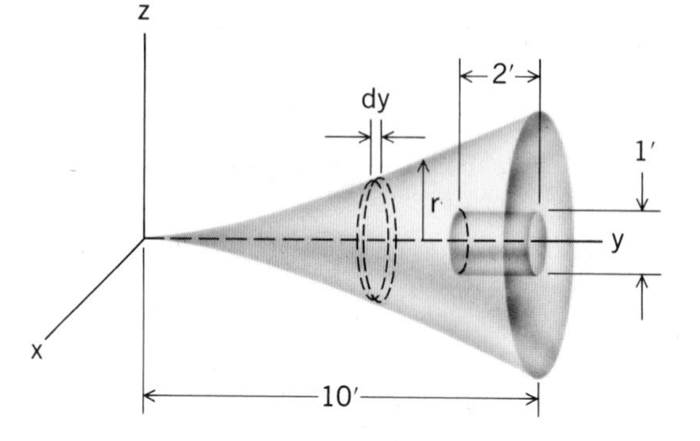


**Figure 4.30.** Body of revolution. Find center of gravity.

We need only compute $\bar{y}$, since it is clear that $\bar{z} = \bar{x} = 0$, owing to symmetry. We first compute the weight of the body. Using slices of thickness $dy$, as shown in the diagram, we sum the weight of all slices in the body assuming it is whole by letting $y$ run from 0 to 10 in an integration. We then subtract the weight of a 2-ft cylinder of diameter 1 ft to take into account the cylindrical cavity inside the body. Thus, we have, noting that the area of a circle is $\pi r^2$ or $\pi D^2/4$

$$W = \int_0^{10} (\pi r^2)\, dy\, \rho g - \frac{\pi(1^2)}{4}(2)(\rho g) \tag{b}$$

Using Eq. (a) to replace $r^2$ in terms of $y$, we get

$$\begin{aligned} W &= \rho g\left(\pi \int_0^{10} \frac{y^4}{400}\, dy - \frac{\pi}{2}\right) = g\rho\pi\left(50 - \frac{1}{2}\right) \\ &= 49.5\, \pi\rho g \text{ lb} \end{aligned} \tag{c}$$

To get $\bar{y}$, we equate the moment of $W$ about the $x$ axis with that of the weight distribution. For the latter, we sum the moments about the $x$ axis of the weight of all slices, assuming first no inside cavity. Then, we subtract from this the moment about the $x$ axis of the weight of a cylinder forming the cavity in the body. Because $\rho$ is constant, the center of gravity of this latter cylinder is at its geometric center so that the moment arm from the $x$ axis for the weight of the cylinder is clearly 9 ft. Thus, we have

$$-(49.5\pi\rho g)\bar{y} = -\rho g\left\{\int_0^{10} y(\pi r^2\, dy) - \left[\frac{\pi(1^2)}{4}(2)\right]9\right\}$$

$$\bar{y} = \frac{1}{49.5}\left(\int_0^{10} \frac{y^5}{400}\, dy - 4.5\right) = 8.33 \text{ ft}$$

Suppose as will be the case in Problem 4.40 that $\gamma\ (= \rho g)$, which is the *specific weight* (giving weight per unit volume), varies with $y$. That is, $\gamma = \gamma(y)$. Then for this problem, note that:

$$W = \int_0^{10} \pi r^2 \gamma \, dy - \int_8^{10} \pi\left(\frac{1}{2}\right)^2 \gamma \, dy$$

also

$$-W\bar{y} = -\left[\int_0^{10} y\pi r^2 \gamma \, dy - \int_8^{10} y\pi\left(\frac{1}{2}\right)^2 \gamma \, dy\right]$$

Note here we cannot take the short cuts used in the original problem where $\gamma$ was a constant.

**EXAMPLE 4.11**

A plate is shown in Fig. 4.31 lying flat on the ground. The plate is 60 mm thick and has a uniform density. The curved edge is that of a parabola with zero slope at the origin. Find the coordinates of the center of gravity.

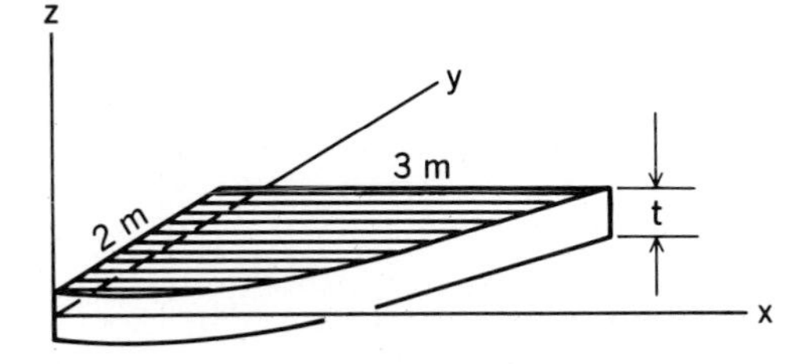


**Figure 4.31.** Find center of gravity of plate.

The equation of a parabola oriented like that of the curved edge of the plate is

$$y = Cx^2 \tag{a}$$

We can determine $C$ by noting that $y = 2m$ when $x = 3m$. Hence,

$$2 = C \cdot 9 \tag{b}$$

Therefore,

$$C = \frac{2}{9}$$

The desired curve then is

$$y = \frac{2}{9}x^2 \tag{c}$$

Therefore,

$$x = \frac{3}{\sqrt{2}}y^{1/2}$$

We shall consider horizontal strips of the plate of width $dy$ (see Fig. 4.32).

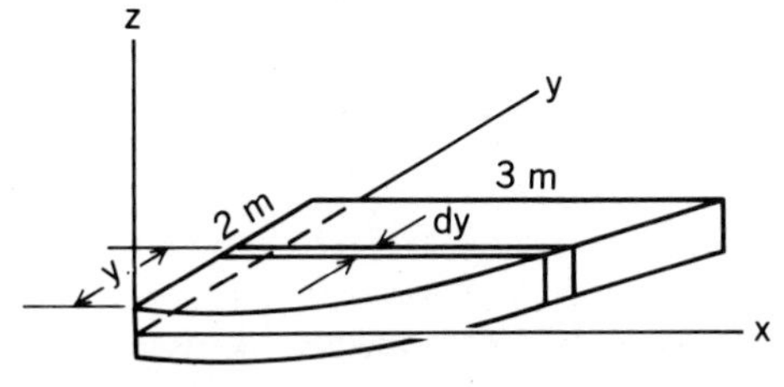


**Figure 4.32.** Use of horizontal strips.

Using the specific weight, $\gamma$, which has units of weight per volume and is equal to $\rho g$, we have for the total weight $W$ of the plate:

$$W = \int_0^2 (dy)(t)(x)\gamma$$

We replace $x$ using Eq. (c) to get

$$W = t\gamma \int_0^2 \left(\frac{3}{\sqrt{2}} y^{1/2}\right) dy$$

where $t$ is the thickness. Integrating, we get

$$W = t\gamma \frac{3}{\sqrt{2}} (y^{3/2}) \left(\frac{2}{3}\right) \Big|_0^2 = t\gamma\sqrt{2}(2)^{3/2} = 4t\gamma \text{ N} \tag{d}$$

We next take moments about the $x$ axis in order to get $\bar{y}$. Thus,

$$\begin{aligned} -W\bar{y} &= -\int_0^2 y(t\, dy\, x)\gamma \\ &= -\gamma t \int_0^2 (y)\left(\frac{3}{\sqrt{2}} y^{1/2}\right) dy \\ &= -\gamma t \frac{3}{\sqrt{2}} (y^{5/2}) \left(\frac{2}{5}\right) \Big|_0^2 \\ &= -\gamma t \left(\frac{3}{\sqrt{2}}\right)\left(\frac{2}{5}\right)[(2^2)(2^{1/2})] \\ &= -\frac{24}{5}\gamma t \end{aligned} \tag{e}$$

Using $4t\gamma$ for $W$ from Eq. (d), we get, for $\bar{y}$:

$$\bar{y} = \tfrac{6}{5} \text{ m} \tag{f}$$

To get $\bar{x}$, we take moments about the $y$ axis, still utilizing the horizontal strips of Fig. 4.32. The center of gravity of a strip is at its center since $\gamma$ is constant and so the moment arm about the $y$ axis is $x/2$.

$$W\bar{x} = \int_0^2 \frac{x}{2}(t\gamma\, dy\, x) \tag{g}$$

Continuing with the calculations, we have

$$\begin{aligned} W\bar{x} &= \frac{t\gamma}{2} \int_0^2 x^2\, dy = \frac{t\gamma}{2} \int_0^2 \left(\frac{9}{2} y\right) dy \\ &= \frac{t\gamma}{2} \frac{9}{2} \frac{y^2}{2} \Big|_0^2 = \frac{9t\gamma}{2} \end{aligned}$$

On replacing $W$ according to Eq. (d), we get, for $\bar{x}$:

$$\bar{x} = \tfrac{9}{8} \text{ m} \tag{h}$$

Finally, is clear that the $\bar{z}$ coordinate is zero for reference $xy$ at the center plane of the plate.

As an exercise (Problem 4.39) you may be asked to solve this problem using vertical strips.

In the previous problems, we used slices of the body having a thickness $dy$. If the specific weight were a function of position, $\gamma(x, y, z)$, we could not readily use such slices, since we cannot easily express the weight of such slices in a simple manner. The reason for this is that in the $x$ and $z$ directions the dimensions of the element are finite, and so $\gamma$ would vary in these directions throughout the element. If, however, we choose an element that is infinitesimal in *all directions*, such as an infinitesimal rectangular parallelepiped, $dx\, dy\, dz$, then $\gamma$ can be assumed to be constant throughout the element. The weight of the element is then easily seen to be $\gamma(dx\, dy\, dz)$, where the coordinates of $\gamma$ correspond to the position of the element. We now illustrate a simple case.

***EXAMPLE 4.12**

Consider a block (see Fig. 4.33) wherein the specific weight $\gamma$ at corner $A$ is 200 lbf/ft³. The specific weight in the block does not change in the $x$ direction. However, it decreases linearly by 50 lbf/ft³ in the $y$ direction, and increases linearly by 50 lbf/ft³ in the $z$ direction, as has been shown in the diagram. What are the coordinates $\bar{x}$, $\bar{y}$ of the center of gravity for this block?

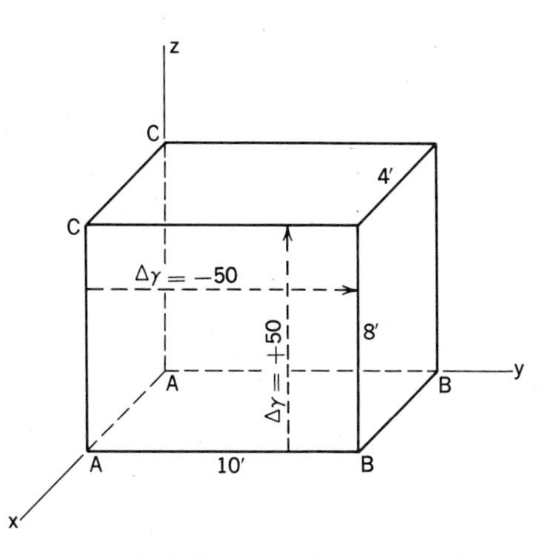

**Figure 4.33.** Block with varying $\gamma$.

We must first express $\gamma$ at any position $P(x, y, z)$. Using simple proportions, we can say

$$\begin{aligned}\gamma &= 200 - \frac{y}{10}(50) + \frac{z}{8}(50) \\ &= 200 - 5y + 6.25z \text{ lbf/ft}^3\end{aligned} \tag{a}$$

We shall first compute the weight of the block (i.e., the resultant force of gravity). We do not use an infinitesimal slice or rectangular rod of the block, as we have done heretofore. With the specific weight varying with both $y$ and $z$, it would not be an easy matter to compute the weight and moment of a slice or a rod. Instead, we shall use an infinitesimal rectangular parallelepiped, $dx\, dy\, dz$, at position $xyz$ as has been shown in Fig. 4.34(a). Because of the vanishingly small size of this element, the

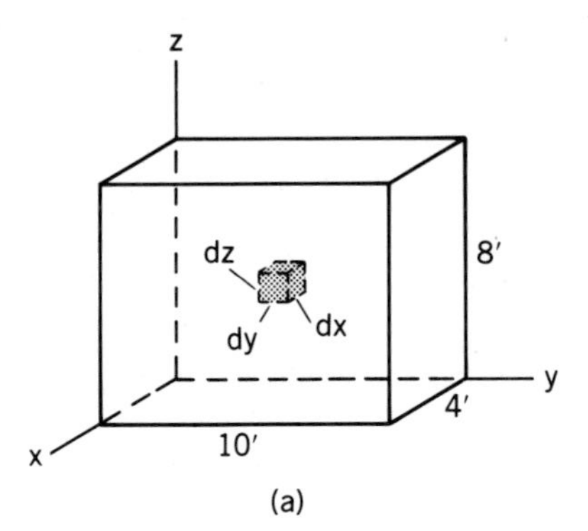


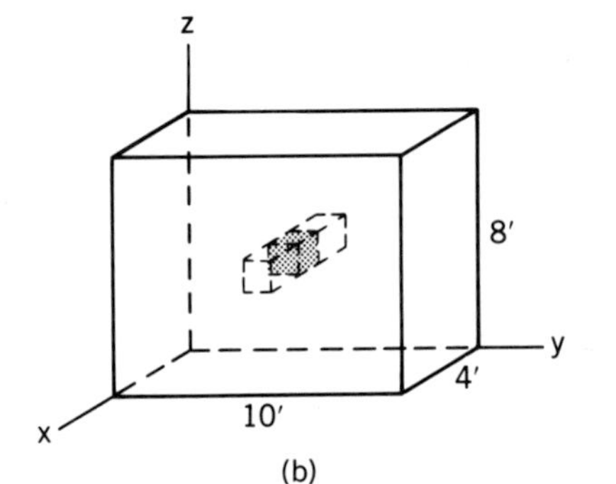


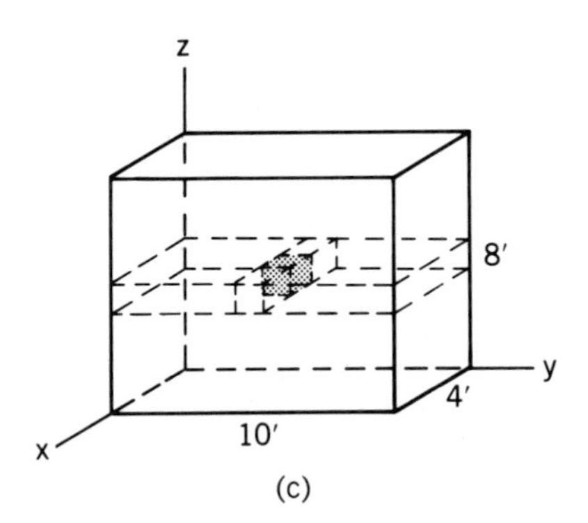


**Figure 4.34.** (a) Element $dx\ dy\ dz$ at $P\ (x, y, z)$; (b) $x$ runs from 0 to 4, while $z$ and $y$ are fixed, to form rectangular rod; (c) $y$ runs from 0 to 10, while holding $z$, to form slice.

specific weight $\gamma$ can be considered constant inside the element, and so the weight $dW$ of the element can be given as

$$dW = \gamma(dx\,dy\,dz) = (200 - 5y + 6.25z)\,dx\,dy\,dz$$

To include the weight of *all* such elements in the block, we first let $x$ "run" from 0 to 4 ft while holding $y$ and $z$ fixed. The rectangular parallelepiped of Fig. 4.34(a) then becomes a rectangular rod as shown in Fig. 4.34(b). Having run its course, $x$ is no longer a variable in this summation process. Next, let $y$ "run" from 0 to 10 while holding $z$ constant. The rectangular rod of Fig. 4.34(b) then becomes an infinitesimal slice, as shown in Fig. 4.34(c). The variable $y$ has thus run its course and is no longer a variable. This leaves only the variable $z$, and now we let $z$ "run" from 0 to 8. Clearly, we cover the entire block by this process.

We can do this mathematically by a process called *multiple integration.* We perform three integrations, paralleling the three steps outlined in the previous paragraph. Thus, we can formulate $W$ as follows:

$$-W\mathbf{k} = \int_0^8 \int_0^{10} \int_0^4 (200 - 5y + 6.25z)\,dx\,dy\,dz\,(-\mathbf{k})$$

We first consider the integration.

$$\int_0^4 (200 - 5y + 6.25z)\,dx$$

As in the first step set forth in the previous paragraph, to go from a rectangular parallelepiped to a rectangular rod, we integrate with respect to $x$ from $x = 0$ to $x = 4$ while holding $y$ and $z$ constant. Thus,

$$\int_0^4 (200 - 5y + 6.25z)\,dx = (200x - 5yx + 6.25zx)\Big|_0^4$$
$$= 800 - 20y + 25z$$

With $x$ no longer a variable (since it has run its course), the equation for $W$ becomes

$$W = \int_0^8 \int_0^{10} (800 - 20y + 25z)\,dy\,dz$$

Now, we hold $z$ constant and integrate with respect to $y$ from 0 to 10. (This takes us from a rectangular rod to a slice.) Thus,

$$\int_0^{10} (800 - 20y + 25z)\,dy = \left(800y - 20\frac{y^2}{2} + 25zy\right)\Big|_0^{10}$$
$$= 8000 - 1000 + 250z$$

Now $y$ has run its course, and we have

$$W = \int_0^8 (7000 + 250z)\, dz$$

By integrating with respect to $z$, we sum up all the slices, and we have covered the entire block. Thus,

$$W = \left(7000z + 250\frac{z^2}{2}\right)\Big|_0^8 = 64{,}000 \text{ lb}$$

To get $\bar{y}$, we equate the moment about the $x$ axis of the resultant force without a couple moment with the moment of the distribution. Thus, using multiple integration as described above:

$$-(64{,}000)\bar{y} = -\int_0^8 \int_0^{10} \int_0^4 y(200 - 5y + 6.25z)\, dx\, dy\, dz$$

Therefore,

$$64{,}000\bar{y} = \int_0^8 \int_0^{10} (200yx - 5y^2x + 6.25yzx)\Big|_0^4 dy\, dz$$

$$= \int_0^8 \int_0^{10} (800y - 20y^2 + 25yz)\, dy\, dz$$

$$= \int_0^8 \left(800\frac{y^2}{2} - \frac{20y^3}{3} + \frac{25y^2}{2}z\right)\Big|_0^{10} dz$$

$$= \int_0^8 (40{,}000 - 6667 + 1250z)\, dz$$

$$= \left(33{,}333z + 1250\frac{z^2}{2}\right)\Big|_0^8 = 307{,}000$$

and

$$\bar{y} = 4.79 \text{ ft}$$

Because $\gamma$ does not depend on $x$, we can directly conclude by inspection that $\bar{x} = 2$.

**Case B. Parallel Force Distribution over a Plane Surface—Center of Pressure.** Let us now consider a normal pressure distribution over a *plane* surface $A$ in the $xy$ plane in Fig. 4.35. The vertical ordinate is taken as a pressure ordinate, so that over the area $A$ we have a pressure distribution $p(x, y)$ represented by the pressure surface. Since in this case there is a parallel force system with one sense of direction, we know that the

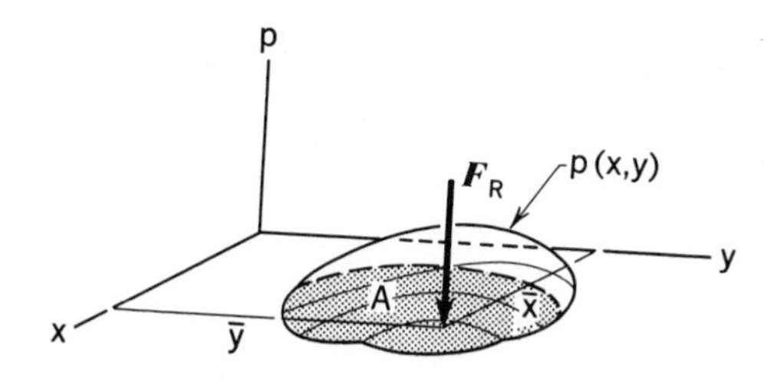


**Figure 4.35.** Pressure distribution.

simplest resultant is a single force, which is given as

$$\boldsymbol{F}_R = -\int p \, d\boldsymbol{A} = -\left(\int p \, dA\right)\boldsymbol{k} \tag{4.12}$$

The position $\bar{x}, \bar{y}$ can be computed by equating the moments about the $x$ and $y$ axes of the resultant force without a couple moment with the corresponding moments of the distribution. Solving for $\bar{x}$ and $\bar{y}$,

$$\bar{x} = \frac{\int px \, dA}{\int p \, dA}$$

$$\bar{y} = \frac{\int py \, dA}{\int p \, dA}$$

Since we know that $p$ is a function of $x$ and $y$ over the surface, we can carry out the preceding integrations either analytically or numerically. The point thus determined is called the *center of pressure*.

(In later chapters, we shall consider distributed frictional forces over plane and curved surfaces. In these cases, the simplest resultant is not necessarily a single force as it was in the special case above.)

**EXAMPLE 4.13**

A plate $ABCD$ on which both distributed and point force systems act is shown in Fig. 4.36. The pressure distribution is given as

$$p = -4y^2 + 100 \text{ psf} \tag{a}$$

Find the simplest resultant for the system.

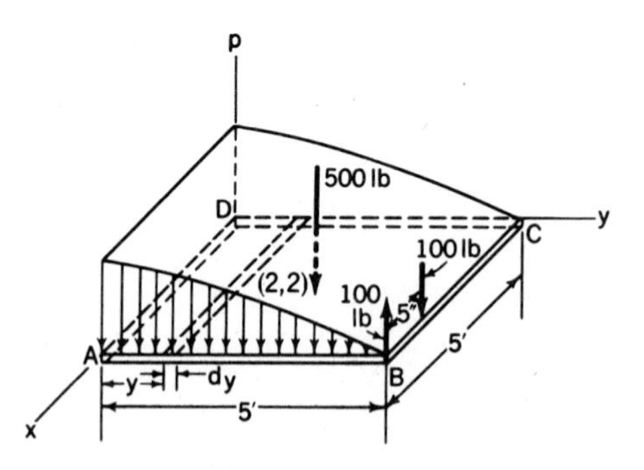


**Figure 4.36.** Find simplest resultant.

To get the resultant force, we consider a strip $dy$ along the plate as shown in Fig. 4.36. The reason for using such a strip is that the pressure $p$ is uniform along this strip, as can be seen from the diagram. Hence, the force from the pressure on the strip is simply $p\,dA = p(dy)(5)$. In summing forces, we simply integrate the forces $p(dy)(5)$ over all the strips of the plate. Thus, we can say that

$$F_R = -\int_0^5 p(5)(dy) - 500$$

$$= -\int_0^5 (-4y^2 + 100)5\,dy - 500 \qquad \text{(b)}$$

$$= \left(20\frac{y^3}{3} - 500y\right)\Big|_0^5 - 500 = -2167 \text{ lb}$$

To get the position $\bar{x}$, $\bar{y}$ of the resultant force $F_R$ without a couple moment, we equate moments of $F_R$ about the $x$ and $y$ axes with that of the original system. Thus, starting with the $x$ axis, we have using strip $dy$ as before:

$$-2167\bar{y} = -\int_0^5 yp(5dy) - (500)(2)$$

$$= -\int_0^5 5y(-4y^2 + 100)\,dy - 1000$$

$$= \left(20\frac{y^4}{4} - 500\frac{y^2}{2}\right)\Big|_0^5 - 1000 = -4125$$

Therefore,

$$\bar{y} = 1.904 \text{ ft}$$

Now, considering the $y$ axis, we still use the strips $dy$ because $p$ is uniform along such strips. However, the force $df = p\,dA = p(5)(dy)$ may be considered acting at the center of the strip, and accordingly has a moment arm about the $y$ axis equal to $\frac{5}{2}$ for each strip. Hence, we can say that

$$2167\bar{x} = \int_0^5 \frac{5}{2}p(5dy) + (500)(2) - \frac{500}{12}$$

$$= \frac{25}{2}\int_0^5 (-4y^2 + 100)\,dy + 1000 - 41.7 = 5125$$

Therefore,

$$\bar{x} = 2.36 \text{ ft}$$

### *EXAMPLE 4.14

What is the simplest resultant and the center of pressure for the pressure distribution shown in Fig. 4.37?

Notice that the pressure varies linearly in the $x$ and $y$ directions. The pressure at any point $x$, $y$ in the distribution can be given as follows with the aid of similar triangles[5]:

$$p = \left(\frac{y}{10}\right)(20) + \left(\frac{x}{5}\right)(30) \qquad \text{(a)}$$

$$= 2y + 6x \text{ Pa}$$

[5]The unit of pressure in SI units is the pascal, where

$$1 \text{ Pa} \equiv 1 \text{ N/m}^2$$

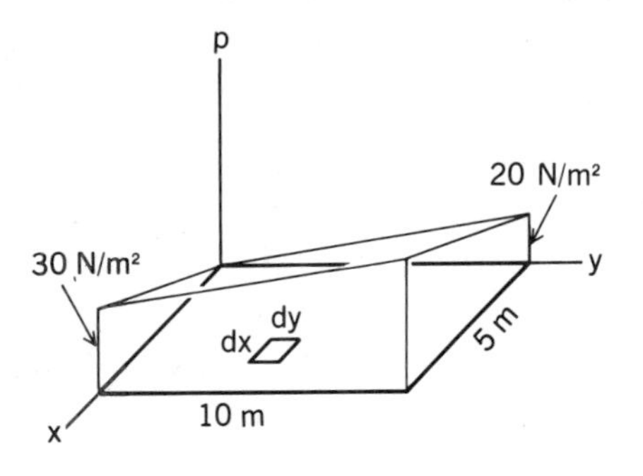


**Figure 4.37.** Nonuniform pressure distribution.

We cannot employ a convenient strip here along which the pressure is uniform, as in Example 4.13. For this reason we consider rectangular area element $dx\,dy$ to work with (see Fig. 4.37). For such small area, we can assume the pressure as constant so that $p\,dx\,dy$ is the force on the element. To find the resultant force, we must integrate over the $10 \times 5$ rectangle. This integration involves two variables and is again a case of *multiple integration*. Thus, we can say that

$$F_R = \int_0^{10} \int_0^5 p\,dx\,dy$$

wherein we first integrate with respect to $x$ while holding $y$ constant and then integrate with respect to $y$ (in this way we cover the entire $10 \times 5$ rectangular area). Thus, we have

$$\begin{aligned} F_R &= \int_0^{10} \int_0^5 (2y + 6x)\,dx\,dy \\ &= \int_0^{10} \left(2yx + \frac{6x^2}{2}\right)\Big|_0^5 dy \\ &= \int_0^{10} (10y + 75)\,dy \\ &= \frac{10y^2}{2} + 75y\Big|_0^{10} = 1250 \text{ N} \end{aligned}$$

To find $\bar{y}$ for $F_R$ without a couple moment, we equate moments of $F_R$ about the $x$ axis with that of the distribution. Thus,

$$-(\bar{y})(1250) = -\int_0^{10} \int_0^5 py\,dx\,dy$$

Therefore,

$$\begin{aligned} \bar{y} &= \frac{1}{1250} \int_0^{10} \int_0^5 (2y + 6x)y\,dx\,dy \\ &= \frac{1}{1250} \int_0^{10} \left(2y^2x + 6y\frac{x^2}{2}\right)\Big|_0^5 dy \\ &= \frac{1}{1250} \int_0^{10} (10y^2 + 75y)\,dy \\ &= \frac{1}{1250}\left(\frac{10y^3}{3} + 75\frac{y^2}{2}\right)\Big|_0^{10} \\ &= 5.67 \text{ m} \end{aligned}$$

As for $\bar{x}$, we proceed as follows:

$$(\bar{x})(1250) = \int_5^{10} \int_0^5 px\,dx\,dy$$

Therefore,

$$\bar{x} = \frac{1}{1250}\int_0^{10}\int_0^{5} (2y + 6x)x\,dx\,dy$$

$$= \frac{1}{1250}\int_0^{10}\left(2y\frac{x^2}{2} + \frac{6x^3}{3}\right)\Bigg|_0^5 dy$$

$$= \frac{1}{1250}\int_0^{10} (25y + 250)\,dy$$

$$= \frac{1}{1250}\left(25\frac{y^2}{2} + 250y\right)\Bigg|_0^{10}$$

$$= 3.00 \text{ m}$$

The center of pressure is thus at (3.00, 5.67) m.

**Case C. Coplanar Parallel Force Distribution.** As we pointed out earlier, this type of loading may be considered for beams loaded symmetrically over the longitudinal midplane of the beam. The loading is represented by an intensity function $w(x)$ as shown in Fig. 4.25. This coplanar parallel force distribution can be replaced by a single force given as

$$\boldsymbol{F}_R = -\int w(x)\,dx\boldsymbol{j}$$

We find the position of $\boldsymbol{F}_R$ without a couple moment by equating moments of $\boldsymbol{F}_R$ and the distribution $w$ about a convenient point of the beam, usually one of the ends. Solving for $\bar{x}$, we get

$$\bar{x} = \frac{\int xw(x)\,dx}{\int w(x)\,dx}$$

**EXAMPLE 4.15**

A simply supported beam is shown in Fig. 4.38 supporting a 1000-lb point force, a 500 lb-ft couple, and a coplanar, parabolic, distributed load $w$ lb/ft. Find the simplest resultant of this force system.

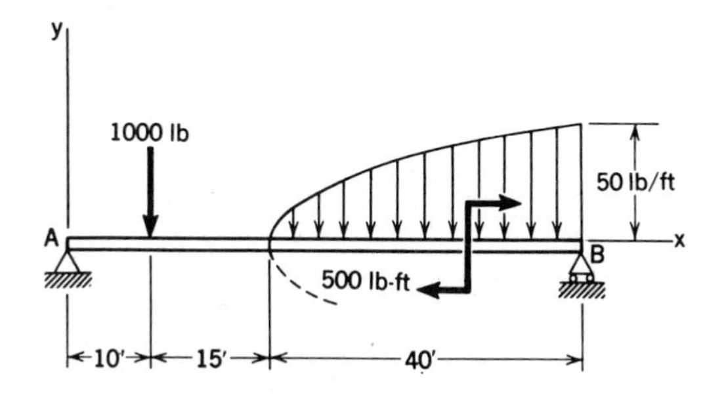


**Figure 4.38.** Find simplest resultant.

To express the intensity of loading for the coordinate system shown in the diagram, we begin with the general formulation

$$w^2 = ax + b \tag{a}$$

Note from the diagram that when $x = 25$ we have $w = 0$, and when $x = 65$, we have $w = 50$. Subjecting Eq. (a) to these conditions, we can determine $a$ and $b$. Thus,

$$0 = a(25) + b \tag{b}$$

$$2500 = a(65) + b \tag{c}$$

Subtracting, we can get $a$ as follows:

$$-2500 = -40a$$

Therefore,

$$a = 62.5$$

From Eq. (b), we get

$$b = -(25)(62.5) = -1562.5$$

Thus, we have

$$w^2 = 62.5x - 1562.5 \text{ lb/ft} \tag{d}$$

Summing forces, we get for $F_R$,

$$F_R = -1000 - \int_{25}^{65} \sqrt{62.5x - 1562.5}\, dx \tag{e}$$

To integrate this, we may change variables as follows:

$$\mu = 62.5x - 1562.5 \tag{f}$$

Therefore,

$$d\mu = 62.5\, dx$$

Substituting into the integral in Eq. (e), we have[6]

$$\begin{aligned} F_R &= -1000 - \int_0^{2500} \mu^{1/2} \frac{d\mu}{62.5} \\ &= -1000 - \frac{1}{62.5}\mu^{3/2}\left(\frac{2}{3}\right)\bigg|_0^{2500} \\ &= -1000 - \frac{1}{62.5}(2500)^{3/2}\left(\frac{2}{3}\right) \\ &= -2333 \text{ lb} \end{aligned}$$

We now compute $\bar{x}$ for the resultant without a couple as follows:

$$-2333\bar{x} = -(10)(1000) - \int_{25}^{65} x\sqrt{62.5x - 1562.5}\, dx - 500 \tag{g}$$

We can evaluate the integral most readily by consulting the mathematical formulas in Appendix I. We find the following formula:

---

[6]Do not forget to change the limits for $\mu$. Thus, from Eq. (f), the upper limit is $(62.5)(65) - 1562.5 = 2500$, whereas the lower limit is $(62.5)(25) - 1562.5 = 0$.

$$\int x\sqrt{a + bx}\,dx = -\frac{2(2a - 3bx)\sqrt{(bx + a)^3}}{15b^2}$$

In our case $b = 62.5$ and $a = -1562.5$, so the indefinite integral for our case is

$$\int x\sqrt{62.5x - 1562.5}\,dx = -\frac{(2)(-3125 - 187.5x)\sqrt{(62.5x - 1562.5)^3}}{(15)(3906)}$$

Putting in limits, we have

$$\int_{25}^{65} x\sqrt{62.5x - 1562.5}\,dx = -\frac{(2)(-3125 - 187.5x)\sqrt{(62.5x - 1562.5)^3}}{(15)(3906)}\Bigg|_{25}^{65}$$

$$= 65{,}333 - 0 = 65{,}333$$

Going back to Eq. (g), we can now solve easily for $\bar{x}$. Thus,

$$\bar{x} = -\frac{1}{2330}[-(10)(1000) - 65{,}300 - 500]$$

$$= 32.5 \text{ ft}$$

Before closing, it will be pointed out that, for a loading function $w(x)$, the resultant, $\int_0^x w\,dx$, equals the *area* under the loading curve. This fact is particularly useful for the case of a triangular loading function such as is shown in Fig. 4.39. Hence, we can say on inspection that the resultant force has the value

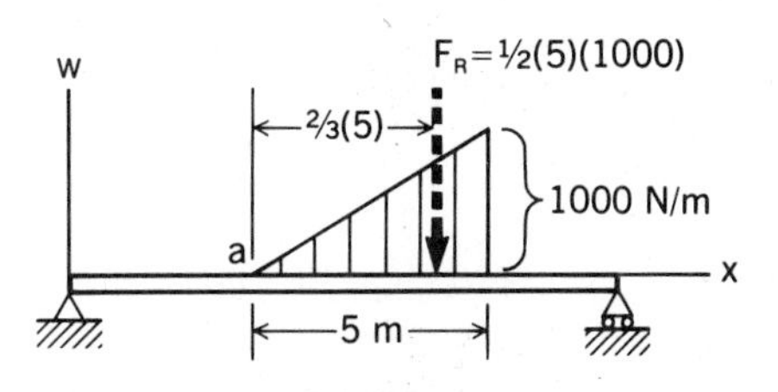


**Figure 4.39.** Triangular loading resultant.

$$F_R = \tfrac{1}{2}(5)(1000) = 2500 \text{ N}$$

Furthermore, you can readily show that the *simplest* resultant has a line of action that is $(\frac{2}{3}) \times$ (length of loading) from the toe of the loading.[7] Thus, $F_R$ without a couple moment is at a position $(\frac{2}{3})(5)$ to the right of $a$ (see Fig. 4.39). You are urged to use this information when needed.

Finally, in the case of a body made up of simple shapes (subbodies) such as cones and cubes, we can find the center of gravity by using the centers of gravity of the known shapes. Thus, we can say on taking moments about the $y$ axis that

$$W_{\text{total}}(x_c) = \sum_i W_i(x_c)_i \tag{4.13}$$

where $W_i$ is the weight of the $i$th subbody and where $(x_c)_i$ is the $x$ coordinate to the center of gravity of the $i$th subbody. Such bodies are called *composite bodies* (see Fig. P.4.51 for an example).

[7] In Chapter 8, you will learn that the simplest resultant force for a distribution $w(x)$ goes through the *centroid* of the area under $w(x)$. The centroid will be carefully defined at that time.

## Problems

**4.37.** A force field is given as

$$F(x, y, z, t) = (10x + 5)\boldsymbol{i} + (16x^2 + 2z)\boldsymbol{j} + 15\boldsymbol{k} \text{ N}$$

What is the force at position (3, 6, 7) m? What is the difference between the force at this position and that at the origin?

**4.38.** A magnetic field is developed such that the body force on the rectangular parallelepiped of metal is given as

$$f = (.01x + \tfrac{1}{8})\boldsymbol{k} \text{ oz/lbm}$$

If the specific weight of the metal is 450 $\text{lb/ft}^3$, what is the *simplest* resultant body force from such a field? Note that at the earth's surface the number of pounds mass equals the number of pounds force.

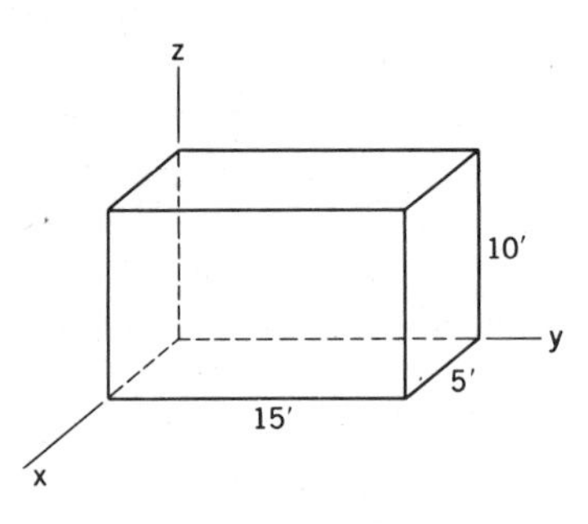


Figure P.4.38

**4.39.** Do the problem given in Example 4.11 using vertical strips.

**4.40.** A body of revolution has a variable specific weight such that $\gamma = (36 + .01x^2) \text{ kN/m}^3$ with $x$ in meters. A hole of diameter 3 m and length 6 m is cut from the body as shown. Where is the center of gravity?

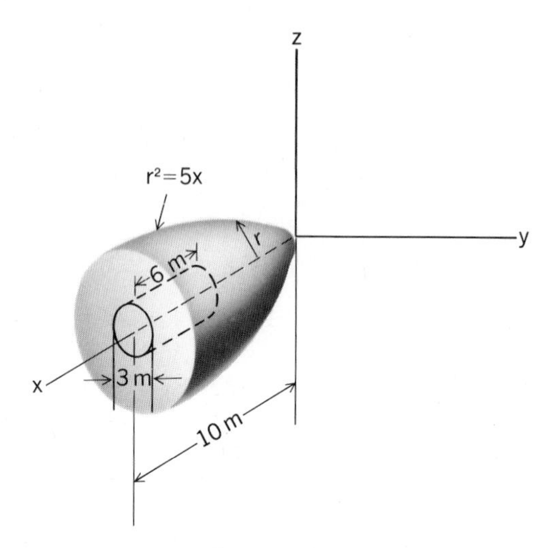

Figure P.4.40

**4.41.** The specific weight $\gamma$ of the material in the solid cylinder varies linearly as one goes from face $A$ to face $B$. If

$$\gamma_A = 400 \text{ lbf/ft}^3, \qquad \gamma_B = 500 \text{ lbf/ft}^3$$

what is the position of the center of gravity of the cylinder?

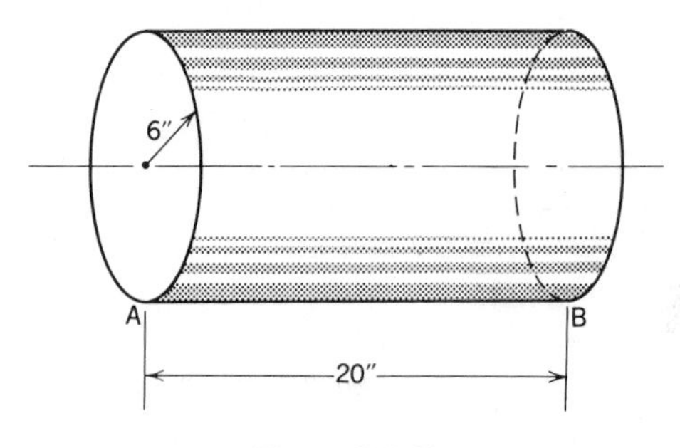


Figure P.4.41

**4.42.** The specific weight of the material in a right circular cone is constant. What is the center of gravity of the cone? *Hint*: Rotate cone 90° so that gravity is perpendicular to the $z$ axis. Use concept of similar triangles to show that $r/R = (h - z)/h$ and solve for $r$ needed for the integration.

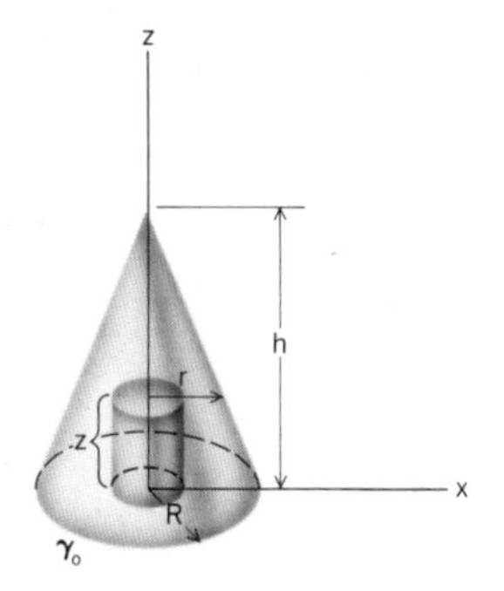


Figure P.4.42

**4.43.** Show that the center of gravity of the right triangular plate of thickness $t$ is at $x = a/3$ and $y = b/3$.

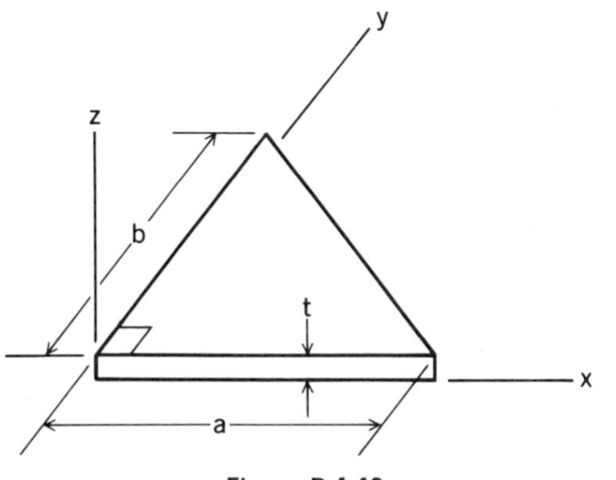

Figure P.4.43

**4.44.** Show that the volume and center of gravity of the conical frustum are, respectively,

$$\frac{\pi h}{3}(r_2^2 + r_1 r_2 + r_1^2)$$

and

$$\frac{h}{4}\,\frac{3r_2^2 + r_1^2 + 2r_1 r_2}{r_2^2 + r_1^2 + r_1 r_2}$$

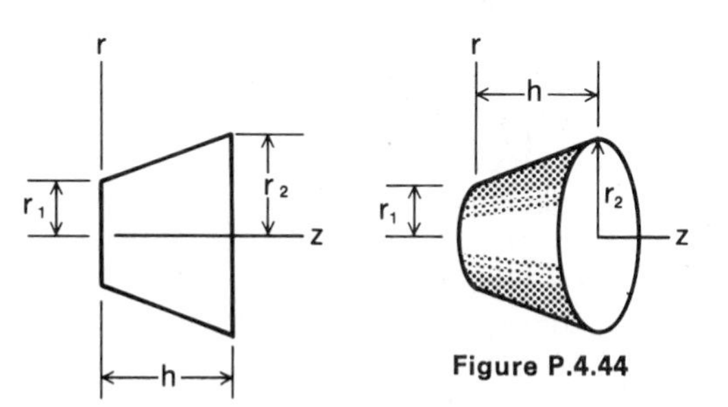


Figure P.4.44

**4.45.** Find the center of gravity of the plate bounded by a straight line and a parabola.

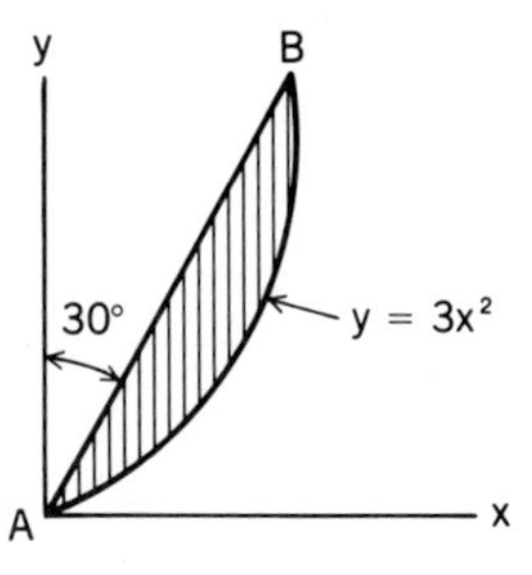

Figure P.4.45

**4.46.** A massive radio-wave antenna for detection of signals from outer space is a body of revolution with a parabolic face (see the diagram). These antennas may be carved from rock in a valley away from other disturbing signals. What would the antenna weigh if made from concrete (23. 6 kN/m³) for location in a remote desert area?

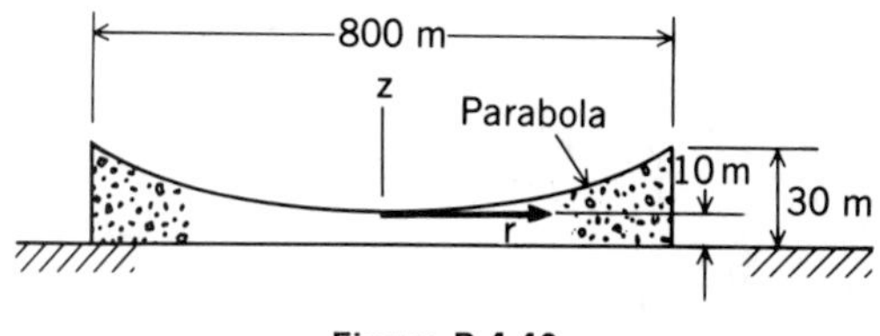

Figure P.4.46

**4.47.** In Problem 4.46, find the distance from the ground to the center of gravity if the total weight is $2.37 \times 10^8$ kN.

***4.48.** A plate of thickness 30 mm has a specific weight $\gamma$ that varies linearly in the $x$ direction from 26 kN/m³ at $A$ to 36 kN/m³ at $B$, and varies as the square of $y$ from 26 kN/m³ at $A$ to 40 kN/m³ at $C$. Where is the center of gravity of the plate?

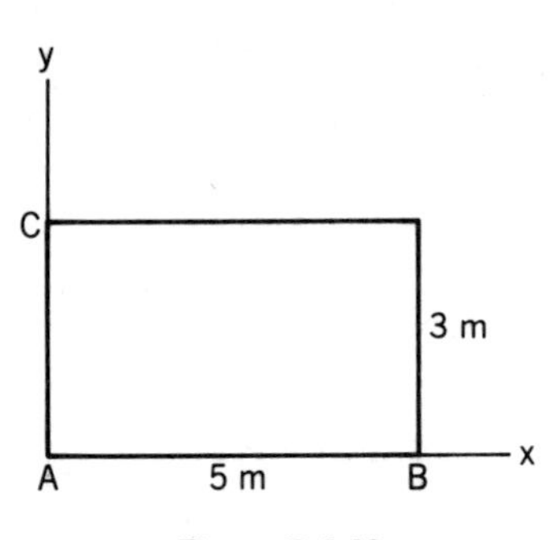


Figure P.4.48

***4.49.** Suppose in Problem 4.38 that

$$f = (.01x + .2y + .3z)\boldsymbol{k} \text{ oz/lbm}$$

Find the *simplest* resultant for $\gamma = 450$ lb/ft³. Find the proper line of action.

**4.50.** After a fast stop and swerve to the left, the load of sand (specific weight = 15 kN/m³) in a dump truck is in the position shown. What is the simplest resultant force on the truck from the sand and where does it act? If the truck was full (with a level top) before the stop, how much sand spilled? Use the results of Problem 4.43.

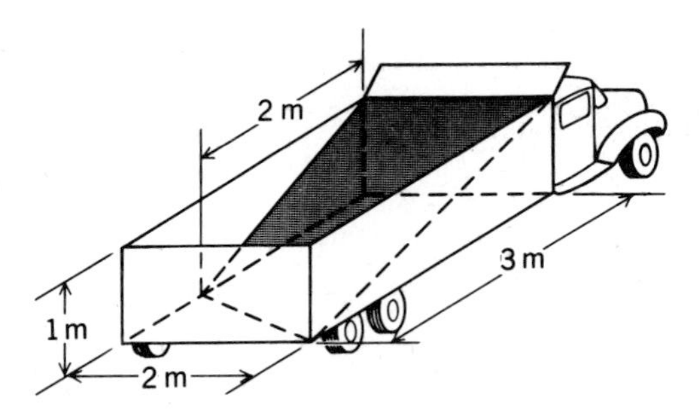


Figure P.4.50

*In Problems 4.51 through 4.53, use the known positions of centers of gravity of simple shapes.*

**4.51.** Find the weight and center of gravity of a large steam turbine for power generation needed for earthquake safety calculations. The specific weights of each turbine component are shown. The big cylinder having a radius $r_2$ of 5 m is 14 m long. Half of this cylinder is embedded in the large block.

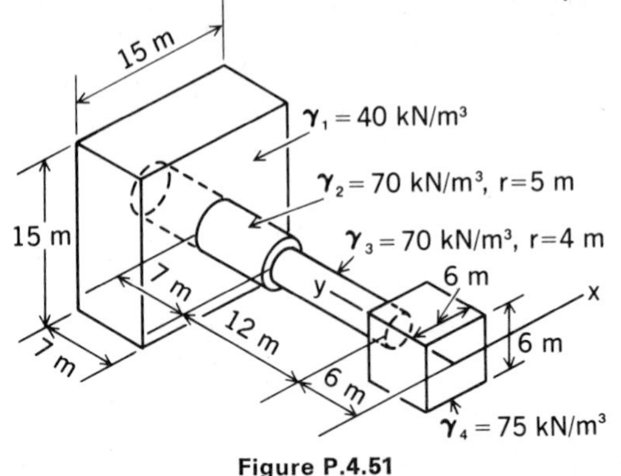


Figure P.4.51

**4.52.** An I-beam cantilevered out from a wall weighs 30 lb/ft and supports a 300-lb hoist. Steel (487 lb/ft³) cover plates 1 in. thick are welded on the beam near the wall to increase the carrying capacity of the beam. What is the moment at the wall due to the weight of the reinforced beam and the hoisted load of 4000 lb at the outermost position of the hoist? What is the simplest resultant force and its location?

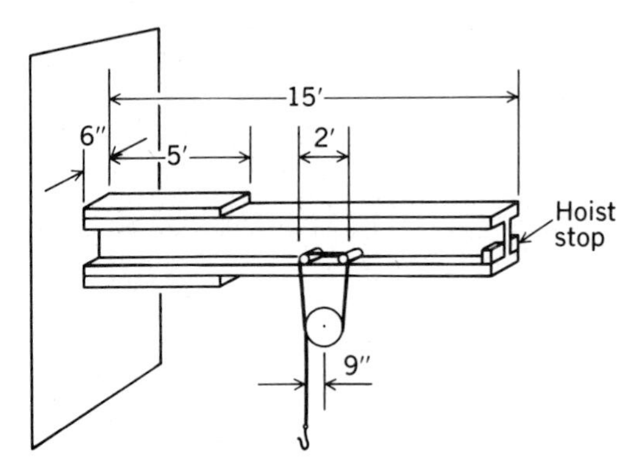


Figure P.4.52

**4.53.** The bulk materials trailer weighs 10,000 lb and is filled with cement ($\gamma = 94$ lb/ft³) in the front compartment (sections 1 and 2), and half-filled with water ($\gamma = 62.5$ lb/ft³) in the rear compartment (sections 3 and 4). What is the simplest resultant force, and where does it act? What is the resultant when the water is drained? Use the center of gravity and volume results from Problem 4.44 (conical frustum).

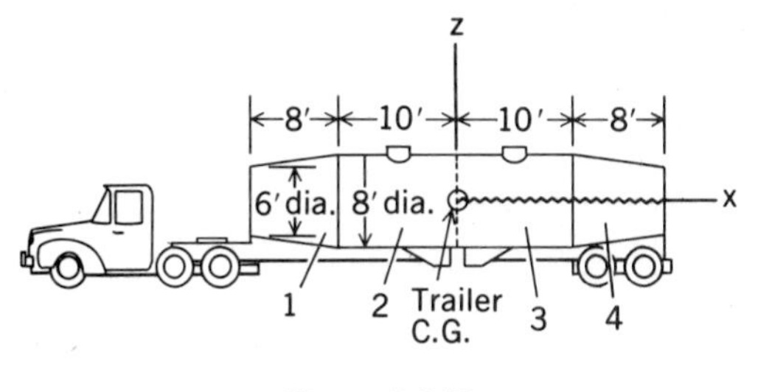


Figure P.4.53

**4.54.** Find the simplest resultant of a normal pressure distribution over the rectangular area with sides $a$ and $b$. Give the coordinates of the center of pressure.

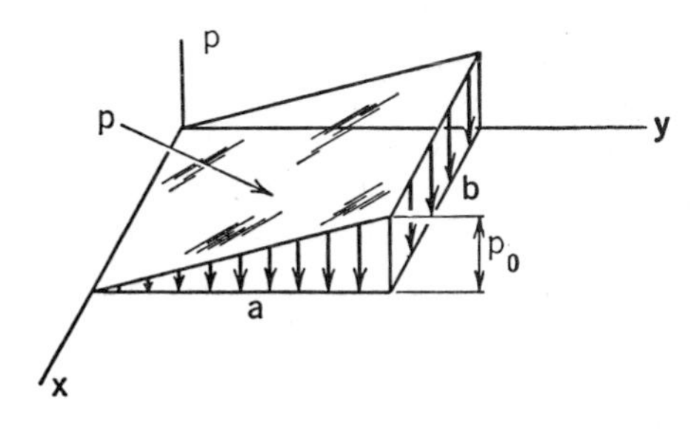


Figure P.4.54

**4.55.** Find the simplest resultant acting on wall $ABCD$. Give the coordinates of the center of pressure. The pressure varies such that $p = A/(y+1) + B$ psi, with $y$ in feet, from 10 psi to 50 psi, as indicated in the diagram.

**4.56.** One floor of a warehouse is divided into four areas. Area 1 is stacked high with TV sets such that the distributed load is $p = 120\ \text{lb/ft}^2$. Area 2 has refrigerators with $p = 65\ \text{lb/ft}^2$. Area 3 has stereos stacked so that $p = 80\ \text{lb/ft}^2$. Area 4 has washing machines with $p = 50\ \text{lb/ft}^2$. What is the simplest resultant force and where does it act?

**4.57.** Consider a pressure distribution $p$ forming a hemispherical surface over a domain of radius 5 m. If the maximum pressure is 5 Pa, what is the *simplest* resultant from this pressure distribution?

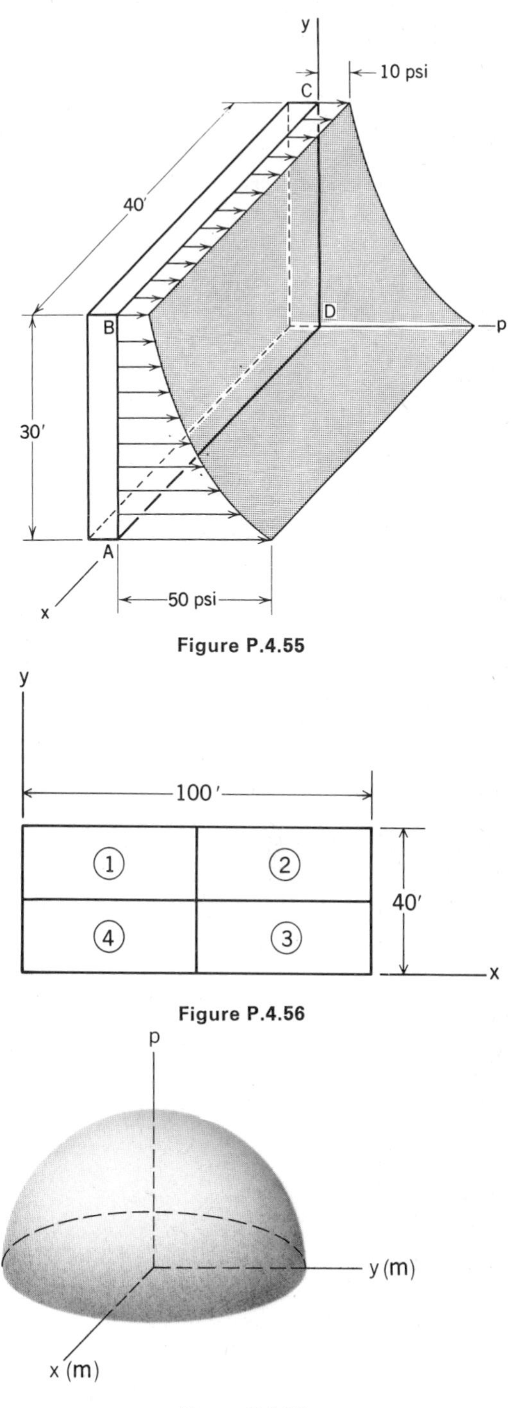


Figure P.4.55

Figure P.4.56

Figure P.4.57

***4.58.** The pressure $p_0$ at the corner $O$ of the plate is 50 Pa and increases linearly in the $y$ direction by 5 Pa/m. In the $x$ direction, it increases parabolically starting with zero slope so that in 20 m the pressure has gone from 50 Pa to 500 Pa. What is the simplest resultant for this distribution? Give the coordinates of the center of pressure.

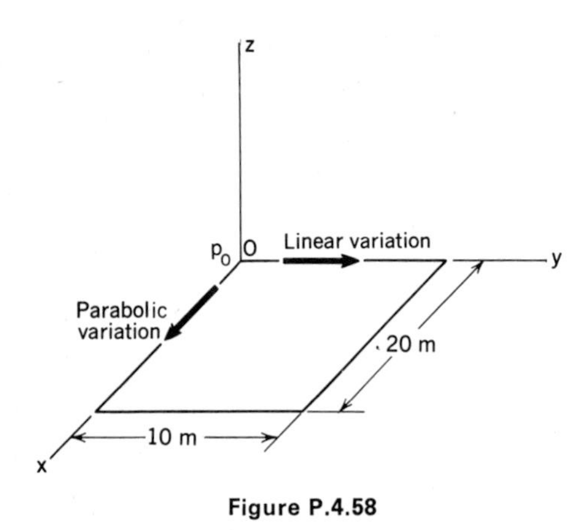


Figure P.4.58

*In Problems 4.59 through 4.63, we are concerned with pressure from water on a submerged surface. These are hydrostatic problems. You learned in physics that the pressure in water is $\gamma d$, where $\gamma$ is the specific weight and d is the vertical depth below the free surface.*

**4.59.** A sluice-gate door in a dam is 3 m wide and 3 m high. The water level in the dam is 4 m above the top of the door. The gate is opened until the water level falls 4 m. What is the simplest resultant force on the closed door at both water levels? Where do the forces act (i.e., where is the "center of pressure" in each case)? Water weighs 9818 N/m³.

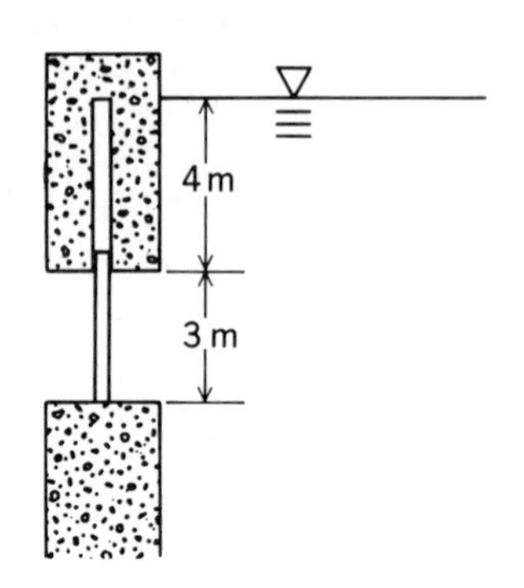


Figure P.4.59

**4.60.** A cylindrical tank of water is rotated at constant angular speed $\omega$ until the water ceases to change shape. The result is a free surface which, from fluid mechanics considerations, is that of a paraboloid. If the pressure varies directly as the depth below the free surface, what is the resultant force on a quadrant of the base of the cylinder? Take $\gamma = 62.4$ lb/ft³. [*Hint:* Use circular strip in quadrant having area $\frac{1}{4}(2\pi r\, dr)$.]

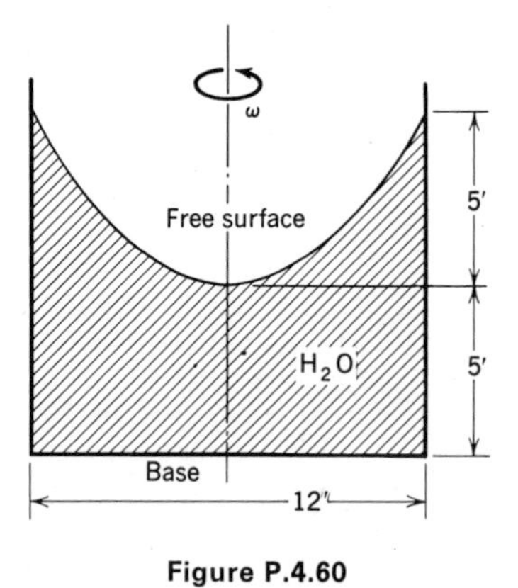


Figure P.4.60

**4.61.** What is the simplest resultant force from the water and where does it act on the 60-m-high 800-m-long straight earthfill dam? (Water weighs 9818 N/m³.)

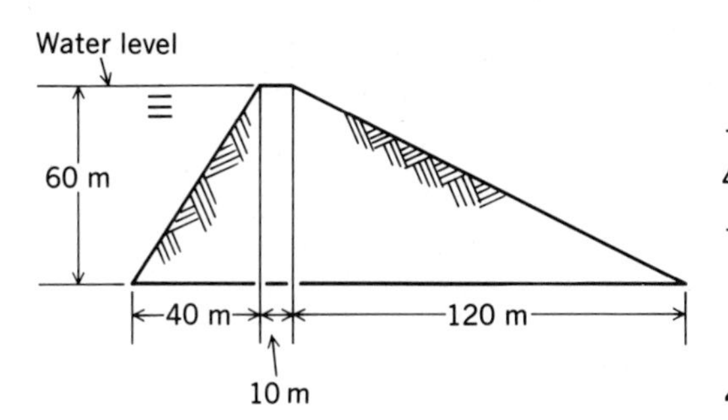


Figure P.4.61

**4.62.** A block 1 ft thick is submerged in water. Compute the simplest resultant force and the center of pressure on the bottom surface. Take $\gamma = 62.4$ lb/ft³.

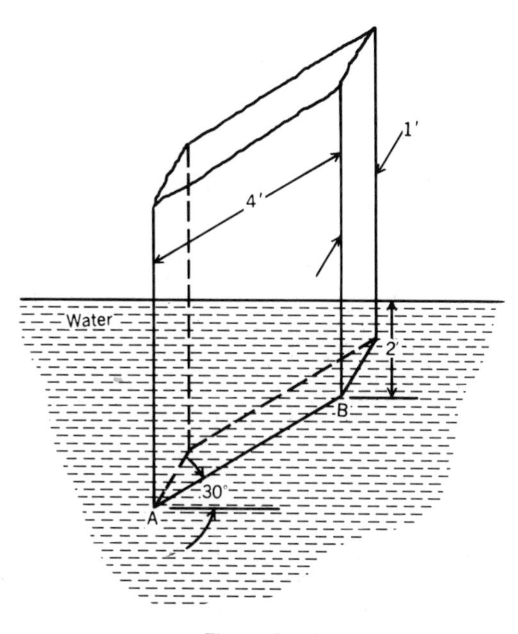


Figure P.4.62

***4.63.** What is the resultant force from water and where does it act on the 40-m-high circular concrete dam between two walls of a rocky gorge? (Water weighs 9818 N/m³.)

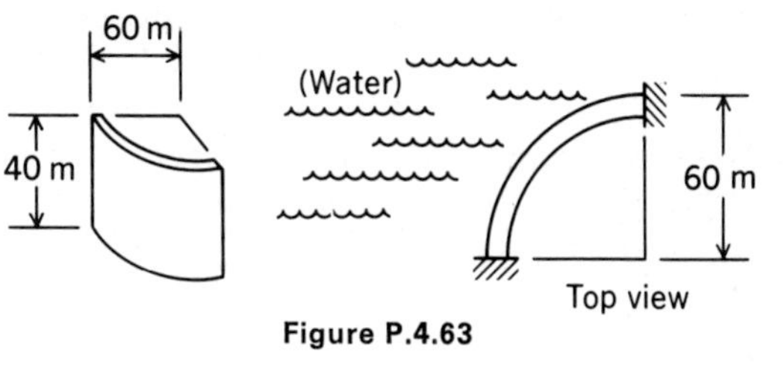


Figure P.4.63

**4.64.** The weight of the wire *ABCD* per unit length, *w*, increases linearly from 4 oz/ft at *A* to 20 oz/ft at *D*. Where is the center of gravity of the wire?

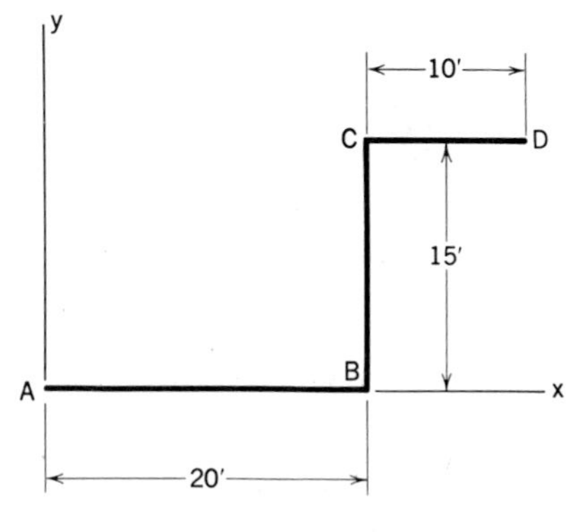


Figure P.4.64

**4.65.** Find the center of gravity of the wire. The weight per unit length increases as the square of the length of wire from a value of 3 oz/ft at *A* until it reaches the value of 8 oz/ft at *C*. It then decreases 1 oz/ft for every 10 ft of length.

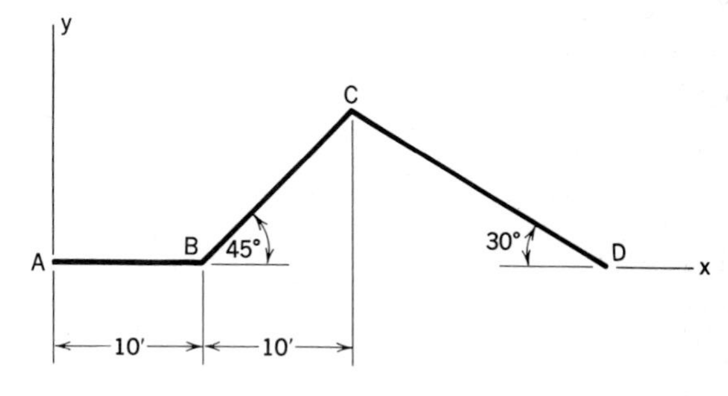


Figure P.4.65

**4.66.** Sandbags are piled on a beam. Each bag is 1 ft wide and weighs 100 lb. What is the simplest resultant force and where does it act? What linear mathematical function of the distributed load can be used to represent the sandbags over the left 3 ft of the beam?

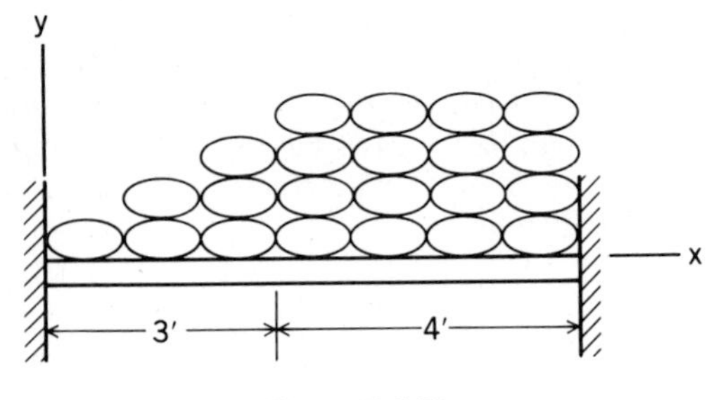


Figure P.4.66

**4.67.** A cantilever beam is subjected to a linearly varying load over part of its length. What is the *simplest* resultant force, and where does it act? What is the moment at the supported end?

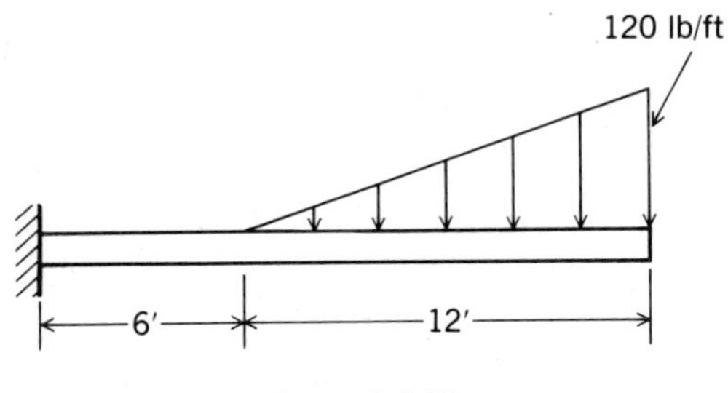


Figure P.4.67

**4.68.** Compute the *simplest* resultant force for the loads acting on the cantilever beam.

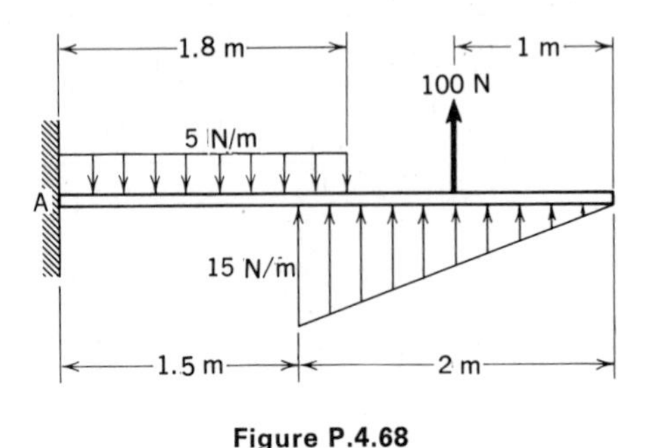


Figure P.4.68

**4.69.** Find the resultant force system at $A$ for the forces on the bent cantilever beam.

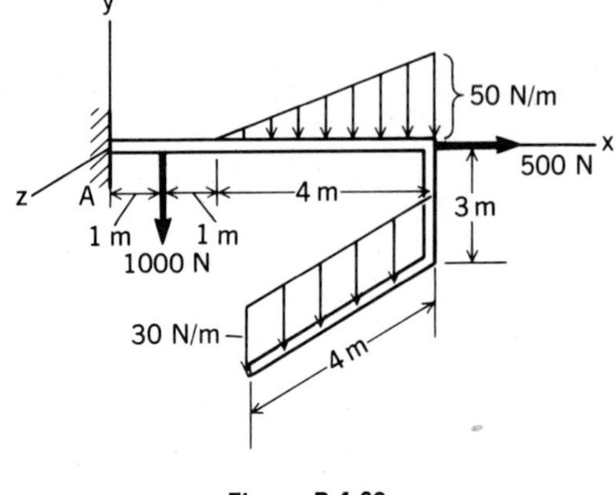

Figure P.4.69

## 4.6 Closure

We now have the tools that enable us to replace, for purposes of rigid-body mechanics, any system of forces by a resultant consisting of a force and a couple moment. These tools will prove very helpful in our computations. More important at this time, however, is the fact that in considering conditions of equilibrium for rigid bodies we need only concern ourselves with this resultant to reach conclusions valid for any force system, no matter how complex. From this viewpoint, we shall develop the fundamental equations of statics in Chapter 5 and then employ them to solve a large variety of problems.

## Review Problems

**4.70.** Replace the force and couples acting on the plate by a single force. Give the intercept of the line of action of this force with the vertical edge *BC* of the plate.

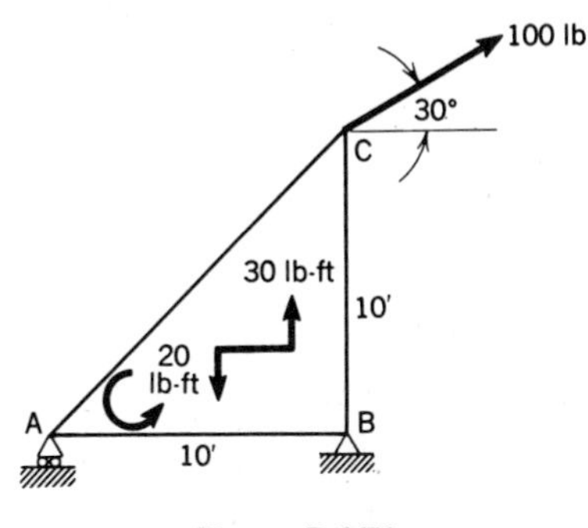


Figure P.4.70

**4.71.** A 100-kN bridge pier supports a 10-m segment of roadway weighing 150 kN and a 150-kN truck. The truck is located at the same position along the roadway as the pier. What is the equivalent force system acting on the base of the bridge pier when the truck is (a) in the center of the outside lane and (b) in the center of the inside lane?

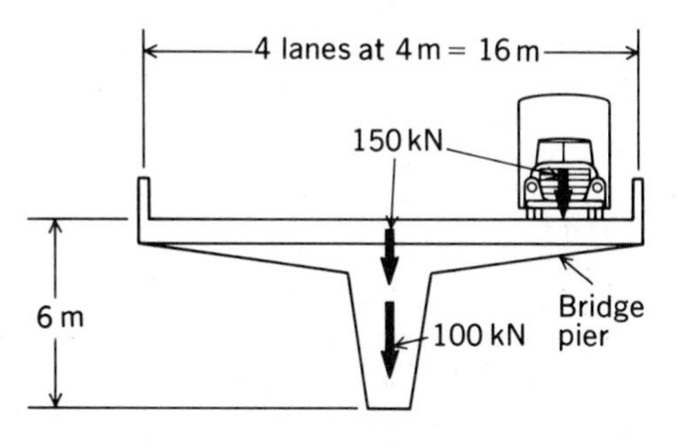


Figure P.4.71

**4.72.** A force $\boldsymbol{F} = 10\boldsymbol{i} + 3\boldsymbol{j} - 2\boldsymbol{k}$ lb goes through a point whose position vector is $\boldsymbol{r} = 6\boldsymbol{i} - 2\boldsymbol{j}$ ft. Find an equivalent system such that the force goes through position $\boldsymbol{r} = 2\boldsymbol{i} + 3\boldsymbol{k}$ ft.

**4.73.** A Jeep weighs 11 kN and has both a front winch and a rear power take-off. The tension in the winch cable is 5 kN. The power take-off develops 300 N-m of torque $T$ about an axis parallel to the $x$ axis. If the driver weighs 800 N, what is the resultant force system at the indicated center of gravity of the Jeep where we can consider the weight of the Jeep to be concentrated?

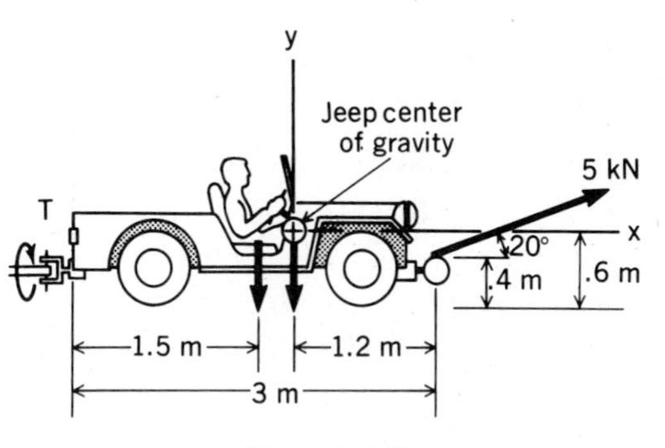


Figure P.4.73

**4.74.** What is the *simplest* resultant for the forces and couple acting on the beam?

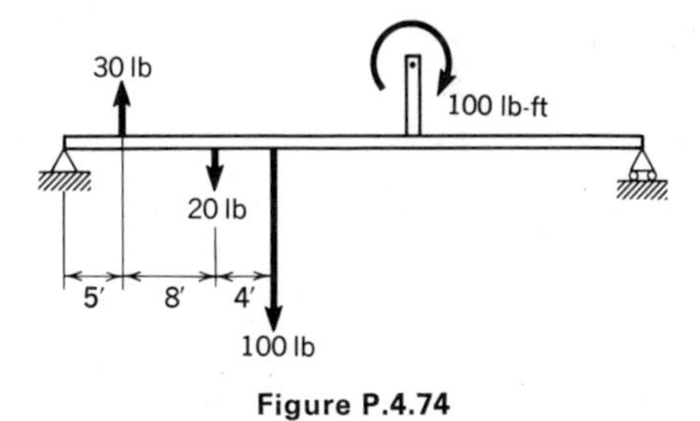


Figure P.4.74

**4.75.** A heavy duty off-the-road dump truck is loaded with iron ore that weighs 51 kN/m³. What is the *simplest* resultant force on the truck and where does it act?

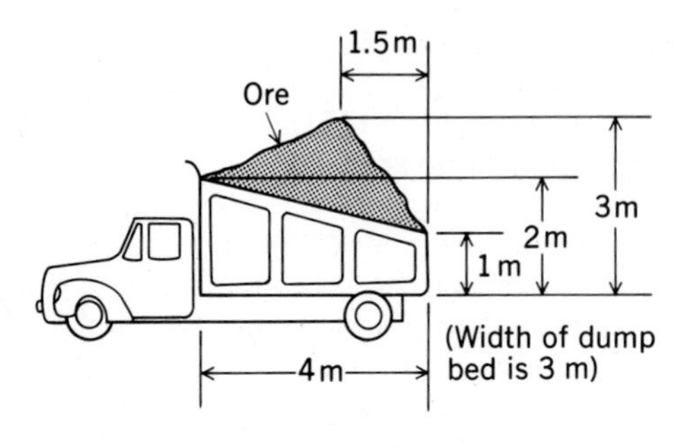


Figure P.4.75

**4.76.** The L-shaped concrete post supports an elevated railroad. The concrete weighs 150 lb/ft³. What is the simplest resultant force from the weight and the load and where does it act?

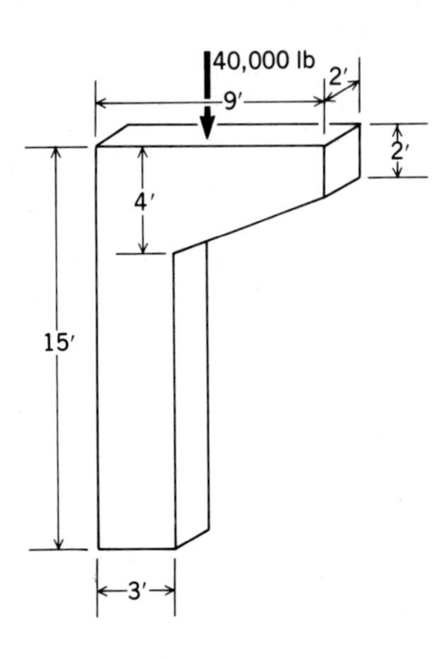


Figure P.4.76

**4.77.** Explain why the system shown can be considered a system of parallel forces. Find the *simplest* resultant for this system. The grid is composed of 1-m squares.

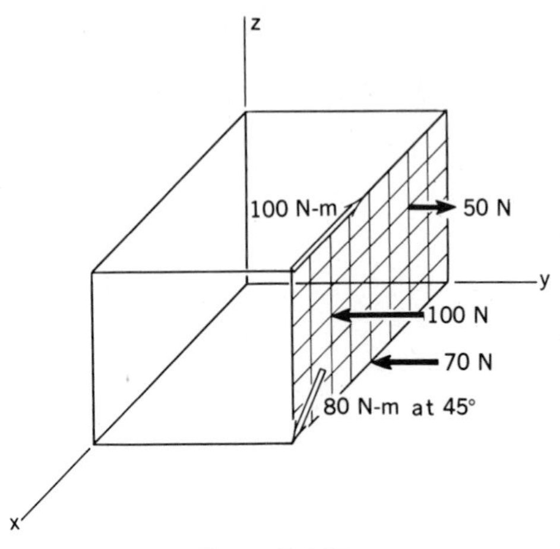

Figure P.4.77

**4.78.** A plate of thickness $t$ has as the upper edge a parabolic curve with infinite slope at the origin. Find the $x, y$ coordinates of the center of gravity for this plate.

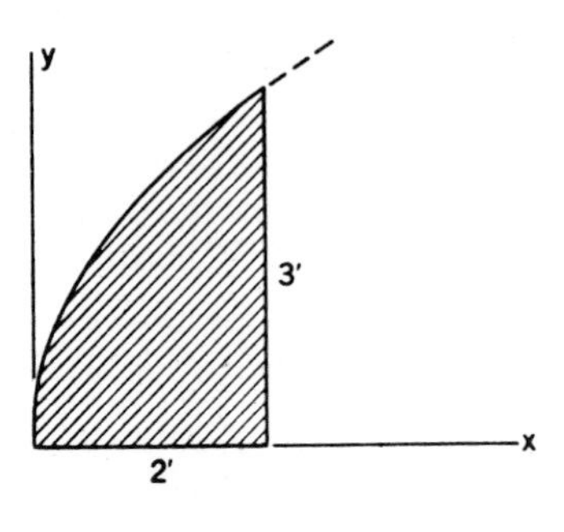


Figure P.4.78

**4.79.** A rectangular tank contains water. At the top of the water there is a pressure of .1380 N/mm² absolute. What is the simplest resultant force in the inside surface of the door $AB$? Where is the center of pressure relative to the bottom of the door? (*Hint:* It is known that the pressure in the water equals $\gamma d$, where $d$ is the distance below the surface of the water, plus the pressure on the surface. For water, $\gamma = 8190$ N/m³.)

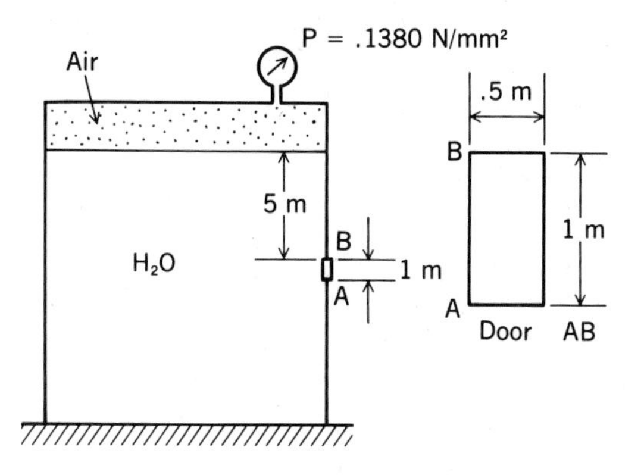


**Figure P.4.79**

**4.80.** The specific weight of the material in a right circular cone varies directly as the square of the distance $y$ from the base. If $\gamma_0 = 50$ lb/ft³ is the specific weight at the base, and if $\gamma' = 70$ lb/ft³ is the specific weight at the tip, where is the center of gravity of the cone? (See hint in Problem 4.42.)

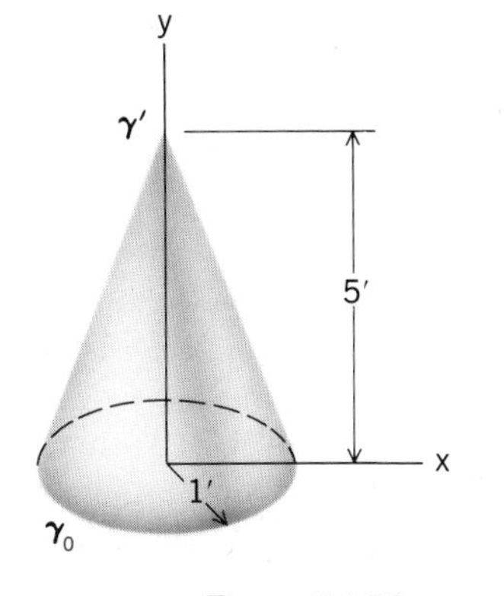


**Figure P.4.80**

***4.81.** A block has a rectangular portion removed (darkened region). If the specific weight is given as

$$\gamma = (2.0x + y + 3xyz) \text{ kN/m}^3$$

find the $\bar{x}$ for the center of gravity.

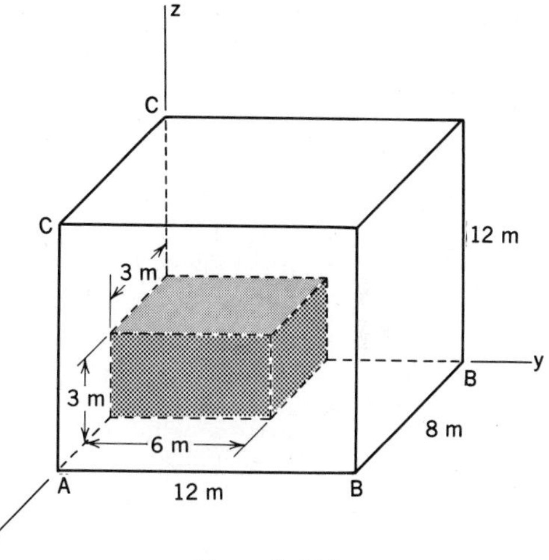

**Figure P.4.81**

**4.82.** Compute the *simplest* resultant for the loads shown acting on the simply supported beam. Give the line of action.

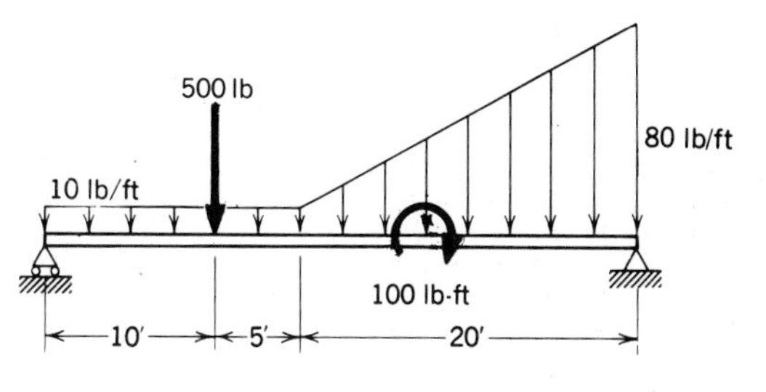


**Figure P.4.82**

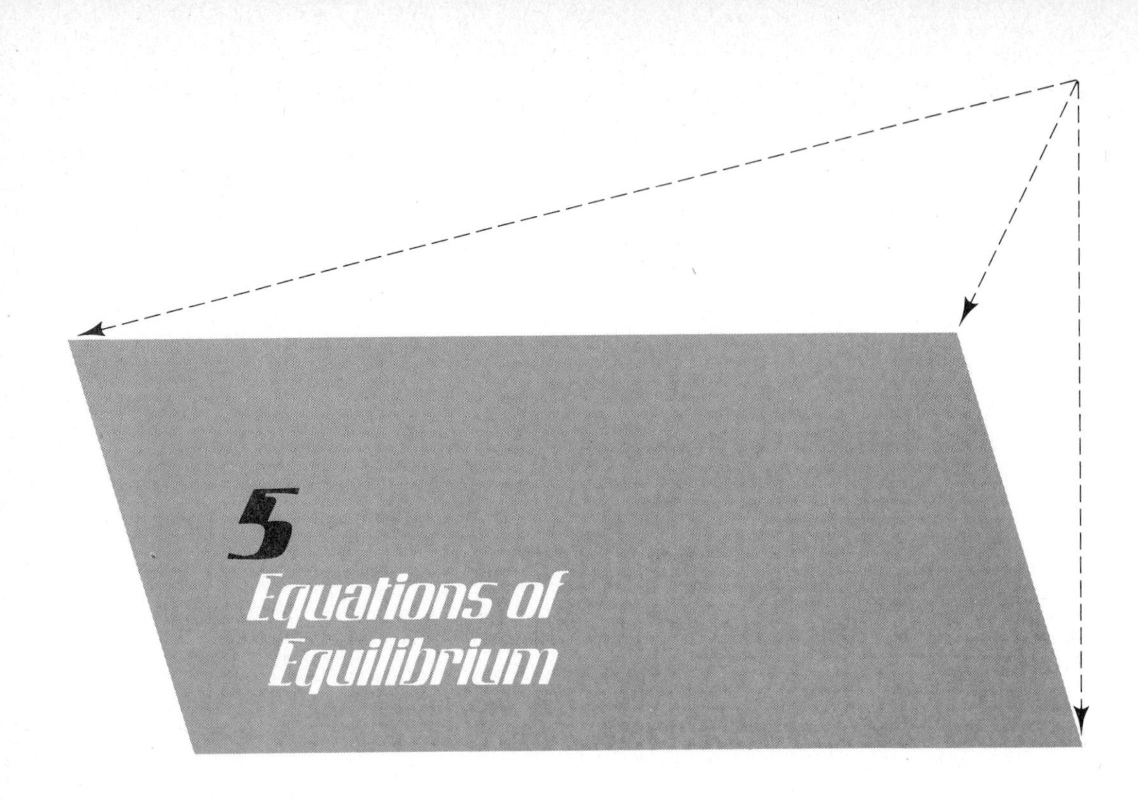


## 5.1 Introduction

You will recall from Section 1.10 that a *particle* in equilibrium is one that is stationary or that moves uniformly relative to an inertial reference. A *body* is in equilibrium if all the particles that may be considered to comprise the body are in equilibrium. It follows, then, that a rigid body in equilibrium cannot be rotating relative to an inertial reference. In this chapter, we shall consider bodies in equilibrium for which the rigid-body model is valid. For these bodies, there are certain simple equations that relate all the surface and body forces, or their equivalents, that act on the body. With these equations, we can sometimes ascertain the value of a certain number of unknown forces. For instance, in the beam shown in Fig. 5.1, we know the loads $F_1$ and $F_2$ and also the weight $W$ of the beam, and we want to determine the forces transmitted to the earth so that we can design a foundation to support the structure properly. Know-

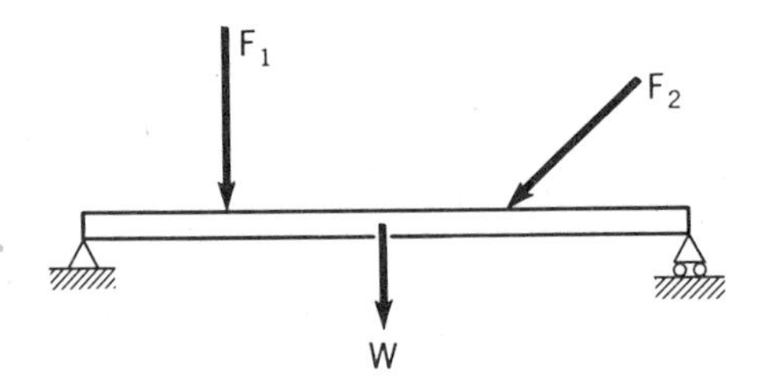


**Figure 5.1.** Loaded beam.

ing that the beam is in equilibrium and that the small deflection of the beam will not appreciably affect the forces transmitted to the earth, we can write rigid-body equations of equilibrium involving the unknown and known forces acting on the beam and thus arrive at the desired information.

Note in the beam problem above that a number of steps are implied. First, there is the singling out of the beam itself for discussion. Then, we express certain equations of equilibrium for the beam, which we take as a rigid body. Finally, there is the evaluation of the unknowns and interpretation of the results. In this chapter, we will carefully examine each of these steps.

Of critical importance is the need to be able to isolate a body or part of a body for analysis. Such a body is called a *free body*. We will first carefully investigate the development of free-body diagrams. We urge you to pay special heed to this topic, since *it is the most important step in the solving of mechanics problems.* An incorrect free-body diagram means that all ensuing work, no matter how brilliant, will lead to wrong results. More than just a means of attacking statics problems, the free-body concept is your first exposure to the overridingly important topic of *engineering analysis* in general.[1] We now examine this critical step.

## 5.2 The Free-Body Diagram

Since the equations of equilibrium for a particular body actually stem from the dynamic considerations of the body, we must be sure to include *all* the forces (or their equivalents) acting *on* this body, because they all affect the motion of the body and must be accounted for. To help identify all the forces and so ensure the correct use of the equations of equilibrium, we isolate the body in a simple diagram and show *all* the forces from the *surroundings* that act *on* the body. Such a diagram is called a *free-body diagram*. When we isolate the beam in our problem from its surroundings, we get Fig. 5.2. On the left end, there is an unknown force from the ground that has a magnitude denoted as $R_1$ and a direction denoted as $\theta$, with a line of action going

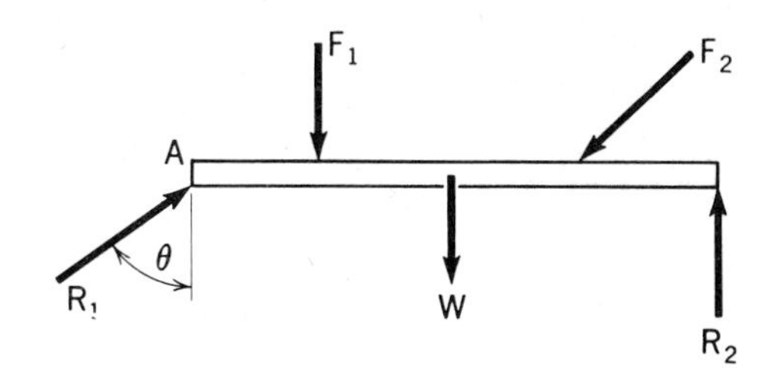


**Figure 5.2.** Free-body diagram of beam.

[1]The author has found through many years of experience that the absence of a free-body diagram in a student's work on a particular problem signifies that:

1. There will most likely be errors in the analysis of the problem, or
2. Even worse, the student does not have a good grasp of the problem.

through a known point $A$. [We may also use components $(R_1)_x$ and $(R_1)_y$ as unknowns rather than $R_1$ and $\theta$.] The right side involves a force in the vertical direction with an unknown magnitude denoted as $R_2$. The direction is vertical because the beam is on rollers to allow for thermal expansion and to relieve stretching of the beam in the axial direction. As a result, the ground exerts a negligibly small horizontal force there. Once all the forces acting on the beam have been identified, including the three unknown quantities $R_1$, $R_2$, and $\theta$, we can, by using three equations of equilibrium, solve for these unknowns.

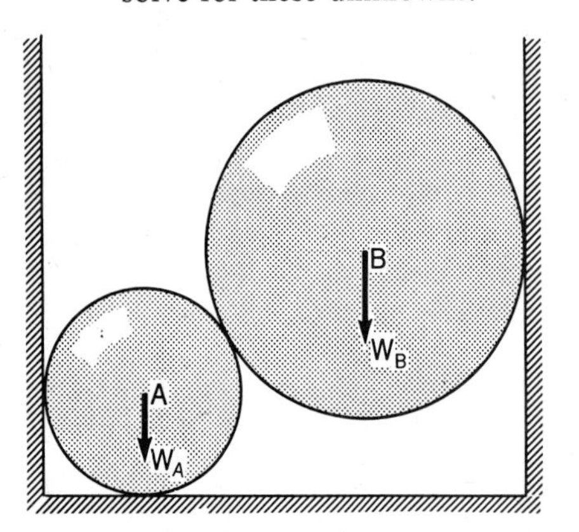


**Figure 5.3.** Smooth spheres in equilibrium.

Consider now the spheres shown in Fig. 5.3 in a condition of equilibrium with surfaces smooth and hard enough to permit us to neglect friction completely. The contact forces thus must be in a direction normal to the surface of contact. The free bodies of the spheres are shown in Fig. 5.4. Notice that $F_3$ is the magnitude of the force from sphere $B$ on sphere $A$, while the reaction, also shown as $F_3$ according to Newton's third law, is the magnitude of the force from sphere $A$ on sphere $B$.

You might be tempted to consider a portion of the container as a free body in the manner shown in Fig. 5.5. But even if this diagram did clearly depict a body (which it does not!), it would not qualify as a free body, since all the forces acting on the body have not been shown.

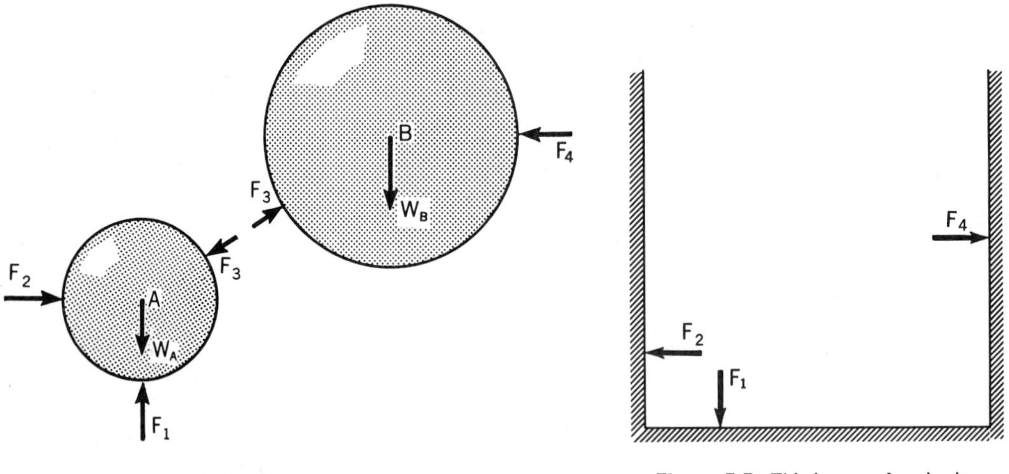


**Figure 5.4.** Free-body diagrams.

**Figure 5.5.** This is *not* a free-body diagram.

In engineering problems, bodies are often in contact in a number of standard ways. In Fig. 5.6, you will find the types of forces transmitted from body $M$ to body

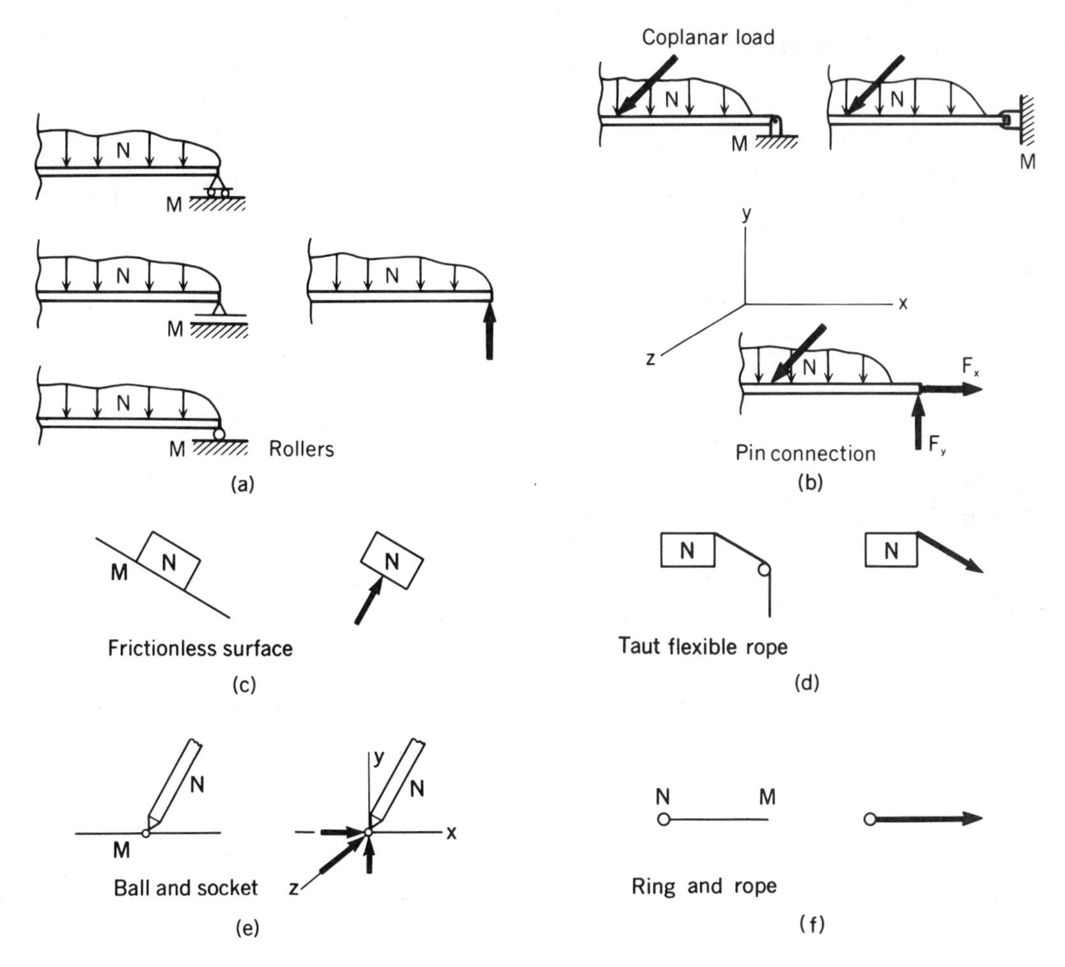


**Figure 5.6.** Standard connections.

$N$ for body connections that are often found in practice. (These are not free-body diagrams, since all the forces on any body have not been shown.)

In general, to ascertain the nature of the force system that a body $M$ is capable of transmitting to a second body $N$ through some connector or support, we may proceed in the following manner. Mentally move the bodies relative to each other in each of three orthogonal directions. In those directions where relative motion is impeded or prevented by the connector or support, there can be a force component at this connector or support in a free-body diagram of either body $M$ or $N$. Next, mentally rotate bodies $M$ and $N$ relative to each other about the orthogonal axes. In each direction about which relative rotation is impeded or prevented by the connector or support, there can be a couple-moment component at this connector or support in a free-body diagram of body $M$ or $N$. Now as a result of equilibrium considerations of body $M$ or $N$, certain force and couple-moment components that are capable of being generated at a support or connector will be zero for the particular loadings at hand. Indeed, one can often readily recognize this by inspection.

For instance, consider the pin-connected beam shown in Fig. 5.7. If we mentally move the beam relative to the ground in the $x$, $y$, and $z$ directions, we get resistance from the pin for each direction, and so the ground at $A$ can transmit force components $A_x$, $A_y$, and $A_z$. However, because the loading is coplanar in the $xy$ plane, the force component $A_z$ must be zero and can be deleted. Next, mentally rotate the beam relative to the ground at $A$ about the three orthogonal axes. Because of the smooth pin connection, there is no resistance about the $z$ axis and so $M_z = 0$. But there is resistance about the $x$ and $y$ axes. However, the coplanar loading in the $xy$ plane cannot exert moments about the $x$ and $y$ axes, and so the couple moments $M_x$ and $M_y$ are zero. All told, then, we just have force components $A_x$ and $A_y$ at the pin connection, as has been shown earlier in Fig. 5.6(b), wherein we relied on physical reasoning for this result.

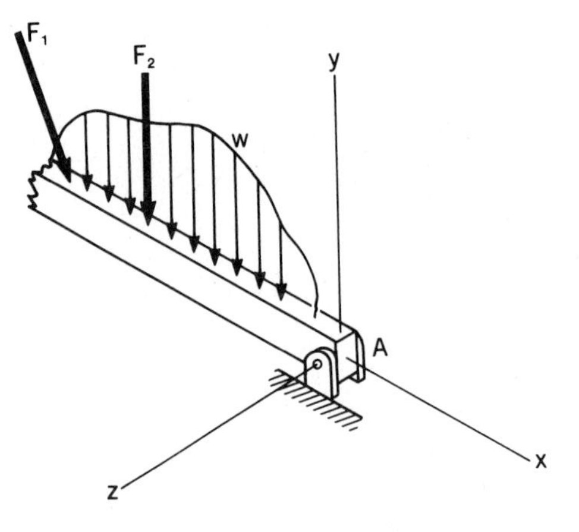


**Figure 5.7.** Pin connection.

### 5.3 Free Bodies Involving Interior Sections

Let us consider a rigid body in equilibrium as shown in Fig. 5.8. Clearly, every portion of this body must also be in equilibrium. If we consider the body as two parts $A$ and $B$, we can present either part in a free-body diagram. To do this, we must include on the portion chosen to be the free body the forces *from the other part* that arise at the common section (Fig. 5.9). The surface between both sections may be any curved or plane surface, and over it there will be a continuous force distribution. In the general case, we know that such a distribution can be replaced by a single force and a single couple moment, and this has been done in the free-body diagram of parts $A$ and $B$ in Fig. 5.9. Notice that Newton's third law has been observed.

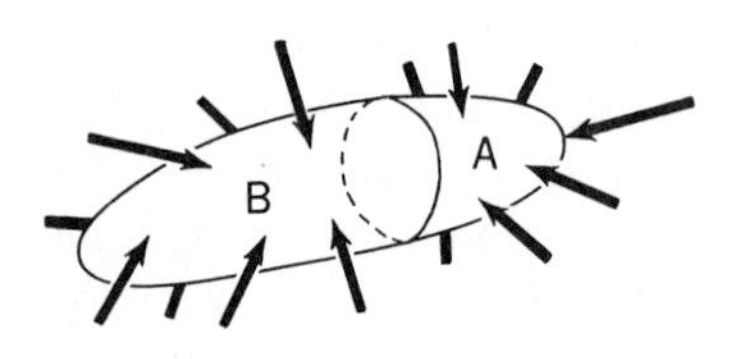


**Figure 5.8.** Rigid body in equilibrium.

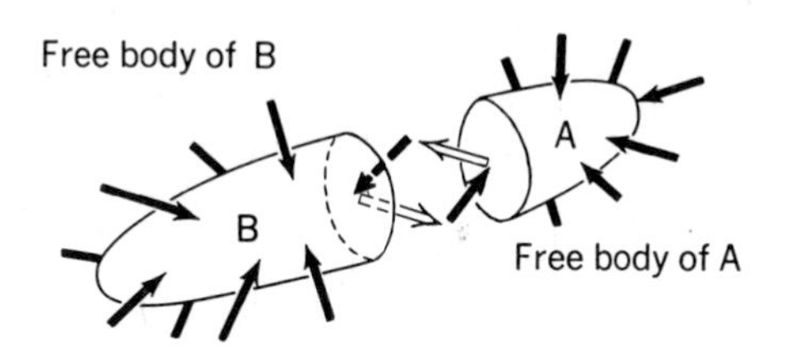


**Figure 5.9.** Free bodies of parts *A* and *B*.

As a special case, consider a beam with one end embedded in a massive wall (cantilever beam) and loaded along the *xy* plane (Fig. 5.10). A free body of the portion of the beam extending from the wall is shown in Fig. 5.11. Because of the geometric symmetry about the *xy* plane and the fact that the loads are in this plane, the exposed

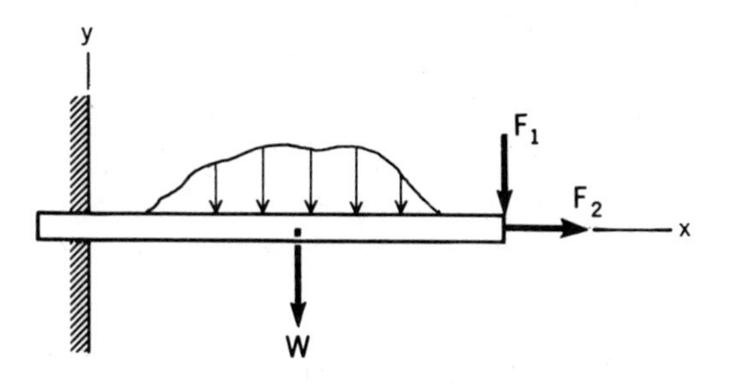


**Figure 5.10.** Cantilever beam.

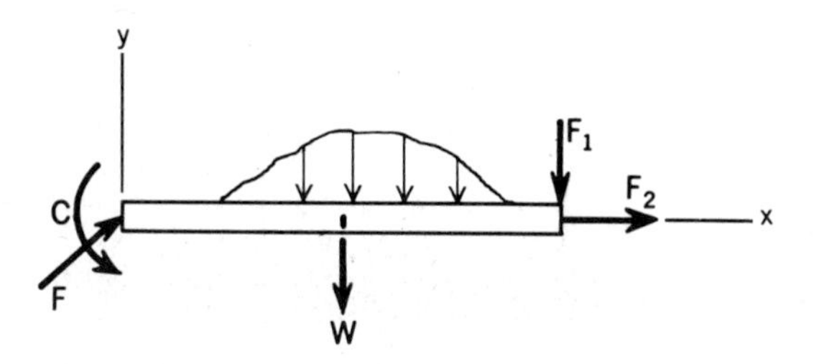


**Figure 5.11.** Free-body diagram of cantilever beam.

forces in the cut section can be considered coplanar. Hence, this distribution can be replaced by a force and a couple moment in the center plane, and it is the usual practice to decompose the force into components $F_y$ and $F_x$. Although a line of action for the force can be found that would enable us to eliminate the couple moment, it is desirable in structural problems to work with an equivalent system that has the force passing through the center of the beam cross section, and thus to have a couple moment. In the next section, we will see how $F$ and $C$ can be ascertained.

**EXAMPLE 5.1**

As a further illustration of a free-body diagram, we shall now consider the frame[2] shown in Fig. 5.12, which consists of members connected by frictionless pins. The force systems acting on the assembly and its parts will be taken as coplanar. We shall now sketch free-body diagrams of the assembly and its parts.

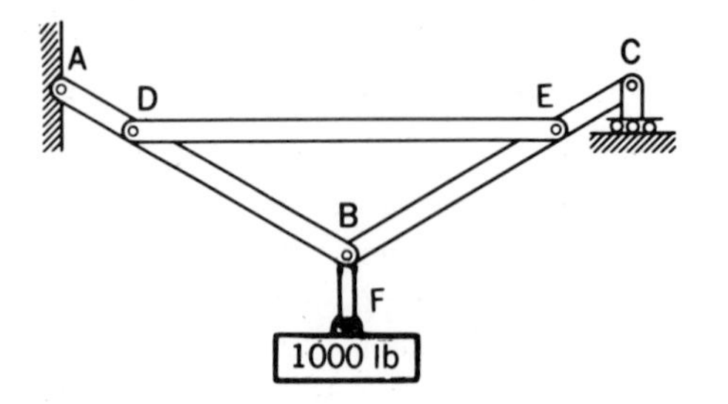


**Figure 5.12.** A frame.

**Free-body diagram of the entire assembly.** The magnitude and direction of the force at $A$ from the wall onto the assembly is not known. However, we know that this force is in the plane of the system. Therefore, two components are shown at this point (Fig. 5.13). Since the direction of the force $C$ is known, there are then three unknown scalar quantities, $A_y$, $A_x$, and $C$, for the free body.

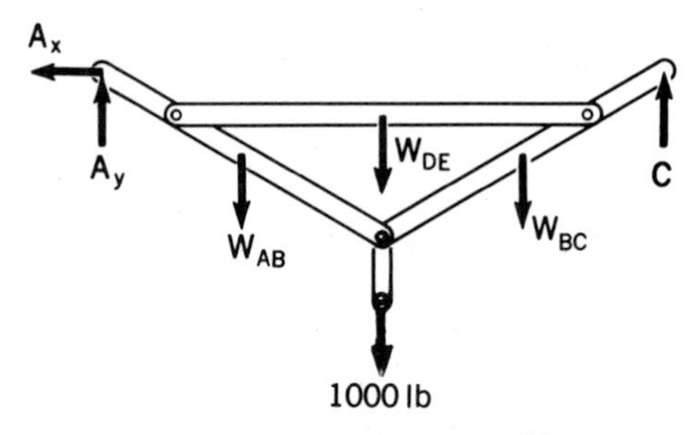


**Figure 5.13.** Free-body diagram of frame.

**Free-body diagram of the component parts.** When two members are pinned together, such as members $DE$ and $AB$ or $DE$ and $BC$, we usually consider the pin to be part of one of the bodies. However, when more than two members are connected at a pin, such as members $AB$, $BC$, and $BF$ at $B$, we often isolate the pin and consider that all members act on the pin rather than directly on each other, as illustrated in Fig. 5.14. Notice the forces that form pairs of reactions have been enclosed with dashed lines.

Do not be concerned about the proper sense of an unknown force component that you draw on the free-body diagram, for you may choose either a positive or negative sense for these components. When the values of these quantities are ascer-

---

[2]A *frame* is a system of connected straight or bent, long, slender members.

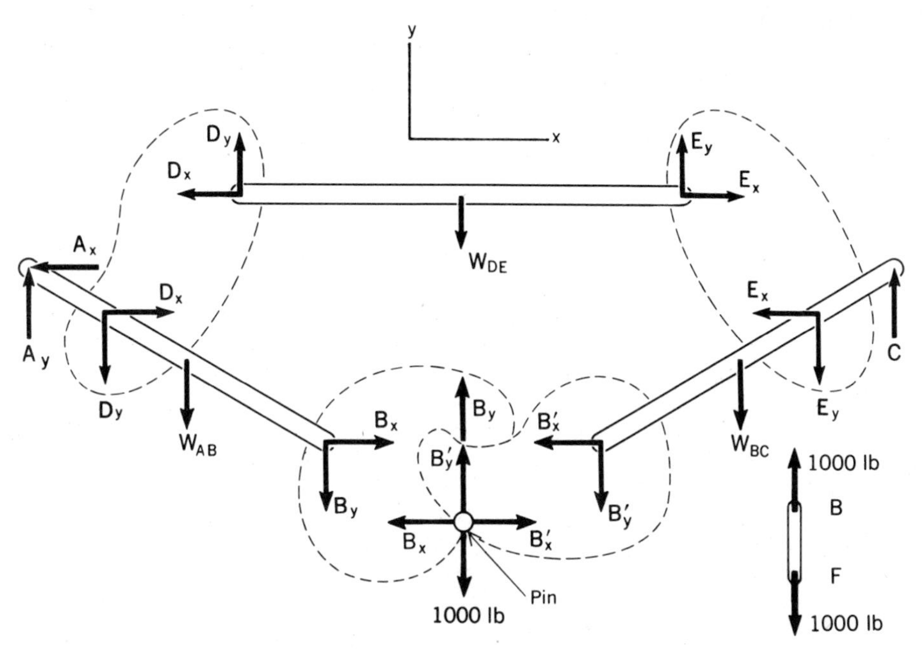


**Figure 5.14.** Free-body diagrams of parts.

tained by methods of statics, the proper sense for each component can then be established; but, having chosen a sense for a component, you must be sure that the *reaction* to this component has the *opposite* sense—else you will violate Newton's third law.

**Free-body diagram of portion of the assembly to the right of *M—M*.** In making a free body of the portion to the right of section *M—M* (see Fig. 5.15), we must remember

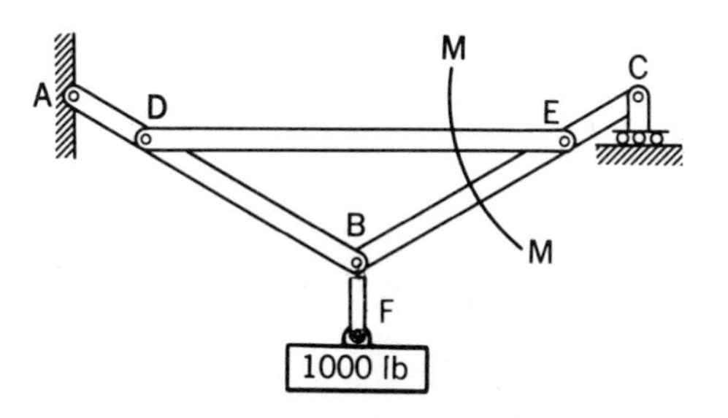


**Figure 5.15.** Cut along *M—M*.

to put in the weight of the portions of the members remaining *after* the cut has been made. At the two cuts made by *M—M* we must replace coplanar force distributions by resultants, as in the case of the previously considered cantilever beam. This is accomplished by inserting two force components and a couple moment as was done for the cantilever beam. Note in Fig. 5.16 that there are seven unknown scalar quantities

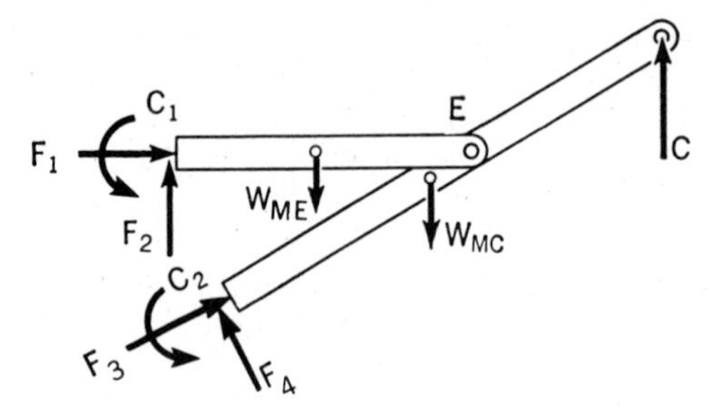


Figure 5.16. Free-body diagram.

for this free-body diagram. They are $C_1$, $C_2$, $F_1$, $F_2$, $F_3$, and $F_4$. Apparently, the number of unknowns varies widely for the various free bodies that may be drawn for the system. For this reason, you must choose the free-body diagram that is suitable for your needs with some discretion in order to effectively solve for the desired unknowns.

**EXAMPLE 5.2**

Draw a free-body diagram of the beam $AB$ and the pulley in Fig. 5.17. The weight of the pulley is $W_D$, and the weight of the beam is $W_{AB}$.

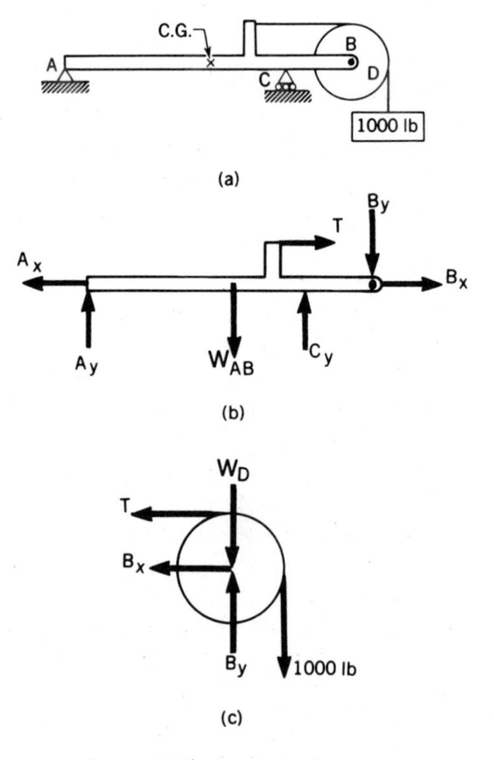


Figure 5.17. Free body diagrams.

The free-body diagram of beam $AB$ is shown in Fig. 5.17(b). The weight of the beam has been shown at the center of gravity. Components $B_x$ and $B_y$ are forces from the pulley $D$ acting on the beam through the pin at $B$. The free-body diagram of the pulley is shown in Fig. 5.17(c).

Some students may be tempted to put the weight of the pulley at $B$ in the free-body diagram of beam $AB$. The argument given is that this weight "goes through $B$." To put the pulley weight at $B$ on free body $AB$ is strictly speaking an error! The fact is that the weight of the pulley is a body force acting throughout the *pulley* and *does not* act on the *beam BD*. It so happens that the simplest resultant of this body force distribution on $D$ goes through a *position* corresponding to pin $B$. This does not alter the fact that this weight acts *on the pulley* and *not on the beam*. The beam can only feel forces $B_x$ and $B_y$ transmitted from the pulley to the beam through pin $B$. These forces are related to the pulley weight as well as the tension in the cord around the pulley through equations of equilibrium for the free body of the pulley itself.

## Problems

**5.1.** Draw the free-body diagram when the gas-grill lid is lifted at the handle to a 45° open position.

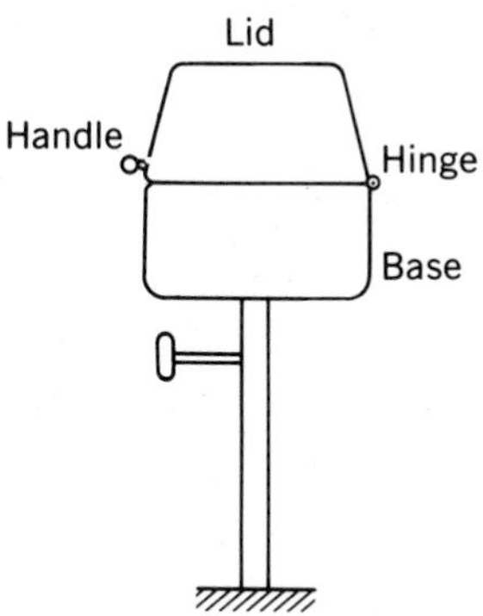


**Figure P.5.1**

**5.2.** A large antenna is supported by three guy wires and rests on a large spherical ball. Draw the free-body diagram of the antenna.

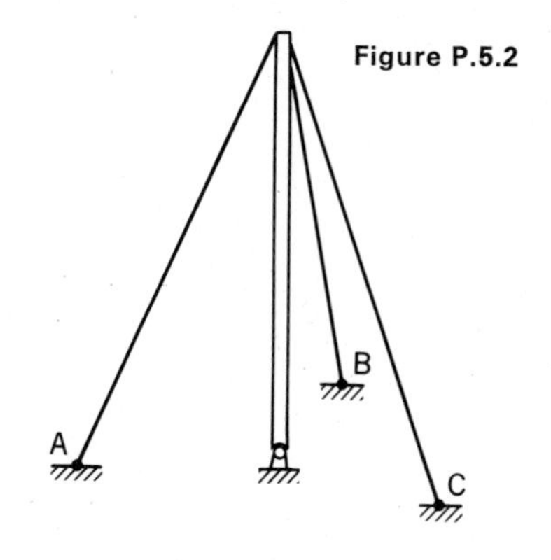


**Figure P.5.2**

**5.3.** Draw a free-body diagram of the A-frame.

A-frame

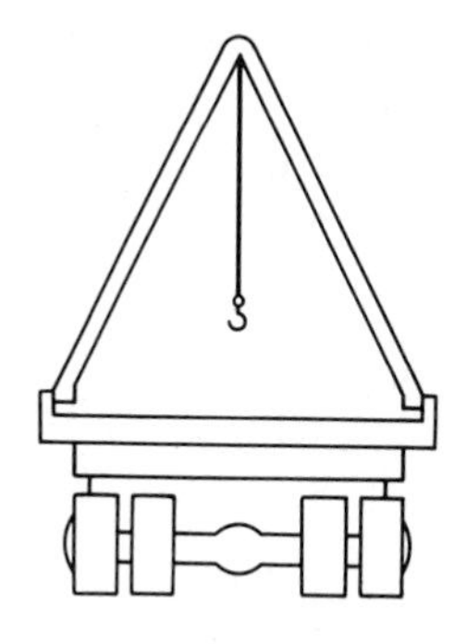

**Figure P.5.3**

**5.4.** Draw complete free-body diagrams for the member $AB$ and for cylinder $D$. Neglect friction at the contact surfaces of the cylinder. The weights of the cylinder and the member are denoted as $W_D$ and $W_{AB}$, respectively.

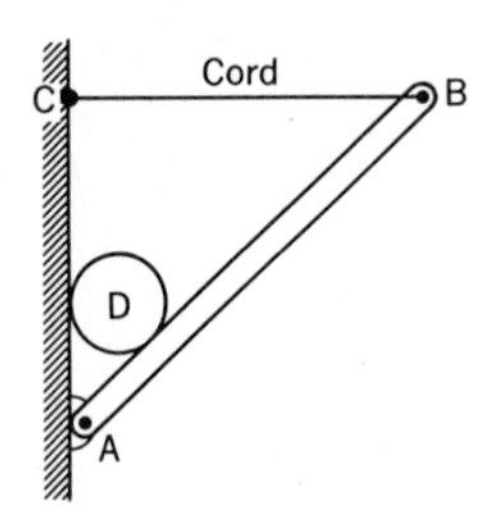


Figure P.5.4

**5.5.** Draw free-body diagrams of the plate $ABCD$ and the bar $EG$. Assume that there is no friction at the pulley $H$ or at the contact surface $C$.

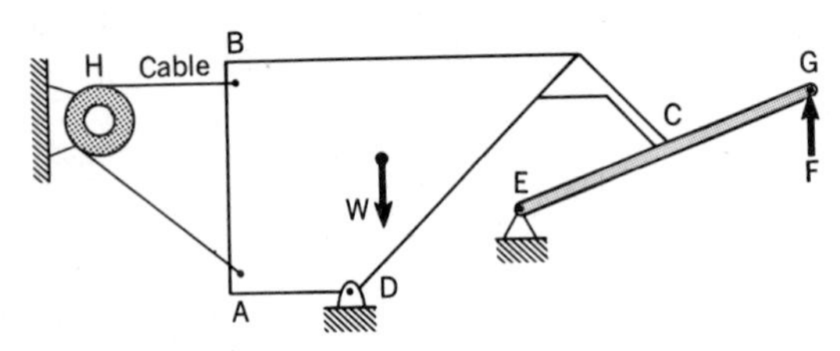


Figure P.5.5

**5.6.** Draw the free-body diagram of one part of the two-piece posthole digger.

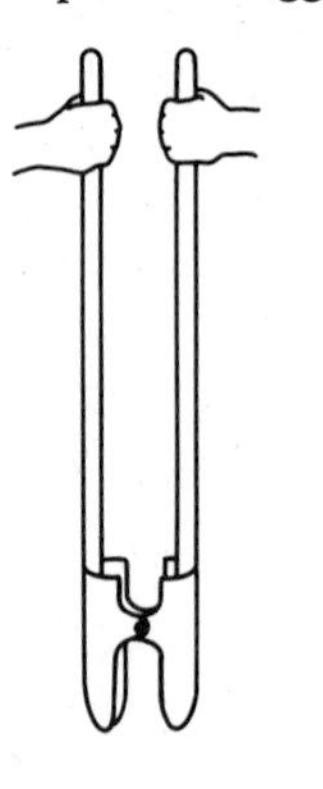

Figure P.5.6

**5.7.** Draw a free-body diagram for each member of the system. Neglect the weights of the members. Replace the distributed load by a resultant.

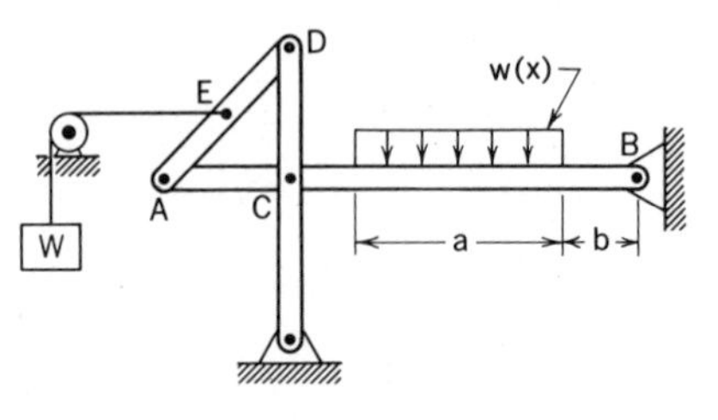


Figure P.5.7

**5.8.** Draw the free-body diagrams for the oars of a rowboat when the rower pushes with one hand and pulls with the other (i.e., turns the boat).

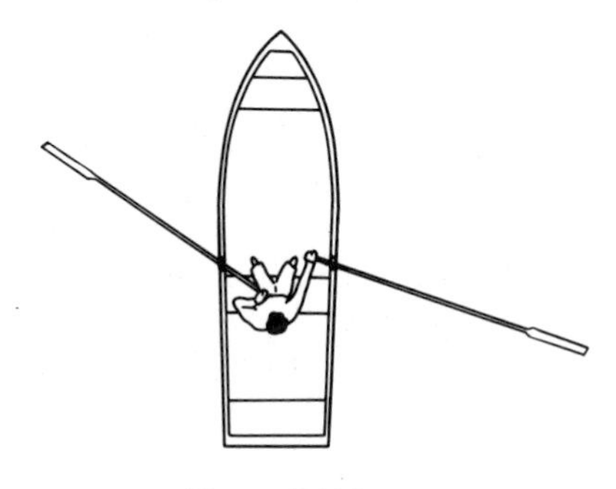

Figure P.5.8

**5.9.** Make a free-body diagram of the portion of the beam that is exposed from the wall. Replace all distributions by simplest equivalent force systems. Neglect the weight of the beam.

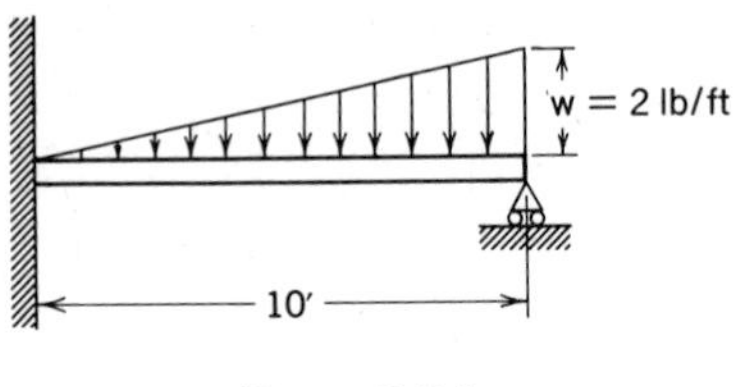


Figure P.5.9

**5.10.** Two cantilever beams are pinned together at $A$. Draw free-body diagrams of each cantilever beam.

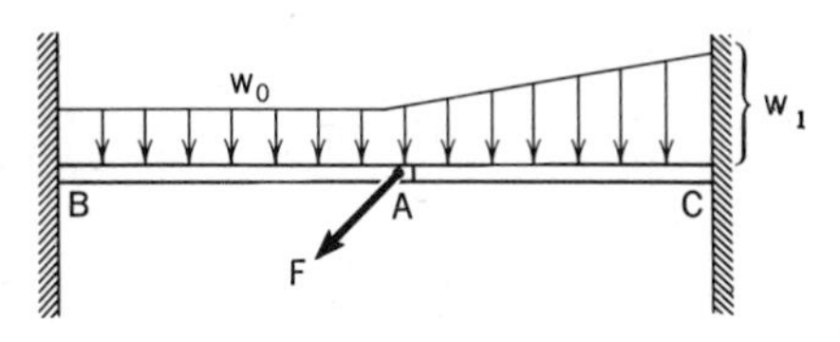


Figure P.5.10

**5.11.** Draw free-body diagrams of each part of the tree-branch trimmer.

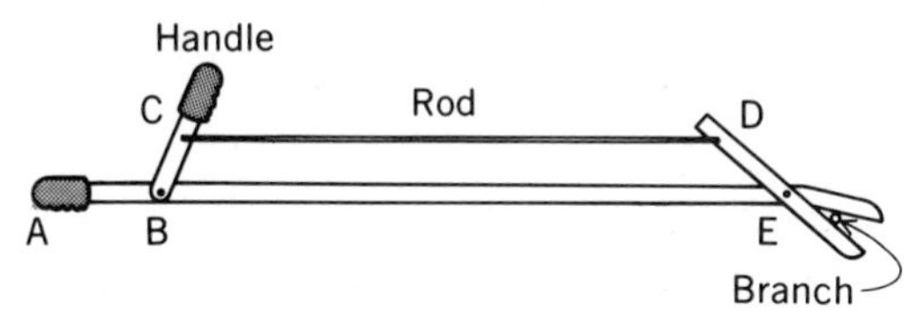


Figure P.5.11

**5.12.** Draw free-body diagrams for the two booms and the body $E$ of the power shovel. Consider the weight of each part to act at a central location. (Regard the shovel and payload as concentrated forces, $W_S$ and $W_{PL}$, respectively.)

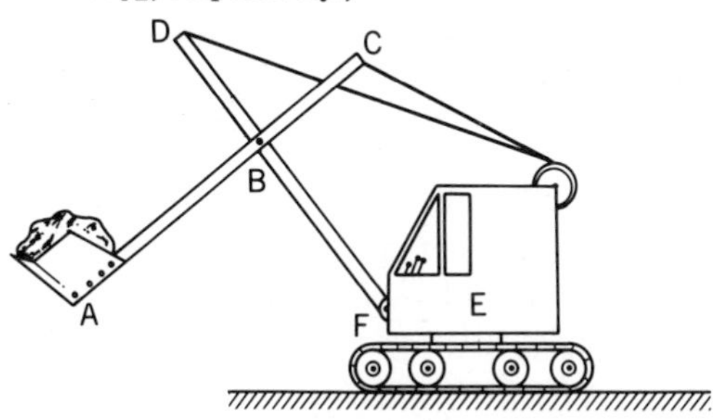


Figure P.5.12

**5.13.** Draw the free-body diagram for the bulldozer, $B$, hydraulic ram, $R$, and tractor, $T$. Consider the weight of each part $B$, $R$, and $T$.

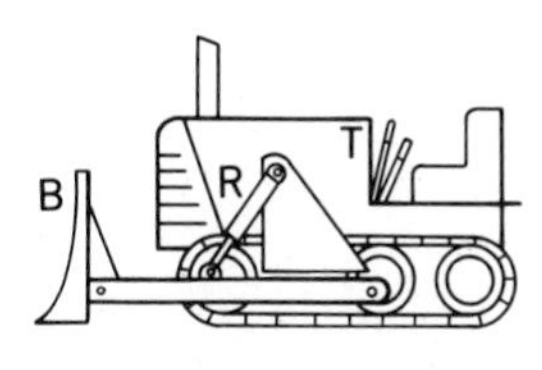


Figure P.5.13

**5.14.** Draw a free-body diagram first of the whole apparatus, then of each of its parts: $AB$, $AC$, $BC$, and $D$. Include the weights of all bodies. Label forces.

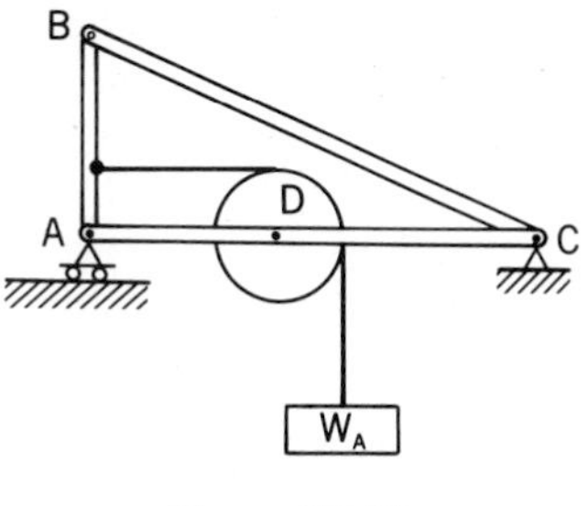


Figure P.5.14

**5.15.** Draw a free-body diagram of members $CG$, $AG$, the disc $B$, and the pin at $G$. Include as the only weight that of disc $B$. Label all forces. (*Hint:* Consider the pin at $G$ as a separate free body.)

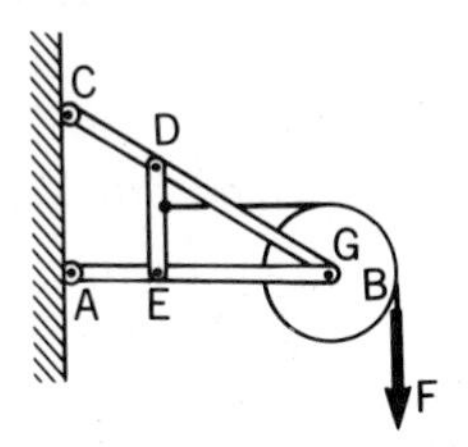


Figure P.5.15

**5.16.** Draw the free-body diagram of the horizontally bent cantilevered beam. Use only $xyz$ components of all vectors drawn.

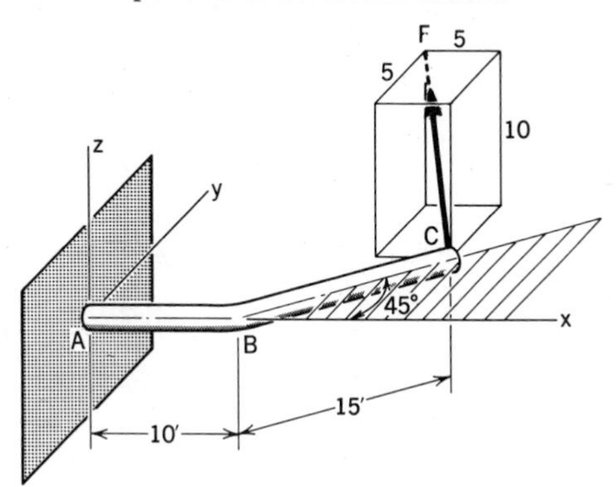


Figure P.5.16

## 5.4 Equations of Equilibrium

For every free-body diagram, we can replace the system of forces and couples acting on the body by a single force and a single couple moment at a point $a$. The force will have the same magnitude and direction, no matter where point $a$ is chosen to move the entire system by methods discussed earlier. However, the couple-moment vector will depend on the point chosen. We will prove in dynamics that:

> *The necessary conditions for a rigid body to be in equilibrium are that the resultant force $\boldsymbol{F}_R$ and the resultant couple moment $\boldsymbol{C}_R$ for any point a be zero vectors.*

That is,

$$\boldsymbol{F}_R = \boldsymbol{0} \tag{5.1a}$$

$$\boldsymbol{C}_R = \boldsymbol{0} \tag{5.1b}$$

We shall prove in dynamics, furthermore, that the conditions above are *sufficient* to maintain an *initially stationary* body in a state of equilibrium. These are the fundamental equations of statics. You will remember from Section 4.3 that the resultant $\boldsymbol{F}_R$ is the sum of the forces moved to the common point, and that the couple moment $\boldsymbol{C}_R$ is equal to the sum of the moments of all the original forces and couples taken about this point. Hence, the equations above can be written

$$\sum_i \boldsymbol{F}_i = \boldsymbol{0} \tag{5.2a}$$

$$\sum_i \boldsymbol{\rho}_i \times \boldsymbol{F}_i + \sum_i \boldsymbol{C}_i = \boldsymbol{0} \tag{5.2b}$$

where the $\boldsymbol{\rho}_i$'s are displacement vectors from the common point $a$ to any point on the lines of action of the respective forces. From this form of the equations of statics, we can conclude that for equilibrium to exist, *the vector sum of the forces must be zero and the moment of the system of forces and couples about any point in space must be zero.*

Now that we have summed forces and have taken moments about a point $a$, we will demonstrate that we cannot find another *independent* equation by taking mo-

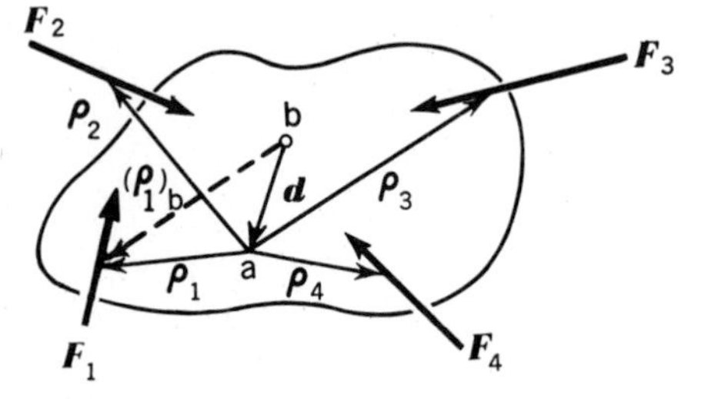


Figure 5.18. Consider moments about point $b$.

ments about a *different* point $b$. For the body in Fig. 5.18, we have initially the following equations of equilibrium using point $a$:

$$\boldsymbol{F}_1 + \boldsymbol{F}_2 + \boldsymbol{F}_3 + \boldsymbol{F}_4 = \boldsymbol{0} \tag{5.3}$$

$$\boldsymbol{\rho}_1 \times \boldsymbol{F}_1 + \boldsymbol{\rho}_2 \times \boldsymbol{F}_2 + \boldsymbol{\rho}_3 \times \boldsymbol{F}_3 + \boldsymbol{\rho}_4 \times \boldsymbol{F}_4 = \boldsymbol{0} \tag{5.4}$$

The new point $b$ is separated from $a$ by the position vector $\boldsymbol{d}$. The position vector (shown dashed) from $b$ to the line of action of the force $\boldsymbol{F}_1$ can be given in terms of $\boldsymbol{d}$ and the displacement vector $\boldsymbol{\rho}_1$ as follows. Similarly for $(\boldsymbol{\rho}_2)_b$, which is not shown, and others.

$$(\boldsymbol{\rho}_1)_b = (\boldsymbol{d} + \boldsymbol{\rho}_1)$$

$$(\boldsymbol{\rho}_2)_b = (\boldsymbol{d} + \boldsymbol{\rho}_2), \quad \text{etc.}$$

The moment equation for point $b$ can then be given as

$$(\boldsymbol{\rho}_1 + \boldsymbol{d}) \times \boldsymbol{F}_1 + (\boldsymbol{\rho}_2 + \boldsymbol{d}) \times \boldsymbol{F}_2 + (\boldsymbol{\rho}_3 + \boldsymbol{d}) \times \boldsymbol{F}_3 + (\boldsymbol{\rho}_4 + \boldsymbol{d}) \times \boldsymbol{F}_4 = \boldsymbol{0}$$

Using the distributive rule for cross products, we can restate this equation as

$$(\boldsymbol{\rho}_1 \times \boldsymbol{F}_1 + \boldsymbol{\rho}_2 \times \boldsymbol{F}_2 + \boldsymbol{\rho}_3 \times \boldsymbol{F}_3 + \boldsymbol{\rho}_4 \times \boldsymbol{F}_4) + \boldsymbol{d} \times (\boldsymbol{F}_1 + \boldsymbol{F}_2 + \boldsymbol{F}_3 + \boldsymbol{F}_4) = \boldsymbol{0} \tag{5.5}$$

Since the expression in the second set of parentheses is zero, in accordance with Eq. 5.3, the remaining portion degenerates to Eq. 5.4, and thus we have not introduced a new equation. Therefore, *there are only two independent vector equations of equilibrium for any single free body.*

We shall now show that instead of using Eqs. 5.3 and 5.4 as the equations of equilibrium, we can instead use Eqs. 5.4 and 5.5. That is, instead of summing forces and then taking moments about a point for equilibrium, we can instead take moments about *two* points. Thus, if Eq. 5.4 is satisfied for point $a$, then for point $b$ we end up in Eq. 5.5 with

$$\boldsymbol{d} \times (\boldsymbol{F}_1 + \boldsymbol{F}_2 + \boldsymbol{F}_3 + \boldsymbol{F}_4) = \boldsymbol{0} \tag{5.6}$$

If point $b$ can be any point in space making $\boldsymbol{d}$ arbitrary, then the above equation indicates that the vector sum of forces is zero. We thus have equilibrium since $\boldsymbol{F}_R = \boldsymbol{0}$ and $\boldsymbol{C}_R = \boldsymbol{0}$.

Using the vector Eqs. 5.2, we can now express the scalar equations of equilibrium. Since, as you will recall, the rectangular components of the moment of a force about a point are the moments of the force about the orthogonal axes at the point, we may state these equations in the following manner:

$$\begin{aligned} &\sum_i (F_x)_i = 0 \quad \text{(a)} \qquad &&\sum_i (M_x)_i = 0 \quad \text{(d)} \\ &\sum_i (F_y)_i = 0 \quad \text{(b)} \qquad &&\sum_i (M_y)_i = 0 \quad \text{(e)} \\ &\sum_i (F_z)_i = 0 \quad \text{(c)} \qquad &&\sum_i (M_z)_i = 0 \quad \text{(f)} \end{aligned} \tag{5.7}$$

From this set of equations, it is clear that *no more than six unknown scalar quantities in the general case can be solved by methods of statics for a single free body.*[3]

[3]Keep in mind that we can also take moments about two sets of axes just as we could take moments about two points for the vector equations of equilibrium.

We can easily express *any number* of scalar equations of equilibrium for a free body by selecting references that have different axis directions, along which we can sum forces and about which we can take moments. However, in choosing six *independent* equations, we will find that the remaining equations will be dependent on these six. That is, these equations will be sums, differences, etc. of the independent set and so will be of no use in solving for desired unknowns.

## 5.5 Special Cases of Equilibrium

The preceding conditions for equilibrium apply to the general case. We shall now consider a number of important *special* cases of equilibrium primarily to ascertain the number of scalar equations that are necessary for equilibrium for these special cases. With this information, we will then know the number of unknown scalar quantities for any free body that can be solved by methods of statics. If there are more such unknowns than available independent equations, no amount of algebraic perseverance will lead to the solution of the unknowns for the chosen free body.

The type of *simplest* resultant for each special system of forces is most useful in determining the number of scalar equations available in a given problem. The procedure is to classify the force system, note what simplest resultant force system is associated with the classification, and then consider the number of scalar equations necessary and sufficient to guarantee this resultant to be zero. The following cases exemplify this procedure.

**Case A. Concurrent System of Forces.** In this case, since the simplest resultant is a single force at the point of concurrency, the only requirement for equilibrium is that this force be zero. We can ensure this condition if the orthogonal components of this force are separately equal to zero. Thus, we have *three* equations of equilibrium of the form

$$\sum_i (F_x)_i = 0, \qquad \sum_i (F_y)_i = 0, \qquad \sum_i (F_z)_i = 0 \tag{5.8}$$

As was pointed out in the general vector discussion, there are other ways of ensuring a zero resultant. Suppose that the moments of the concurrent force system are zero about three nonparallel axes: $\alpha$, $\beta$, and $\gamma$. That is,

$$\sum_i (M_\alpha)_i = 0, \qquad \sum_i (M_\beta)_i = 0, \qquad \sum_i (M_\gamma)_i = 0 \tag{5.9}$$

Any one of the following three conditions must then be true:

1. The resultant force $F_R$ is zero.
2. $F_R$ cuts all three axes (see Fig. 5.19).
3. $F_R$ cuts two axes and is parallel to the third (see Fig. 5.20).

We can guarantee condition 1 and thus equilibrium if we select axes $\alpha$, $\beta$, and $\gamma$ so that no straight line can intersect all three axes or can cut two axes and be parallel to

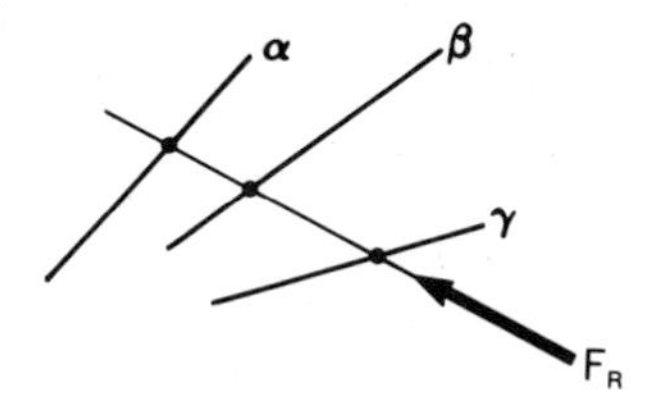


**Figure 5.19.** $F_R$ cuts three axes.

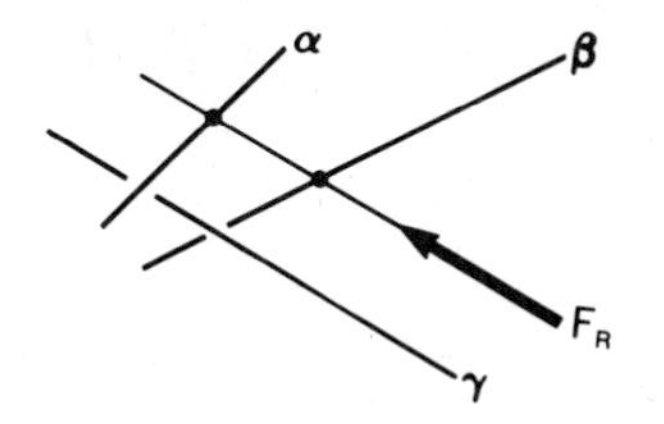


**Figure 5.20.** $F_R$ cuts two axes and is parallel to third.

the third. Then we can use Eqs. 5.9 as the equations of equilibrium under the aforestated conditions rather than using Eqs. 5.8. What happens if an axis used violates these conditions? The resulting equation will either be an *identity* $0 = 0$ or will be dependent on a previous independent equation of equilibrium for one of the axes. No harm is done. One should use other axes until three independent equations are found.

Similarly, one can sum forces in one direction and take moments about two axes. Setting these equal to zero can yield three independent equations of equilibrium. If not, use other axes.

The essential conclusion to be drawn is that *there are three independent equations of equilibrium for a concurrent force system.*

**Case B. Coplanar Force System.** We have shown that the simplest resultant for a coplanar force system is a single force or a single couple moment. Thus, to ensure that the resultant force is zero, we require for a coplanar system in the $xy$ plane:

$$\sum_i (F_x)_i = 0, \qquad \sum_i (F_y)_i = 0 \tag{5.10}$$

To ensure that the resultant couple moment is zero, we require for moments about any axis parallel to the $z$ axis:

$$\sum_i (M_z)_i = 0 \tag{5.11}$$

We conclude that there are *three* scalar equations of equilibrium for a coplanar force system. Other combinations, such as two moment equations for two axes parallel to the $z$ axis and a single force summation, if properly chosen, may be employed to give the three independent scalar equations of equilibrium, as was discussed in case A.

**Case C. Parallel Forces in Space.** In the case of parallel forces in space, we already know that the simplest resultant can be either a single force or a couple moment. If the forces are in the $z$ direction, then

$$\sum_i (F_z)_i = 0 \tag{5.12}$$

ensures that the resultant force is zero. Also,

$$\sum_i (M_x)_i = 0, \qquad \sum_i (M_y)_i = 0 \tag{5.13}$$

guarantees that the resultant couple moment is zero, where the $x$ and $y$ axes may be chosen in any plane perpendicular to the direction of the forces.[4] Thus, three independent scalar equations are available for equilibrium of parallel forces in space.

A summary of the special cases discussed in this section is given below. For even simpler systems such as the concurrent-coplanar and the parallel-coplanar systems, clearly, there is one less equation of equilibrium.

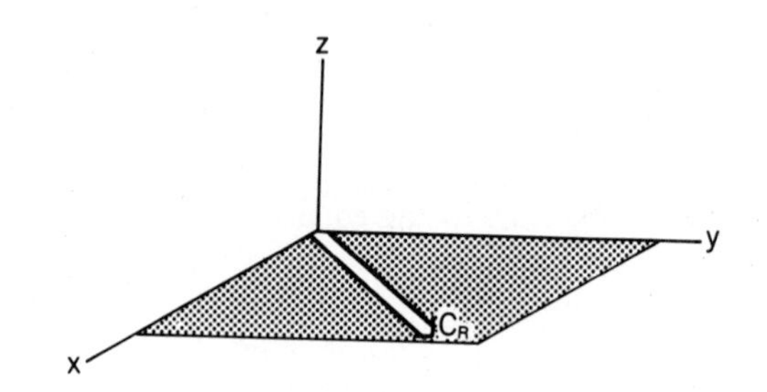


**Figure 5.21.** Resultant couple moment put in *xy* plane.

SUMMARY FOR SPECIAL CASES

| *System* | *Simplest Resultant* | *Number of Equations for Equilibrium* |
|---|---|---|
| Concurrent (three-dimensional) | Single force | 3 |
| Coplanar | Single force or single couple moment | 3 |
| Parallel (three-dimensional) | Single force or single couple moment | 3 |

## 5.6 Problems of Equilibrium

We shall now examine problems of equilibrium in which the rigid-body assumption is valid. To solve such problems, we must find the value of certain unknown forces and couple moments. We first draw a free-body diagram of the entire system or portions thereof to clearly *expose* pertinent unknowns for analysis. We then write the equilibrium equations in terms of the unknowns along with the known forces and geometry. As we have seen, for any free body there is a limited number of independent scalar equations of equilibrium. Thus, at times we must employ several free-body diagrams for portions of the system to produce enough independent equations to solve all the unknowns.

[4]For parallel forces in the $z$ direction, a simplest resultant consisting of a couple moment only must have this couple moment parallel to the $xy$ plane (see Fig. 5.21). Recall from Chapter 3 that the orthogonal $xyz$ components of $\mathbf{C}_R$ equals the torques of the system about these axes. Hence, by setting $\sum_i (M_x)_i = \sum_i (M_y)_i = 0$, we are ensuring that $\mathbf{C}_R = \mathbf{0}$.

For any free body, we may proceed by expressing two basic vector equations of statics. After carrying out such vector operations as cross products and additions in the equations, we form scalar equations. These scalar equations are then solved simultaneously (together with scalar equations from other free-body diagrams that may be needed) to find the unknown forces and couple moments. We can also express the scalar equations immediately by using the alternative scalar equilibrium relations that we formulated in previous sections. In the first case, we start with more compact vector equations and arrive at the expanded scalar equations by the formal procedures of vector algebra. In the latter case, we evaluate the expanded scalar equations by carrying out arithmetic operations on the free-body diagram as we write the equations. Which procedure is more desirable? It all depends on the problem and the investigator's skill in vector manipulation. It is true that many statics problems submit easily to a direct scalar approach, but the more challenging problems of statics and dynamics definitely favor an initial vector approach. In this text, we shall employ the particular procedure that the occasion warrants.

In statics problems, we must assign a sense to each component of an unknown force or couple moment in order to write the equations. If, on solving the equations, *we obtain a negative sign for a component, then we have guessed the wrong sense for that component.* Nothing need be redone should this occur. Continue with the remainder of the problem, retaining the minus sign (or signs). At the end of the problem, report the correct sense of your force components and couple-moment components.

We shall now solve and discuss a number of problems of equilibrium. These problems are divided into four classes of force systems:

1. Concurrent.
2. Coplanar.
3. Parallel.
4. General.

**Concurrent Force Systems.** Recall that there are three independent equations of equilibrium for a concurrent force system acting on a body. If the concurrent forces are coplanar, there are but two independent equations. We shall illustrate both cases, starting for simplicity with the coplanar case.

**EXAMPLE 5.3**

A 500-lb weight is suspended by flexible cables as shown in Fig. 5.22. Determine the tension in the cables.

A suitable free body that exposes the desired unknown quantities is the ring $C$, which may be considered as a particle for this computation because of its comparatively small size (Fig. 5.23). Physical intuition indicates that the cables should be in tension and hence pulling away from $C$, and we have so indicated in the diagram. The force system acting on a particle must always be a concurrent system. Here we have the additional fact that it is coplanar as well, and therefore we may solve for the two unknowns. We shall proceed directly to the scalar equations of equilibrium.

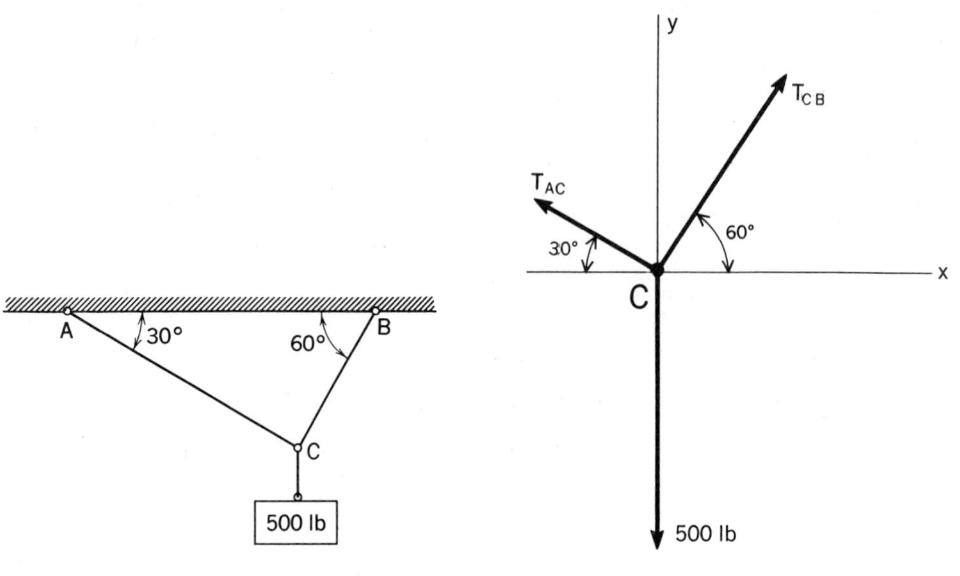


Figure 5.22. Find tensions in cables.

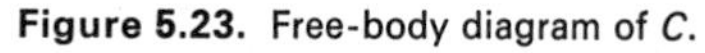
Figure 5.23. Free-body diagram of *C*.

Thus,

$\underline{\sum F_y = 0:}$

$$-500 + T_{CB} \sin 60^\circ + T_{AC} \sin 30^\circ = 0 \tag{a}$$

$\underline{\sum F_x = 0:}$

$$-T_{AC} \cos 30^\circ + T_{CB} \cos 60^\circ = 0 \tag{b}$$

By solving these equations simultaneously, we get the desired results:

$$T_{CB} = 433 \text{ lb}, \qquad T_{AC} = 250 \text{ lb}$$

Since the signs for $T_{CB}$ and $T_{AC}$ are positive, we have chosen the correct senses for the forces in the free-body diagram. Thus, the cables are in tension, confirming the physical intuition that guided our initial choices.

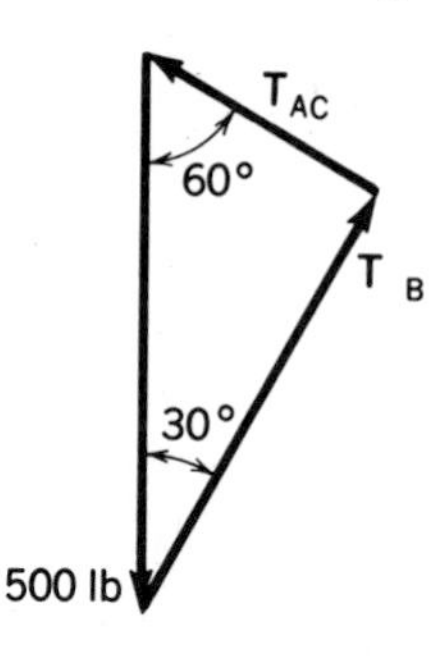


Figure 5.24. Force polygon.

Another way of arriving at the solution is to consider the *force polygon* that was discussed in Section 2.3. Because the forces are in equilibrium, the polygon must close; that is, the head of the final force must coincide with the tail of the initial force. In this case, we have a right triangle, as shown in Fig. 5.24, drawn approximately to scale.

From trigonometric considerations of this right triangle, we can state:

$$T_{CB} = 500 \cos 30^\circ = 433 \text{ lb}$$

$$T_{AC} = 500 \sin 30^\circ = 250 \text{ lb}$$

The force polygon may thus be used to good advantage when three concurrent coplanar forces are in equilibrium.

As a final alternative, let us now initiate the computations for the unknown tensions directly from the basic *vector* equations of statics. First, we must express all forces in vector notation:

$$\boldsymbol{T}_{CB} = T_{CB}(.500\boldsymbol{i} + .866\boldsymbol{j})$$
$$\boldsymbol{T}_{AC} = T_{AC}(-.866\boldsymbol{i} + .500\boldsymbol{j})$$

We get the following equation when the vector sum of the forces is set equal to zero:

$$T_{CB}(.500\boldsymbol{i} + .866\boldsymbol{j}) + T_{AC}(-.866\boldsymbol{i} + .500\boldsymbol{j}) - 500\boldsymbol{j} = \boldsymbol{0}$$

Choosing point $C$, the point of concurrency, we see clearly that the sum of moments of the forces about this point is zero, so the second basic equation of equilibrium is intrinsically satisfied. We now regroup the preceding equation in the following manner:

$$(.500T_{CB} - .866T_{AC})\boldsymbol{i} + (.866T_{CB} + .500T_{AC} - 500)\boldsymbol{j} = \boldsymbol{0}$$

To satisfy this equation, each of the quantities in parentheses must be zero. This gives the scalar equations (a) and (b) stated earlier, from which the scalar quantities $T_{CB}$ and $T_{AC}$ can be solved.

The three alternative methods of solution are apparently of equal usefulness in this simple problem. However, the force polygon is only practicably useful for three concurrent coplanar forces, where the trigonometric properties of the triangle can be directly used. The other two methods can be readily extended to more complex concurrent problems.

**EXAMPLE 5.4**

What are the forces in the cables shown in Fig. 5.25 supporting a 50-kg mass? Note that cable $BD$ lies in the $zy$ plane.

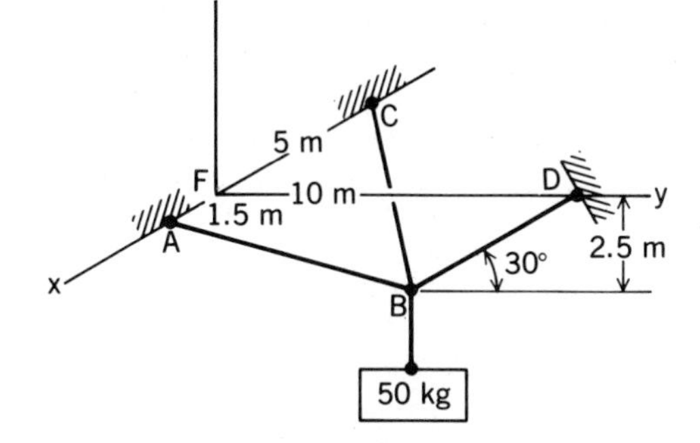


**Figure 5.25.** Find forces in cables.

If we take the connecting point $B$ of the cables as a free body (see Fig. 5.26), we have three unknown concurrent forces which can be solved from the equations of equilibrium. The 50-kg mass has a weight of (9.81)(50) = 490.5 N, as shown in the diagram.

We shall first express the forces vectorially. For this purpose, note from Fig. 5.26 that

$$\frac{2.5}{ED} = \tan 30^\circ$$

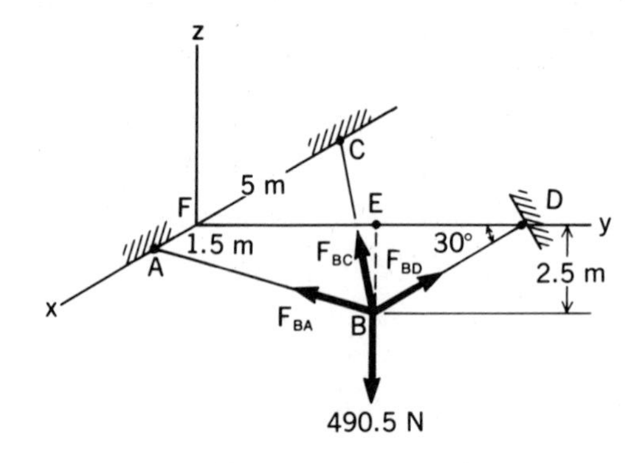


**Figure 5.26.** Free-body diagram of *B*.

Therefore,

$$ED = \frac{2.5}{\tan 30^\circ} = 4.33 \text{ m} \tag{a}$$

Accordingly, we have

$$FE = 10 - 4.33 = 5.67 \text{ m} \tag{b}$$

We can now give the forces as follows using unit vectors for the first two:

$$\begin{aligned}
\boldsymbol{F}_{BA} &= F_{BA}\left(\frac{1.5\boldsymbol{i} - 5.67\boldsymbol{j} + 2.5\boldsymbol{k}}{\sqrt{1.5^2 + 5.67^2 + 2.5^2}}\right) \\
&= F_{BA}(.235\,\boldsymbol{i} - .889\boldsymbol{j} + .392\boldsymbol{k}) \text{ N} \\
\boldsymbol{F}_{BC} &= F_{BC}\left(\frac{-5\boldsymbol{i} - 5.67\boldsymbol{j} + 2.5\boldsymbol{k}}{\sqrt{5^2 + 5.67^2 + 2.5^2}}\right) \\
&= F_{BC}(-.628\boldsymbol{i} - .712\boldsymbol{j} + .314\boldsymbol{k}) \text{ N} \\
\boldsymbol{F}_{BD} &= F_{BD}(\cos 30^\circ\boldsymbol{j} + \sin 30^\circ\boldsymbol{k}) \\
&= F_{BD}(.866\boldsymbol{j} + .500\boldsymbol{k}) \text{ N}
\end{aligned}$$

Now, summing the forces acting on $B$ equal to zero, we get

$$\begin{aligned}
&F_{BA}(.235\boldsymbol{i} - .889\boldsymbol{j} + .392\boldsymbol{k}) + F_{BC}(-.628\boldsymbol{i} - .712\boldsymbol{j} + .314\boldsymbol{k}) \\
&\qquad + F_{BD}(.866\boldsymbol{j} + .500\boldsymbol{k}) - 490.5\boldsymbol{k} = \boldsymbol{0}
\end{aligned} \tag{c}$$

The scalar equations are

$$.235F_{BA} - .628F_{BC} = 0 \tag{d}$$

$$-.889F_{BA} - .712F_{BC} + .866F_{BD} = 0 \tag{e}$$

$$.392F_{BA} + .314F_{BC} + .500F_{BD} - 490.5 = 0 \tag{f}$$

From Eq. (d), we may solve for $F_{BC}$ in terms of $F_{AB}$. Thus,

$$F_{BC} = \frac{.235}{.628}F_{BA} = .374F_{BA} \tag{g}$$

Substituting Eq. (g) into Eqs. (e) and (f), we get

$$-1.155F_{BA} + .866F_{BD} = 0 \tag{h}$$

$$.509F_{BA} + .500F_{BD} = 490.5 \tag{i}$$

Now multiplying Eq. (i) by 1.155/.509 = 2.269, we have for the equations above,

$$-1.155F_{BA} + .866F_{BD} = 0 \qquad \text{(j)}$$

$$1.155F_{BA} + 1.135F_{BD} = 1113 \qquad \text{(k)}$$

Adding these equations, we may directly determine $F_{BD}$. Thus,

$$F_{BD} = 556 \text{ N}$$

From Eq. (h), we then may determine $F_{BA}$ to be

$$F_{BA} = \frac{.866}{1.155}(556) = 417 \text{ N}$$

Finally, from Eq. (g) we get $F_{BC}$:

$$F_{BC} = (.374)(417) = 156.0 \text{ N}$$

**Coplanar Force System.** The simplest resultant for a coplanar force distribution is a single force or a single couple moment. Hence, there are three independent equations of equilibrium for a given free body. We shall first examine a problem for which only one free-body diagram is needed, and then we shall consider a problem involving several free-body diagrams.

**EXAMPLE 5.5**

A crane weighing 3000 lb supports a 10,000-lb load as shown in Fig. 5.27. Determine the supporting forces at $A$, which is a pinned connection, and at $B$, which is a roller.

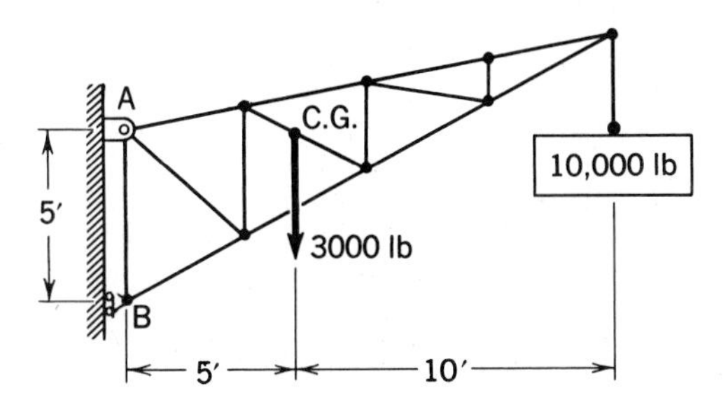


**Figure 5.27.** Loaded crane.

A free-body diagram of the main structure exposes the desired unknowns (Fig. 5.28). Note that since the system of forces may be taken as coplanar, we have three equations of equilibrium available and we may accordingly solve for the three unknown quantities from this single free-body diagram. Hence:

$\underline{\sum F_x = 0:}$

$$A_x + B = 0$$

Therefore,

$$A_x = -B$$

$\underline{\sum F_y = 0:}$

$$A_y - 3000 - 10{,}000 = 0$$

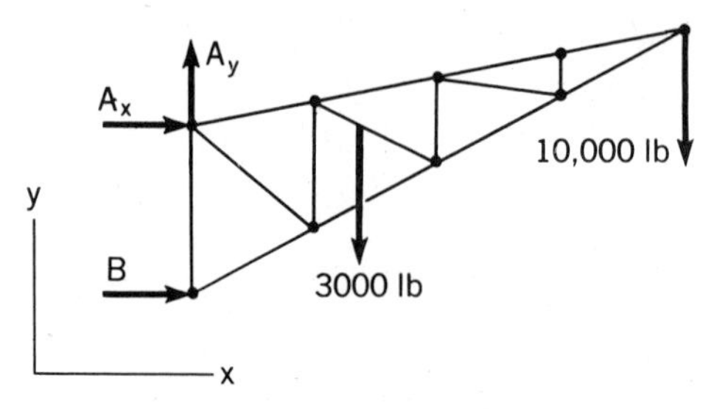


**Figure 5.28.** Free-body diagram of crane.

Therefore,

$$A_y = 13{,}000 \text{ lb}$$

$\underline{\sum M_B = 0:}$

$$-(5)(3000) - (15)(10{,}000) - 5A_x = 0$$

Therefore,

$$A_x = -33{,}000 \text{ lb}$$

The results are

$$A_x = -33{,}000 \text{ lb}, \qquad A_y = 13{,}000 \text{ lb}, \qquad B = 33{,}000 \text{ lb}$$

Note that $A_x$ has a negative sign. As noted earlier, this is the result of having chosen the wrong sense for $A_x$ at the outset of the computations. All we need do now is recognize that $A_x$ has a sense opposite to what is shown in Fig. 5.28.

We have now solved the forces from the *wall onto the structure.* The forces from the *structure onto the wall* are the *reactions* to these forces and are equal and opposite.

**EXAMPLE 5.6**

In Fig. 5.29 is shown a frame where the pulley at $D$ has a mass of 200 kg. Neglecting the weights of the bars, find the force transmitted from one bar to another at $C$.

To expose force components $C_x$ and $C_y$, we form the free body of bar $BD$. This is shown in F.B.D. I in Fig. 5.30. It is clear that for this free body we have six unknowns and only three independent equations of equilibrium.[5] The free-body diagram of the bent bar $AC$ is then drawn (F.B.D. II in Fig. 5.30). Here, we have three more equations but we bring in three more unknowns. Finally, the free-body diagram of the pulley (F.B.D. III in Fig. 5.30) gives three more equations with no additional unknowns. We now have nine equations available and nine unknowns and can proceed with confidence. Since only two of the unknowns are desired, we

[5]It should be noted that it is possible to have situations wherein there are more unknowns than independent equations of equilibrium for a given free body, but wherein some of the unknowns—perhaps the desired ones—can be still determined by the equations available. However, not all the unknowns of the free body can be solved. Accordingly, be alert for such situations, so as to minimize the work involved. In this case, we must consider other free-body diagrams.

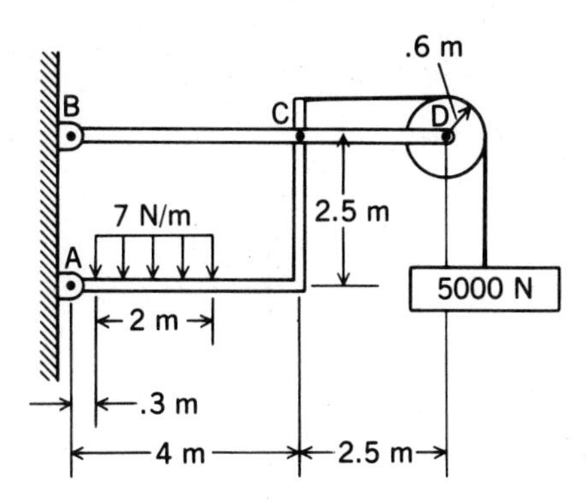


Figure 5.29. Loaded frame.

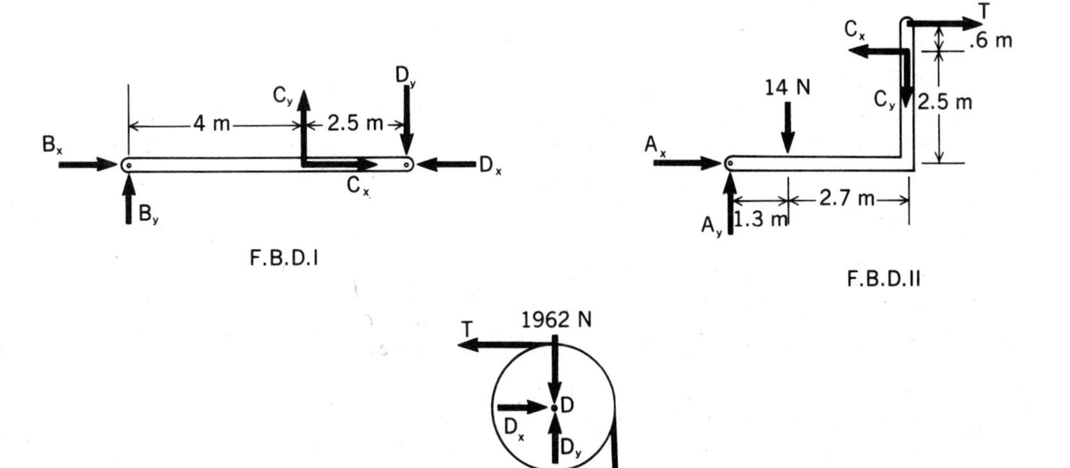

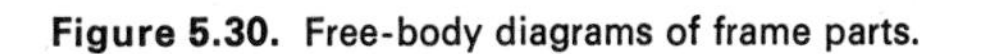

Figure 5.30. Free-body diagrams of frame parts.

shall take select scalar equations from each of the free-body diagrams to arrive at the components $C_x$ and $C_y$ most quickly.

*From F.B.D. III:*

$\underline{\sum M_D = 0:}$

$$(T)(.6) - (5000)(.6) = 0$$

Therefore,

$$T = 5000 \text{ N}$$

$\underline{\sum F_x = 0:}$

$$-T + D_x = 0$$

Therefore,

$$D_x = 5000 \text{ N}$$

$\underline{\sum F_y = 0:}$

$$-1962 - 5000 + D_y = 0$$

Therefore,

$$D_y = 6962 \text{ N}$$

*From F.B.D. I:*
$\underline{\sum M_B = 0:}$

$$(4)(C_y) - (6.5)(D_y) = 0$$

Therefore,

$$C_y = 11{,}313 \text{ N}$$

*From F.B.D. II:*
$\underline{\sum M_A = 0:}$

$$-(1.3)(14) - (T)(3.1) - C_y(4) + C_x(2.5) = 0$$

Therefore,

$$C_x = 24{,}300 \text{ N}$$

We can give the force at $C$ as

$$\boldsymbol{C} = 24{,}300\boldsymbol{i} + 11.313\boldsymbol{j} \text{ N}$$

## Problems

**5.17.** In a tug of war, when team $B$ pulls with 400-lb force, how much force must team $C$ exert for a draw? With what force does team $A$ pull?

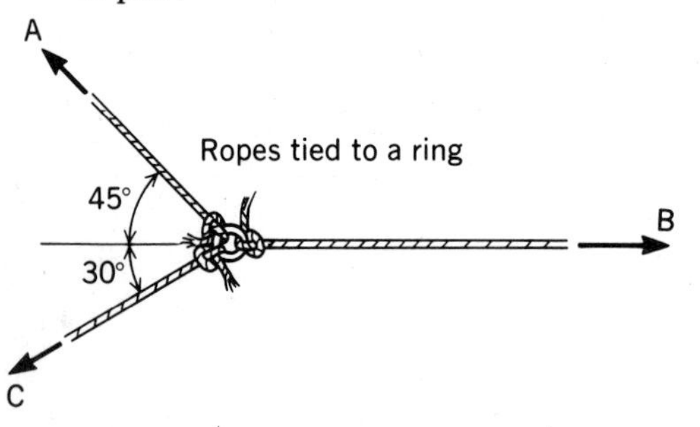


Figure P.5.17

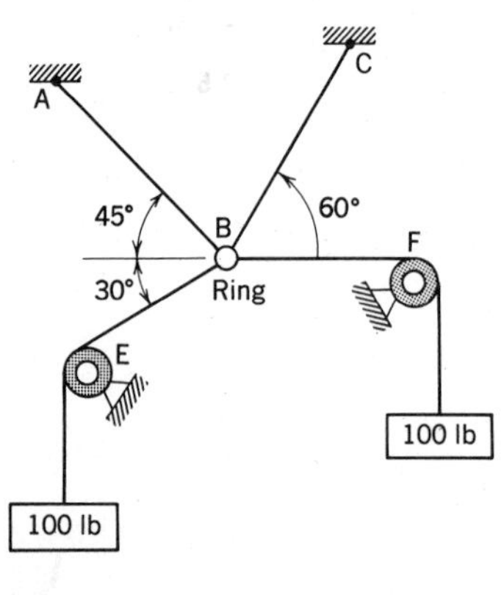

Figure P.5.18

**5.18.** Find the tensile force in cables $AB$ and $CB$. The remaining cables ride over frictionless pulleys $E$ and $F$.

**5.19.** Find the force transmitted by wire $BC$. The pulley $E$ can be assumed to be frictionless in this problem.

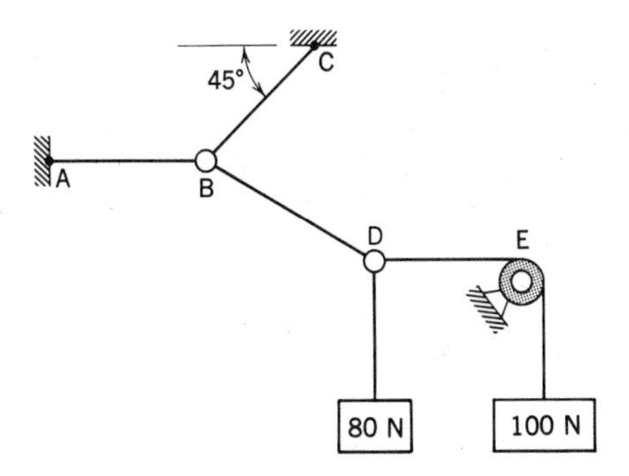


Figure P.5.19

**5.20.** A 700-N circus performer causes a .15-m sag in the middle of a 12-m tightrope with a 5000-N initial tension. What additional tension is induced in the cable? What is the cable tension when the performer is 3 m from the end and the sag is .12 m?

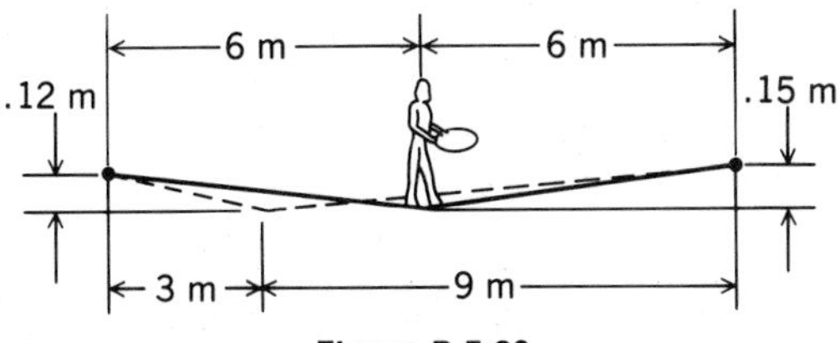

Figure P.5.20

**5.21.** A 27-lb mirror is held up by a wire fastened to two hooks on the mirror frame. (a) What is the force on the wall hook and the tension in the wire? (b) If the wire will break at a tension of 32 lb, must the wall hook be moved (i.e., the wire lengthened or shortened and the 4 in. rise distance changed)? If so, to what point?

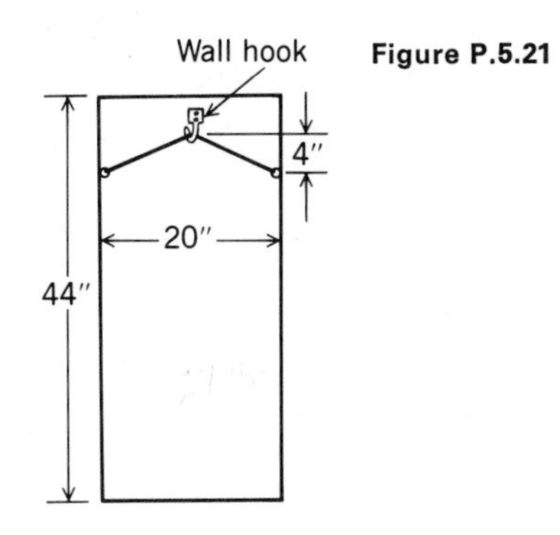


Figure P.5.21

**5.22.** Explain why equilibrium of a concurrent force system is guaranteed by having $\sum_i (F_y)_i = 0$, $\sum_i (M_d)_i = 0$, and $\sum_i (M_e)_i = 0$. Axes $d$ and $e$ are not parallel to the $xz$ plane. Moreover, the axes are oriented so that the line of action of the resultant force cannot intersect both axes.

**5.23.** Cylinders $A$ and $B$ weigh 500 N each and cylinder $C$ weighs 1000 N. Compute all contact forces.

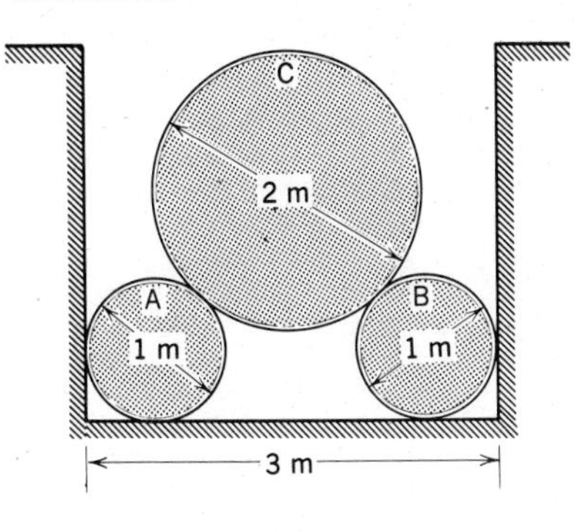


Figure P.5.23

**5.24.** A block having a mass of 500 kg is held by five cables. What are the tensions in these cables?

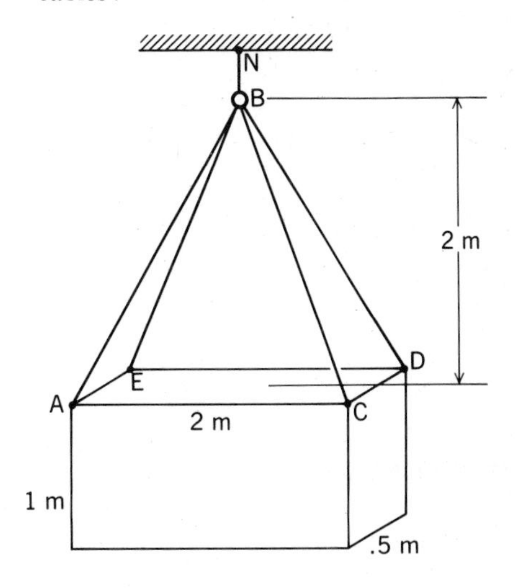


Figure P.5.24

**5.25.** What are the tensions in cables $AB$, $BC$, and $BD$? Points $A$ and $B$ are in the $yz$ plane.

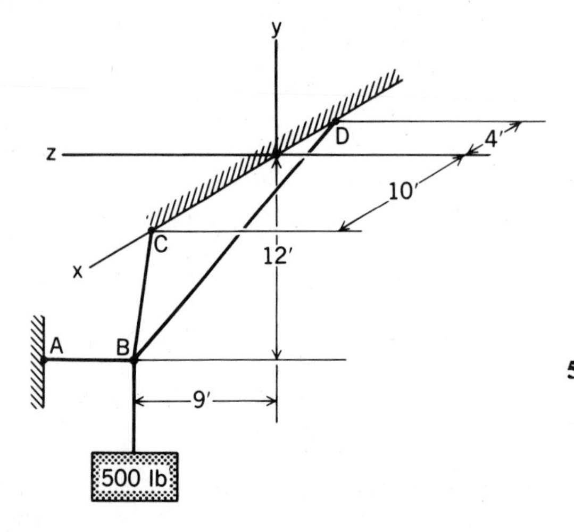


Figure P.5.25

**5.26.** An elastic cord $AB$ is just taut before the 1000-N force is applied. If it takes 5.0 N/mm of elongation of the cord, what is the tension $T$ in the cord after the 1000-N force is applied? Set up the equation for $T$ but do not solve.

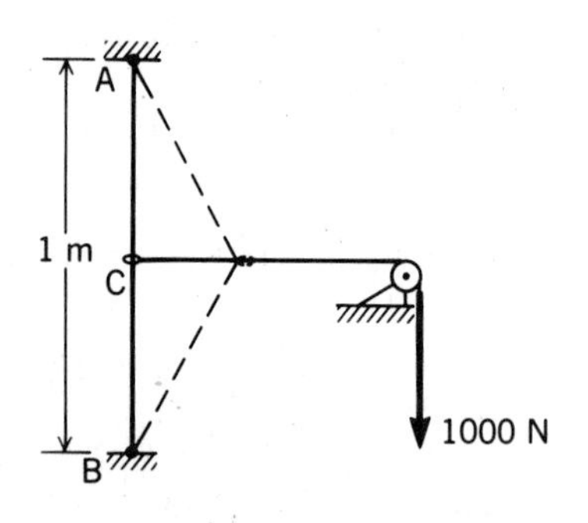


Figure P.5.26

**5.27.** A thin-walled cylinder of outside radius 1 m and weight 500 N rests on an incline. What friction force $f$ at $A$ is needed for this configuration? What is the tension in wire $CB$?

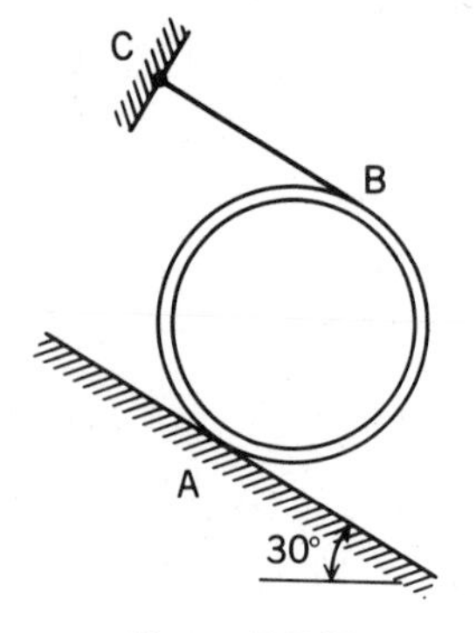


Figure P.5.27

**5.28.** A stepped cylinder is pulled down an incline by a force $F$ which is increased from zero to 20 N very very slowly while always maintaining the 30° inclination shown. If the cylinder is in equilibrium when $F = 0$, how far does $O$ move as a result of $F$ after equilibrium has been established with $F = 20$ N. The stepped cylinder has a mass of 10 kg. There is no slipping at the base. The force from the spring is $K$ times the extension of the spring. For this spring, $K = 5$ N/mm.

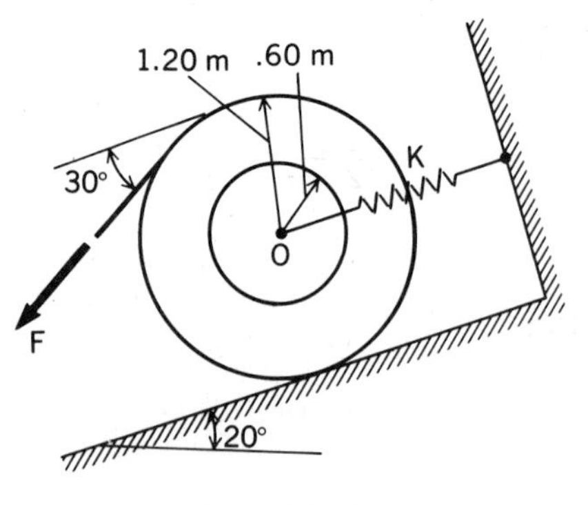


Figure P.5.28

**5.29.** A 10-kg ring is supported by a smooth surface $E$ and a wire $AB$. A body $D$ having a mass of 3 kg is fixed to the ring at the orientation shown. What is the tension in the wire $AB$? What is its orientation $\alpha$?

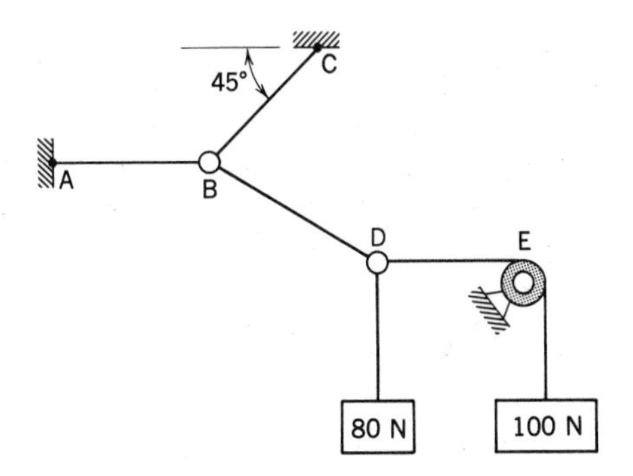


Figure P.5.19

**5.20.** A 700-N circus performer causes a .15-m sag in the middle of a 12-m tightrope with a 5000-N initial tension. What additional tension is induced in the cable? What is the cable tension when the performer is 3 m from the end and the sag is .12 m?

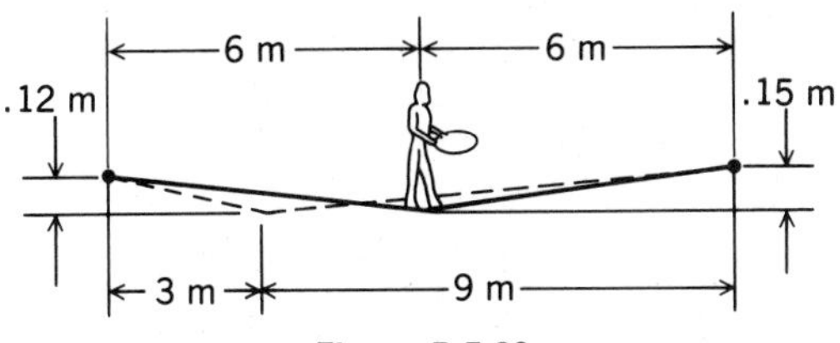


Figure P.5.20

**5.21.** A 27-lb mirror is held up by a wire fastened to two hooks on the mirror frame. (a) What is the force on the wall hook and the tension in the wire? (b) If the wire will break at a tension of 32 lb, must the wall hook be moved (i.e., the wire lengthened or shortened and the 4 in. rise distance changed)? If so, to what point?

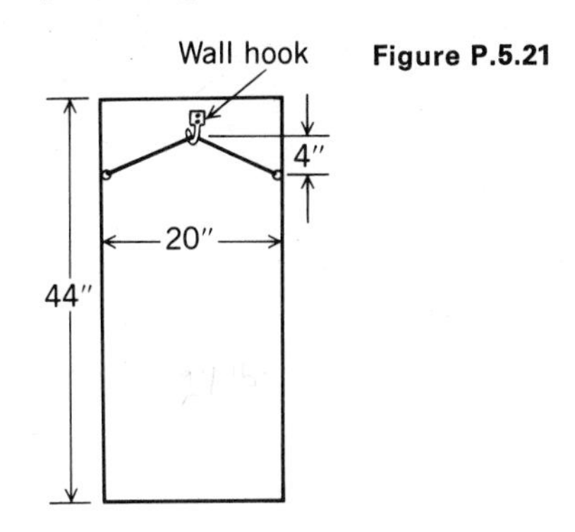


Figure P.5.21

**5.22.** Explain why equilibrium of a concurrent force system is guaranteed by having $\sum_i (F_y)_i = 0$, $\sum_i (M_d)_i = 0$, and $\sum_i (M_e)_i = 0$. Axes $d$ and $e$ are not parallel to the $xz$ plane. Moreover, the axes are oriented so that the line of action of the resultant force cannot intersect both axes.

**5.23.** Cylinders $A$ and $B$ weigh 500 N each and cylinder $C$ weighs 1000 N. Compute all contact forces.

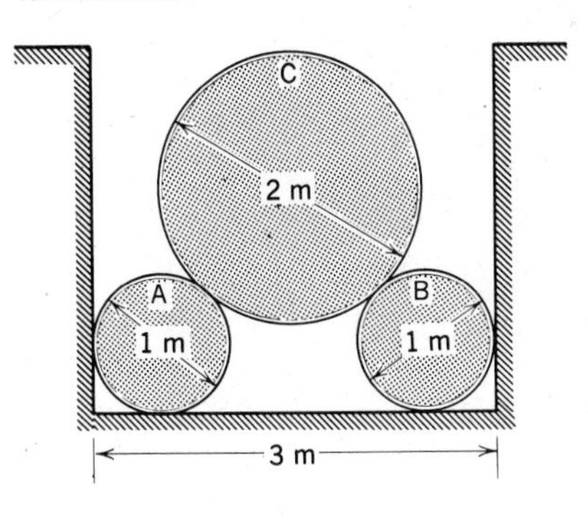


Figure P.5.23

**5.24.** A block having a mass of 500 kg is held by five cables. What are the tensions in these cables?

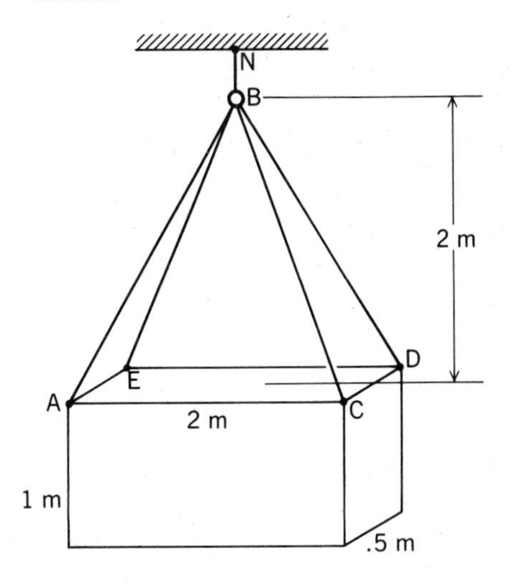


Figure P.5.24

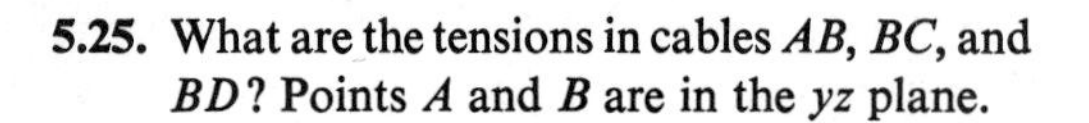
**5.25.** What are the tensions in cables $AB$, $BC$, and $BD$? Points $A$ and $B$ are in the $yz$ plane.

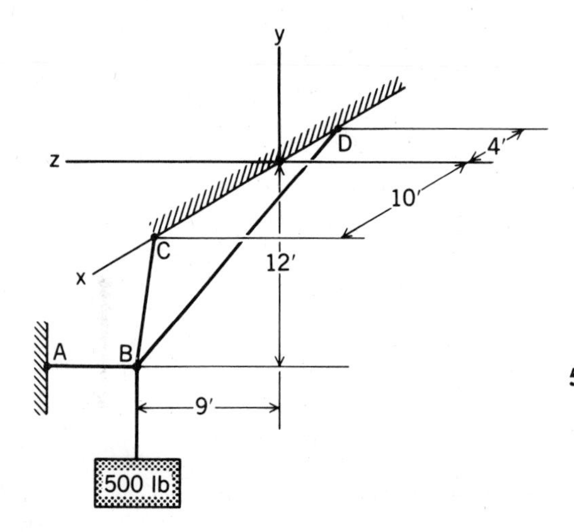


**Figure P.5.25**

**5.26.** An elastic cord $AB$ is just taut before the 1000-N force is applied. If it takes 5.0 N/mm of elongation of the cord, what is the tension $T$ in the cord after the 1000-N force is applied? Set up the equation for $T$ but do not solve.

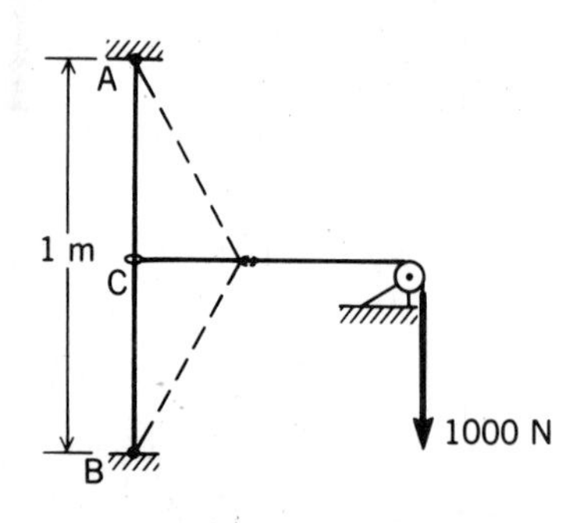


**Figure P.5.26**

**5.27.** A thin-walled cylinder of outside radius 1 m and weight 500 N rests on an incline. What friction force $f$ at $A$ is needed for this configuration? What is the tension in wire $CB$?

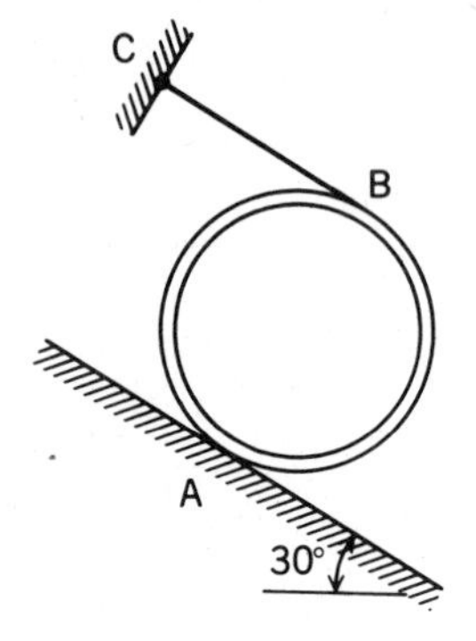


**Figure P.5.27**

**5.28.** A stepped cylinder is pulled down an incline by a force $F$ which is increased from zero to 20 N very very slowly while always maintaining the 30° inclination shown. If the cylinder is in equilibrium when $F = 0$, how far does $O$ move as a result of $F$ after equilibrium has been established with $F =$ 20 N. The stepped cylinder has a mass of 10 kg. There is no slipping at the base. The force from the spring is $K$ times the extension of the spring. For this spring, $K =$ 5 N/mm.

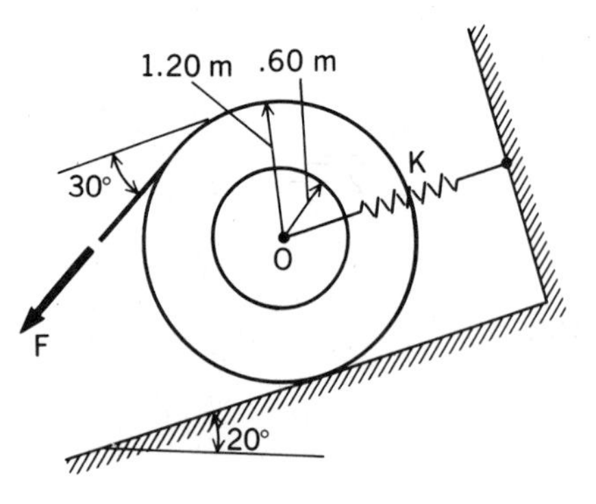


**Figure P.5.28**

**5.29.** A 10-kg ring is supported by a smooth surface $E$ and a wire $AB$. A body $D$ having a mass of 3 kg is fixed to the ring at the orientation shown. What is the tension in the wire $AB$? What is its orientation $\alpha$?

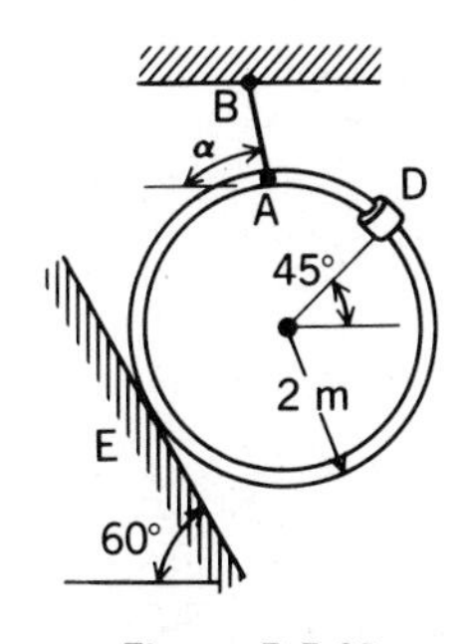


**Figure P.5.29**

**5.30.** What is the tension in the cables of a 10-ft-wide 12-ft-long 6000-lb castle drawbridge when the bridge is first raised? When the bridge is at 45°? What are the reactions at the hinge pin?

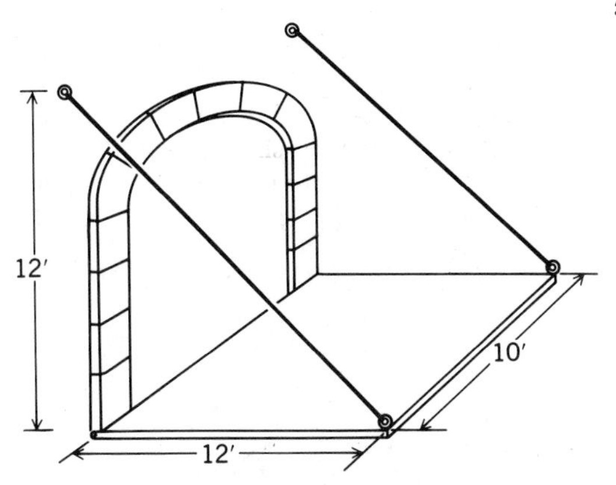


**Figure P.5.30**

**5.31.** A small hoist has a lifting capacity of 20 kN. What is the maximum cable tension and the corresponding reactions at *C*? Do not consider weight of beam.

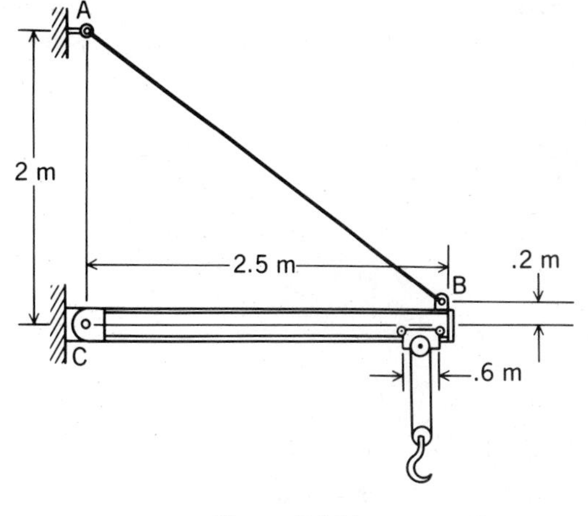


**Figure P.5.31**

**5.32.** A uniform block weighing 500 lb is constrained by three wires. What are the tensions in these wires?

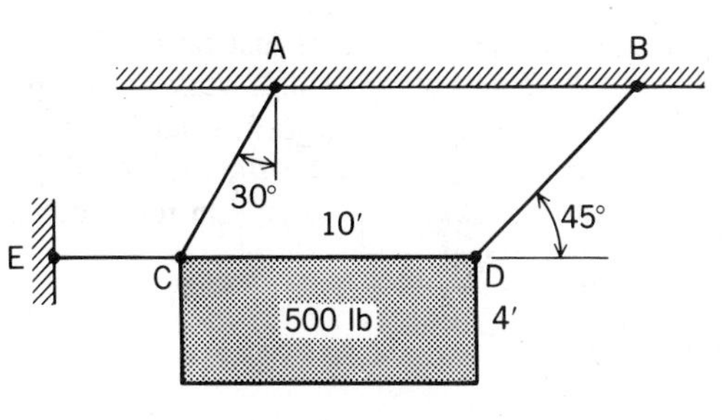


**Figure P.5.32**

**5.33.** Find the supporting forces for the frame shown.

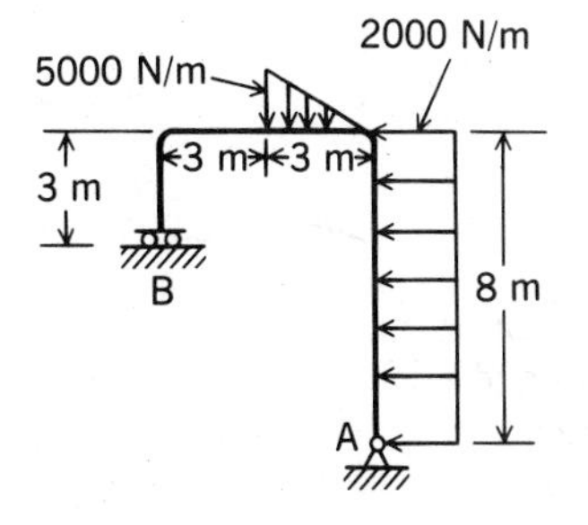


**Figure P.5.33**

**5.34.** A 300-kN tank is climbing up a 30° incline at constant speed. What is the torque developed on the rear drive wheels to accomplish this? Assume that all other wheels are free-turning.

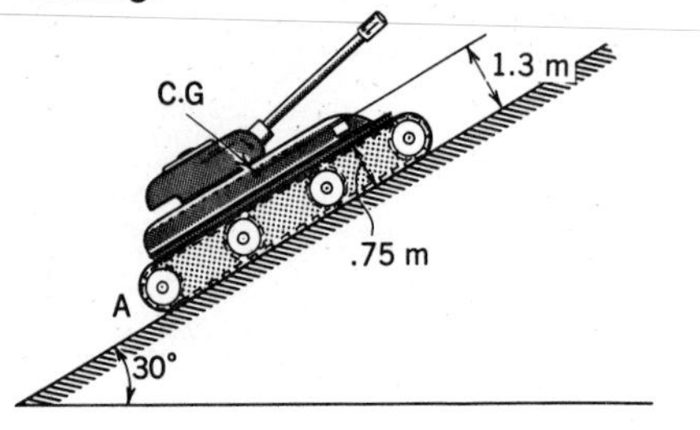


**Figure P.5.34**

**5.35.** Find the components of the forces acting on pins *A*, *B*, and *C* connecting and supporting the blocks shown. Block I weighs 10 kN, and block II weighs 30 kN.

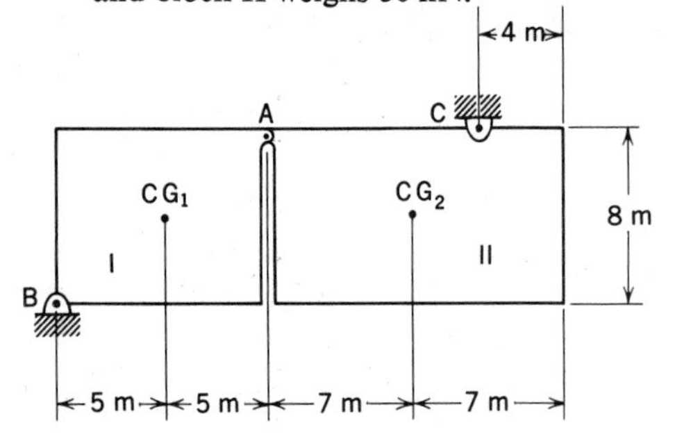


**Figure P.5.35**

**5.36.** What force *F* do the pliers develop on the pipe section *D*? Neglect friction.

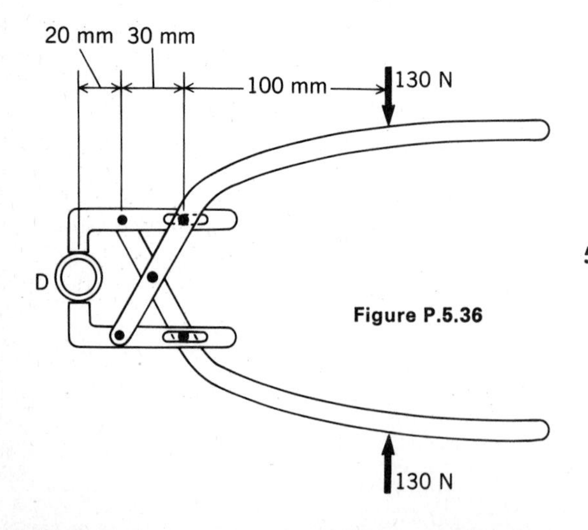


**Figure P.5.36**

**5.37.** What are the supporting forces for the frame? Neglect all weights except the 10-kN weight.

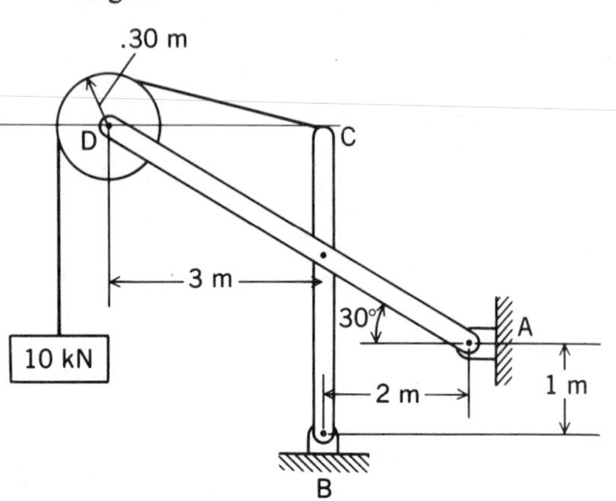


**Figure P.5.37**

**5.38.** A 20-m circular arch must withstand a wind load given for $0 < \theta < \pi/2$ as

$$f = 5000\left(1 - \frac{\theta}{\pi/2}\right) \text{ N/m}$$

where $\theta$ is measured in radians. Note that for $\theta > \pi/2$, there is no loading. What are the supporting forces? (*Hint:* What is the point for which taking moments is simplest?)

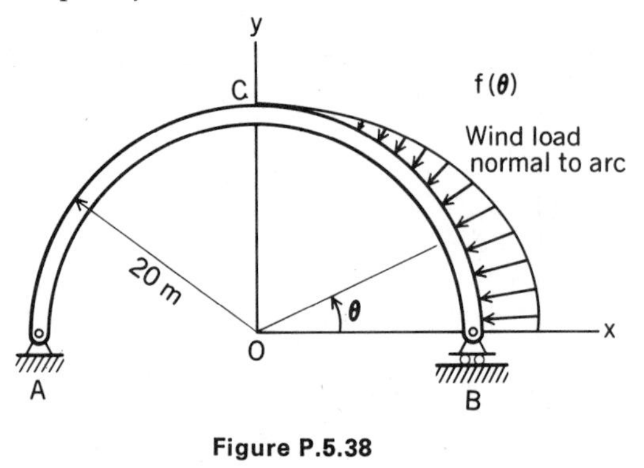


**Figure P.5.38**

**5.39.** An arch is formed by uniform plates *A* and *B*. Plate *A* weighs 5 kN and plate *B* weighs 2 kN. What are the supporting forces at *C*, *D*, and *E*?

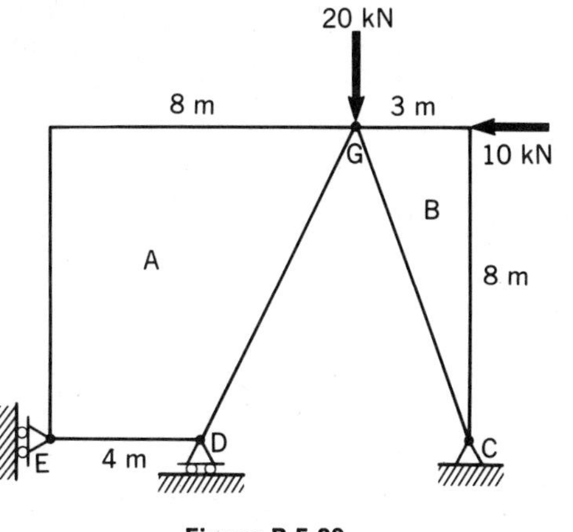


Figure P.5.39

**5.40.** Find the supporting forces on the beam *EF* and the supporting forces at *A*, *B*, *C*, and *D*.

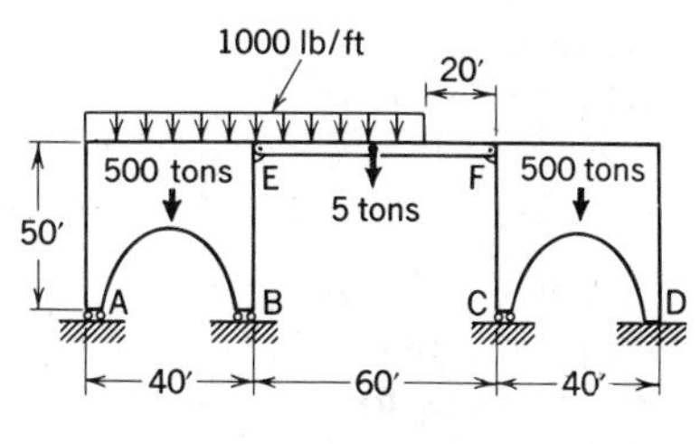


Figure P.5.40

**5.41.** What is the supporting force system at *A* for the cantilever beam? Neglect the weight of the beam.

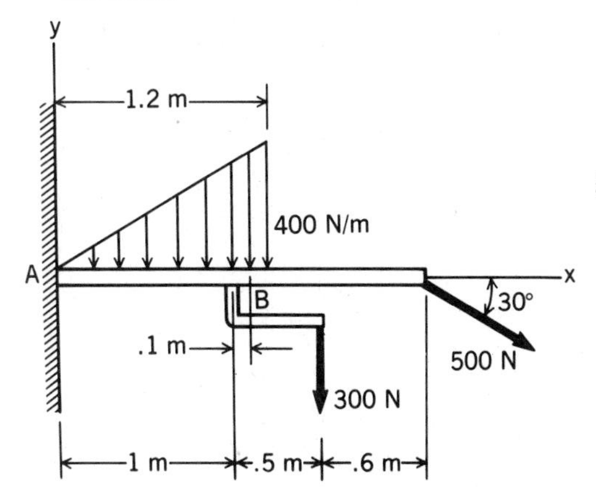


Figure P.5.41

**5.42.** In Problem 5.41, find the force system transmitted through the cross section at *B*.

**5.43.** A cantilever beam *AB* is pinned at *B* to a simply supported beam *BC*. For the loads given, find the supporting force system at *A*. Determine force components normal and axial to the beam *AB*. Neglect the weights of the beams.

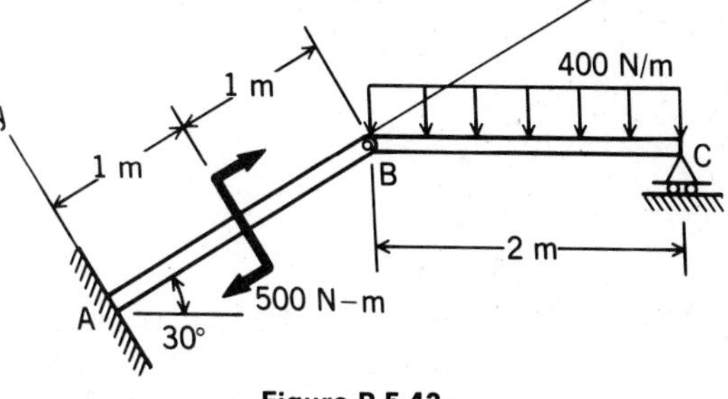


Figure P.5.43

**5.44.** Find the supporting force system at *A*.

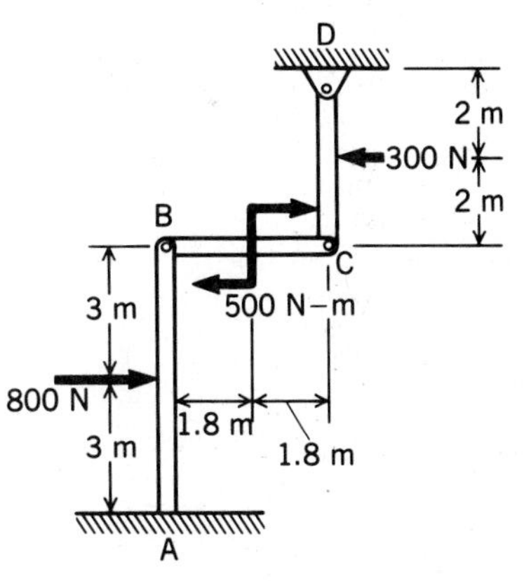


Figure P.5.44

**5.45.** A light bent rod *AD* is pinned to a straight light rod *CB* at *C*. The bent rod supports a uniform load. A spring is stretched to connect the two rods. The spring has a spring constant of $10^4$ N/m, and its unstretched length is .8 m. Find the supporting forces at *A* and *B*. The force in the spring is $10^4$ times the elongation in meters.

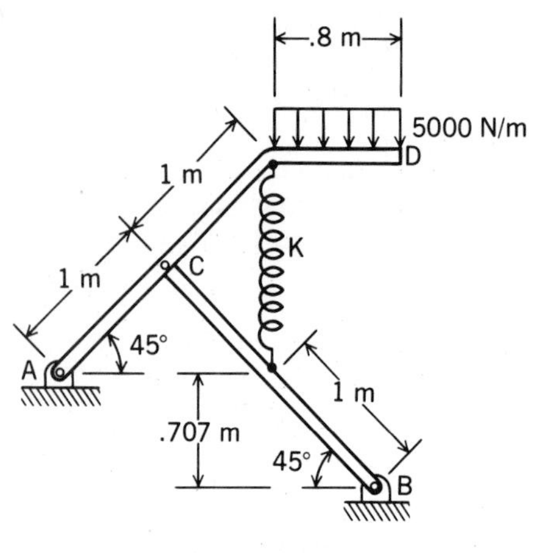


Figure P.5.45

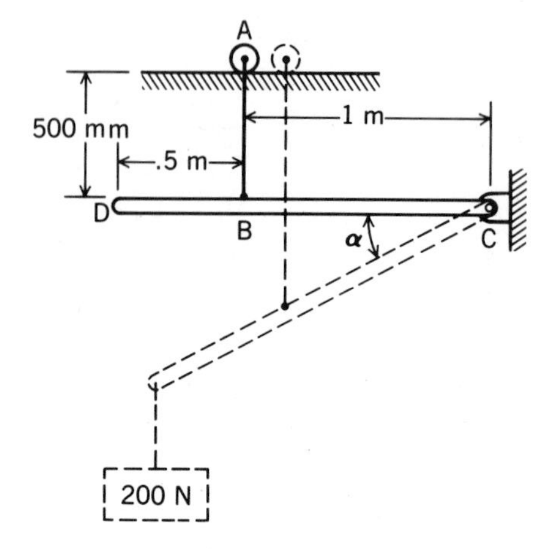


Figure P.5.47

**5.46.** Light rods $AD$ and $AC$ are pinned together at $C$ and support a 300-N and a 100-N load. What are the supporting forces at $A$ and $B$?

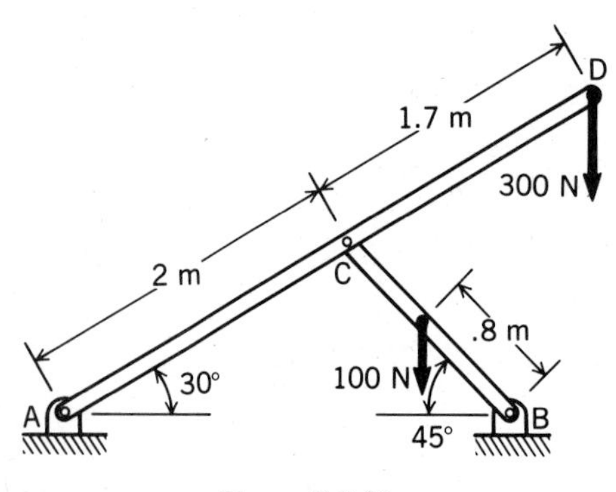


Figure P.5.46

**5.47.** A light rod $CD$ is held in a horizontal position by a strong elastic band $AB$ (shock cord) which acts like a spring in that it takes $10^3$ N per meter of elongation of the band. The upper part of the band is connected to a small wheel free to roll on a horizontal surface. What is the angle $\alpha$ needed to support a 200-N load as shown?

**5.48.** Solve for the supporting forces at $A$ and $C$. $AB$ weighs 100 lb, and $BC$ weighs 150 lb.

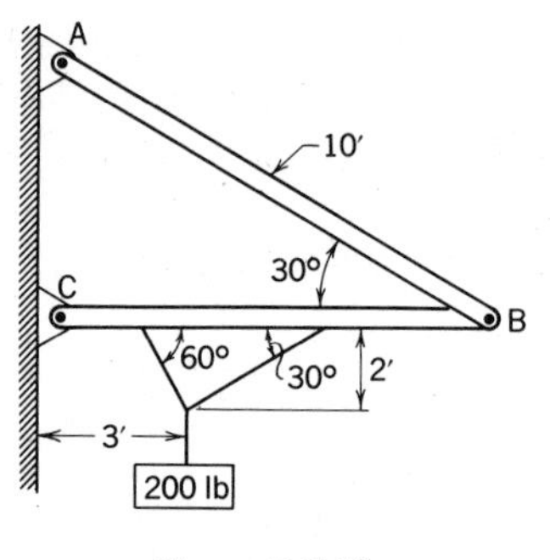


Figure P.5.48

**5.49.** What torque $T$ is needed to maintain the configuration shown for the compressor if $p_1 = 5$ psig? The system lies horizontally.

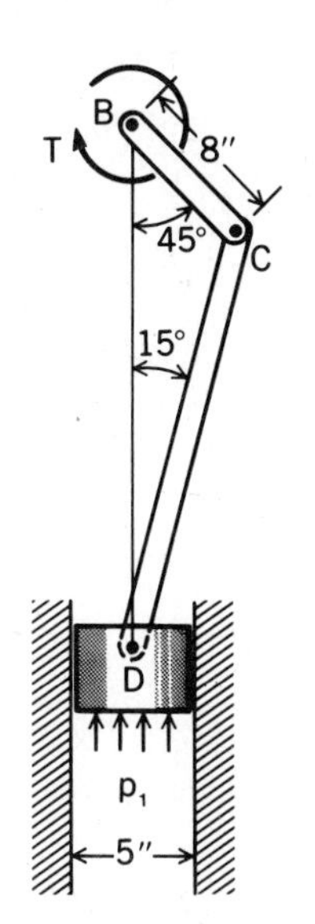


Figure P.5.49

**5.50.** Work Problem 5.49 for the system oriented vertically with $BC$ weighing 3 lb and $CD$ weighing 5 lb.

**5.51.** Neglecting friction, find the angle $\beta$ of line $AB$ for equilibrium in terms of $\alpha_1$, $\alpha_2$, $W_1$, and $W_2$.

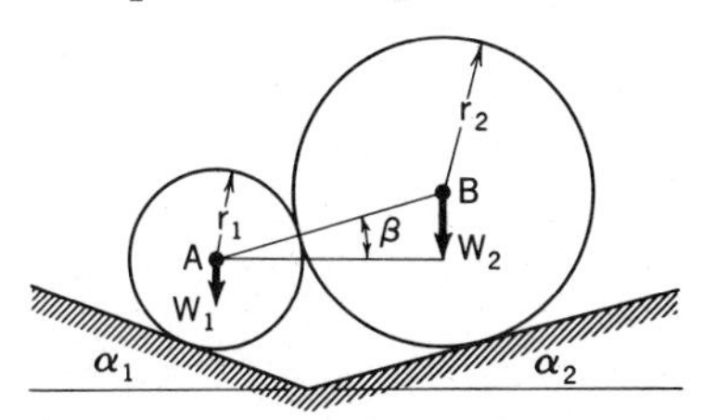


Figure P.5.51

**5.52.** If the rod $CD$ weighs 20 lb, what torque $T$ is needed to maintain equilibrium? The system is in a vertical plane. Cylinder $A$ weighs 10 lb and cylinder $B$ weighs 5 lb. Disregard friction. At $D$ there is a slot.

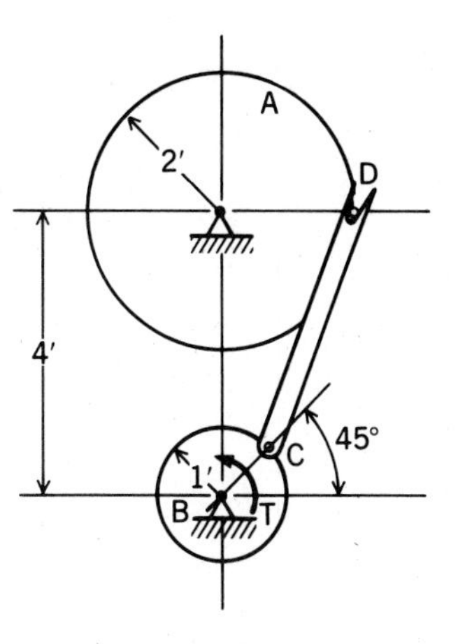


Figure P.5.52

**5.53.** Find the supporting forces at $A$ and $G$. The weight of $W$ is 500 N and the weight of $C$ is 200 N. Neglect all other weights. The cord connecting $C$ and $D$ is vertical.

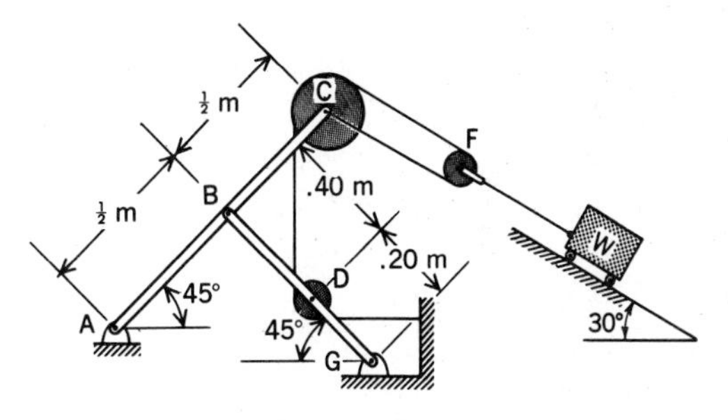


Figure P.5.53

**5.54.** What torque $T$ is needed for equilibrium if cylinder $B$ weighs 500 N and $CD$ weighs 300 N?

Figure P.5.54

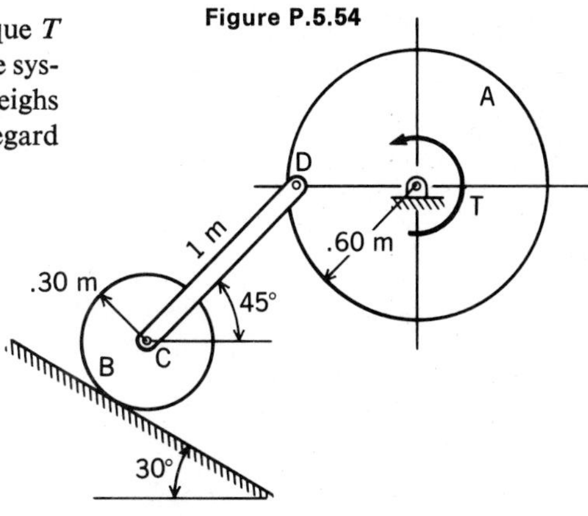

**5.55.** A bar $AB$ is pinned to two identical planetary gears each of diameter .30 m. Gear $E$ is pinned to bar $AB$ and meshes with the two planetary gears, which in turn mesh with stationary gear $D$. If a torque $T$ of 100 N-m is applied to the bar $AB$, what external torque is needed to be applied to the upper gear $D$ to maintain equilibrium? The system is horizontal.

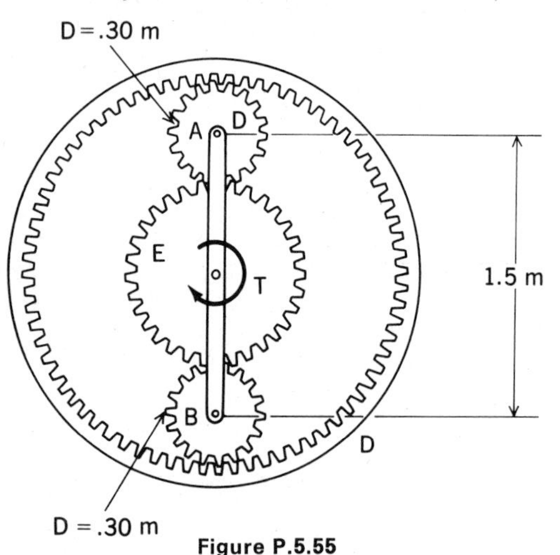


Figure P.5.55

**5.56.** In Problem 5.55, equilibrium is maintained by applying a torque on gear $E$ rather than gear $D$. What is this torque?

**5.57.** A Bucyrus–Erie transit crane is holding a chimney having a weight of 20 kN. The chimney is held by a cable that goes over a pulley at $A$, then goes over a second pulley at $D$, and then to a winch at $K$. The position of boom $AH$ (on top) is maintained by two separate cables, one from $A$ to $B$, and the other from $B$ to pulley $C$. Find the tensions in cables $AB$ and $BC$. Note that $BC$ is vertical for the setup shown. Consider only the weight of the load and neglect friction.

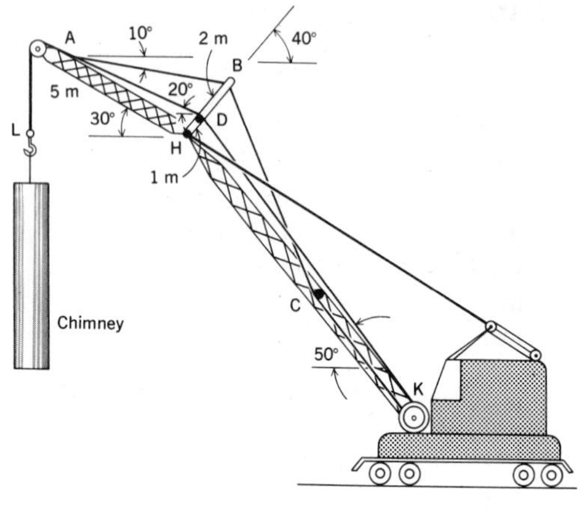

Figure P.5.57

**5.58.** What are the forces at the arm connections at $B$ and the cable tensions when the power shovel is in the position shown? Arm $AC$ weighs 13,000 N, arm $DF$ weighs 11,000 N, and the shovel and payload together weigh 9000 N and act at the center of gravity as shown. $B$ is at the same elevation as $G$.

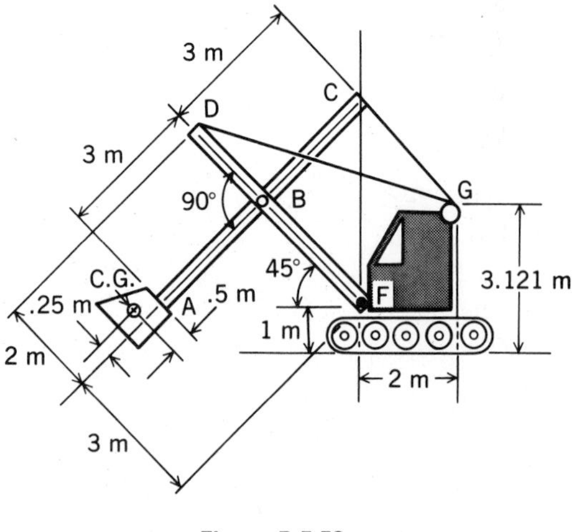


Figure P.5.58

**Parallel Force Systems.** The simplest resultant is a single force or a single couple moment. There are then three independent equations of equilibrium. If the forces are also coplanar, then we have only two independent equations of equilibrium. We shall examine the latter case in the following example.

**EXAMPLE 5.7**

Determine the forces required to support the uniform beam in Fig. 5.31 shown loaded with a couple, a point force, and a parabolic distribution of load. The weight of the beam is 100 lb.

Since a couple can be rotated without affecting the equilibrium of the body, we can orient the couple so that the forces are vertical. Accordingly, we have here a beam loaded by a system of parallel coplanar loads. Clearly, the supporting forces must be vertical, as shown in Fig. 5.32, where we have a free-body diagram of the beam. Since there are only two unknown quantities, we can handle the problem by statical consideration of this free body.

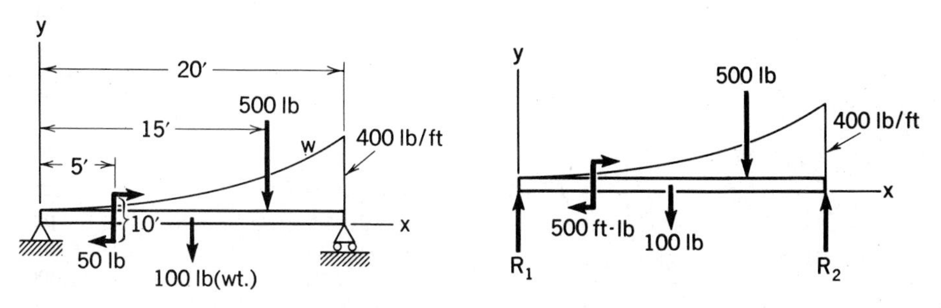


**Figure 5.31.** Find supporting forces.

**Figure 5.32.** Free-body diagram.

The equation for the loading curve must be $w = ax^2 + b$, where $a$ and $b$ are to be determined from the loading data and the choice of reference. With an $xy$ reference at the left end, as shown, we then have the conditions:

1. When $x = 0$, $w = 0$.
2. When $x = 20$, $w = 400$.

To satisfy these conditions, $b$ must be zero and $a$ must be unity; the loading function is thus given as

$$w = x^2 \text{ lb/ft}$$

In this problem, we shall again work directly with the scalar equations. By summing moments about the left and right ends of the beam, we can then solve the unknowns directly:

$\underline{\sum M_1 = 0:}$

$$-500 - (10)(100) - (15)(500) - \int_0^{20} x^3\,dx + 20R_2 = 0$$

Integrating and canceling terms, we get

$$-9000 - \frac{x^4}{4}\bigg|_0^{20} + 20R_2 = 0$$

By inserting limits and solving, we get one of the unknowns:

$$R_2 = 2450 \text{ lb}$$

Next,

$\underline{\sum M_2 = 0:}$

$$-20(R_1) - 500 + (10)(100) + (5)(500) + \int_0^{20} (20 - x)x^2\,dx = 0$$

Solving for $R_1$, we have

$$R_1 = 817 \text{ lb}$$

As a *check* on these computations, we can sum forces in the vertical direction. The result must be zero (or as close to zero as the accuracy of our calculations permits):

$\underline{\sum F_y = 0:}$

$$R_1 + R_2 - 100 - 500 - \int_0^{20} x^2\,dx = 0$$

$$3267 - 600 - \frac{x^3}{3}\bigg|_0^{20} = 0$$

Therefore,

$$2667 - 2667 = 0$$

Always take the opportunity to check a solution in this manner (i.e., by using a redundant equilibrium equation). In later problems, we shall rely heavily on calculated reactions; thus, we must make sure they are correct.

**General Force Systems.** The simplest resultant in the general case is a force and a couple moment. Six equations of equilibrium can be given for each free-body diagram. We now examine two examples for this case.

**EXAMPLE 5.8**

A derrick is shown in Fig. 5.33 supporting a 1000-lb load. The vertical beam has a ball-and-socket connection into the ground at $d$ and is held by guy wires. Neglect the weight of the members and guy wires, and find the tensions in the guy wires $ac$, $bc$, and $ce$.

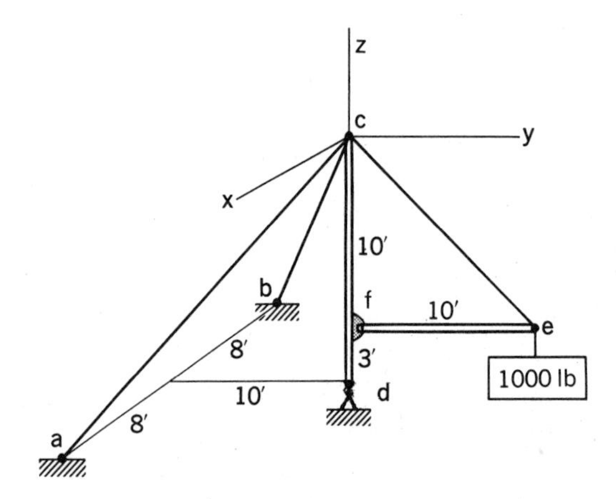


**Figure 5.33.** Loaded derrick.

If we select as a free body both members and the interconnecting guy wire *ce*, we shall expose two of the desired unknowns (Fig. 5.34). Note that this is a general three-dimensional force system with only five unknowns. Although all these unknowns can be solved by statical considerations of this free body, you will notice that, if we take moments about point *d*, we will involve in a vector equation only the desired unknowns $T_{bc}$ and $T_{ac}$. Accordingly, all unknown forces need not be computed for this free-body diagram. You should always look for such short cuts in situations such as these.

To determine the unknown tension $T_{ce}$, we must employ another free-body diagram. Either the vertical or horizontal member will expose this unknown in a manner susceptible to solution. The latter has been selected and is shown in Fig. 5.35. Note that we have here a coplanar force system with three unknowns. Again, you can see that, by taking moments about point *f*, we will involve only the desired unknown.

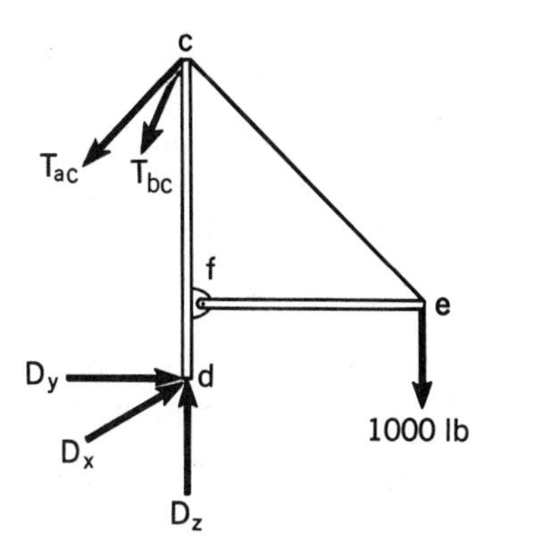


Figure 5.34. Free-body diagram 1.

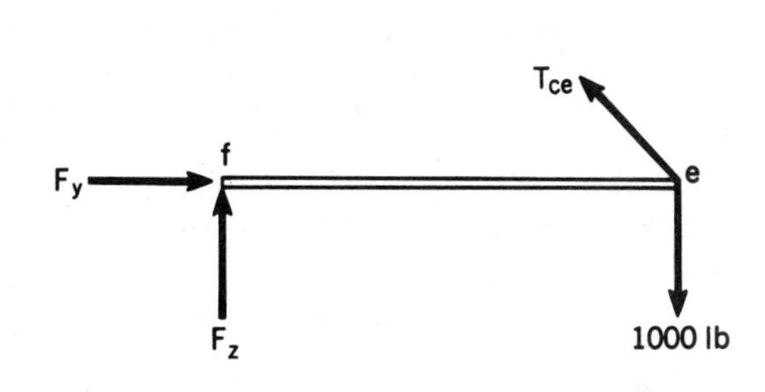


Figure 5.35. Free-body diagram 2.

The vector $\boldsymbol{T}_{ac}$ may then be given as

$$\boldsymbol{T}_{ac} = T_{ac}\left[\frac{1}{\sqrt{333}}(8\boldsymbol{i} - 10\boldsymbol{j} - 13\boldsymbol{k})\right] \quad \text{(a)}$$

Similarly, we have for $\boldsymbol{T}_{bc}$,

$$\boldsymbol{T}_{bc} = T_{bc}\left[\frac{1}{\sqrt{333}}(-8\boldsymbol{i} - 10\boldsymbol{j} - 13\boldsymbol{k})\right] \quad \text{(b)}$$

Using the free-body diagram in Fig. 5.34, we now set the sum of moments about point *d* equal to zero. Thus, employing the relations above, we get

$$13\boldsymbol{k} \times \frac{T_{ac}}{\sqrt{333}}(8\boldsymbol{i} - 10\boldsymbol{j} - 13\boldsymbol{k}) + 13\boldsymbol{k} \times \frac{T_{bc}}{\sqrt{333}}(-8\boldsymbol{i} - 10\boldsymbol{j} - 13\boldsymbol{k}) + 10\boldsymbol{j} \times (-1000\boldsymbol{k}) = \boldsymbol{0} \quad \text{(c)}$$

When we make the substitution of variable $t_1 = T_{ac}/\sqrt{333}$ and $t_2 = T_{bc}/\sqrt{333}$, the preceding equation becomes

$$[130(t_1 + t_2) - 10{,}000]\boldsymbol{i} + [104(t_1 - t_2)]\boldsymbol{j} = \boldsymbol{0} \quad \text{(d)}$$

The scalar equations,

$$130(t_1 + t_2) - 10{,}000 = 0$$
$$104(t_1 - t_2) = 0$$

can now be readily solved to give $t_1 = t_2 = 38.5$. Hence, we get $T_{ac} = 38.5\sqrt{333} = 702$ lb and $T_{bc} = 38.5\sqrt{333} = 702$ lb.[6]

Turning finally to the free-body diagram 2 in Fig. 5.35, we see that, in summing moments about $f$, the horizontal component of the tension $T_{ce}$ has a zero-moment arm. Thus,

$$(10)(0.707)T_{ce} - (10)(1000) = 0$$

Hence,

$$T_{ce} = 1414 \text{ lb}$$

**EXAMPLE 5.9**

A blimp is shown in Fig. 5.36 fixed at the mooring tower $D$ by a ball-joint connection, and held by cables $AB$ and $AC$. The blimp has a mass of 1500 kg. The simplest resultant force $\boldsymbol{F}$ from air pressure (including the effects of wind) is

$$\boldsymbol{F} = 17{,}500\boldsymbol{i} + 1000\boldsymbol{j} + 1500\boldsymbol{k} \text{ N}$$

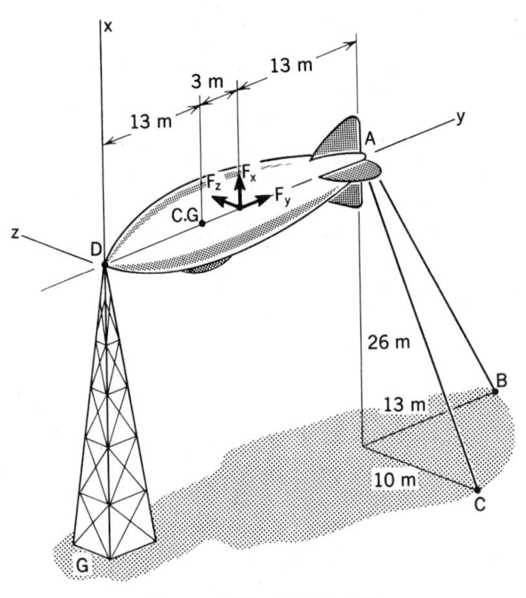


**Figure 5.36.** Tethered blimp.

at a position shown in the diagram. Compute the tension in the cables as well as the force transmitted to the ball joint at the top of the tower at $D$. Also, what force sys-

[6] By taking moments about the line connecting points $a$ and $d$, we could get $T_{bc}$ directly using the scalar triple product. We suggest that you try this.

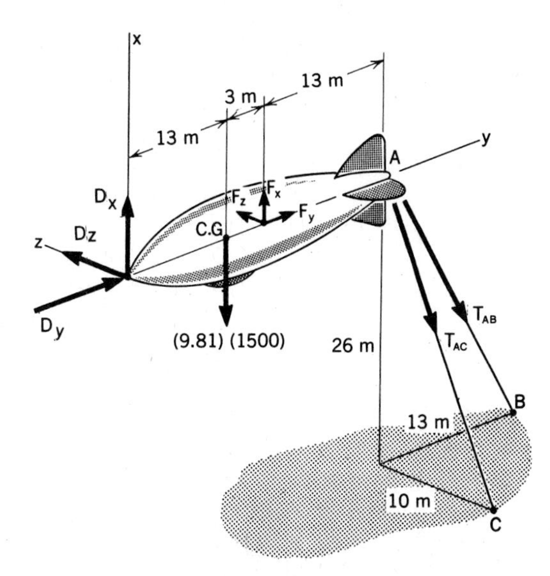


**Figure 5.37.** Free-body diagram of blimp.

tem is transmitted to the ground at $G$ through the mooring tower? The tower weighs 5000 N.

We shall first consider a free-body diagram of the blimp, as shown in Fig. 5.37. We have five unknown forces here, and we can solve all of them by using equations of equilibrium for this free body. As a first step, we express the cable tensions vectorially. That is,

$$\boldsymbol{T}_{AC} = T_{AC}\left(\frac{-26\boldsymbol{i} - 10\boldsymbol{k}}{\sqrt{26^2 + 10^2}}\right) = T_{AC}(-.933\boldsymbol{i} - .359\boldsymbol{k})$$

$$\boldsymbol{T}_{AB} = T_{AB}\left(\frac{-26\boldsymbol{i} + 13\boldsymbol{j}}{\sqrt{26^2 + 13^2}}\right) = T_{AB}(-.894\boldsymbol{i} + .447\boldsymbol{j})$$

We now go back to the basic vector equations of equilibrium. Thus:

$\underline{\sum \boldsymbol{F}_i = \boldsymbol{0}}$:

$$D_x\boldsymbol{i} + D_y\boldsymbol{j} + D_z\boldsymbol{k} - (1500)(9.81)\boldsymbol{i} + 17{,}500\boldsymbol{i} + 1000\boldsymbol{j} + 1500\boldsymbol{k} + T_{AC}(-.933\boldsymbol{i} - .359\boldsymbol{k}) + T_{AB}(-.894\boldsymbol{i} + .447\boldsymbol{j}) = \boldsymbol{0}$$

The scalar equations are:

$$D_x + 2785 - .933T_{AC} - .894T_{AB} = 0 \quad \text{(a)}$$

$$D_y + 1000 + .447T_{AB} = 0 \quad \text{(b)}$$

$$D_z + 1500 - .359T_{AC} = 0 \quad \text{(c)}$$

Next take moments about point $D$.

$\underline{\sum (\boldsymbol{M}_i)_D = \boldsymbol{0}}$:

$$13\boldsymbol{j} \times (9.81)(1500)(-\boldsymbol{i}) + 16\boldsymbol{j} \times (17{,}500\boldsymbol{i} + 1000\boldsymbol{j} + 1500\boldsymbol{k}) + 29\boldsymbol{j} \times T_{AC}(-.933\boldsymbol{i} - .359\boldsymbol{k}) + 29\boldsymbol{j} \times T_{AB}(-.894\boldsymbol{i} + .447\boldsymbol{j}) = \boldsymbol{0}$$

Carrying out the various cross products, we end up only with $\boldsymbol{k}$ and $\boldsymbol{i}$ components, thus generating two scalar equations.[7] They are:

$$-10.41T_{AC} + 24{,}000 = 0 \tag{d}$$

$$25.9T_{AB} + 27.1T_{AC} - 88{,}700 = 0 \tag{e}$$

We now have five independent equations for five unknowns. We can thus solve these equations simultaneously. From Eq. (d), we have

$$T_{AC} = 2305 \text{ N}$$

From Eq. (e), we have

$$T_{AB} = 1012 \text{ N}$$

From Eq. (c), we have

$$D_z = -673 \text{ N}$$

From Eq. (b), we have

$$D_y = -1452 \text{ N}$$

From Eq. (a), we have

$$D_x = 270 \text{ N}$$

Next, consider the mooring tower as a free body (Fig. 5.38). Notice that in showing the forces at the ball joint $D$, we have taken into account both of the negative signs shown above for $D_y$ and $D_z$ as well as Newton's third law.

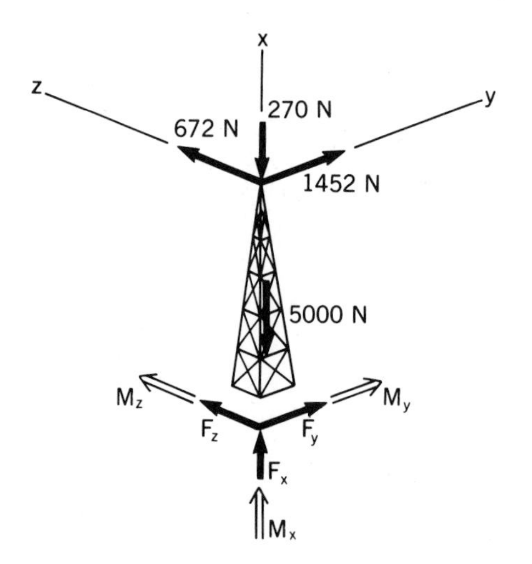


**Figure 5.38.** Free-body diagram of mooring tower.

Again, using the basic vector equations of statics, we have

$$-270\boldsymbol{i} + 1452\boldsymbol{j} + 672\boldsymbol{k} + F_x\boldsymbol{i} + F_y\boldsymbol{j} + F_z\boldsymbol{k} - 5000\boldsymbol{i} = \boldsymbol{0}$$

---

[7]The third equation is $0 = 0$. That is, there are no moments about the $y$ axis, because all forces pass through the $y$ axis.

Hence,

$$F_x = 5270 \text{ N}$$
$$F_y = -1452 \text{ N}$$
$$F_z = -672 \text{ N}$$

Now take moments about the base at *F*. We get

$$26\boldsymbol{i} \times (-270\boldsymbol{i} + 1452\boldsymbol{j} + 672\boldsymbol{k}) + M_x\boldsymbol{i} + M_y\boldsymbol{j} + M_z\boldsymbol{k} = \boldsymbol{0}$$

From this, we get

$$M_x = 0$$
$$M_y = 17{,}470 \text{ N-m}$$
$$M_z = -37{,}800 \text{ N-m}$$

We can conclude that, at the center of the base, the force system from the ground is

$$\boldsymbol{F} = 5270\boldsymbol{i} - 1452\boldsymbol{j} - 672\boldsymbol{k} \text{ N}$$
$$\boldsymbol{C} = 17{,}470\boldsymbol{j} - 37{,}800\boldsymbol{k} \text{ N-m}$$

The force system acting on the ground at the center of the base is the reaction to the system above. Thus,

$$\boldsymbol{F}_{\text{ground}} = -5270\boldsymbol{i} + 1452\boldsymbol{j} + 672\boldsymbol{k} \text{ N}$$
$$\boldsymbol{C}_{\text{ground}} = -17{,}470\boldsymbol{j} + 37{,}800\boldsymbol{k} \text{ N-m}$$

## Problems

**5.59.** The triple pulley sheave and the double pulley sheave weigh 15 lb and 10 lb, respectively. What rope force is necessary to lift a 350-lb engine? What is the force on the ceiling hook?

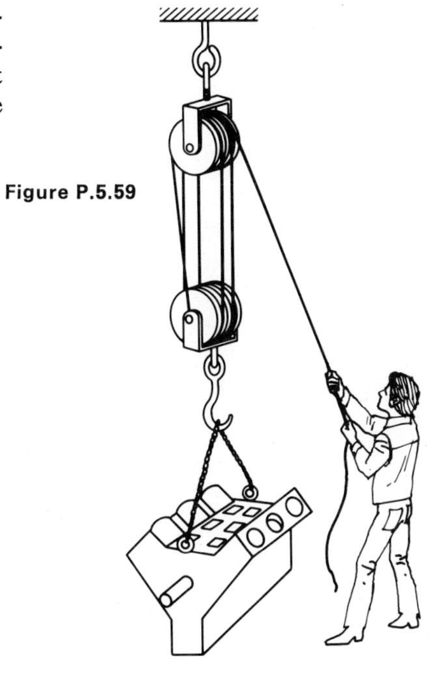

Figure P.5.59

**5.60.** A multipurpose pry bar can be used to pull nails in the three positions. If a force of 400 lb is required to remove a nail and a carpenter can exert 50 lb, which position(s) must he use?

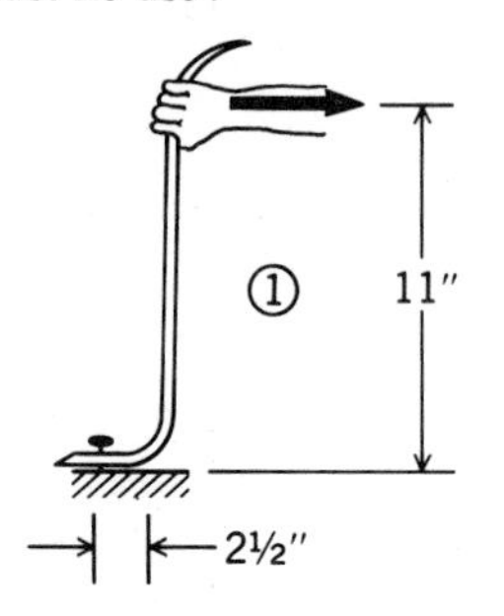


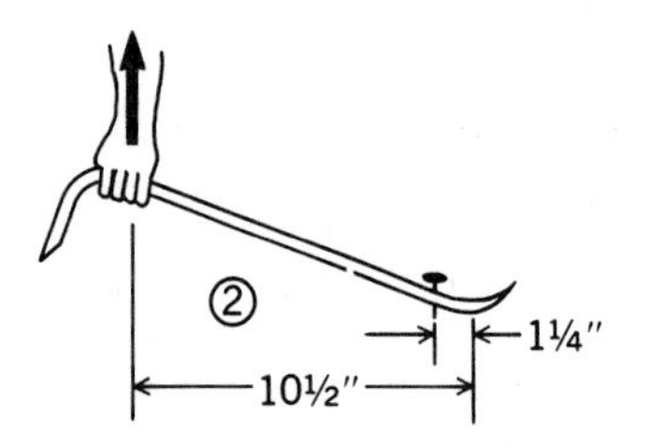


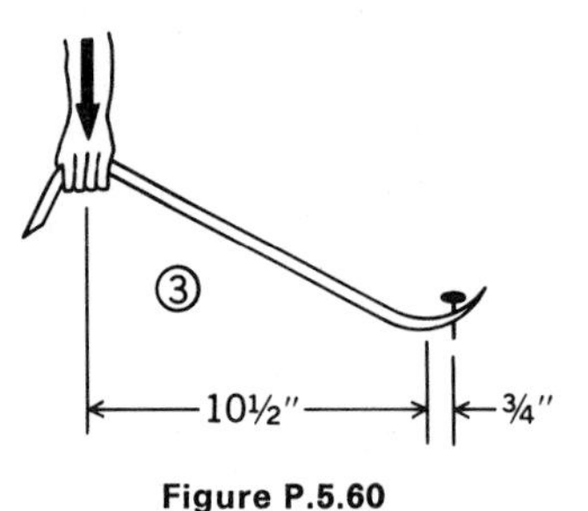


Figure P.5.60

**5.61.** At what position must the operator of the counterweight crane locate the 50-kN counterweight when he lifts a 10-kN load of steel?

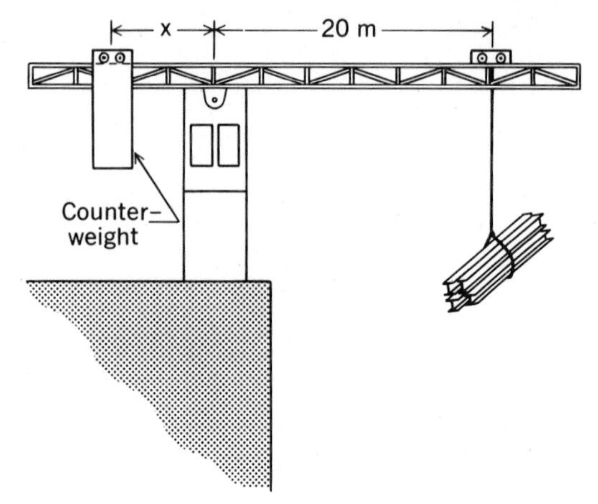


Figure P.5.61

**5.62.** A Jeep winch is used to raise itself by a force of 2 kN. What are the reactions at the Jeep tires with and without the winch load? The driver weighs 800 N, and the Jeep weighs 11 kN. The center of gravity of the Jeep is shown.

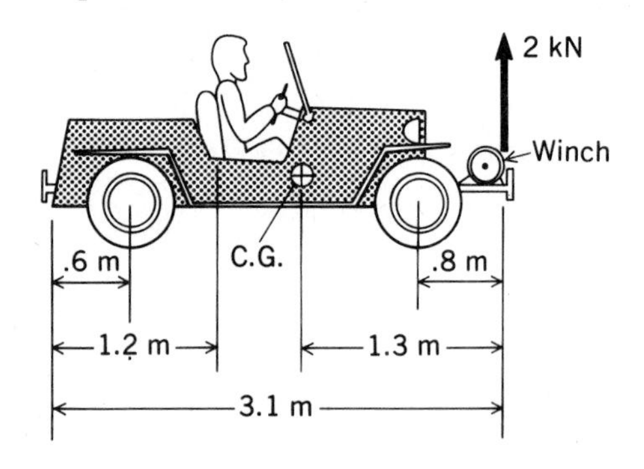


Figure P.5.62

**5.63.** A *differential pulley* is shown. Compute $F$ in terms of $W$, $r_1$, and $r_2$.

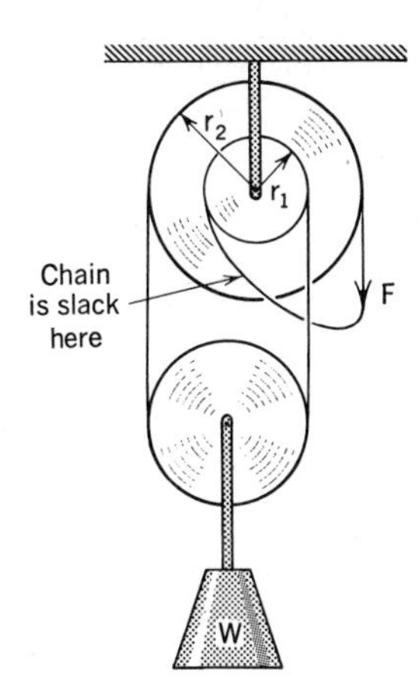


Figure P.5.63

**5.64.** What is the longest portion of pipe weighing 400 lb/ft that can be lifted without tipping the 12,000-lb tractor? Take the center of gravity of the tractor at the geometric center.

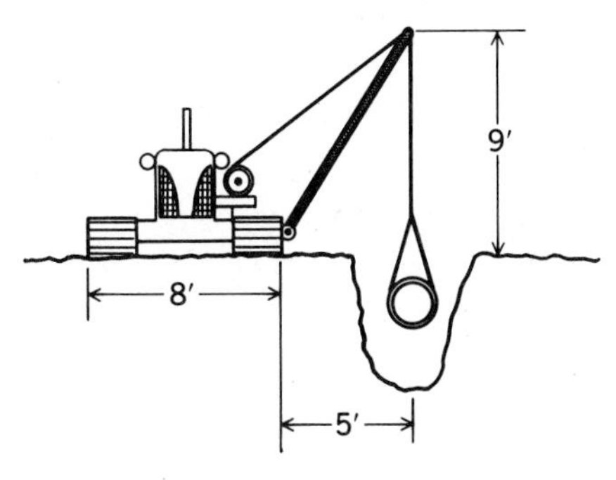


Figure P.5.64

**5.65.** The L-shaped concrete post supports an elevated railroad. The concrete weighs 150 lb/ft³. What are the reactions at the base of the post?

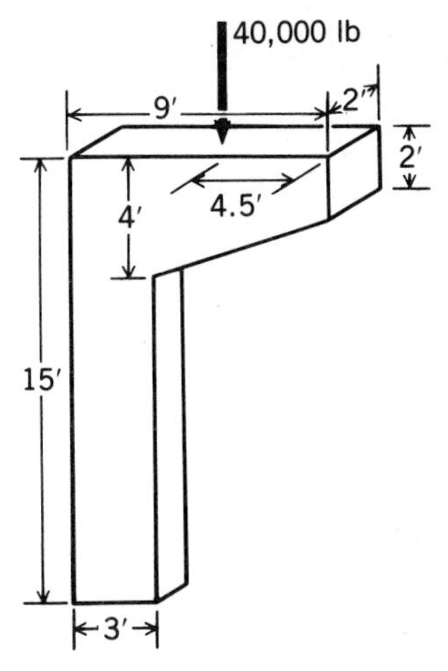


Figure P.5.65

**5.66.** Two hoists are operated on the same overhead track. Hoist *A* has a 3000-lb load, and hoist *B* has a 4000-lb load. What are the reactions at the ends of the track when the hoists are in the position shown?

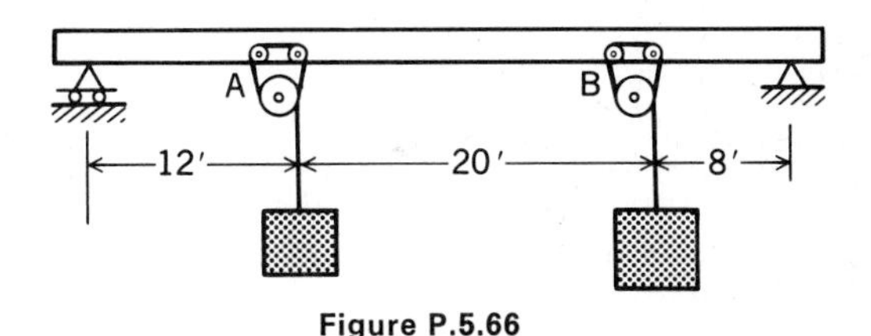


Figure P.5.66

**5.67.** An I-beam cantilevered out from a wall weighs 30 lb/ft and supports a 300-lb hoist. Steel (487 lb/ft³) cover plates 1 in. thick are welded on the beam near the wall to increase the moment-carrying capacity of the beam. What are the reactions at the wall when a 400-lb load is hoisted at the outermost position of the hoist?

Figure P.5.67

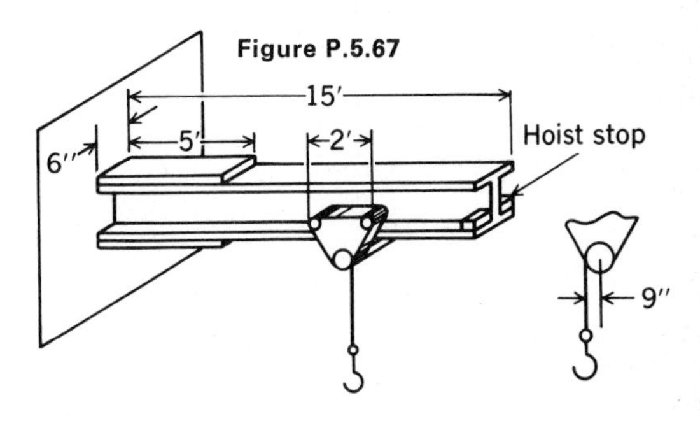

**5.68.** Find the supporting force system for the cantilever beam shown pinned at $C$.

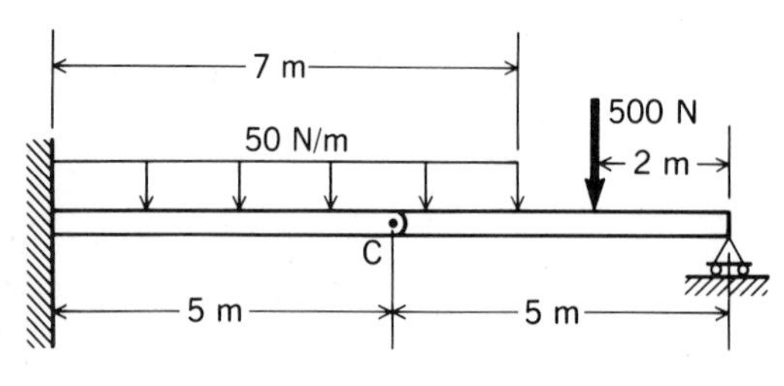


Figure P.5.68

**5.69.** Find the supporting force system for the cantilever beams connected to bar $AB$ by pins.

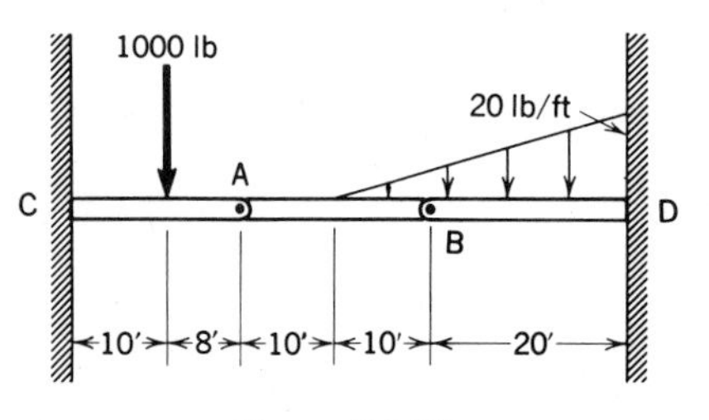


Figure P.5.69

**5.70.** The trailer weighs 50 kN and is loaded with crates weighing 90 kN and 40 kN. What are the reactions at the rear wheel and on the tractor at $A$?

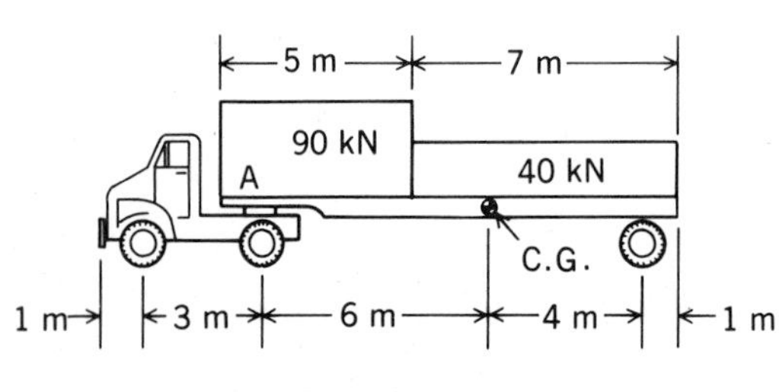

Figure P.5.70

**5.71.** What load $W$ will a pull $P$ of 100 lb lift in the pulley system? Sheaves $A$, $B$, and $C$ weigh 20 lb, 15 lb, and 30 lb, respectively. Assume first that the three sheaves are frictionless and find $W$. Then, calculate $W$ that can be raised at constant speed for the case where the resisting torque in each of sheaves $A$ and $B$ is .01 times the total force at the bearing of each of sheaves $A$ and $B$.

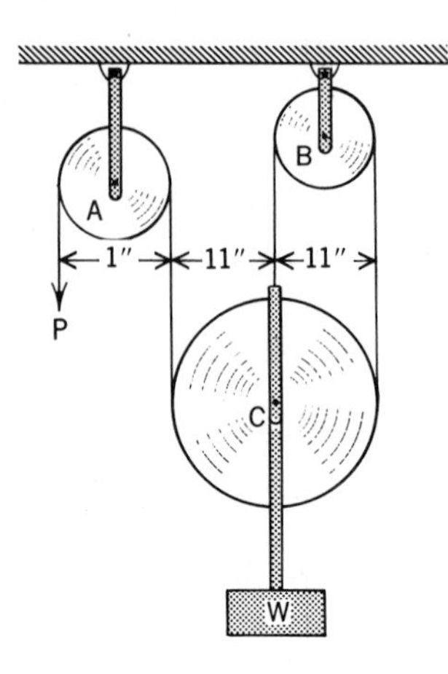


Figure P.5.71

**5.72.** A piece of pop art is being developed. The weight of the body enclosed by the full lines is 2 kN. What is the smallest distance $d$ that the artist can use for cutting a .5-m diameter hole and still avoid tipping? The body is uniform in thickness.

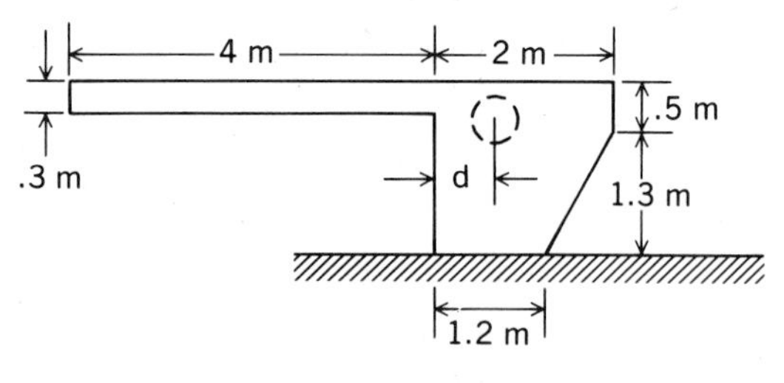


Figure P.5.72

**5.73.** What is the largest weight $W$ that the crane can lift without tipping? What are the supporting forces when the crane lifts this load? What is the force and couple-moment system transmitted through section $C$ of the beam? Compute the force and couple-moment system transmitted through section $D$. The crane weighs 10 tons, having a center of gravity as shown in the diagram.

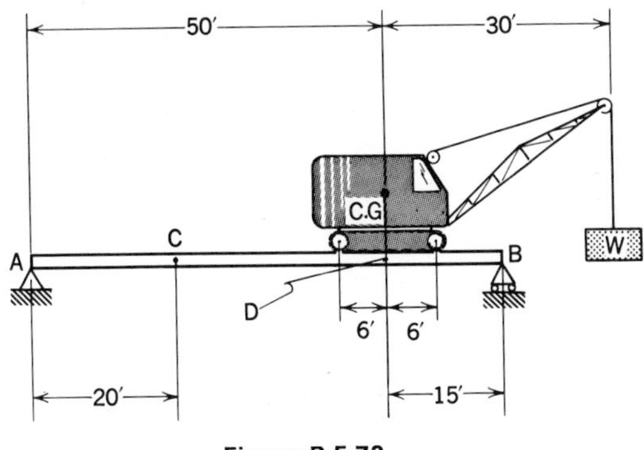


Figure P.5.73

**5.74.** A 20-kN block is being raised at constant speed. If there is no friction in the three pulleys, what are forces $F_1, F_2$, and $F_3$ needed for the job? The block is not rotating in any way. The line of action of the weight vector passes through point $C$ as shown.

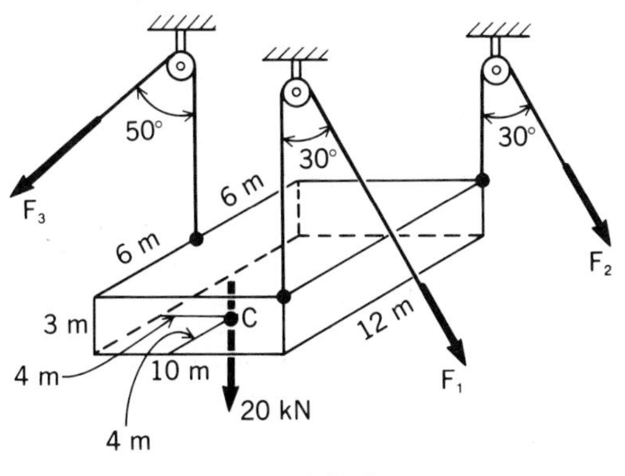


Figure P.5.74

**5.75.** A 10-ton sounding rocket (used for exploring outer space) has a center of gravity shown as C.G.$_1$. It is mounted on a launcher whose weight is 50 tons with a center of gravity at C.G.$_2$. The launcher has three identical legs separated 120° from each other. Leg $AB$ is in the same plane as the rocket and supporting arms $CDE$. What are the supporting forces from the ground? What torque is transmitted from the horizontal arm $CD$ to the ramp $ED$ by the rack and pinion at hinge $D$ to counteract the weight of the rocket?

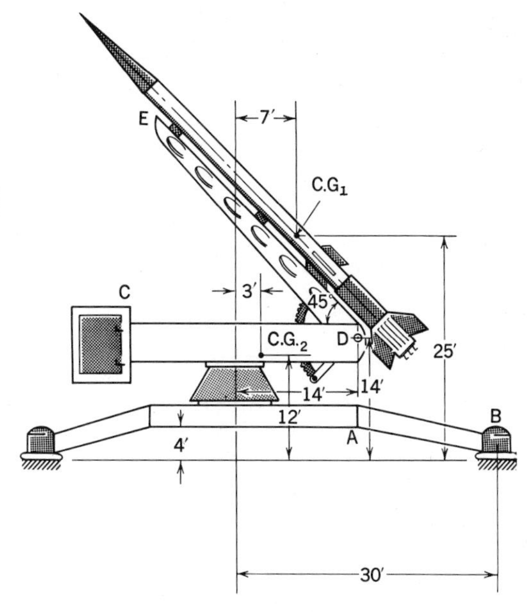


Figure P.5.75

**5.76.** A door is hinged at $A$ and $B$ and contains water whose specific weight $\gamma$ is 62.5 lb/ft³. A force $F$ normal to the door keeps the door closed. What are the forces on the hinges $A$ and $B$ and the force $F$ to counteract the water? As noted in Chapter 4, the pressure in the water above atmosphere is given as $\gamma d$, where $d$ is the perpendicular distance from the free surface of the water.

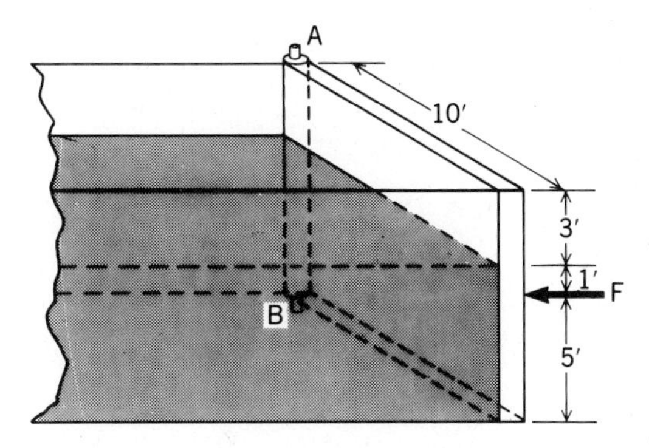


Figure P.5.76

**5.77.** A row of books of length 800 mm and weighing 200 N sits on a three-legged table as shown. The legs are equidistant from each other with one leg coinciding with the $y$ axis. If the table weighs 400 N, will it tip? If not, what are the forces on the legs?

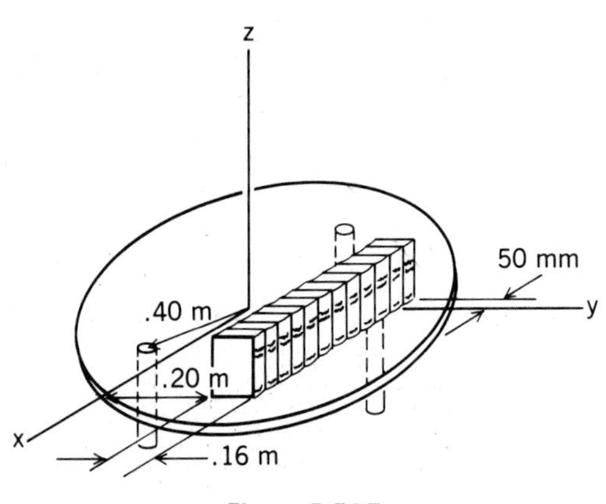


Figure P.5.77

**5.78.** A small helicopter is in a hovering maneuver. The helicopter rotor blades give a lifting force $F_1$ but there results from the air forces on the blades a torque $C_1$. The rear rotor prevents the helicopter from rotating about the $z$ axis but develops a torque $C_2$. Compute the force $F_1$ and couple $C_2$ in terms of the weight $W$. How are $F_3$ and $C_1$ related?

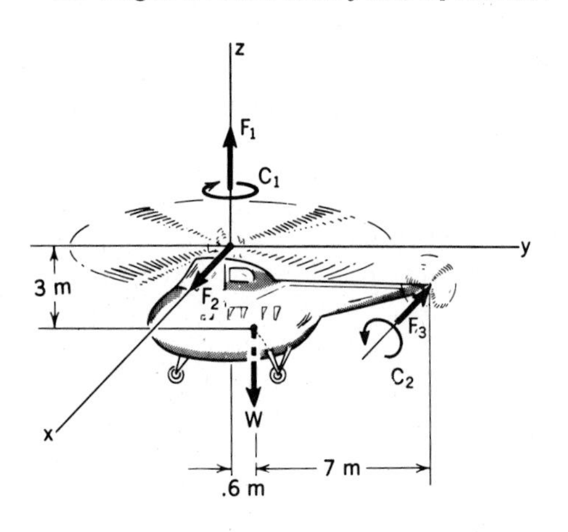


Figure P.5.78

**5.79.** Find the supporting force and couple-moment system for the cantilever beam. What is the force and couple-moment system transmitted through a cross section of the beam at $B$?

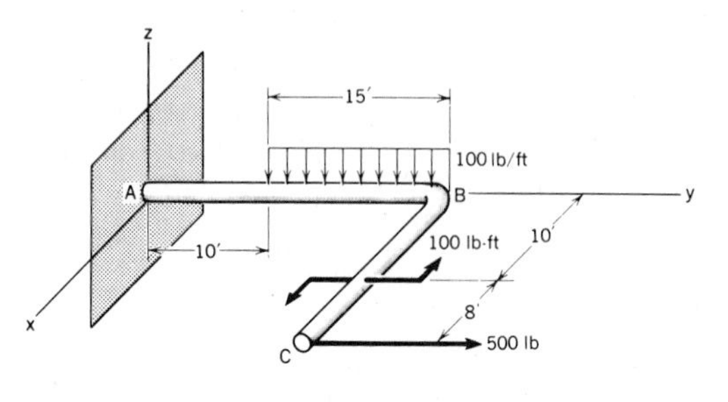


Figure P.5.79

**5.80.** A structure is supported by a ball-and-socket joint at $A$, a pin connection at $B$ offering no resistance in the direction $AB$, and a simple roller support at $C$. What are the supporting forces for the loads shown?

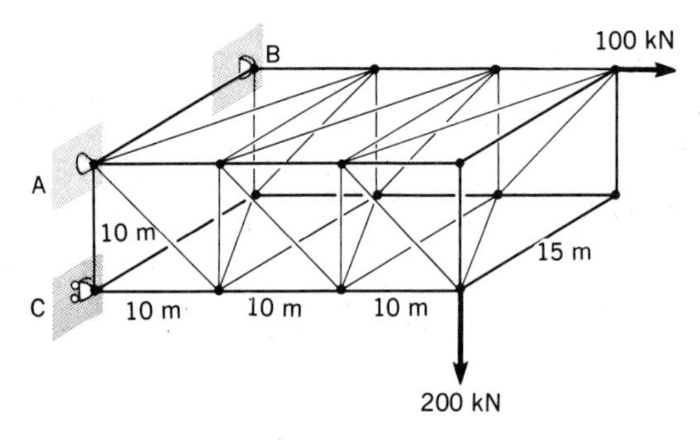


Figure P.5.80

**5.81.** Compute the value of $F$ to maintain the 200-lb weight shown. Assume that the bearings are frictionless, and determine the forces from the bearings on the shaft at $A$ and $B$.

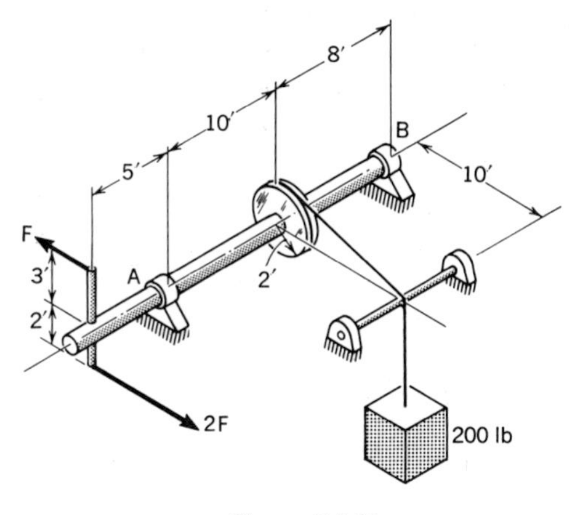


Figure P.5.81

**5.82.** A bar with two right-angle bends supports a force $\boldsymbol{F}$ given as

$$\boldsymbol{F} = 10\boldsymbol{i} + 3\boldsymbol{j} + 100\boldsymbol{k} \text{ N}$$

If the bar has a weight of 10 N/m, what is the supporting force system at $A$?

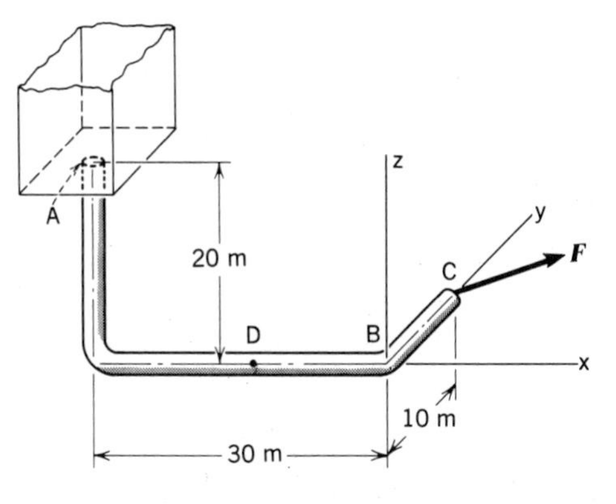


Figure P.5.82

**5.83.** What is the resultant of the force system transmitted across the section at $A$? The couple is parallel to plane $M$.

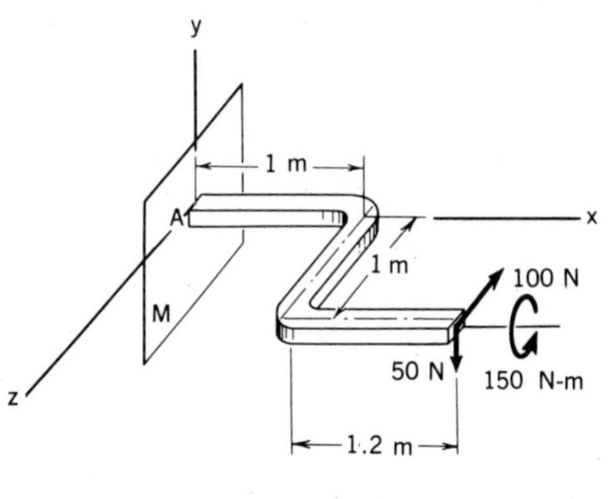

Figure P.5.83

**5.84.** Determine the vertical force $F$ that must be applied to the windlass to maintain the 100-lb weight. Also, determine the supporting forces from the bearings onto the shaft. The handle $DE$ on which the force is applied is in the indicated $xz$ plane.

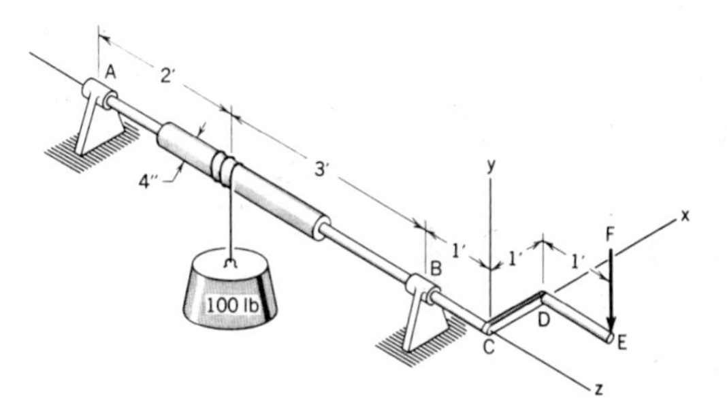


Figure P.5.84

**5.85.** A transport plane has a gross weight of 70,000 lb with a center of gravity as shown. Wheels $A$ and $B$ are locked by the braking system while an engine is being tested under load prior to take off. A thrust $T$ of 3000 lb is developed by this engine. What are the supporting forces?

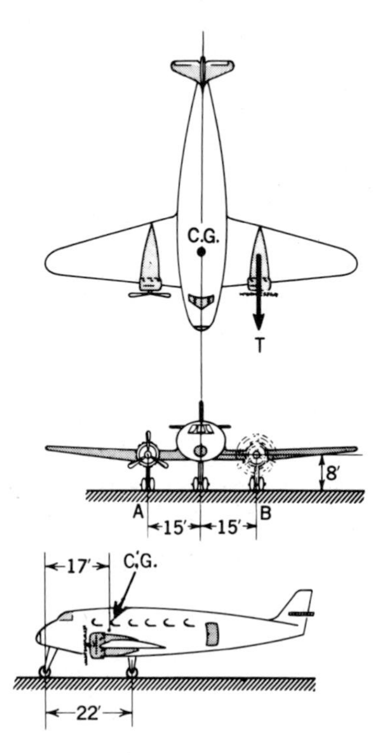


Figure P.5.85

**5.86.** Two cables *GH* and *KN* support a rod *AB* which connects to a ball-and-socket joint support at *A* and supports a 500-kg body *C* at *B*. What are the tensions in the cable and the supporting forces at *A*?

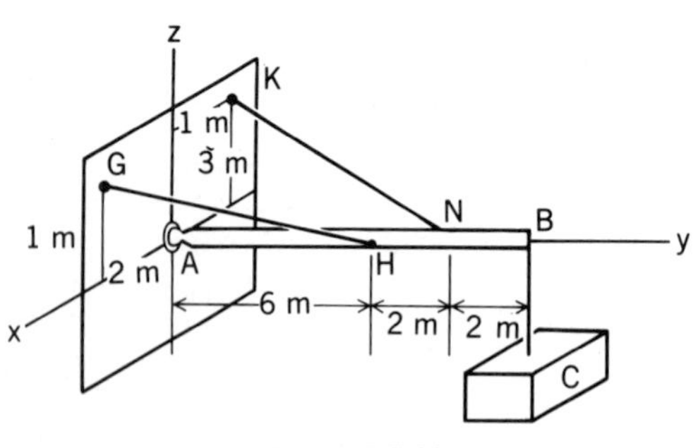


Figure P.5.86

**5.87.** What change in elevation for the 100-lb weight will a couple of 300 lb-ft support if we neglect friction in the bearings at *A* and *B*? Also, determine the supporting force components at the bearings for this configuration.

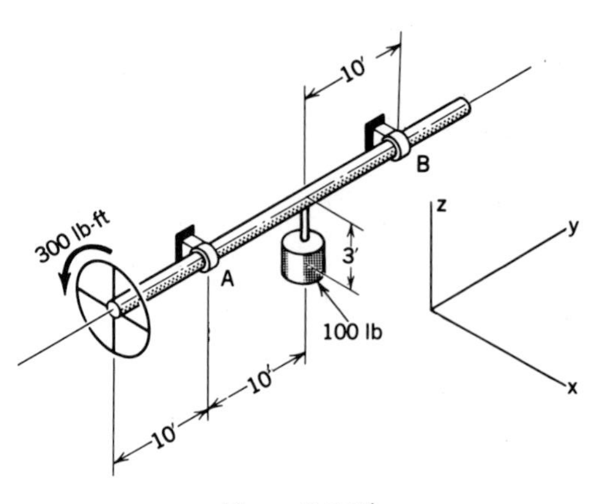


Figure P.5.87

**5.88.** Determine the force *P* required to keep the 150-N door of an airplane open 30° while in flight. The force *P* is exerted in a direction normal to the fuselage. There is a net pressure increase on the outside surface of .02 N/mm². Also, determine the supporting forces at the hinges. Consider that the top hinge supports any vertical force on the door.

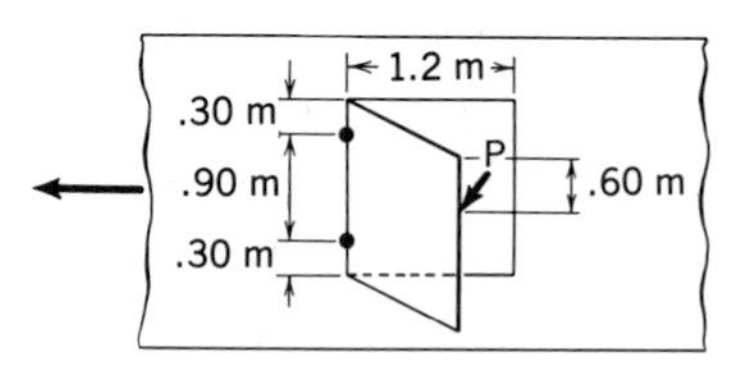


Figure P.5.88

**5.89.** What force *P* is needed to hold the door in a horizontal position? The door weighs 50 lb. Determine the supporting forces at *A* and *B*. At *A* there is a pin and at *B* there is a ball-and-socket joint.

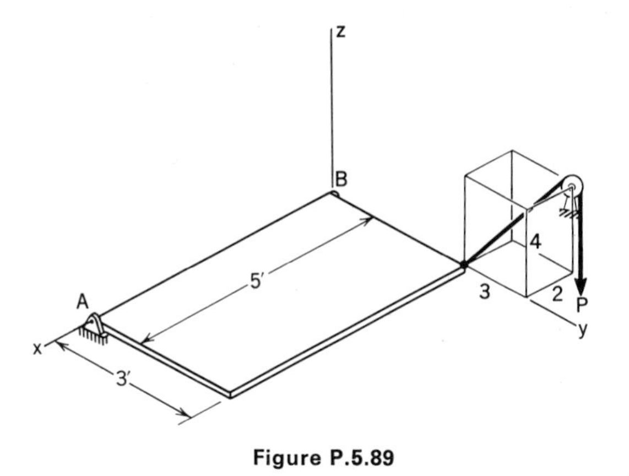


Figure P.5.89

**5.90.** A rod $AB$ is held by a ball-and-socket joint at $A$ and supports a 100-kg mass $C$ at $B$. This rod is in the $zy$ plane and is inclined to the $y$ axis by an angle of 15°. The rod is 16 m long and $F$ is at its midpoint. Find the forces in cables $DF$ and $EB$.

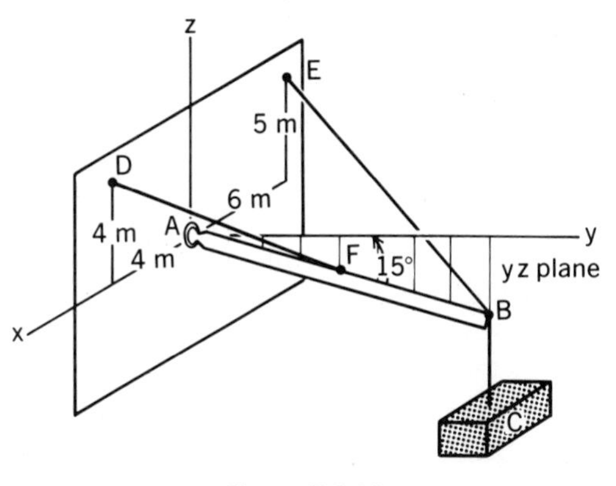


Figure P.5.90

**5.91.** A bent rod $ADGB$ supports two weights—one at the center of $AD$ and one at the center of $DG$. There are ball-and-socket joint supports at $A$ and $B$. With one scalar equation using the triple scalar product, determine the tension in cable $DC$.

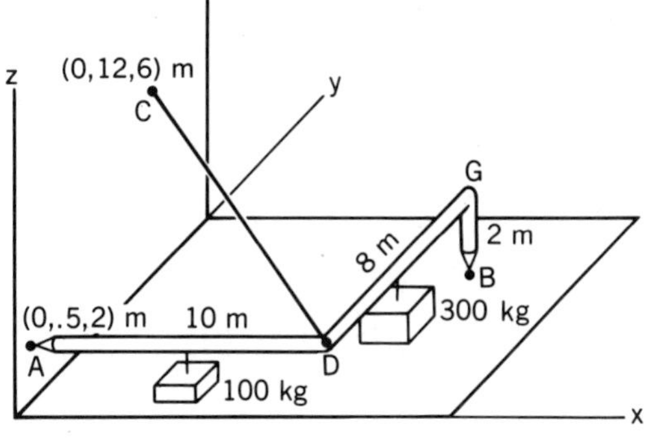

Figure P.5.91

**5.92.** Four cables support a block of weight 5000 N. The edges of the block are parallel to the coordinate axes. Point $B$ is at $(7, 7, -15)$. What are the forces in the cables and the direction cosines for cable

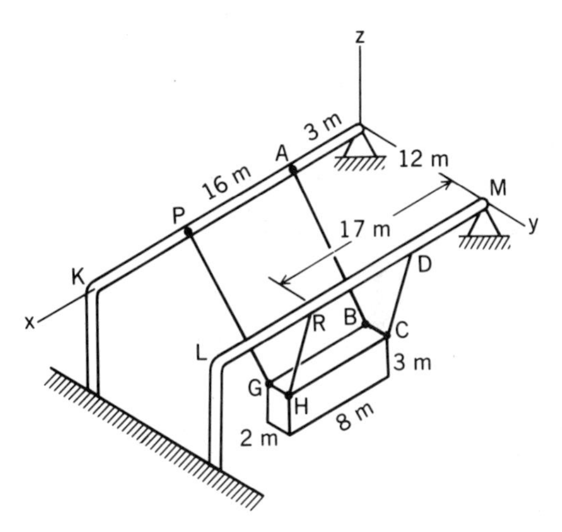


Figure P.5.92

**5.93.** A bar can rotate parallel to plane $A$ about an axis of rotation normal to the plane at $O$. A weight $W$ is held by a cord that is attached to the bar over a small pulley that can rotate freely as the bar rotates. Find the value of $C$ for equilibrium if $h = 300$ mm, $W = 30$ N, $\phi = 30°$, $l = 700$ mm, and $d = 500$ mm.

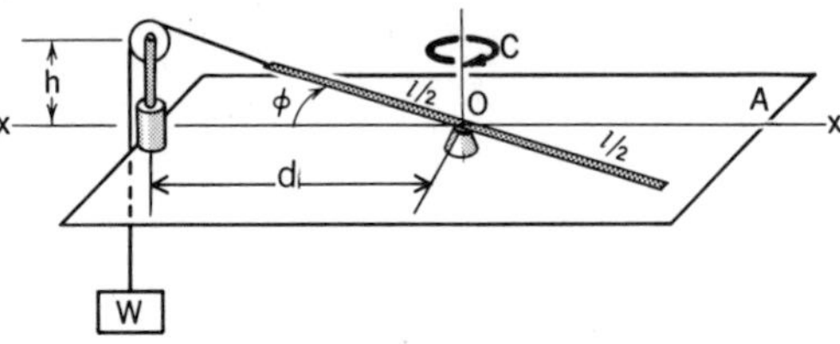


Figure P.5.93

**5.94.** A uniform bar of length $l$ and weight $W$ is connected to the ground by a ball-and-socket joint, and rests on a semicylinder from which it is not allowed to slip down by a wall at $B$. If we consider the wall and cylinder to be frictionless, determine the supporting forces at $A$ for the following data:

$$l = 1 \text{ m}$$
$$c = .30 \text{ m}$$
$$r = .20 \text{ m}$$
$$b = .40 \text{ m}$$
$$W = 100 \text{ N}$$

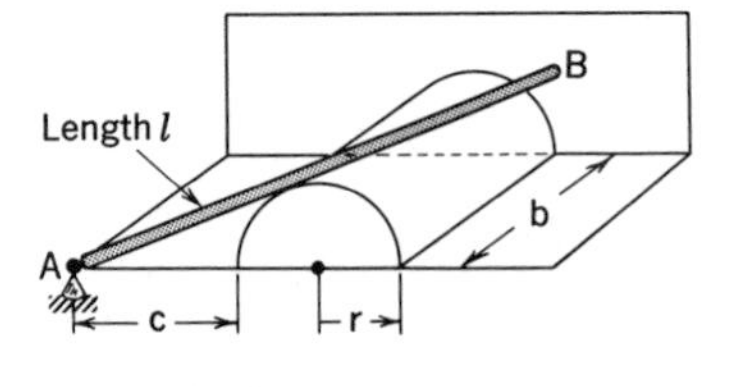


Figure P.5.94

## 5.7 Two-Force Members

We shall now consider a simple case of equilibrium that occurs quite often and from which simple, useful conclusions may readily be drawn.

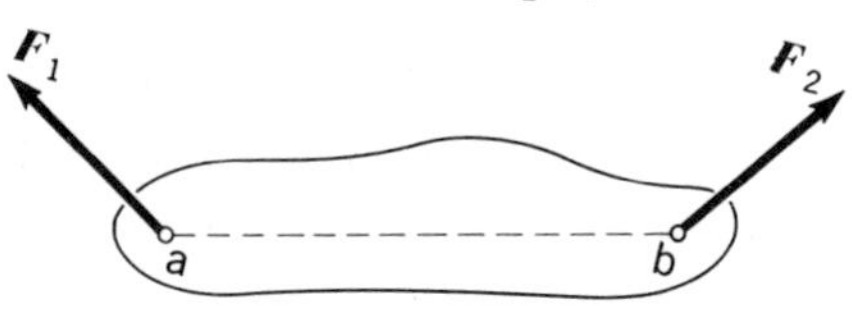


**Figure 5.39.** Two-force member.

Consider a rigid body on which *two forces* are, respectively, acting at points $a$ and $b$ (see Fig. 5.39). If the body is in equilibrium, the first basic equation of statics, 5.1(a), stipulates that $\boldsymbol{F}_1 = -\boldsymbol{F}_2$; that is, the forces must be *equal* and *opposite*. The second fundamental equation of statics, 5.1(b), requires that $\boldsymbol{M} = \boldsymbol{0}$, indicating that the forces be *collinear* so as not to form a nonzero couple. With points $a$ and $b$ given as points of application for the two forces in Fig. 5.39, clearly the *common line of action for the forces must coincide with the line segment ab*. Such bodies, where there are only *two points of loading*, are called "two-force" members.

We often have to deal with pin-connected structural members with loads applied at the pins. If we neglect friction at the pins and also the weight of the members, we can conclude that only two forces act on each member. These forces, then, must be equal and opposite and must have lines of action that are collinear, with the line joining the points of application of the forces. If the member is straight (see Fig. 5.40), the common line of action of the two forces coincides with the centerline of the member.[8] The top member in Fig. 5.40 is a *compression* member, the one below a *tensile* member.

Before considering an example, it should be pointed out that the forces $F_1$ and $F_2$ in Fig. 5.39 may be the resultants of systems of concurrent forces at $a$ and $b$ respec-

[8]Note that the bent member in Fig. 5.41, if weightless, is also a two-force member. The line of action of the forces must coincide with the line $ab$ connecting the points of application of the two forces.

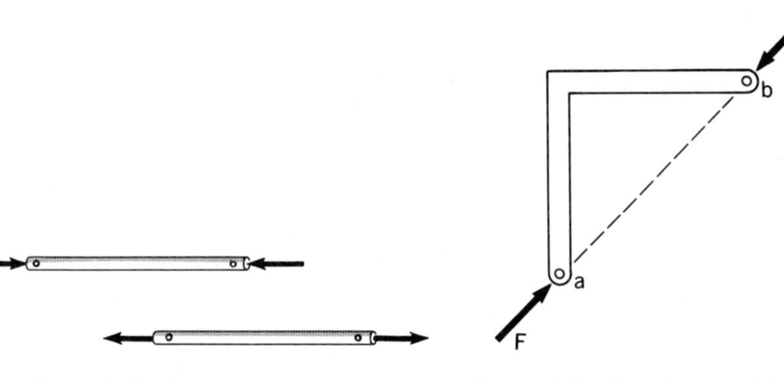


**Figure 5.40.** Compression and tension members.

**Figure 5.41.** Line of action of $F$ colinear with $ab$.

tively. Since concurrent forces are always equivalent to their resultant at the point of concurrency, the member in Fig. 5.39 is still a two-force member with the resulting restrictions on the resultants $F_1$ and $F_2$.

**EXAMPLE 5.10**

A device for crushing rocks is shown in Fig. 5.42. A piston $D$ having an 8-in. diameter is activated by a pressure $p$ of 50 psig (above that of the atmosphere). Rods $AB$, $BC$, and $BD$ can be considered weightless for this problem. What is the horizontal force transmitted at $A$ to the trapped rock shown in the diagram?

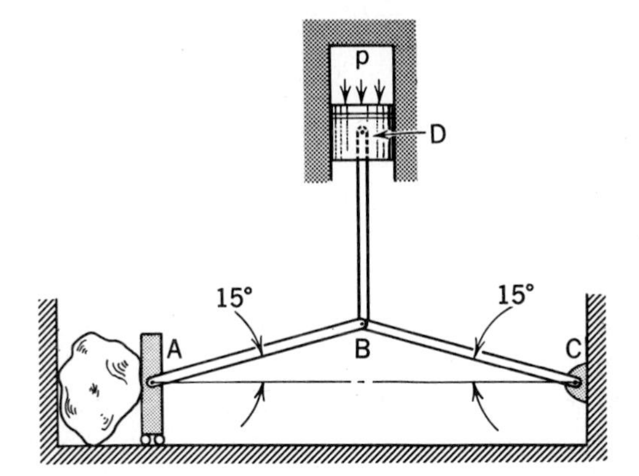


**Figure 5.42.** Rock crusher.

We have here three two-force members coming together at $B$. Accordingly, if we isolate pin $B$ as a free body, we will have three forces acting on the pin. These forces must be collinear with the centerlines of the respective members, as explained earlier (Fig. 5.43).

The force $F_D$ is easily computed by considering the action of the piston. Thus, we get

$$F_D = (50)\frac{\pi 8^2}{4} = 2510 \text{ lb}$$

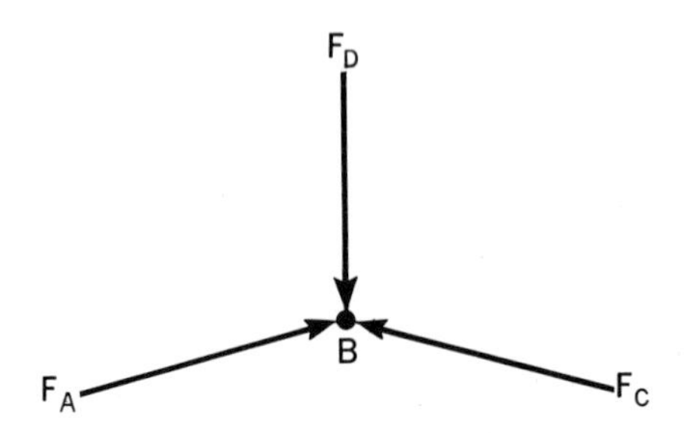


**Figure 5.43.** Free-body diagram of *B*.

Summing forces at pin *B*:

$\sum F_x = 0$:

$$F_A \cos 15° - F_C \cos 15° = 0$$

$$F_A = F_C$$

$\sum F_y = 0$:

$$2F_A \sin 15° - 2510 \text{ lb} = 0$$

$$F_A = \frac{2510}{(2)(0.259)} = 4850 \text{ lb}$$

The force transmitted to the rock in the horizontal direction is then 4850 cos 15° = 4690 lb.

Of less direct use is the *three-force* theorem. It states that a system of three forces in equilibrium must be *coplanar* and either be *concurrent* or be *parallel*.[9]

## 5.8 Static Indeterminacy

Examine the simple beam in Fig. 5.44, with known external loads and weight. If the deformation of the beam is small, and the final positions of the external loads after deformation differ only slightly from their initial positions, we can assume the beam to be rigid and, using the undeformed geometry, we can solve for the supporting forces $A$, $B_x$, and $B_y$. This is possible since we have three equations of equilibrium available. Suppose, now, that an additional support is made available to the beam, as indicated in Fig. 5.45. The beam can still be considered a rigid body, since the applied loads will shift even less because of deformation. Therefore, the equivalent force coming from the ground to counteract the applied loads and weight of the beam must be the *same* as before. In the first case, in which two supports were given, however, a unique set of values for the forces $A$, $B_x$, and $B_y$ gave us the required resistance. In other words,

[9]To prove this, assume that two of the forces intersect at a point *A*. Show from the basic equations of equilibrium that the forces must be coplanar and concurrent. Now assume the forces do *not* intersect. Setting moments of the system equal to zero about two points along the line of action of one of the forces, show that the system must be coplanar. Now since the forces do not intersect, they must be parallel. The theorem will thus have been proved.

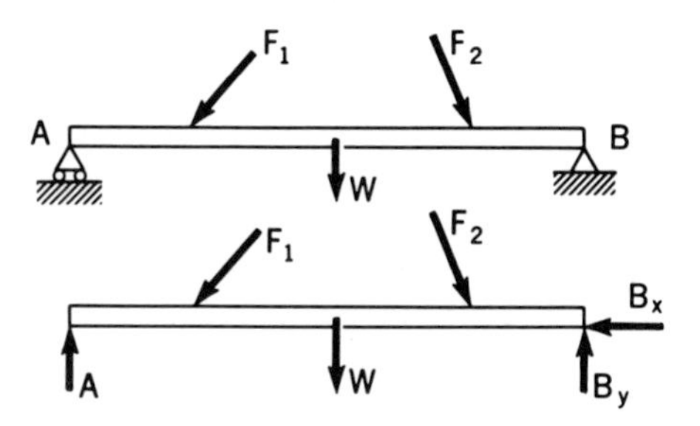


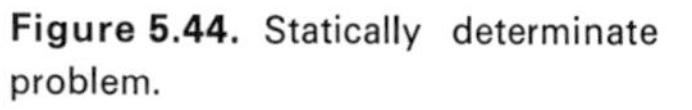
**Figure 5.44.** Statically determinate problem.

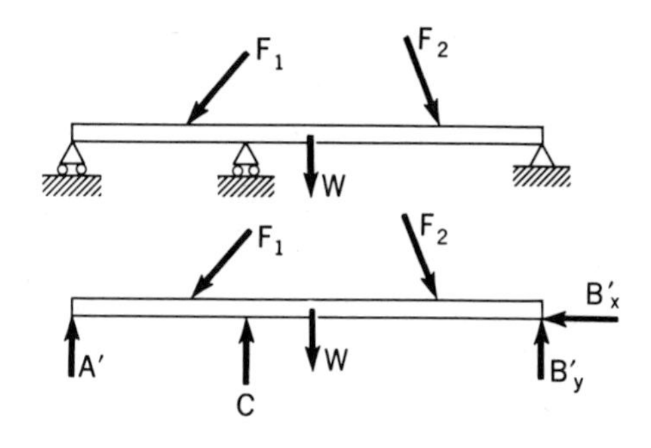


**Figure 5.45.** Statically indeterminate problem.

we were able to solve for these forces by statics alone, without further considerations. In the second case, rigid body statics will give the required *same* equivalent supporting force system, but now there are an infinite number of possible combinations of values of the supporting forces that will give us this equivalent system demanded by equilibrium of rigid bodies. To decide on the proper combination of supporting forces requires additional computation. Although the deformation properties of the beam were unimportant up to this point, they now become the all-important criterion in apportioning the supporting forces. These problems are termed *statically indeterminate*, in contrast to the statically determinate type, in which statics and the rigid-body assumption suffice. For a given system of loads and masses, two models—the rigid-body model and models taught in other courses involving elastic behavior—are accordingly both employed to achieve a desired end. In summary:

> *In statically indeterminate problems, we must satisfy both the equations of equilibrium for rigid bodies and the equations that stem from deformation considerations. In statically determinate problems, we need only satisfy the equations of equilibrium.*

In the discussion thus far, we used a beam as the rigid body and discussed the statical determinacy of the supporting system. Clearly, the same conclusions apply to any structure that, without the aid of the external constraints, can be taken as a rigid body. If, for such a structure as a free body, there are as many unknown supporting force and couple-moment components as there are equations of equilibrium, and if these equations can be solved for these unknowns, we say that the structure is *externally statically determinate.*

On the other hand, should we desire to know the forces transmitted *between* internal members of this kind of structure (i.e., one that does not depend on the external constraints for rigidity), we then examine free bodies of these members. When all the unknown force and couple-moment components can be found by the equations of equilibrium for these free bodies, we then say that the structure is *internally statically determinate.*

There are structures that depend on the external constraints for rigidity (see the

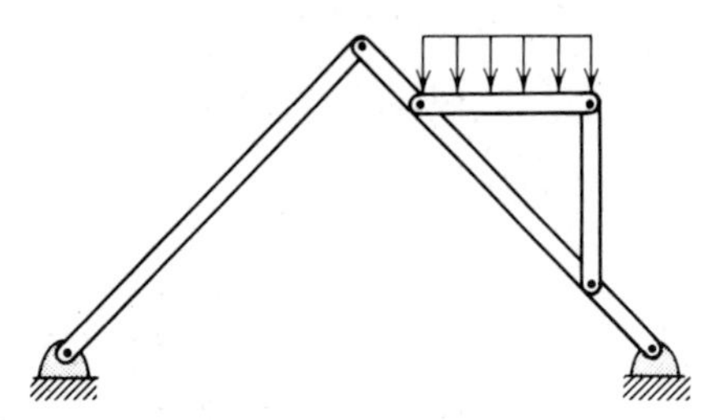
Figure 5.46. Nonrigid structure.

structure shown in Fig. 5.46). Mathematically speaking, we can say for such structures that the supporting force system always depends on both the internal forces and the external loads. (This is in contrast to the previous case, where the supporting forces could, for the externally statically determinate case, be related directly with the external loads without consideration of the internal forces.) In this case, we do not distinguish between internal and external statical determinacy, since the evaluation of supporting forces will involve free bodies of some or all of the internal members of the structure; hence, some or all of the internal forces and moments will be involved. For such cases, we simply state that the structure is statically determinate if, for all the unknown force and couple-moment components, we have enough equations of equilibrium that can be solved for these unknowns.

## Problems

**5.95.** Draw free-body diagrams for the hoe, arms, and tractor of the backhoe. Consider the weight of each part to act at a central location. The backhoe is not digging at the instant shown. Neglect the weights of the hydraulic systems *CE*, *AB*, and *FH*.

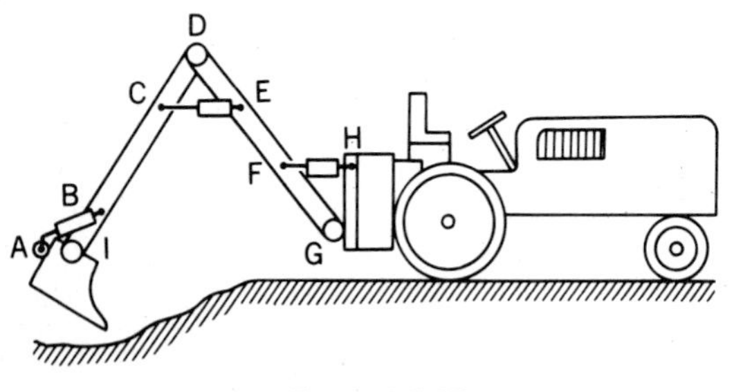

Figure P.5.95

**5.96.** A parking-lot gate arm weighs 150 N. Because of the taper, the weight can be regarded as concentrated at a point 1.25 m from the pivot point. What force must be exerted by the solenoid to lift the gate? What solenoid force is necessary if a 300-N counterweight is placed .25 m to the left of the pivot point?

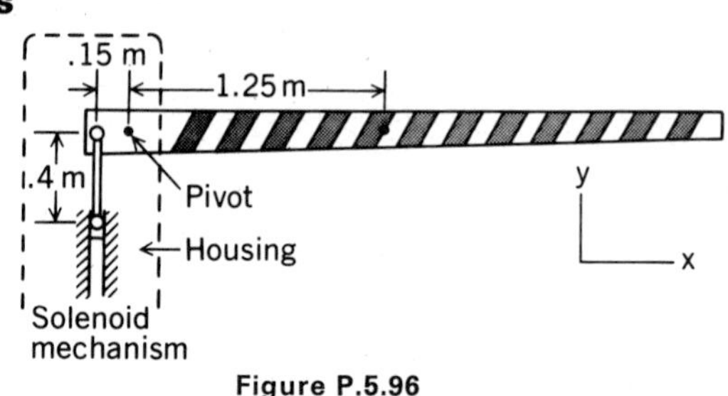

Figure P.5.96

**5.97.** Find the force delivered at *C* in a horizontal direction to crush the rock. Pressure $p_1 = 100$ psig and $p_2 = 60$ psig (pressures measured above atmospheric pressure). The diameters of the pistons are 6 in. each. Neglect the weight of the rods.

Figure P.5.97

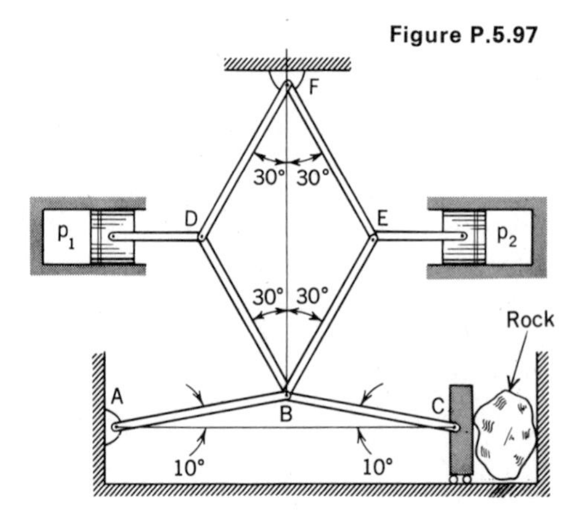

**5.98.** A Broyt X-20 digger carries a 20-kN load as shown. If hydraulic ram *CB* is normal to *BA*, where *A* is the axis of rotation for member *E*, find the force needed by ram *CB*. Do not consider the weights of the members.

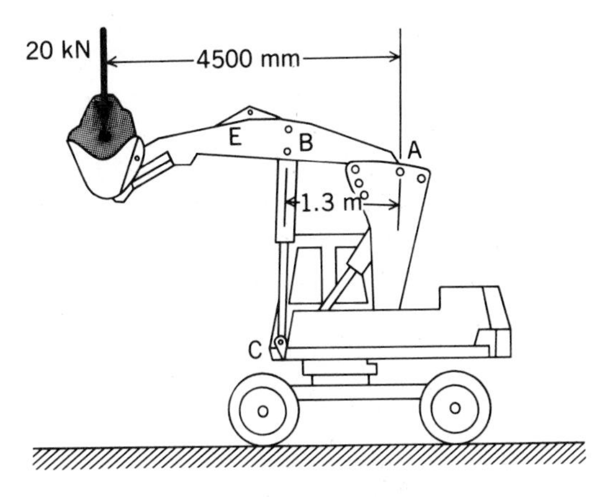

Figure P.5.98

**5.99.** What force $F_1$ will be developed by the 500-lb load? Neglect friction. The design is symmetrical.

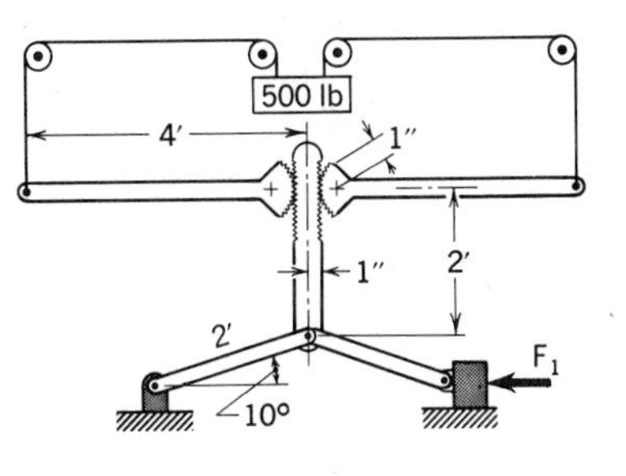


Figure P.5.99

**5.100.** Find the forces in the cables *DB* and *CB* as well as the compression member *AB*. The 500-N force is parallel to the *y* axis.

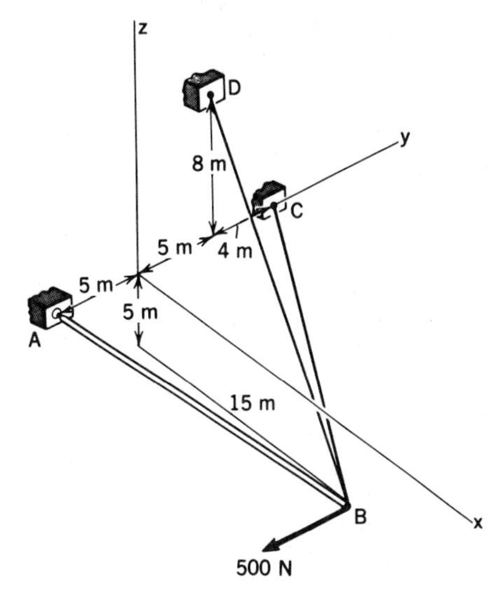


Figure P.5.100

**5.101.** Find the values of *F* and *C* so that members *AB* and *CD* fail simultaneously. The maximum load for *AB* is 15 kN and for *CD* is 22 kN. Neglect the weight of the members.

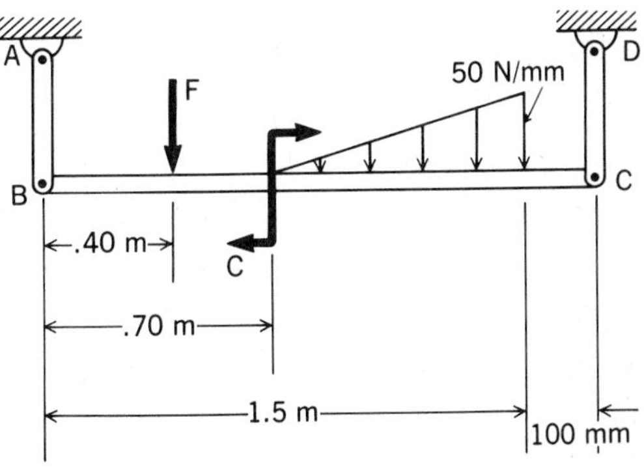


Figure P.5.101

**5.102.** The landing carriage of a transport plane supports a stationary vertical load of 50 kN per wheel. There are two wheels on each side of shock strut *AB*. Find the force in member *EC*, and the forces transmitted to the fuselage at *A*, if the brakes are locked and the engines are tested resulting in a thrust of 5 kN, 40% of which is resisted by this landing gear.

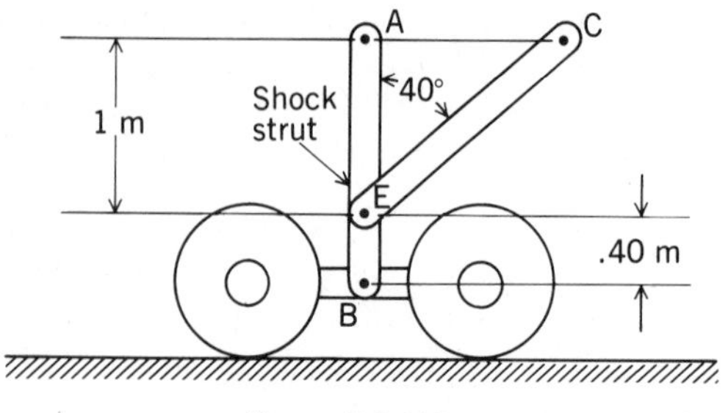


**Figure P.5.102**

**5.103.** Find the supporting forces at $A$ and $B$ in the frame. Neglect weights of members.

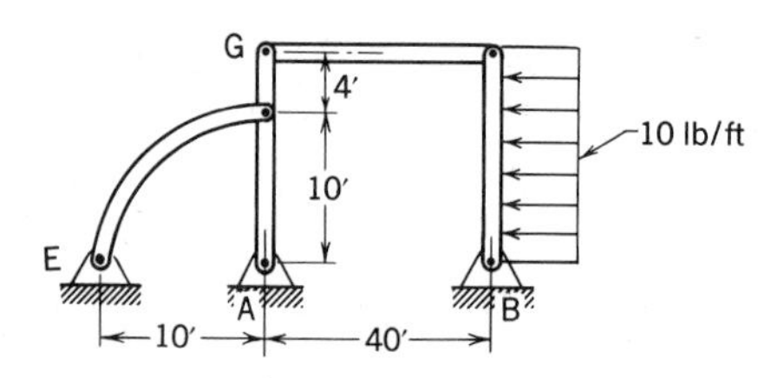


**Figure P.5.103**

**5.104.** A bolt cutter has a force of 130 N applied at each handle. What is the force on the bolt from the cutter edge?

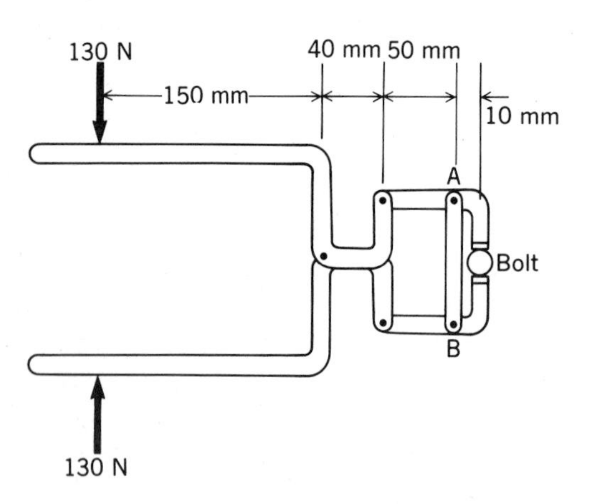


**Figure P.5.104**

**5.105.** A steam locomotive is developing a pressure of .20 N/mm² gage. If the train is stationary, what is the total traction force from the two wheels shown? Neglect the weight of the various connecting rods. Neglect friction in piston system and connecting rod pins.

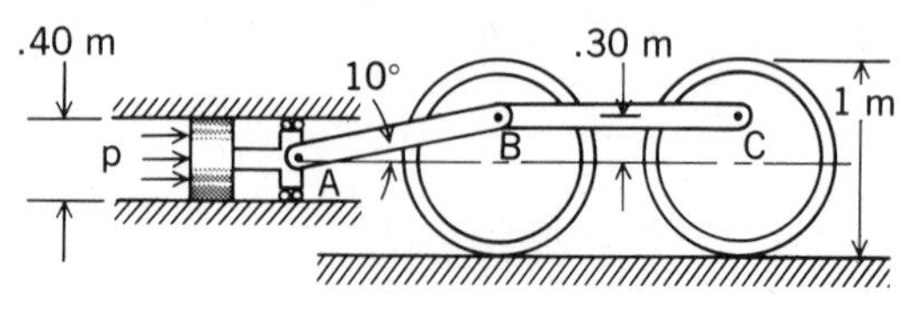

**Figure P.5.105**

**5.106.** Find the supporting forces at $A$, $B$, and $C$. Neglect the weight of the rod.

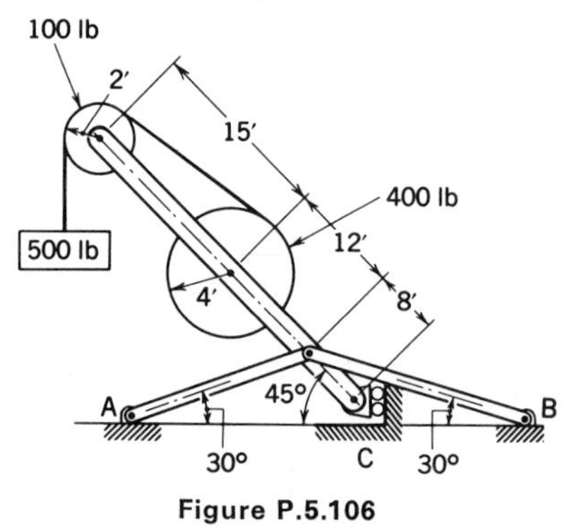


**Figure P.5.106**

**5.107.** The 5000-lb van of an airline food-catering truck rises straight up until its floor is level with the airplane floor. What forces exist at each joint of the scissors assembly? What force must be exerted by the hydraulic ram?

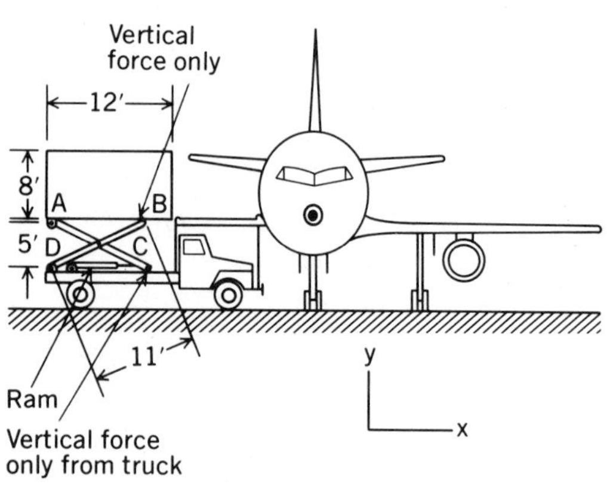


**Figure P.5.107**

**5.108.** The pavement exerts a force of 1000 lb on the tire. The tire, brakes, and so on, weigh 100 lb; the center of gravity is taken at the center plane of the tire. Determine the force from the spring and the compression force in $CD$.

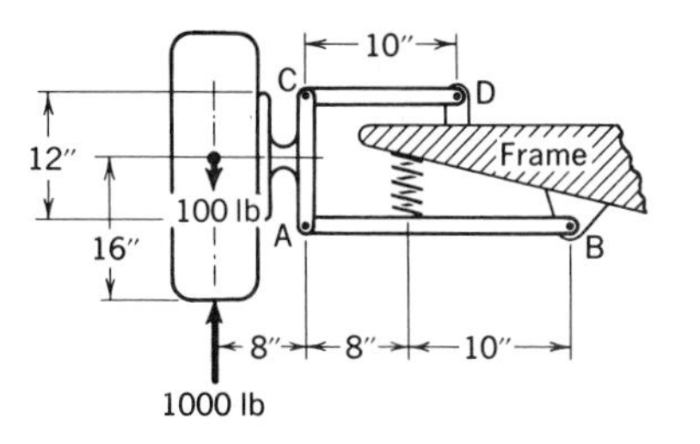


Figure P.5.108

**5.109.** A *flyball governor* is shown rotating at a constant speed $\omega$ of 500 rpm. The weights $C$ and $D$ are each of mass 500 g and are pin-connected to light rods. The centrifugal force on the weights, you will recall from physics, is given as $mr\omega^2$, where $r$ is the radial distance to the particle from the axis of rotation and $\omega$ is in rad/sec. Using this centrifugal force, and, imagining that we are rotating with the system, we can consider that we have equilibrium. (This is the D'Alembert principle that you learned in physics.) What is the tension in the rods and the downward force $F$ at $B$ needed to maintain the configuration shown for the given $\omega$?

Figure P.5.109

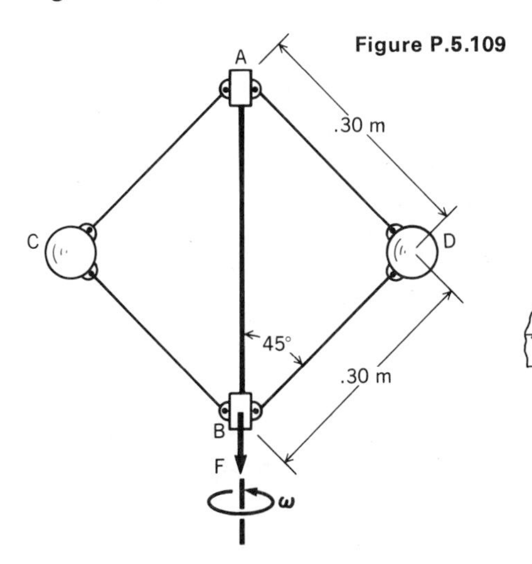


**5.110.** Another kind of flyball governor is shown. If $\omega_1 = 3$ rad/sec, compute the tension in $AG$ and $AE$. Neglect the weights of the members but assume that $HC$ and $HB$ are stiff. The weights $C$ and $B$ each have a mass of 200 g. What is the force $F$ needed to maintain the configuration? (*Hint:* Read the discussion of centrifugal force and D'Alembert's principle in Problem 5.109.)

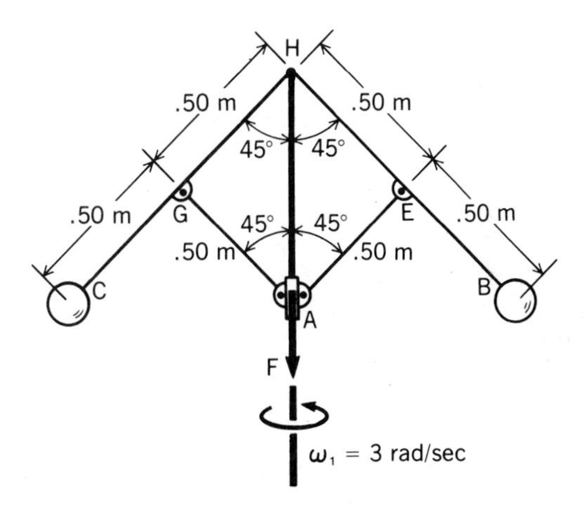


Figure P.5.110

**5.111.** Determine the supporting forces at $A$, $C$, $D$, $G$, $F$, and $H$ for the structure.

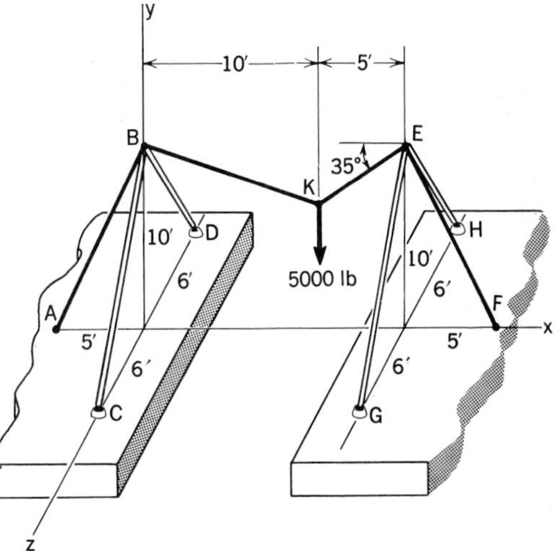


Figure P.5.111

**5.112.** Find the supporting forces at the ball-and-socket connections $A$, $D$, and $C$. Members $AB$ and $DB$ are pinned together through member $EC$ at $B$.

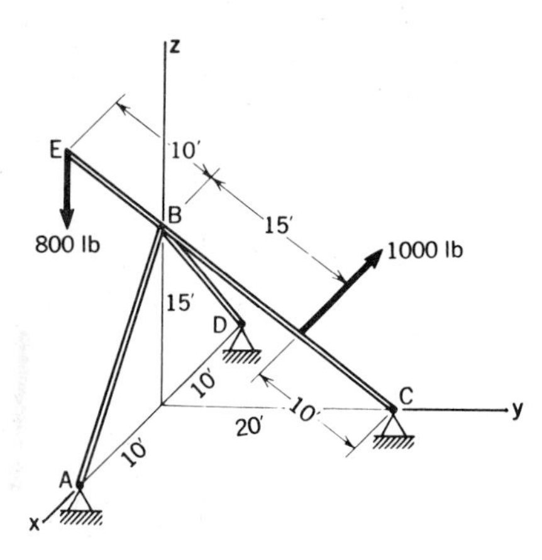


Figure P.5.112

**5.113.** A tractor with a bulldozer is used to push an earthmover picking up dirt. If the tractor force on the earthmover is 150 kN, what are the reactions of the bulldozer on the tractor at $B$ and $A$?

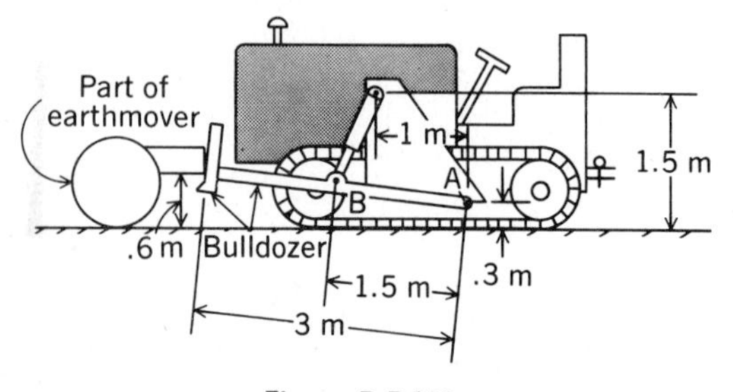


Figure P.5.113

**5.114.** A hydraulic-lift platform for loadingt ucks supports a weight $W$ of 5000 lb. Only one side of the system has been shown; the other side is identical. If the diameter of the pistons in the cylinders is 4 in., what pressure $p$ is needed to support $W$ when $\theta = 60°$. The following data apply.

$l = 24$ in., $\quad d = 60$ in., $\quad e = 10$ in.

Neglect friction everywhere. (*Hint:* Only two free-body diagrams need be drawn.)

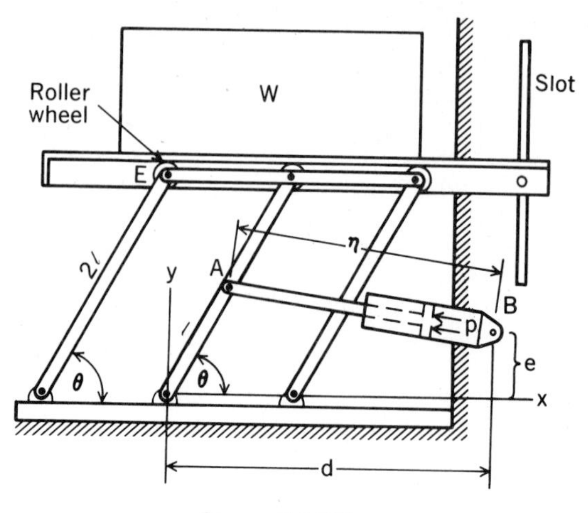


Figure P.5.114

**5.115.** A Bucyrus–Erie Dynahoe digger is partially shown. To develop the indicated forces in the bucket, what forces must hydraulic cylinders $HB$ and $CD$ develop? Consider only the 3-kN and 5-kN loads and not the weights of the members.

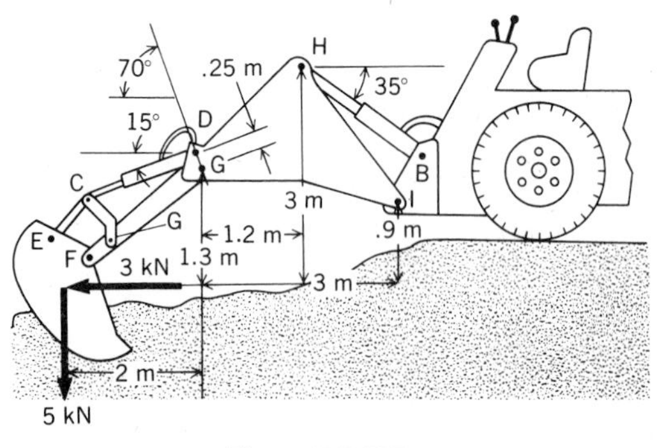


Figure P.5.115

**5.116.** A block of material weighing 200 lb is supported by members $KC$ and $HB$, whose weight we neglect, a ball-and-socket-joint support at $A$, and a smooth, frictionless support at $E$. Members $KC$ and $HB$ have directions collinear with diagonals of the block as shown. What are the supporting forces for this block?

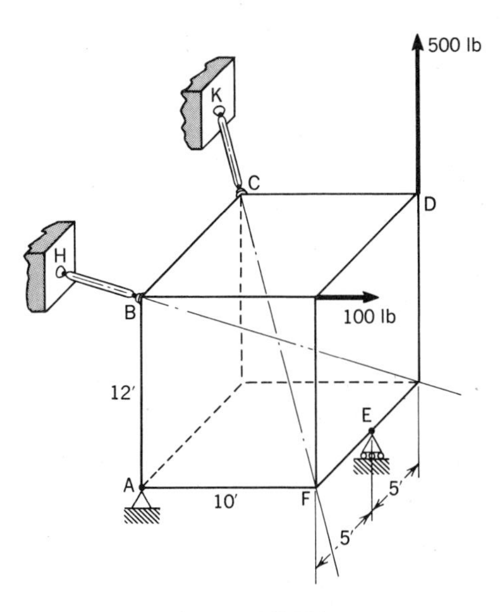


Figure P.5.116

**5.117.** A trap door is kept open by a rod *CD*, whose weight we shall neglect. The door has hinges at *A* and *B* and has a weight of 200 lb. A wind blowing against the outside surface of the door creates a pressure increase of 2 lb/ft². Find the force in the rod, assuming that it cannot slip from the position shown. Also determine the forces transmitted to the hinges. Only hinge *B* can resist motion along direction *AB*.

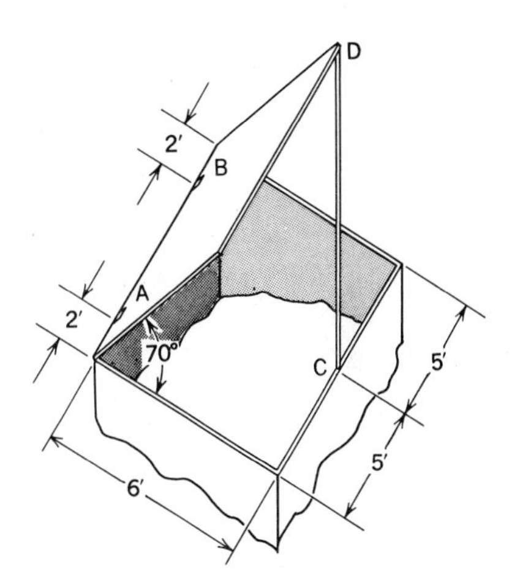


Figure P.5.117

## 5.9 Closure

We can now draw a free-body diagram that exposes a system of forces which, for equilibrium, must satisfy certain equations. By noting the type of simplest resultant for the system, we can readily deduce how many unknown quantities can be found for the free body. We thus have the direct means of solving statically determinate problems. And we also have available some of the conditions (the rigid-body equations of equilibrium) that must be satisfied in certain statically indeterminate problems. However, considerations beyond the scope of this text are necessary for the solution of these latter problems.[10]

In Chapter 6, we shall consider certain types of bodies that are of great engineering interest. The problems will be statically determinate and will involve nothing that is fundamentally new. We devote a separate chapter to these problems because they contain sign conventions and techniques that are important and complex enough to warrant such a study. We therefore proceed to an introduction of statically determinate structural mechanics problems.

---

[10]For such problems, see I. H. Shames, *Introduction to Solid Mechanics*, Prentice-Hall, Inc., Englewood Cliffs, N.J., 1975.

## Review Problems

**5.118.** Determine the tensions in all the cables. Block $A$ has a mass of 600 kg. Note that $GH$ is in the $yz$ plane.

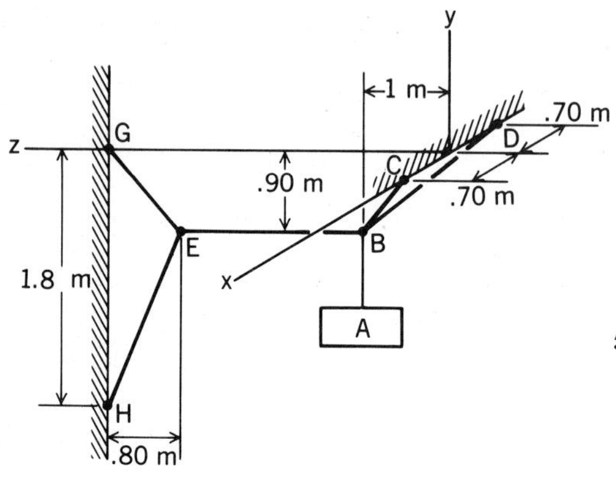

Figure P.5.118

**5.119.** Determine the force components at $G$. $E$ weighs 300 lb.

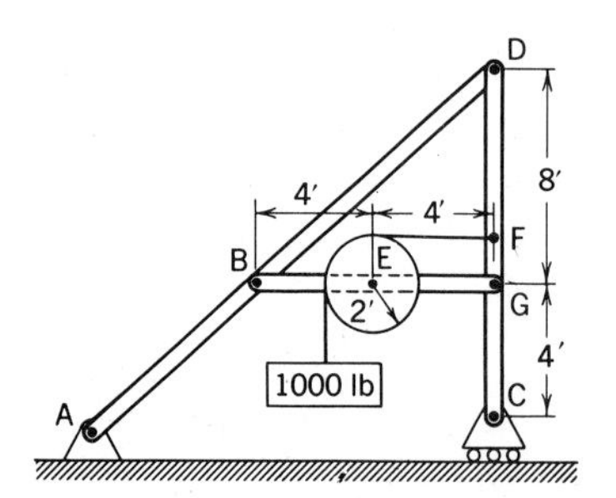

Figure P.5.119

**5.120.** A scenic excursion train with cog wheels for steep inclines weighs with load 30 tons. If the cog wheels have a mean radius to the contact points of the teeth of 2 ft, what torque must be applied to the driver wheels $A$ if wheels $B$ run free? What force do wheels $B$ transmit to the ground?

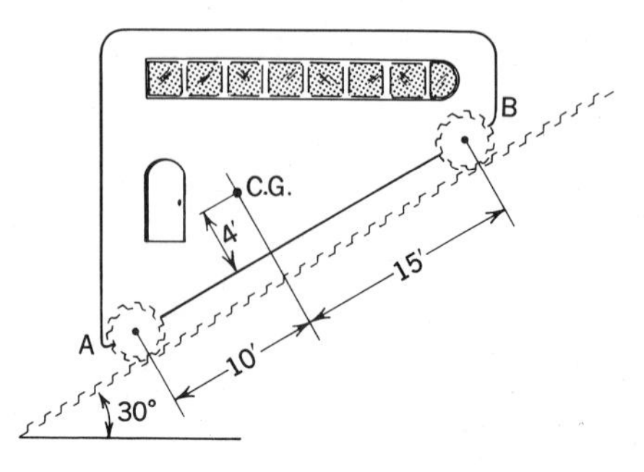

Figure P.5.120

**5.121.** Find the forces on the block of ice from the hooks at $A$ and $F$.

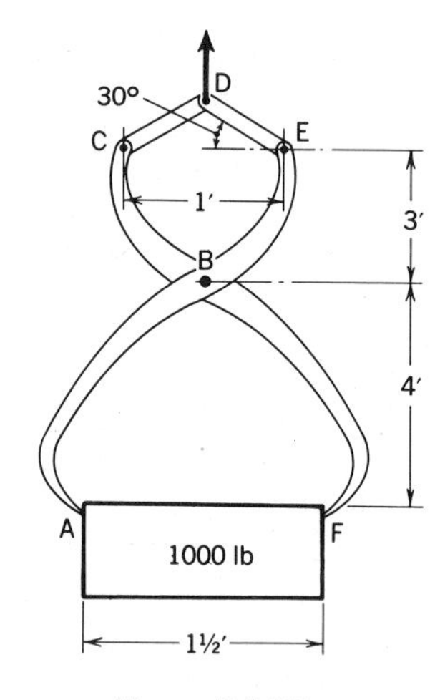


Figure P.5.121

**5.122.** Members $AB$ and $BC$ weighing, respectively, 50 N and 200 N are connected to each other by a pin. $BC$ connects to a disc $K$ on which a torque $T_K = 200$ N-m is applied. What torque $T$ is needed on $AB$ to keep the system in equilibrium at the configuration shown?

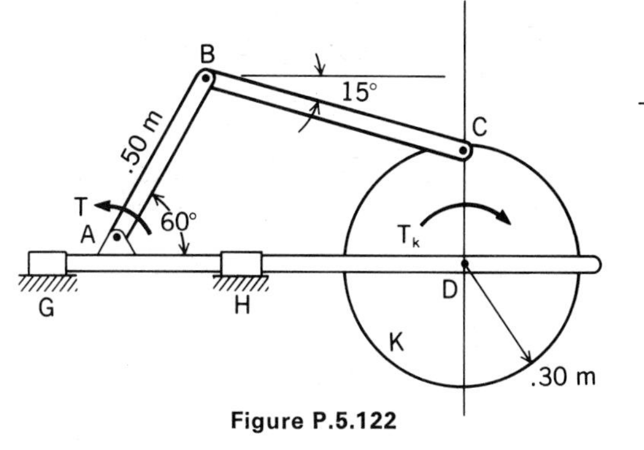


Figure P.5.122

**5.123.** A transport jet plane has a weight without fuel of 220 kN. If one wing is loaded with 50 kN of fuel, what are the forces in each of the three landing gear?

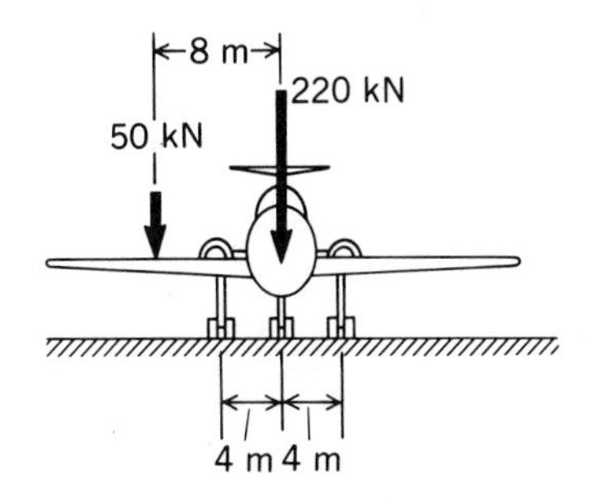


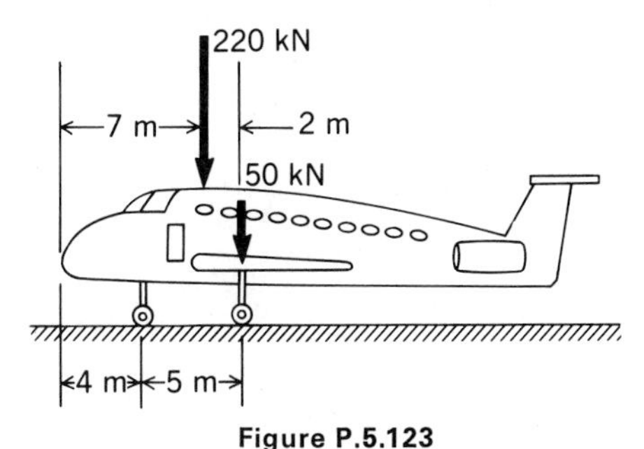


Figure P.5.123

**5.124.** A rod $AB$ is connected by a ball-and-socket joint to a frictionless sleeve at $A$, and by a ball-and-socket joint to a fixed position at $B$. What are the supporting forces at $B$ and at $A$ if we neglect the weight of $AB$? The 100-lb load is connected to the center of $AB$.

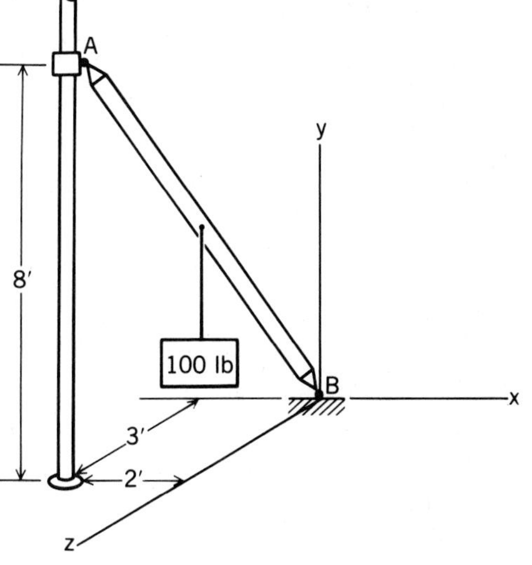


Figure P.5.124

**5.125.** A beam weighing 400 lb is held by a ball-and-socket joint at $A$ and by two cables $CD$ and $EF$. Find the tension in the cables. They are attached at opposite ends of the beam as shown.

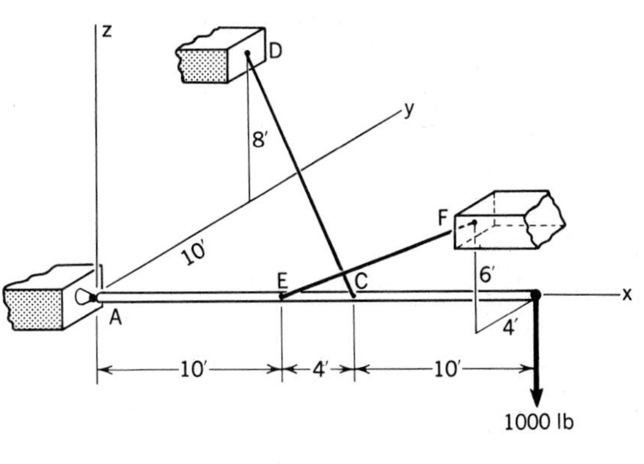

Figure P.5.125

**5.126.** What should the values of $R$ and $C$ be if the supporting rods $AB$ and $CD$ are to fail simultaneously? Rod $AB$ can withstand a 5000-lb force, and rod $DC$ can withstand an 8000-lb force. Neglect the weights of the members.

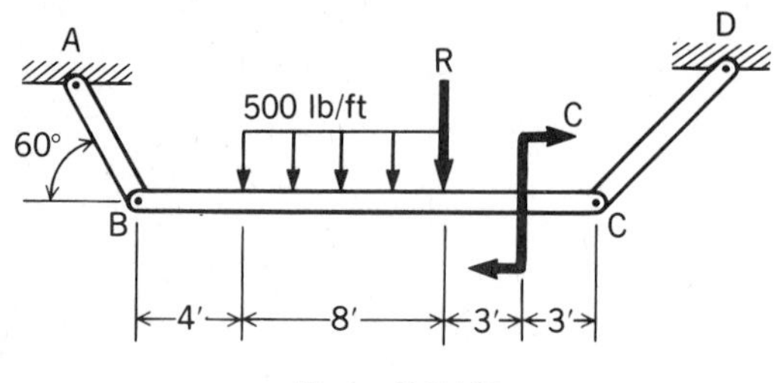


Figure P.5.126

**5.127.** A mechanism consists of two weights $W$ each of weight 50 N, four light linkage rods each of length $a$ equal to 200 mm, and a spring $K$ whose spring constant is 8 N/mm. The spring is unextended when $\theta = 45°$. If held vertically, what is the angle $\theta$ for equilibrium? Neglect friction. The force from the spring equals $K$ times the compression of the spring.

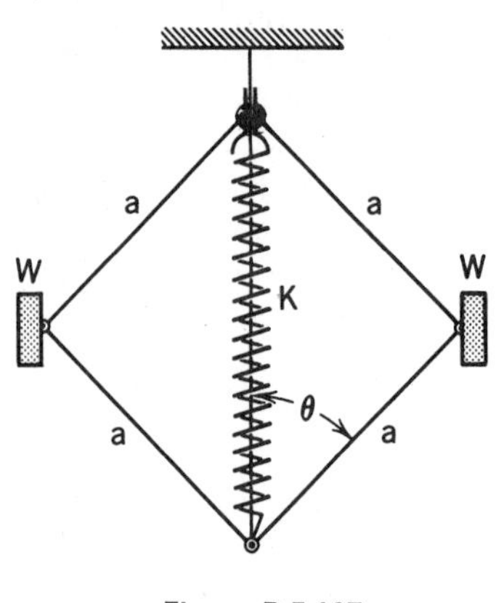

Figure P.5.127

**5.128.** Find the compressive force in pawl $AB$. What is the resultant supporting force system at $E$?

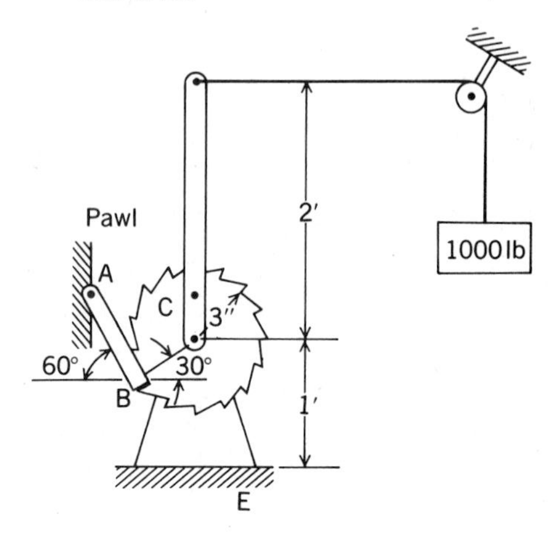


Figure P.5.128

**5.129.** A 10-kN load is lifted in the front loader bucket. What are the forces at the connections to the bucket and to arm $AE$? Hydraulic ram $DF$ is perpendicular to arm $AE$, and $BC$ is horizontal. Points $A$ and $F$ are at the same height above the ground.

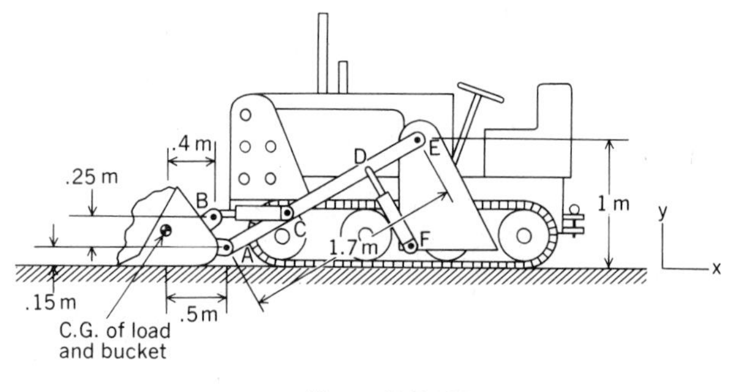


Figure P.5.129

# 6 Introduction to Structural Mechanics

## Part A Trusses

### 6.1 The Structural Model

A *truss* is a system of members that are fastened together to support stationary and moving loads.[1] Everyday examples of trusses are shown in Figs. 6.1 and 6.2. Each member of a truss is usually of uniform cross section along its length; however, the various members typically have different cross-sectional areas because they must transmit different forces. Our purpose in Part A of this chapter is to set forth methods for determining forces in members of an elementary class of trusses.

As a first step, we shall divide trusses into two main categories according to geometry. A truss consisting of a coplanar system of members is called a *plane truss*. Examples of plane trusses are the sides of a bridge (see Fig. 6.1) and a roof truss (see Fig. 6.2). A three-dimensional system of members, on the other hand, is called a *space truss*. A common example of a space truss is the tower from an electric power transmission system (see Fig. 6.3). Both plane trusses and space trusses consist of members having cross sections resembling the letters H, I, and L. Such members are commonly

[1] A truss is different than a frame (see the footnote on p. 140) in that the members of a truss are always connected together at the ends of the members, as will soon become evident.

**Figure 6.1.** Foot bridge near author's home. Sides of structure are plane trusses.

**Figure 6.2.** Roof trusses that are simple plane trusses.

**Figure 6.3.** Space trusses supporting transmission lines sending power into the northeast grid of the U.S.

used in many structural applications. These members are fastened together to form a truss by being welded, riveted, or bolted to intermediate structural elements called *gusset plates* such as has been shown in Fig. 6.4(a) for the case of a plane truss. The analysis of forces and moments in such connections is clearly quite complicated. Fortunately, there is a way of simplifying these connections such as to incur very little loss in accuracy in determining forces in the members. Specifically, if the centerlines of the members are *concurrent* at the connections, such as is shown in Fig. 6.4(a) for the coplanar case, then we can replace the complex connection at the points of concurrency by a simple pin connection in the coplanar truss and a simple bali-and-socket connection for the space truss. Such a replacement is called an *idealization* of the system. This is illustrated for a plane truss in Fig. 6.4, where the actual connector or joint is shown in (a) and the idealization as a pinned joint is shown in (b).

In order to maximize the load-carrying capacity of a truss, the external loads must be applied at the joints. The prime reason for this rule is the fact that the members

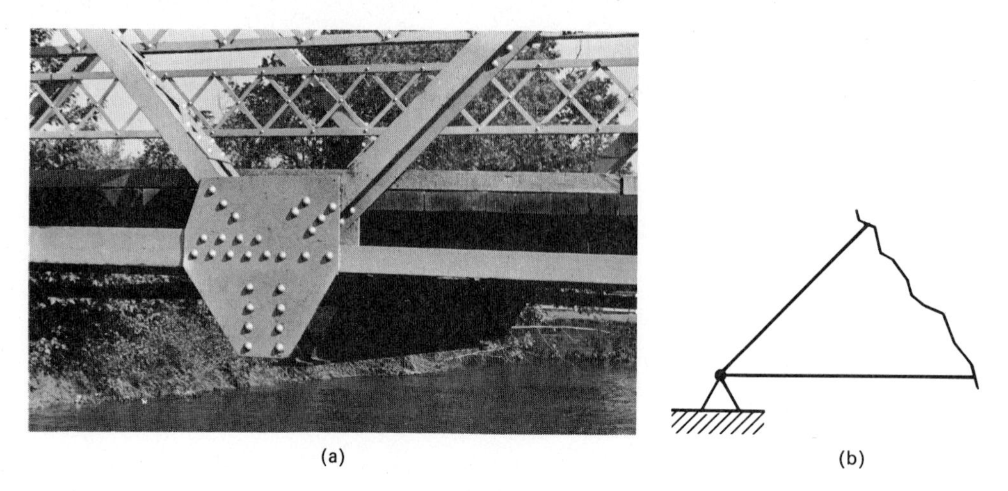

(a) (b)

**Figure 6.4.** (a) Gusset plate; (b) idealization.

of a truss are long and slender, thus rendering them less able to carry loads transverse to their centerlines away from the ends.[2] If the weights of the members are neglected, as is sometimes the case, it should be apparent that each member is a *two-force member*, and accordingly is either a tensile member or a compression member. If the weight is not negligible, the common practice as an approximation is to apply half the weight of a member to each of its two joints. Thus, the idealization of a member as a two-force member is still valid.

## 6.2 The Simple Truss

An idealized truss as described in Section 6.1 is termed *just-rigid* if the removal of any of its members destroys its rigidity. If removing a member does not destroy rigidity, the structure is said to be *over-rigid.* We shall be concerned with just-rigid trusses in Part A of this chapter.[3]

The most elementary just-rigid truss is one with three members connected to form a triangle. Just-rigid space trusses may be built up from this triangle by adding for each new joint three new members, as is shown in Fig. 6.5.[4] Trusses constructed in this manner are called *simple space trusses.* The *simple plane truss* is built up from an elementary triangle by adding two new members for each new pin as shown in Fig. 6.6. Clearly, the simple plane truss is just-rigid.

[2] You will understand these limitations more clearly when you study buckling in your strength of materials course.

[3] Over-rigid structures are studied in courses of strength of materials and structural mechanics. They are internally statically indeterminate and deformation must be taken into account when computing forces in the members.

[4] To ensure that the space truss is just-rigid, no set of three new members can be coplanar. Why?

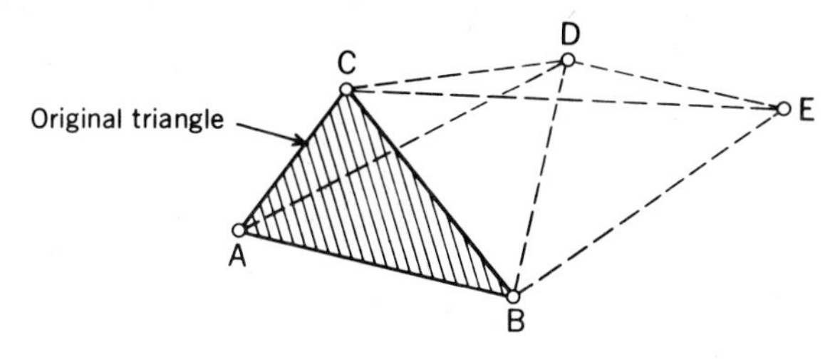


**Figure 6.5.** Simple space truss.

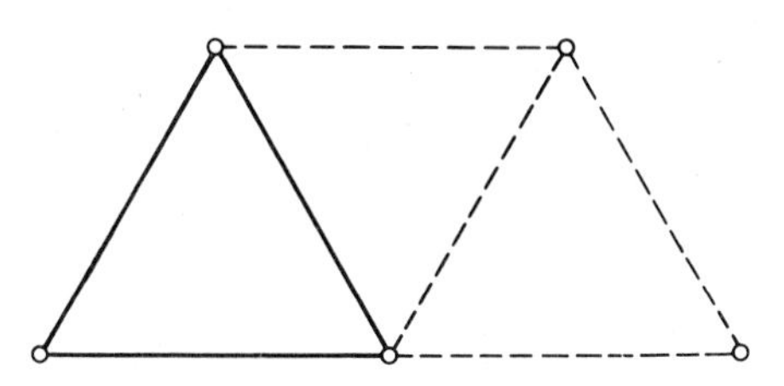

**Figure 6.6.** Simple plane truss.

A simple relationship exists between the number of joints $j$ and the number of members $m$ in a simple truss. You can directly verify by examining the simple space truss in Fig. 6.5 that $m$ is related to $j$ as follows:

$$m = 3j - 6 \tag{6.1}$$

Similarly, for the simple plane truss in Fig. 6.6 you can verify that

$$m = 2j - 3 \tag{6.2}$$

You will learn in more advanced structures courses that Eqs. 6.1 and 6.2 hold generally for just-rigid space trusses and for just-rigid plane trusses, respectively.

We now show that if the supporting force system is statically determinate, we can compute the forces in all the members of simple trusses. Specifically, in examining the ball joints of simple space trusses, we can see that in the general three-dimensional case, a ball joint with only three unknown forces acting on it from the members can always be found. (One such joint is the last joint formed.) Each unknown force from a member onto this joint must have a direction collinear with that member, and hence has a known direction. There are, then, only three unknown scalars, and since we have a concurrent force system they can be determined by statics alone. We then find another joint with only three unknowns and so carry on the computations until the forces in the entire structure have been evaluated. For the simple plane truss, a similar procedure can be followed. The free body of at least one joint has only two unknown forces. We have a concurrent, coplanar force system, and we accordingly can solve the corresponding two equilibrium equations in two unknowns at that joint. We then proceed to the other joints, thereby evaluating all member forces by the use of statics alone.

## 6.3 Solution of Simple Trusses

Generally, the first step in a truss analysis is to compute the supporting forces in the overall truss. This calculation of the external forces or reactions that must exist to keep the truss in equilibrium is independent of whether the truss, internally, is statically determinate or statically indeterminate. Simply regard the truss as a rigid body to which forces are applied, some known (given applied forces) and some unknown (reactions),[5] and solve for the reactions as we did in Chapter 5. We have shown a simple plane truss in Fig. 6.7(a) and have shown the features of the truss in Fig. 6.7(b) that are essential for the calculations of the reactions. Note that members *CB*, *DB*, and *DE* need not be shown in the free body since they provide *internal* forces for the body.

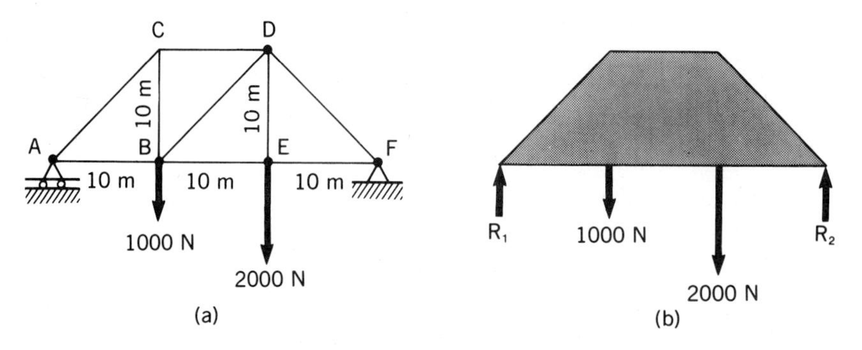


**Figure 6.7.** Free-body analysis of truss.

Once the free-body diagram has been carefully drawn, use three equations of equilibrium to determine the reactions of a plane truss (six equations for a space truss). It is highly advisable to then check your results by using another (dependent) equation of equilibrium. You will be using the computed reactions for many subsequent calculations involving forces in internal members. Accordingly, with much work at stake, it is important to start off with a correct set of reactions.

We shall present two methods for determining the forces in the members of the truss. One is called the *method of joints* and the other is called the *method of sections*. As will be seen in the following sections, the prime difference between these methods lies in the choice of the free bodies to be used.

## 6.4 Method of Joints

In the method of joints, the free-body diagrams to be used, once the reactions are determined, are the pins or ball joints and the forces applied to them by the attached members and external loads. Note that we have already alluded to this method in

[5]Supporting forces are often called *reactions* in structural mechanics.

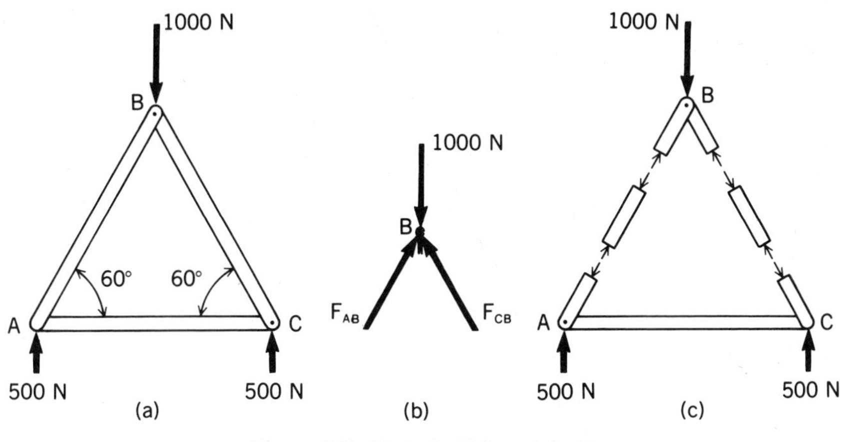


**Figure 6.8.** Method of joints—joint *B*.

Section 6.2. Consider first the triangular plane truss shown in Fig. 6.8(a). Notice we have already determined the reactions.

Next, consider the free body of pin *B* [Fig. 6.8(b)]. The unknown forces from the members are shown collinear with these members since they are two-force members. We can solve for these forces by setting the sum of forces equal to zero in the horizontal and vertical directions, to get

$$F_{BA} = F_{CB} = 577 \text{ N}$$

Because both forces are *pushing* against pin *B*, the corresponding members are *compressive* rather than tension members. We can most readily see this fact by considering Fig. 6.8(c), where members *AB* and *CB* have been cut at various places. Notice that *AB* is also pushing against pin *A* as does *BC* against pin *C*. Thus, once having decided that the members are compressive members as a result of considerations at a pin at one end of the member, we can conclude that the member is pushing with equal force against the pin at the other end. To make for speed and accuracy as we go from one joint to another, we recommend that, once the nature of the loading in a member has been established by considerations at a pin, we mark down this value using a T for tension or a C for compression after it on the truss diagram, as shown in Fig. 6.9(a). Note also that appropriate arrows are drawn in the members. These arrows represent forces developed by the members on the pins. Hence, for *compression* the arrows point *toward* the pins, and for *tension* they point *away* from the pins. Accordingly, if we now consider the free body of pin *A* as shown in Fig. 6.9(b), we know the direction and value of the force on *A* from member *BA*.

If a negative value is found for a force at a pin, the sense of the force has been taken incorrectly at the outset. With this in mind, we decide whether the member associated with the force is a tension or compression member. And we label the member accordingly, as shown in Fig. 6.9(a) for use later in examining the pin at the other end of the member as a free body.

We now consider the solution of a plane truss problem by the method of joints in greater detail.

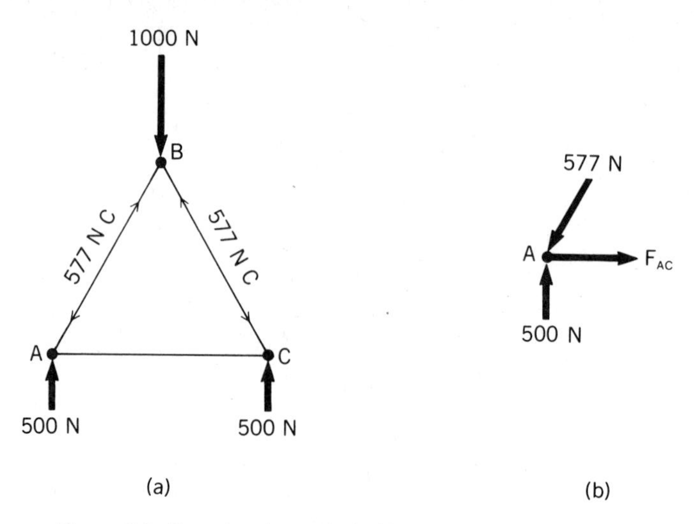


**Figure 6.9.** Procedure for method of joints. (a) Notation for members *AB* and *BC*; (b) free body diagram of *A*.

**EXAMPLE 6.1**

A simple plane truss is shown in Fig. 6.10. Two 1000-lb loads are shown acting on pins *C* and *E*. We are to determine the force transmitted by each member. Neglect the weight of the members.

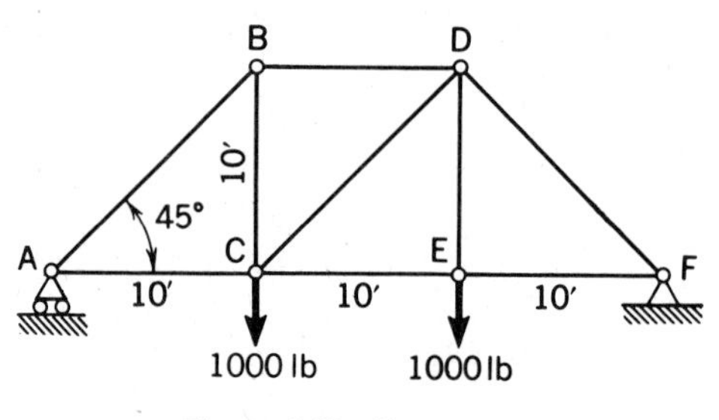


**Figure 6.10.** Plane truss.

In this simple loading, we see by inspection that there are 1000-lb vertical forces at each support. We shall begin, then, by studying pin *A*, for which there are only two unknowns.

**Pin *A*.** The forces on pin *A* are the known 1000-lb supporting force and two unknown forces from the members *AB* and *AC*. The orientation of these forces is known from the geometry of the truss, but the magnitude and sense must be determined. To help in interpreting the results, put the forces in the same position as the corresponding bars in the space diagram (Fig. 6.11). That is, avoid the force diagram in Fig. 6.12, which is equivalent to the one in Fig. 6.11 but which may lead to errors in interpretation. There are two unknowns for the concurrent coplanar force sys-

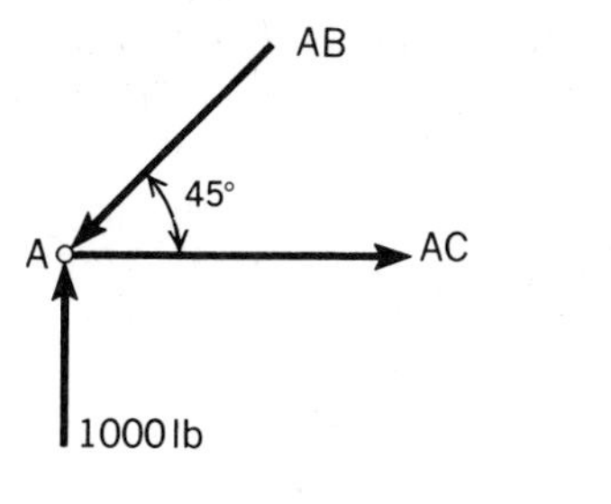


Figure 6.11. Pin $A$.

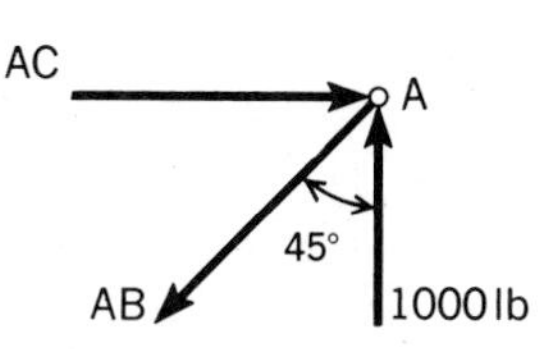


Figure 6.12. Pin $A$—avoid this diagram.

tem in Fig. 6.11 and thus, if we use the scalar equations of equilibrium, we may evaluate $AB$ and $AC$:

$\underline{\sum F_x = 0:}$

$$AC - 0.707AB = 0$$

$\underline{\sum F_y = 0:}$

$$-0.707AB + 1000 = 0$$

Therefore,

$$AB = 1414 \text{ lb}; \quad AC = 1000 \text{ lb}$$

Since both results are positive, we have chosen the proper senses for the forces. We can then conclude on examining Fig. 6.11 that $AB$ is a compression member, whereas $AC$ is a tension member.[6] In Fig. 6.13, we have labeled the members accordingly.

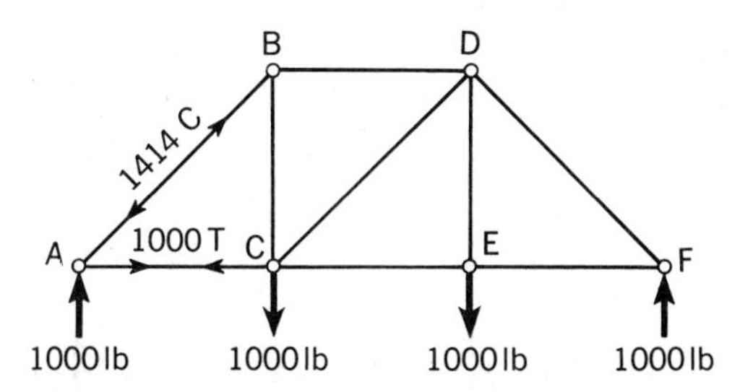


Figure 6.13. Notation for members $AB$ and $AC$.

If we next examine pin $C$, clearly since there are three unknowns involved for this pin, we cannot solve the forces by equilibrium equations at this time. However, pin $B$ can be handled, and once $BC$ is known, the forces on pin $C$ can be solved.

**Pin *B*.** Since $AB$ is a compression member (see Fig. 6.13) we know that it exerts a force of 1414 lb directed against pin $B$ as has been shown in Fig. 6.14. As for members $BC$ and $BD$, we assign senses as shown.

[6]Had we used Fig. 6.12 as a free body, the state of loading in the members (i.e., tension or compression) would not be clear. Therefore, we strongly recommend putting forces representing members in positions coinciding with the members.

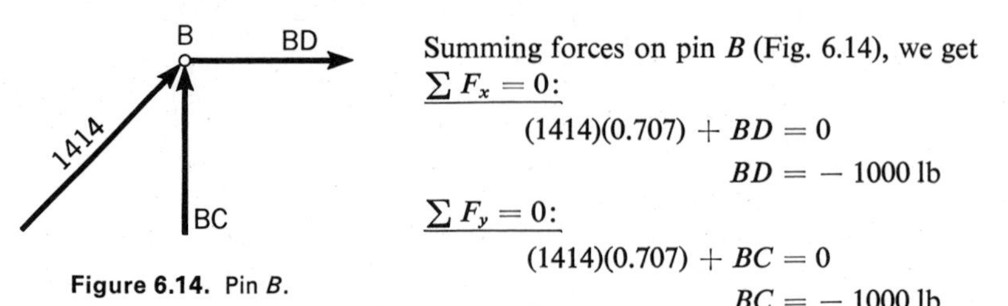


Figure 6.14. Pin *B*.

Summing forces on pin *B* (Fig. 6.14), we get

$\underline{\sum F_x = 0:}$

$$(1414)(0.707) + BD = 0$$
$$BD = -1000 \text{ lb}$$

$\underline{\sum F_y = 0:}$

$$(1414)(0.707) + BC = 0$$
$$BC = -1000 \text{ lb}$$

Here we have obtained two negative quantities, indicating that we have made incorrect choices of sense. Keeping this in mind, we can conclude that member *BD* is a compression member, whereas member *BC* is a tension member.

We can proceed in this manner from joint to joint. At the last joint all the forces will have been computed without using it as a free body. Thus, it is available to be

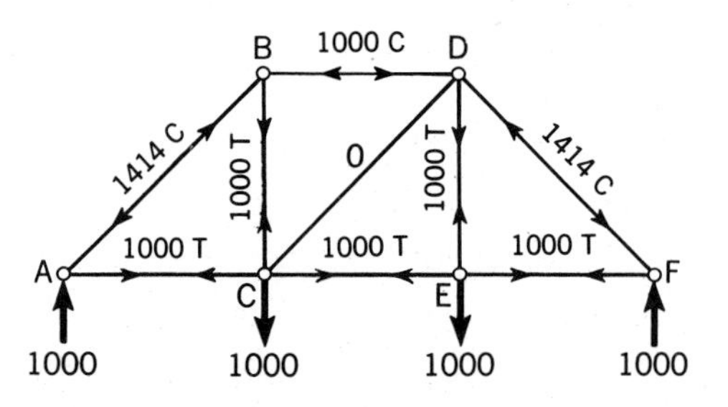


Figure 6.15. Solution for truss.

used as a check on the solution. That is, the sum of the known forces for the last joint in the $x$ and $y$ directions should be zero or close to zero, depending on the accuracy of your calculations. We urge you to take advantage of this check. The final solution is shown in Fig. 6.15. Notice that member *CD* has zero load. This does not mean that we can get rid of this member. Other loadings expected for the truss will result in nonzero force for *CD*. Furthermore, without *CD* the truss will not be rigid.

### *EXAMPLE 6.2

Ascertain the forces transmitted by each member of the three-dimensional truss [Fig. 6.16(a)].

We can readily find the supporting forces for this simple structure by considering the whole structure as a free body and by making use of the symmetry of the loading and geometry. The results are shown in Fig. 6.16(b).

**Joint *F*.** It is clear, on an inspection of the forces in the $x$ direction acting on joint *F*, that the force *FE* must be zero, since all other forces are in a plane at right angles to it. These other forces are shown in Fig. 6.17. Summing forces in the $y$ and $z$ directions, we get

$\underline{\sum F_y = 0:}$

$$-FD\frac{20}{\sqrt{20^2 + 10^2}} + 2000 = 0$$

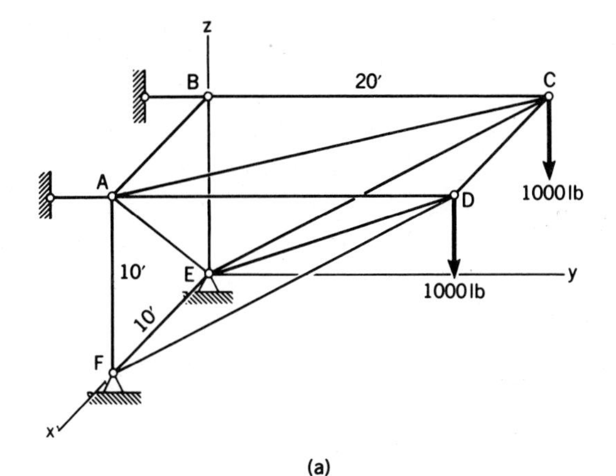


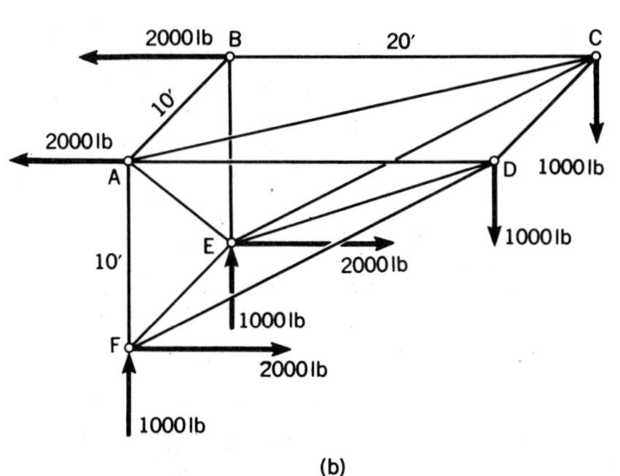


**Figure 6.16.** (a) Space truss and (b) free-body diagram.

Therefore,

$$FD = 2240 \text{ lb of compression}$$

$\underline{\sum F_z = 0:}$

$$-AF + 1000 - 2240 \frac{10}{\sqrt{500}} = 0$$

Therefore,

$$AF = 1000 - 1000 = 0$$

**Figure 6.17.** Pin *F*.

**Joint *B*.** Going to joint *B*, we see [Fig. 6.16(b)] that $AB = 0$ and $BE = 0$, since there are no other force components on pin *B* in the directions of these members. Finally, $BC = 2000$ lb of tension.

**Joint *A*.** Let us next consider joint *A* (Fig. 6.18). We can express force $\overrightarrow{AC}$ and $\overrightarrow{AE}$ vectorially. Thus,

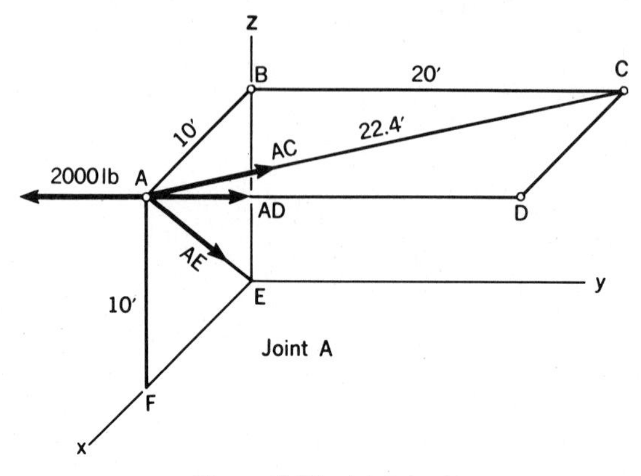


**Figure 6.18.** Joint *A*.

$$\vec{AC} = AC\frac{-10\boldsymbol{i} + 20\boldsymbol{j}}{\sqrt{10^2 + 20^2}} = AC\,(-.447\boldsymbol{i} + .894\boldsymbol{j})\text{ lb}$$

$$\vec{AE} = AE\frac{-10\boldsymbol{i} - 10\boldsymbol{k}}{\sqrt{10^2 + 10^2}} = AE\,(-.707\boldsymbol{i} - .707\boldsymbol{k})\text{ lb}$$

Summing forces, we have

$$-2000\boldsymbol{j} + AD\boldsymbol{j} + AC(-.447\boldsymbol{i} + .894\boldsymbol{j}) + AE(-.707\boldsymbol{i} - .707\boldsymbol{k}) = \boldsymbol{0}$$

Hence,

$$.894AC + AD = 2000 \quad \text{(a)}$$

$$-.447AC - .707AE = 0 \quad \text{(b)}$$

$$-.707AE = 0 \quad \text{(c)}$$

We see that $AE = AC = 0$ and $AD = 2000$ lb of tension.

**Joint *D*.** We now consider joint *D* (Fig. 6.19). Forces $\vec{FD}$ and $\vec{ED}$ are expressed as follows:

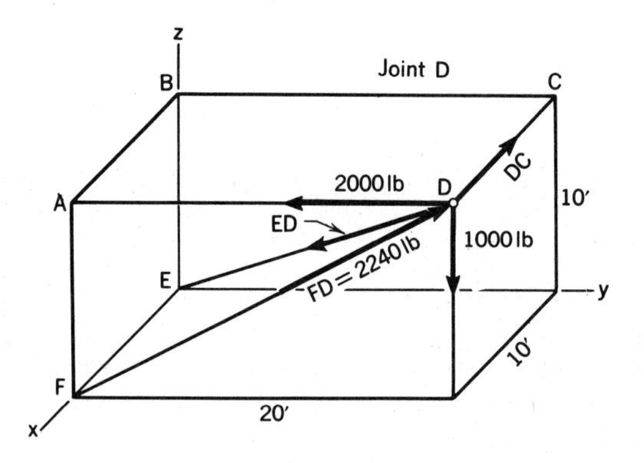


**Figure 6.19.** Joint *D*.

$$\overrightarrow{ED} = ED\frac{-10\boldsymbol{i} - 20\boldsymbol{j} - 10\boldsymbol{k}}{\sqrt{10^2 + 20^2 + 10^2}} = ED(-.408\boldsymbol{i} - .816\boldsymbol{j} - .408\boldsymbol{k}) \text{ lb}$$

$$\overrightarrow{FD} = FD\frac{20\boldsymbol{j} + 10\boldsymbol{k}}{\sqrt{20^2 + 10^2}} = 2240(.894\boldsymbol{j} + .447\boldsymbol{k}) \text{ lb}$$

Hence, summing forces, we get

$$\begin{aligned} -2000\boldsymbol{j} - 1000\boldsymbol{k} - DC\boldsymbol{i} + 2240(.894\boldsymbol{j} + .447\boldsymbol{k}) \\ + ED(-.408\boldsymbol{i} - .816\boldsymbol{j} - .408\boldsymbol{k}) = \boldsymbol{0} \end{aligned} \quad \text{(d)}$$

Thus,

$$-2000 + 2000 - .816ED = 0 \quad \text{(e)}$$

$$DC + .408ED = 0 \quad \text{(f)}$$

$$-1000 + 1000 - .408ED = 0 \quad \text{(g)}$$

We see here that $ED = 0$ and $DC = 0$.

**Joint *E*.** The only nonzero forces on joint $E$ are the supporting forces and $CD$, as shown in Fig. 6.20(a). We may solve for $CE$ directly and get 2240 lb of compression.

**Joint *C*.** As a check on our problem, we can examine joint $C$. The only nonzero forces are shown on the joint [Fig. 6.20(b)]. The reader may readily verify that the solution checks.

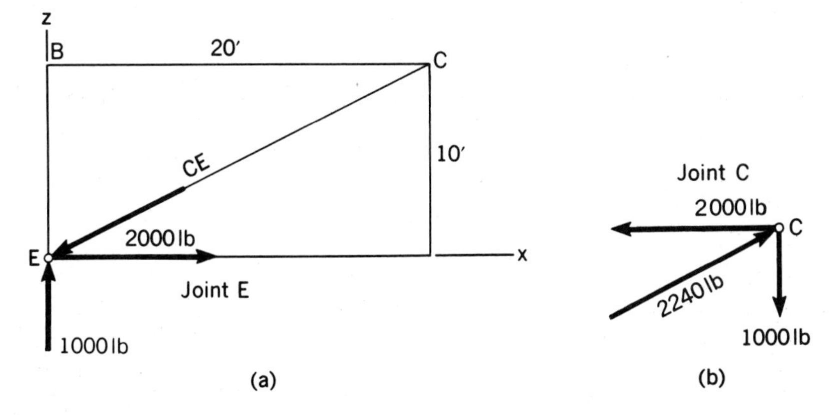


**Figure 6.20.** (a) Joint $E$; (b) check at joint $C$.

Before proceeding with the problems, it will be well to comment on the loading of plane roof trusses. Usually there will be a series of separated parallel trusses supporting the loading from the roof such as is shown in Fig. 6.21, where a wind pressure is shown on a roof as $p$. Now the inside truss can be considered to support the loading over a region extending halfway to each neighboring truss (shown as distance $d$). Furthermore, pins $A$ and $B$ support the force exerted on area $lhmk$ while pins $B$ and $C$ support the forces exerted on area $lrvh$. When dealing with the entire inside truss as a free body, you can use the resultant force from pressure over $krvm$. However, when dealing with the pins as a free body you must use the forces coming on to each pin as

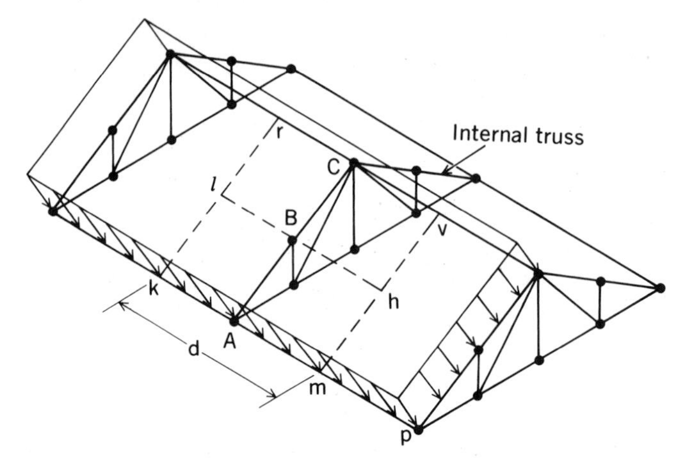


Figure 6.21. Roof trusses supporting a wind load.

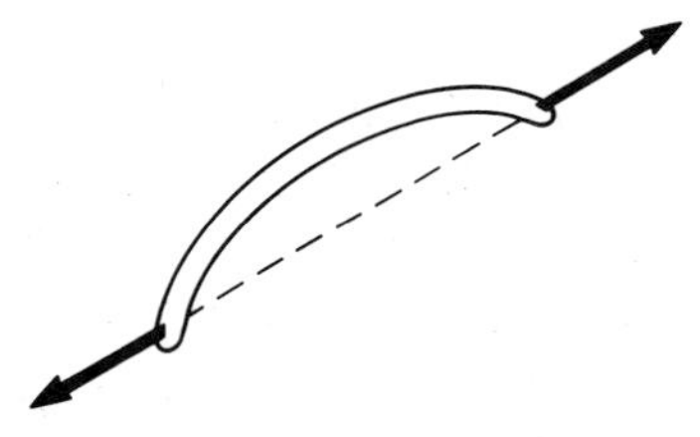

Figure 6.22. Curved two-force member.

described above and *not* the total resultant which was used for the free body of the entire internal truss.

Finally, we wish to remind you that a curved member in a truss, such as appears in Problems 6.5 and 6.9, is a two-force member with forces coming only from the pins. Recall that, for such members, the force transmitted to the pins must be collinear with the line connecting the points of application of the pins, such as is shown in Fig. 6.22.

## Problems

**6.1.** State which of the trusses shown are simple trusses and which are not.

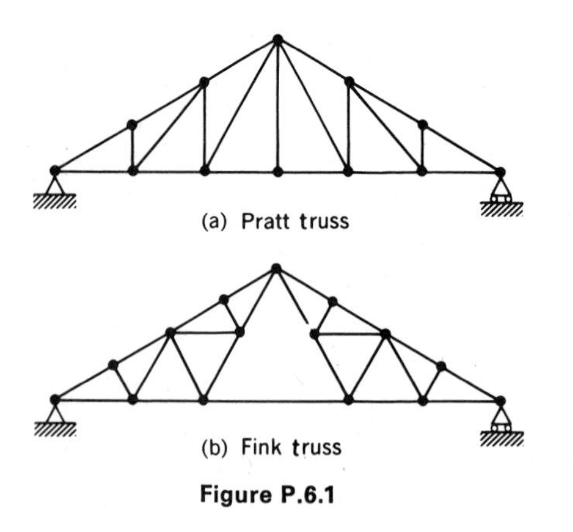


Figure P.6.1

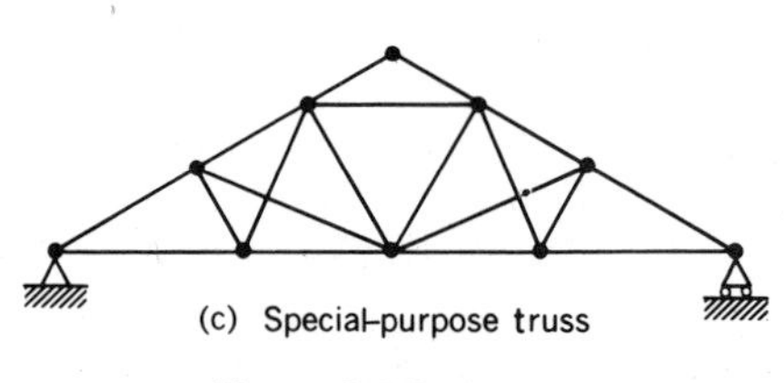


Figure P.6.1. (cont.)

**6.2.** Find the forces transmitted by each member of the truss.

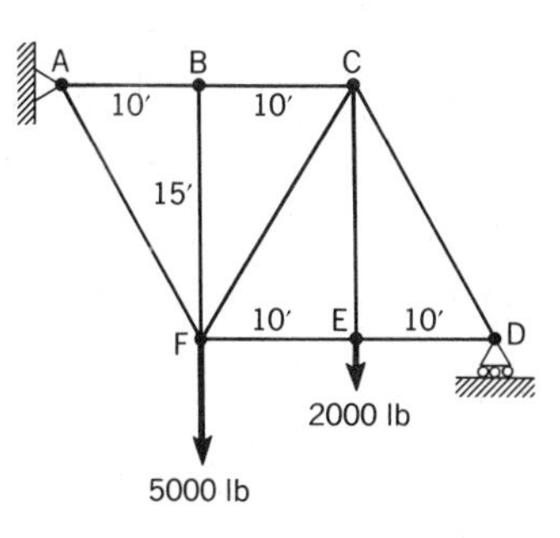


Figure P.6.2

**6.3.** The simple country-road bridge has floor beams to carry vehicle loads to the truss joints. Find the forces in all members for a truck-loaded weight of 160 kN. Floor beams 1 are supported by pins $A$ and $B$, while floor beams 2 are supported by pins $B$ and $C$.

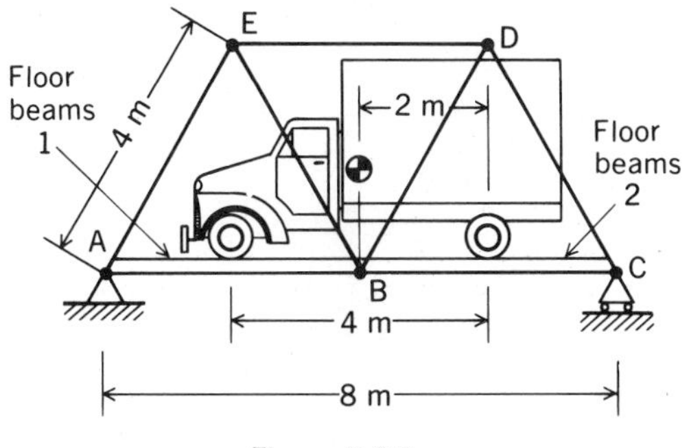

Figure P.6.3

**6.4.** A rooftop pond is filled with cooling water from an air conditioner and is supported by a series of parallel plane trusses. What are the forces in each member of an inside truss? The roof trusses are spaced at 10 ft apart. Water weighs 62.4 lb/ft³.

Figure P.6.4

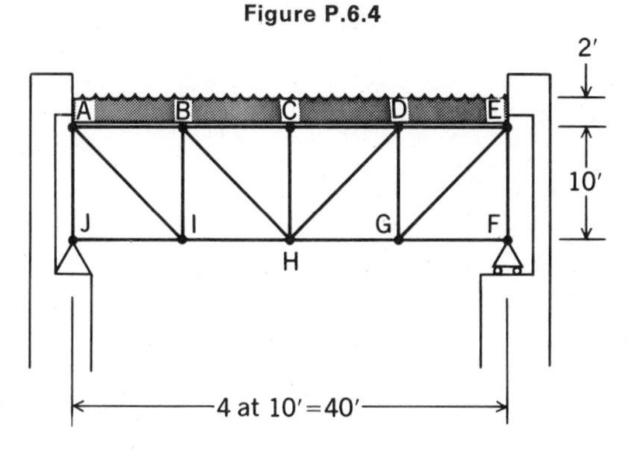


**6.5.** Find the forces transmitted by the straight members of the truss. DC is circular.

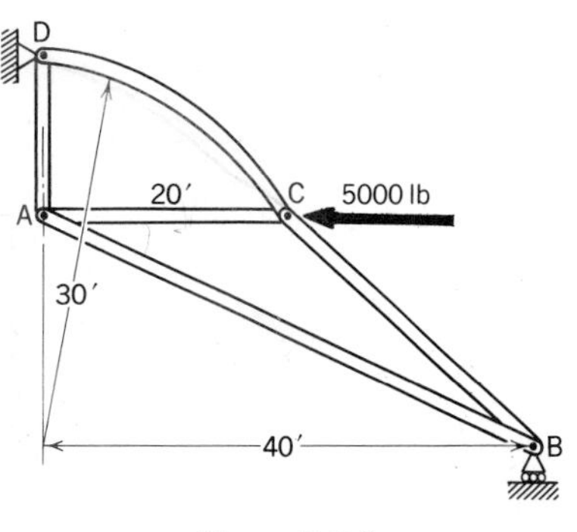


Figure P.6.5

**6.6.** Roof trusses such as the one shown are spaced 6 m apart in a long, rectangular building. During the winter, snow loads of up to 1 kN/m² (or 1 kPa) accumulate on the central portion of the roof. Find the force in each member for a truss not at the ends of the building.

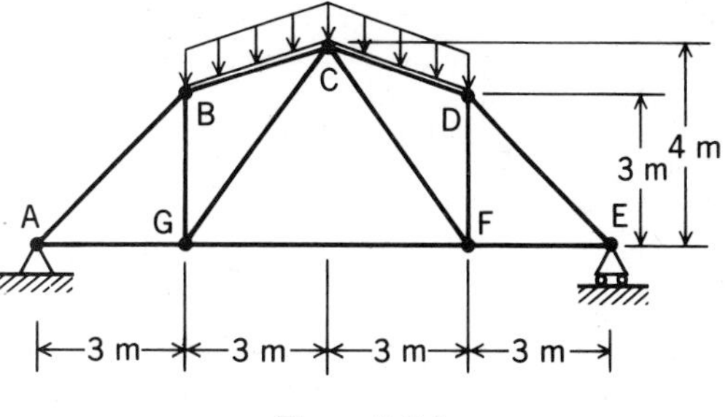


Figure P.6.6

**6.7.** The bridge supports a roadway load of 1000 lb/ft for each of the two trusses. Each member weighs 30 lb/ft. Compute the forces in the members, accounting approximately for the weight of the members.

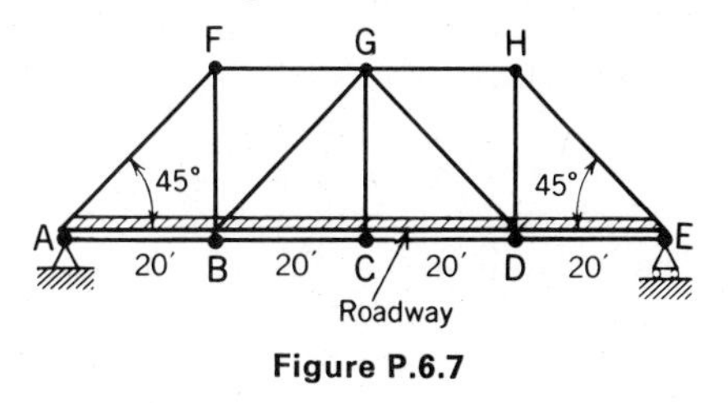


Figure P.6.7

**6.8.** Roadway and vehicle loads are transmitted to the highway bridge truss as the idealized forces shown. What are the forces in members?

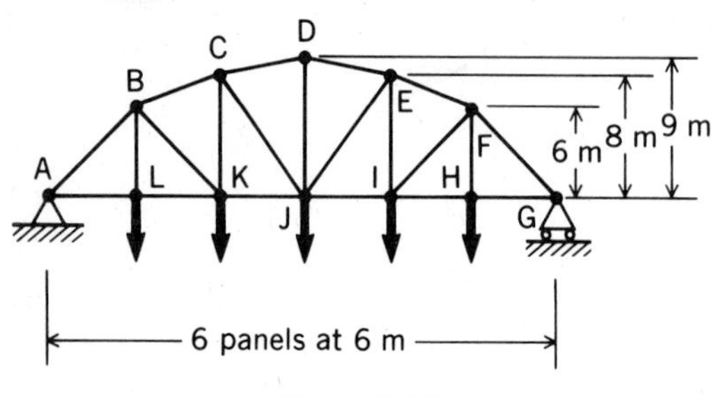


Figure P.6.8

**6.9.** A 5-kN hoist to lift railroad cars for truck repair has a 150-kN capacity and hangs from a truss with an L-shaped member to clear boxcars. What are the forces in the straight members?

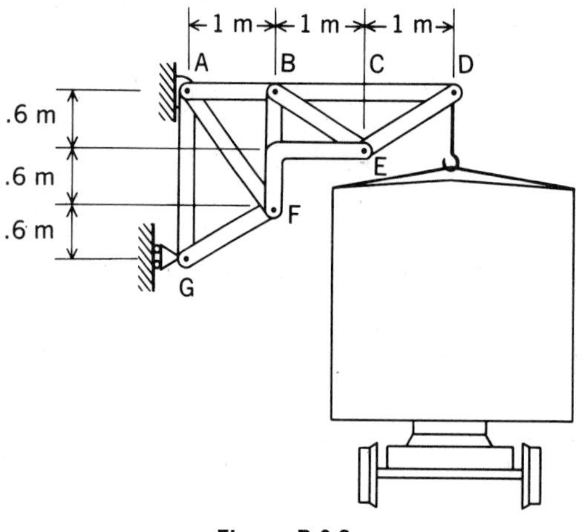


Figure P.6.9

**6.10.** A 5-kN traveling hoist has a 50-kN capacity and is suspended from a beam weighing 1 kN/m, which, in turn, is fastened to the roof truss at $I$ and $G$ as shown. In addition, wind pressures of up to 2 kN/m² (or 2 kPa) act on the side of the roof. The resulting force is transmitted to pins $A$ and $J$. If the trusses are spaced 10 m apart, what are the forces in each member for an internal truss when the hoist is in the middle of the span?

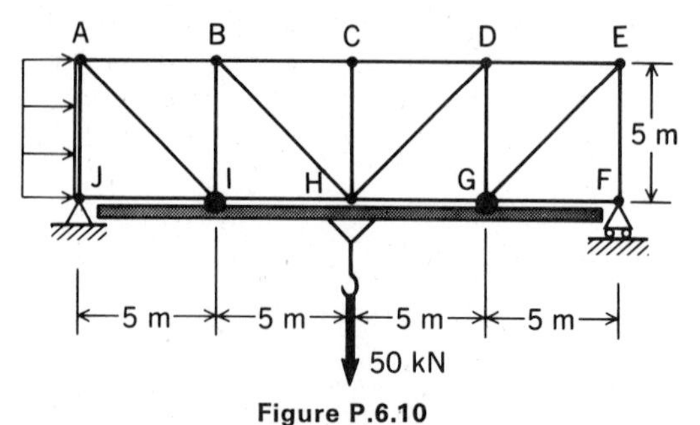

Figure P.6.10

**6.11.** Find the forces in the members of the truss. The 1000-lb force is parallel to $y$, and the 500-lb force is parallel to $z$.

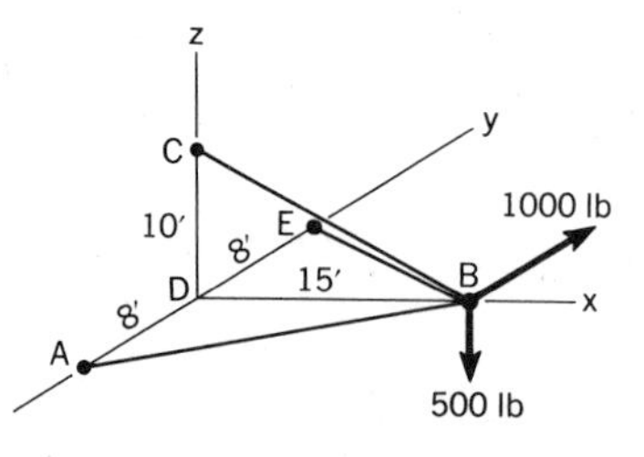


Figure P.6.11

**6.12.** Find the forces in the members and the supporting forces for the space truss $ABCD$. Note that $BDC$ is in the $xz$ plane.

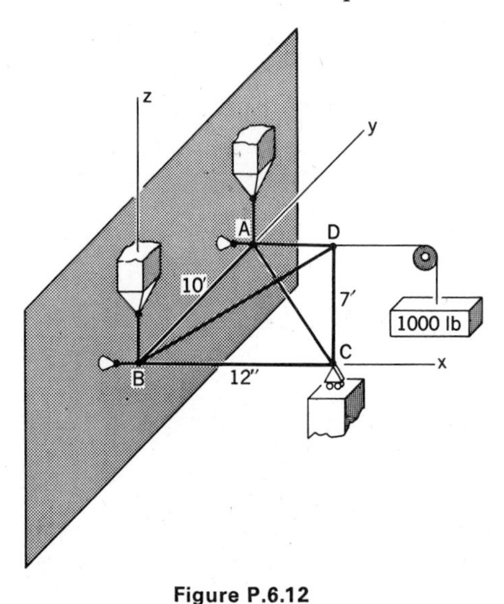


Figure P.6.12

**6.13.** A space truss *ABCDE* supports a 50-kN vertical load as well as a 10-kN horizontal load and rests on smooth, mutually perpendicular surfaces. Assume that the contact between the space truss and the smooth surface is at the ball joints. What are the forces in the members?

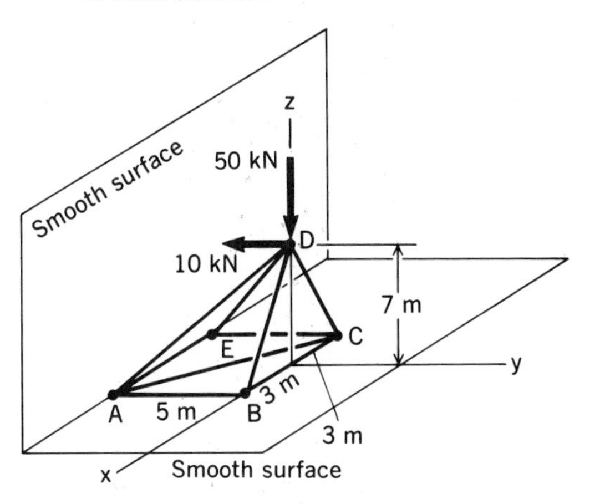


Figure P.6.13

**6.14.** Find the forces in the members of the space truss under the action of a force $\boldsymbol{F}$ given as

$$\boldsymbol{F} = 10\boldsymbol{i} - 6\boldsymbol{j} - 12\boldsymbol{k} \text{ kN}$$

Note that $C$ is a ball-and-socket joint while $A$, $F$, and $E$ are on rollers.

**6.15.** The plane of ball-and-socket joints *CDHE* of the space truss is in the *zy* plane, while the plane of *FGDE* is parallel to the *xz* plane. Note that this is *not* a simple space truss. Nevertheless, the forces in the members can be ascertained by choosing a desirable starting joint and proceeding by statics from joint to joint. Determine the forces in all the members and then determine the supporting forces.

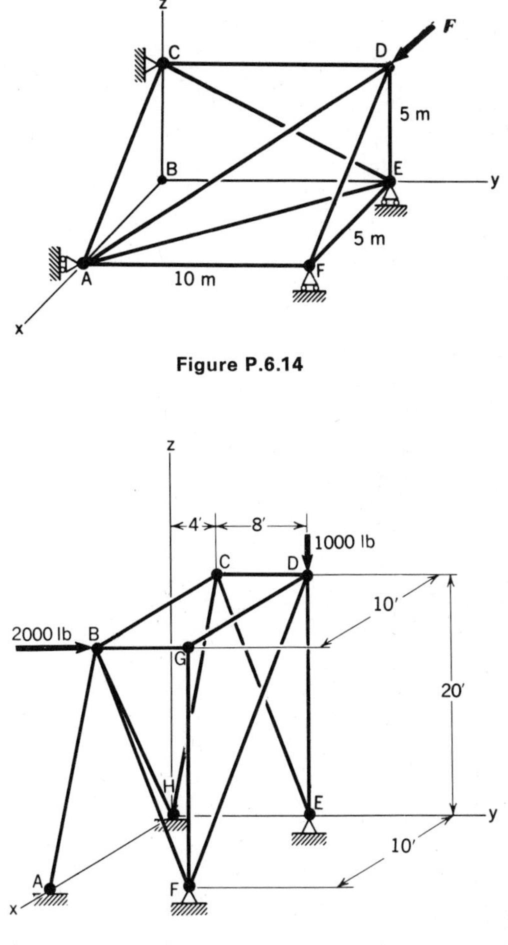


Figure P.6.14



Figure P.6.15

## 6.5 Method of Sections

In the method of sections that we shall use for plane trusses, we employ free-body diagrams that are generally different than that of the method of joints, as was pointed out earlier. *A free body in this method is formed by cutting away a portion of a truss and including at the cut sections the forces that are transmitted across these sections.* We then use the equations of equilibrium for these free bodies. In this way, we can expose for calculation individual members well inside a truss and avoid the laborious process of proceeding joint by joint until reaching a joint on which the desired unknown force acts.

Generally, a free body is created by passing a section (or cut) through the truss such as section *A–A* or section *B–B* in Fig. 6.23(a). Note that the section can be straight or curved. The corresponding free-body diagrams [see Fig. 6.23(b) for cut *A–A* and Fig. 6.23(c) for cut *B–B*] involve coplanar force systems. We have, accordingly, three equations of equilibrium available for each free body. Note that in contrast to the method

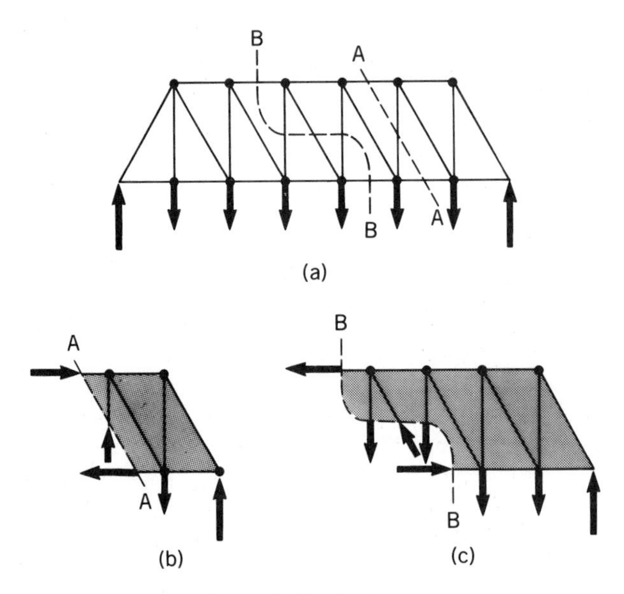


**Figure 6.23.** Section cuts.

of joints, one or more equilibrium equations can most profitably be moment equations. The choice of the section (or sections) to find the desired unknowns inside a truss involves ingenuity on the part of the engineer. He will want the fewest and simplest sections to find desired forces for one or more members inside the truss. The method of sections is used for finding limited information. The method of joints for such problems is by contrast one of "brute force."

We now illustrate the method of sections in the following examples.

**EXAMPLE 6.3**

In Example 6.1 suppose that we wish to know the force in member *CE* only. To avoid the laborious joint-by-joint procedure, we employ a portion of the truss to the left of cut *K–K*, as shown in Fig. 6.24. Notice that the forces from the other part of the truss acting on this part through the cut members have been included, and in this way the desired force has been exposed. The sense of these exposed forces is not known, but we do know the orientations, as explained in our earlier discussions. Using the equations of equilibrium and taking advantage of the fact that the lines of action of some of the exposed unknown forces are concurrent at certain joints, we may readily solve for the unknowns if they number three or less. To determine $CE$, we take moments about a point corresponding to joint $D$ through which the lines of action of forces $BD$ and $CD$ pass:

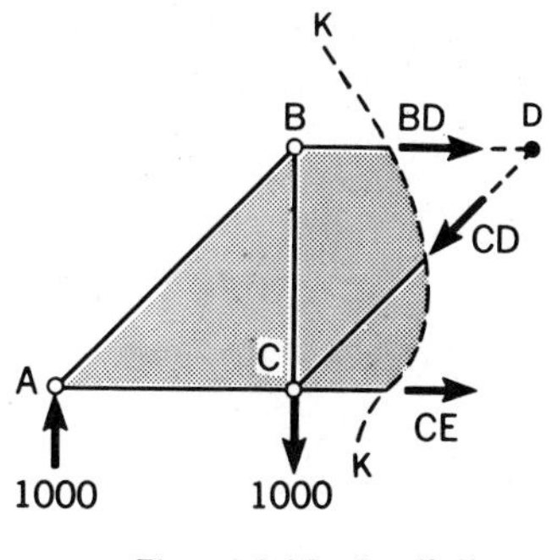


**Figure 6.24.** Cut *K–K*.

$\underline{\sum M_D = 0}$:

$$-(1000)(20) + (1000)(10) + 10CE = 0$$

Therefore,

$$CE = 1000 \text{ lb}$$

Our ingenuity here has led us to one equation with only one unknown, the desired force $CE$. By observing the free-body diagram in Fig. 6.24, we can clearly see that $CE$ is a tension member.

If we desire $BD$ also, we can take moments about point $C$ through which the lines of action of $CE$ and $CD$ pass. However, $BD$ now comes out negative, indicating that we have made an incorrect choice of sense. With this in mind, we can conclude that $BD$ is in compression.

Perhaps a suitable single section with sufficient unknowns for a solution cannot be found. We may then have to take several sections before we can expose the desired force in a free body with enough simultaneous equations to effect a solution. These problems are no different from the ones we studied in Chapter 5, where several free-body diagrams were needed to generate a complete set of equations containing the unknown quantity. We now consider such a problem.

**EXAMPLE 6.4**

A plane truss is shown in Fig. 6.25 for which only the force in member $AB$ is desired. The supporting forces have been determined and are shown in the diagram.

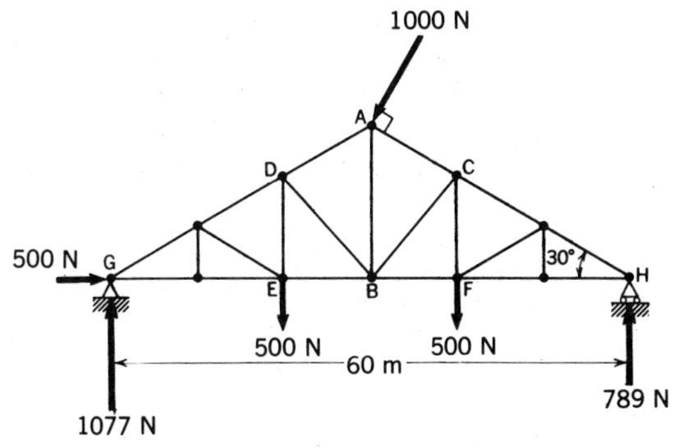


**Figure 6.25.** Plane truss.

In Fig. 6.26, we have shown a cut *J–J* of the truss exposing force *AB*. (This is the same force diagram as that which results from the free-body diagram of pin *A*.) We have here three unknown forces for which only two equations of equilibrium are available. We must use an additional free body.

Thus, in Fig. 6.27 we have shown a second cut *K–K*. Note that by taking moments about joint *B*, we can solve for *AC* directly. With this information, we can

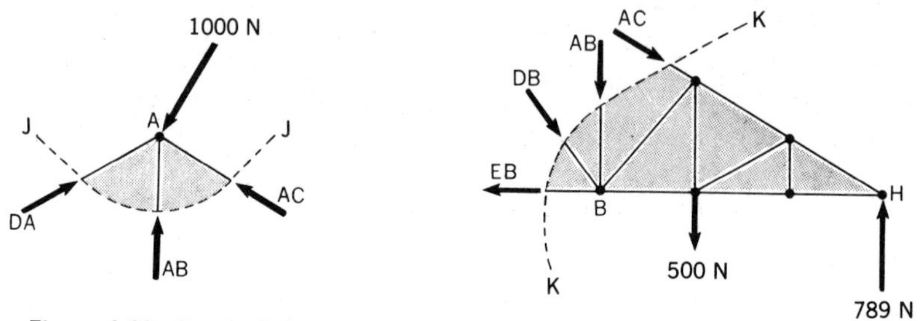


**Figure 6.26.** Free body I from cut *J–J*.

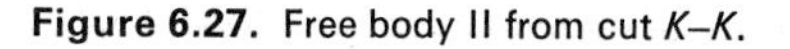

**Figure 6.27.** Free body II from cut *K–K*.

then return to the first cut to get the desired unknown *AB*. Accordingly, we have, for free body II:

$\underline{\sum M_B = 0:}$

$$-(10)(500) + (30)(789) - (AC)(\sin 30°)(30) = 0$$

(Note we have transmitted *AC* to joint *H* in evaluating its moment contribution.) Solving for *AC*, we get

$$AC = 1245 \text{ N}$$

Summing forces for free body I, we have

$\underline{\sum F_x = 0:}$

$$DA \cos 30° - AC \cos 30° - 1000 \sin 30° = 0$$

Therefore,

$$DA = 1822 \text{ N}$$

$\underline{\sum F_y = 0:}$

$$DA \sin 30° + AC \sin 30° + AB - 1000 \cos 30° = 0$$

Therefore,

$$AB = -667 \text{ N}$$

We see that member *AB* is a tension member rather than a compression member as was our initial guess in drawing the free-body diagrams.

In retrospect, you will note that, in the method of joints, errors made early will of necessity propagate through the calculations. There is, on the other hand, much less likelihood of this occurring in the method of sections, since many of the equations will be independent. However, for simple trusses with many members, we may profitably use the method of joints in conjunction with a computer for which the brute-force approach of the method of joints is ideally suited.

## Problems

**6.16.** Find the forces in members *CB* and *BE* of the plane truss.

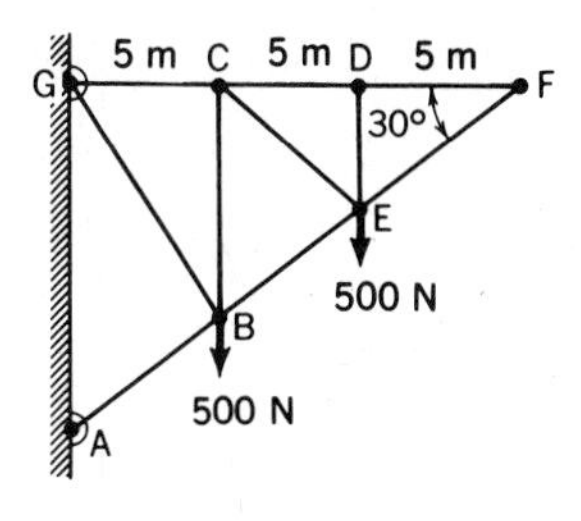


**Figure P.6.16**

**6.17.** For the roof truss: (a) Find the forces transmitted by member *DC*. (b) What is the force transmitted by *DE*?

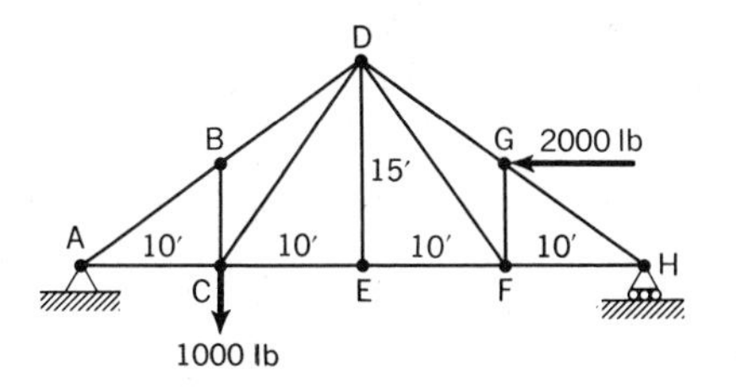


**Figure P.6.17**

**6.18.** Determine the force transmitted by member *KU* in the plane truss.

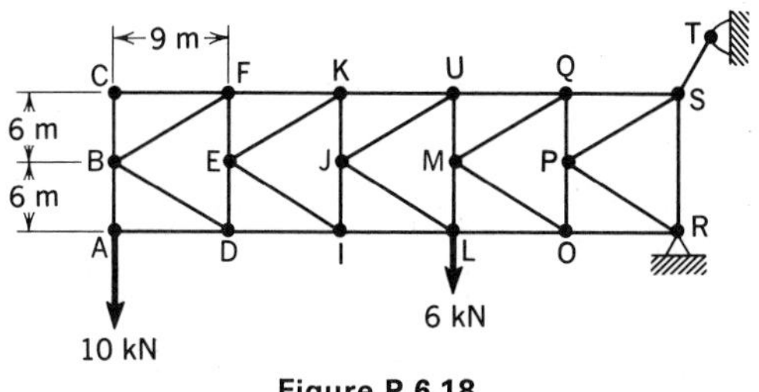


**Figure P.6.18**

**6.19.** In the roof truss of Problem 6.6, find the force in member *GF*. Remember loads are applied to the pins.

**6.20.** Find the forces in members *CD*, *DG*, and *HG* in the plane truss.

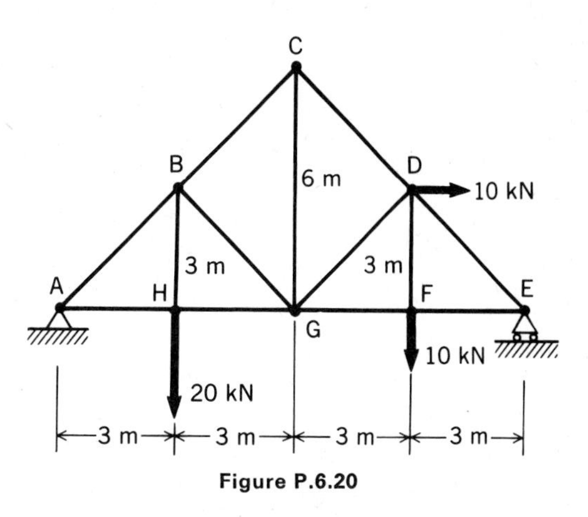


**Figure P.6.20**

**6.21.** In Problem 6.9, find the force in members *BF* and *AB*.

**6.22.** The roof is subjected to a wind loading of 20 lb/ft². Find the forces in members *LK* and *KJ* if the trusses are spaced 10 ft apart.

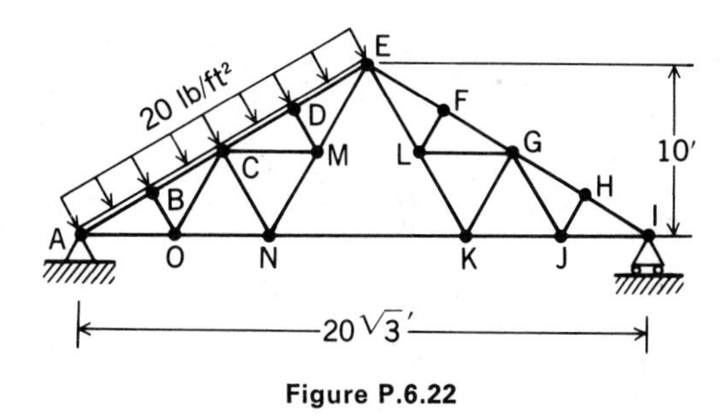


**Figure P.6.22**

**6.23.** The guideways for a large overhanging crane are suspended from certain joints of the truss (*M*, *K*, *J*, and *G*). Find the forces in members *BC*, *BK*, *DE*, *DI*, and *EF*. Neglect the truss and guideway weights. Guideways only transmit supported loads to pins and are not considered part of the truss structurally.

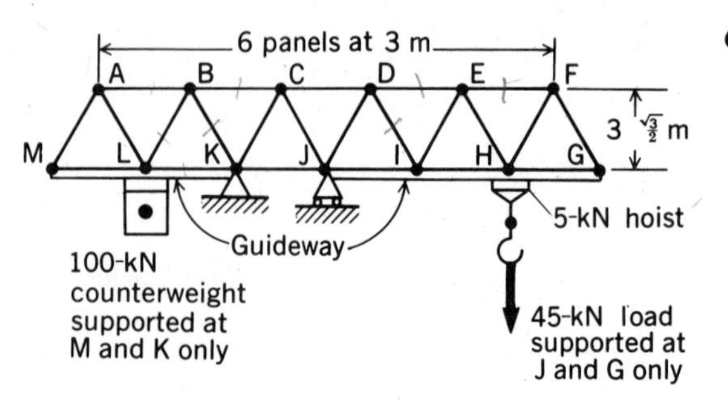


Figure P.6.23

**6.24.** Find the force in members *HE*, *FH*, *FE*, and *FC* of the truss.

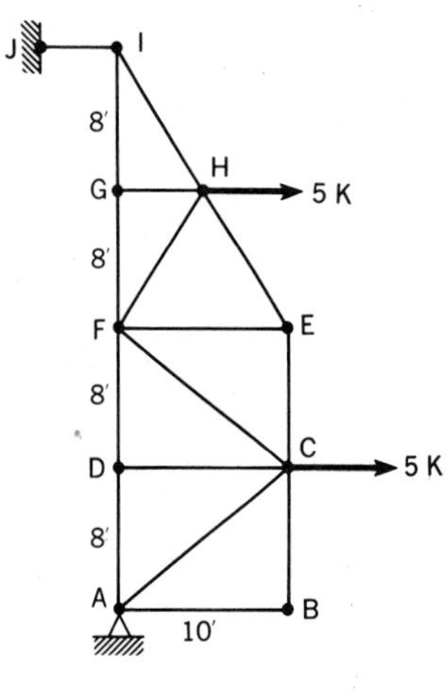


Figure P.6.24

**6.25.** Find the forces in member *JF* in the truss.

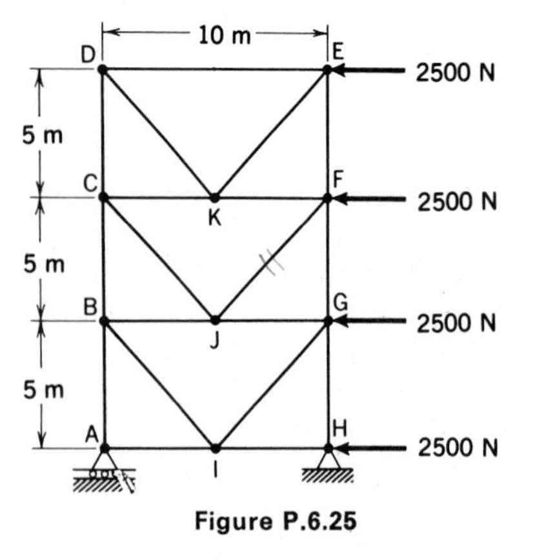


Figure P.6.25

**6.26.** Find the force in members *FI*, *EF*, and *DH* in the truss. Neglect the weight of the pulleys.

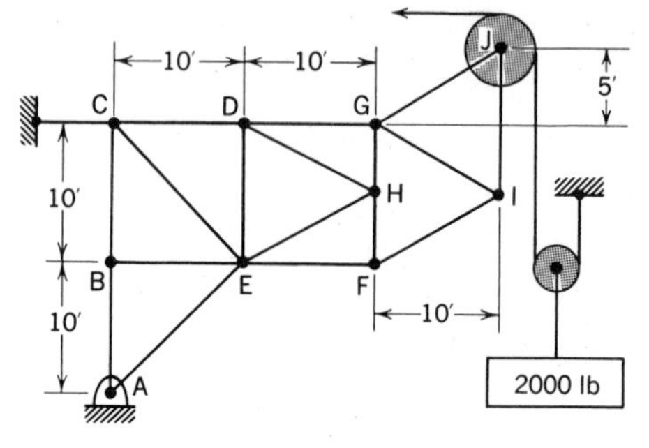


Figure P.6.26

**6.27.** A railroad engine is starting to cross the deck-type truss bridge shown. If the weight of the engine is idealized by the four 50-kip loads,[7] find the forces in members *AB*, *BL*, *CK*, *CL*, *LK*, *DK*, *KJ*, and *DJ*.

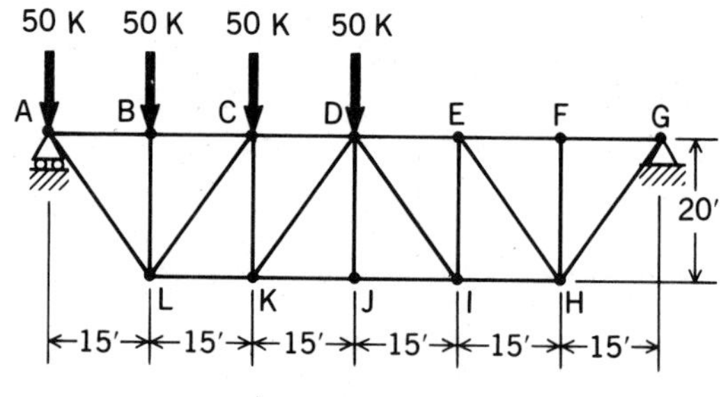


Figure P.6.27

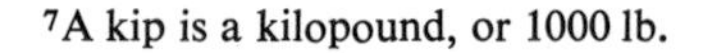

[7]A kip is a kilopound, or 1000 lb.

**6.28.** A truss supports a roadway load of 800 lb/ft per truss. Concentrated loads have been shown representing approximations of vehicle loading for each truss at some instant of time. The bridge has six 20-ft panels. Determine the forces in members *EG*, *FH*, and *IJ*.

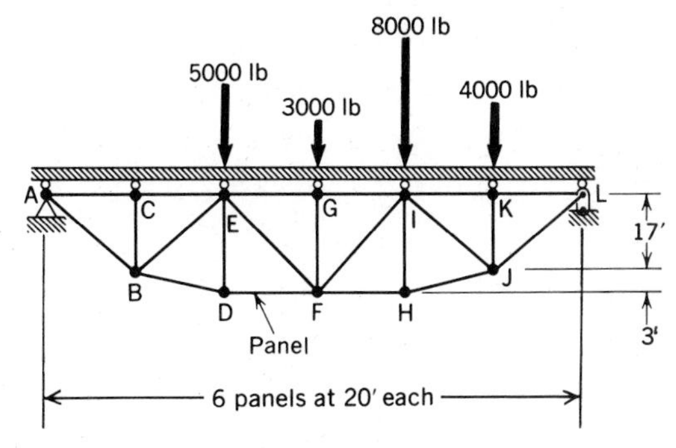

**Figure P.6.28**

# Part B
# Section Forces in Beams

## 6.6 Introduction

In Part A, we considered a number of problems involving members loaded axially along the axes of the members. The resultant force at any section was easily established as a single axial force. We shall now consider thin prismatic members that are loaded *transversely* as well as axially. Generally, when such members are loaded transversely, we call them *beams*. Of considerable use will be certain components of the resultant force system acting on *cross sections* of the beams. We shall set forth methods in this section for computing these quantities. We consider beams with a vertical plane of symmetry along the axis of the beam.

## 6.7 Shear Force, Axial Force, and Bending Moment

Consider first a beam with an arbitrary intensity of loading $w(x)$ in the plane of symmetry and a load $P$ along the direction of the beam applied at the end $A$ as shown in Fig. 6.28(a). It will be assumed that the supporting forces have been determined. To find the force transmitted across the cross-sectional interface at position $x$, we take a portion of the beam as a free body so as to "expose" the section of the beam at $x$ as shown in Fig. 6.28(b). Since we have a coplanar-loading distribution, we know from rigid-body mechanics that, depending on the problem, we can replace the distribution at section $x$ most simply by a single force or a single couple moment in the plane of the external loads. If the resultant is most simply a single force, we know that it must have a particular line of action. This line of action does not usually go through the center of the cross section. Since the actual position of the intersection of this force with the

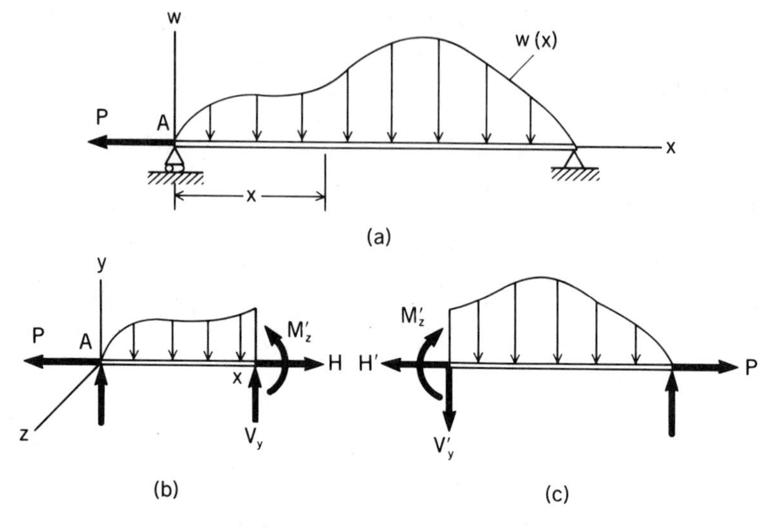


**Figure 6.28.** Resultant at a section.

cross section is of little interest in beam theory, we deliberately take the position of the resultant force to be at the center of the cross section at all times and include the proper couple moment $M_z$ to accompany the force. Furthermore, we decompose the force into orthogonal components—in this case, a vertical force $V_y$ and a horizontal force $H$. These quantities are shown in Fig. 6.28(b). Since these quantities are used to such a great extent in structural work, we have associated names with them. They are $V_y \equiv$ *shear-force* component, $H \equiv$ *axial-force* component, $M_z \equiv$ *bending-moment* component.[8] If we had a three-dimensional load, there would have been one additional shear

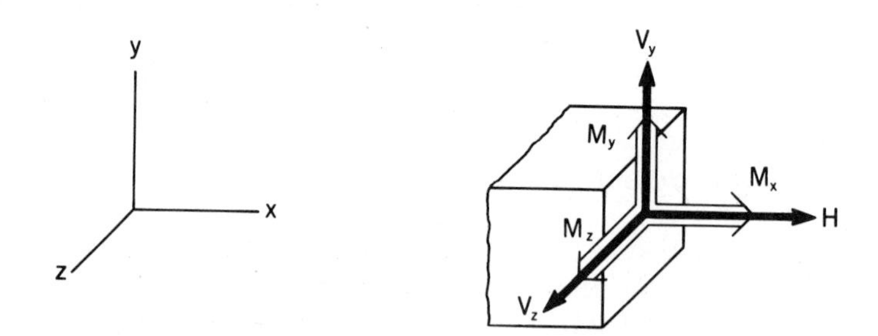


**Figure 6.29.** Section resultant for three-dimensional loading.

component $V_z$ (see Fig. 6.29), one additional bending-moment component $M_y$, and a couple moment along the axis of the beam $M_x$, which we shall call the *twisting moment*.

Notice in Fig. 6.28(c) that a second free-body diagram has been drawn which

---

[8]For curved beams, shear forces $V$ are always tangent to the cross section, whereas axial force $H$ is always normal to the cross section.

exposes the "other side" of the cross section at position $x$. The shear force, axial force, and bending moment for this section have been primed in the diagram. We know from Newton's third law that they should be equal and opposite to the corresponding unprimed quantities in part (b) of the diagram. We can thus choose for our computations either a left-hand or a right-hand free-body diagram. But this poses somewhat of a problem for us when we come to reporting the signs of the transmitted forces and couple moments at a section. We cannot use the direction of a force or couple moment at the section. Clearly, this would be inadequate since the sense of the force or couple moment at a section would depend on whether a left-hand or a right-hand free-body diagram was used. To associate an unambiguous sign for shear force, axial force, and bending moment at a section, we adopt the following convention:

*A force component for a section is positive if the area vector of the cross section and the force component both have senses either in the positive or in the negative directions of any one or two of the reference axes.*[9]

The same is true for the bending moment.

Thus, consider Fig. 6.28. For the left-hand free-body diagram, the area vector for section $x$ points in the positive $x$ direction. Note also that $H$, $V_y$, and the vectorial representation of $M_z$ also point in positive directions of the $xyz$ axes. Hence, according to our convention we have drawn a positive shear force, a positive axial force, and a positive bending moment at the section at $x$. For the right-hand free-body diagram, the cross-sectional area vector points in the negative $x$ direction. And, since $H'$, $V'_y$, and $M'_z$ point in negative directions of the $x$, $y$, and $z$ axes, these components are again positive for the section at $x$ according to our convention. Clearly, by employing this convention, we can easily and effectively specify the force system at a section without the danger of ambiguity.

As pointed out earlier, we can solve for $V_y$, $H$, and $M_z$ at section $x$ using rigid-body mechanics for either a left-hand or a right-hand free-body diagram provided that we know all the external forces. The quantities $V_y$, $H$, and $M_z$ will depend on $x$, and for this reason, it is the practice to sketch shear-force and bending-moment diagrams to convey this information for the entire beam.

We now illustrate the computation of $V$ and $M$.

### EXAMPLE 6.5

We shall express the shear-force and bending-moment equations for the simply supported beam shown in Fig. 6.30(a), whose weight we shall neglect. The support forces obtained from equilibrium are 500 N each.

To get the shear force at a section $x$, we isolate either the left or right side of the beam at $x$ and employ the equations of equilibrium on the resulting free body. If $x$

---

[9]Some authors employ the reverse convention for shear force from the one that we have proposed. Our convention is consistent with the usual convention used in the theory of elasticity for the sign of stress at a point, and it is for this reason that we have employed this convention rather than the other one.

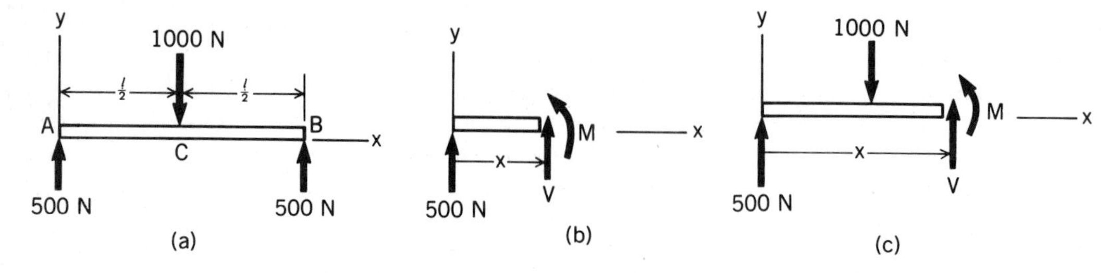


**Figure 6.30.** Simply supported beam.

lies between $A$ and $C$ of the beam, the only noninternal force present for a left-hand free body is the left supporting force [see Fig. 6.30(b)]. Notice that we have used directions for $V$ and $M$ (there is no need for subscripts in the simple problem) corresponding to the *positive* states from the point of view of our convention. Clearly, the *algebraic* sign we get for these quantities from equilibrium calculations will then correspond to the *convention* sign. If $x$ is between $C$ and $B$ for such a free body, two external forces appear [see Fig. 6.30(c)]. Therefore, if the shear force is to be expressed as a function of $x$, clearly separate equations covering the two ranges, $0 < x < l/2$ and $l/2 < x < l$, are necessary.[10] Summing forces we then get

$\underline{0 < x < l/2:}$

$$500 + V = 0; \qquad \text{therefore, } V = -500 \text{ N} \tag{a}$$

$\underline{l/2 < x < l:}$

$$500 - 1000 + V = 0; \qquad \text{therefore, } V = 500 \text{ N} \tag{b}$$

Now let us turn to the bending-moment equations. Again, we must consider two discrete regions. Taking moments about position $x$, we get

$\underline{0 \leq x \leq l/2:}$

$$-500x + M = 0; \qquad \text{therefore, } M = 500x \text{ N-m} \tag{c}$$

$\underline{l/2 \leq x \leq l:}$

$$-500x + 1000\left(x - \frac{l}{2}\right) + M = 0; \qquad \text{therefore, } M = 500(l - x) \text{ N-m} \tag{d}$$

**EXAMPLE 6.6**

Determine the shear-force and bending-moment equations for the simply supported beam shown in Fig. 6.31. Neglect the weight of the beam.

We must first find the supporting forces for the beam. Hence, we have:

$\underline{\sum M_B = 0:}$

$$-R_1(22) + (50)(8)(14) + (1000)(14) - 500 = 0$$

Therefore,

$$R_1 = 868 \text{ lb}$$

$\underline{\sum M_A = 0:}$

$$R_2(22) - 500 - (50)(8)(8) - (1000)(8) = 0$$

---

[10]We exclude points $A$, $C$, and $B$ because, as you will soon see in the shear-force diagrams, $V$ is *indeterminate* at locations of point forces. Bending moment, however, will be continuous except at a point couple moment.

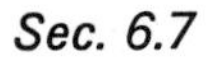

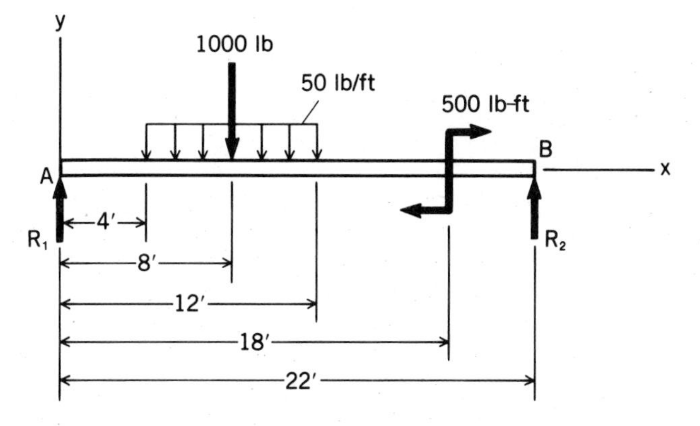


**Figure 6.31.** Simply supported beam.

Therefore,

$$R_2 = 532 \text{ lb}$$

In Fig. 6.32(a) we have shown a free-body diagram exposing sections between the left support and the uniform load. Summing forces and taking moments about a

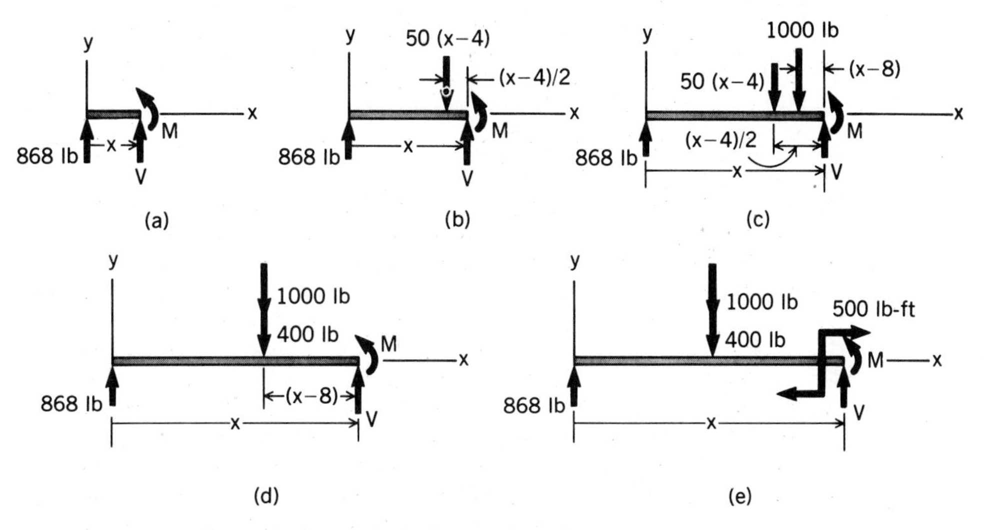


**Figure 6.32.** Free-body diagrams for various ranges.

point in the section, where we have drawn $V$ and $M$ as positive according to our convention, we get

$\underline{0 < x \leq 4:}$

$$868 + V = 0; \qquad \text{therefore, } V = -868 \text{ lb}$$

$$-868x + M = 0; \qquad \text{therefore, } M = 868x \text{ ft-lb}$$

The next interval is between the beginning of the uniform load and the point force. Thus, observing Fig. 6.32(b):

$\underline{4 < x \leq 8:}$

$$868 - 50(x - 4) + V = 0$$

Therefore,

$$V = 50x - 1068 \text{ lb}$$

$$-868x + \frac{50(x - 4)^2}{2} + M = 0$$

Therefore,

$$M = -25x^2 + 1068x - 400 \text{ ft-lb}$$

We now consider the interval between the point force and the end of the uniform load. Thus, observing Fig. 6.32(c):

$\underline{8 \leq x < 12:}$

$$868 - 50(x - 4) - 1000 + V = 0$$

Therefore,

$$V = 50x - 68 \text{ lb}$$

$$-868x + \frac{50(x - 4)^2}{2} + 1000(x - 8) + M = 0$$

Therefore,

$$M = -25x^2 + 68x + 7600 \text{ ft-lb}$$

The next interval is between the end of uniform loading and the point couple. We can now replace the uniform loading by its resultant of 400 lb, as shown in Fig. 6.32(d). Thus,

$\underline{12 < x \leq 18:}$

$$868 - 400 - 1000 + V = 0$$

Therefore,

$$V = 532 \text{ lb}$$

$$-868x + 1400(x - 8) + M = 0$$

Therefore,

$$M = -532x + 11{,}200 \text{ ft-lb}$$

The last interval goes from the point couple to the right support. It is to be pointed out that the point couple *does not* contribute directly to the shear force and we could have used the above formulation for $V$ for interval $18 < x \leq 22$. However, the couple *does* contribute directly to the bending moment, thus requiring the additional interval. Accordingly, using Fig. 6.32(e), we get

$$V = 532 \text{ lb} \quad \text{(as in previous interval)}$$

Whereas for $M$ we have

$$-868x + 1400(x - 8) - 500 + M = 0$$

Therefore,

$$M = -532x + 11{,}700 \text{ ft-lb}$$

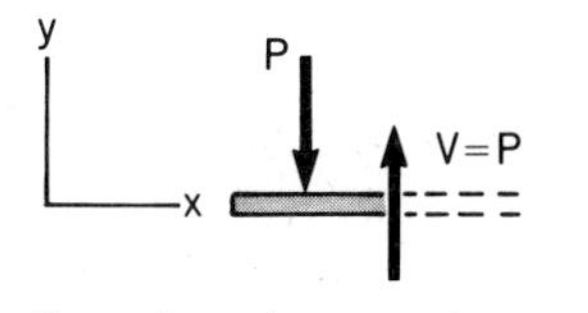


Shear-force increment induced by P on sections to right of it.

(a)

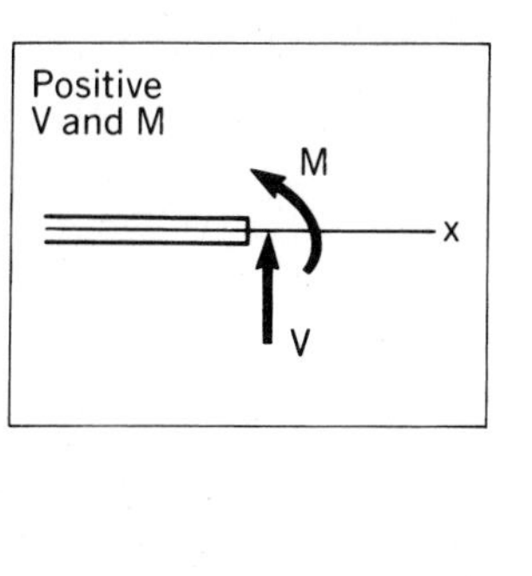


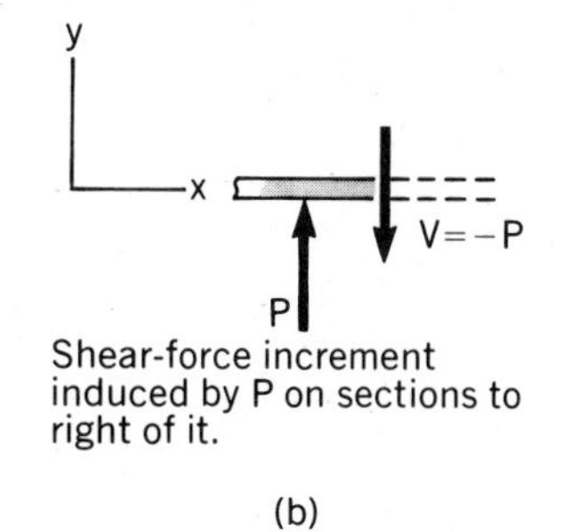


(b)

**Figure 6.33.** Shear induced by *P*.

We wish to point out now that we can determine shear-force and bending-moment equations in a less formal manner than what has been shown thus far. In this connection, it will be useful to note that a downward force $P$, as shown in Fig. 6.33(a), induces on sections to the right a positive shear force (see insert) of value $+P$, whereas an upward force of $P$ induces on sections to the right of it a negative shear force $-P$ [see Fig. 6.33(b)]. Also, an upward force $P$ induces on sections at a distance $\xi$ to the right of it a positive bending moment $P\xi$ [see Fig. 6.34(a)], whereas a downward force

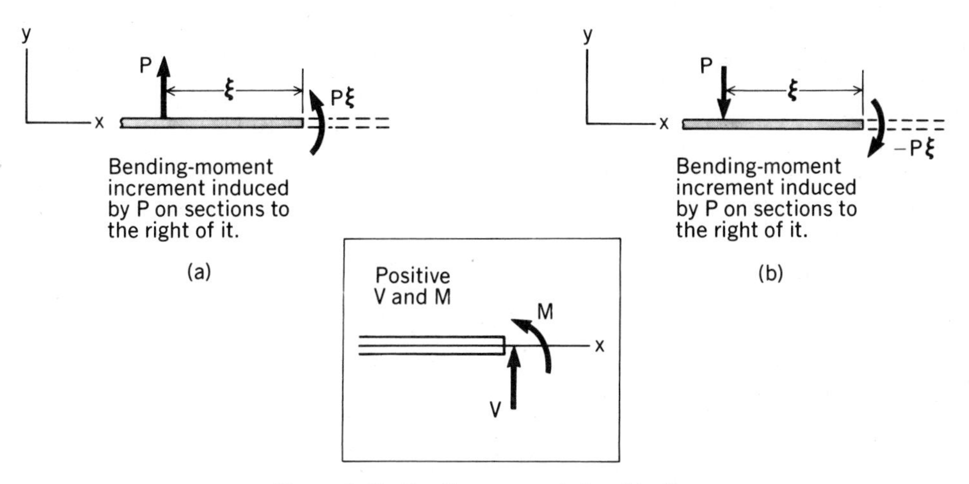


**Figure 6.34.** Bending moment induced by *P*.

$P$ induces on sections $\xi$ to the right of it a negative bending moment $-P\xi$ [Fig. 6.34(b)]. Finally, as can be seen in Fig. 6.35(a), a clockwise couple moment $C$ induces a positive bending moment $+C$ on the sections to the right of it (it does not induce a shear force), whereas a counterclockwise couple moment $C$ [Fig. 6.35(b)] induces a negative bending moment $-C$ on sections to the right of it. In the following example, we shall show how by this reasoning we may more directly formulate the shear-force and bending-moment equations.

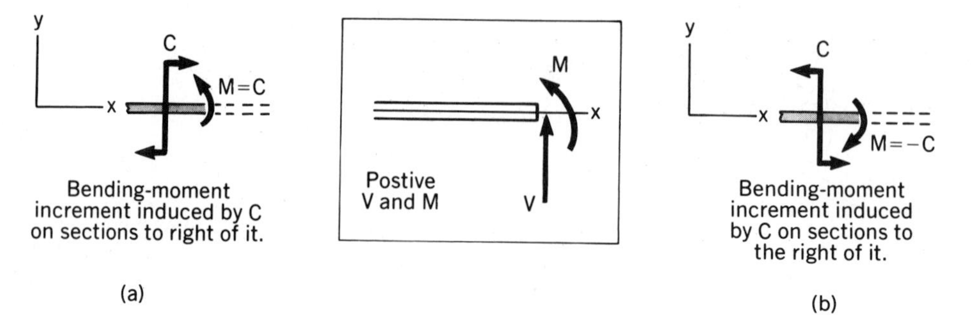


**Figure 6.35.** Bending moment induced by *C*.

**EXAMPLE 6.7**

Evaluate the shear-force and bending-moment equations for the beam shown in Fig. 6.36.

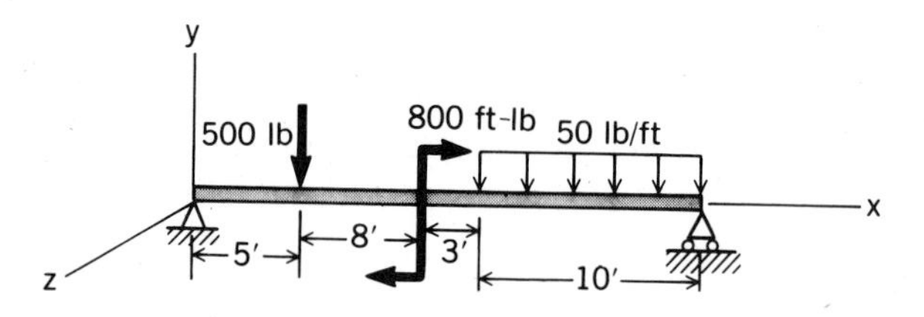


**Figure 6.36.** Simply supported beam.

A free-body diagram of the beam is shown in Fig. 6.37. We can immediately compute the supporting forces as follows:

$\underline{\sum M_2 = 0:}$

$$-R_1(26) + (500)(21) - 800 + (500)(5) = 0$$

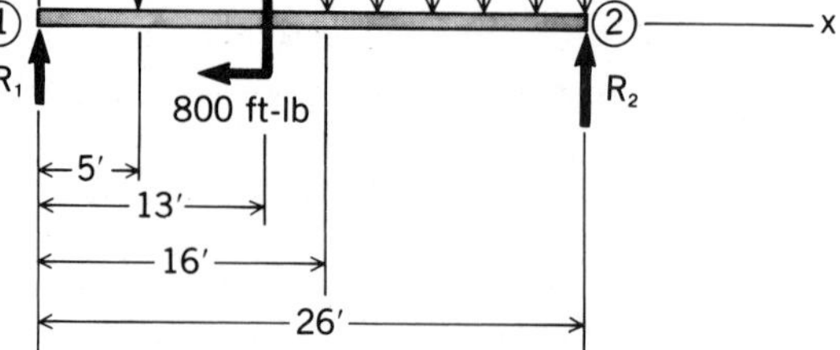


**Figure 6.37.** Free-body diagram of beam.

Therefore,

$$R_1 = 469 \text{ lb}$$

$\underline{\sum M_1 = 0:}$

$$R_2(26) - (500)(21) - 800 - (500)(5) = 0$$

Therefore,

$$R_2 = 531 \text{ lb}$$

We shall now directly give the shear force $V$ and bending moment $M$ while viewing Fig. 6.37. Thus,

$\underline{0 < x < 5:}$

$$V = -469 \text{ lb}$$

$$M = 469x \text{ ft-lb}$$

$\underline{5 < x < 13:}$

$$V = -469 + 500 = 31 \text{ lb}$$

$$M = 469x - 500(x - 5) = -31x + 2500 \text{ ft-lb}$$

$\underline{13 < x \leq 16:}$

$$V = 31 \text{ lb} \quad \text{(same as previous interval)}$$

$$M = 469x - 500(x - 5) + 800 = -31x + 3300 \text{ ft-lb}$$

$\underline{16 \leq x < 26:}$

$$V = -469 + 500 + 50(x - 16) = -769 + 50x \text{ lb}$$

$$M = 469x - 500(x - 5) + 800 - \frac{50(x - 16)^2}{2} = -25x^2 + 769x - 3100 \text{ ft-lb}$$

We shall present effective methods of sketching the shear-force and bending-moment diagrams in Section 6.7.

Before we proceed further, it must be carefully pointed out that the replacement of a distributed load by a single resultant force is only meaningful for the particular free body on which the force distribution acts. Thus, to compute the reactions for the entire beam taken as a free body (Fig. 6.38), we can replace the weight distribution $w_0$ by the total weight at position $L/2$ (Fig. 6.39). For the bending moment at $x$, the resultant of the loading for the free body shown in Fig. 6.40 becomes $wx$ and is midway at position $x/2$. In other words, *in making shear-force and bending-moment equations and diagrams, we cannot replace loading distributions over the entire beam by a resultant and then proceed*; there is inherent in these equations an infinite number of free bodies, each smaller than the beam itself, which makes the abovementioned replacements invalid for shear-force and bending-moment considerations.

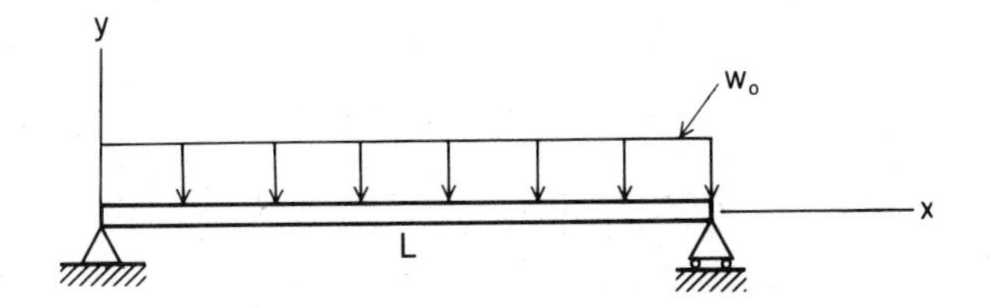


**Figure 6.38.** Uniform loading.

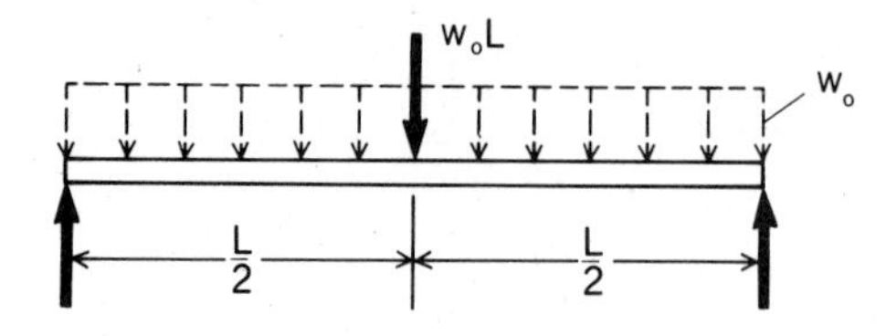


**Figure 6.39.** Resultant for $w_0$ for entire beam.

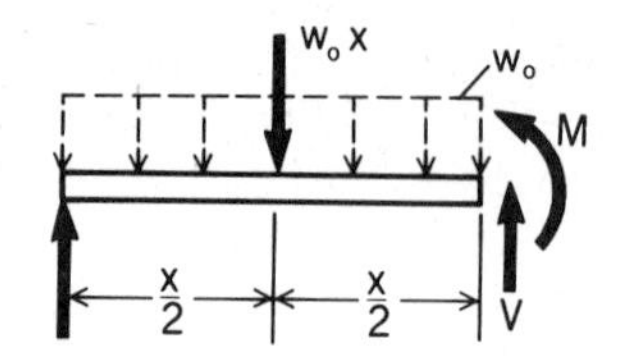


**Figure 6.40.** Resultant for $w_0$ for portion $x$ of beam.

## Problems

*In Problems 6.29 through 6.40 make use of free-body diagrams.*

**6.29.** Formulate the shear-force and bending-moment equations for the simply supported beam. Do not include the weight of the beam.

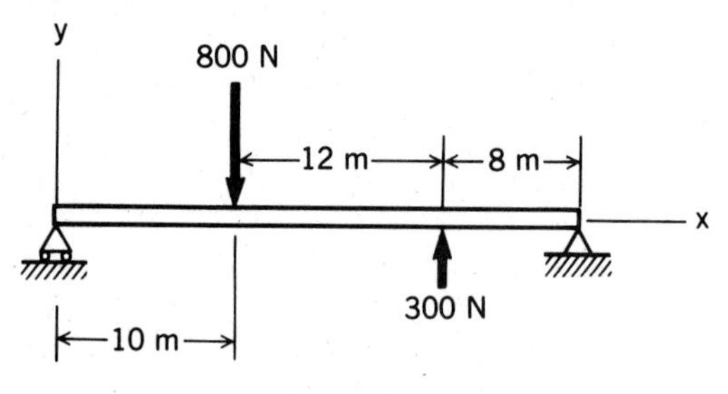


**Figure P.6.29**

**6.30.** Formulate the shear-force and bending-moment equations for the cantilever beam. Do not include the weight of the beam.

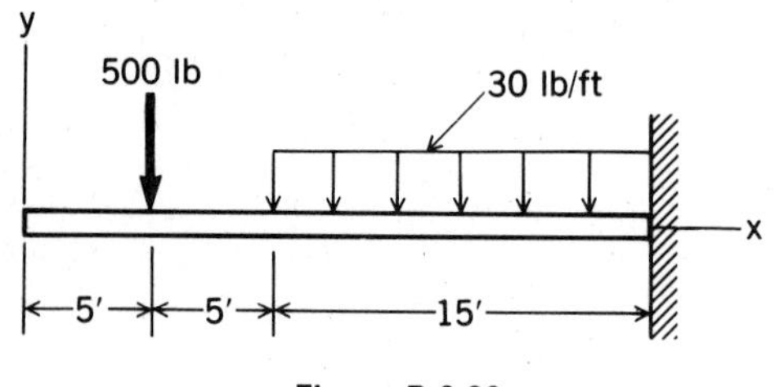


**Figure P.6.30**

**6.31.** Determine the shear-force and bending-moment equations for the simply supported beam.

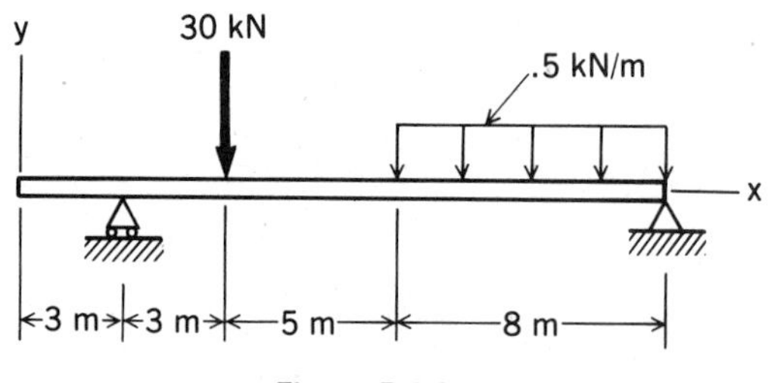


**Figure P.6.31**

**6.32.** For the beam shown, what is the shear force and bending-moment at the following positions?

(a) 5 ft from the left end
(b) 12 ft from the left end
(c) 5 ft from the right end

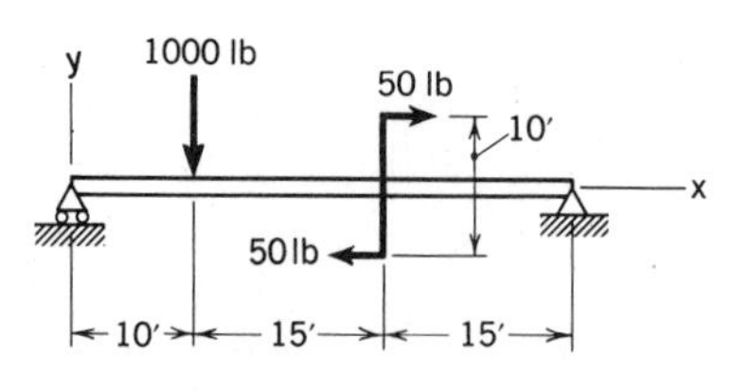


**Figure P.6.32**

**6.33.** Formulate the shear-force and b nding-moment equations for the simply supported beam.

**Figure P.6.33**

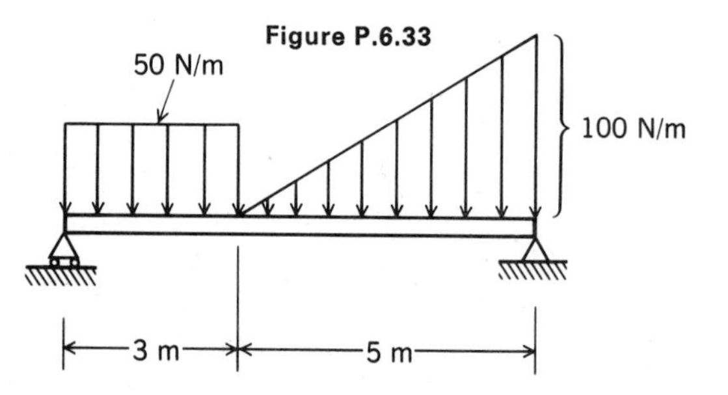

**6.34.** Compute shear force and bending moments for the bent beam as functions of $s$ along the centerline of the beam.

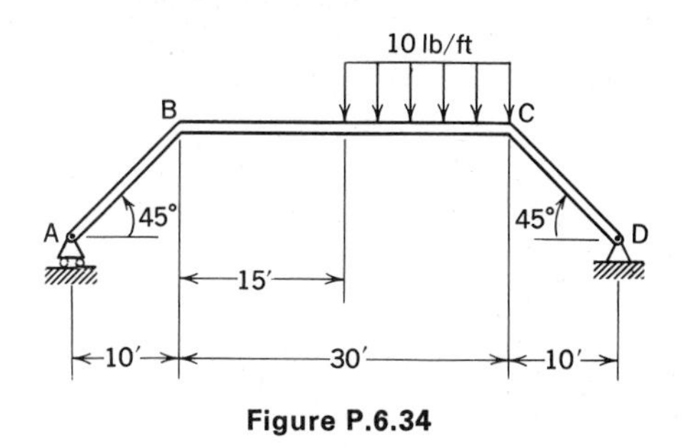


Figure P.6.34

**6.35.** A simply supported beam is loaded in two planes. This means there will be shear-force components $V_y$ and $V_z$ and bending-moment components $M_z$ and $M_y$. Compute these as functions of $x$. The beam is 40 ft in length.

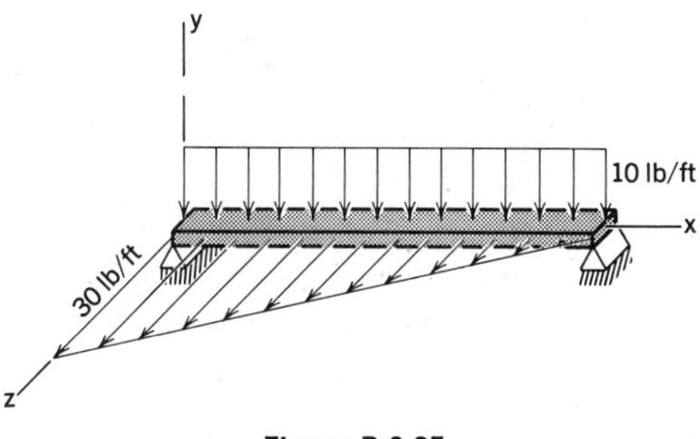

Figure P.6.35

**6.36.** What are the shear force, bending moment, and axial force for the three-dimensional cantilever beam? Give your results separately for the three portions $AB$, $BC$, and $CD$. Neglect the weight of the member. Use $s$ as the distance along the centerline from $D$.

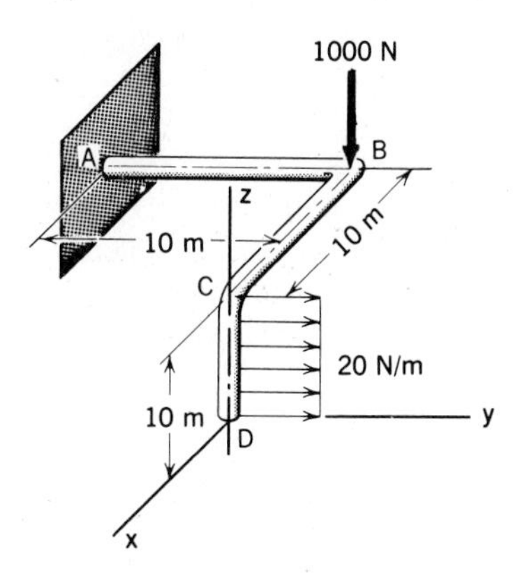

Figure P.6.36

**6.37.** Oil flows from a tank through a pipe $AB$. The oil weighs 40 lb/ft³ and, in flowing, develops a drag on the pipe of 1 lb/ft. The pipe has an inside diameter of 3 in. and a length of 20 ft. Flow conditions are assumed to be the same along the entire length of the pipe. What are the shear force, bending moment, and axial force along the pipe? The pipe weighs 10 lb/ft.

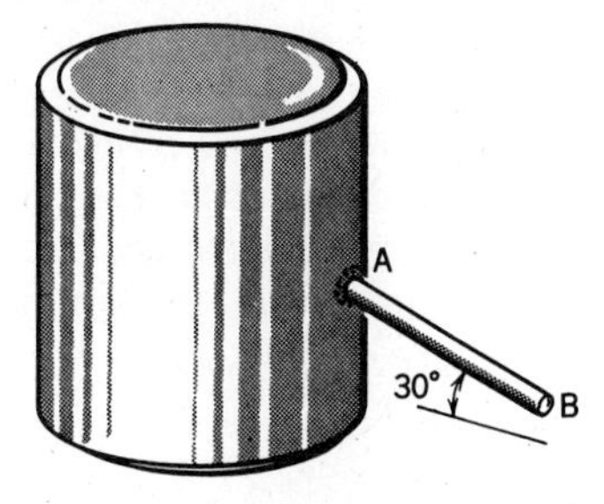


Figure P.6.37

**6.38.** Determine the shear force, bending moment, and axial force as functions of $\theta$ for the circular beam.

Figure P.6.38

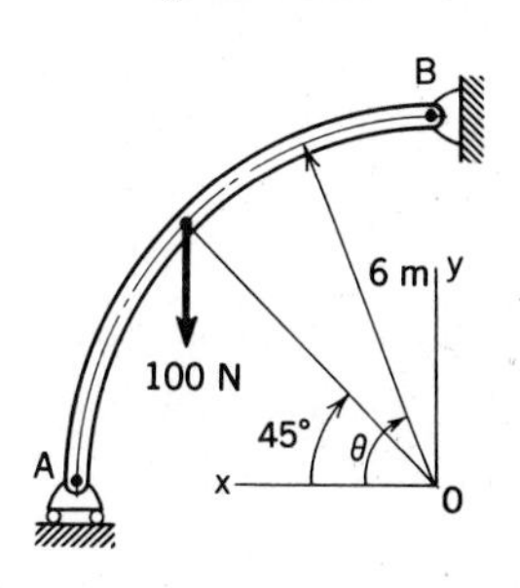

**6.39.** A hoist can move along a beam while supporting a 10,000-lb load. If the hoist starts at the left and moves from $\bar{x} = 3$ to $\bar{x} = 12$, determine the shear force and bending moment at $A$ in terms of $\bar{x}$. At what position $\bar{x}$ do we get the maximum shear force at $A$ and the maximum bending moment at $A$? What are their values?

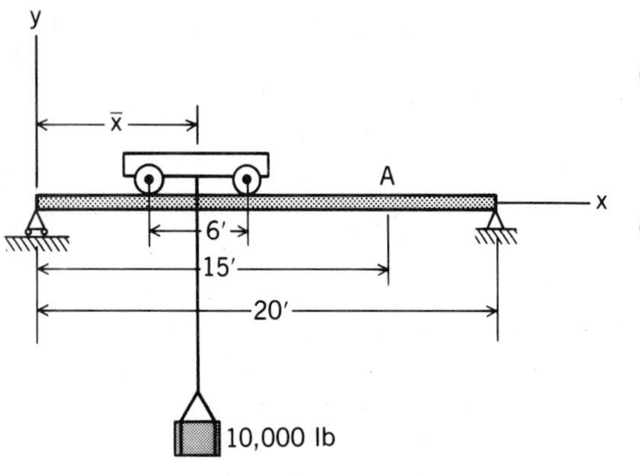


Figure P.6.39

**6.40.** A pipe weighs 10 lb/ft and has an inside diameter of 2 in. If it is full of water and the pressure of the water is that of the atmosphere at the entrance $A$, compute the shear force, axial force, and bending moment of the pipe from $A$ to $D$. Use coordinate $s$ measured from $A$ along the centerline of the pipe.

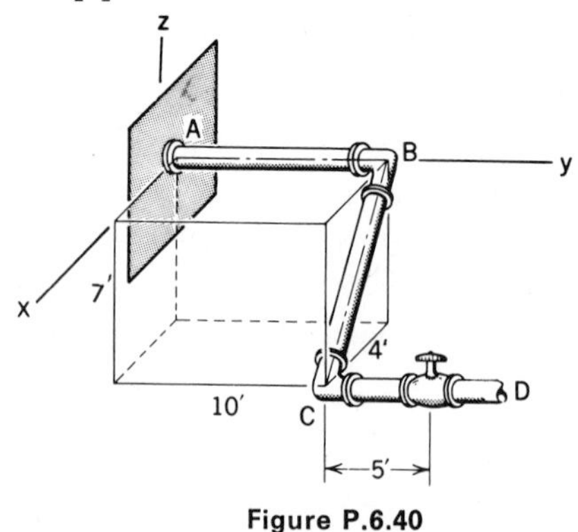


Figure P.6.40

**6.41.** After finding the supporting forces, determine for Problem 6.29 the shear-force and bending-moment equations without the further aid of free-body diagrams.

**6.42.** Determine the shear-force and bending-moment equations for Problem 6.30 without the aid of free-body diagrams.

**6.43.** In Problem 6.31, after determining the supporting forces, determine the shear-force and bending-moment equations without the aid of free-body diagrams.

**6.44.** In Problem 6.32, after finding the supporting forces, write the shear force and bending moment as a function of $x$ for the beam without the aid of free-body diagrams.

**6.45.** Give the shear-force and bending-moment equations for the cantilever beam. Except for determining the supporting forces, do not use free-body diagrams.

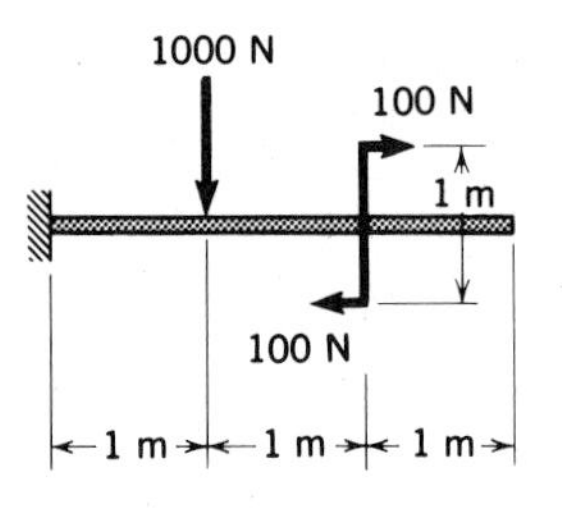


Figure P.6.45

**6.46.** Formulate the shear-force and bending-moment equations for the simply supported beam. (*Suggestion:* For the domain $5 < x < 10$, it is simplest to replace the indicated downward triangular load, going from 400 N/m to zero, by a uniform 400-N/m uniform downward load from $x = 5$ to $x = 10$ plus a triangular upward load going from zero to 400 N/m in the interval.)

Figure P.6.46

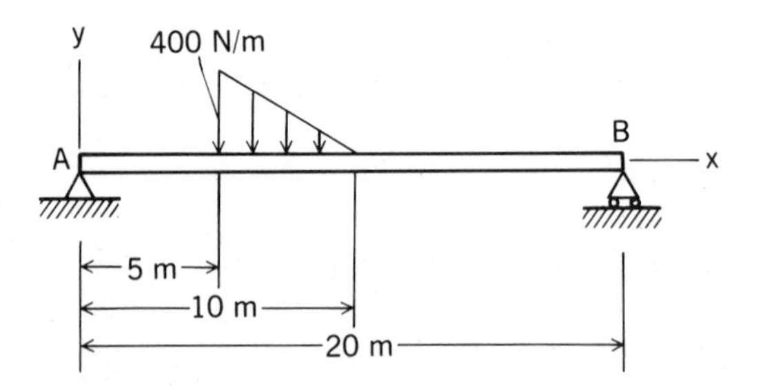

**6.47.** After finding the supporting forces for the simply supported beam $AB$, express the shear-force and bending-moment equations without the aid of free-body diagrams. The 10-kN load is applied to a bracket welded to the beam $AB$.

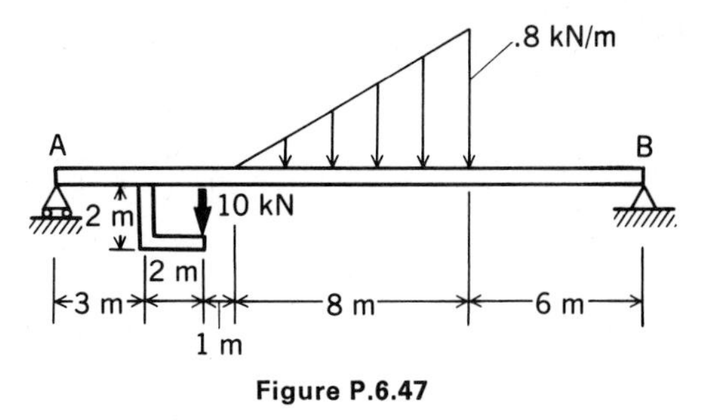


Figure P.6.47

## 6.8 Differential Relations for Equilibrium

In Section 6.7, we considered free bodies of *finite size* comprising variable portions of a beam in order to ascertain the resultant force system at sections along the beam. We shall now proceed in a different manner by examining an *infinitesimal slice* of the beam. Equations of equilibrium for this slice will then yield *differential equations* rather than algebraic equations for the variables $V$ and $M$.

Consider a slice $\Delta x$ of the beam shown in Fig. 6.41. We adopt the convention that intensity of loading $w$ in the positive coordinate direction is positive. We shall assume here that the weight of the beam has been included in the intensity of loading so that all forces acting on the element have been shown on the free-body diagram of the element in Fig. 6.42. Note we have employed positive shear-force and bending-moment

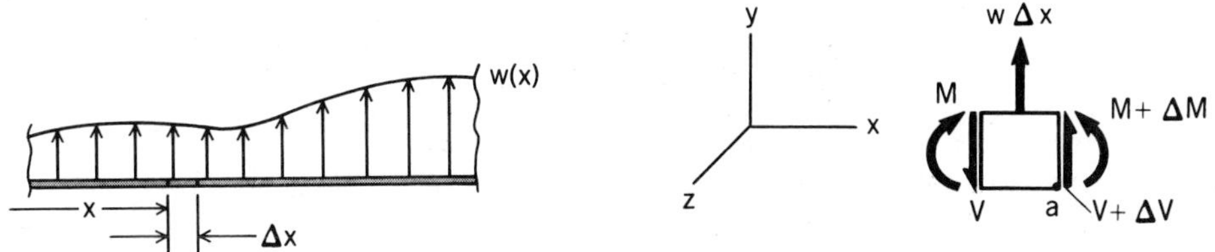


**Figure 6.41.** Element $\Delta x$ of beam.

**Figure 6.42.** Free-body diagram of element.

conventionwise in the free-body diagram for reasons explained earlier. We now apply the equations of equilibrium. Thus, summing forces:

$\underline{\sum F_y = 0:}$

$$-V + (V + \Delta V) + w\,\Delta x = 0$$

Taking moments about edge $a$ of the element, we get

$\underline{\sum M_a = 0:}$

$$-M + V\,\Delta x - (w\,\Delta x)(\beta\,\Delta x) + (M + \Delta M) = 0$$

where $\beta$ is some fraction which, when multiplied by $\Delta x$, gives the proper moment arm of the force $w\,\Delta x$ about edge $a$. These equations can be written in the following manner after we cancel terms and divide through by $\Delta x$:

$$\frac{\Delta V}{\Delta x} = -w$$

$$\frac{\Delta M}{\Delta x} = -V + w\beta\, \Delta x$$

In the limit as $\Delta x \longrightarrow 0$, we get the following differential equations:

$$\boxed{\begin{aligned} \frac{dV}{dx} &= -w \qquad \text{(a)} \\ \frac{dM}{dx} &= -V \qquad \text{(b)} \end{aligned}} \tag{6.3}$$

We may next integrate Eqs. 6.3(a) and 6.3(b) from position 1 along the beam to position 2. Thus, we have

$$(V)_2 - (V)_1 = -\int_1^2 w\, dx$$

Therefore,

$$\boxed{(V)_2 = (V)_1 - \int_1^2 w\, dx} \tag{6.4}$$

$$(M)_2 - (M)_1 = -\int_1^2 V\, dx$$

Therefore,

$$\boxed{(M)_2 = (M)_1 - \int_1^2 V\, dx} \tag{6.5}$$

Equation (6.4) means that the change in the shear force between two points on a beam equals minus the area under the loading curve between these points provided that there is no point force present in the interval.[11] Note that, if $w(x)$ is positive in an interval, the area under this curve is positive in this interval; if $w(x)$ is negative in an interval, the area under this curve is negative in this interval. Similarly, Eq. 6.5 indicates that the change in bending moment between two points on a beam equals minus the area of the shear-force diagram between these points provided that there are no point couple moments applied in the interval. If $V(x)$ is positive in an interval, the area under this curve is positive in this interval; if $V(x)$ is negative in an interval, the area under the curve is negative for this interval. In sketching the diagram, we shall make use of Eqs. 6.4 and 6.5 as well as the differential equations 6.3.

**EXAMPLE 6.8**

Sketch the shear-force and bending-moment distributions for the simply supported beam shown in Fig. 6.43 and label the key points.

[11]The differential equation 6.3(a) is only meaningful with a continuous loading present, while Eq. 6.3(b) is only valid in the absence of point couple moments.

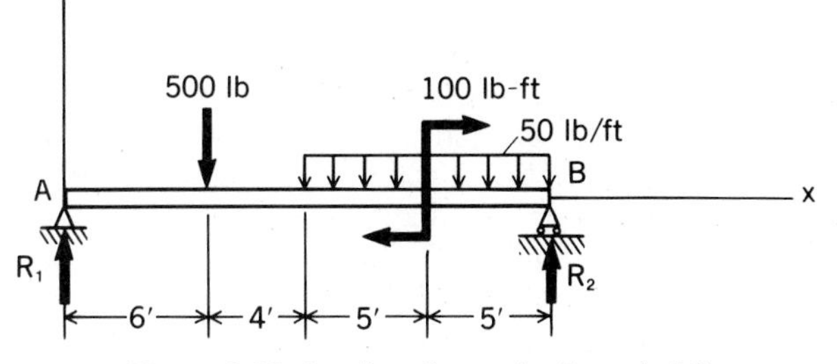


**Figure 6.43.** Loading diagram for Example 6.8.

The supporting forces $R_1$ and $R_2$ are found by rules of statics. Thus,

$\sum M_B = 0$:

$$-R_1(20) + (500)(14) + (50)(10)(10/2) - 100 = 0$$

Therefore,

$$R_1 = 470 \text{ lb}$$

$\sum M_A = 0$:

$$R_2(20) - (500)(6) - (50)(10)(15) - 100 = 0$$

Therefore,

$$R_2 = 530 \text{ lb}$$

In sketching the diagrams, we shall employ Eqs. 6.3, 6.4, and 6.5—i.e., the differential equations of equilibrium and their integrals. Accordingly, we first draw the loading diagram in Fig. 6.44(a), and we shall then sketch the shear-force and

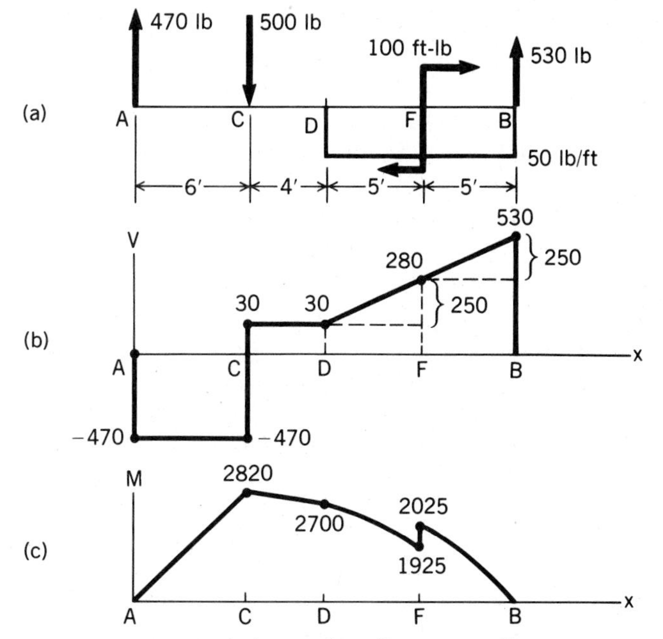


**Figure 6.44.** Shear-force and bending-moment diagrams.

bending-moment diagrams directly below without the aid of the shear-force and bending-moment equations, evaluating key points as we go.

Note as we start on the shear diagram that the 470-lb supporting force induces a negative shear of $-470$ lb just to the right of the support. Now from $A$ to $C$, the area under the loading curve is zero and so, in accordance with Eq. 6.4, there is no change in the value of shear between $A$ and $C$. Hence, $V_C = -470$ lb, as shown in Fig. 6.44(b). Also, since $w = 0$ between $A$ and $C$, the slope of the shear curve should be zero, in accordance with Eq. 6.3(a). And so we have a horizontal line for $V$ between $A$ and $C$. Now as we cross $C$, the 500-lb downward force will induce a positive increment of shear of value 500 on sections to the right of it. Accordingly, $V$ *jumps* from $-470$ lb to $+30$ lb as we cross $C$. Between $C$ and $D$ there is no loading $w$, so $V_D = V_C$ and we have a 30-lb shear force at point $D$. Again, since $w = 0$ in this interval, the slope of the shear curve is zero and we have a horizontal line for the shear curve between $C$ and $D$. Since there is no concentrated load at $D$, there is no sudden change in shear as we cross this point. Next, the change in shear between $D$ and $B$ is minus the area of the loading curve[12] in this interval in accordance with Eq. 6.4. But this area is $(-50)(10) = -500$. Hence, from Eq. 6.4 the value of $V_B$ (just to the left of the support) is $V_D - (-500) = 530$ lb. Also, since $w$ is negative and constant between $D$ and $B$, the slope of the shear curve should be positive and constant, in accordance with Eq. 6.3(a). Hence, we can draw a straight line between $V_D = 30$ lb and $V_B = 530$ lb. As we now cross the right support force, we see that it induces a negative shear of 530 lb on sections to the right of the support, and so at $B$ the shear curve comes back to zero.

We now proceed with the bending-moment curve. With no point couple moment present at $A$, the value of $M_A$ must be zero. The change in moment between $A$ and $C$ is then minus the area underneath the shear curve in this interval. We can then say from Eq. 6.5 that $M_C = M_A - (-470)(6) = 0 + 2820 = 2820$ ft-lb, and we denote this in the moment diagram. Furthermore the value of $V$ is a negative constant in the interval and, accordingly [see Eq. 6.3(b)], the slope of the moment curve is positive and constant. We can then draw a straight line between $M_A$ and $M_C$. Between $C$ and $D$ the area for the shear diagram is 120 lb-ft, and so we can say that $M_D = M_C - (120) = 2820 - 120 = 2700$ ft-lb. Again, with $V$ constant and positive in the interval, the slope of the moment curve must be negative and constant in the interval and has been so drawn. Between $D$ and $F$ the area under the shear curve is readily seen to be $(30)(5) + \frac{1}{2}(5)(250) = 775$ ft-lb. Hence, the bending moment goes from 2700 ft-lb at $D$ to 1925 ft-lb at $F$. Now the shear curve is positive and *increasing* in value as we go from $D$ to $F$. This means that the slope of the bending-moment curve is negative and becoming *steeper* as we go from $D$ to $F$. As we go by $F$ we encounter the 100-ft-lb point couple moment and we can say that this point couple moment induces a positive 100-ft-lb moment on sections to the right of point $F$. Accordingly, there is a sudden increase in bending moment of 100 ft-lb at $F$, as has been shown in the diagram. The area of the shear diagram between $F$ and $B$ is readily seen from Fig. 6.44(b) to be $(280)(5) + \frac{1}{2}(5)(250) = 2025$ ft-lb. We see then that the

[12]Note that the point couple moment has a zero net force and so need not be of concern in the interval from $D$ to $B$ as far as shear is concerned. However, it will be a point where sudden change occurs in the bending-moment diagram.

bending moment goes to zero at $B$. Since the shear force is positive and *increasing* between $F$ and $B$, we conclude that the slope of the bending-moment curve is negative and becoming *steeper* as we approach $B$. We have thus drawn the shear-force and bending-moment diagram and have labeled all key points.

Note that to be correct both the shear-force and bending-moment curves must go to zero at the end of the beam to the right of the right support. This serves as a check on the correctness of the calculations.

In Example 6.8, we can get equations and diagrams of shear force and bending moment independently of each other. With simple loadings such as point forces, point couples, and uniform distributions, this can readily be done. Indeed, this covers many problems that occur in practice. Usually, all that is needed are the labeled diagrams of the kind that we set forth in the previous problem. In problems with more complex loadings, we usually set forth the equations in the customary manner and then sketch the curves using the *equations* to give key values of $V$ and $M$ (the areas for the various curves are no longer the simple familiar ones, thus precluding advantageous use of Eqs. 6.4 and 6.5); the key points are then connected by curves sketched by making use of the slope relations as in Example 6.8.

It will be helpful to remember that if a curve has *increasing* negative[13] or positive values, the subsequent curve must have a *steepening* slope over the corresponding range. On the other hand, if a curve has *decreasing* negative or positive values, the subsequent curve must have a *flattening* slope over the corresponding range.

You will note in the preceding examples that the key points of the shear-force and bending-moment diagrams were evaluated and marked. The maximum value of both the shear force and bending moment were easily depicted from these diagrams. We wish to note, in this regard, that at points on shear-force and bending-moment curves where there is zero value of slope, there may be possible maximum values of shear

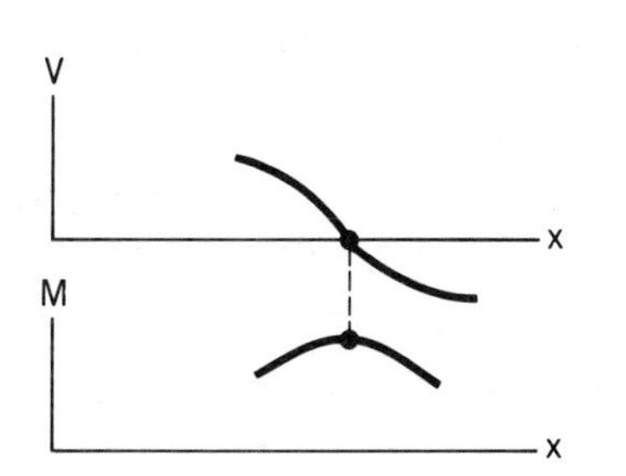


**Figure 6.45.** At $w = 0$, possible maximum for $V$.

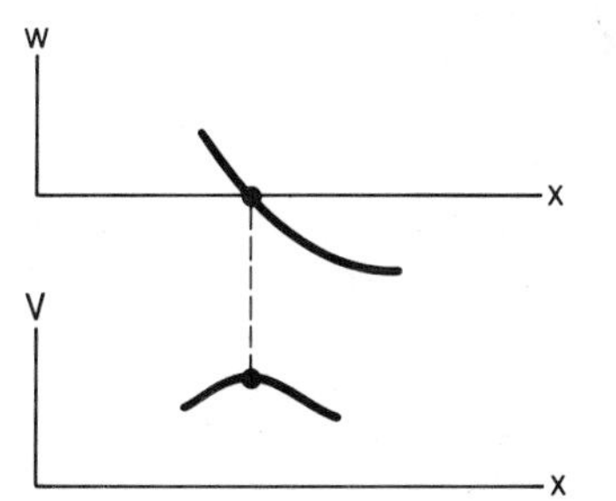


**Figure 6.46.** At $V = 0$, possible maximum for $M$.

force and bending moment, respectively, for the beam. This is illustrated for shear force in Fig. 6.45 and for bending moment in Fig. 6.46. Note that where the loading

[13]By an increasing negative value, we mean here an increasing absolute value. That is, $-200$ is considered larger negatively than $-100$.

curve $w$ crosses the $x$ axis, we accordingly have the position of a possible maximum shear force; similarly, where the shear curve $V$ crosses the $x$ axis, we accordingly have the position of a possible maximum bending moment. These respective positions and corresponding values of shear force and bending moment should be evaluated and marked in the diagram.

## Problems

**6.48.** After finding the supporting forces of the cantilever beam, sketch the shear-force and bending-moment diagrams labeling key points.

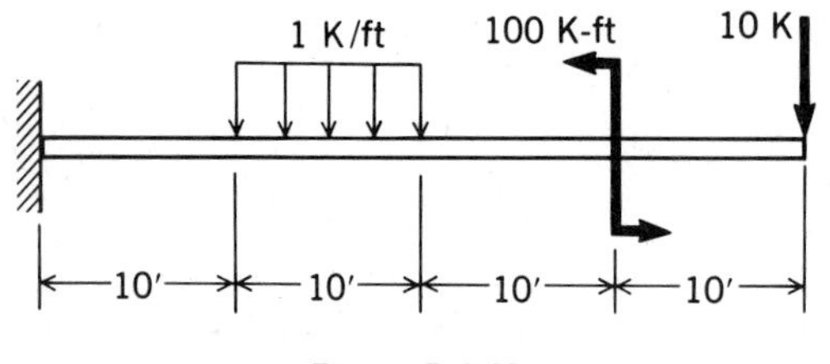


Figure P.6.48

**6.49.** What is the maximum negative bending moment in the region between the supports for the simply supported beam?

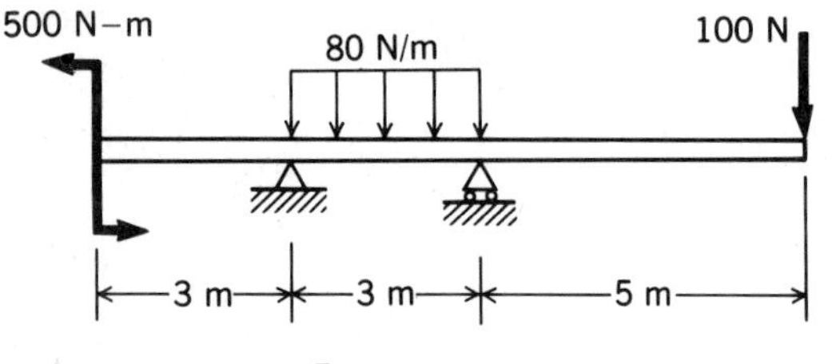


Figure P.6.49

**6.50.** Find the supporting forces for the simply supported beam. Then sketch the shear-force and bending-moment diagrams, labeling key points.

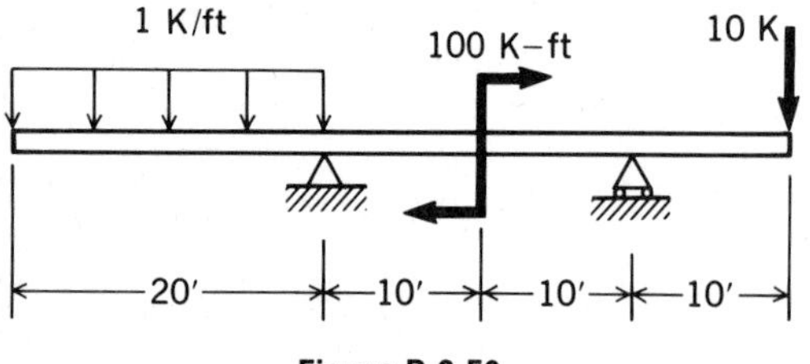


Figure P.6.50

**6.51.** Sketch the shear-force and bending-moment diagrams and compute key points for the overhanging beam.

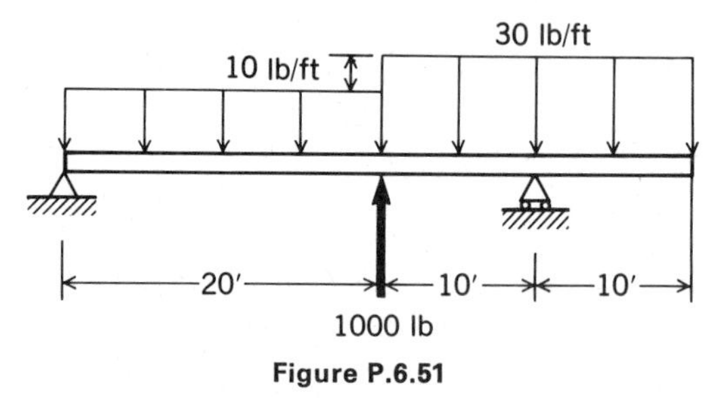


Figure P.6.51

**6.52.** A simply supported beam $AB$ is shown. A bar $CD$ is welded to the beam. After determining the supporting forces, sketch the shear-force and bending-moment diagram and determine the maximum bending moment. (*Hint:* Find the position for $V = 0$ using similar triangles.)

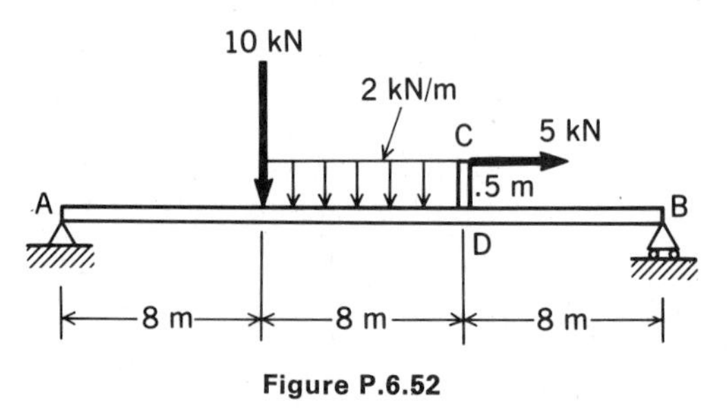


Figure P.6.52

**6.53.** Show the shear-force and bending-moment diagrams and evaluate key points only for the cantilever beam.

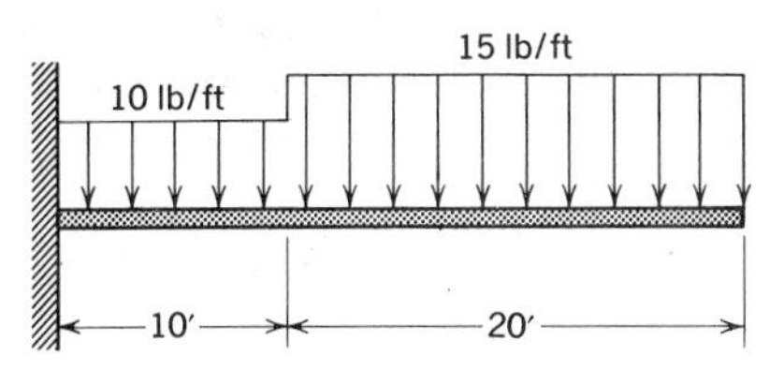


Figure P.6.53

**6.54.** Sketch the shear-force and bending-moment diagrams for the sinusoidally loaded beam. What is the maximum bending moment?

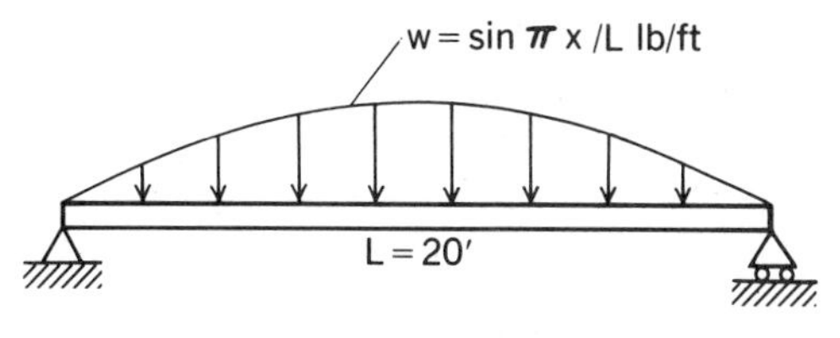


Figure P.6.54

**6.55.** Formulate the shear-force and bending-moment equations for the beam. Sketch the shear and moment diagrams.

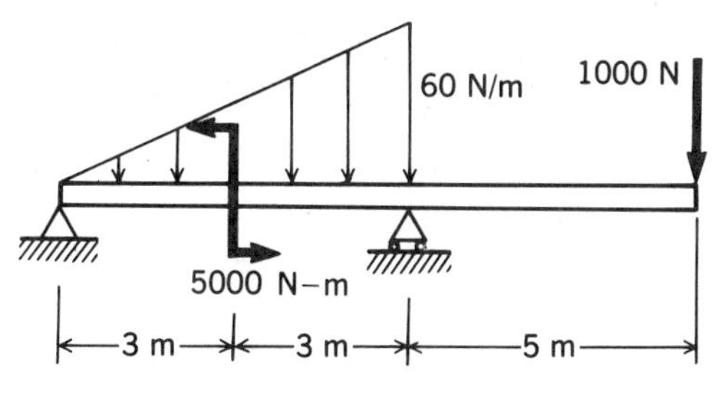


Figure P.6.55

**6.56.** A simply supported I-beam is shown. A hole must be cut through the web to allow passage of a pipe that runs horizontally at right angles to the beam.

(a) Where, within the marked 24-ft section, would the hole least affect the moment-carrying capacity of the beam?

(b) In the same marked section, where should the hole go to least affect the shear-carrying capacity of the beam?

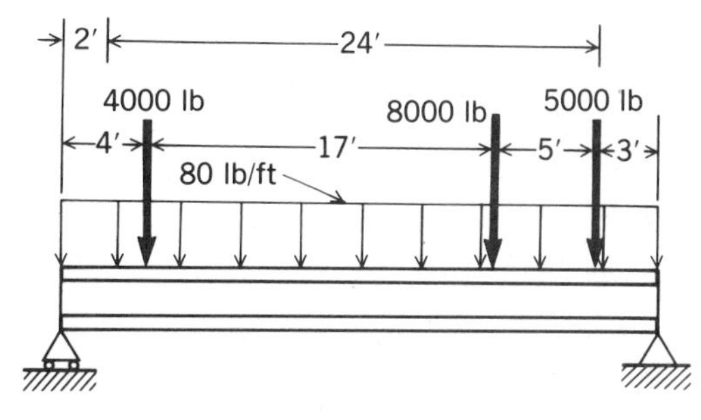


Figure P.6.56

***6.57.** A cantilever beam supports a parabolic and a triangular load. What are the shear-force and bending-moment equations? Sketch the shear-force and bending-moment diagrams. See the suggestion in Problem 6.46 regarding the triangular load.

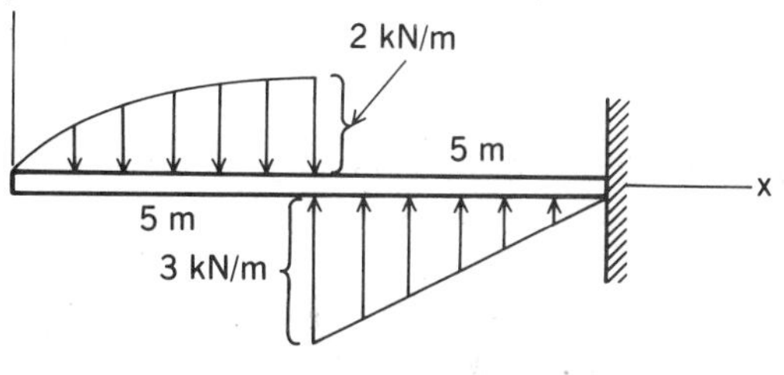


Figure P.6.57

**6.58.** Determine the shear-force and bending-moment equations for the beam. Then sketch the diagrams using the aforementioned equations if necessary to ascertain key points in the diagrams, such as the position between the supports where $V = 0$. What is the bending moment there?

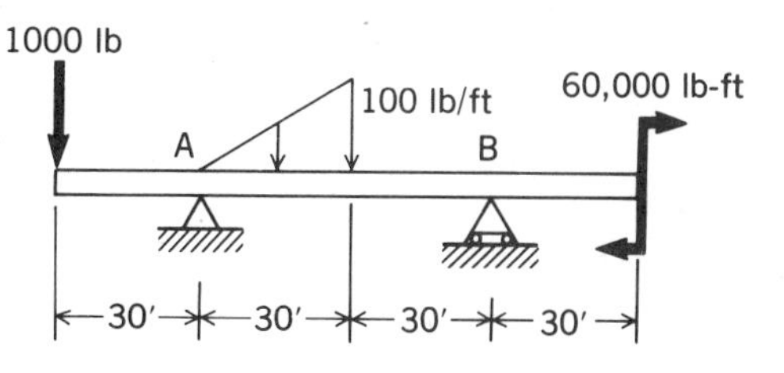


Figure P.6.58

# Part C
# Chains and Cables

## 6.9 Introduction

We often encounter relatively flexible cables or chains that are used to support loads. In suspension bridges, for example, we find a coplanar arrangement in which a cable supports a large load. The weight of the cable itself in such cases may often be considered negligible. In transmission lines, on the other hand, the principal force is the weight of the cable itself. In this section, we shall evaluate the shape of and the tension in the cables for both these cases.

To facilitate computations, the model of the structural system will be assumed to be perfectly flexible and inextensible. The flexibility assumption means that at the center of any cross section of the cable only a tensile force is transmitted and there can be no bending moment there. The force transmitted through the cable must, under these conditions, be tangent to the cable at all positions along the cable. The inextensibility assumption means that the length of the cable is constant.

## 6.10 Coplanar Cables

We shall now consider the case of a cable suspended between two rigid supports $A$ and $B$ under the action of a loading function $w(x)$ given per unit length as measured in the *horizontal* direction. This loading will be considered to be coplanar with the cable and directed vertically, as shown in Fig. 6.47. Consider an element of the cable of length

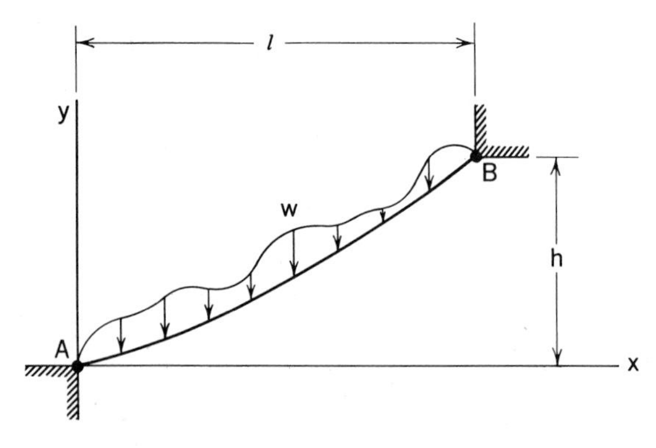


**Figure 6.47.** Coplanar cable; $w = w(x)$.

$\Delta s$ as a free body (Fig. 6.48). Summing forces in the $x$ and $y$ directions, respectively, we get

$$-T\cos\theta + (T+\Delta T)\cos(\theta+\Delta\theta) = 0 \quad (6.6a)$$

$$-T\sin\theta + (T+\Delta T)\sin(\theta+\Delta\theta) - w_{av}\,\Delta x = 0 \quad (6.6b)$$

where $w_{av}$ is the average loading. Dividing by $\Delta x$ and taking the limit as $\Delta x \longrightarrow 0$, we have

$$\lim_{\Delta x\to 0}\left[\frac{(T+\Delta T)\cos(\theta+\Delta\theta) - T\cos\theta}{\Delta x}\right] = 0$$

$$\lim_{\Delta x\to 0}\left[\frac{(T+\Delta T)\sin(\theta+\Delta\theta) - T\sin\theta}{\Delta x}\right] = w$$

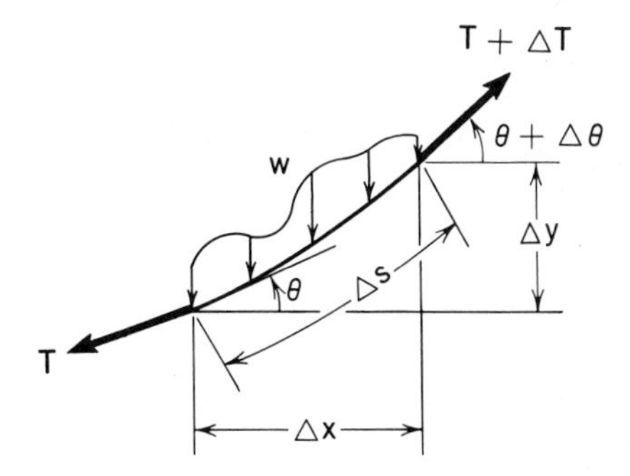


**Figure 6.48.** Element of cable.

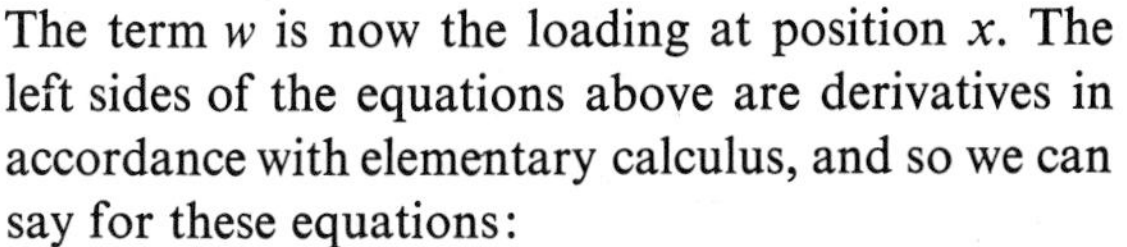

The term $w$ is now the loading at position $x$. The left sides of the equations above are derivatives in accordance with elementary calculus, and so we can say for these equations:

$$\frac{d(T\cos\theta)}{dx} = 0 \quad (6.7a)$$

$$\frac{d(T\sin\theta)}{dx} = w \quad (6.7b)$$

From Eq. 6.7(a), we conclude that

$$T\cos\theta = \text{constant} = H \quad (6.8)$$

where clearly the constant $H$ represents the *horizontal* component of the tensile force anywhere along the cable. Integrating Eq. 6.7(b), we get

$$T\sin\theta = \int w(x)\,dx + C_1' \quad (6.9)$$

where $C_1'$ is a constant of integration. Solving for $T$ in Eq. 6.8 and substituting into Eq. 6.9, we get

$$\frac{\sin\theta}{\cos\theta} = \frac{1}{H}\int w(x)\,dx + C_1$$

Noting that $\sin\theta/\cos\theta = \tan\theta = dy/dx$, we have, on carrying out a second integration:

$$\boxed{y = \frac{1}{H}\int\left[\int w(x)\,dx\right]dx + C_1x + C_2} \quad (6.10)$$

Equation 6.10 is the deflection curve for the cable in terms of $H$, $w(x)$, and the constants of integration. The constants of integration must be determined by the boundary conditions at the supports $A$ and $B$.

**EXAMPLE 6.9**

A cable is shown in Fig. 6.49 terminating at points at the same elevation. The loading distribution is uniform, given by constant $w$. Other known data are the span,

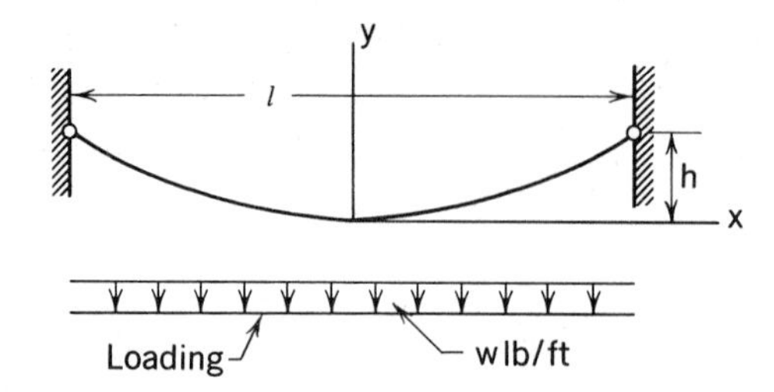


**Figure 6.49.** Cable with sag $h$.

$l$, and the sag, $h$. The maximum force in the cable, the shape of the cable, and the length of the cable are desired. Neglect the weight of the cable itself.

We have placed the reference at the center of the cable for simplicity as shown in the diagram. Noting that $w(x) = w =$ constant for this problem, we can proceed directly with the integrations in Eq. 6.10. Thus, we have

$$y = \frac{1}{H}\int\left[\int w\,dx\right]dx + C_1 x + C_2 = \frac{1}{H}\int wx\,dx + C_1 x + C_2$$

Therefore,

$$y = \frac{1}{H}\frac{wx^2}{2} + C_1 x + C_2 \tag{a}$$

The deflection curve is thus a *parabola.* We now require that $y = dy/dx = 0$, when $x = 0$. Thus, the constants $C_1 = C_2 = 0$. The deflection curve then is simply

$$y = \frac{w}{2H}x^2 \tag{b}$$

To get the constant $H$, we set $y = h$ for $x = l/2$. Thus,

$$h = \frac{w}{2H}\frac{l^2}{4}$$

Therefore,

$$H = \frac{wl^2}{8h} \tag{c}$$

The deflection curve is now fully established in terms of the data of the problem in the form

$$y = \frac{w}{2(wl^2/8h)}x^2 = 4\frac{hx^2}{l^2} \tag{d}$$

We next compute the *maximum tension* in the cable. Equation 6.8 can be used for this purpose. Solving for $T$, we get

$$T = \frac{H}{\cos\theta} \tag{e}$$

from which is apparent that the maximum value of $T$ occurs where $\theta$ is greatest. Examining the slope of the deflection curve,

$$\frac{dy}{dx} = \frac{w}{H}x \tag{f}$$

it is apparent that the largest $\theta$ occurs at $x = l/2$ (i.e., at the supports). Hence, from above we have, for $\theta_{max}$:

$$\theta_{max} = \tan^{-1}\left(\frac{dy}{dx}\right)_{x=l/2} = \tan^{-1}\left(\frac{w}{H}\frac{l}{2}\right) \quad \text{(g)}$$

Consequently, we get for $T_{max}$:

$$T_{max} = \frac{H}{\cos\left[\tan^{-1}(wl/2H)\right]} \quad \text{(h)}$$

From trigonometric consideration of the denominator,

$$T_{max} = \frac{H(4H^2 + w^2l^2)^{1/2}}{2H} = H\left[1 + \left(\frac{wl}{2H}\right)^2\right]^{1/2} \quad \text{(i)}$$

Substituting for $H$ using Eq. (c), we then get, on rearranging the terms,

$$T_{max} = \frac{wl}{2}\sqrt{1 + \left(\frac{l}{4h}\right)^2} \quad \text{(j)}$$

Finally, to determine the *length of the cable* for the given conditions, we must perform the following integration:

$$L = 2\int_0^{s_{max}} ds = 2\int_0^{s_{max}} \sqrt{dx^2 + dy^2} = 2\int_0^{l/2} \sqrt{1 + \left(\frac{dy}{dx}\right)^2}\, dx \quad \text{(k)}$$

Now the slope, $dy/dx$, equals $wx/H$ [see Eq. (f)], which on substituting for $H$ [see Eq.(c)] becomes $8hx/l^2$. Therefore,

$$L = 2\int_0^{l/2} \sqrt{1 + \left(\frac{8hx}{l^2}\right)^2}\, dx$$

This may be integrated using a formula to be found in Appendix I to give

$$L = \left[x\sqrt{1 + \left(\frac{8hx}{l^2}\right)^2} + \frac{l^2}{8h}\sinh^{-1}\frac{8hx}{l^2}\right]_0^{l/2}$$

Substituting limits, we have

$$L = \left[\frac{l}{2}\sqrt{1 + \left(\frac{4h}{l}\right)^2} + \frac{l^2}{8h}\sinh^{-1}\frac{4h}{l}\right]$$

Rearranging so that the result is given as a function of the sag ratio $h/l$ and the span $l$, we get finally

$$L = \frac{l}{2}\left[\sqrt{1 + 16\left(\frac{h}{l}\right)^2} + \frac{1}{4h/l}\sinh^{-1}\frac{4h}{l}\right] \quad \text{(l)}$$

Another possible approach to determining the length of the cable is to expand the integrand in Eq. (k) as a power series using the *binomial theorem*. Thus, we have

$$L = 2\int_0^{l/2} \left[1 + \frac{1}{2}\left(\frac{dy}{dx}\right)^2 - \frac{1}{8}\left(\frac{dy}{dx}\right)^4 + \dots\right] dx \quad \text{(m)}$$

provided that $|dy/dx| < 1$ at all positions along the interval.[14] Now, employ Eq.

[14]Otherwise, the series diverges. Hence, this approach is limited to cases where the slope of the cable is less than 45°.

(f) to replace $dy/dx$ in Eq. (m) to get

$$L = 2\int_0^{l/2} \left(1 + \frac{1}{2}\frac{w^2}{H^2}x^2 - \frac{1}{8}\frac{w^4}{H^4}x^4 + \dots\right) dx \tag{n}$$

We can integrate a power series term by term and so we have, for $L$:

$$L = l\left[1 + \frac{1}{24}\left(\frac{w^2}{H^2}\right)l^2 - \frac{1}{640}\left(\frac{w^4}{H^4}\right)l^4 + \dots\right] \tag{o}$$

For cables having small slopes, the series converges rapidly and only the first few terms need generally be employed.

In Example 6.9, the supports are at the same level and consequently the position of zero slope is known (i.e., it is at the midpoint). We found it simplest to set our reference $xy$ at this point. In problems at the end of this section, the supports may not be at the same level. For such cases the reference is best taken at one of the supports. Also, the slope of the cable is often known at some point, and the problem may then be solved in much the same way as Example 6.9.

In the previous development, the loading was given as a function of $x$. Let us now consider the case of a cable *loaded only by its own weight*. The loading function is now most easily expressed as a function of $s$, the position along the cable. Equations 6.6 apply to this case provided that we replace $\Delta x$ by $\Delta s$. Dividing through by $\Delta s$ and taking the limit as $\Delta s \longrightarrow$ zero, we get equations analogous to Eqs. 6.7.

$$\frac{d(T\cos\theta)}{ds} = 0$$

$$\frac{d(T\sin\theta)}{ds} = w(s)$$

Integrating, we have

$$T\cos\theta = H \tag{6.11a}$$

$$T\sin\theta = \int w(s)\,ds + C_1' \tag{6.11b}$$

Eliminating $T$ from Eqs. 6.11, we get, as in the previous development:

$$\frac{dy}{dx} = \frac{1}{H}\int w(s)\,ds + C_1 \tag{6.12}$$

The right side of the equation is a function of $s$. Thus, we cannot directly integrate as a next step. Accordingly, note that

$$dy = (ds^2 - dx^2)^{1/2}$$

Hence, from this equation,

$$\frac{dy}{dx} = \left[\left(\frac{ds}{dx}\right)^2 - 1\right]^{1/2} \tag{6.13}$$

Substituting for $dy/dx$ in Eq. 6.12 using the preceding result, we get

$$\left[\left(\frac{ds}{dx}\right)^2 - 1\right]^{1/2} = \frac{1}{H}\int w(s)\,ds + C_1$$

Solving for $ds/dx$, we have

$$\frac{ds}{dx} = \left\{1 + \left[\frac{1}{H}\int w(s)\,ds + C_1\right]^2\right\}^{1/2}$$

Separating variables and integrating, we get

$$x = \int \frac{ds}{\left\{1 + [(1/H)\int w(s)\,ds + C_1]^2\right\}^{1/2}} + C_2 \tag{6.14}$$

As a first step, determine if possible the constant $C_1$ by applying a slope-boundary condition to Eq. 6.12. With this $C_1$ in Eq. 6.14, solve for $s$ as a function of $x$. Next, substitute for $s$ in Eq. 6.12 using this relation. Finally, integrate Eq. 6.12 with respect to $x$ to get $y$ as a function of $x$. Boundary conditions must then be used to determine $H$ as well as the remaining constant of the integration. The following examples will illustrate how these steps are carried out.

**EXAMPLE 6.10**

Consider a uniform cable having a span $l$ and a sag $h$ as shown in Fig. 6.50. The weight per unit length $w$ of the cable is a constant.

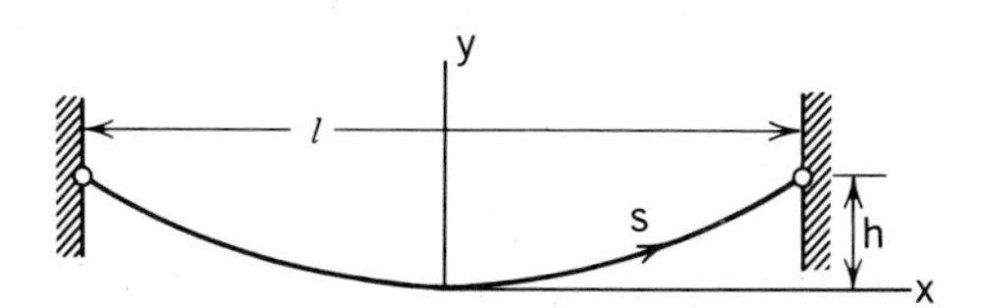


**Figure 6.50.** Uniform cable loaded by its own weight.

For simplicity, we have placed a reference at the center of the span where the slope of the cable is zero. Accordingly, consider Eq. 6.12 for this case:

$$\frac{dy}{dx} = \frac{1}{H}\int w\,ds + C_1 = \frac{w}{H}s + C_1 \tag{a}$$

When $s = 0$ we require that $dy/dx = 0$, whereupon $C_1$ is zero. Now consider Eq. 6.14:

$$\begin{aligned} x &= \int \frac{ds}{\left\{1 + [(1/H)\int w\,ds]^2\right\}^{1/2}} + C_2 \\ &= \int \frac{ds}{\{1 + [(w/H)s]^2\}^{1/2}} + C_2 \end{aligned} \tag{b}$$

Integrating the right side of the equation using an integration formula from Appendix I, we get

$$x = \frac{H}{w} \sinh^{-1} \frac{sw}{H} + C_2 \tag{c}$$

The constant $C_2$ must also be zero, since $x = 0$ at $s = 0$. Solving for $s$ from Eq. (c), we get

$$s = \frac{H}{w} \sinh \frac{xw}{H} \tag{d}$$

Substituting for $s$ in Eq. (a) using the preceding result, we have

$$\frac{dy}{dx} = \sinh \frac{w}{H} x \tag{e}$$

Integrating, we get

$$y = \frac{H}{w} \cosh \frac{w}{H} x + C_3$$

Since $y = 0$ at $x = 0$, the constant $C_3$ becomes $-H/w$. We then have the deflection curve:

$$y = \frac{H}{w}\left(\cosh \frac{w}{H} x - 1\right) \tag{f}$$

This curve is called a *catenary curve*.[15]

To determine $H$, we set $y = h$ when $x = l/2$. Thus,

$$h = \frac{H}{w}\left(\cosh \frac{wl}{2H} - 1\right) \tag{g}$$

This equation can be solved by trial and error. We may then proceed to determine the maximum force in the cable as well as the length of the cable in the manner followed in Example 6.9.

**EXAMPLE 6.11**

A water skier is shown in Fig. 6.51 dangling from a kite that is towed by a powerboat at a speed of 30 mph. The boat develops a thrust of 200 lb. The drag on

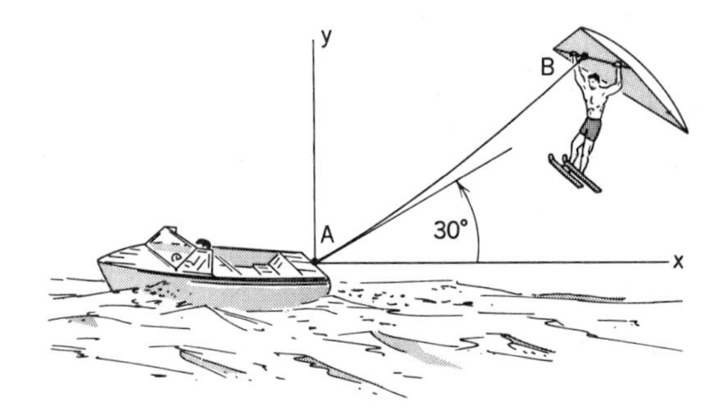


**Figure 6.51.** Analyze tow rope *AB*.

the boat from the water is estimated as 100 lb. At the support $A$, the rope has a tangent of 30°. If the man weighs 150 lb, find the height and the lift of the kite as

[15]The Latin for chain is *catena*.

well as the maximum tension in the rope. The kite weighs 25 lb. The uniform rope is 50 ft long and weighs .5 lb/ft. Neglect aerodynamic effects on the rope.

We start with Eq. 6.12, which becomes for this case:

$$\frac{dy}{dx} = \frac{w}{H}s + C_1 \tag{a}$$

Using a reference at $A$ as shown in the diagram, we know that $dy/dx = \tan 30° = .577$ when $s = 0$. Thus, we get for $C_1$,

$$C_1 = .577$$

Equation 6.14 is considered next. We have

$$x = \int \frac{ds}{\{1 + [(w/H)s + .577]^2\}^{1/2}} + C_2$$

Integrating by making a change in variable and using the proper integration formula in Appendix I, we get

$$x = \frac{H}{w} \sinh^{-1}\left(\frac{w}{H}s + .577\right) + C_2 \tag{b}$$

Solving for $s$, we have

$$s = \frac{H}{w}\left\{\sinh\left[(x - C_2)\frac{w}{H}\right] - .577\right\} \tag{c}$$

Substituting for $s$ in Eq. (a) using Eq. (c), we get

$$\frac{dy}{dx} = \sinh\left[(x - C_2)\frac{w}{H}\right] \tag{d}$$

Integrating again, we have

$$y = \frac{H}{w}\cosh\left[(x - C_2)\frac{w}{H}\right] + C_3 \tag{e}$$

We must now evaluate the unknown constants $C_2$, $C_3$, and $H$ using the boundary conditions and data of the problem. First, since $H$ is the horizontal component of force transmitted by the rope, we know that $H$ is the thrust of the boat minus the drag of the water. Thus,

$$H = 100 \text{ lb}$$

Also, $x = 0$ when $s = 0$, so that from Eq. (c), we get $C_2$ as follows:

$$\sinh\left(-\frac{.5}{100}C_2\right) = .577$$

Therefore,

$$-\frac{.5}{100}C_2 = \sinh^{-1} .577 = .549$$

Hence,

$$C_2 = -109.8$$

Finally, note that $x = 0$ when $y = 0$. From Eq. (e), we can then get constant $C_3$ in the following manner:

$$C_3 = -\frac{100}{.5}\cosh\left[\frac{-.5}{100}(-109.8)\right]$$
$$= -200 \cosh .548 = -231$$

We may now evaluate the position $x'$, $y'$ of point $B$ of the kite. To get $x'$, we insert for $s$ in Eq. (b) the value of 50 ft. Thus,

$$x' = \frac{100}{.5}\sinh^{-1}\left(\frac{.5}{100}50 + .577\right) - 109.8 = 40.9 \text{ ft}$$

Now from Eq. (e) we can get $y'$ and consequently the desired height.

$$y' = \frac{100}{.5}\cosh\left[(40.9 + 109.8)\frac{.5}{100}\right] - 231$$
$$= 28.6 \text{ ft} \qquad \text{(f)}$$

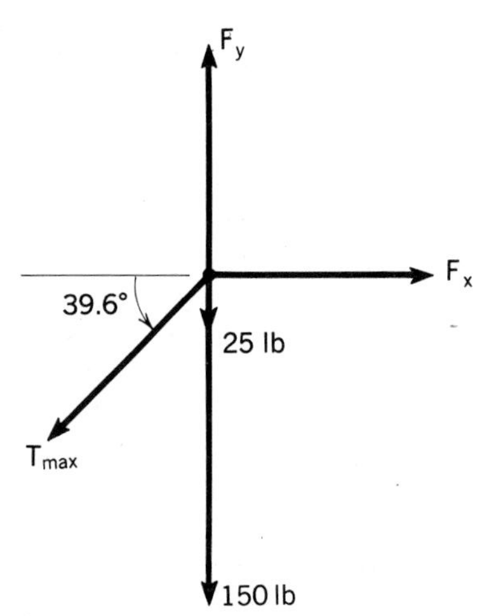


**Figure 6.52.** Free-body diagram of kite support.

The maximum tension in the rope occurs at point $B$, where $\theta$ is greatest. To get $\theta_{max}$, we go back to Eq. (a). Thus,

$$\left(\frac{dy}{dx}\right)_{max} = \tan\theta_{max} = \frac{.5}{100}(50) + .577 = .827$$

Therefore,

$$\theta_{max} = 39.6° \qquad \text{(g)}$$

Hence, from Eq. 6.11(a) we have for $T_{max}$:

$$T_{max} = \frac{100}{\cos 39.6°} = 130 \text{ lb} \qquad \text{(h)}$$

To get the lifting force of the kite, we draw a free-body diagram of point $B$ of the kite as shown in Fig. 6.52. Note that $F_y$ and $F_x$ are, respectively, the aerodynamic lift and drag forces on the kite. The lift force $F_y$ of the kite then becomes

$$F_y = 175 + T_{max}\sin 39.6° = 258 \text{ lb} \qquad \text{(i)}$$

## Problems

**6.59.** Find the length of a cable stretched between two supports at the same elevation with span $l = 200$ ft and sag $h = 50$ ft, if it is subjected to a vertical load of 4 lb/ft uniformly distributed in the horizontal direction. (Assume that the weight of the cable is either negligible or included in the 4-lb/ft distribution.) Find the maximum tension.

**6.60.** A cable supports a 8000-km uniform bar. What is the equation of the cable and what is the maximum tension in the cable?

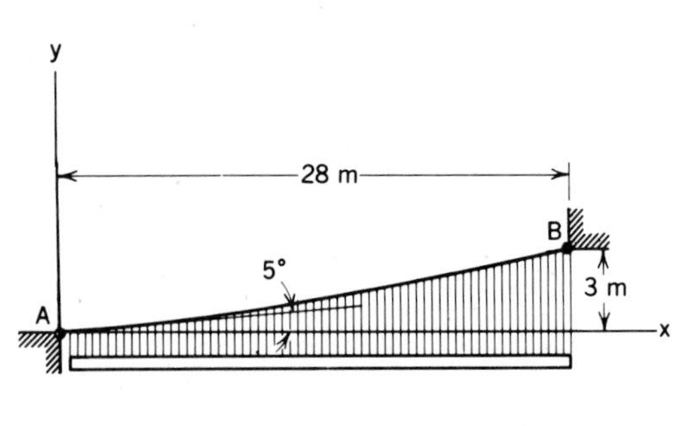


**Figure P.6.60**

**6.61.** A cable supports a uniform loading of 100 lb/ft. If the lowest point of the cable occurs 20 ft from point *A* as shown, what is the maximum tension in the cable and its length? Use *A* as the origin of reference.

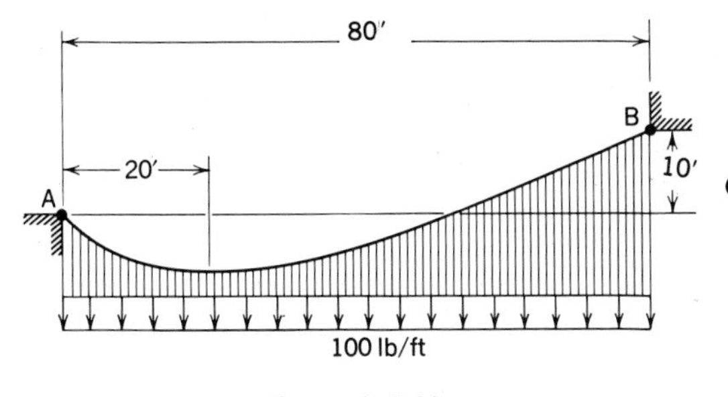


**Figure P.6.61**

**6.62.** A uniform cable is shown whose weight we shall neglect. If a loading given as $5x$ N/m is imposed on the cable, what is the deflection curve of the cable if there is a zero slope of the curve at point *A*? What is the maximum tension?

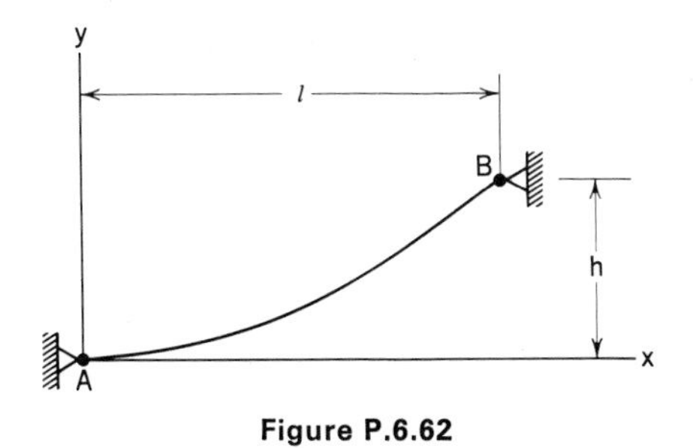


**Figure P.6.62**

**6.63.** The left side of a cable is mounted at an elevation 7 m below the right side. The sag, measured from the left support, is 7 m. Find the maximum tension if the cable has a *uniform* loading in the vertical direction of 1500 N/m. (*Suggestion:* Place reference at position of zero slope and determine the location of this point from the boundary conditions.)

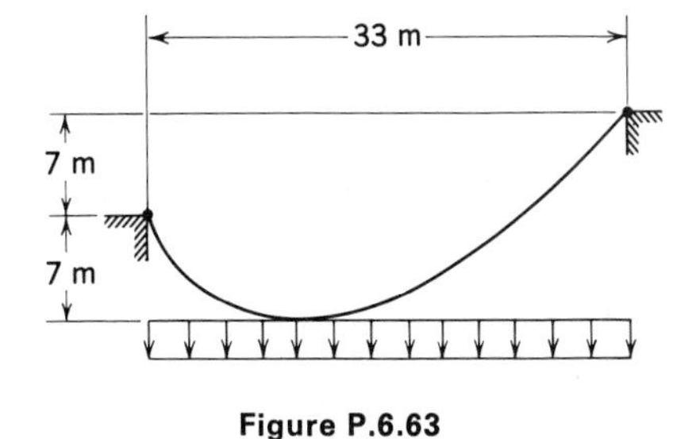


**Figure P.6.63**

**6.64.** A blimp is dragging a chain of length 500 ft and weight 10 lb/ft. A thrust of 300 lb is developed by the blimp as it moves against an air resistance of 200 lb. How much chain is on the ground and how high is the blimp? The vertical lift of the blimp on the cable is taken as 1000 lb.

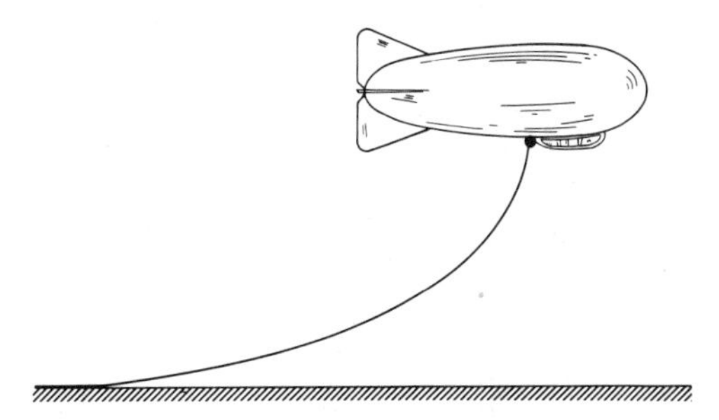

**Figure P.6.64**

**6.65.** A large balloon has a buoyant force of 100 lb. It is held by a 150-ft cable whose weight is .5 lb/ft. What is the height *h* of the balloon above the ground when a steady wind causes it to assume the position shown? What is the maximum tension on the cable?

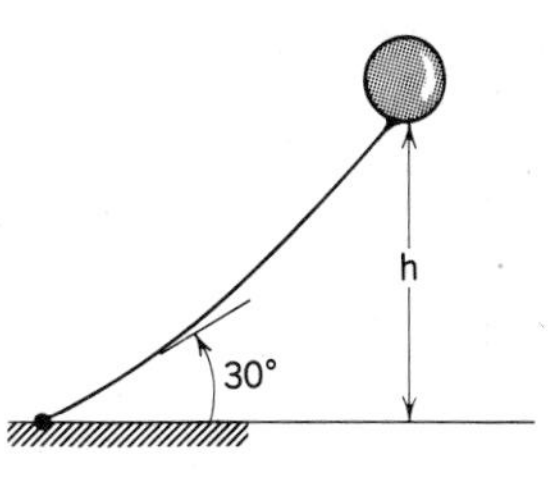


**Figure P.6.65**

**6.66.** What is the deflection curve for the uniform cable shown weighing 30 N/m? Find the maximum tension. Compute the height $h$ of the support $B$.

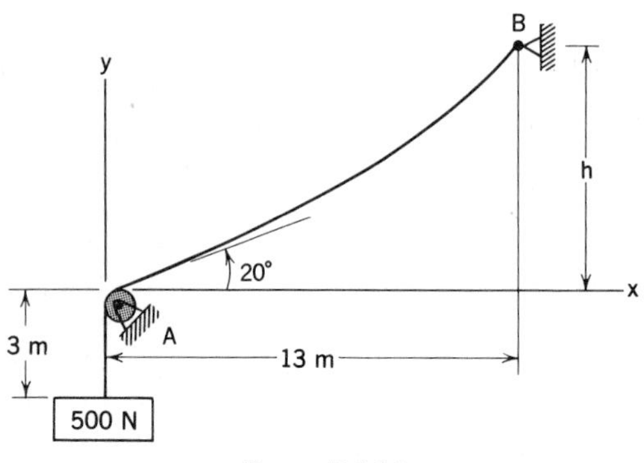


Figure P.6.66

**6.67.** A search boat is dragging the lake floor for stolen merchandise using a 100-m chain weighing 100 N/m. The tension of the chain at support point $B$ is 5000 N and the chain makes an angle of 50° there. What is the height of point $B$ above the lake bed? Also, what length of chain is dragging along on the bottom? Do not consider buoyant effects.

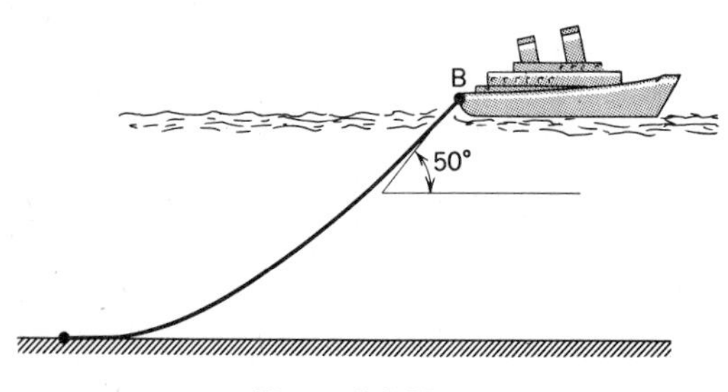


Figure P.6.67

**6.68.** A cable weighing 3 lb/ft is stretched between two points on the same level. If the length of the cable is 450 ft and the tension at the points of support is 1500 lb, find the sag and the distance between the points of support. Put reference at left support.

**6.69.** A flexible, inextensible cable is loaded by concentrated forces. If we neglect the weight of the cable, what are the supporting forces at $A$ and $B$? What are the tensions in the chord $AC$ and the angle $\alpha$? (*Hint:* Proceed by using finite free bodies and working from first principles.)

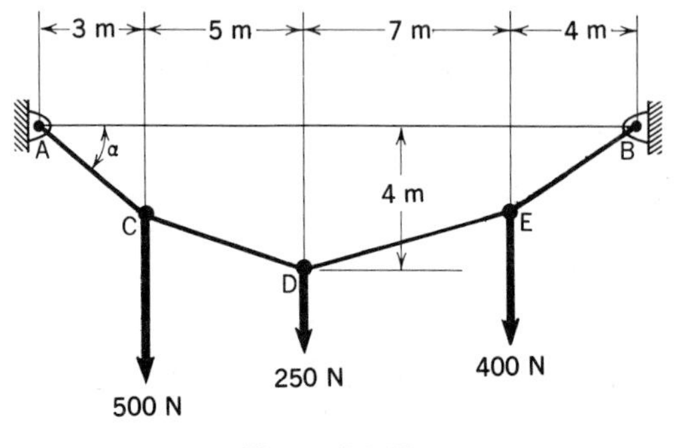


Figure P.6.69

**6.70.** A system of two inextensible, flexible cables is shown supporting a 2000-lb platform in a horizontal position. What are the inclinations of the cable segments $AB$, $BC$, and $DE$ to accomplish this and what lengths should they be? Neglect the weight of the segments and note the hint in Problem 6.69.

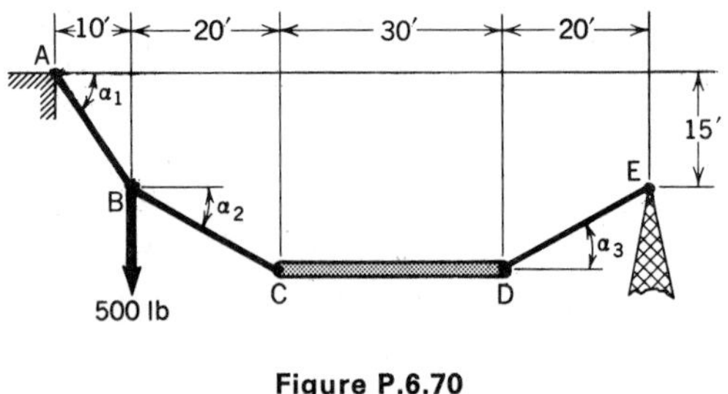


Figure P.6.70

## 6.11 Closure

Essentially what we have done in this chapter is to apply previously developed material to situations of singular importance in engineering. Further information on structures can be found in books on strength of materials and structural mechanics.[16] We turn again to new material in Chapter 7, where we will discuss the Coulomb laws of friction.

## Review Problems

**6.71.** A 3-kN traveling hoist has a 27-kN capacity and is suspended from a beam weighing .5 kN/m. The beam is fastened to several trusses spaced 8 m apart. What are the forces in each truss member when the fully loaded hoist is located at point $C$ directly under the truss shown? Assume that the hoist acts on pin $C$ and that pin $C$ also supports half of the I-beam $G$ between each of the adjacent trusses.

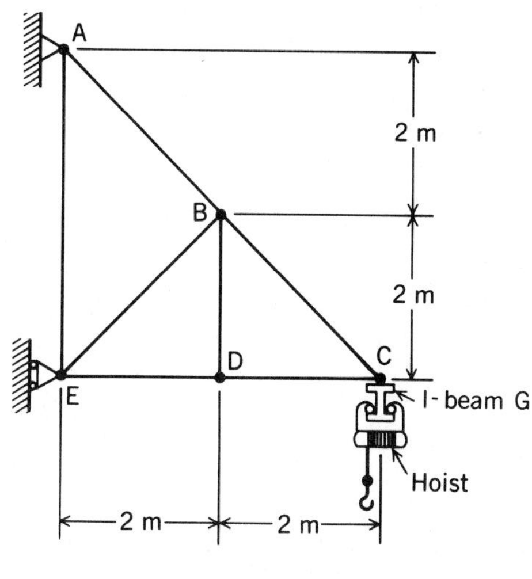

Figure P.6.71

**6.72.** The truss is used to support the roof of a low-clearance train-car repair shed (hence the curved members). The roof is subjected to a snow load of 1 kN/m². What are the forces in the straight members if the trusses are 10 m apart?

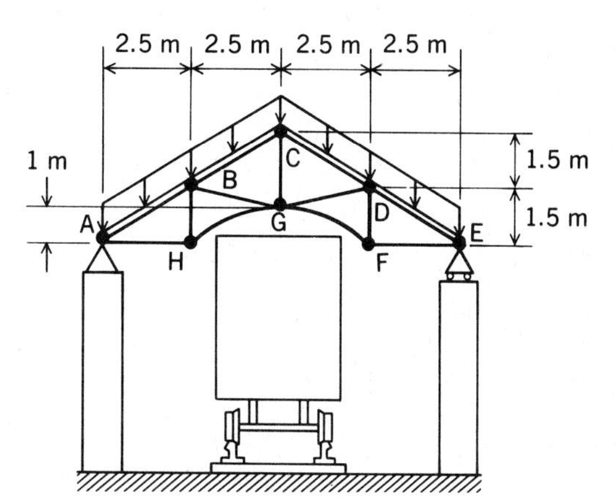


Figure P.6.72

**6.73.** Find forces in all the members of the space truss. Note that $ACE$ is in the $xz$ plane.

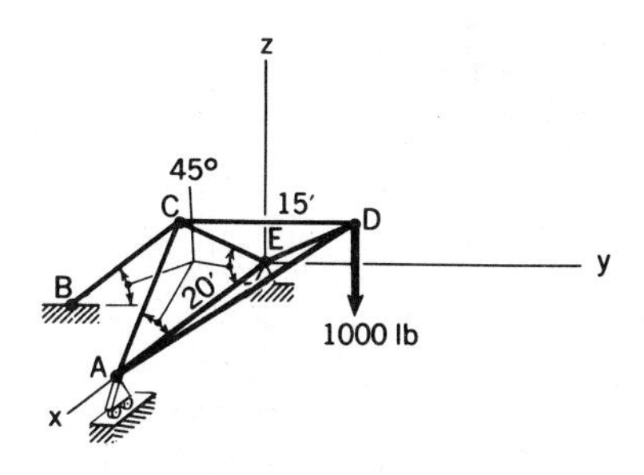


Figure P.6.73

**6.74.** Determine the forces in members $BG$, $BF$, and $CE$ for the plane truss.

---

[16]See I. H. Shames, *Introduction to Solid Mechanics*, Prentice-Hall, Inc., Englewood Cliffs, N.J., 1975.

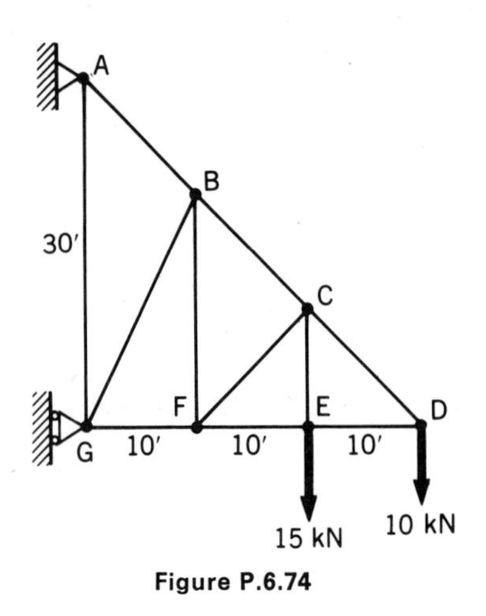


Figure P.6.74

**6.75.** Express the shear-force and bending-moment equations with the aid of free-body diagrams. Then express $V$ and $M$ without the diagrams.

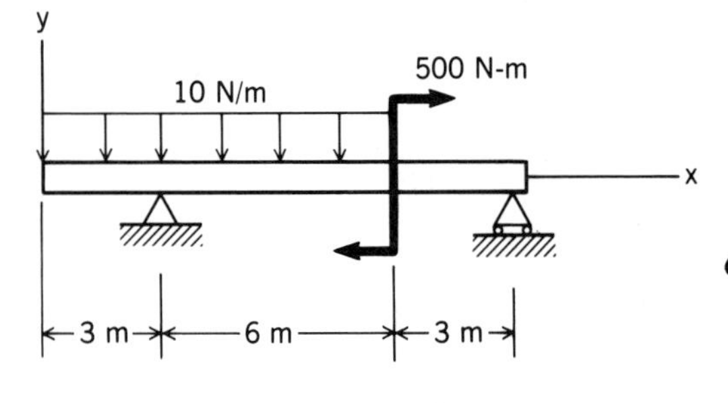


Figure P.6.75

**6.76.** Express the shear-force and bending-moment equations without the aid of free-body diagrams.

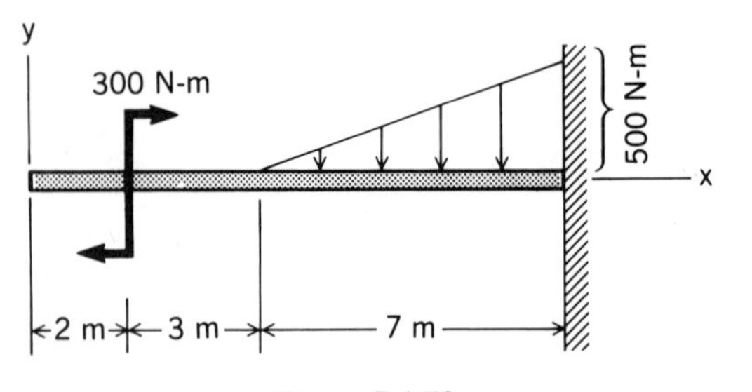


Figure P.6.76

**6.77.** Give the shear-force and bending-moment equations for the beam, and sketch shear-force and bending-moment diagrams. At what position between supports is the bending moment equal to zero?

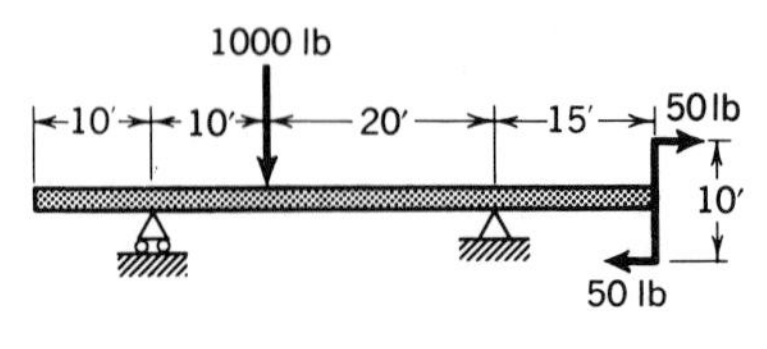


Figure P.6.77

**6.78.** Sketch the shear-force and bending-moment diagrams labeling key points.

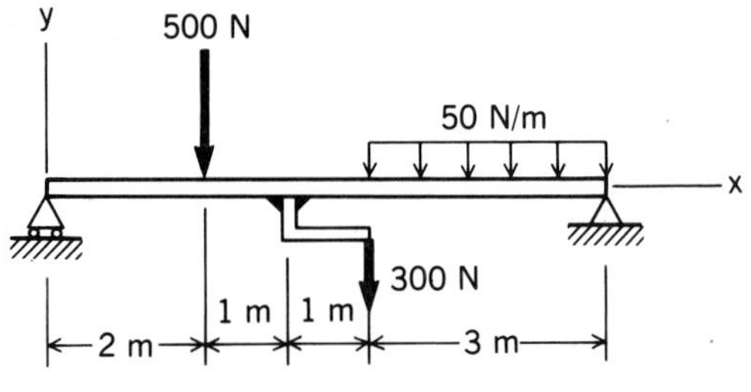


Figure P.6.78

**6.79.** Find the shape of a cable stretched between two points on the same level, $l$ units apart with sag $h$, and subjected to a vertical loading of

$$w(x) = 5 \cos \frac{\pi x}{l} \text{ N/m}$$

distributed in the horizontal direction. The coordinate $x$ is measured from the zero slope position of the cable.

**6.80.** A uniform cable weighs 1 lb/ft. It is connected to a uniform rod at *B*. This rod is free to swing about hinge *C*. If a force component $F_x$ of $-200$ lb is exerted at *A* as shown, what is the resulting angle of inclination $\alpha$? The cable is 50 ft long and the rod is 20 ft long. What is the weight of the rod?

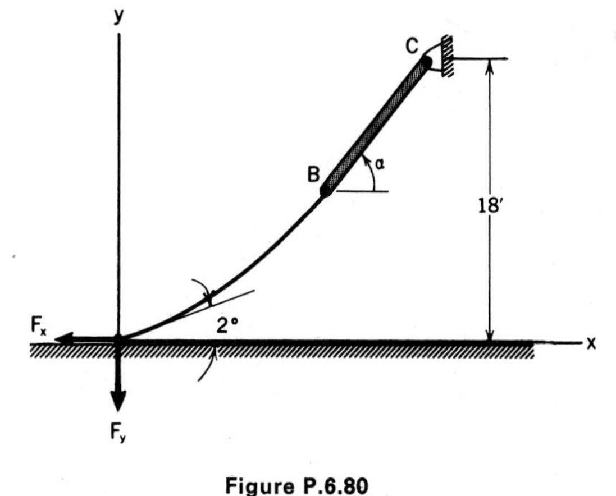


**Figure P.6.80**

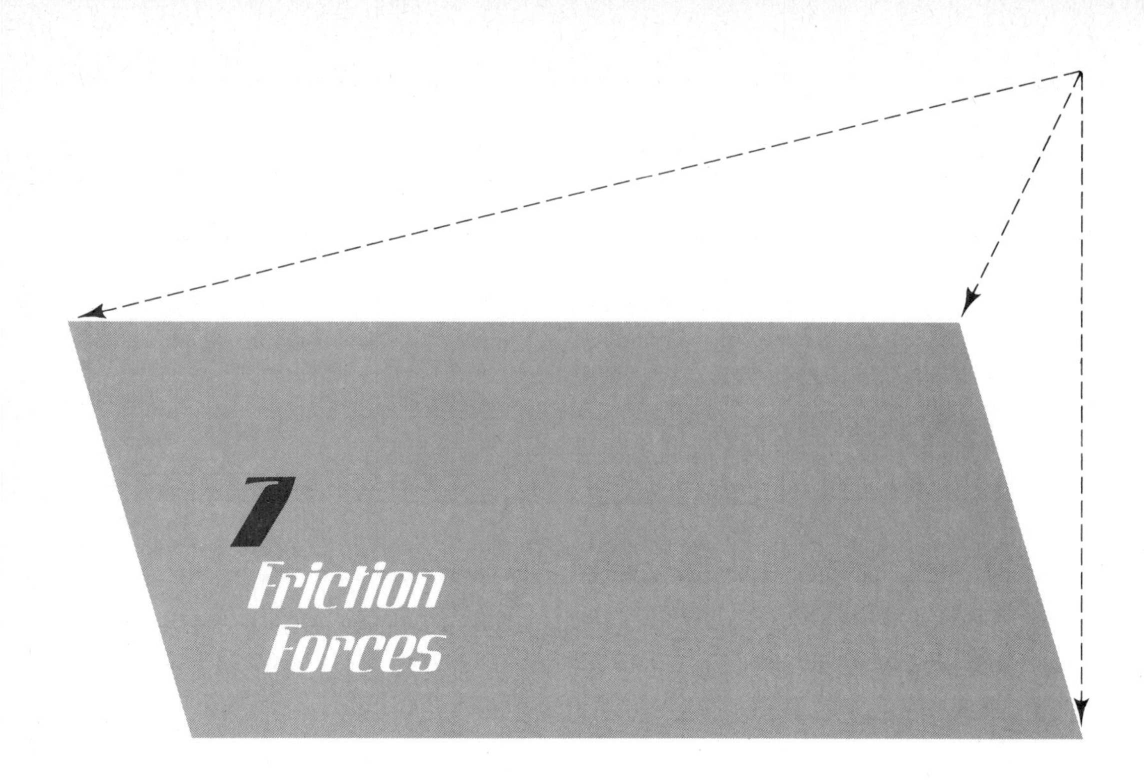


## 7.1 Introduction

*Friction* is the force distribution at the surface of contact between two bodies that prevents or impedes sliding motion of one body relative to the other. This force distribution is tangent to the contact surface and has, for the body under consideration, a direction at every point in the contact surface that is in opposition to the possible or existing slipping motion of the body at that point.

Frictional effects are associated with energy dissipation and are therefore sometimes considered undesirable. At other times, however, this means of changing mechanical energy to heat is a beneficial one, as for example in brakes, where the kinetic energy of a vehicle is dissipated into heat. In statics applications, frictional forces are often necessary to maintain equilibrium.

*Coulomb friction* is that friction which occurs between bodies having dry contact surfaces, and is not to be confused with the action of one body on another separated by a film of fluid such as oil. These latter problems are termed *lubrication problems* and are studied in the fluid mechanics courses. Coulomb, or *dry*, friction is a complicated phenomenon, and actually not much is known about its true nature.[1] The major cause of dry friction is believed to be the microscopic roughness of the surfaces of contact. Interlocking microscopic protuberances oppose the relative motion between the surfaces. When sliding is present between the surfaces, some of these protuberances

---

[1]For a more complete discussion of friction, see F. P. Bowden and D. Tabor, *The Friction and Lubrication of Solids*, Oxford University Press, New York, 1950.

either are sheared off or are melted by high local temperatures. This is the reason for the high rate of "wear" for dry-body contact and indicates why it is desirable to separate the surfaces by a film of fluid.

We have previously employed the terms "smooth" and "rough" surfaces of contact. A "smooth" surface can only support a normal force. On the other hand, a "rough" surface in addition can support a force tangent to the contact surface (i.e., a friction force). In this chapter, we shall consider situations whereby the friction force can be directly related to the normal force at a surface of contact. Other than including this new relationship, we use only the usual static equilibrium equations.

## 7.2 Laws of Coulomb Friction

Everybody has gone through the experience of sliding furniture along a floor. We exert a continuously increasing force which is completely resisted by friction until the object begins to move—usually with a lurch. The lurch occurs because once the object begins to move, there is a decrease in frictional force from the maximum force attained under static conditions. An idealized plot of this force as a function of time is shown in Fig. 7.1. There the force $P$ applied to the furniture, idealized as a block in Fig. 7.2, is shown to drop from the highest or limiting value to a lower value which is constant with time. This latter constant value is independent of the velocity of the object. The condition corresponding to the maximum value is termed the condition of *impending motion* or *impending slippage*.

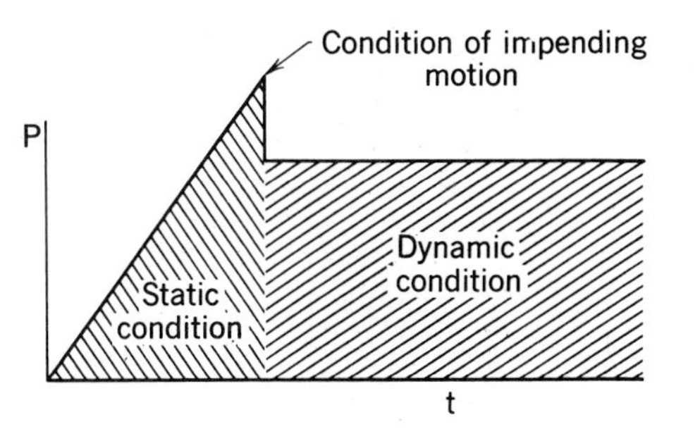


**Figure 7.1.** Idealized plot of friction force $P$.

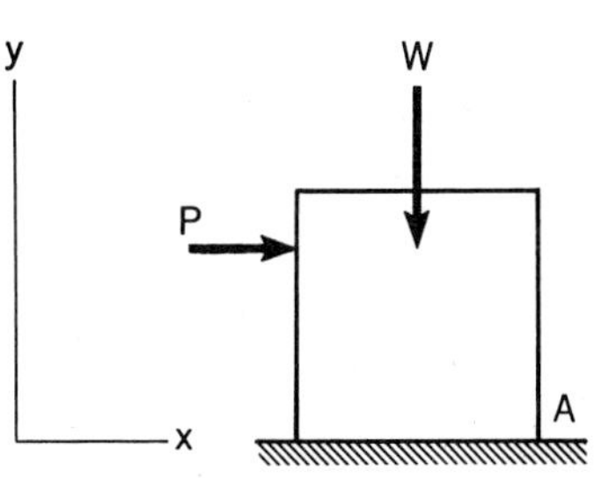


**Figure 7.2.** Idealization of furniture.

By carrying out experiments on blocks tending to move without rotation or actually moving without rotation on flat surfaces, Coulomb in 1781 presented certain conclusions which are applicable at the condition of *impending slippage* or once *slippage has begun*. These have since become known as Coulomb's laws of friction. For block problems, he reported that:

1. The total force of friction that can be developed is independent of the magnitude of the area of contact.

2. For low relative velocities between sliding objects, the frictional force is practically independent of velocity. However, the sliding frictional force is less than the frictional force corresponding to impending slippage.
3. The total frictional force that can be developed is proportional to the normal force transmitted across the surface of contact.

Conclusions 1 and 2 may come as a surprise to most of you and be contrary to your "intuition." Nevertheless, they are accurate enough statements for many engineering applications. More precise studies of friction, as was pointed out earlier, are complicated and involved. We can express conclusion 3 mathematically as:

$$f \propto N$$

Therefore,

$$f = \mu N \tag{7.1}$$

where $\mu$ is called the *coefficient of friction.*

Equation 7.1 is valid *only at conditions of impending slippage or while the body is slipping*. Since the limiting static friction force exceeds the dynamic force friction, we differentiate between coefficients of friction for those conditions. Thus, we have coefficients of *static* friction and coefficients of *dynamic* friction, $\mu_s$ and $\mu_d$, respectively. The accompanying table is a small list of static coefficients. The corresponding coefficients of friction for dynamic conditions are about 25% less.

STATIC COEFFICIENTS OF FRICTION[2]

| | |
|---|---|
| Steel on cast iron | .40 |
| Copper on steel | .36 |
| Hard steel on hard steel | .42 |
| Mild steel on mild steel | .57 |
| Rope on wood | .70 |
| Wood on wood | .20–.75 |

Let us consider carefully the simple block problem used to develop the laws of Coulomb. Note that we have:

1. A plane surface of contact.
2. An impending or actual motion which is in the same direction for all area elements of the contact surface. Thus, there is no impending or actual rotation between the bodies in contact.
3. The further implication that the properties of the respective bodies are uniform at the contact surface. Thus, the coefficient of friction $\mu$ is constant for all area elements of the contact surface.

[2]F. P. Bowden and D. Tabor, *The Friction and Lubrication of Solids*, Oxford University Press, New York, 1950.

What do we do if any of these conditions is violated? We can always choose an *infinitesimal* part of area of contact between the bodies. Such an infinitesimal area can be considered plane even though the general surface of contact of which it is an infinitesimal part is not. Furthermore, the relative motion at this infinitesimal contact surface may be considered as along a straight line even though the finite surface of which it is a part may not have such a simple straight motion. Finally, for the infinitesimal area of contact, we may consider the materials to be uniform even though the properties of the material vary over the finite area of contact. In short, when conditions 1 through 3 do not prevail, we can still use Coulomb's law *in the small* (i.e., at infinitesimal contact areas) and then integrate the results. We shall call such problems *complex* surface contact problems and we shall examine a series of such problems in Section 7.4.

We now examine simple contact problems where Coulomb's laws apply to the contact surface *as a whole* without requiring integration procedures. We shall thus consider uniform blocklike bodies akin to those used by Coulomb. Also, we shall consider bodies which however complex have very *small* contact surfaces, such as in Fig. 7.3(a). Clearly, the whole contact surface can be considered an infinitesimal plane area and for reasons set forth earlier, we shall directly use Coulomb's laws when appropriate as has been shown in Fig. 7.3(b).

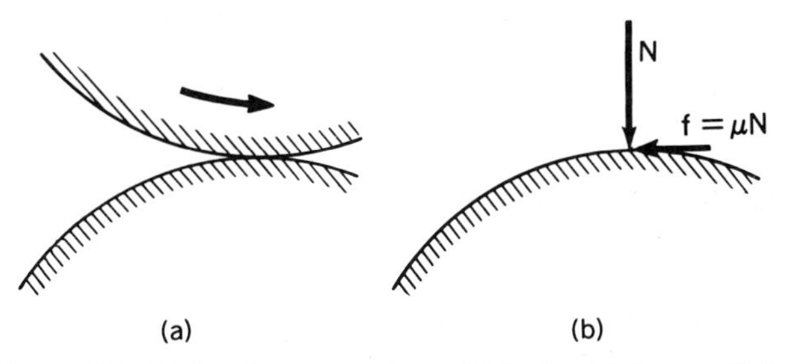


**Figure 7.3.** (a) Small contact surface; (b) Coulomb's laws applied.

Before proceeding to the problems, we have one additional point to make. For a finite simple surface of contact, such as the block shown in Fig. 7.2, we must note that we do not generally know the line of action for the simplest resultant supporting force $N$, since we do not generally know the normal force distribution between the two bodies. Hence, we cannot take moments for such free-body diagrams without introducing additional unknown distances in the equation. Consequently, for such problems we limit ourselves to summing forces only. This is not true, however, when we have a *point* contact surface such as in Fig. 7.3(a). The line of action of the supporting force must be at the point of contact, and we can thus take moments without introducing additional unknown distances.

## 7.3 Simple Contact Friction Problems

Two common classes of statics problems involve dry friction. In one class, we know that motion is impending, or has been established and is uniform, and we desire information about certain forces that are present. We can then express friction forces at surfaces of contact where there is impending or actual slippage as $\mu N$ according to Coulomb's law and, using $f_i$ for other friction forces, proceed by methods of statics. However, the proper direction must be given to *all* friction forces. That is, *they must oppose possible, impending, or actual relative motion at the contact surfaces.* In the second class of problems, external loads on a body are given, and we desire to determine whether the friction forces present are sufficient to maintain equilibrium. One way to attack this latter type of problem is to assume that impending motion exists in the various possible directions, and to solve for the external forces required for such conditions. By comparing the actual external forces present with those required for the various impending motions, we can then deduce whether the body can be restrained by frictional forces from sliding.

The following examples are used to illustrate the two classes of problems.

**EXAMPLE 7.1**

In Fig. 7.4(a) is shown an automobile on a roadway inclined at an angle $\theta$ with the horizontal. If the coefficients of static and dynamic friction between the tires and the road are taken as 0.6 and 0.5, respectively, what is the maximum inclination $\theta_{max}$ that the car can climb at uniform speed? It has rear-wheel drive and has a total loaded weight of 3600 lb. The center of gravity for this loaded condition has been shown in the diagram.

Let us assume that the drive wheels do not "spin"; that is, there is zero relative velocity between the tire surface and the road surface at the point of contact. Then, clearly, the maximum friction force possible is $\mu_s$ times the normal force at this contact surface, as has been indicated in Fig. 7.4(b).

We can consider this to be a coplanar problem with three unknowns, $N_1$, $N_2$, and $\theta_{max}$. Accordingly, since the friction force is restricted to a point, three equations of equilibrium are available. Using the reference $xy$ shown in the diagram, we have:

$\underline{\sum F_x = 0:}$

$$.6N_1 - 3600 \sin \theta_{max} = 0 \tag{a}$$

$\underline{\sum F_y = 0:}$

$$N_1 + N_2 - 3600 \cos \theta_{max} = 0 \tag{b}$$

$\underline{\sum M_A = 0:}$

$$10N_2 - (3600 \cos \theta_{max})(5) + (3600 \sin \theta_{max})(1) = 0 \tag{c}$$

To solve for $\theta_{max}$, we eliminate $N_1$ from Eqs. (a) and (b), getting as a result the equation

$$N_2 = 3600 \cos \theta_{max} - 6000 \sin \theta_{max} \tag{d}$$

Now, eliminating $N_2$ from Eq. (c) using Eq. (d), we get

$$18{,}000 \cos \theta_{max} - 56{,}400 \sin \theta_{max} = 0$$

Therefore,

$$\tan \theta_{max} = .320 \tag{e}$$

Hence,

$$\theta_{max} = 17.7° \tag{f}$$

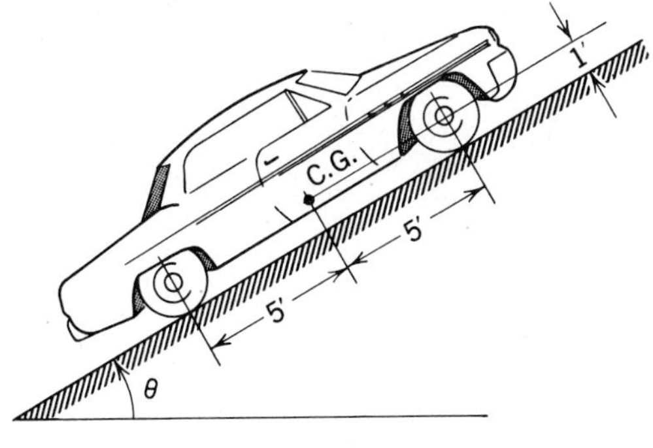


(a)

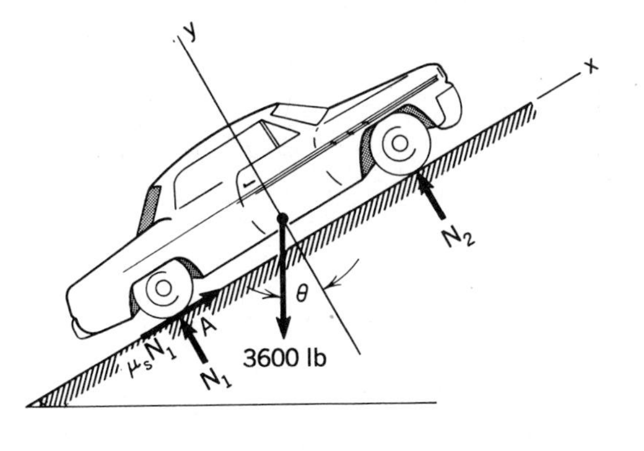

(b)

**Figure 7.4.** (a) Find maximum $\theta$; (b) free-body diagram using Coulomb's law.

If the drive wheels were caused to spin, we would have to use $\mu_d$ in place of $\mu_s$ for this problem. We would then arrive at a smaller $\theta_{max}$, which for this problem would be 14.7°.

**EXAMPLE 7.2**

Using the data of Example 7.1, compute the torque needed by the drive wheels to move the car at a uniform speed up an incline where $\theta = 15°$. Also, assume that the brakes have "locked" while the car is in a parked position on the incline. What force is then needed to tow the car either up the incline or down the incline with the brakes in this condition? The diameter of the tire is 25 in.

A free-body diagram for the first part of the problem is shown in Fig. 7.5(a). Note that the friction force $f$ will now be determined by Newton's law and not by Coulomb's law, since we do not have impending slippage between the wheel and the road for this case. Accordingly, we have, for $f$:

$\underline{\sum F_x = 0:}$

$$f - 3600 \sin 15° = 0$$

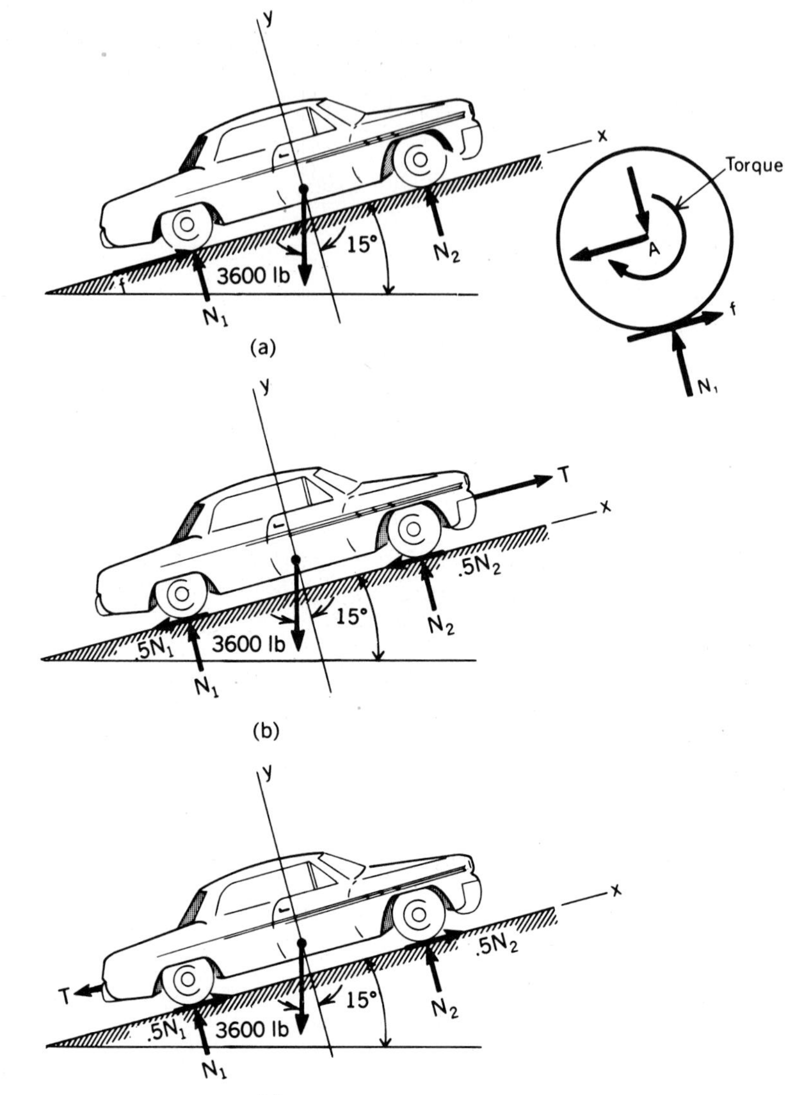


**Figure 7.5.** Free-body diagrams: (a) climbing at uniform speed; (b) under tow upward; (c) under tow downward.

Therefore,

$$f = 932 \text{ lb}$$

The torque needed is then computed using the rear wheels as a free body [see Fig. 7.5(a)]. Taking moments about $A$, we have

$$\text{torque} = (f)(r) = (932)\left(\frac{25/2}{12}\right) = 971 \text{ ft-lb}$$

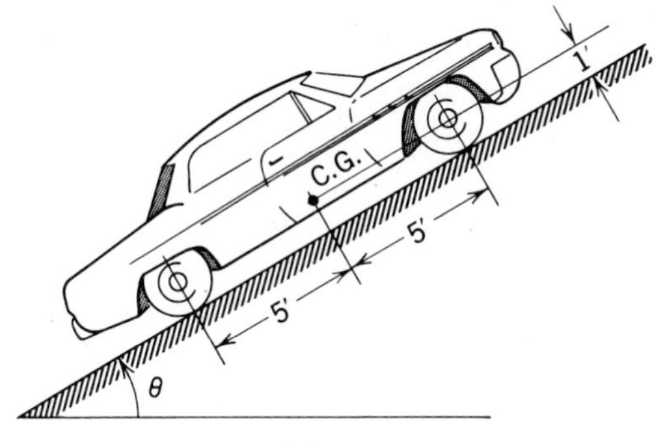


(a)

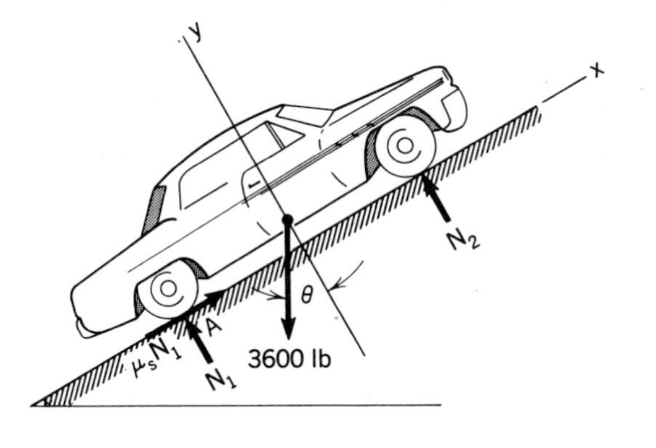

(b)

**Figure 7.4.** (a) Find maximum $\theta$; (b) free-body diagram using Coulomb's law.

If the drive wheels were caused to spin, we would have to use $\mu_d$ in place of $\mu_s$ for this problem. We would then arrive at a smaller $\theta_{\max}$, which for this problem would be 14.7°.

**EXAMPLE 7.2**

Using the data of Example 7.1, compute the torque needed by the drive wheels to move the car at a uniform speed up an incline where $\theta = 15°$. Also, assume that the brakes have "locked" while the car is in a parked position on the incline. What force is then needed to tow the car either up the incline or down the incline with the brakes in this condition? The diameter of the tire is 25 in.

A free-body diagram for the first part of the problem is shown in Fig. 7.5(a). Note that the friction force $f$ will now be determined by Newton's law and not by Coulomb's law, since we do not have impending slippage between the wheel and the road for this case. Accordingly, we have, for $f$:

$\underline{\sum F_x = 0:}$

$$f - 3600 \sin 15° = 0$$

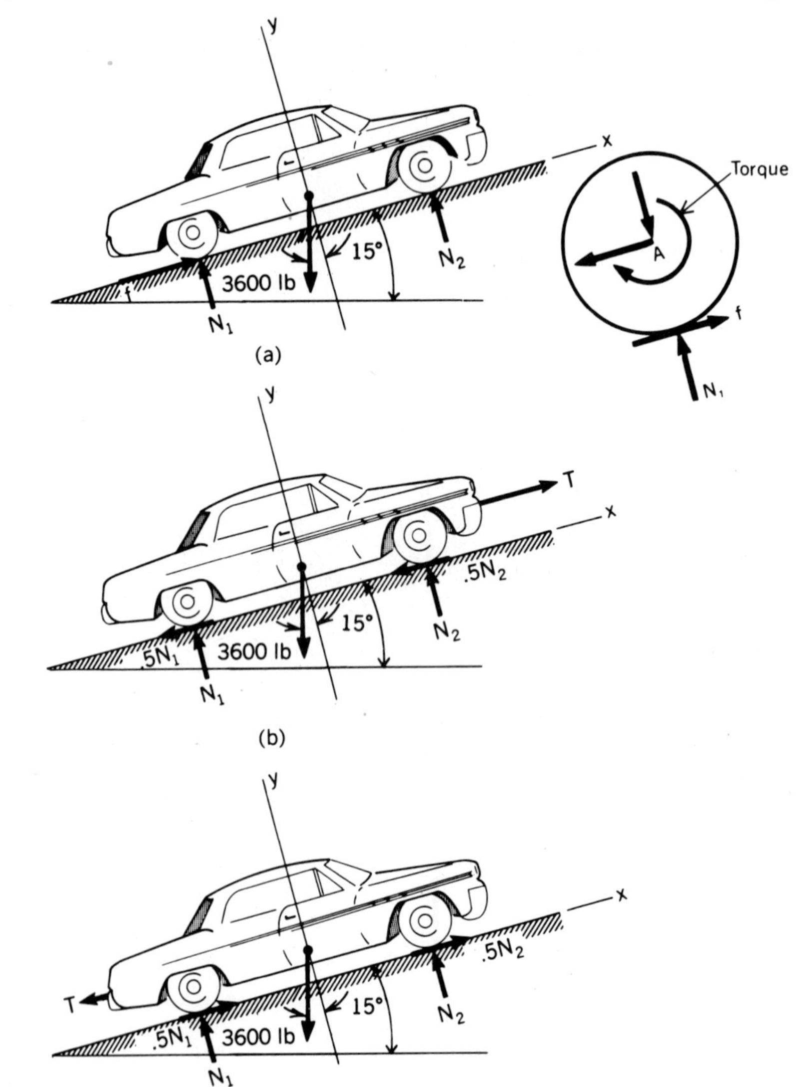


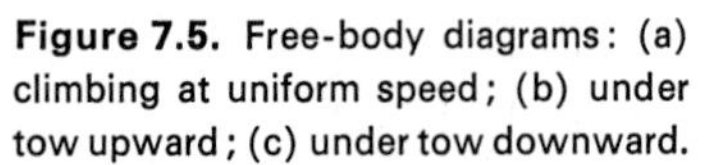
**Figure 7.5.** Free-body diagrams: (a) climbing at uniform speed; (b) under tow upward; (c) under tow downward.

Therefore,

$$f = 932 \text{ lb}$$

The torque needed is then computed using the rear wheels as a free body [see Fig. 7.5(a)]. Taking moments about $A$, we have

$$\text{torque} = (f)(r) = (932)\left(\frac{25/2}{12}\right) = 971 \text{ ft-lb}$$

For the second part of the problem, we have shown the required free body in Fig. 7.5(b). Note that we have used Coulomb's law for the friction forces with the *dynamic* friction coefficient $\mu_d$. We now write the equations of equilibrium for this free body.

$\sum F_x = 0$:

$$T - .5(N_1 + N_2) - 3600 \sin 15° = 0 \tag{a}$$

$\sum F_y = 0$:

$$(N_1 + N_2) - 3600 \cos 15° = 0 \tag{b}$$

Solving for $N_1 + N_2$ from Eq. (b), and substituting into Eq. (a), we can now solve for $T$. Hence,

$$T = (.5)(3600)(.966) + 932 = 2670 \text{ lb} \tag{c}$$

For towing the car down the incline we must reverse the direction of the friction force as shown in Fig. 7.5(c). Solving for $T$ as in the previous calculation, we get

$$T = (.5)(3600)(.966) - 932 = 807 \text{ lb} \tag{d}$$

**EXAMPLE 7.3**

In Fig. 7.6, a strongbox of mass 75 kg rests on a floor. The static coefficient for the contact surface is .20. What is the highest position $h$ for a horizontal load $P$ at which the box will move without tipping?

The free-body diagram for the strongbox is shown in Fig. 7.7. The condition of impending motion has been recognized by the use of Coulomb's law. Furthermore, by concentrating the supporting and friction forces at the left corner, we are stipulating *impending tipping* for the problem. This latter condition imposes the desired *largest* possible value of $h$ for equilibrium without tipping.

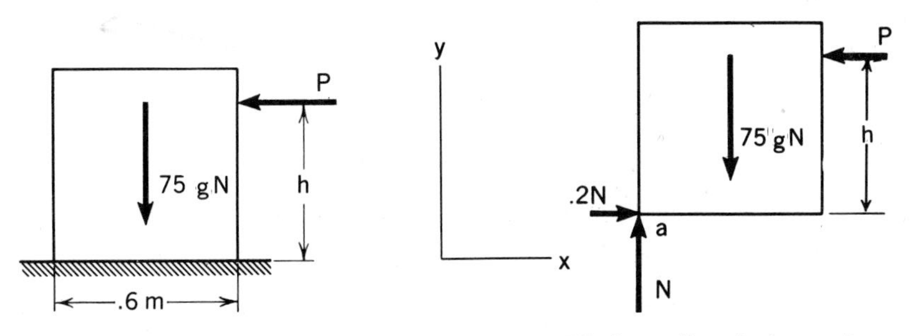


Figure 7.6. Strongbox being pushed.

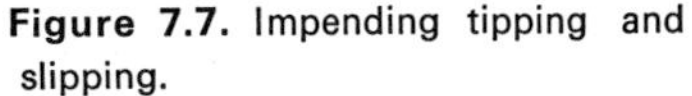

Figure 7.7. Impending tipping and slipping.

The pertinent forces constitute a coplanar system of forces at the midplane of the strongbox. We proceed with the scalar equations of equilibrium:

$\sum F_y = 0$:

$$N = 75\,g = 736 \text{ N}$$

$\sum F_x = 0$:

$$P = .2\,N = 147.15 \text{ N}$$

$\sum M_a = 0$:

$$-(75\,g)(.3) + (147.15)h = 0$$

Therefore,

$$h = 1.500 \text{ m}$$

Thus, the height of the applied load must be less than 1.50 m in order to avoid tipping.

The three examples presented illustrated the *first* type of friction problem wherein we know the nature of the motion or impending motion present in the system and we determine certain forces or positions of certain forces. In the last example of this series, we illustrate the *second* type of friction problem set forth earlier—namely the problem of deciding whether bodies will move or not move under prescribed external forces.

**EXAMPLE 7.4**

The coefficient of static friction for all contact surfaces in Fig. 7.8 is .2. Does the 50-lb force move the block $A$ up, hold it in equilibrium, or is it too small to prevent $A$ from coming down and $B$ from moving out? The 50-lb force is exerted at the midplane of the blocks so that we can consider this a coplanar problem.

We can compute a force $P$ in place of the 50-lb force to cause impending motion of block $B$ to the left, and a force $P$ for impending motion of block $B$ to the right. In this way, we can judge by comparison the action that the 50-lb force will cause.

The free-body diagrams for impending motion of block $B$ to the left have been shown in Fig. 7.9, which contains the unknown force $P$ mentioned above. We need not be concerned about the correct location of the centers of gravity of the blocks, since we shall only add forces in the analysis. (We do not know the line of action of

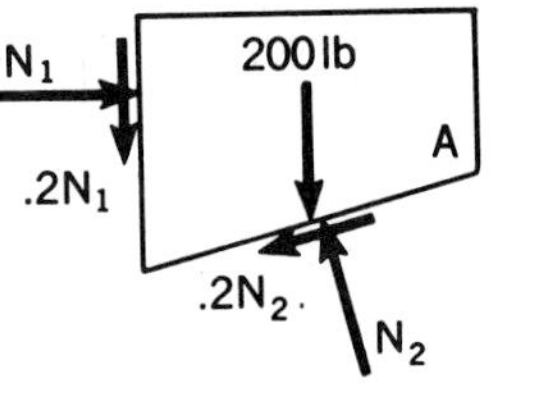


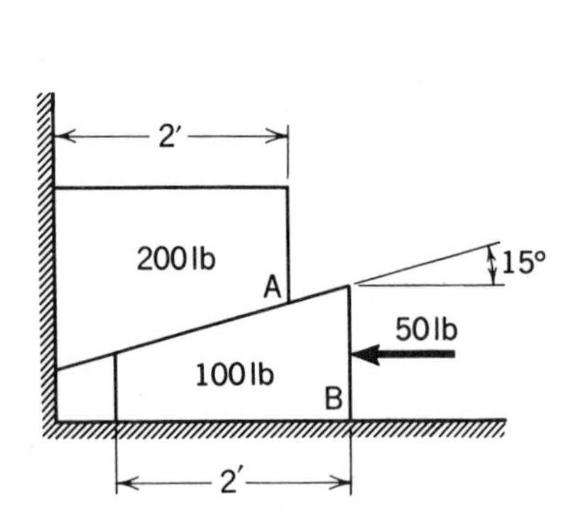


Figure 7.8. Do blocks move?

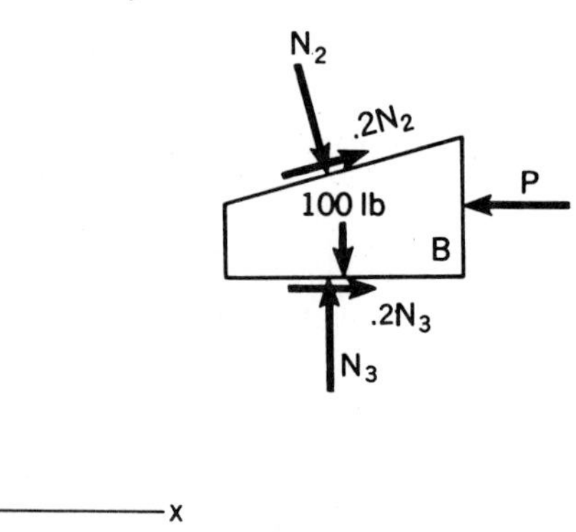


Figure 7.9. Impending motion of $B$ to the left.

the normal forces at the contact surfaces and therefore cannot take moments.) Summing forces on block $A$, we get

$$N_2 \cos 15°\boldsymbol{j} - N_2 \sin 15°\boldsymbol{i} - .2N_1\boldsymbol{j} - 200\boldsymbol{j}$$
$$+ N_1\boldsymbol{i} - .2N_2 \cos 15°\boldsymbol{i} - .2N_2 \sin 15°\boldsymbol{j} = \boldsymbol{0}$$

The scalar equations are:

$$N_1 - .259N_2 - .1932N_2 = 0$$
$$.966N_2 - .2N_1 - 200 - .0518N_2 = 0$$

Solving simultaneously, we get

$$N_2 = 243 \text{ lb}, \qquad N_1 = 109.8 \text{ lb}$$

For the free-body diagram of $B$, we have, on summing forces,

$$-N_2 \cos 15°\boldsymbol{j} + N_2 \sin 15°\boldsymbol{i} - P\boldsymbol{i} + .2N_3\boldsymbol{i} + N_3\boldsymbol{j}$$
$$- 100\boldsymbol{j} + .2N_2 \cos 15°\boldsymbol{i} + .2N_2 \sin 15°\boldsymbol{j} = \boldsymbol{0}$$

This yields the following scalar equations:

$$-P + 62.9 + .2N_3 + 46.9 = 0$$
$$-235 + N_3 - 100 + 12.6 = 0$$

Solving simultaneously, we have

$$P = 174 \text{ lb}$$

Clearly, the stipulated force of 50 lb is insufficient to induce a motion on block $B$ to the left, so further computation is necessary.

Next, we reverse the direction of force $P$ and compute what its value must be to move the block $B$ to the right. The frictional forces in Fig. 7.9 are all reversed, and the vector equation of equilibrium for block $A$ becomes

$$N_2 \cos 15°\boldsymbol{j} - N_2 \sin 15°\boldsymbol{i} + .2N_1\boldsymbol{j} - 200\boldsymbol{j}$$
$$+ N_1\boldsymbol{i} + .2N_2 \cos 15°\boldsymbol{i} + .2N_2 \sin 15°\boldsymbol{j} = \boldsymbol{0}$$

The scalar equations are:

$$N_1 - .259N_2 + .1932N_2 = 0$$
$$.966N_2 + .2N_1 - 200 + .0518N_2 = 0$$

Solving simultaneously, we get

$$N_2 = 194.1 \text{ lb}, \qquad N_1 = 12.80 \text{ lb}$$

For free body $B$ we have, on summing forces,

$$-N_2 \cos 15°\boldsymbol{j} + N_2 \sin 15°\boldsymbol{i} + P\boldsymbol{i} - .2N_3\boldsymbol{i} + N_3\boldsymbol{j}$$
$$-100\boldsymbol{j} - .2N_2 \cos 15°\boldsymbol{i} - .2N_2 \sin 15°\boldsymbol{j} = \boldsymbol{0}$$

The following are the scalar equations:

$$P - .2N_3 + 50.3 - 37.5 = 0$$
$$-100 + N_3 - 187.5 - 10.05 = 0$$

Solving, we get $P = 46.7$ lb. Thus, we would have to *pull to the right* to get block $B$ to move in this direction. We can now conclude from this study that the blocks are in equilibrium.

## Problems

**7.1.** A block has a force $F$ applied to it. If this force has a time variation as shown in the diagram, draw a simple sketch showing the friction force variation with time. Take $\mu_s = .3$ and $\mu_d = .2$ for the problem.

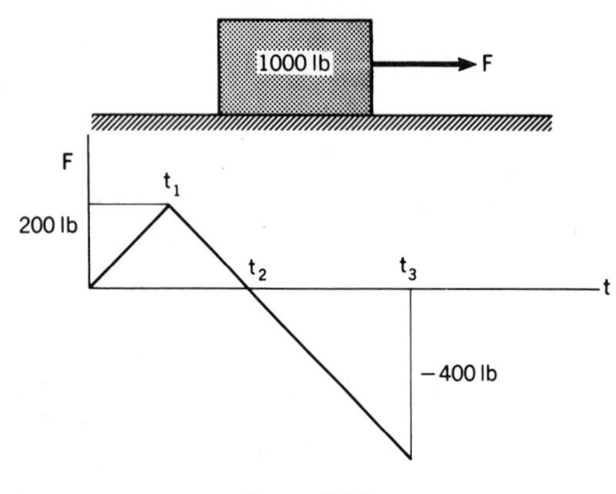


Figure P.7.1

**7.2.** Show that by increasing the inclination $\phi$ on an inclined surface until there is impending slippage of supported bodies, we reach the *angle of repose* $\phi_s$, so that $\tan \phi_s = \mu_s$.

**7.3.** To what angle must the driver elevate the dump bed of the truck to cause the wooden crate of weight $W$ to slide out? For wood on steel, $\mu_s = .6$ and $\mu_d = .4$.

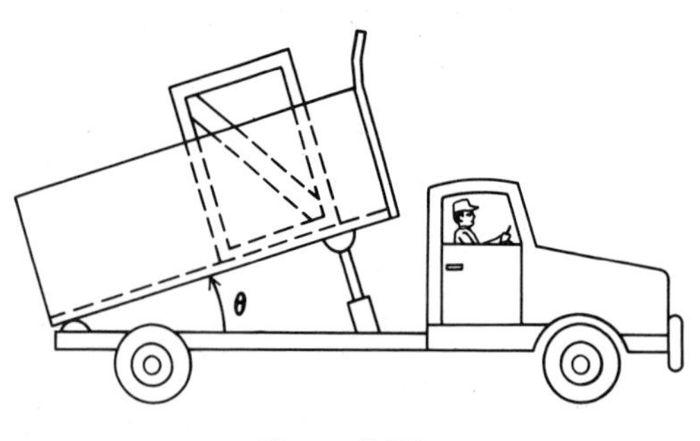


Figure P.7.3

**7.4.** A platform is suspended by two ropes which are attached to blocks that can slide horizontally. At what value of $W$ does the platform begin to descend? Will $W$ start tipping?

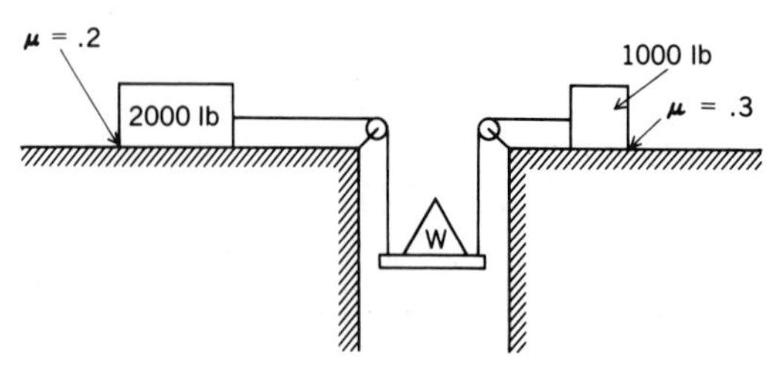

Figure P.7.4

**7.5.** Explain how a violin bow, when drawn over a string, maintains the vibration of the string. Do this in terms of friction forces and the difference in static and dynamic coefficients of friction.

**7.6.** What is the value of the force $F$, inclined at 30° to the horizontal, needed to get the block just started up the incline? What is the force $F$ needed to keep it just moving up at a constant speed? The coefficients of static and dynamic friction are .3 and .275, respectively.

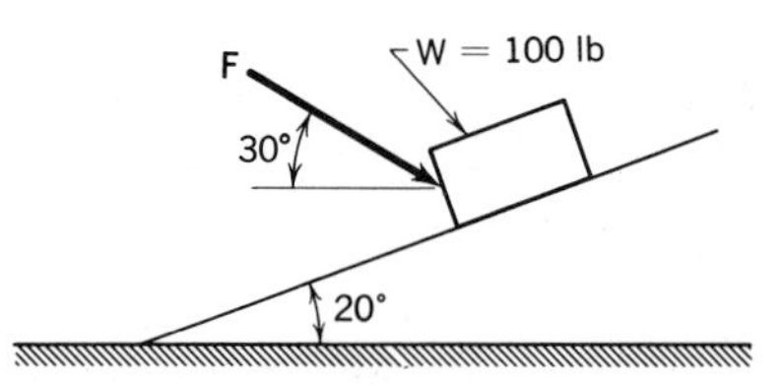


Figure P.7.6

**7.7.** Bodies $A$ and $B$ weigh 500 N and 300 N, respectively. The platform on which they are placed is raised from the horizontal position to an angle $\theta$. What is the *maximum* angle that can be reached before the bodies slip down the incline? Take $\mu_s$ for body $B$ and the plane as .2 and $\mu_s$ for body $A$ and the plane as .3.

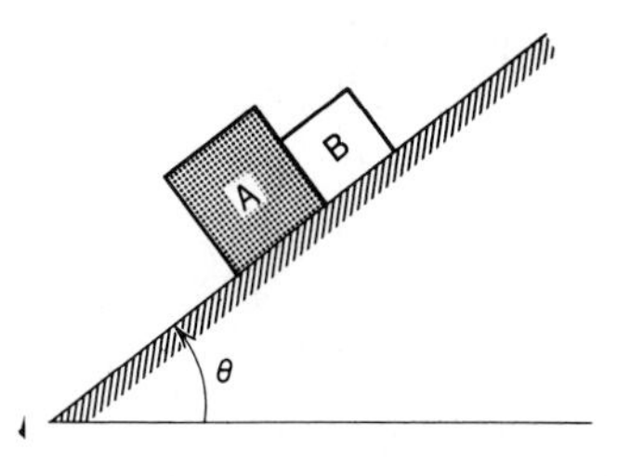


Figure P.7.7

**7.8.** A 30-ton tank is moving up a 30° incline. If $\mu_s = .6$ for the contact surface between tread and ground, what *maximum* torque can be developed at the rear drive sprocket with no slipping? What maximum towing force $F$ can the tank develop? Take the mean diameter of the rear sprocket as 2 ft.

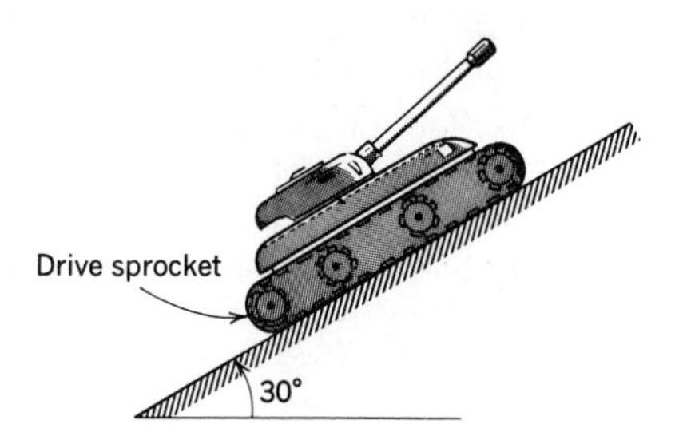


Figure P.7.8

**7.9.** A 500-lb crate $A$ rests on a 1000-lb crate $B$. The centers of gravity of the crates are at the geometric centers. The coefficients of static friction between contact surfaces are shown in the diagram. The force $T$ is increased from zero. What is the first action to occur?

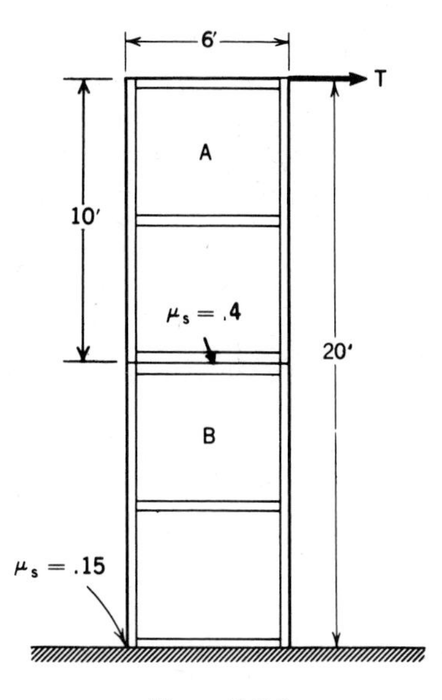


Figure P.7.9

**7.10.** What force $F$ is needed to get the 300-kg block moving to the right? The coefficient of static friction for all surfaces is .3.

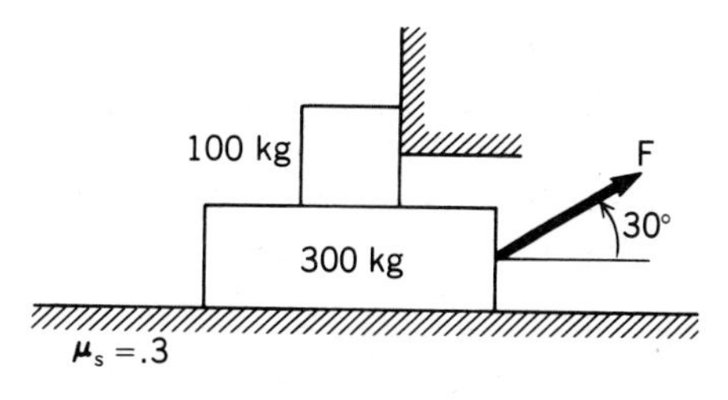


Figure P.7.10

**7.11.** A chute is shown having sides that are at right angles to each other. The chute is 30 ft in length with end $A$ 10 ft higher than end $B$. Cylinders weighing 200 lb are to slide down the chute. What is the *maximum* allowable coefficient of friction so there cannot be sticking of the cylinders along the chute?

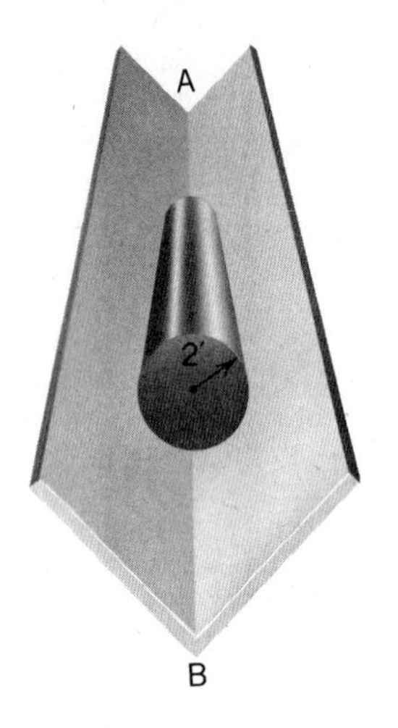


Figure P.7.11

**7.12.** A block rests on a surface for which there is a coefficient of friction $\mu_s = .2$. Over what range of angle $\beta$ will there be no movement of the block for the 150-N force? (You will have to solve an equation by trial and error.)

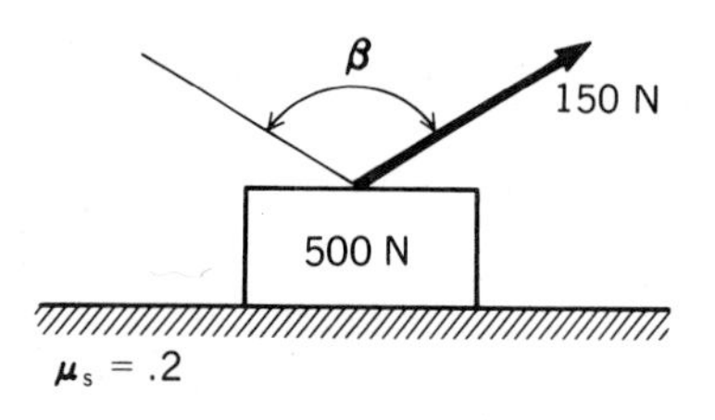

Figure P.7.12

**7.13.** What is the *largest* load that can be suspended without moving blocks *A* and *B*? The coefficient of friction for all plane surfaces of contact is .3. Block *A* weighs 500 N and block *B* weighs 700 N. Neglect friction in the pulley system.

Figure P.7.13

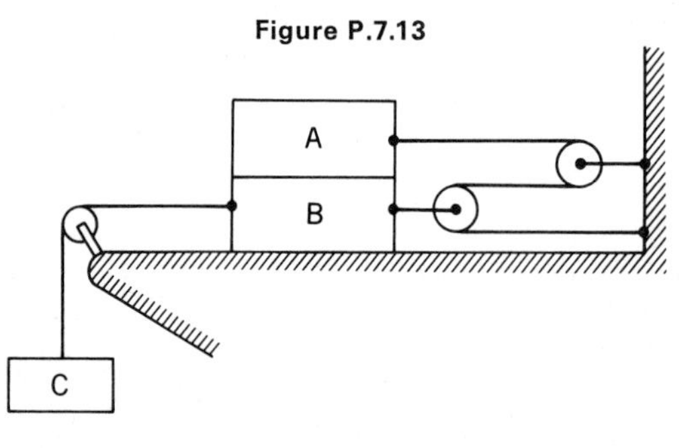


**7.14.** What is the *minimum* force *F* to hold the cylinders, each weighing 100 lb? Take $\mu_s = .2$ for all surfaces of contact.

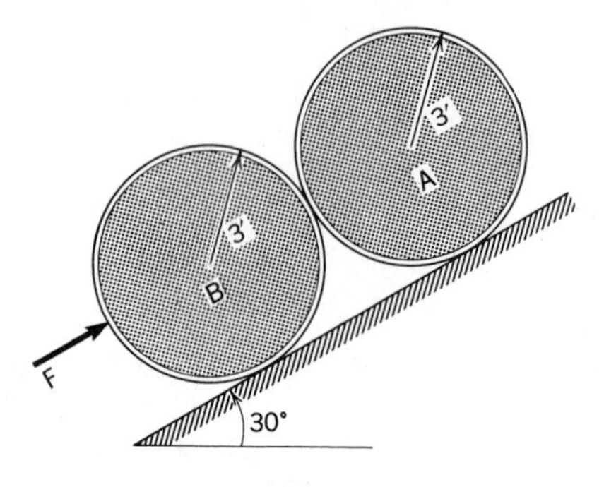


Figure P.7.14

**7.15.** A compressor is shown. If the pressure in the cylinder is 1.40 N/mm² above atmosphere (gage), what *minimum* torque *T* is needed to start the system to move? Neglect the weight of the crank and connecting rod as far as their contribution toward moving the system. Consider friction only on the piston where with the cylinder walls there is a coefficient of friction $\mu = .15$.

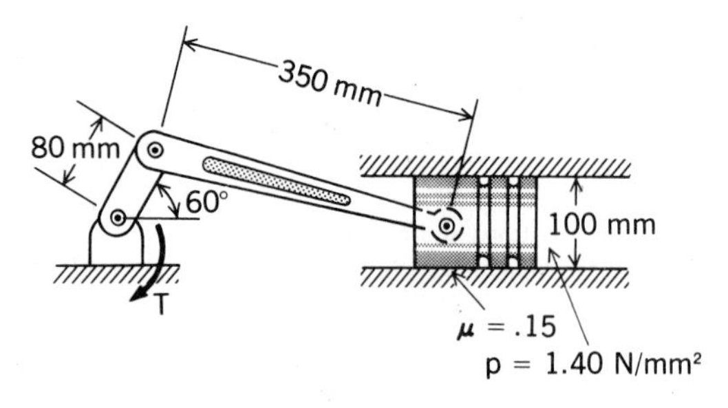

Figure P.7.15

**7.16.** Find the cord tension, the friction forces, and whether the blocks move if $\mu = .2$.

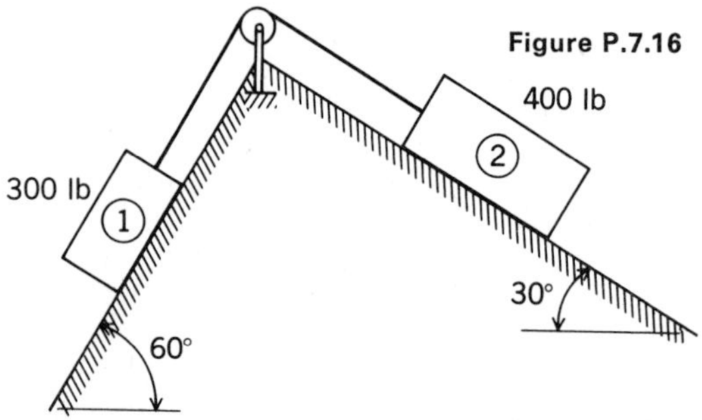


Figure P.7.16

**7.17.** Given that $\mu_s = .2$ for all surfaces, find the force $P$ needed to start the block $A$ to the right.

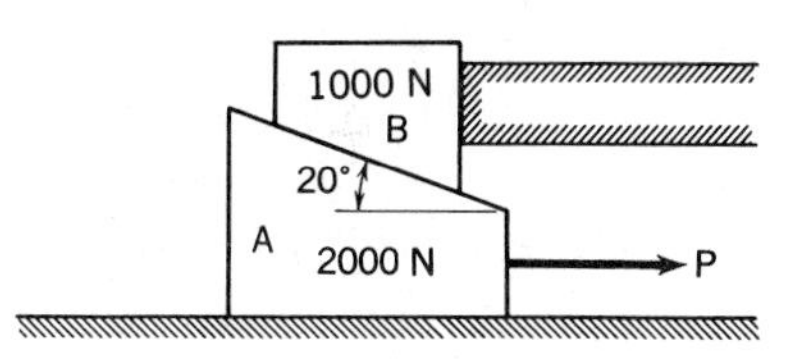

Figure P.7.17

**7.18.** The cylinder shown weighs 200 N and is at rest. What is the friction force at $A$? If there is impending slippage, what is the friction coefficient? The supporting plane is inclined at 60° to the horizontal.

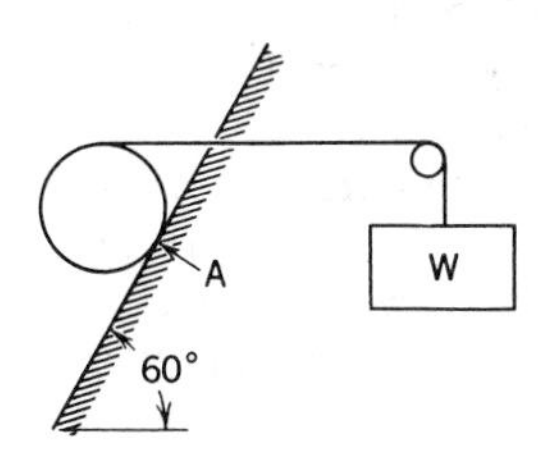


Figure P.7.18

**7.19.** Armature $B$ is stationary while rotor $A$ rotates with angular speed $\omega$. In armature $B$ there is a braking system. If $\mu_d = .4$, what is the braking torque on $A$ for a force $F$ of 300 N? Note that the rod on which $F$ is applied is pinned at $C$ to the armature $B$. Neglect friction between $B$ and the brake pads $G$ and $H$.

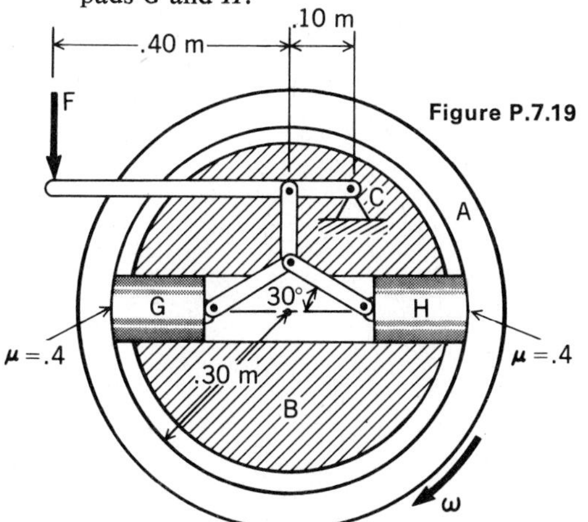


Figure P.7.19

**7.20.** A 200-lb load is placed on the luggage rack of the 4500-lb station wagon. Will the station wagon climb the hill easier or harder with the luggage than without? Explain. The coefficient of friction is .55.

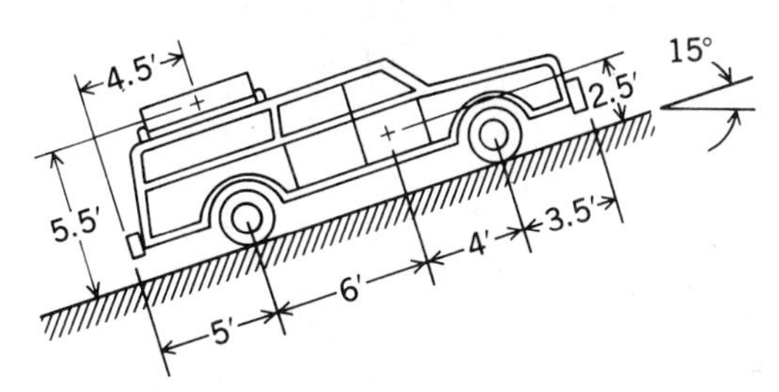


Figure P.7.20

**7.21.** An insect tries to climb out of a hemispherical bowl of radius 600 mm. If the coefficient of friction between insect and bowl is .4, how high up does the insect go? If the bowl is spun about a vertical axis, the bug gets pushed out in a radial direction by the force $mr\omega^2$, as you learned in physics. At what speed $\omega$ will the bug just be able to get out of the bowl?

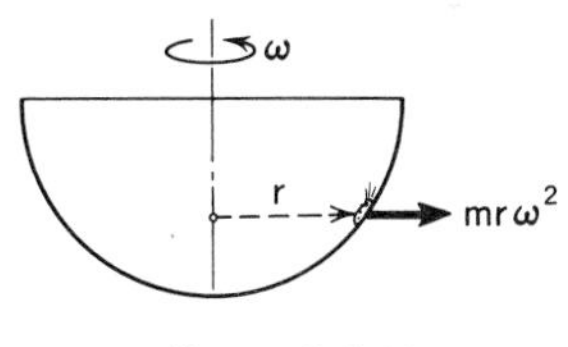


Figure P.7.21

**7.22.** A block $A$ of mass 500 kg rests on a stationary support $B$ where the static coefficient of friction $\mu_s = .4$. On the right side, support $C$ is on rollers. The dynamic coefficient of friction $\mu_d$ of the support $C$ with body $A$ is .2. If $C$ is moved at constant speed to the left, how far does it move before body $A$ begins to move?

Figure P.7.22

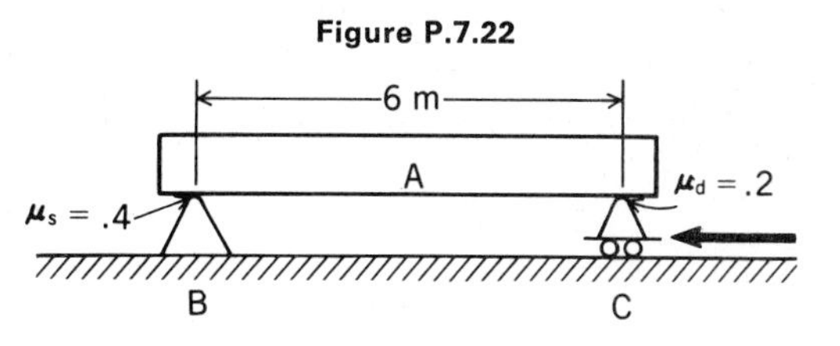

**7.23.** In a preliminary grinding operation for a 1500-N car engine block, the grinding wheel is pushed against the block with a 500-N force. What force must be exerted by the hydraulic ram to move the block to the right if (a) the wheel rotates clockwise and (b) the wheel rotates counterclockwise? The coefficient of friction between the grinding wheel and the block is .7 and between the table and the block is .2.

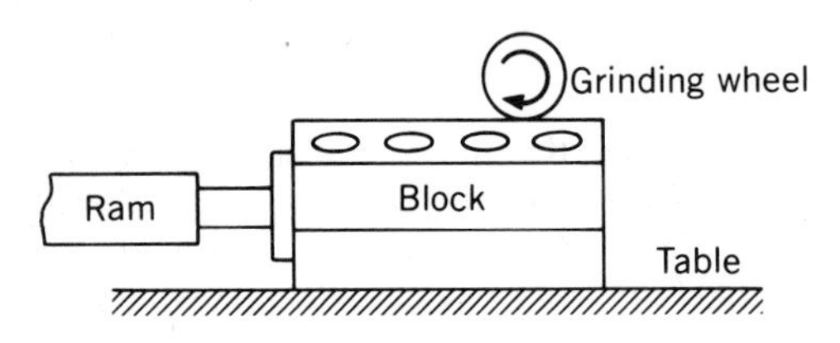


Figure P.7.23

**7.24.** An 8000-lb tow truck with four-wheel drive develops a torque of 750 lb-ft at each axle. What is the *heaviest* car that can be towed up a 10° slope if $\mu_s = .3$?

truck tire radius = 18 in.

**7.25.** A 7-m ladder weighing 250 N is being pushed by force *F*. What is the *minimum* force needed to get the ladder to move? The coefficient of friction for all contact surfaces is .4.

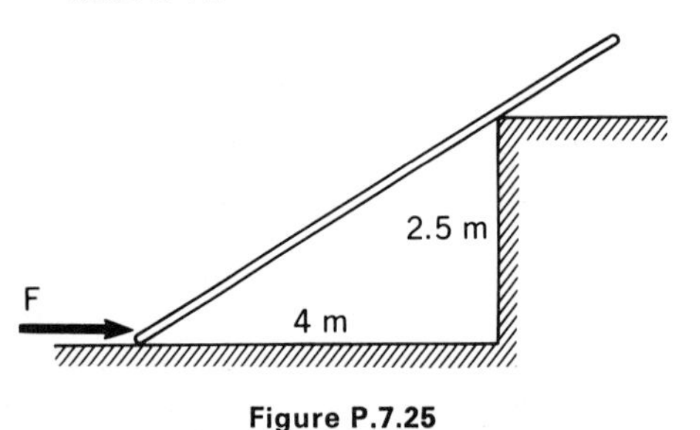


Figure P.7.25

**7.26.** In Problem 7.25 if *F* is released will the ladder begin to slide down?

**7.27.** Can a force *P* roll the 50-lb cylinder over the step? The coefficient of static friction is .4. What is the value of *P* if this can be done?

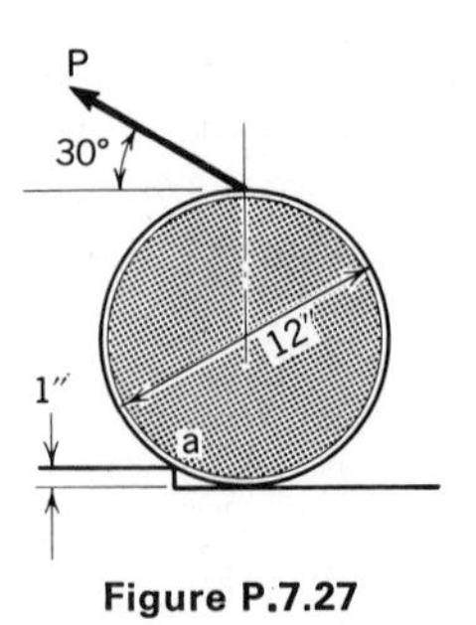


Figure P.7.27

**7.28.** The block of weight *W* is to be moved up an inclined plane. A rod of length *c* with negligible weight is attached to the block and the force *F* is applied to the top of this rod. If the coefficient of starting friction is $\mu_s$, determine in terms of *a*, *d*, and $\mu_s$ the *maximum* length *c* for which the block will begin to slide rather than tip.

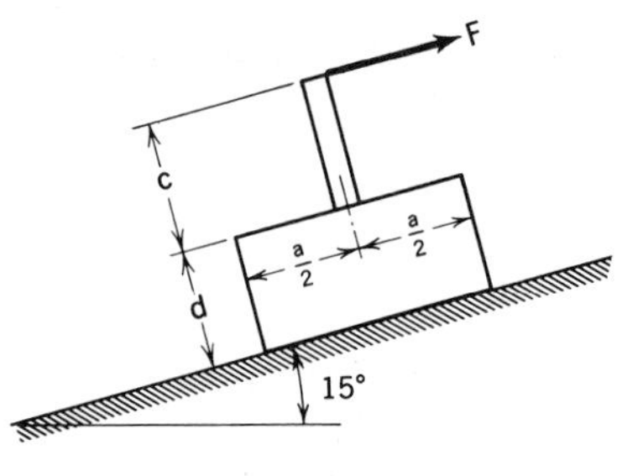


Figure P.7.28

**7.29.** Determine the range of values of $W_1$ for which the block will either slide up the plane or slide down the plane. At what value of $W_1$ is the friction force zero? $W_2 = 100$ lb.

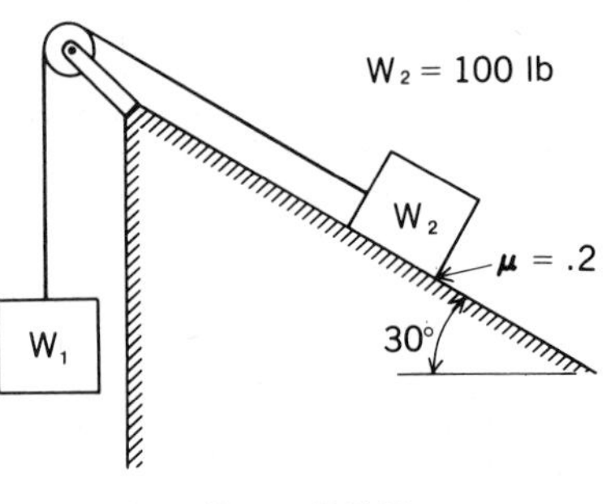


Figure P.7.29

**7.30.** A 200-kN tractor is to push a 60-kN concrete beam up a 15° incline at a construction site. If $\mu_d = .5$ between beam and dirt, and if $\mu_s$ is .6 between tractor tread and dirt, can the tractor do the job? If so, what torque must be developed on the tractor drive sprocket which is .8 m in diameter? What force $P$ is then developed to push the beam?

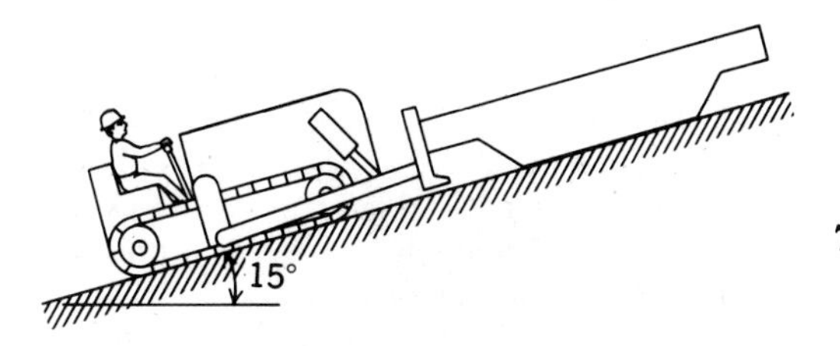


**Figure P.7.30**

**7.31.** What is the *minimum* coefficient of friction required just to maintain the bracket and its 500-lb load in a static position? (Assume point contacts at the horizontal centerlines of the arms.) The center of gravity is 7 in. from the shaft centerline.

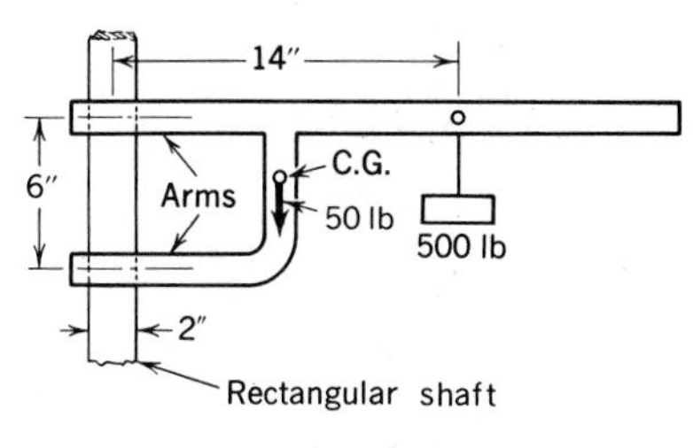

**Figure P.7.31**

**7.32.** If the coefficient of friction in Problem 7.31 is .2, at what *minimum* distance from the centerline of the vertical shaft can we support the 500-lb load without slipping?

**7.33.** A rod is held by a cord at one end. If the force $F = 200$ N, and if the rod weighs 450 N, what is the *maximum* angle $\alpha$ that the rod can be placed for $\mu_s$ between the rod and the floor equal to .4? The rod is 1 m in length.

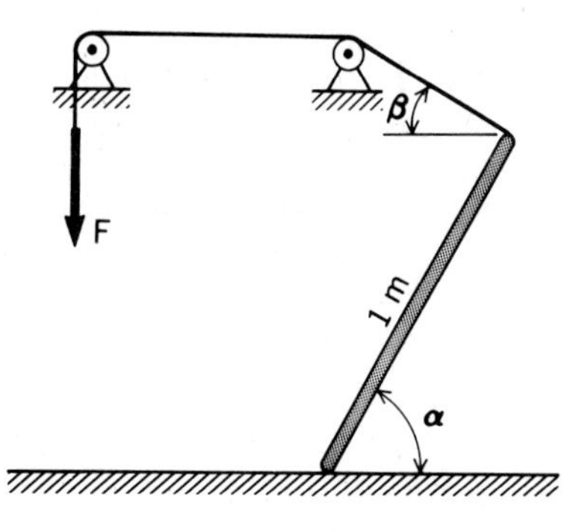


**Figure P.7.33**

**7.34.** Suppose that the ice lifter is used to support a hard block of material by friction only. What is the *minimum* coefficient of static friction, $\mu_s$, to accomplish this for any weight $W$ and for the geometry shown in the diagram?

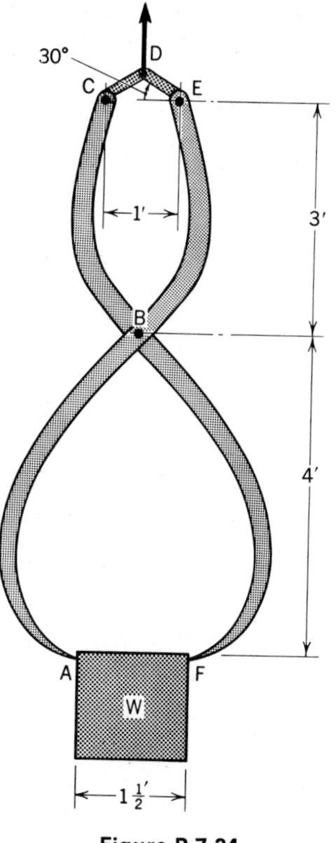


**Figure P.7.34**

**7.35.** A rectangular case is loaded with uniform vertical thin rods such that when it is full, as shown in (a), the case has a total weight of 1000 lb. The case weighs 100 lb when empty and has a coefficient of static friction of .3 with the floor as shown in the diagram. A force $T$ of 200 lb is maintained on the case. If the rods are unloaded as shown in (b), what is the limiting value of $x$ for equilibrium to be maintained?

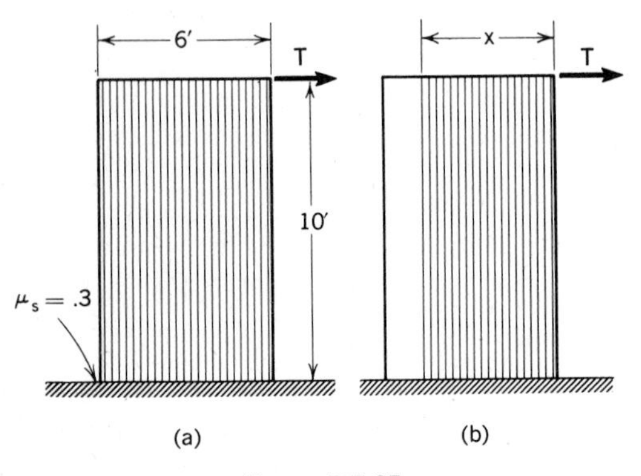


Figure P.7.35

**7.36.** A beam supports load $C$ weighing 500 N. At supports $A$ and $B$, the coefficient of frictions is .2. At the contact surface between load $C$ and the beam, the coefficient of friction is .75. If force $F$ moves $C$ steadily to the left, how far does it move before the beam begins to move? The beam weighs 200 N. Neglect the height $t$ of the beam in your calculations.

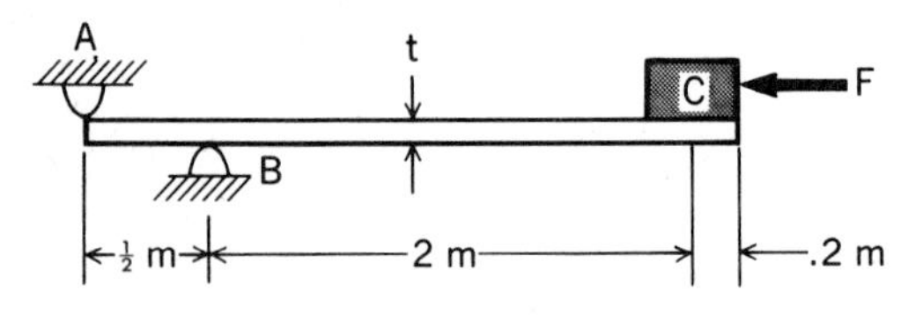


Figure P.7.36

**7.37.** Do Problem 7.36 for the case where the height $t$ is taken into account. Take $t$ = 120 mm.

**7.38.** A rod is supported by two wheels spinning in opposite directions. If the wheels were horizontal, the rod would be placed centrally over the wheels for equilibrium. However, the wheels have an inclination of 20° as shown, and the rod must be placed at a position off center for equilibrium. If the coefficient of friction is $\mu_d = .8$, how many feet off center must the rod be placed?

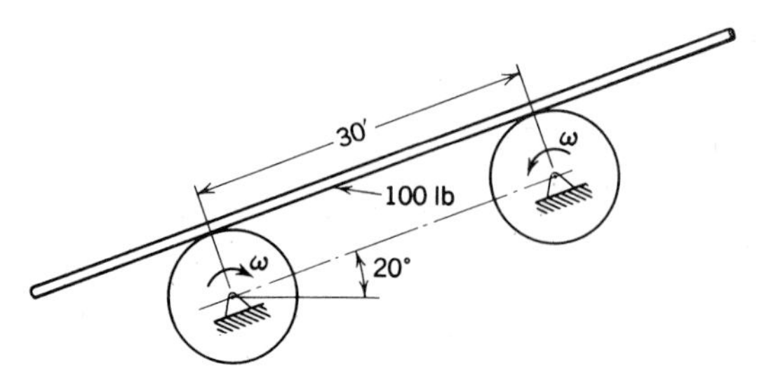


Figure P.7.38

**7.39.** How much force $F$ must be applied to the wedge to begin to raise the crate? Neglect changes in geometry. What force must the stopper block provide to prevent the crate from moving to the left? The coefficient of friction between all surfaces is .3.

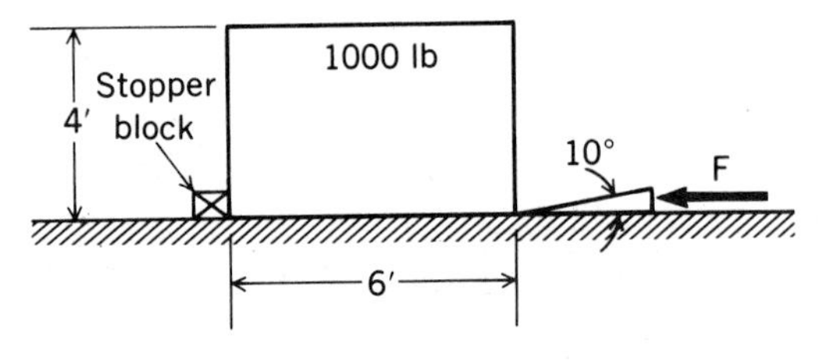


Figure P.7.39

**7.40.** What is the *maximum* height $x$ of a step so that the force $P$ will roll the 50-lb cylinder over the step with no slipping at $a$? Take $\mu_s = .3$.

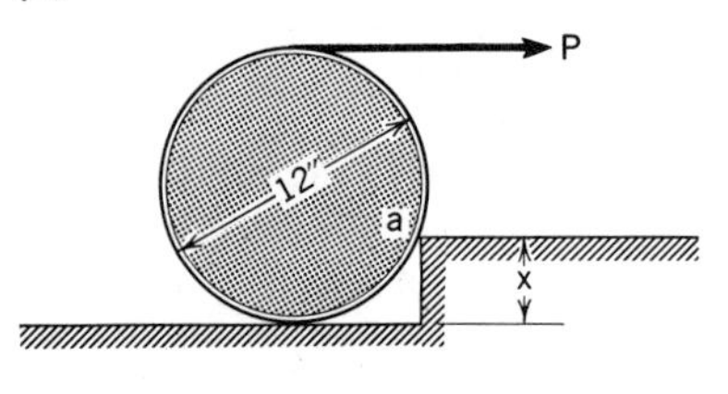


Figure P.7.40

**7.41.** Two identical light rods are pinned together at $B$. End $C$ of rod $BC$ is pinned while end $A$ of rod $AB$ rests on a rough floor having a coefficient of friction with the rod of $\mu_d = .5$. The spring requires a force of 5 N/mm of stretch. A load $F = 300$ N is applied slowly at $B$ and then maintained constant. What is the angle $\theta$ when the system ceases to move? The spring is unstretched when $\theta = 45°$. (*Hint:* You will have to solve an equation by trial and error.)

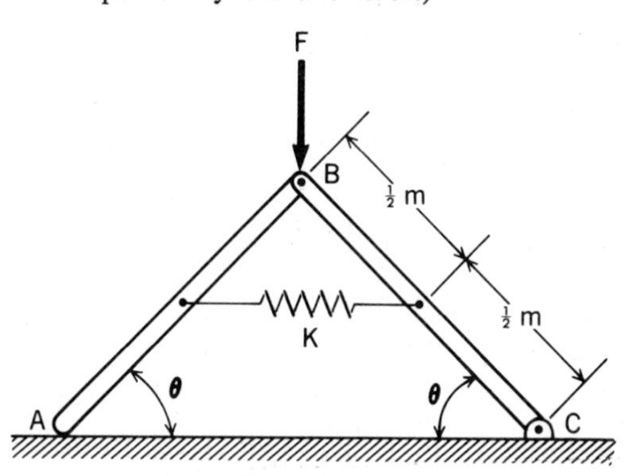


Figure P.7.41

**7.42.** A device for "throwing" baseballs is shown. This device is to be found in amusement parks for batting practice. The two wheels are inflated automobile tires that rotate as shown with a constant speed $\omega$. A baseball is fed to a point where it is touching the wheels. What is the *minimum* separation $d$ of the wheels if the ball is to be drawn into the slot and then ejected on the other side as a pitched ball? The coefficient of friction at the contact surfaces is .4.

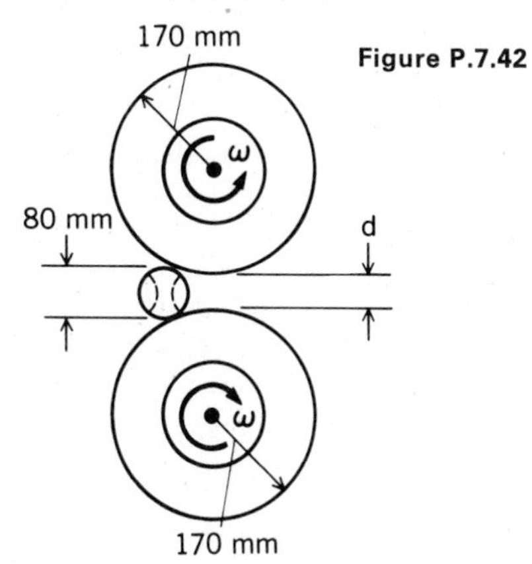


Figure P.7.42

**7.43.** What is the *maximum* angle $\alpha$ for which there will be equilibrium if $A$ has a mass of 50 kg. The coefficient of friction at the supports is equal to .3. What is the force in each of the supporting members?

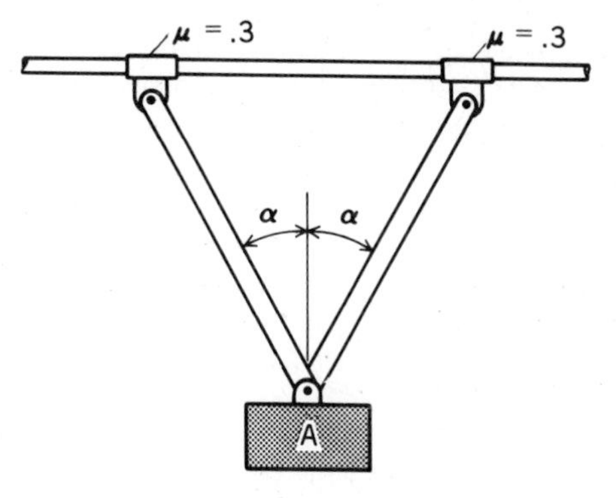


Figure P.7.43

**7.44.** What is the *maximum* angle $\alpha$ for which there will be equilibrium if $A$ weighs 1000 N and if $\mu$ at the supporting surface is .3? The rods are each 1.3 m long. You will have to solve a transcendental equation by trial and error.

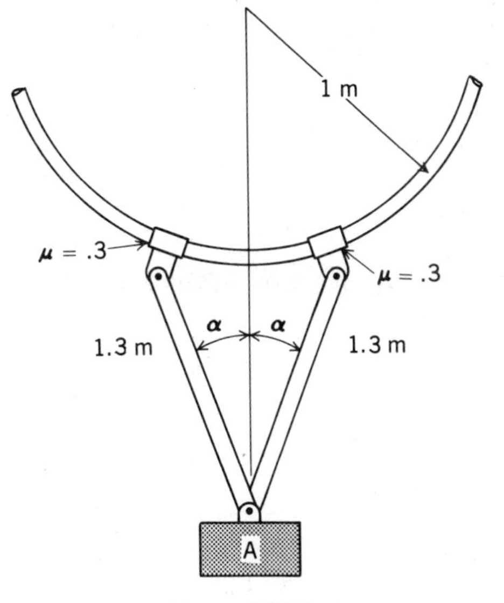


Figure P.7.44

**7.45.** The rod $AD$ is pulled at $A$ and it moves to the left. If the coefficient of dynamic friction for the road at $A$ and $B$ is .4, what must the *minimum* value of $W_2$ be to prevent the block from tipping when $\alpha = 20°$? With this value of $W_2$, determine the minimum coefficient of static friction between the block and the supporting plane needed to just prevent the block from sliding. $W_1$ is 100 N.

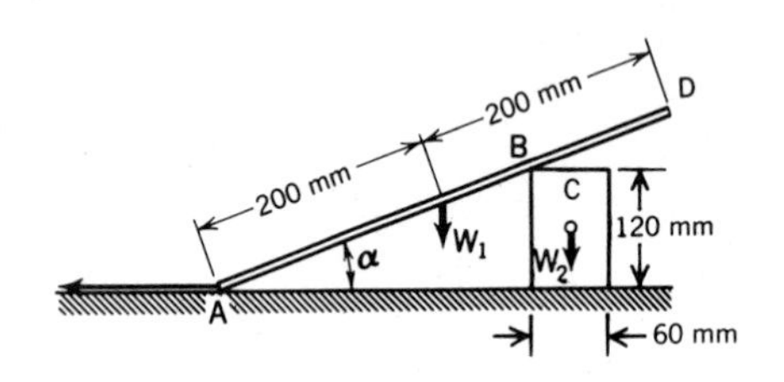


Figure P.7.45

**7.46.** If we neglect friction at the rollers, and if the coefficient of static friction is .2 for all surfaces, ascertain whether the 5000-lb weight will go up, go down, or stay stationary.

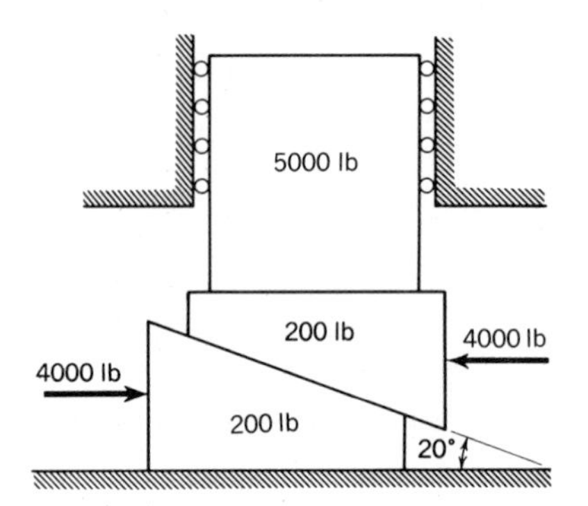

Figure P.7.46

## 7.4 Complex Surface Contact Friction Problems

In the examples undertaken heretofore, the nature of the relative impending or actual motion between the plane surfaces of contact was quite simple—that of motion without rotation. We shall now examine more general types of contacts between bodies. In Example 7.5, we have a plane contact surface but with varying direction of impending or slipping motion for the area elements as a result of rotation. In such problems we shall have to apply Coulomb's laws locally to infinitesimal areas of contact and to integrate the results, for reasons explained in Section 7.2. To do this, we must ascertain the distribution of the normal force at the contact surface, an undertaking that is usually difficult and well beyond the capabilities of rigid-body statics, as explained in Chapter 5. However, we can at times *approximately* compute frictional effects by *estimating* the manner of distribution of the normal force at the surface of contact. We now illustrate this.

**EXAMPLE 7.5**

Compute the frictional resistance to rotation of a rotating solid cylinder with an attached pad $A$ pressing against a flat dry surface with a force $P$ (see Fig. 7.10). The pad $A$ and the stationary flat dry surface consitute a dry *thrust bearing*.

The direction of the frictional forces distributed over the contact surface is no longer simple. We therefore take an infinitesimal area for examination. This area is shown in Fig. 7.10, where the element has been formed from polar-coordinate differentials so as to be related simply to the boundaries. The area $dA$ is equal to

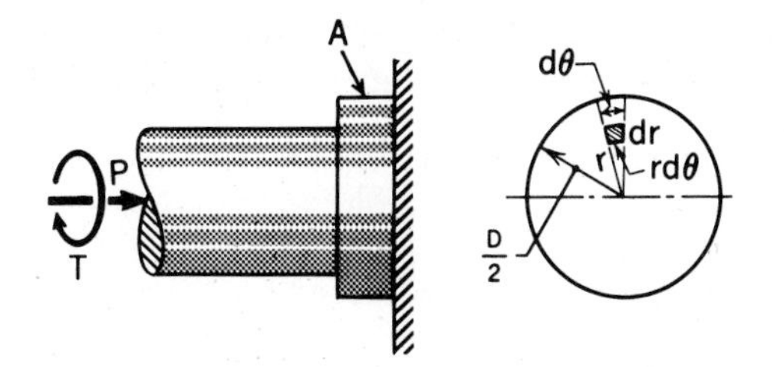


**Figure 7.10.** Dry thrust bearing.

$r\,d\theta\,dr$. We shall *assume* that the normal force $P$ is uniformly distributed over the entire area of contact. The normal force on the area element is then

$$dN = \frac{P}{\pi D^2/4} r\,d\theta\,dr \tag{a}$$

The friction force associated with this force during motion is

$$df = \mu_d \frac{P}{\pi D^2/4} r\,d\theta\,dr \tag{b}$$

The direction of $df$ must oppose the relative motion between the surfaces. The relative motion is rotation of concentric circles about the centerline, so the direction of a force $df_1$ (Fig. 7.11) must lie tangent to a circle of radius $r$. At 180° from the position of the area element for $df_1$, we may carry out a similar calculation for a force $df_2$, which for the same $r$ must be equal and opposite to $df_1$, thus forming a couple. Since the entire area may be decomposed in this way, we can conclude that there are only couples in the plane of contact. If we take moments of all infinitesimal forces about the center, we get the magnitude of the total frictional couple moment. The direction of the couple moment is along the shaft axis. First, consider area elements on the ring of radius $r$:

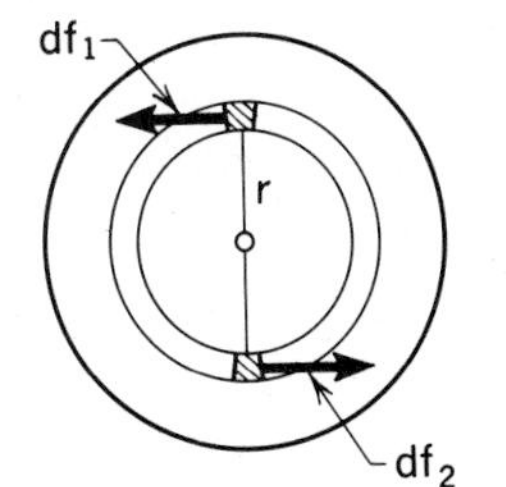


**Figure 7.11.** Friction forces form couples.

$$dM = \int_0^{2\pi} r \mu_d \frac{P}{\pi D^2/4} r\,d\theta\,dr \tag{c}$$

Taking $\mu_d$ as constant and holding $r$ constant, we have on integration with respect to $\theta$:

$$dM = \mu_d \frac{P}{\pi D^2/4} 2\pi r^2\,dr$$

We thus account for all area elements on the ring of radius $r$. To account for all the rings of the contact surface, we next integrate with respect to $r$ from zero to $D/2$. Clearly, this gives us the total resisting torque $M$. Thus,

$$M = \mu_d \frac{8P}{D^2} \int_0^{D/2} r^2\,dr = \frac{PD\mu_d}{3} \tag{d}$$

What we have performed in the last three steps is *multiple integration*, which we introduced in Chapter 4 when dealing with rectangular coordinates.

## 7.5 Belt Friction

A flexible belt is shown in Fig. 7.12 wrapped around a portion of a drum, with the amount of wrap indicated by angle $\beta$. The angle $\beta$ is called the *angle of wrap*. Assume that the drum is stationary and tensions $T_1$ and $T_2$ are such that motion is impending between the belt and the drum. We shall take the impending motion of the belt to be clockwise relative to the drum, and therefore the tension $T_1$ exceeds tension $T_2$.

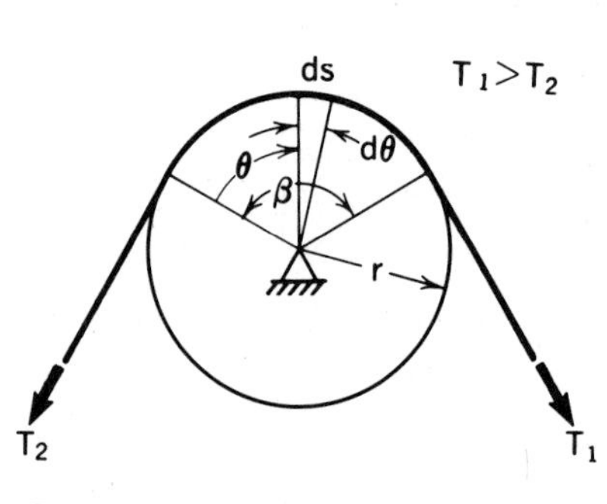


**Figure 7.12.** Belt wrapped around drum.

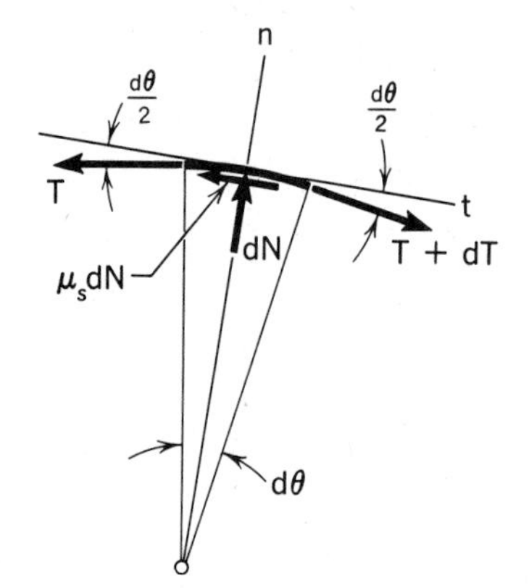


**Figure 7.13.** Free-body diagram of segment of belt; impending slippage.

Consider an infinitesimal segment of the belt as a free body. This segment subtends an angle $d\theta$ at the drum center as shown in Fig. 7.13. Summing force components in the radial and transverse directions and equating them to zero, we get the following scalar equations:

$\underline{\sum F_t = 0:}$

$$-T\cos\frac{d\theta}{2} + (T + dT)\cos\frac{d\theta}{2} - \mu_s\, dN = 0$$

Therefore,

$$dT\cos\frac{d\theta}{2} = \mu_s\, dN$$

$\underline{\sum F_n = 0:}$

$$-T\sin\frac{d\theta}{2} - (T + dT)\sin\frac{d\theta}{2} + dN = 0$$

Therefore,

$$-2T\sin\frac{d\theta}{2} - dT\sin\frac{d\theta}{2} + dN = 0$$

The sine of a very small angle approximately equals the angle itself in radians. Furthermore, to the same degree of accuracy, the cosine of a small angle approaches unity. (That these relations are true may be seen by expanding the sine and cosine in a power series and then retaining only the first terms.) The preceding equilibrium equations then become

$$dT = \mu_s \, dN \tag{7.2a}$$

$$-T \, d\theta - dT \frac{d\theta}{2} + dN = 0 \tag{7.2b}$$

In the last equation, we have an expression involving the product of two infinitesimals. This quantity may be considered negligible compared to the other terms of the equation involving only one differential. Thus, we have for this equation:

$$T \, d\theta = dN \tag{7.3}$$

From Eqs. 7.2a and 7.3, we may form an equation involving $T$ and $\theta$. Thus, by eliminating $dN$ from the equations, we have

$$dT = \mu_s T \, d\theta$$

Hence,

$$\frac{dT}{T} = \mu_s \, d\theta$$

Integrating both sides around the portion of the belt in contact with the drum,

$$\int_{T_2}^{T_1} \frac{dT}{T} = \int_0^{\beta} \mu_s \, d\theta$$

we get

$$\ln \frac{T_1}{T_2} = \mu_s \beta$$

or

$$\frac{T_1}{T_2} = e^{\mu_s \beta} \tag{7.4}$$

We therefore have established a relation between the tensions on each part of the belt at a condition of impending motion between the belt and the drum. The same relation can be reached for a *rotating* drum with impending slippage between the belt and the drum *if we neglect centrifugal effects on the belt*. Furthermore, by using the *dynamic* coefficient of friction in the formula above, we have the case of the belt slipping over either a rotating or stationary drum (again neglecting centrifugal effects on the belt). Thus, for all such cases, we have

$$\boxed{\frac{T_1}{T_2} = e^{\mu \beta}} \tag{7.5}$$

where the proper coefficient of friction must be used to suit the problem, and the angle $\beta$ must be expressed in *radians*. Note that *the ratio of tensions depends only on the angle*

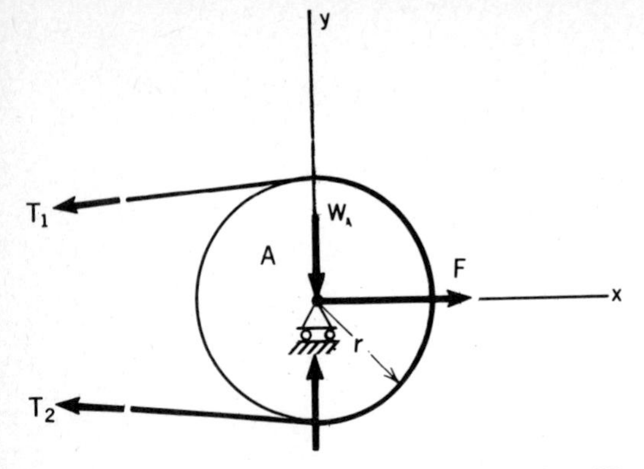


**Figure 7.14.** Force $F$ affects $T_1$ and $T_2$ but not $T_1/T_2$.

*of wrap $\beta$ and the coefficient of friction $\mu$.* Thus, if the drum $A$ is forced to the right, as shown in Fig. 7.14, the tensions will increase, but if $\beta$ is not affected by the action, the ratio of $T_1/T_2$ for impending or actual slippage is not affected by this action. However, the *torque* developed by belt on drum as a result of friction *is* affected by the force $F$. The torque is easily determined by using the drum and the portion of the belt in contact with the drum as a free body, as is shown in Fig. 7.14. Thus,

$$\text{torque} = T_1 r - T_2 r = (T_1 - T_2)r \tag{7.6}$$

If we pull the drum to the right without disturbing the angle of wrap, we can see from Eq. 7.5 that the tensions $T_1$ and $T_2$ must increase by the same factor. And if we call this factor $H$, the new tensions become $HT_1$ and $HT_2$, respectively. Substituting into Eq. 7.6, we see that the frictional torque is also increased by the same factor:

$$\text{torque} = H(T_1 - T_2)r = H(\text{torque})_{\text{original}}$$

Sometimes we know the force $F$ (Fig. 7.14) which acts on the drum support to maintain the belt tension. Summing forces on the free-body diagram in Fig. 7.14, we get in the $x$ direction:

$$(T_1)_x + (T_2)_x = F_{\text{known}} \tag{7.7}$$

At impending slippage or at slippage, Eq. 7.5 is valid, and with Eq. 7.7, we can solve for $T_1$ and $T_2$. With Eq. 7.6, the torque that the belt is capable of developing on the drum now becomes a simple computation.

**EXAMPLE 7.6**

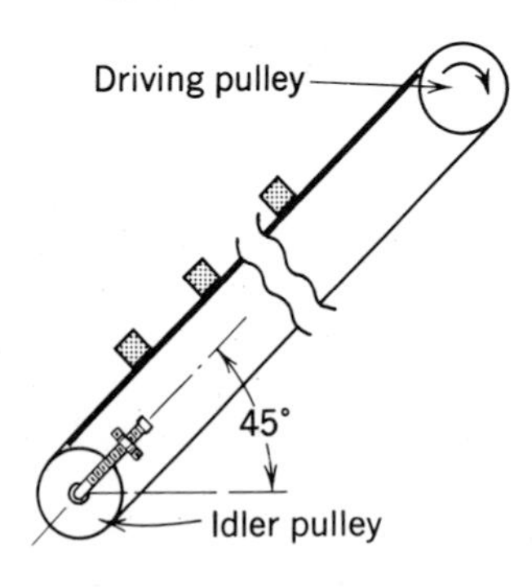


**Figure 7.15.** Conveyor.

A conveyor is moving ten 50-lb boxes at a 45° setting (Fig. 7.15). The coefficient of friction between the belt and the bed of the coveyor is .05. Furthermore, the coefficient of friction between the driving pulley and the belt is .4. The idler pulley is moved along the direction of the conveyor by a crank mechanism so that the idler pulley is subject to a force $F$ of 500 lb. Compute the maximum tension found in the belt and ascertain if there will be slipping on the driving pulley. Neglect the weight of the belt.

In Fig. 7.16, we have shown free-body diagrams of various parts of the conveyor.[3] For the portion of the belt on the conveyor frame, we can sum forces normal and tangent to the belt:

$\underline{\sum F_n = 0:}$

$$N - (10)(50)(.707) = 0$$

[3]The weights of the pulleys have been counteracted by supporting forces at the axles and have not been shown.

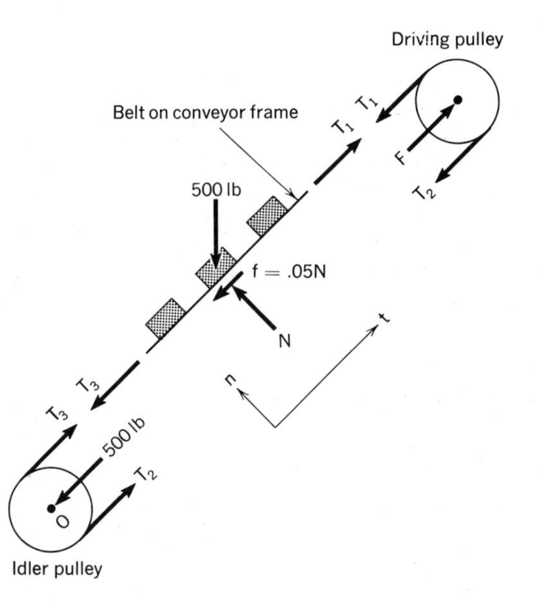


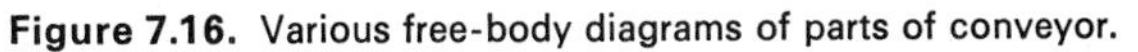

**Figure 7.16.** Various free-body diagrams of parts of conveyor.

Therefore,

$$N = 354 \text{ lb} \tag{a}$$

$\underline{\sum F_t = 0:}$

$$T_1 - T_3 - (10)(50)(.707) - (.05)(354) = 0$$

Therefore,

$$T_1 - T_3 = 371 \text{ lb} \tag{b}$$

For the idler pulley, we have

$\underline{\sum M_0 = 0:}$

$$T_3 = T_2 \tag{c}$$

$\underline{\sum F_t = 0:}$

$$T_3 + T_2 = 500 \tag{d}$$

From Eqs. (c) and (d), we conclude that

$$T_2 = T_3 = 250 \text{ lb} \tag{e}$$

From Eq. (b) we now get for the maximum tension $T_1$:

$$T_1 = T_3 + 371 = 621 \text{ lb} \tag{f}$$

We must next check the driving pulley to ensure that there is no slippage occurring. For the condition of impending slippage, we have, using as $T_2$ the value of 250 lb and solving for $T_1$

$$T_1 = T_2 e^{.4\pi} = (250)(3.51) = 878 \text{ lb}$$

Clearly, since the $T_1$ needed is only 621 lb, we do not have slippage at the driving pulley, and we conclude that the maximum tension is indeed 621 lb.

**EXAMPLE 7.7**

An electric motor (not shown) in Fig. 7.17 drives at constant speed the pulley $B$, which connects to pulley $A$ by a belt. Pulley $A$ is connected to a compressor (not shown) which requires 700 N-m torque to drive it at constant speed $\omega_A$. If $\mu_s$ for the belt and either pulley is .4, what minimum value of the indicated force $F$ is required to have no slipping anywhere?

As a first step, we determine the angles of wrap $\beta$ for the respective pulleys. For this purpose, we first compute $\alpha$ (Fig. 7.18). Note that the radii $O_AD$ and $O_BE$, being perpendicular to the same line $DE$, are therefore parallel to each other. Drawing $EC$ parallel to $O_AO_B$, we then form $\alpha$ in the cross-hatched triangle. Hence, we can say:

$$\alpha = \sin^{-1}\frac{CD}{CE} = \sin^{-1}\frac{r_A - r_B}{O_A O_B} = \sin^{-1}\frac{.50 - .30}{2} = 5.74°$$

Hence,

$$\beta_A = 180° + 2(5.74) = 191.5°$$
$$\beta_B = 180° - 2(5.74) = 168.5°$$

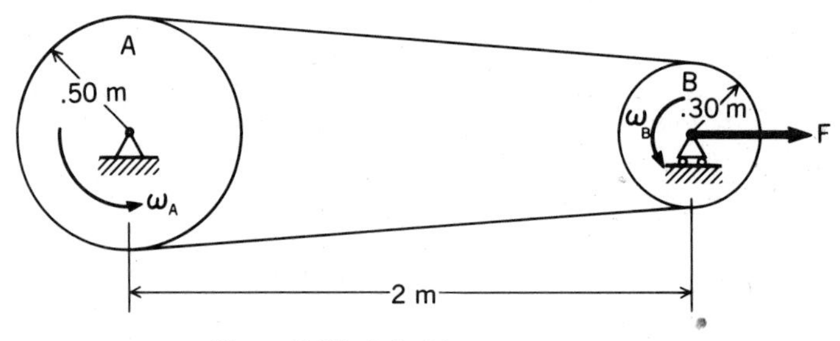


Figure 7.17. Belt-driven compressor.

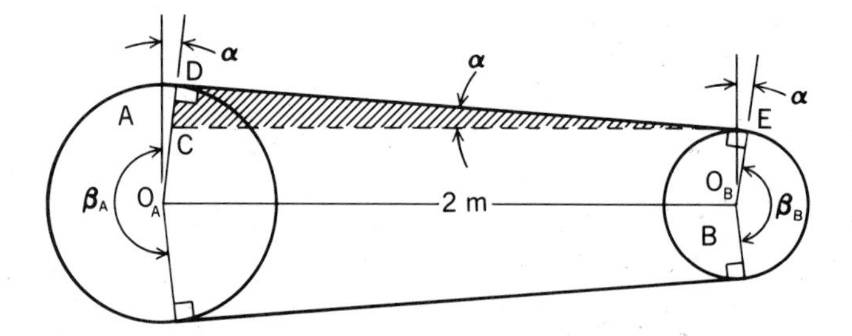


Figure 7.18. Find angles of wrap.

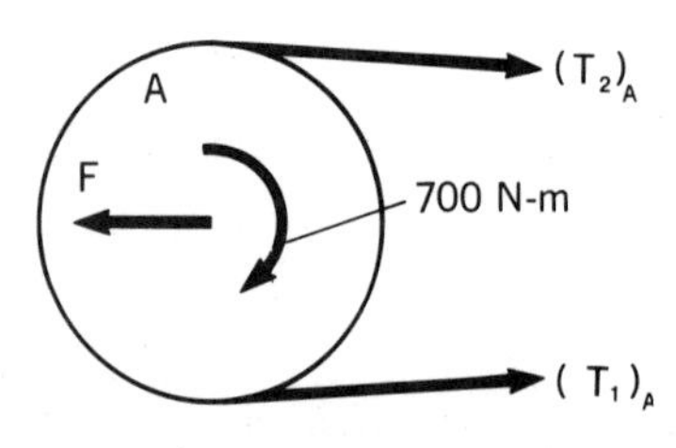


Figure 7.19. Free-body diagram of $A$.

Now consider pulley $A$ as a free body in Fig. 7.19. Note that the minimum force $F$ corresponds to the condition of impending slippage. Accordingly, for this condition at $A$, we have

$$\frac{(T_1)_A}{(T_2)_A} = e^{\mu_s \beta_A} = e^{(.4)[(191.5/360)2\pi]} = 3.81 \qquad \text{(a)}$$

Also, summing moments about the center of the pulley, we have

$$[(T_1)_A - (T_2)_A](.50) - 700 = 0 \tag{b}$$

Therefore,

$$(T_1)_A - (T_2)_A = 1400$$

Solving Eqs. (a) and (b) simultaneously, we get

$$(T_1)_A = 1898 \text{ N}; \qquad (T_2)_A = 498 \text{ N}$$

From equilibrium, we can compute force $F$ as follows:

$$(1898 + 498) \cos 5.74° - F = 0 \tag{c}$$

Therefore,

$$F = 2384 \text{ N}$$

Now go to pulley $B$ to see what minimum force $F$ is needed so that the belt does not slip on it during operations. Consider in Fig. 7.20 the free-body diagram of pulley $B$. For impending slipping on pulley $B$, we have

$$\frac{(T_1)_B}{(T_2)_B} = e^{\mu_s \beta_B} = e^{(.4)[(168.5/360)2\pi]} = 3.24 \tag{d}$$

The torque for pulley $B$ needed to develop 700 N-m on pulley $A$ is next computed. Thus,[4]

$$M_B = \frac{r_B}{r_A} M_A = \frac{.30}{.50} M_A = \frac{.30}{.50}(700)$$

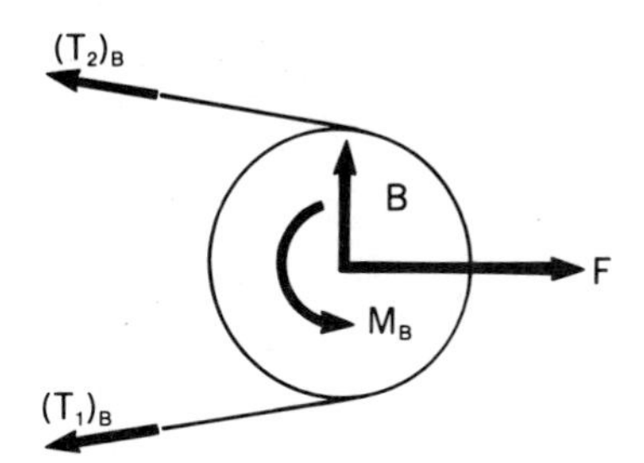


**Figure 7.20.** Free-body diagram of *B*.

Therefore,

$$M_B = 420 \text{ N-m}$$

Summing moments in Fig. 7.20 about the center of $B$, we then have on using the above result

$$-[(T_1)_B - (T_2)_B](.30) + 420 = 0$$

Therefore,

$$(T_1)_B - (T_2)_B = 1400 \tag{e}$$

Solving Eqs. (d) and (e) simultaneously, we get

$$(T_1)_B = 2025 \text{ N}; \qquad (T_2)_B = 625 \text{ N}$$

Hence, the minimum $F$ needed for pulley $B$ is

$$F = (2025 + 625) \cos 5.74° = 2637 \text{ N}$$

Thus, for no slipping on *either* pulley we require $F = 2637$ N as a minimum value.

---

[4]Note that the ratio of transmitted torques $M_2/M_1$ between directly connected pulleys and gears will equal $r_2/r_1$ or $D_2/D_1$ of the pulleys or gears. Can you verify this yourself?

## Problems

**7.47.** Compute the frictional resisting torque for the concentric dry thrust bearing. The coefficient of friction is taken as $\mu_d$.

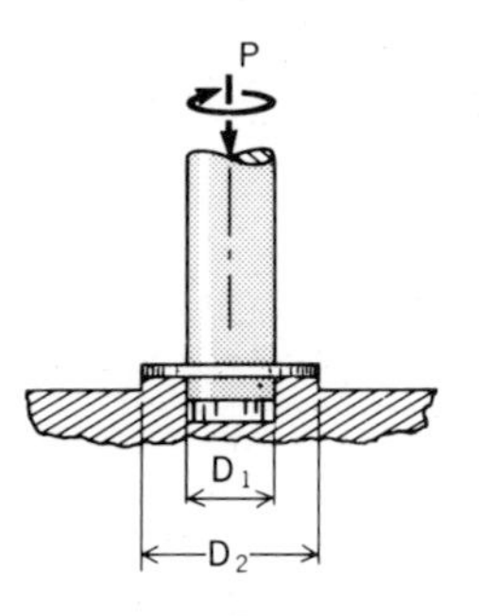


Figure P.7.47

**7.48.** The support end of a dry thrust bearing is shown. Four pads form the contact surface. If a shaft creates a 100-N thrust uniformly distributed over the pads, what is the resisting torque for a coefficient of friction of .1?

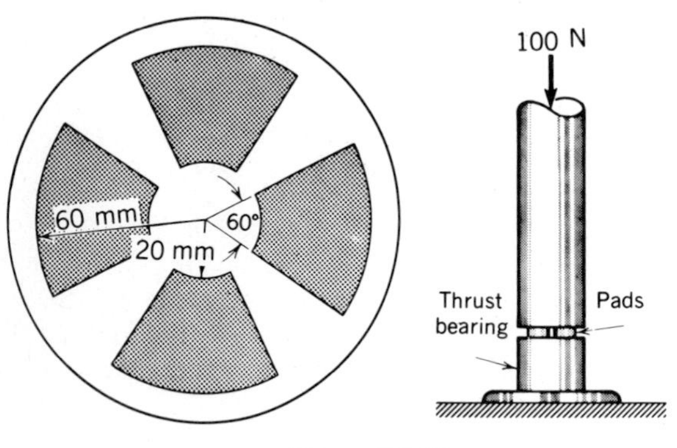


Figure P.7.48

**7.49.** In Example 7.5, the normal force distribution at the contact surface is not uniform but, as a result of wear, is inversely proportional to the radius $r$. What, then, is the resisting torque $M$?

**7.50.** Compute the frictional torque needed to rotate the truncated cone relative to the fixed member. The cone has a 20-mm-diameter base and a 60° cone angle and is cut off 3 mm from the cone tip. The coefficient of dynamic friction is .2.

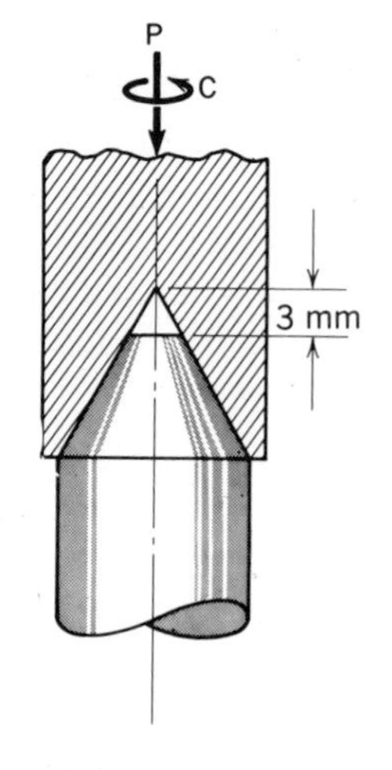


Figure P.7.50

**7.51.** A 1000-N block is being lowered down an inclined surface. The block is pinned to the incline at $C$, and at $B$ a cord is played out so as to cause the body to rotate at uniform speed about $C$. Taking $\mu_d$ to be .3 and assuming the contact pressure is uniform along the base of the block, compute $T$ for the configuration shown in the diagram.

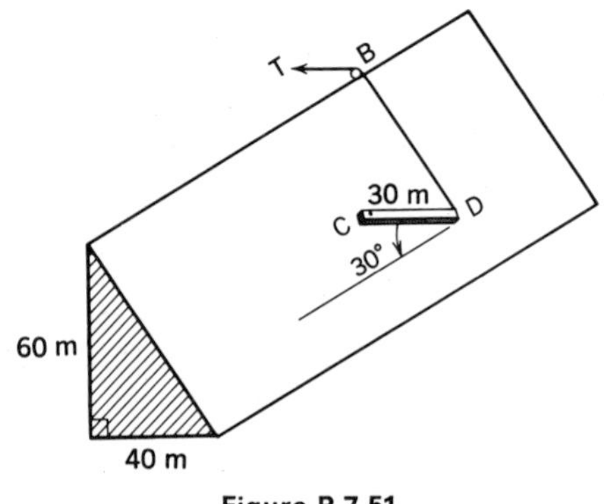


Figure P.7.51

**7.52.** A pulley requires 200 N-m torque to get it rotating. The angle of wrap is $\pi$ radians, and $\mu_s$ is known to be 5.2. What is the *minimum* horizontal force $F$ required to create enough tension in the belt so that it can rotate the pulley?

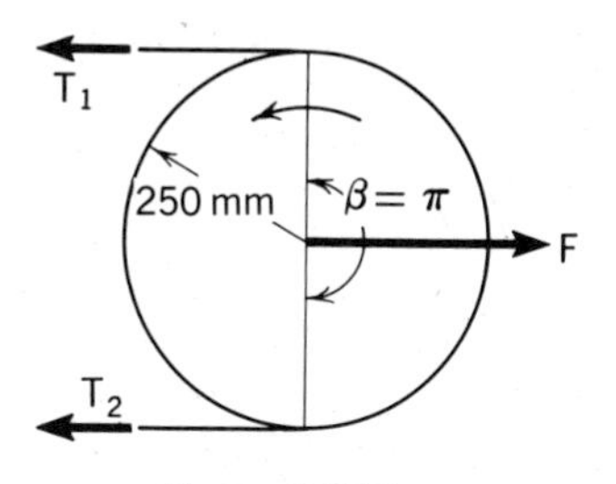


Figure P.7.52

**7.53.** If in Problem 7.52 the belt is wrapped $2\frac{1}{2}$ times around the pulley, what is the *minimum* horizontal force $F$ needed to rotate the pulley?

**7.54.** The seaman pulls with 100-N force and wants to stop the motorboat from moving away from the dock under power. How few wraps $n$ of the rope must he make around the post if the motorboat develops 3500 N of thrust and the coefficient of friction between the rope and the post is .2?

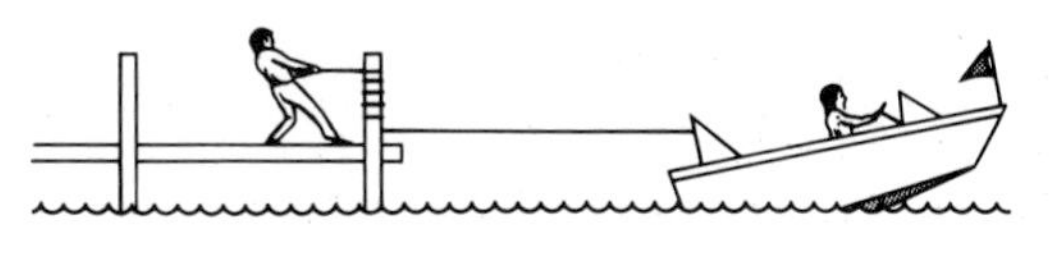

Figure P.7.54

**7.55.** A length of belt rests on a flat surface and runs over a quarter of the drum. A load $W$ rests on the horizontal portion of the belt, which in turn is supported by a table. If the coefficient of friction for all surfaces is .3, compute the *maximum* weight $W$ that can be moved by rotating the drum.

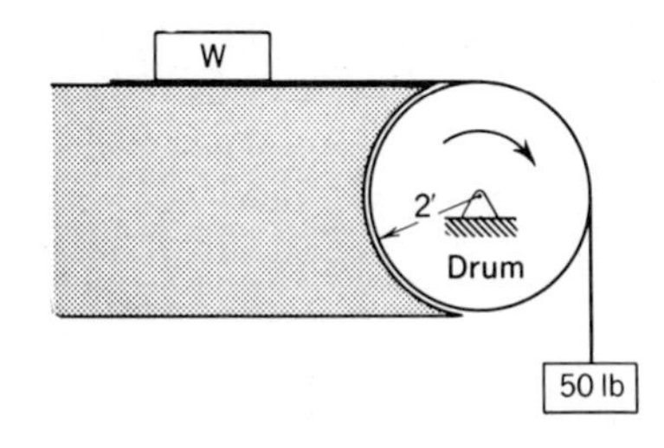


Figure P.7.55

**7.56.** The rope holding the 50-lb weight $E$ passes over the drum and is attached at $A$. The weight of $C$ is 60 lb. What is the *minimum* coefficient of friction between the rope and the drum to maintain equilibrium?

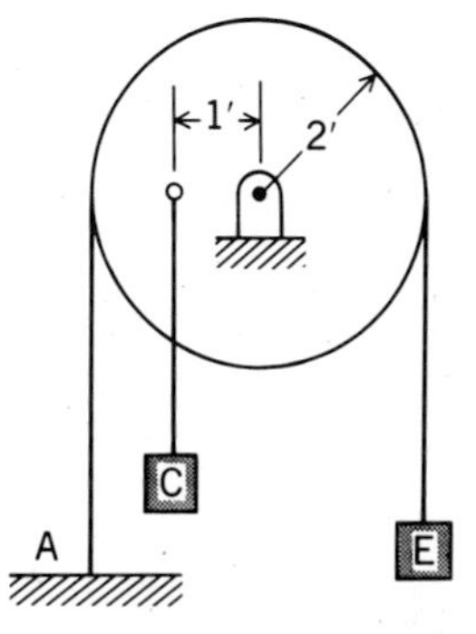


Figure P.7.56

**7.57.** What is the *maximum* weight that can be supported by the system in the position shown? Pulley $B$ *cannot* turn. Bar $AC$ is fixed to cylinder $A$, which weighs 500 N. The coefficient of static friction for all contact surfaces is .3.

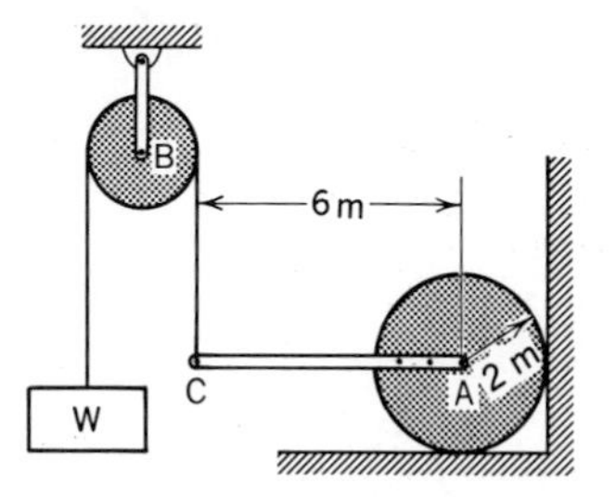


Figure P.7.57

**7.58.** A mountain climber of weight $W$ hangs freely suspended by one rope that is fastened at one end to his waist, wrapped one-half turn about a rock with $\mu_s = .2$, and held at the other end in his hand. What *minimum* force in terms of $W$ must he pull with to maintain his position? What minimum force must he pull with to gain altitude?

Figure P.7.58

**7.59.** Pulley $B$ is turned by a diesel engine and drives pulley $A$ connected to a generator. If the torque that $A$ must transmit to the generator is 500 N-m, what is the *minimum* coefficient of friction between the belt and pulleys for the case where the force $F$ is 2000 N?

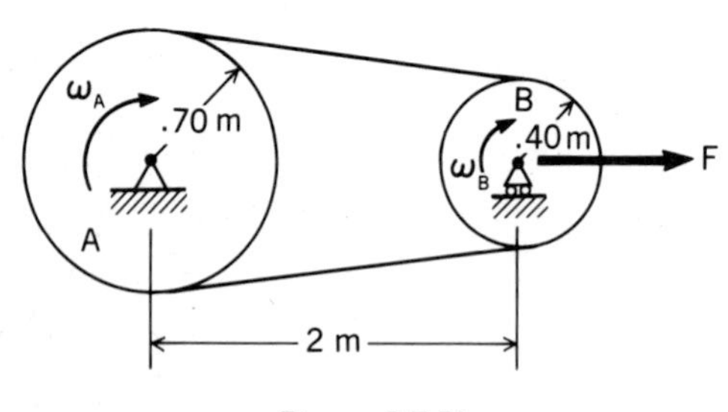


Figure P.7.59

**7.60.** A hand brake is shown. If $\mu_d = .4$, what is the resisting torque when the shaft is rotating?

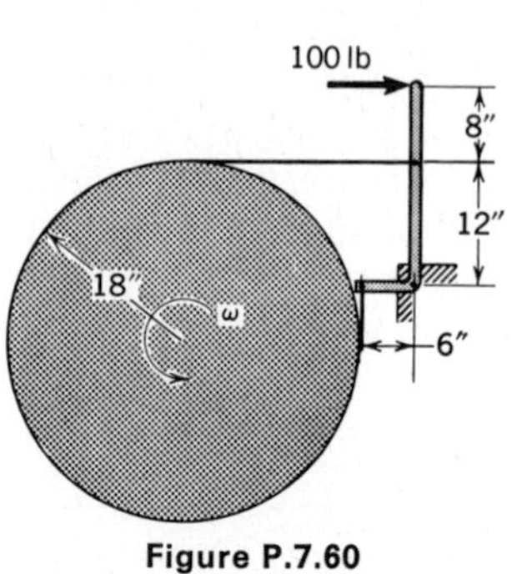


Figure P.7.60

**7.61.** A conveyor is shown with two driving pulleys $A$ and $B$. Driver $A$ has an angle of wrap of 330°, whereas $B$ has a wrap of 180°. If the coefficient of friction between the belt and the bed of the conveyor is .1, and the weight to be transported is 10,000 N, what is the *smallest* coefficient of friction between the belt and the driving pulleys? One-fifth of the load can be assumed to be between the active pulleys at all times, and the tension in the slack side (underneath) is 2000 N. There is a free-wheeling pulley at the left end of the conveyor. You will have to solve an equation by trial and error.

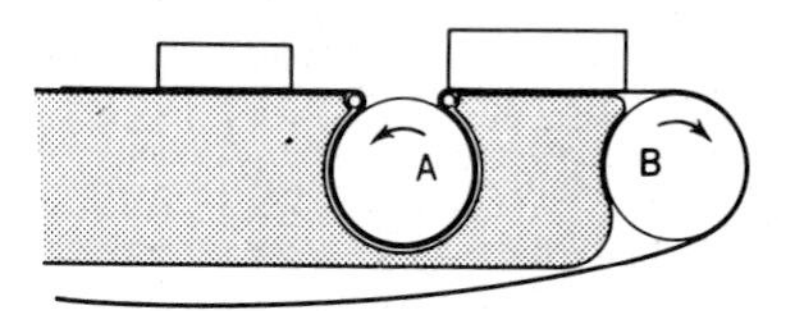


Figure P.7.61

**7.62.** A freely turning idler pulley is used to increase the angle of wrap for the pulleys shown. If the tension in the slack side below is 200 lb, find the *maximum* torque that can be transmitted by the pulleys for a coefficient of friction of .3.

Figure P.7.62

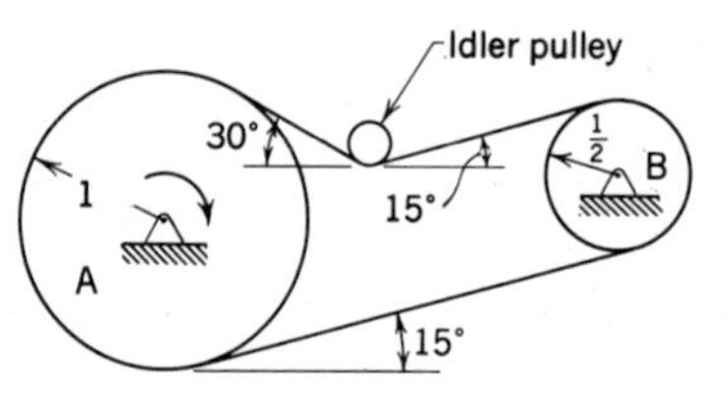

**7.63.** Rod $AB$ weighing 200 N is supported by a cable wrapped around a semicylinder having a coefficient of friction $\mu_s$ equal to .2. A weight $A$ having a mass of 10 kg can slide on rod $AB$. What is the maximum range $x$ from the centerline that the center of $A$ can be placed without causing slippage?

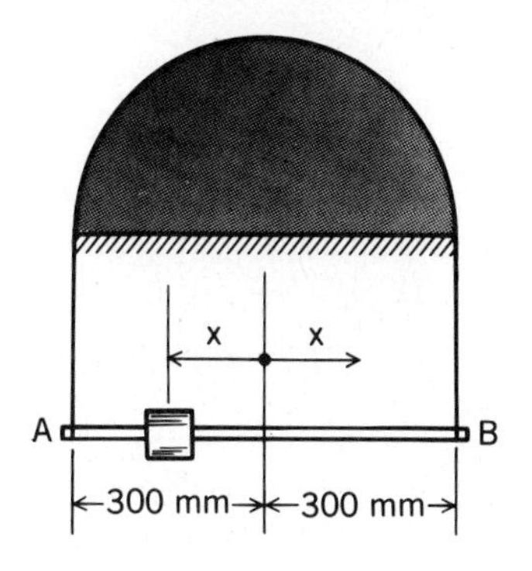


Figure P.7.63

**7.64.** The cable mechanism shown is similar to that used to move the station indicator on a radio. If the indicator jams, what force is developed at the indicator base to free the jam when the torque applied to the handle is 10 lb-in.? Also, what are the forces in the various regions of the cable? The coefficient of friction is .15.

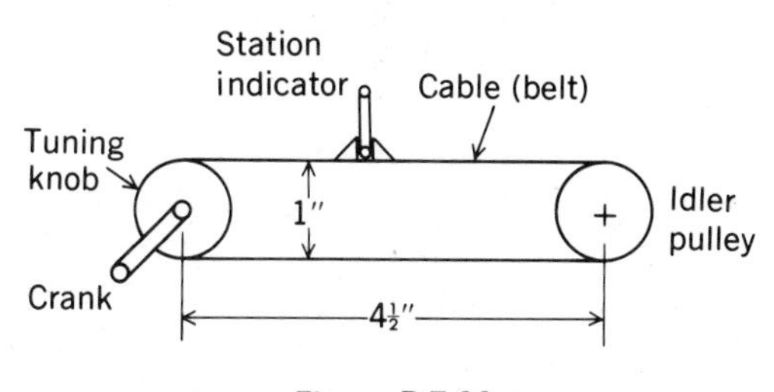


Figure P.7.64

**7.65.** What are the *minimum* possible supporting force components needed for pulley $B$ as a result of the action of the belt? The coefficient of friction for the belt and pulley $B$ is .3 and for the belt and pulley $A$ is .4. The torque that the belt delivers to $A$ is 200 N-m.

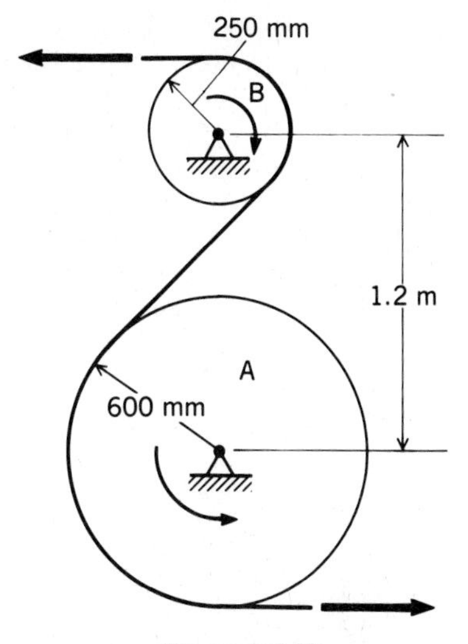


Figure P.7.65

**7.66.** From first principles, show that the normal force per unit length, $w$, acting on a drum from a belt is given as

$$w = \frac{T_2}{r} e^{\mu\theta}$$

Use the indicated diagram as an aid. [*Hint:* Start with Eq. 7.2(a) and use Eq. 7.4 for any point $a$.]

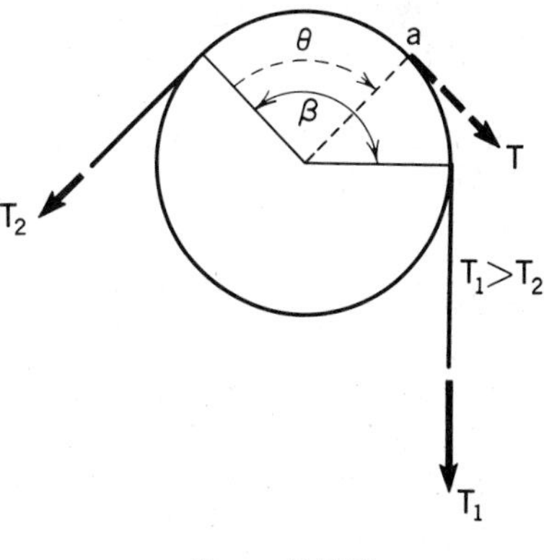


Figure P.7.66

**7.67.** What *minimum* force $F$ is needed so that drum $A$ can transmit a clockwise torque of 500 N-m without slipping? The coefficient of friction, $\mu_s$, for $A$ and the belt is .4. What *minimum* coefficient of friction is needed for $B$ for no slipping?

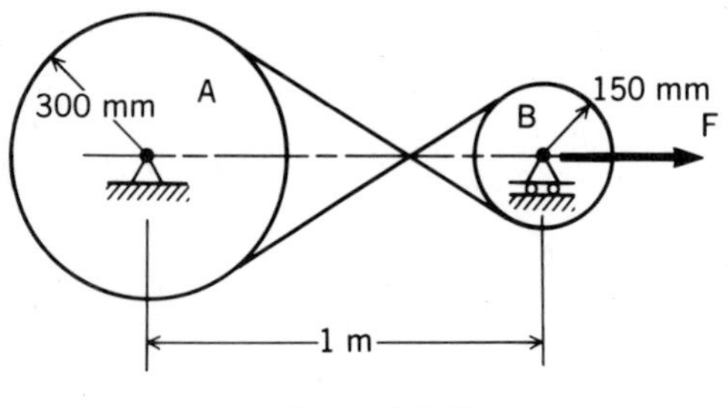


Figure P.7.67

**7.68.** What is the *minimum* weight $B$ that will prevent rotation induced by body $C$ weighing 500 N? The weight of $A$ is 100 N. The coefficient of friction between the belts and $A$ is .4, and between $A$ and the walls is .1. Neglect friction at pulley $G$.

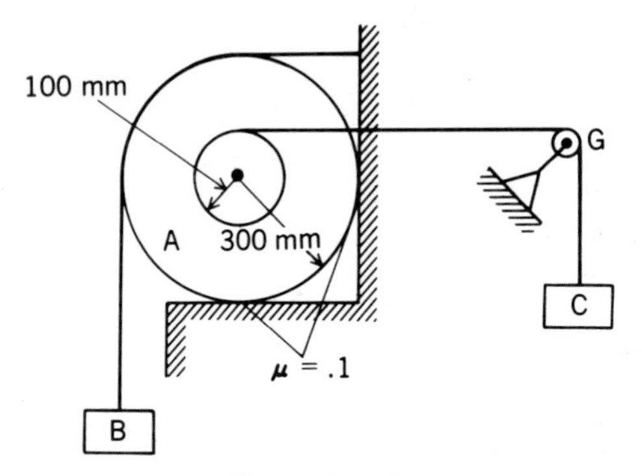


Figure P.7.68

**7.69.** A V-belt is shown. Show that

$$\frac{T_1}{T_2} = e^{\mu_s \beta / \sin(\alpha/2)}$$

for impending slippage. Use a development analogous to that of the flat belt in Section 7.5.

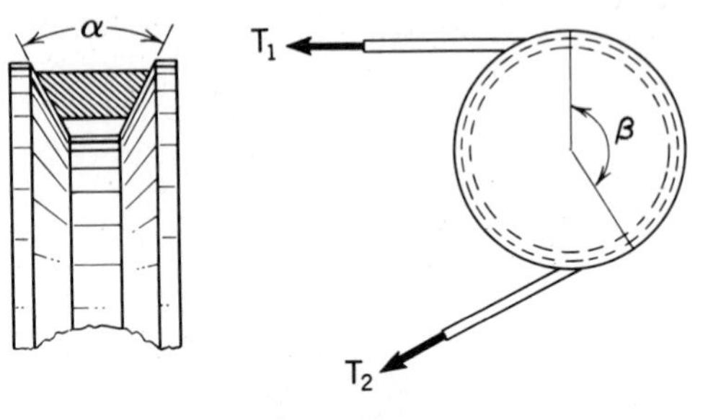


Figure P.7.69

**7.70.** An electric motor drives a pulley $B$ which drives three V-belts having the cross section shown. These V-belts then drive a compressor through pulley $A$. If the torque needed to drive the compressor is 1000 N-m, what *minimum* force $F$ is needed to do the job? The coefficient of friction between belts and pulleys is .5. See Problem 7.69 before doing this problem.

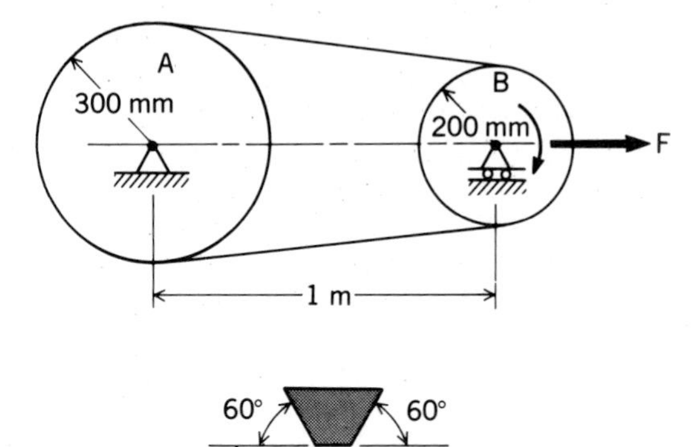

Figure P.7.70

**7.71.** A drum of radius $r$ is rotated by a belt with a constant speed $\omega$ rad/sec. What is the relation between $T_1$ and $T_2$ for the case of impending slipping between drum and belt if *centrifugal* effects are counted? The belt has a mass per unit length of $m$ kg/m. Recall from physics that the centrifugal force of a particle of mass $M$ is $Mr\omega^2$, where $r$ is the distance from the axis of rotation. Assume the belt is thin compared to the radius of the drum. The desired result is

$$\frac{T_1 - r^2\omega^2 m}{T_2 - r^2\omega^2 m} = e^{\mu\beta}$$

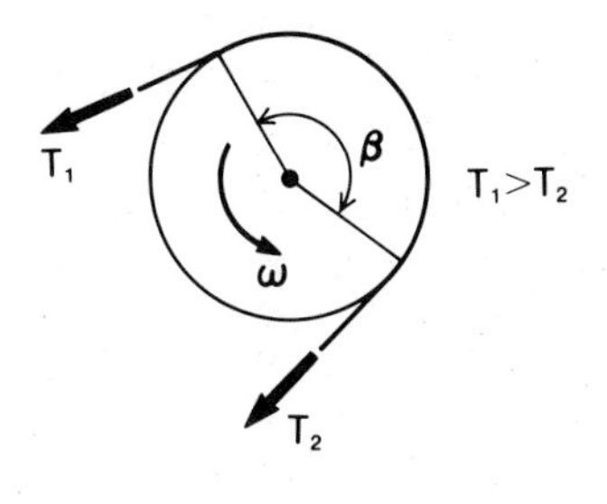


Figure P.7.71

**7.72.** A pulley $A$ is driven by an outside agent at a speed $\omega$ of 100 rpm. A belt weighing 30 N/m is driven by the pulley. If $T_2 = 200$ N, what is the *maximum* possible tension $T_1$ computed without considering centrifugal effects? Compute $T_1$ counting centrifugal effects, and give the percentage error incurred by not including centrifugal effects. The coefficient of friction between the belt and the pulley is .3. See Problem 7.71 before doing this problem.

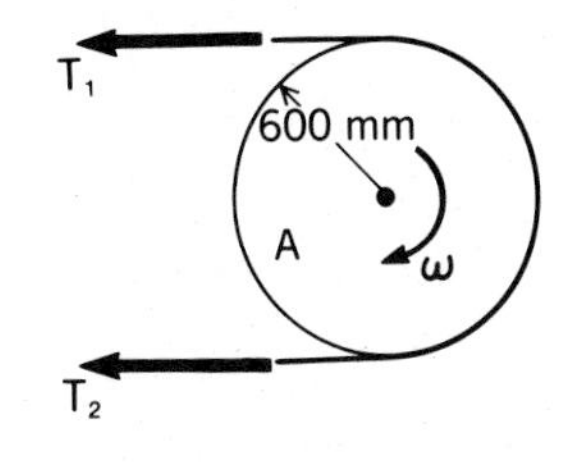


Figure P.7.72

## 7.6 The Square Screw Thread

We shall now consider the action of a nut on a screw that has square threads (Fig. 7.21). Let us take $r$ as the mean radius from the centerline of the screw to the thread. The *pitch*, $p$, is the distance along the screw between adjacent threads, and the *lead*, $L$, is the distance that a nut will advance in the direction of the axis of the screw in one revolution. For screw threads that are single-threaded, $L$ equals $p$. For an $n$-threaded screw, the lead $L$ is $np$.

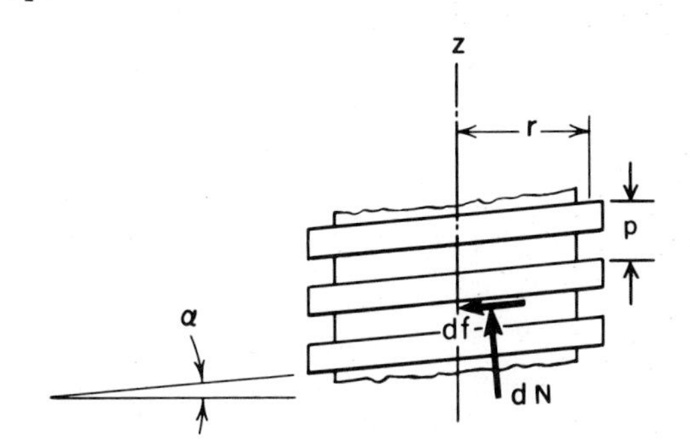


**Figure 7.21.** Square screw thread.

Forces are transmitted from screw to nut over several revolutions of thread, and hence we have a distribution of normal and friction forces. However, because of the narrow width of the thread, we may consider the distribution to be confined at a distance $r$ from the centerline, thus forming a "loading" strip winding around the centerline of the screw. Figure 7.21 illustrates infinitesimal normal and frictional forces on an infinitesimal part of the strip. The local slope $\tan \alpha$ as one looks in radially is determined by considering the definition of $L$, the lead. Thus,

$$\text{slope} = \tan\alpha = \frac{L}{2\pi r} = \frac{np}{2\pi r}$$

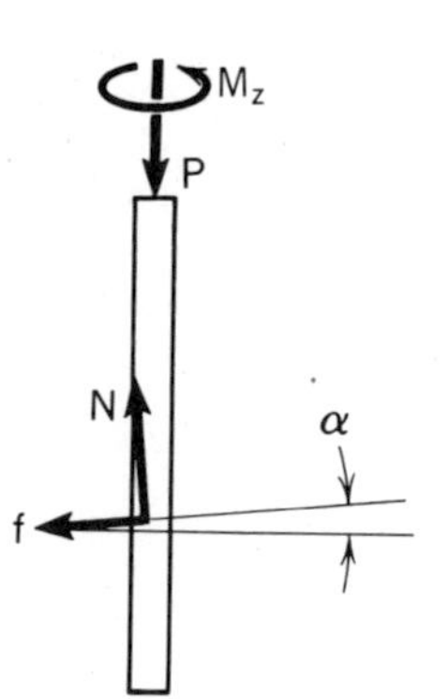

**Figure 7.22.** Free-body diagram.

All elements of the proposed distribution have the same inclination (direction cosine) relative to the $z$ direction. In the summation of forces in this direction, therefore, we can consider the distribution to be replaced by a single normal force $N$ and a single friction force $f$ at the inclinations shown in Fig. 7.22 at a position anywhere along the thread. And, since the elements of the distribution have the same moment arm about the centerline in addition to the common inclination, we may use the concentrated forces mentioned above in taking moments about the centerline. There is thus a "limited equivalence" between $N$ and $f$ and the force distribution from the nut onto the screw. The other forces on the screw will be considered as an axial load $P$ and a torque $M_z$ collinear with $P$ (Fig. 7.22). For equilibrium at a condition of impending motion to raise the screw, we then have the following scalar equations:[5]

$\underline{\sum F_z = 0:}$

$$-P + N\cos\alpha - \mu N \sin\alpha = 0 \tag{a}$$

$\underline{\sum M_z = 0:}$

$$-\mu N \cos\alpha\, r - N \sin\alpha\, r + M_z = 0 \tag{b}$$

These equations may be used to eliminate the force $N$ and so get a relation between $P$ and $M_z$ that will be of practical significance. This may readily be done by solving for $N$ in (a) and substituting into (b). The result is

$$\boxed{M_z = \frac{Pr\,(\mu\cos\alpha + \sin\alpha)}{\cos\alpha - \mu\sin\alpha}} \tag{7.8}$$

An important question arises when we employ the screw and nut in the form of a jack as shown in Fig. 7.23. Once having raised a load $P$ by applying the torque $M_z$ to the jackscrew, does the device maintain the load at the raised position when the applied torque is released, or does the screw unwind under the action of the load and thus lower the load? In other words, is this a *self-locking* device? To examine this, we go back to the equations of equilibrium. Setting $M_z = 0$ and changing the direction of the friction forces, we have the condition for impending "unwinding" of the screw. Eliminating $N$ from the equations, we get

[5]The equations also apply to *steady rotation* of the nut on the screw, in which case one uses the dynamic coefficient of friction $\mu_d$ in the equations.

$$\frac{Pr\,(-\mu_s \cos\alpha + \sin\alpha)}{\cos\alpha + \mu_s \sin\alpha} = 0$$

This requires that

$$-\mu_s \cos\alpha + \sin\alpha = 0$$

Therefore,

$$\mu_s = \tan\alpha \qquad (7.9)$$

We can conclude that, if the coefficient of friction $\mu_s$ equals or exceeds $\tan\alpha$, we will have a self-locking condition. If $\mu_s$ is less than $\tan\alpha$, the screw will unwind and will not support a load $P$ without the proper external torque.

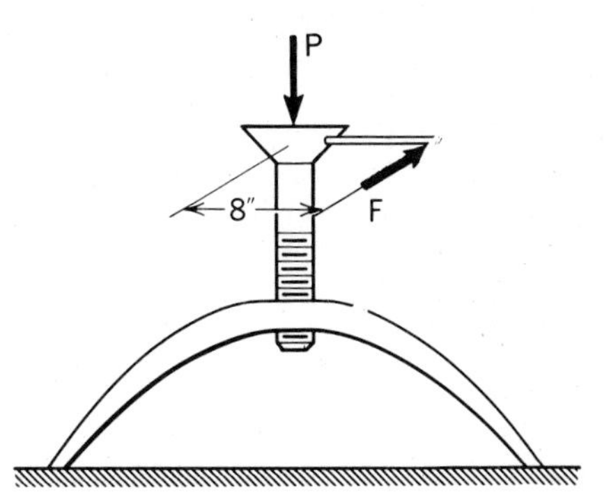

Figure 7.23. Jackscrew.

**EXAMPLE 7.8**

A jackscrew with a double thread of mean diameter 2 in. is shown in Fig. 7.23. The pitch is .2 in. If a force $F$ of 40 lb is applied to the device, what load $W$ can be raised? With this load on the device, what will happen if the applied force $F$ is released? Take $\mu_s = .3$ for the surfaces of contact.

The applied torque $M_z$ is clearly:

$$M_z = \tfrac{8}{12}(40) = 26.7 \text{ lb-ft} \qquad \text{(a)}$$

The angle $\alpha$ for this screw is given as

$$\tan\alpha = \frac{(2)(.2)}{(2\pi)(1)} = .0636 \qquad \text{(b)}$$

Therefore,

$$\alpha = 3.64°$$

Using Eq. 7.8 we can solve for $P$. Thus,

$$\begin{aligned} P &= \frac{M_z(\cos\alpha - \mu_s \sin\alpha)}{r(\mu_s \cos\alpha + \sin\alpha)} \\ &= \frac{(26.7)[.998 - (.3)(.0635)]}{\frac{1}{12}[(.3)(.998) + .0635]} = 864 \text{ lb} \end{aligned} \qquad \text{(c)}$$

The load $W$ is 864 lb. The device is self-locking since $\mu_s$ exceeds $\tan\alpha = .0636$. To lower the load requires a reverse torque. We may readily compute this torque by using Eq. 7.8 with the friction forces reversed. Thus,

$$\begin{aligned} (M_z)_{\text{down}} &= \frac{864(\frac{1}{12})[-(.3)(.998) + .0635]}{.998 + (.3)(.0635)} \\ &= -16.71 \text{ lb-ft} \end{aligned} \qquad \text{(d)}$$

## *7.7 Rolling Resistance

Let us now consider the situation where a hard roller moves without slipping along a horizontal surface while supporting a load $W$ at the center. Since we know from experience that a horizontal force $P$ is required to maintain uniform motion, some sort of

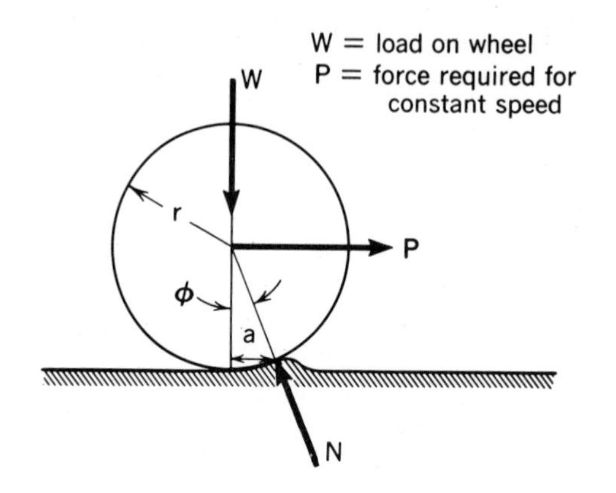


**Figure 7.24.** Rolling resistance model.

resistance must be present. We can understand this resistance if we examine the deformation shown in an exaggerated manner in Fig. 7.24. If force $P$ is along the centerline as shown, the equivalent force system coming onto the roller from the region of contact must be that of a force $N$ whose line of action also goes through the center of the roller since, you will recall from Chapter 5, three nonparallel forces must be concurrent for equilibrium. In order to develop a resistance to motion, clearly $N$ must be oriented at an angle $\phi$ with the vertical direction, as is shown in Fig. 7.24. The scalar equations of equilibrium become

$$W = N\cos\phi; \qquad P = N\sin\phi$$

Therefore,

$$\frac{P}{W} = \tan\phi \tag{7.10}$$

Since the area of contact is small, we note that $\phi$ is a small angle and that $\tan\phi \approx \sin\phi$. The $\sin\phi$ is seen to be $a/r$ from Fig. 7.24. Therefore, we may say that

$$\frac{P}{W} = \frac{a}{r} \tag{7.11a}$$

Solving for $P$, we get

$$P = \frac{Wa}{r} \tag{7.11b}$$

The distance $a$ in these equations is called the *coefficient of rolling resistance.*

Coulomb suggested that for *variable* loads $W$ the ratio $P/W$ is constant for given materials and a given geometry ($r$ = constant). Looking at Eq. 7.11a, we see that $a$ must then be a constant for given geometry and materials. Coulomb added that, for given materials and *variable* radius, the ratio $P/W$ varies inversely as $r$; that is, as the radius of the cylinder is increased, the resistance to uniform motion for a given load $W$ decreases. Thus, considering Eq. 7.11a again, we may conclude that, for given materials, $a$ is also constant for all sizes of rollers and loads. However, other investigators have contested both statements, particularly the latter one, and there is a need for further investigation in this area. Lacking better data, we present the following list of rolling coefficients for your use, but we must caution that you should not expect great accuracy from this general procedure.

COEFFICIENTS OF ROLLING RESISTANCE

| | $a$ (in.) |
|---|---|
| Steel on steel | .007 –.015 |
| Steel on wood | .06 –.10 |
| Pneumatic tires on smooth road | .02 –.03 |
| Pneumatic tires on mud road | .04 –.06 |
| Hardened steel on hardened steel | .0002–.0005 |

**EXAMPLE 7.9**

What is the rolling resistance of a railroad freight car weighing 100 tons? The wheels have a diameter of 30 in. The coefficient of rolling resistance between wheel and track is .001 in. Compare the resistance to that of a truck and trailer having the same total weight and with tires having a diameter of 4 ft. The coefficient of rolling resistance $a$ for the truck tires and road is .025 in.

We can use Eq. 7.11b directly for the desired results. Thus, for the railroad freight car, we have[6]

$$P_1 = \frac{(100)(2000)(.001)}{15} = 13.33 \text{ lb} \qquad \text{(a)}$$

For the truck, we get

$$P_2 = \frac{(100)(2000)(.025)}{24} = 208 \text{ lb} \qquad \text{(b)}$$

We see a decided difference between the two vehicles, with clear advantage toward the railroad freight car.

## Problems

**7.73.** A simple C-clamp is used to hold two pieces of metal together. The clamp has a single square thread with a pitch of .12 in. and a mean diameter of .75 in. The coefficient of friction is .30. Find the torque required if a 1000-lb compressive load is required on the blocks. If the thread is a double thread, what is the required torque?

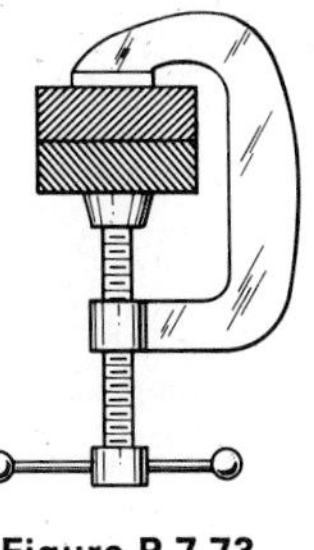

Figure P.7.73

[6]The number of wheels $n$ plays no role here since we divide the load by $n$ to get the load per wheel and later multiply by $n$ to get the total resistance.

**7.74.** The mast of a sailboat is held by wires called shrouds, as shown in the diagram. Racing sailors are careful to get the proper tension in the shrouds by adjusting the turnbuckle at the bottom of the shrouds. When we do this we say we are "tuning" the boat. If a tension of 150 N exists in the shroud, what torque is needed to start tightening further by turning the turnbuckle? The pitch of the single threaded screw is 1.5 mm and the mean diameter is 8.0 mm. The coefficient of friction is .2.

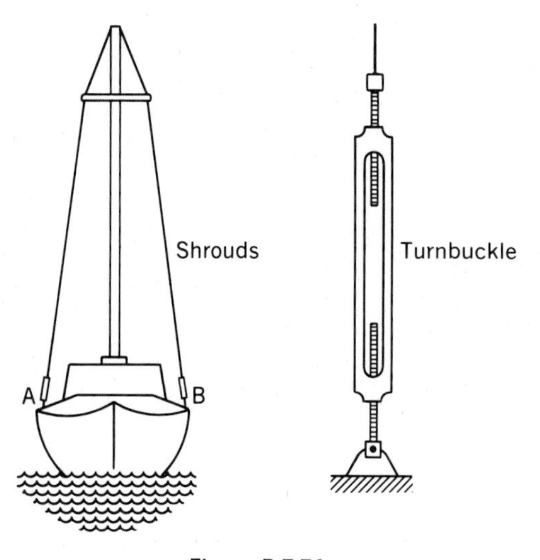


**Figure P.7.74**

**7.75.** Forces $F$ of 50 lb are applied to the jackscrew shown. The thread diameter is 2 in. and the pitch is $\frac{1}{2}$ in. The coefficient of friction for the thread is .05. The weight $W$ is not permitted to rotate and so the collar must rotate on the shaft of the screw. If the coefficient of friction for the collar and shaft is .1, determine the weight $W$ that can be lifted by this system.

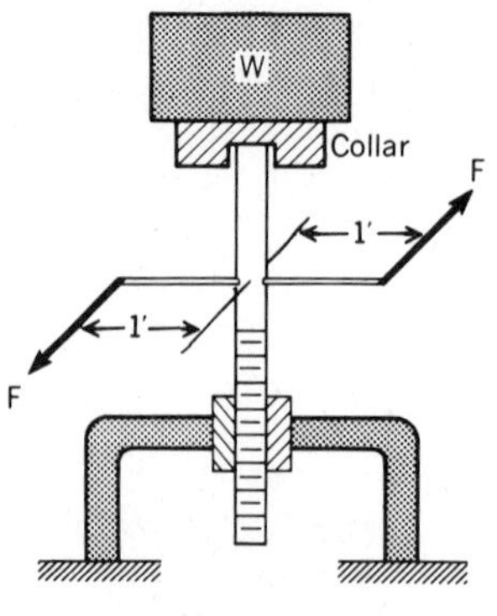

**Figure P.7.75**

**7.76.** A brake is shown. Force is developed at the brake shoes by turning $A$, which has a single right-handed square screw thread at $B$ and a single left-handed screw thread at $C$. The diameter of the screw thread is $1\frac{1}{2}$ in. and the pitch is .3 in. If the coefficient of friction is .1 for the thread and .4 for the brake shoes, what resisting torque is developed on the wheel by a 100 in.-lb torque at $A$?

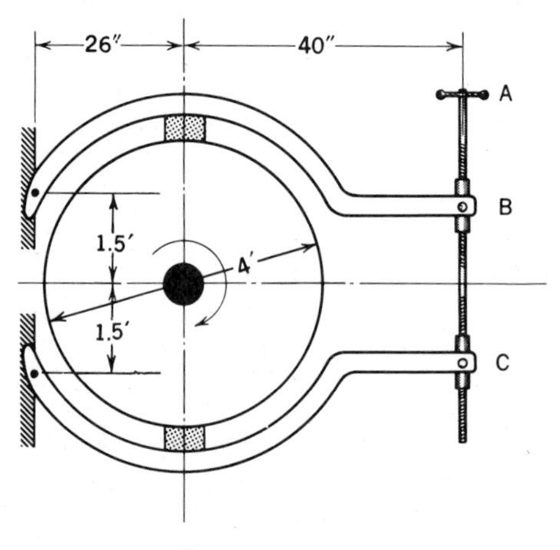


**Figure P.7.76**

**7.77.** A triangular-threaded screw is shown. In a manner paralleling the development of the square-thread formulation in Section 7.6, show that

$$M_z = \frac{rP(\mu \cos \alpha + \cos \theta \tan \alpha)}{\cos \theta - \mu \sin \alpha}$$

where

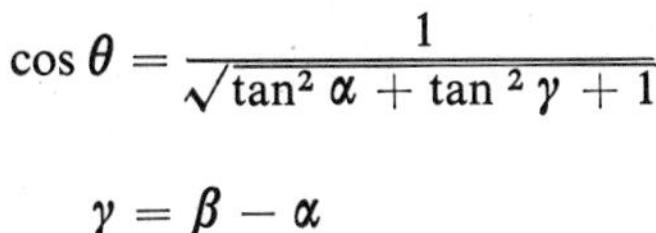

$$\cos\theta = \frac{1}{\sqrt{\tan^2\alpha + \tan^2\gamma + 1}}$$

and

$$\gamma = \beta - \alpha$$

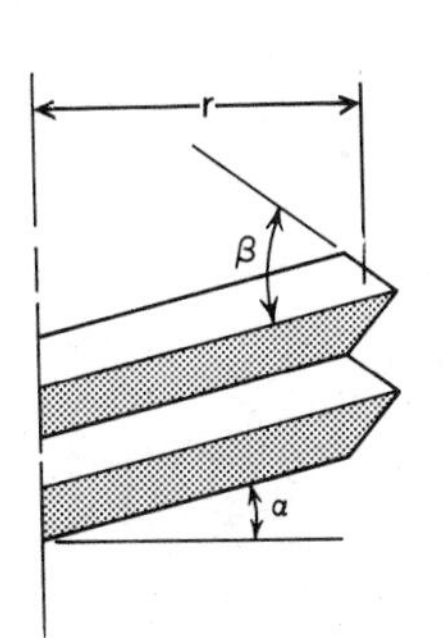


**Figure P.7.77**

**7.78.** Consider a single-threaded screw where the pitch $p = 4.5$ mm and the mean radius is 20 mm. For a coefficient of friction $\mu = .3$, what torque is needed on the nut for it to turn under a load of 1000 N? Compute this for a square thread and then do it for a triangular thread where the angle $\beta$ is 30°. See Problem 7.77 before doing this problem.

**7.79.** If the coefficient of rolling friction of a cylinder on a flat surface is .05 in., at what inclination of the surface will the cylinder of radius $r = 1$ ft roll with uniform velocity?

**7.80.** A 65-kN vehicle designed for polar expeditions is on a very slippery ice surface for which the coefficient of friction between tires and ice is .005. Also, the coefficient of rolling friction is known to be .8 mm. Will the vehicle be able to move? The vehicle has four-wheel drive.

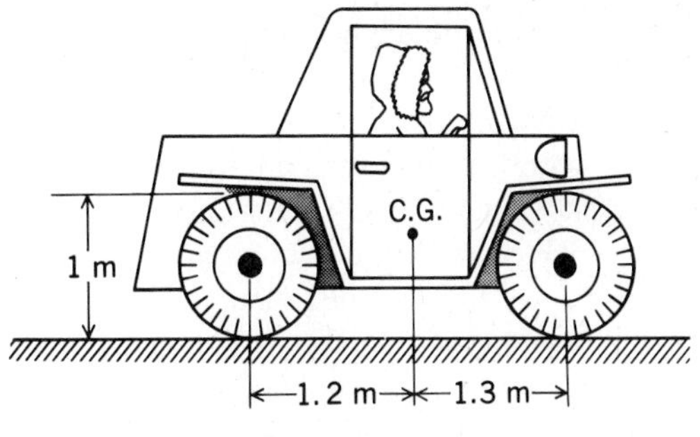

**Figure P.7.80**

**7.81.** In Problem 7.80, suppose there is only rear-wheel drive available. What is the minimum coefficient of friction needed between tires and ground for the vehicle to move?

**7.82.** A roller thrust bearing is shown supporting a force $P$ of 2.5 kN. What torque $T$ is needed to turn the shaft $A$ at constant speed if the only resistance is that from the ball bearings? The coefficient of rolling for the balls and the bearing surfaces is .01270 mm. The mean radius from the centerline of the shaft and the balls is 30 mm.

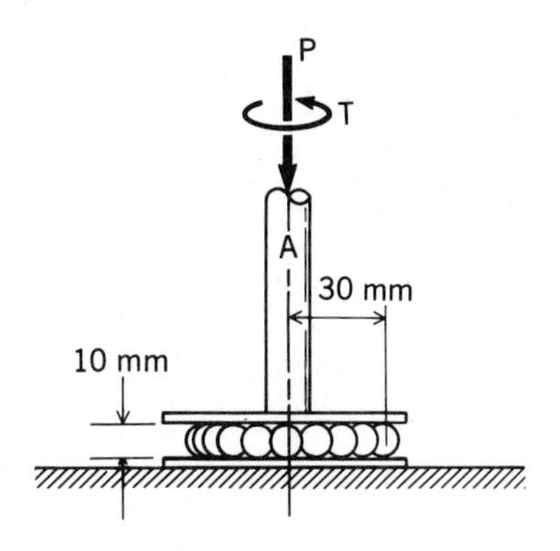


**Figure P.7.82**

## 7.8 Closure

In this chapter, we have examined the results of two independent experiments: that of impending or actual sliding of one body over another and that of a cylinder or sphere rolling at constant speed over a flat surface. Without any theoretical basis, the results of such experiments must be used in situations that closely parallel the experiments themselves.

In the case of a rolling cylinder, both rolling resistance and sliding resistance are present. However, for a cylinder accelerating with any appreciable magnitude, only sliding friction need be accounted for. With no acceleration on a horizontal surface, only rolling resistance need be considered. Most situations fall into these categories. For very small accelerations, both effects are present and must be taken into account. We can then expect only a crude result for such computations.

Before going further, we must carefully define certain properties of plane surfaces in order to facilitate later computations in mechanics where such properties are most useful. This will be done in Chapter 8.

## Review Problems

**7.83.** Find the force $F$ needed to start the 200-N weight moving to the right, if the coefficient of friction is $\mu_s = .35$.

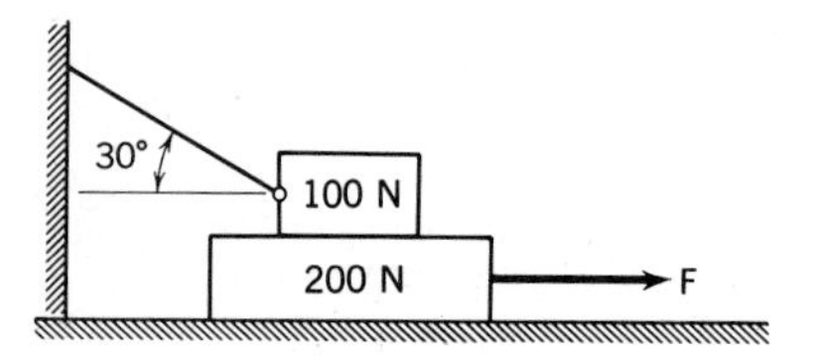


Figure P.7.83

**7.84.** A loaded crate is shown. The crate weighs 500 lb with a center of gravity at its geometric center. The contact surface between crate and floor has a static coefficient of friction of .2. If $\theta = 90°$, show that the crate will slide before one can increase $T$ large enough for tipping to occur. If a stop is to be inserted in the floor at $A$ to prevent slipping so that the crate could be tipped, what *minimum* horizontal force will be exerted on the stop?

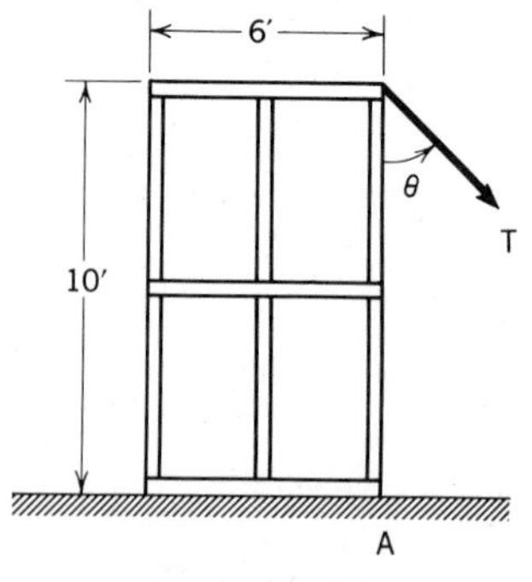


Figure P.7.84

**7.85.** In Problem 7.84, compute a value of $\theta$ and $T$ where slipping and tipping will occur simultaneously. If the actual angle $\theta$ is smaller than this value of $\theta$, is there any further need of the stop at $A$ to prevent slipping?

**7.86.** A friction drive is shown with $A$ the driver disc and $B$ the driven disc. If force $F$ pressing $B$ onto $A$ is 150 N, what is the *maximum* torque $M_2$ that can be developed? For this torque, what is the torque $M_1$ needed for the drive disc $A$? The coefficient of friction between $A$ and $B$ is .7. What vertical force must rod $G$ withstand for the action above?

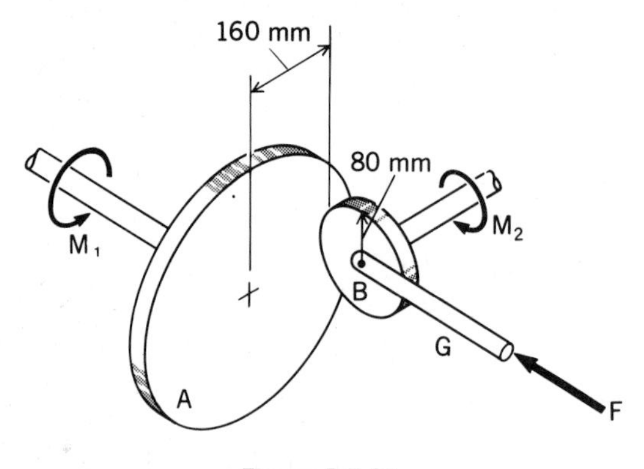


Figure P.7.86

**7.87.** A tug is pushing a barge into a berth. After the barge turns and touches the sides of the pilings, what thrust must the tug develop to move it at uniform speed of 2 knots farther into the berth? The coefficient of friction between the barge and the sides of the berth is .4. The drag from the water is 3000 N along the centerline of the barge.

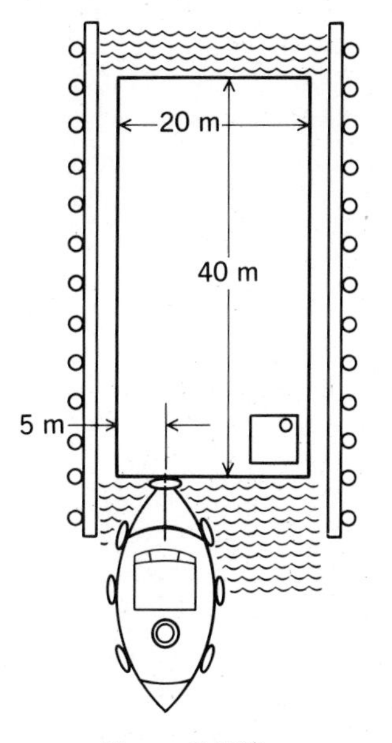


Figure P.7.87

**7.88.** The static and dynamic coefficients of friction for the upper surface of contact $A$ of the cylinder are $\mu_s = .4$, $\mu_d = .3$, and for the lower surface of contact $B$ are $\mu_s = .1$ and $\mu_d = .08$. What is the *minimum* force $P$ needed to just get the cylinder moving?

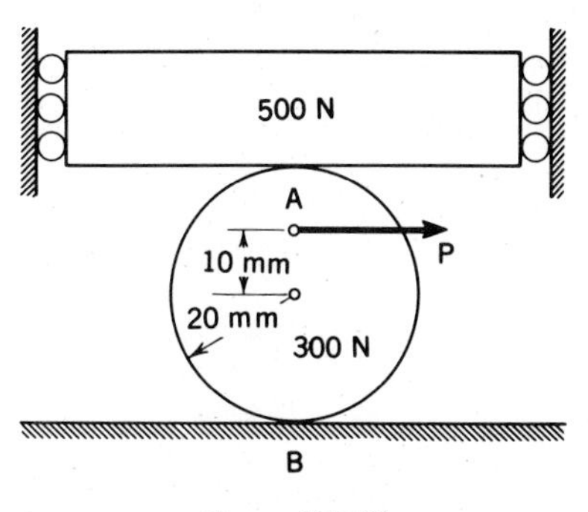

Figure P.7.88

**7.89.** A hot rectangular metal ingot is to be flattened by passing through cylindrical rollers. If the ingot is to be drawn into the rollers by friction once it touches the rollers, what is the *minimum* thickness $t$ of the ingot that can be achieved by this process on one pass? The coefficient of friction for the contact between ingot and cylinder is .3. The cylinders rotate as shown with angular speed $\omega$.

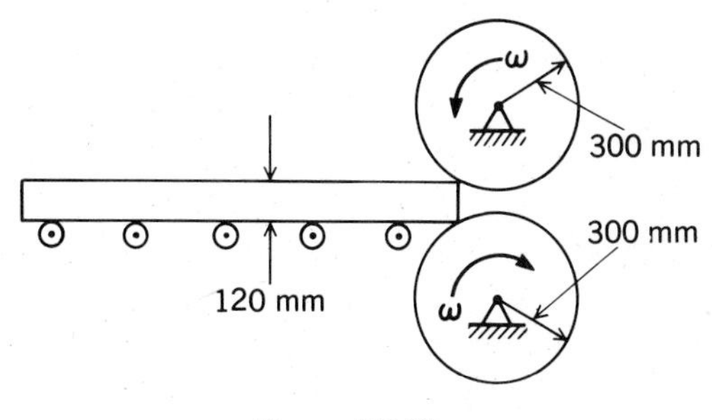


Figure P.7.89

**7.90.** A cone clutch is shown. Assuming that uniform pressures exist between the contact surfaces, compute the *maximum* torque that can be transmitted. The coefficient of friction is .30 and the activating force $F$ is 100 lb. (*Hint:* Assume that the moving cone transmits its 100-lb axial force to the stationary cone by pressure primarily. That is, we will neglect the friction-force component on the cone surface in the axial direction.)

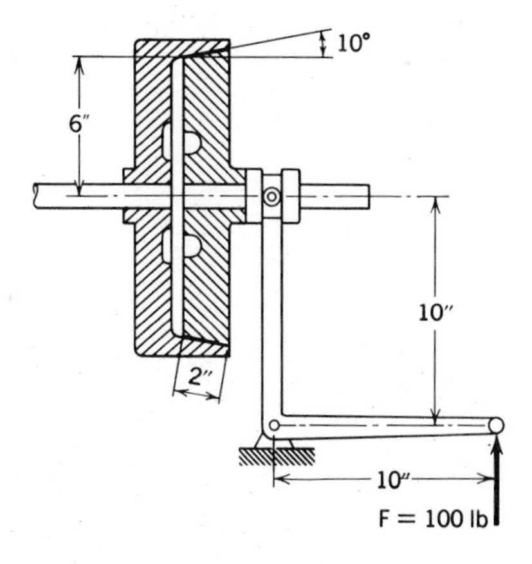


Figure P.7.90

**7.91.** The drum is driven by a motor with a maximum torque capability of 500 lb-ft. The coefficient of friction between the drum and the braking strap (belt) is .4. How much force $P$ must an operator exert to stop the drum if it rotates (1) clockwise and (2) counterclockwise? What are the belt forces in each case?

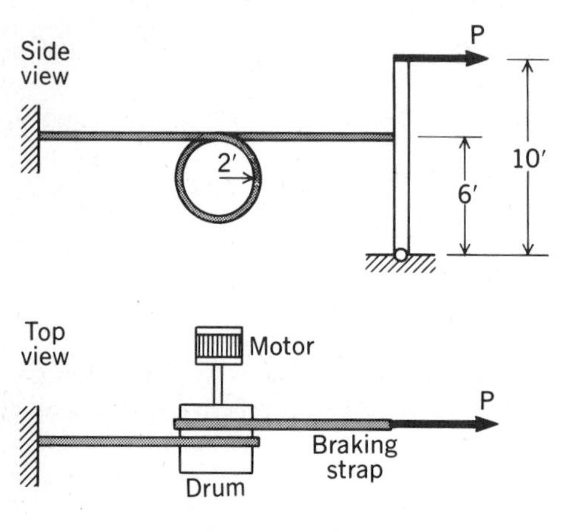


Figure P.7.91

**7.92.** The four drive pulleys shown are used to transmit a torque from pulley $A$ to pulley $D$ on an electric typewriter. If the coefficient of friction between the belts and the pulleys is .3, what is the torque available at pulley $D$ if 10 lb-in. of torque is input to the shaft of pulley $A$? What are the belt forces?

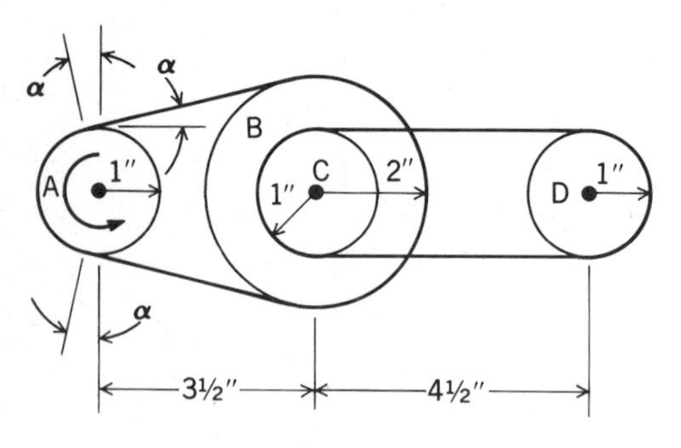


Figure P.7.92

**7.93.** A scissors jack is shown lifting the end of a car so that $R = 6.67$ kN. What torque $T$ is needed for this operation? Note that $A$ is merely a bearing and at $B$ we have a nut. The screw is single-threaded with a pitch of 3 mm and a mean diameter of 20 mm. The coefficient of friction is .3. Neglect the weight of the members and evaluate $T$ for $\theta = 45°$ and for $\theta = 60°$.

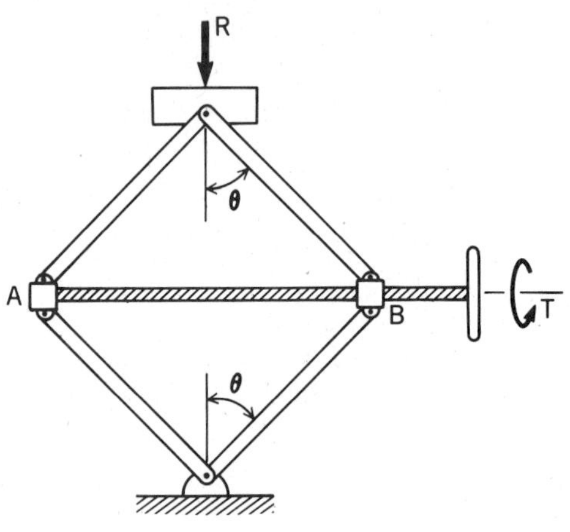


Figure P.7.93

**7.94.** A block $C$ weighing 10 kN is being moved on rollers $A$ and $B$ each weighing 1 kN. What force $P$ is needed to maintain steady motion? Take the coefficient of rolling resistance between the rollers and the ground to be .6 mm and between block $C$ and the rollers to be .4 mm.

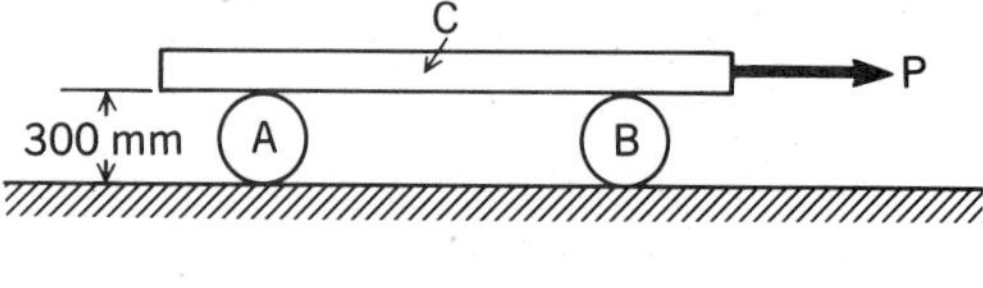

Figure P.7.94

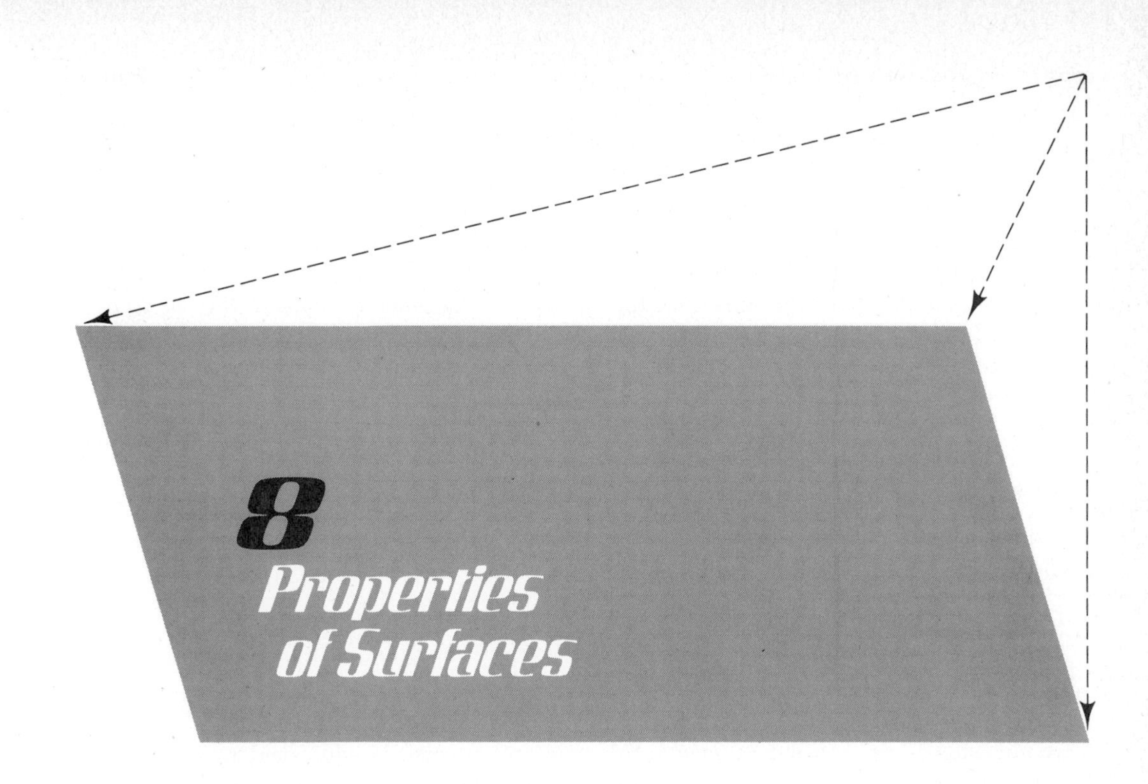


## 8.1 Introduction

If we are buying a tract of land, we certainly want to consider the size and, with equal interest, the shape and orientation of the earth's surface, and possibly its agricultural, geological, or aesthetic potentials. The size of a surface (i.e., the area) is a familiar concept and has been used in the previous sections. Certain aspects of the shape and orientation of a surface will be examined in this chapter. There are a number of formulations that convey meaning about the shape and disposition of a surface relative to some reference. To be sure, these formulations are not used by real estate people, but in engineering work, where a variety of quantitative descriptions are necessary, they will prove most useful. In general, we shall restrict our attention to coplanar surfaces.

## 8.2 First Moment of an Area and the Centroid

A coplanar surface of area $A$ and a reference $xy$ in the plane of the surface are shown in Fig. 8.1. We define the *first moment* of area $A$ about the $x$ axis as

$$M_x = \int_A y \, dA \tag{8.1}$$

and the first moment about the $y$ axis as

$$M_y = \int_A x \, dA \tag{8.2}$$

These two quantities convey a certain knowledge of the shape, size, and orientation of the area which we can use in many analyses of mechanics.

You will no doubt notice the similarity of the preceding integrals to those which would occur for computing moments about the $x$ and $y$ axes from a parallel force distribution oriented normal to the area $A$ in Fig. 8.1. The moment of such a force distribution has been shown for the purposes of rigid-body calculations to be equivalent to that of a single resultant force located at a particular point $\bar{x}, \bar{y}$. Similarly, we can concentrate the entire area $A$ at a position $x_c, y_c$, called the *centroid*,[1] where, for

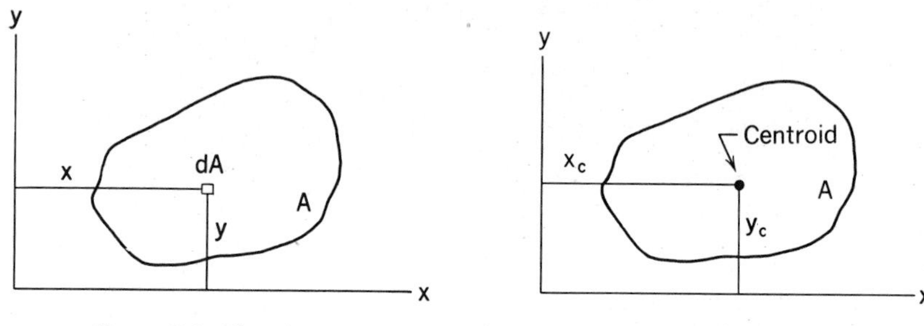

**Figure 8.1.** Plane area.

**Figure 8.2.** Centroidal coordinates.

computations of first moments, this new arrangement is equivalent to the original distribution (Fig. 8.2). The coordinates $x_c$ and $y_c$ are usually called the *centroidal coordinates*. To compute these coordinates, we simply equate moments of the distributed area with that of the concentrated area about both axes:

$$Ay_c = \int_A y\,dA; \qquad \text{therefore, } y_c = \frac{\int_A y\,dA}{A} \tag{8.3a}$$

$$Ax_c = \int_A x\,dA; \qquad \text{therefore, } x_c = \frac{\int_A x\,dA}{A} \tag{8.3b}$$

The location of the centroid of an area can readily be shown to be independent of the reference axes employed. That is, the centroid is a property only of the area itself. We have asked the reader to prove this in Problem 8.1.

If the axes $xy$ have their origin at the centroid, then these axes are called *centroidal axes* and clearly the first moments about these axes must be zero.

Finally, we point out that all axes going through the centroid of an area are called *centroidal axes* for that area. Clearly, the *first moments of an area about any of its centroidal axes must be zero.*

**EXAMPLE 8.1**

A plane surface is shown in Fig. 8.3 bounded by the $x$ axis, the curve $y^2 = 25x$, and a line parallel to the $y$ axis.

---

[1]The concept of the centroid can be used for any geometric quantity. In the next section, we shall consider centroids of volumes and arcs.

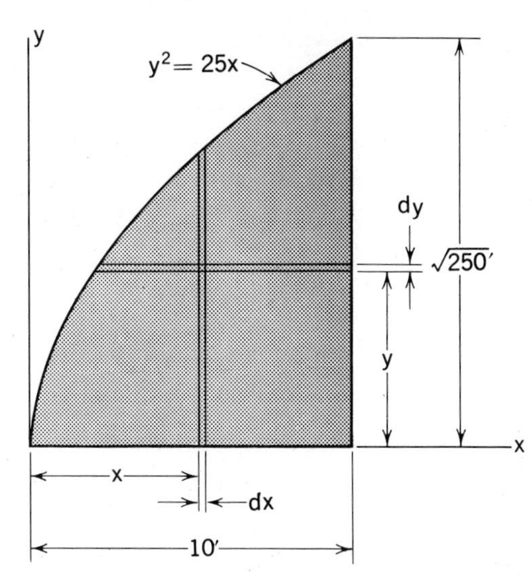


**Figure 8.3.** Find centroid.

We shall first compute $M_x$ and $M_y$ for this area. Using vertical infinitesimal area elements of width $dx$ and height $y$, we have

$$M_y = \int_0^{10} x(y\,dx) = \int_0^{10} x(5\sqrt{x})\,dx$$

$$= \left.\frac{5x^{5/2}}{\frac{5}{2}}\right|_0^{10} = 632 \text{ ft}^3$$

To compute $M_x$, we use horizontal area elements of width $dy$ as shown in the diagram. Thus,

$$M_x = \int_0^{\sqrt{250}} y[(10 - x)dy]$$

$$= \int_0^{\sqrt{250}} \left(10y - \frac{y^3}{25}\right)dy$$

$$= \left.\left(5y^2 - \frac{y^4}{100}\right)\right|_0^{\sqrt{250}} = 625 \text{ ft}^3$$

We could also have used vertical strips for computing $M_x$ as follows:

$$M_x = \int_0^{10} \frac{y}{2}(y\,dx) = \int_0^{10} \frac{25x}{2}\,dx$$

$$= \left.(12.5)\left(\frac{x^2}{2}\right)\right|_0^{10} = 625 \text{ ft}^3$$

To compute the position of the centroid $(x_c, y_c)$, we will need the area $A$ of the surface. Thus, using vertical strips:

$$A = \int_0^{10} y\,dx = \int_0^{10} 5\sqrt{x}\,dx = \left.\frac{5x^{3/2}}{\frac{3}{2}}\right|_0^{10}$$

$$= 105.4 \text{ ft}^2$$

The centroidal coordinates are, accordingly,

$$x_c = \frac{M_y}{A} = \frac{632}{105.4} = 6.00 \text{ ft}$$

$$y_c = \frac{M_x}{A} = \frac{625}{105.4} = 5.93 \text{ ft}$$

To get the moment of the area about an axis $y'$, which is 15 ft to the left of the $y$ axis, we simply proceed as follows:

$$M_{y'} = (A)(x_c + 15) = 105.4(6.00 + 15) = 2213 \text{ ft}^3$$

Consider now a plane area with an *axis of symmetry* such as is shown in Fig. 8.4, where the $y$ axis is collinear with the axis of symmetry. In computing $x_c$ for this area, we have

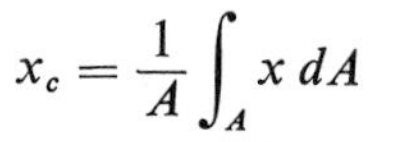

$$x_c = \frac{1}{A}\int_A x\,dA$$

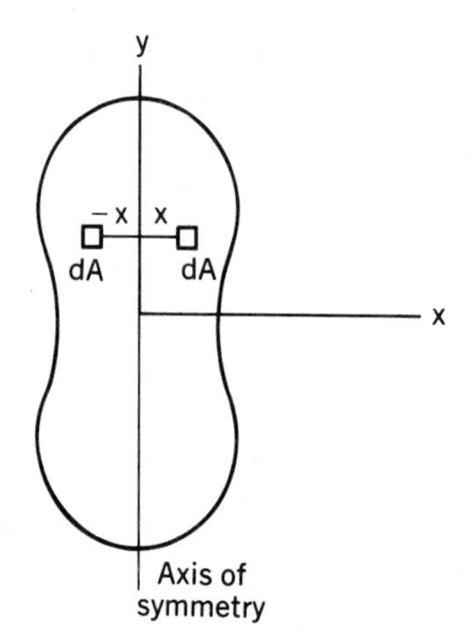


**Figure 8.4.** Area with one axis of symmetry.

In evaluating the integral above, we can consider area elements in symmetric pairs such as shown in Fig. 8.4, where we have shown a pair of area elements which are mirror images of each other about the axis of symmetry. Clearly, the first moment of such a pair about the axis of symmetry is zero. And, since the entire area can be considered as composed of such pairs, we can conclude that $x_c = 0$. Thus, the centroid of an area with one axis of symmetry must therefore lie somewhere along this axis of symmetry. With two orthogonal axes of symmetry, the centroid must lie at the intersection of these axes. Thus, for such areas as circles and rectangles, the centroid is easily determined by inspection.

In many problems, the area of interest can be considered formed by the addition or subtraction of simple familiar areas whose centroids are known by inspection as well as by other familiar areas, such as triangles and sectors of circles whose centroids and areas are given in handbooks. We call areas made up of such simple areas *composite* areas. (A listing of familiar areas is given for your convenience on the inside covers of this text.) For such problems, we can say that

$$x_c = \frac{\sum_i A_i\bar{x}_i}{A}$$

$$y_c = \frac{\sum_i A_i\bar{y}_i}{A}$$

where $\bar{x}_i$ and $\bar{y}_i$ (with proper signs) are the centroidal coordinates to simple area $A_i$ and where $A$ is the total area.

We now illustrate how we can use the composite-area approach for finding the centroid of an area composed of familiar parts as described above.

**EXAMPLE 8.2**

Find the centroid of the shaded section shown in Fig. 8.5.

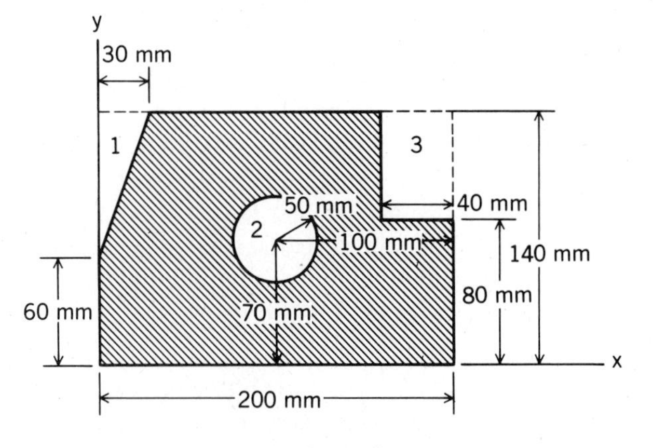


**Figure 8.5.** Composite area.

We may consider four separate areas. These are the triangle (1), the circle (2), and the rectangle (3) all cut from an original rectangular 200 × 140 mm² area which we denote as area (4). In composite-area problems, we urge you to set up a format of the kind we shall now illustrate. Using the positions of the centroid of a right triangle as given in the inside covers of this text, we have:

| $A_i$ | $\bar{x}_i$ | $A_i\bar{x}_i$ | $\bar{y}_i$ | $A_i\bar{y}_i$ |
|---|---|---|---|---|
| $A_1 = -\frac{1}{2}(30)(80) = -1{,}200$ | 10 | $-12{,}000$ | 113.3 | $-136{,}000$ |
| $A_2 = -\pi 50^2 = -7{,}850$ | 100 | $-785{,}000$ | 70 | $-549{,}780$ |
| $A_3 = -(40)(60) = -2{,}400$ | 180 | $-432{,}000$ | 110 | $-264{,}000$ |
| $A_4 = (200)(140) = 28{,}000$ | 100 | $2{,}800{,}000$ | 70 | $1{,}960{,}000$ |
| $A = 16{,}550 \text{ mm}^2$ | | $\sum_i A_i\bar{x}_i = 1.571 \times 10^6 \text{ mm}^3$ | | $\sum_i A_i\bar{y}_i = 1.011 \times 10^6 \text{ mm}^3$ |

Therefore,

$$x_c = \frac{\sum A_i\bar{x}_i}{A} = \frac{1.571 \times 10^6}{16{,}550} = 94.9 \text{ mm}$$

$$y_c = \frac{\sum A_i\bar{y}_i}{A} = \frac{1.011 \times 10^6}{16{,}550} = 61.1 \text{ mm}$$

In closing, we would like to point out that the centroid concept can be of use in finding the simplest resultant of a distributed loading. Thus, consider the distributed loading $w(x)$ shown in Fig. 8.6. The resultant force $F_R$ of this loading, also shown in the diagram, is given as

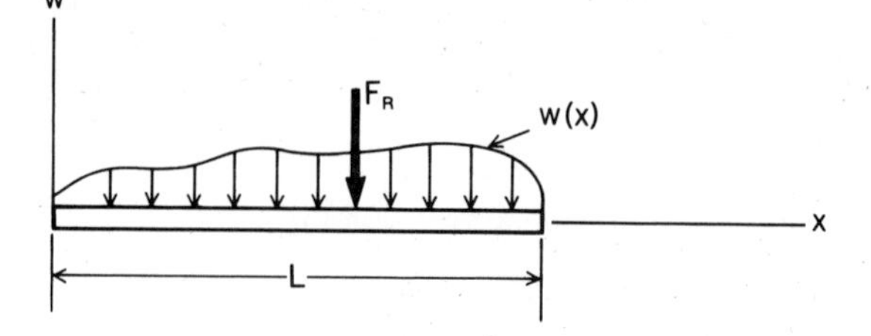


**Figure 8.6.** Loading curve $w(x)$ and its resultant $F_R$.

$$F_R = \int_0^L w(x)\,dx \tag{8.4}$$

From the equation above, we can readily see that the *resultant force equals the area under the loading curve.* To get the position of the *simplest* resultant for the loading, we then say that

$$F_R\bar{x} = \int_0^L xw(x)\,dx$$

Therefore,

$$\bar{x} = \frac{\int_0^L xw(x)\,dx}{F_R} \tag{8.5}$$

The preceding result shows that $\bar{x}$ is actually the centroidal coordinate of the loading curve area from reference $xy$. Thus, the *simplest resultant force of a distributed load acts at the centroid of the area under the loading curve.* Accordingly, for a triangular load such as is shown in Fig. 8.7, we can replace the loading for free bodies on which the entire loading acts by a force $F$ equal to $(w_0)(b - a)(\frac{1}{2})$ at a position $\frac{2}{3}(b - a)$ from the left end of the loading. You will recall that we pointed this out in Chapter 4.

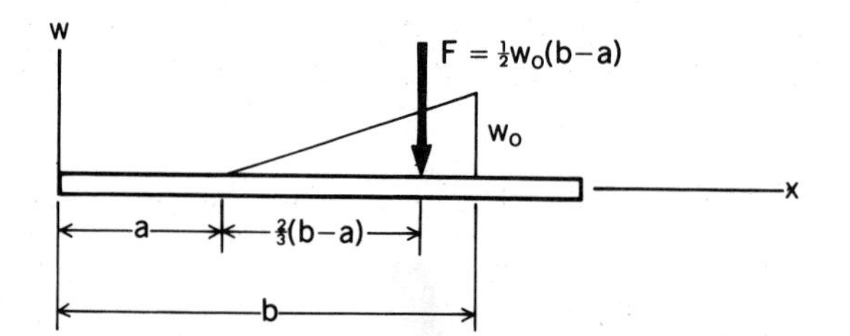


**Figure 8.7.** Triangular loading with simplest resultant.

## Problems

**8.1.** Show that the centroid of area $A$ is the same point for axes $xy$ and $x'y'$. Thus, the position of the centroid of an area is a property only of the area.

**8.2.** Show that the centroid of the right triangle is $x_c = 2a/3$, $y_c = b/3$.

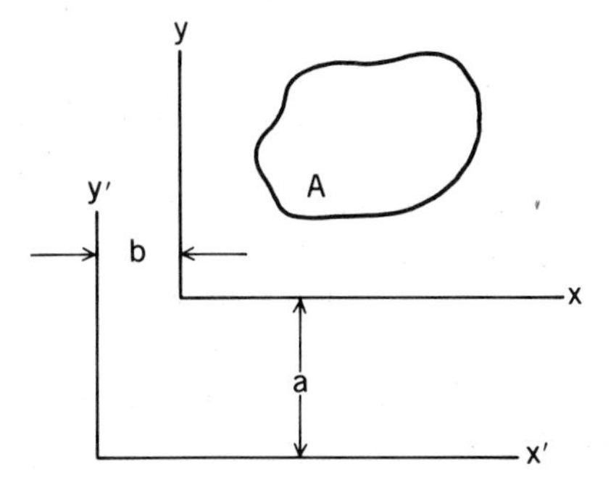

Figure P.8.1

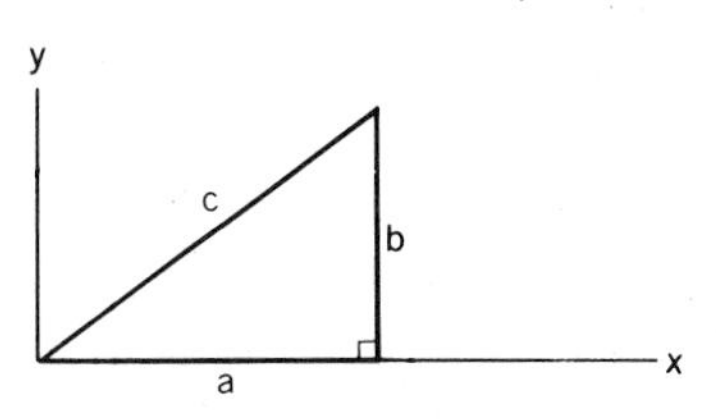

Figure P.8.2

**8.3.** Find the centroid of the area under the half-sine wave. What is the first moment of this area about axis $A$–$A$?

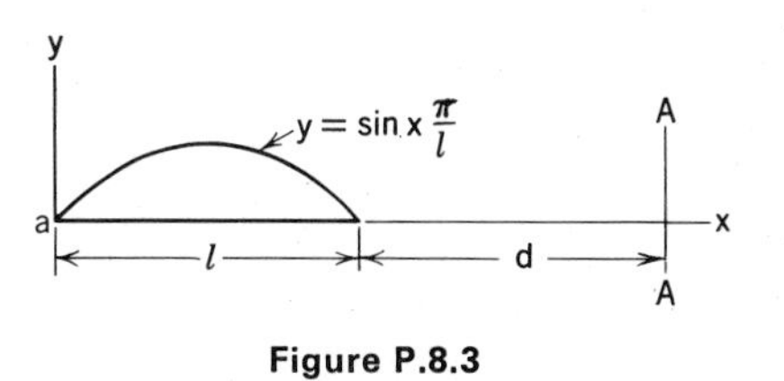


Figure P.8.3

**8.4.** What are the first moments of the area about the $x$ and $y$ axes? The curved boundary is that of a parabola. (*Hint:* The general equation for parabolas of the shape shown is $y^2 = ax + b$.)

Figure P.8.4

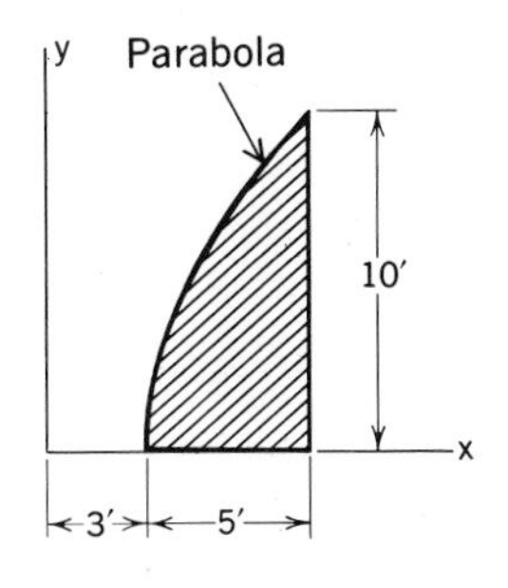


**8.5.** What are the centroidal coordinates for the shaded area? The curved boundary is that of a parabola. (*Hint:* The general equation for parabolas of the shape shown is $y = ax^2 + b$.)

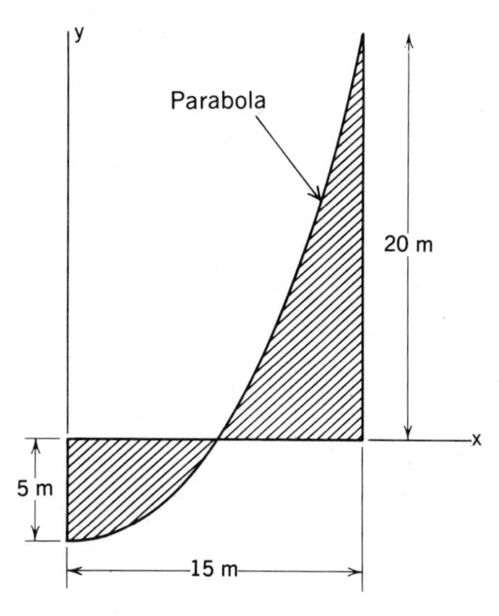


Figure P.8.5

**8.6.** Show that the centroid of the area under a semicircle is as shown in the diagram.

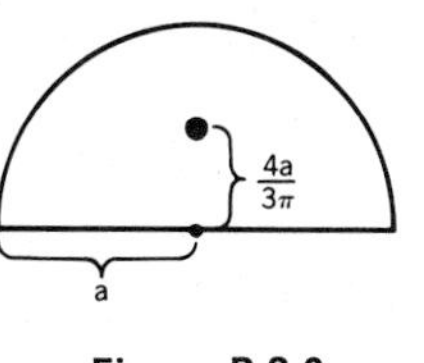

Figure P.8.6

**8.7.** What is the first moment of the area under the parabola about an axis through the origin and going through point $\boldsymbol{r} = 6\boldsymbol{i} + 7\boldsymbol{j}$ m. Take $l = 10$ m.

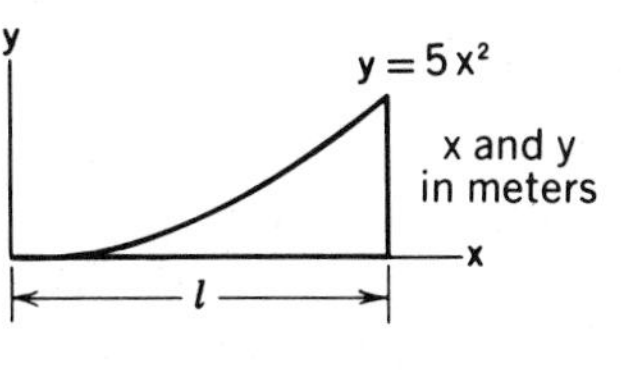


Figure P.8.7

**8.8.** Show that the centroid of the triangle is at $x_c = (a + b)/3$, $y_c = h/3$. (*Hint:* Break the triangle into two right triangles for which the centroids are known from Problem 8.2.)

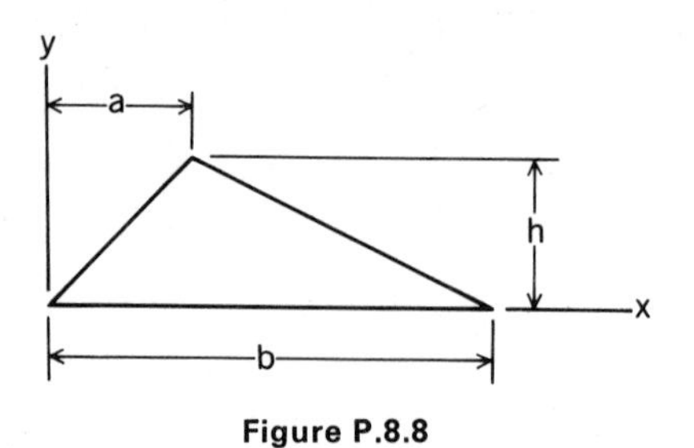


Figure P.8.8

**8.9.** What are the centroidal coordinates for the shaded area? The outer boundary is that of a circle having a radius of 1 m.

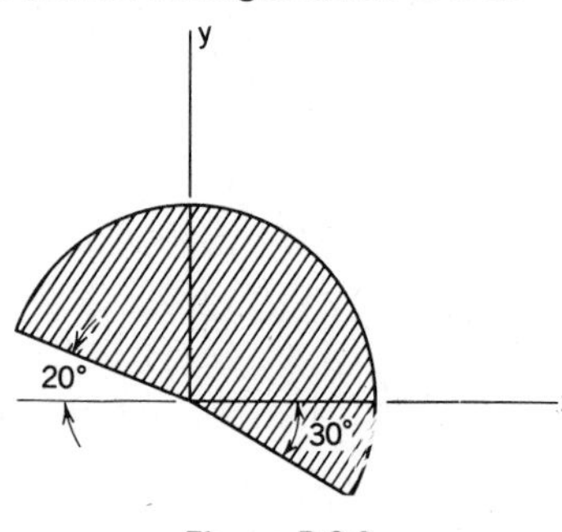


Figure P.8.9

**8.10.** What are the coordinates of the centroid of the shaded area? The parabola is given as $y^2 = 2x$ with $y$ and $x$ in millimeters.

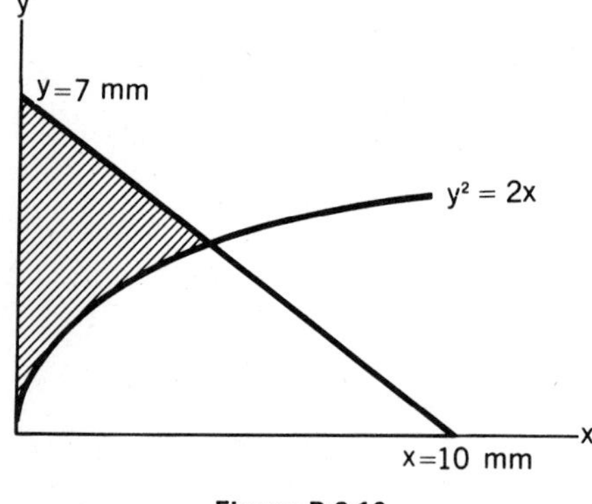


Figure P.8.10

**8.11.** Find the centroid of the shaded area. The equation of the curve is $y = 5x^2$ with $x$ and $y$ in millimeters. What is the first moment of the area about line $AB$?

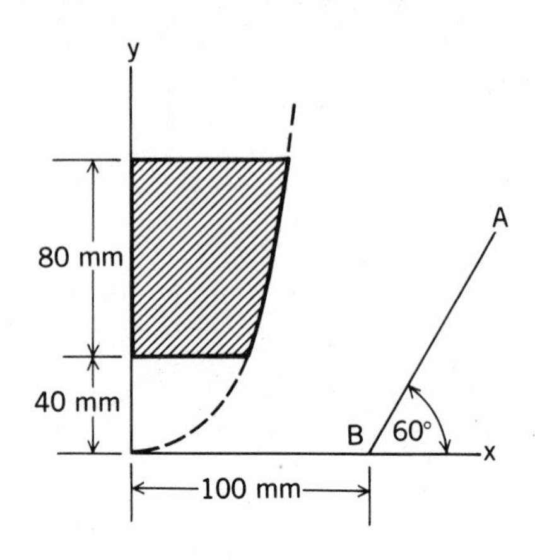


Figure P.8.11

**8.12.** Find the centroid of the shaded area. What is the first moment of this area about line $A$–$A$. The upper boundary is a parabola $y^2 = 3x$ with $x$ and $y$ in millimeters.

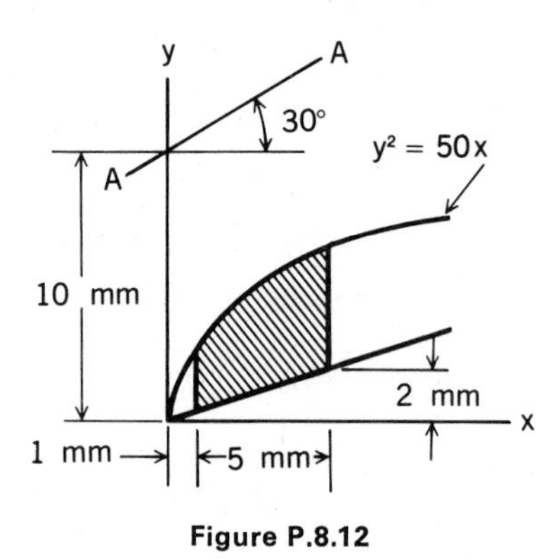


Figure P.8.12

*In the remaining problems of this section, use centroidal positions of simple areas as found in the inside covers.*

**8.13.** Find the centroid of the end shield of a bulldozer blade.

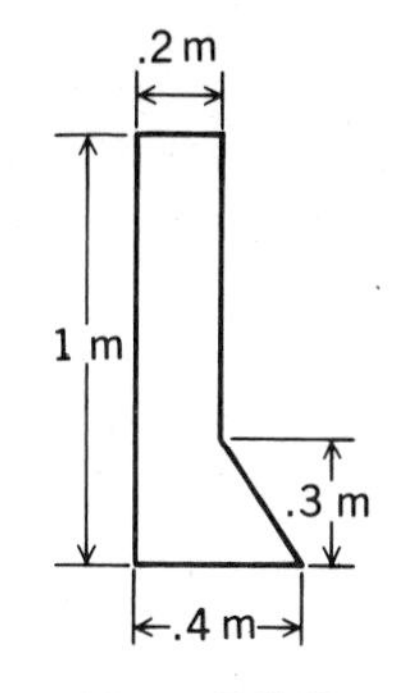


Figure P.8.13

**8.14.** Find the centroid of the truss gusset plate.

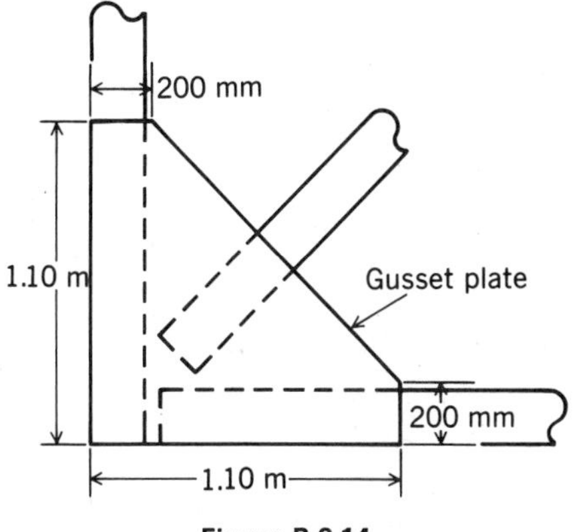


Figure P.8.14

**8.15.** Find the centroid of the indicated area.

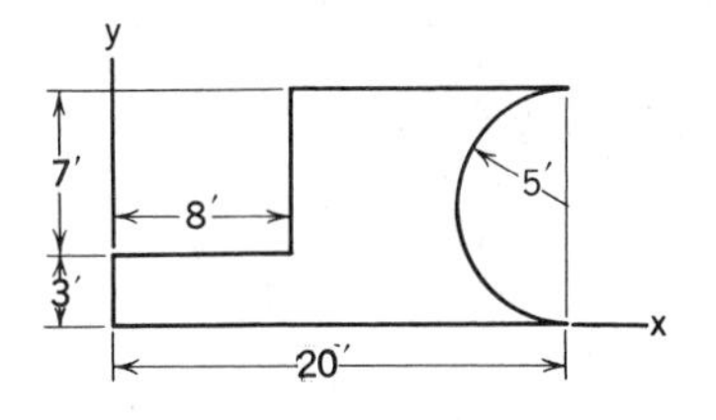


Figure P.8.15

**8.16.** Find the centroidal coordinates for the shaded area shown. Give the results in meters. (*Hint:* See Fig. P.8.6.)

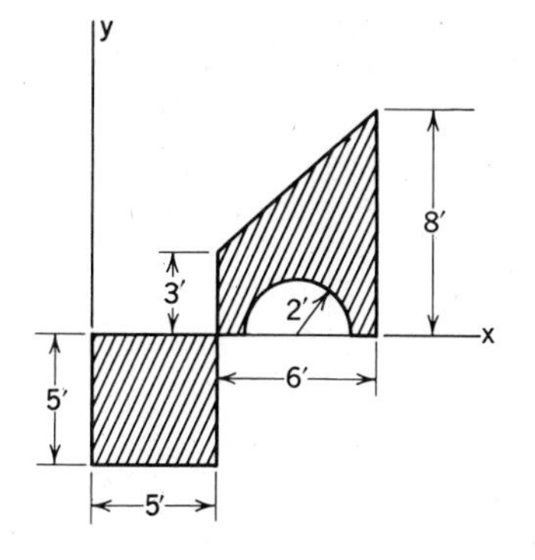

Figure P.8.16

**8.17.** Find the centroid of the end of the bucket of a small front-end loader.

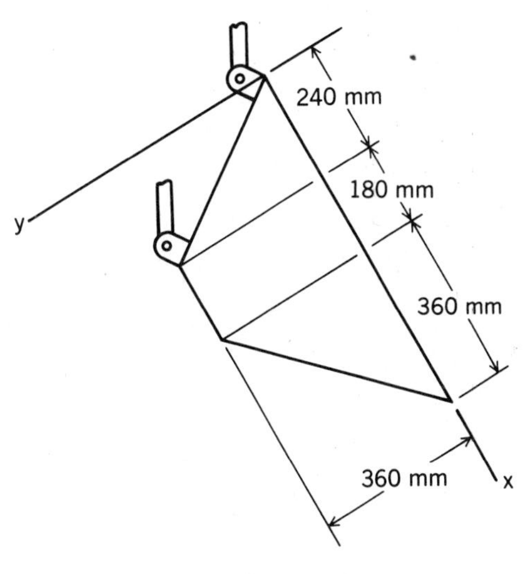


Figure P.8.17

**8.18.** Where is the centroid of the airplane vertical stabilizer (whole area)?

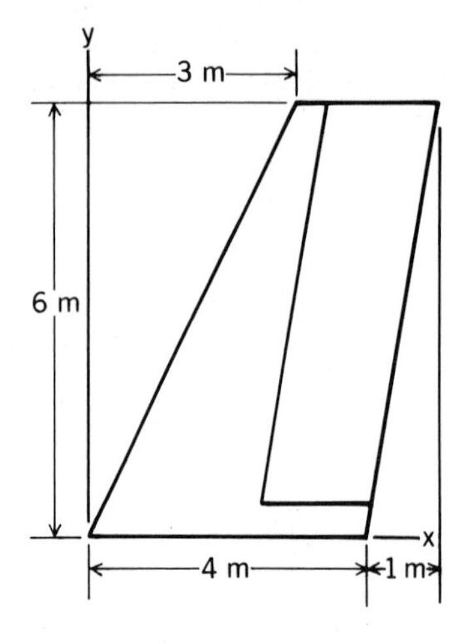


Figure P.8.18

**8.19.** What is the first moment of the shaded area about the diagonal $A$–$A$? (*Hint:* Consider symmetry.)

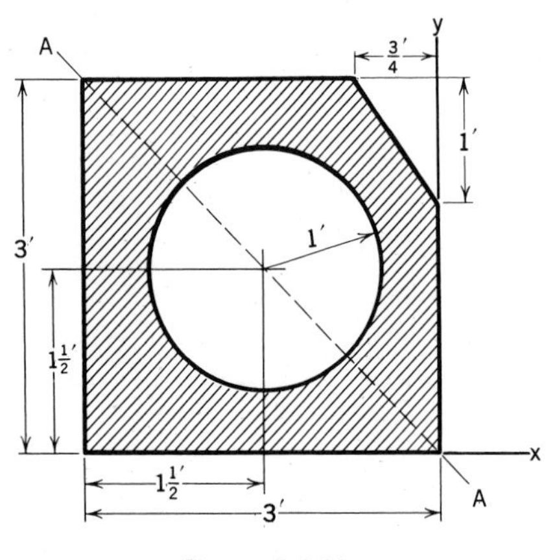


Figure P.8.19

**8.20.** A built-up beam is shown with four 120-mm by 120-mm by 20-mm angles. Find the vertical distance above the base for the centroid of the cross section.

**8.21.** A wide-flange I-beam (identified as 14 WF 202 I-beam) is shown with two reinforcing plates on top. At what height above the bottom is the centroid located?

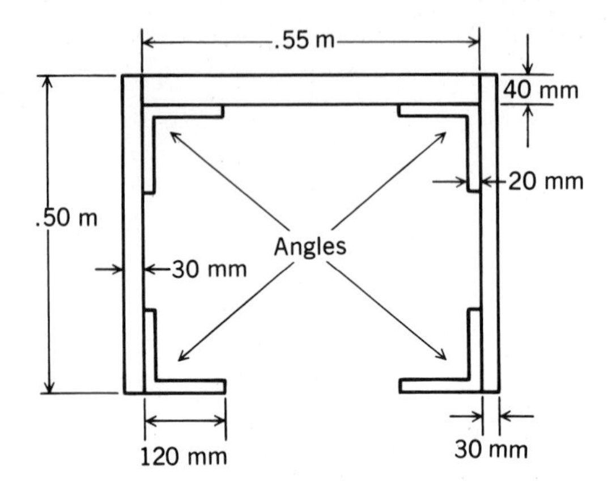


Figure P.8.20

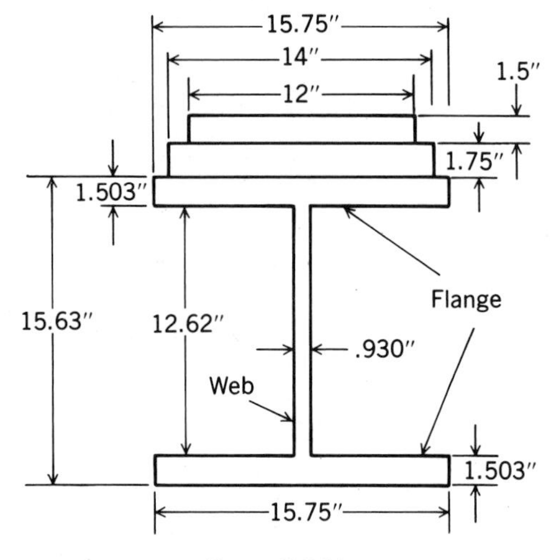


Figure P.8.21

**8.22.** Compute the position of the centroid of the shaded area. (*Hint:* See Fig. P.8.6.)

Figure P.8.22

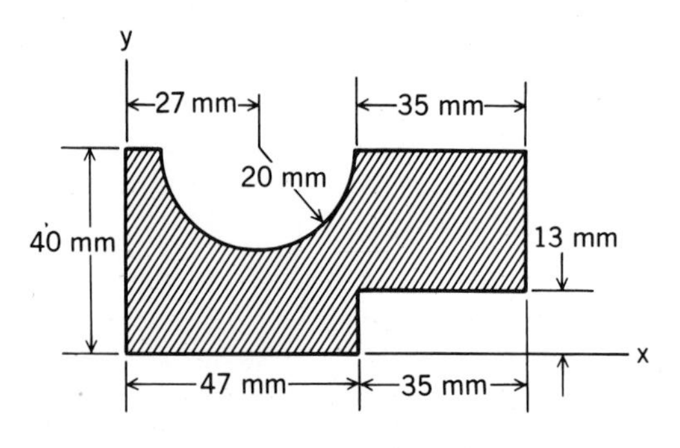

**8.23.** Find the centroid of the sheet metal cover of a centrifugal blower (shown shaded).

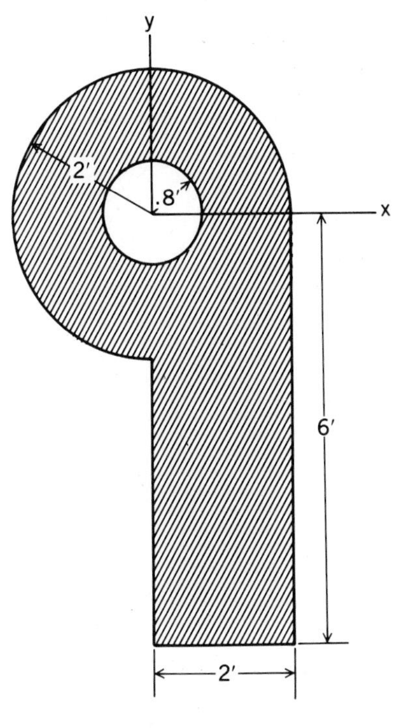


Figure P.8.23

**8.24.** What is the position from the left end of the simplest resultant force of the distribution shown?

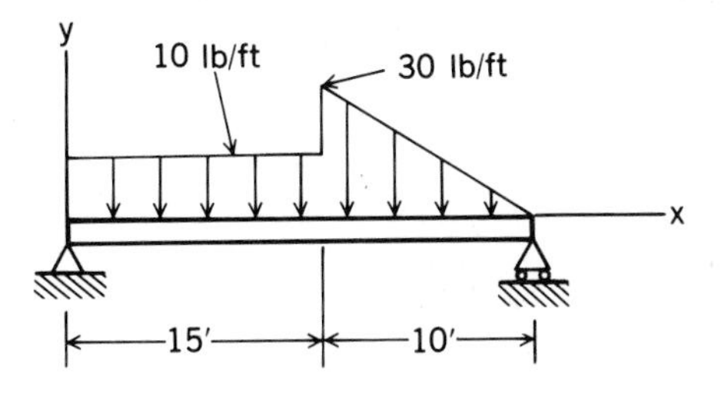


Figure P.8.24

## 8.3 Other Centers

We employ the concepts of moments and centroids in mechanics for three-dimensional bodies as well as for plane areas. Thus, we introduce now the first moment of a volume, $V$, of a body (see Fig. 8.8) about a point $O$ where we have shown a reference $xyz$. We say that the first *moment of volume* $V$ about $O$ is

$$\text{moment vector of volume} \equiv \iiint_V \mathbf{r}\, dv \tag{8.6}$$

The *center of volume*, $\mathbf{r}_c$, is then defined as follows:

$$V\mathbf{r}_c = \iiint_V \mathbf{r}\, dv$$

Therefore,

$$\mathbf{r}_c = \frac{1}{V}\iiint_V \mathbf{r}\, dv \tag{8.7}$$

We see that the center of volume is the point where we could hypothetically concentrate the entire volume of a body for purposes of computing the first moment of the

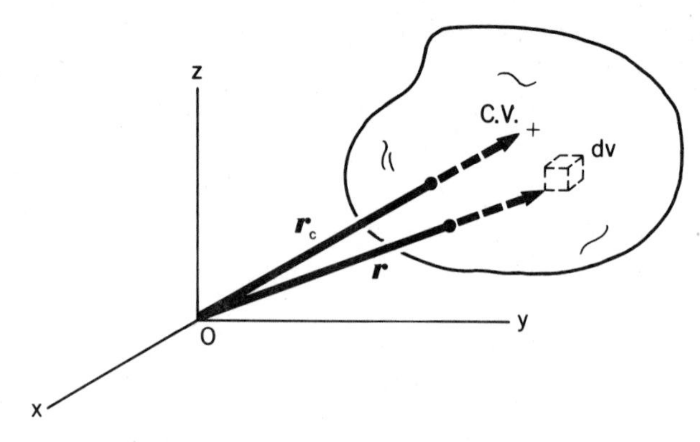


**Figure 8.8.** Center of volume, C.V., of a body.

volume of the body about some point $O$. The components of Eq. 8.7 give the *centroid distances* of volume $x_c$, $y_c$, and $z_c$. Thus, we have

$$x_c = \frac{\iiint x\,dv}{\iiint dv}, \qquad y_c = \frac{\iiint y\,dv}{\iiint dv}, \qquad z_c = \frac{\iiint z\,dv}{\iiint dv} \tag{8.8}$$

The integral $\iiint x\,dv$, it should be noted, gives the first moment of volume about the $yz$ plane, etc.

If we replace $dv$ by $dm = \rho\,dv$ in Eq. 8.6, where $\rho$ is the mass *density*, we get the *first moment of mass* about $O$. That is,

$$\text{moment vector of mass} \equiv \iiint_V \mathbf{r}\,\rho\,dv \tag{8.9}$$

The *center of mass* $\mathbf{r}_c$ is then given as

$$\mathbf{r}_c = \frac{1}{M}\iiint_V \mathbf{r}\rho\,dv \tag{8.10}$$

where $M$ is the total mass of the body. The center of mass is the point in space where hypothetically we could concentrate the entire mass for purposes of computing the first moment of mass about a point $O$. Using components of Eq. 8.10, we can say that

$$x_c = \frac{\iiint x\rho\,dv}{\iiint \rho\,dv}, \qquad y_c = \frac{\iiint y\rho\,dv}{\iiint \rho\,dv}, \qquad z_c = \frac{\iiint z\rho\,dv}{\iiint \rho\,dv}$$

In our work in dynamics, we shall consider the center of mass of a system of $n$ particles (see Fig. 8.9). We will then say:

$$\left(\sum_{i=1}^{n} m_i\right)\mathbf{r}_c = \sum_{i=1}^{n} m_i\mathbf{r}_i$$

Therefore,

$$\boldsymbol{r}_c = \frac{\sum_{i=1}^{n} m_i \boldsymbol{r}_i}{M} \tag{8.11}$$

where $M$ is the total mass of the system. Clearly, if the particles are of infinitesimal mass and constitute a continuous body, we get back Eq. 8.10.

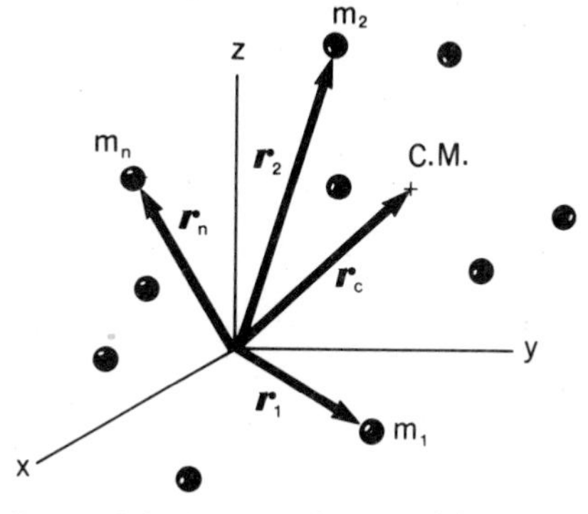

**Figure 8.9.** System of $n$ particles showing center of mass, C.M.

Finally, if we replace $dv$ by $\gamma\ dv$, where $\gamma$ $(= \rho g)$ is the *specific weight*, we arrive at the concept of *center of gravity* discussed in Chapter 4. We have used the center of gravity of a body in many calculations thus far as a point to concentrate the entire weight of a body.

You should have no trouble in concluding from Eq. 8.10 that if $\rho$ is constant throughout a body, the center of mass coincides with the center of volume. Furthermore, if $\gamma$ $(= \rho g)$ is constant throughout a body, the center of gravity of the body corresponds to the center of volume of the body. If, finally, $\rho$ and $g$ are each constant for a body, all three points coincide for the body.

We now illustrate the computation of the center of volume. Computation for the center of mass follows similar lines, and we have already computed centers of gravity in Chapter 4.

**EXAMPLE 8.3**

Consider a volume of revolution formed by revolving the area shown in Fig. 8.3 about the $x$ axis. This volume has been shown in Fig. 8.10. Clearly, the centroid of this volume must lie somewhere along the $x$ axis. We therefore need only com-

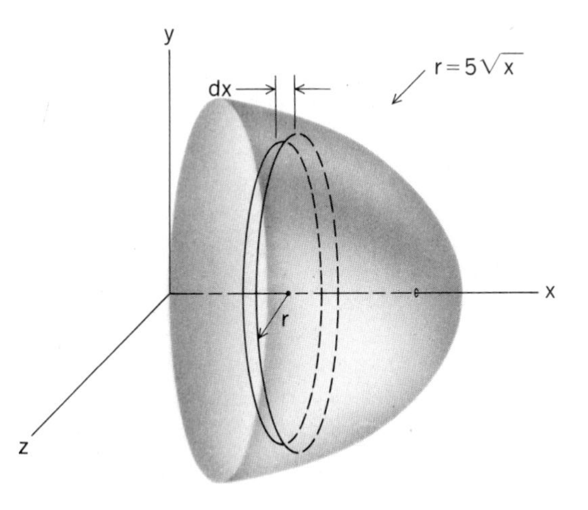

**Figure 8.10.** Body of revolution.

ponent $x_c$. Using $r$, $\theta$, and $x$ as coordinates (cylindrical coordinates), we then have, using slices of thickness $dx$ as volume elements:

$$V = \int_0^{10} (\pi r^2)\, dx = \int_0^{10} (\pi)(25x)\, dx$$

where we have replaced $r^2$ by $25x$ according to the equation for the boundary of the generating area. Integrating, we get

$$V = 25\pi \frac{x^2}{2}\bigg|_0^{10} = 3927 \text{ ft}^3$$

Now we compute $x_c$ by using infinitesimal slices of the body of the kind employed for the computation of $V$. The centroid of each such slice is at the intercept of the slice with the $x$ axis. Thus, we have

$$x_c = \frac{1}{V}\int_0^{10} x(\pi r^2 dx) = \frac{1}{3927}\int_0^{10} x(\pi)(25x)\, dx$$

$$= \frac{25\pi}{3927}\frac{x^3}{3}\bigg|_0^{10} = 6.67 \text{ ft}$$

Many volumes are composed of a number of simple familiar shapes whose centers of volume are either known by inspection or can be found in handbooks (also see the inside cover pages). Such volumes may be called *composite volumes*. To find the centroid of such a volume, we use the known centroids of the composite parts. Thus, for $x_c$ of the composite body whose total volume is $V$, we have

$$x_c = \frac{\sum_i \bar{x}_i V_i}{V}$$

where $\bar{x}_i$ is the $x$ coordinate to the centroid of the $i$th composite body of volume $V_i$. Similarly,

$$y_c = \frac{\sum_i \bar{y}_i V_i}{V}$$

$$z_c = \frac{\sum_i \bar{z}_i V_i}{V}$$

We now illustrate the use of these formulas.

#### EXAMPLE 8.4

What is the coordinate $x_c$ for the center of volume of the body of revolution shown in Fig. 8.11? Note that a cone has been cut away from the left end while, at the right end, we have a hemispherical region.

We have a composite body consisting of three simple domains—a cone (body 1), a cylinder (body 2), and a hemisphere (body 3). Using formulas from the inside covers, we have:

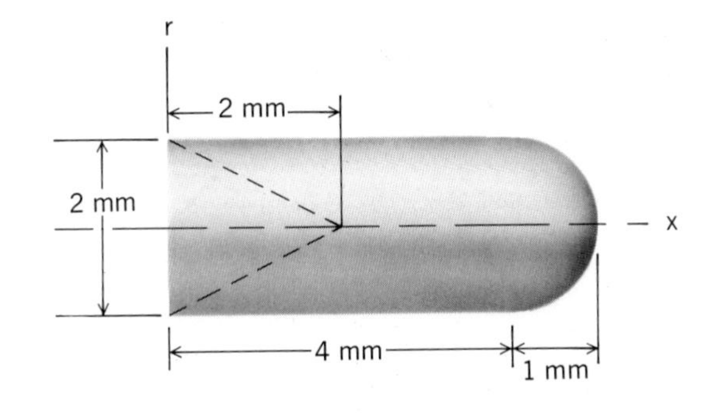


**Figure 8.11.** Composite volume.

| | $V_i$ (mm³) | $\bar{x}_i$ (mm) | $V_i\bar{x}_i$ (mm⁴) |
|---|---|---|---|
| 1. | $-(\frac{1}{3})(\pi)(1^2)(2) = -2.09$ | $\frac{2}{4}$ | $-1.047$ |
| 2. | $(\pi)(1^2)(4) = 12.57$ | $2$ | $25.14$ |
| 3. | $\frac{2}{3}(\pi)(1^3) = 2.09$ | $4 + \frac{3}{8}(1) = 4.38$ | $9.15$ |
| | $V = 12.57$ | | $\sum_i V_i\bar{x}_i = 33.24$ |

Therefore,

$$x_c = \frac{\sum_i V_i\bar{x}_i}{V} = \frac{33.24}{12.57} = 2.64 \text{ mm}$$

In closing, we wish to point out further that curved surfaces and lines have centroids. Since we shall have occasion in the next section to consider the centroid of a line, we simply point out now (see Fig. 8.12) that

$$x_c = \frac{\int x\, dl}{L} \tag{8.12a}$$

$$y_c = \frac{\int y\, dl}{L} \tag{8.12b}$$

where $L$ is the length of the line. Note that the centroid $C$ will not generally lie along the line.

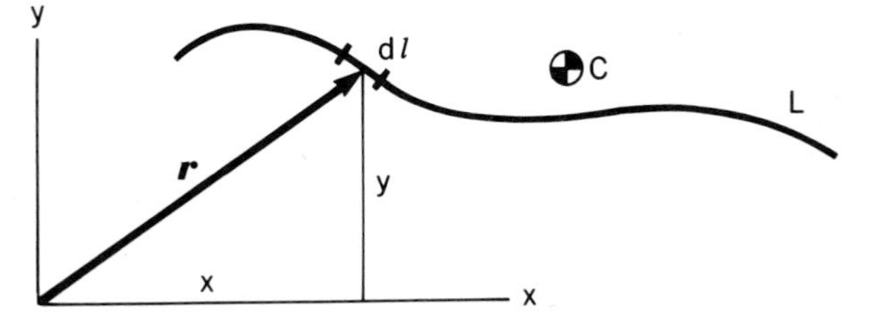


**Figure 8.12.** Centroid for curved line.

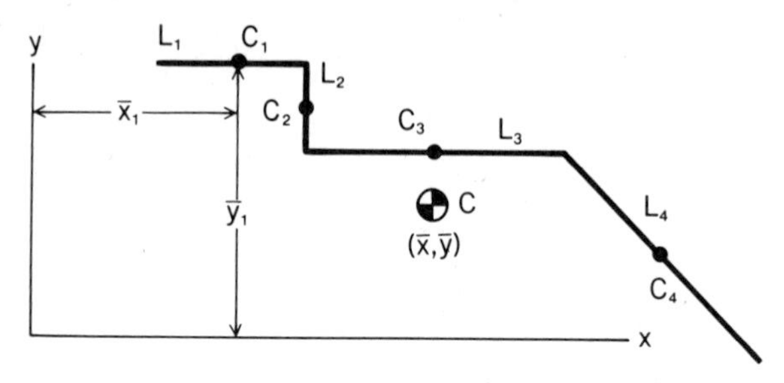


Figure 8.13. Centroid for composite line.

Consider next a curve made up of simple curves each of whose centroids is known. Such is the case shown in Fig. 8.13, made up of straight lines. The line segment $L_1$ has centroid $C_1$ with coordinates $\bar{x}_1\bar{y}_1$, as has been shown in the diagram. We can then say for the entire curve that

$$x_c = \frac{\sum_i \bar{x}_i L_i}{L}$$
$$y_c = \frac{\sum_i \bar{y}_i L_i}{L} \qquad (8.13)$$

## *8.4 Theorems of Pappus–Guldinus

The theorems of Pappus–Guldinus were first set forth by Pappus about 300 A.D. and then restated by the Swiss mathematician Paul Guldin about 1640. These theorems are concerned with the relation of a surface of revolution to its generating curve, and the relation of a volume of revolution to its generating area.

The first of the theorems may be stated as follows:

*Consider a coplanar generating curve and an axis of revolution in the plane of this curve (see Fig. 8.14). The generating curve can touch but must not cross the axis of revolution. The surface of revolution developed by revolving the generating curve about the axis of revolution has an* area *equal to the product of the* length *of the* generating curve *times the* circumference *of the* circle *formed by the* centroid *of the* generating curve *in the process of generating a surface of revolution.*

To prove this theorem, consider first an element $dl$ of the generating curve shown in Fig. 8.14. For a single revolution of the generating curve about the $x$ axis, the line segment $dl$ traces an area

$$dA = 2\pi y\, dl$$

For the entire curve this area becomes the surface of revolution given as

$$A = 2\pi \int y\, dl = 2\pi y_c L \qquad (8.14)$$

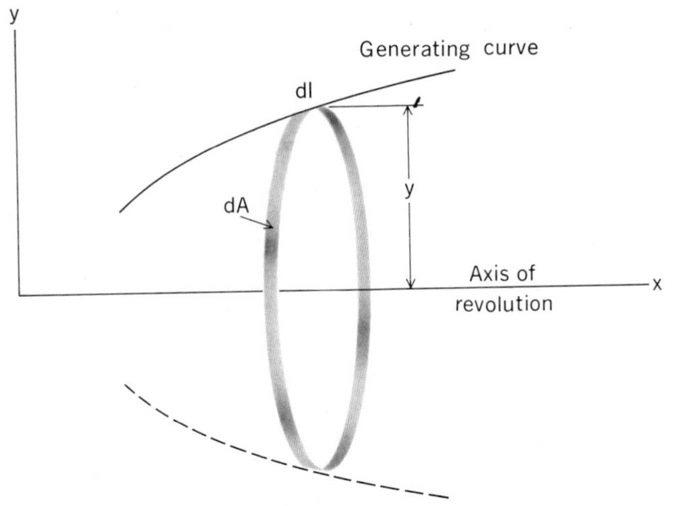


**Figure 8.14.** Coplanar generating curve.

where $L$ is the length of the curve and $y_c$ is the centroidal coordinate of the curve. But $2\pi y_c$ is the circumferential length of the circle formed by having the centroid of the curve rotate about the $x$ axis. The first theorem is thus proved.

Another way of interpreting Eq. 8.14 is to note that the area of the body of revolution is equal to $2\pi$ times the *first moment* of the generating curve about the axis of revolution. If the generating curve is composed of simple curves, $L_i$, whose centroids are known, such as the case shown in Fig. 8.13, then we can express $A$ as follows:

$$A = 2\pi(\sum_i L_i \bar{y}_i) \tag{8.15}$$

where $\bar{y}_i$ is the centroidal coordinate to the $i$th line segment $L_i$.

The second theorem may be stated as follows:

> *Consider a plane surface and an axis of revolution coplanar with the surface but oriented such that the axis can intersect the surface only as a tangent at the boundary or have no intersection at all. The volume of the body of revolution developed by rotating the plane surface about the axis of revolution equals the product of the* area *of the surface times the* circumference *of the* circle *formed by the* centroid *of the surface in the process of generating the body of revolution.*

To prove the second theorem, consider a plane surface $A$ as shown in Fig. 8.15. The volume generated by revolving $dA$ of this surface about the $x$ axis is

$$dV = 2\pi y\, dA$$

The volume of the body of revolution formed from $A$ is then

$$V = 2\pi \int_A y\, dA = 2\pi y_c A \tag{8.16}$$

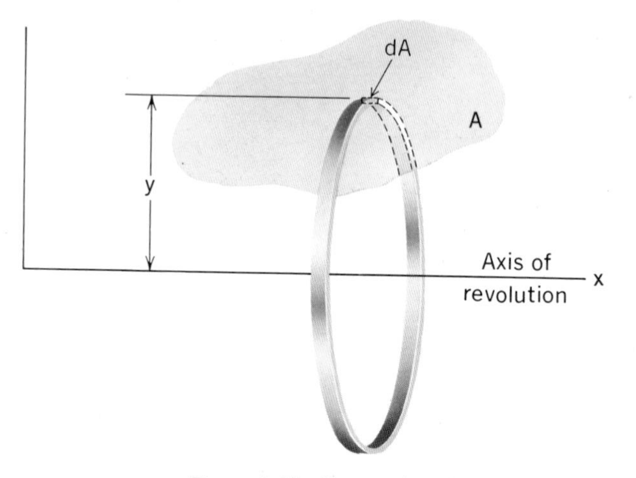


**Figure 8.15.** Plane surface $A$ coplanar with $x$.

Thus, the volume $V$ equals the area of the generating surface $A$ times the circumferential length of the circle of radius $y_c$. The second theorem is thus also proved.[2]

Another way to interpret Eq. 8.16 is to note that $V$ equals $2\pi$ times the *first moment* of the generating area $A$ about the axis of revolution. If this area $A$ is made up of simple areas $A_i$, we can say that

$$V = 2\pi(\sum_i A_i \bar{y}_i) \tag{8.17}$$

where $\bar{y}_i$ is the centroidal coordinate to the $i$th area $A_i$.

We now illustrate the use of theorems of Pappus and Guldinus. As we proceed, it will be helpful to remember the theorems by noting that you multiply a length (or area) of the generator by the distance moved by the centroid of the generator.

#### EXAMPLE 8.5

Determine the surface area and volume of the bulk materials trailer shown in Fig. 8.16.

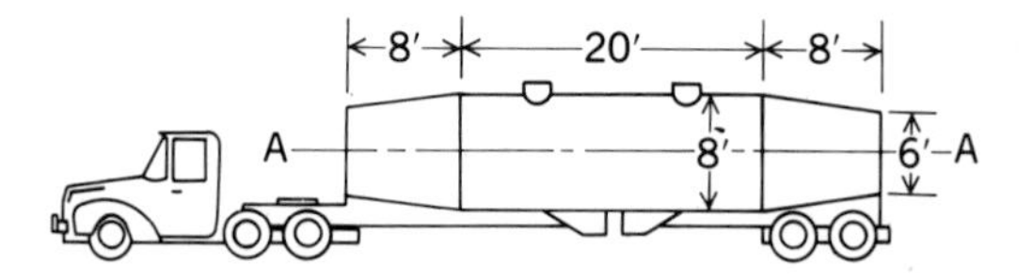


**Figure 8.16.** Bulk materials trailer.

We shall first determine the surface area by considering the first moment about the centerline $A$–$A$ (see Fig. 8.17) of the generating curve of the surface of revolution. This curve is a set of 5 straight lines each of whose centroids is easily known by

[2]It is to be pointed out that the centroid of a volume of revolution will not be coincident with the centroid of a longitudinal cross section taken along the axis of the volume. Example: a cone and its triangular, longitudinal cross section.

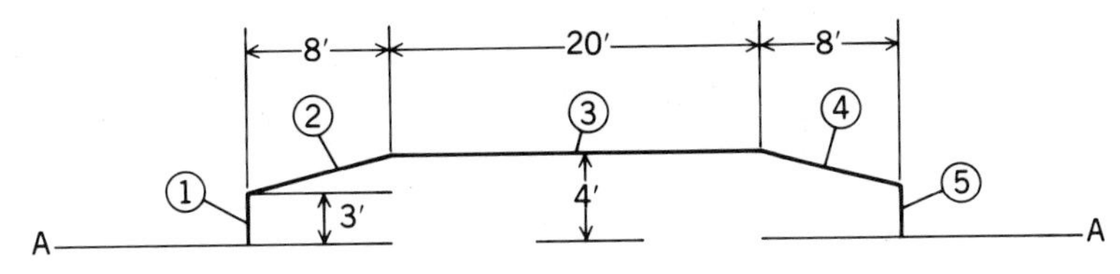


**Figure 8.17.** Generating curve for surface of revolution.

inspection. Accordingly we may use Eq. 8.15. For clarity, we use a column format for the data as follows:

| | $L_i$ (ft) | $\bar{y}_i$ (ft) | $L_i\bar{y}_i$ (ft$^2$) |
|---|---|---|---|
| 1. | 3 | 1.5 | 4.5 |
| 2. | $\sqrt{8^2 + 1^2} = 8.06$ | 3.5 | 28.21 |
| 3. | 20 | 4 | 80 |
| 4. | 8.06 | 3.5 | 28.21 |
| 5. | 3 | 1.5 | 4.5 |
| | | $\sum_i L_i\bar{y}_i =$ | 145.43 |

Therefore,

$$A = (2\pi)(145.43) = 914 \text{ ft}^2$$

To get the volume, we next show in Fig. 8.18 the generating area for the body of revolution. Notice it has been decomposed into simple composite areas. We shall

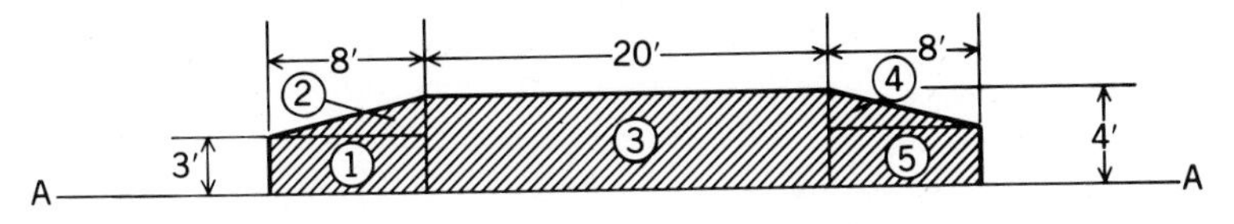


**Figure 8.18.** Generating area for body of revolution.

employ Eq. 8.17 and hence we shall need the first moment about the axis $A$–$A$ of the composite areas. Again, we shall employ a column format for the data.

| | $A_i$ (ft$^2$) | $\bar{y}_i$ (ft) | $A_i\bar{y}_i$ (ft$^3$) |
|---|---|---|---|
| 1. | 24 | 1.5 | 26 |
| 2. | $(\frac{1}{2})(8)(1) = 4$ | $3 + \frac{1}{3} = 3.33$ | 13.33 |
| 3. | 80 | 2 | 160 |
| 4. | 4 | 3.333 | 13.33 |
| 5. | 24 | 1.5 | 36 |
| | | $\sum_i A_i\bar{y}_i =$ | 258.7 |

Therefore,

$$V = 2\pi \sum_i A_i\bar{y}_i = (2\pi)(258.7) = 1625 \text{ ft}^3$$

The theorems of Pappus and Guldinus have enabled us to compute the surface area and the volume of bulk materials trailer quickly and easily.

## Problems

**8.25.** If $r^2 = ax$ in the body of revolution shown, compute the centroidal distance $x_c$ of the body.

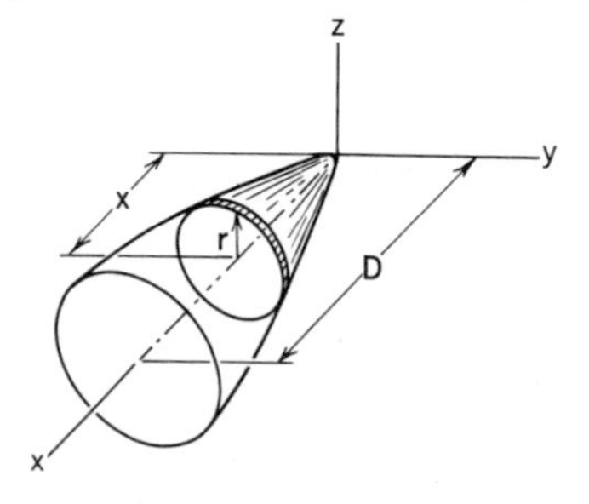


Figure P.8.25

**8.26.** Using vertical elements of volume as shown, compute the centroidal coordinates $x_c$, $y_c$ of the body. Then, using horizontal elements, compute $z_c$.

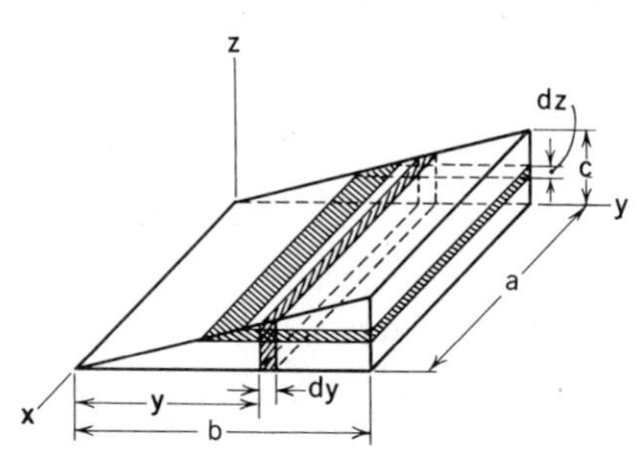


Figure P.8.26

**8.27.** Compute the center of volume of a right circular cylinder of height $h$ and radius at the base $r$.

Figure P.8.27

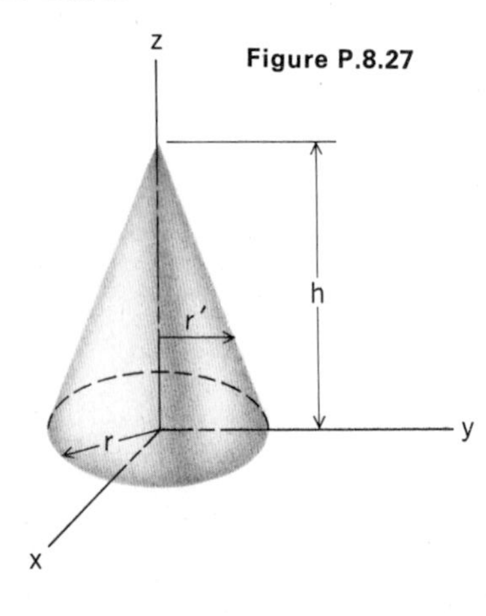


**8.28.** Determine the position of the center of mass of the solid hemisphere having a uniform mass density $\rho$ and with a radius $a$.

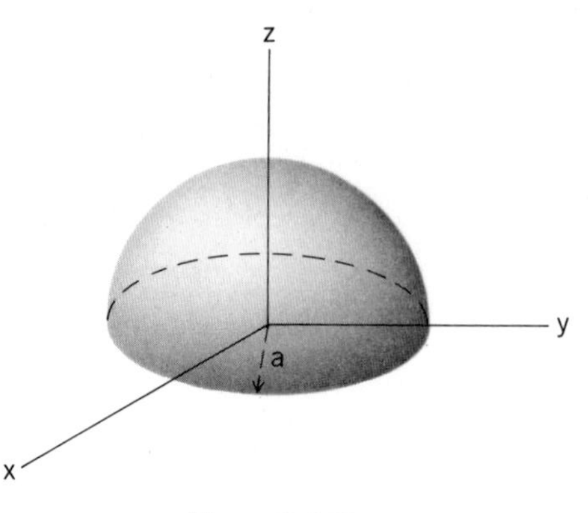


Figure P.8.28

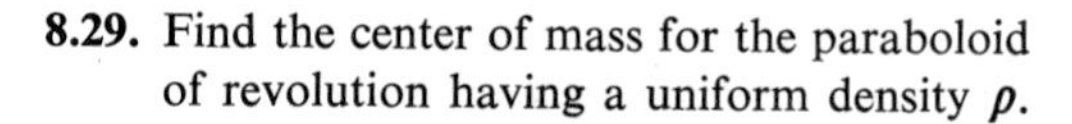

**8.29.** Find the center of mass for the paraboloid of revolution having a uniform density $\rho$.

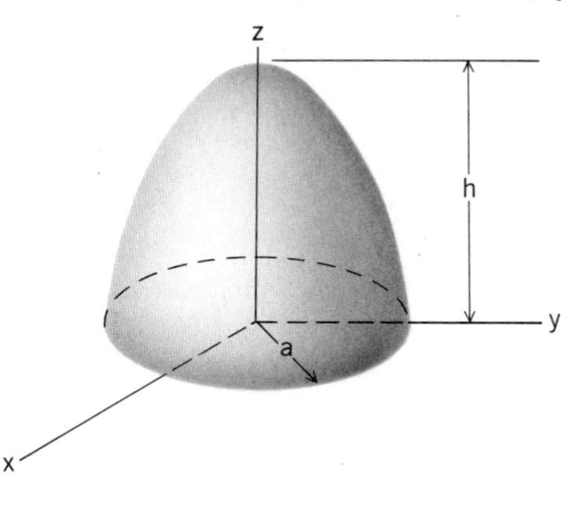


Figure P.8.29

**8.30.** A small bomb has exploded at position $O$. Four pieces of the bomb move off at high speed. At $t = 3$ sec, the following data apply:

| | $m$ (kg) | $\boldsymbol{r}$ (m) |
|---|---|---|
| 1. | .2 | $2\boldsymbol{i} + 3\boldsymbol{j} + 4\boldsymbol{k}$ |
| 2. | .1 | $4\boldsymbol{i} + 4\boldsymbol{j} - 6\boldsymbol{k}$ |
| 3. | .15 | $-3\boldsymbol{i} + 2\boldsymbol{j} - 3\boldsymbol{k}$ |
| 4. | .22 | $2\boldsymbol{i} - 3\boldsymbol{j} + 2\boldsymbol{k}$ |

What is the position of the center of mass?

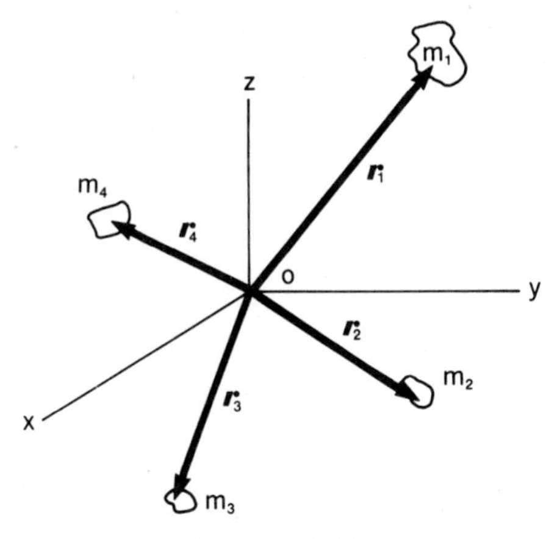


Figure P.8.30

**8.31.** A plate of uniform thickness and density has for its curved edge a rectangular hyperbola ($xy$ = constant). Find the centroid of the upper surface. Find the centers of mass, volume, and gravity for the plate.

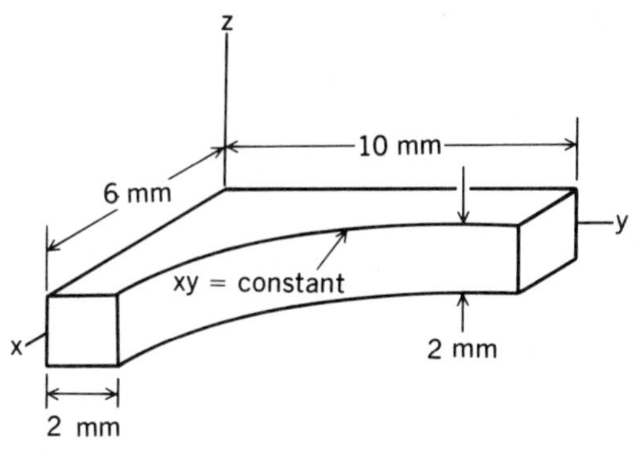

Figure P.8.31

*In Problems 8.32 through 8.38, use the formulas on the inside covers for simple shapes.*

**8.32.** Where must a lifting hook be cast in a tapered concrete beam so that the beam always stays horizontal when lifted?

Figure P.8.32

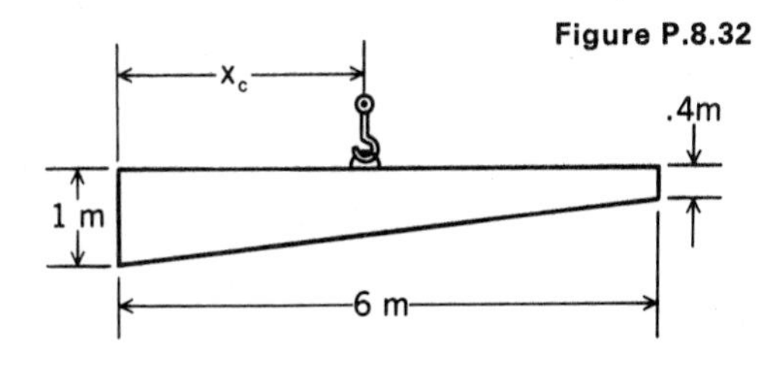


**8.33.** Two solid semicylinders are glued together. Body $A$ has a uniform mass density of 6.54 kN/m$^3$, while body $B$ has a uniform mass density of 10 kN/m$^3$. Determine:

(a) Center of volume
(b) Center of mass
(c) Center of gravity

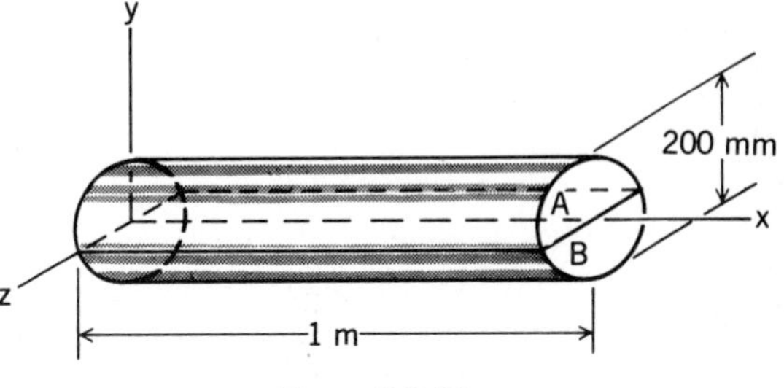


Figure P.8.33

**8.34.** What is the centroid of the body shown? It consists of a cylinder $A$ of length 2 m and diameter 6 m, a shaft $B$ of diameter 2 m and length 8 m, and a block $C$ of length 4 m and height and width of 7 m. The $x$ axis is a centerline for the arrangement. Origin $O$ is at the geometric center of cylinder $A$.

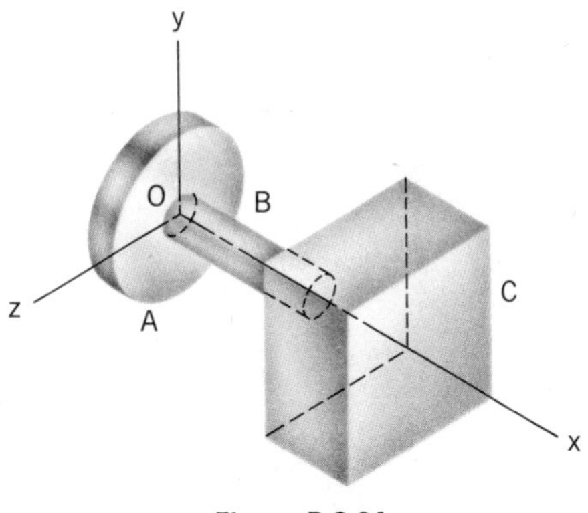


Figure P.8.34

**8.35.** Find the center of volume for the cone–cylinder shown. Note that there is a cylindrical hole of length 16 ft and diameter 4 ft cut into the body.

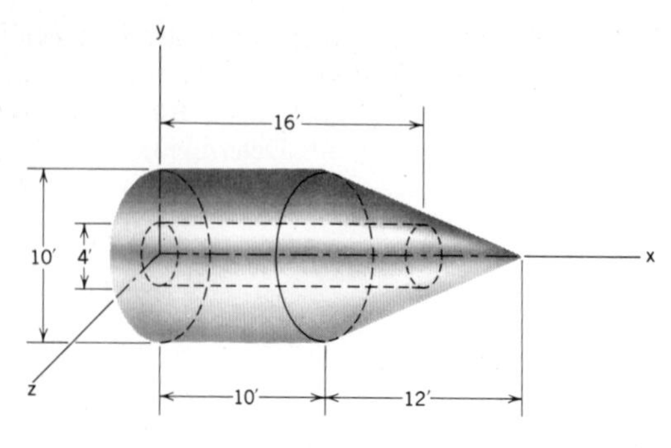

Figure P.8.35

**8.36.** A bent aluminum rod weighing 30 N/m is fitted into a plastic cylinder weighing 200 N, as shown. What are the centers of volume, mass, and gravity?

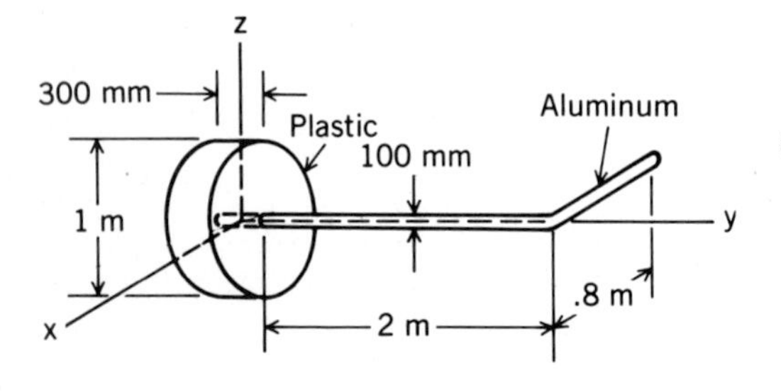


Figure P.8.36

**8.37.** A brass cylinder fits snugly into an aluminum block. The brass weighs 43.2 kN/m³ and the aluminum weighs 30 kN/m³. Find the center of volume, the center of mass, and the center of gravity.

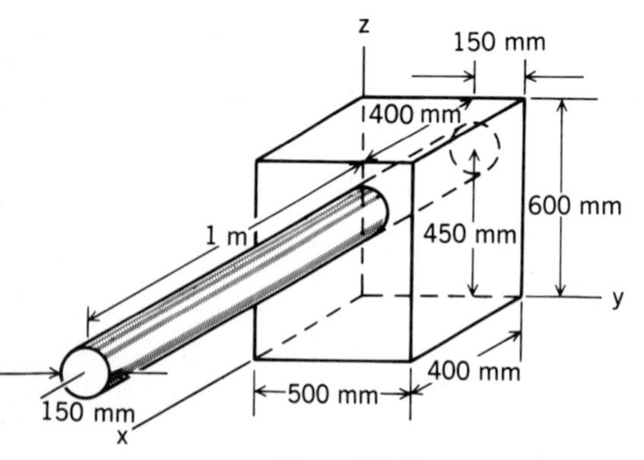


Figure P.8.37

**8.38.** Two thin plates are welded together. One has circle of radius 200 mm cut out as shown. If each plate weighs 450 N/m², what is the position of the center of mass?

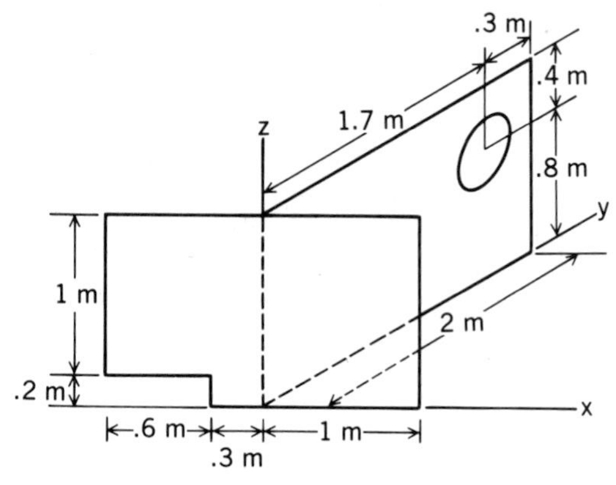


Figure P.8.38

**8.39.** What is the center of mass of the bent wire if it weighs 10 N/m?

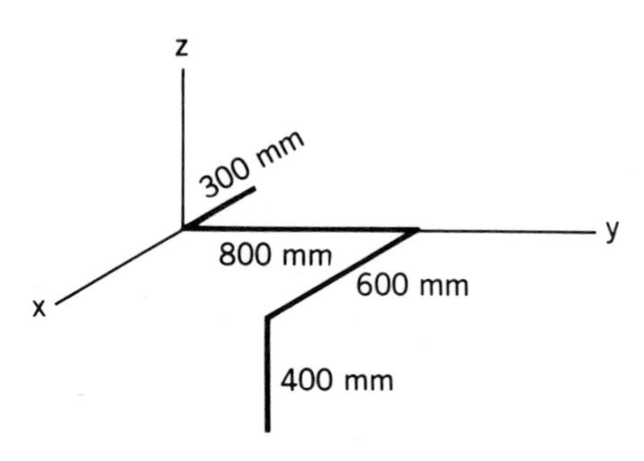


Figure P.8.39

**8.40.** Find the center of mass of the bent wire shown in the *zy* plane. The wire weighs 15 N/m.

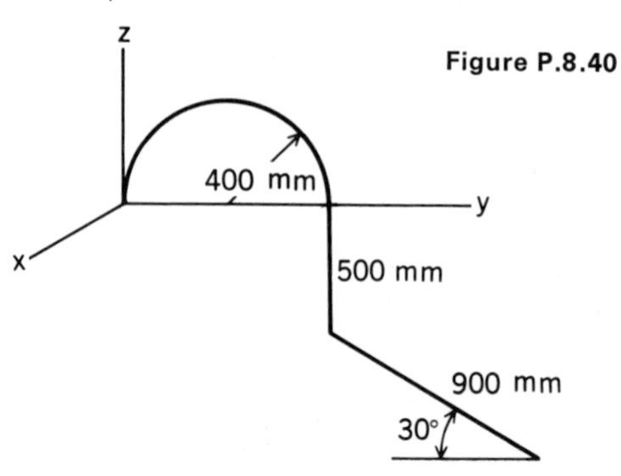


Figure P.8.40

**8.41.** In Problem 8.35, involving a wooden cone–cylinder with a cylindrical hole, find the center of mass for the case where the cylinder has a density of 46.0 lbm/ft³ and the cone has a density of 30.0 lbm/ft³.

**8.42.** The volume of an ellipsoidal body of revolution is known from the calculus to be $\frac{1}{6}\pi ab^2$. If the area of an ellipse is $\pi ab/4$, find the centroid of the area for a semiellipse.

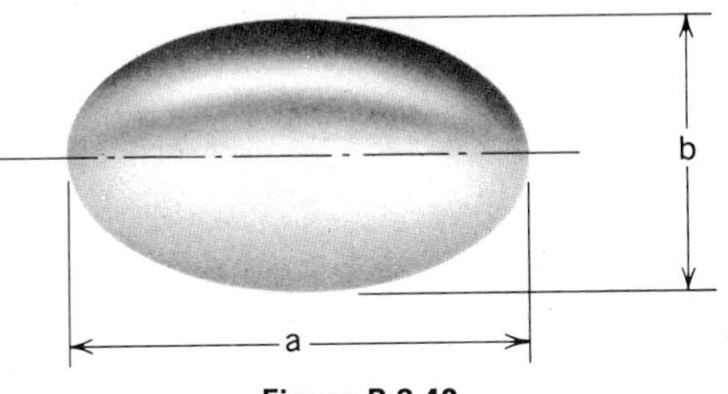


Figure P.8.42

**8.43.** Find the centroidal coordinate $y_c$ of the shaded area shown, using the theorems of Pappus and Guldinus.

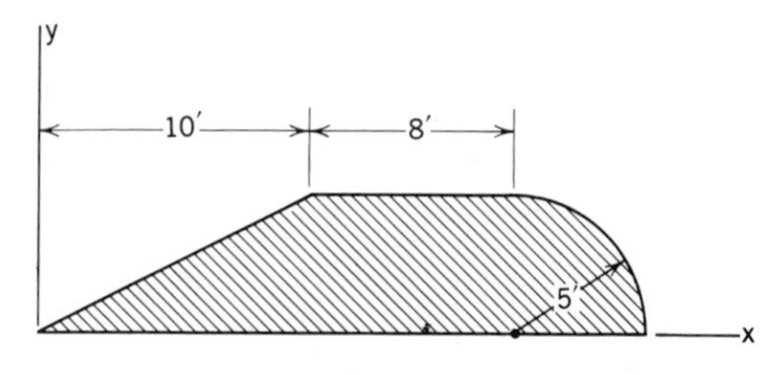


Figure P.8.43

**8.44.** A cutting tool of the lathe is programmed to cut along the dashed line as shown. What are the volume and the area of the body of revolution formed on the lathe?

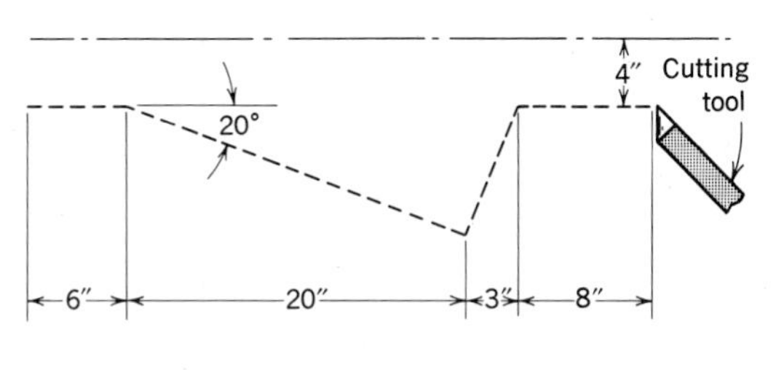


Figure P.8.44

**8.45.** Find the surface area and volume of the right conical frustum.

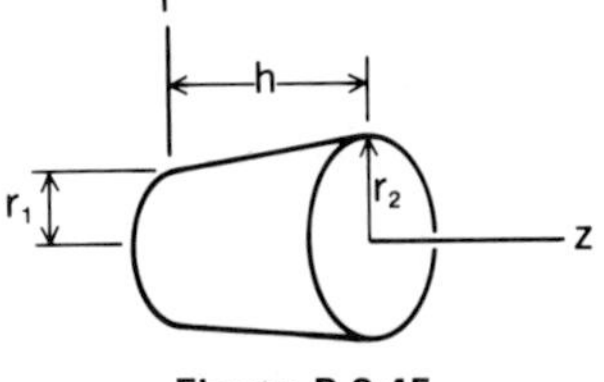


Figure P.8.45

**8.46.** Find the surface area and volume of the earth entry capsule for an unmanned Mars sampling mission. Approximate the rounded nose with a pointed nose as shown with the dashed lines.

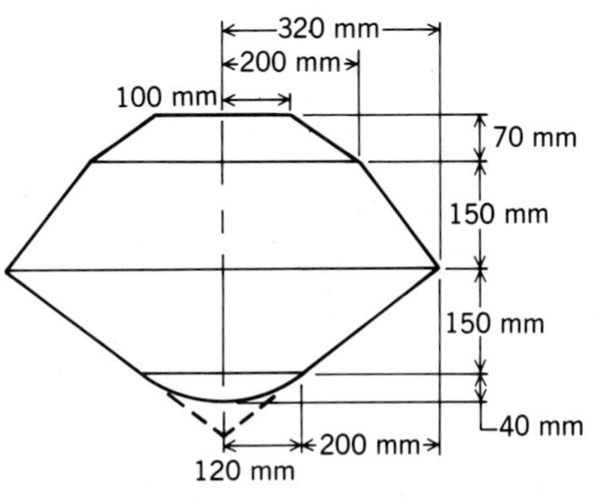

Figure P.8.46

**8.47.** Find the volume and surface area of the Apollo spaceship used for lunar exploration.

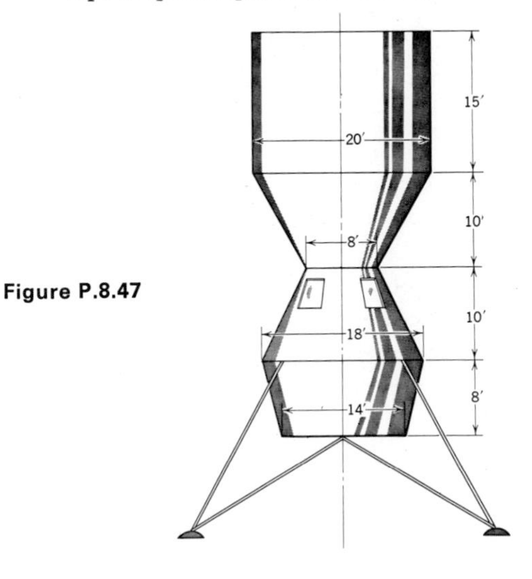


Figure P.8.47

## 8.5 Second Moments and the Product of Area[3] of a Plane Area

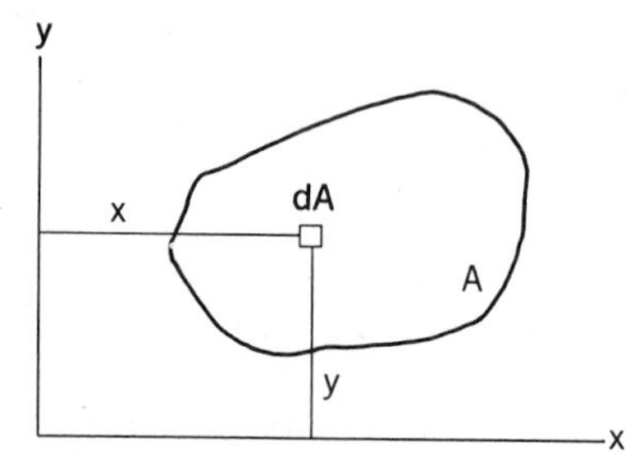


Figure 8.19. Plane surface.

We shall now consider other properties of a plane area relative to a given reference. The *second moments* of the area $A$ about the $x$ and $y$ axes (Fig. 8.19), denoted as $I_{xx}$ and $I_{yy}$, respectively, are defined as:

$$I_{xx} = \int_A y^2\, dA \tag{8.18a}$$

$$I_{yy} = \int_A x^2\, dA \tag{8.18b}$$

The second moment of area cannot be negative, in contrast to the first moment. Furthermore, because the square of the distance from the axis is used, elements of area that are farthest from the axis contribute most to the second moment of area.

In an analogy to the centroid, the entire area may be concentrated at a single point $(k_x, k_y)$ to give the same second moment of area for a given reference. Thus,

$$\begin{aligned} Ak_x^2 = I_{xx} = \int_A y^2\, dA; \qquad \text{therefore, } k_x^2 = \frac{\int_A y^2\, dA}{A} \\ Ak_y^2 = I_{yy} = \int_A x^2\, dA; \qquad \text{therefore, } k_y^2 = \frac{\int_A x^2\, dA}{A} \end{aligned} \tag{8.19}$$

The distances $k_x$ and $k_y$ are called the *radii of gyration. This point will have a position that depends not only on the shape of the area but also on the position of the reference.* This situation is unlike the centroid, whose location is independent of the reference position.

The *product of area* relates an area directly to a set of axes and is defined as

$$I_{xy} = \int_A xy\, dA \tag{8.20}$$

This quantity may be negative. We shall soon show that second moments and products of area are related for a given reference.

If the area under consideration has an axis of symmetry, the product of area for this axis and any axis orthogonal to this axis must be zero. You can readily reach this conclusion by considering the area in Fig. 8.20 which is symmetrical about the axis $A$–$A$. Notice that the centroid is somewhere along this axis. (Why?) The axis of symmetry has been indicated as the $y$ axis, and an arbitrary $x$ axis coplanar with the area has been shown. Also indicated are two elemental areas that are positioned as mirror images about the $y$ axis. The contribution to the product of area of each element is

[3] We often use the expressions *moment* and *product of inertia* for second moment and product of area, respectively. However, we shall also use the former expressions in Chapter 9 in connection with mass distributions.

$xy\,dA$, but with opposite signs, and so the net result is zero. Since the entire area can be considered to be composed of such pairs, it becomes evident that the product of area for such cases is zero. This *should not* be taken to mean that a nonsymmetric area cannot have a zero product of area about a set of axes. We shall discuss this last condition in more detail later.

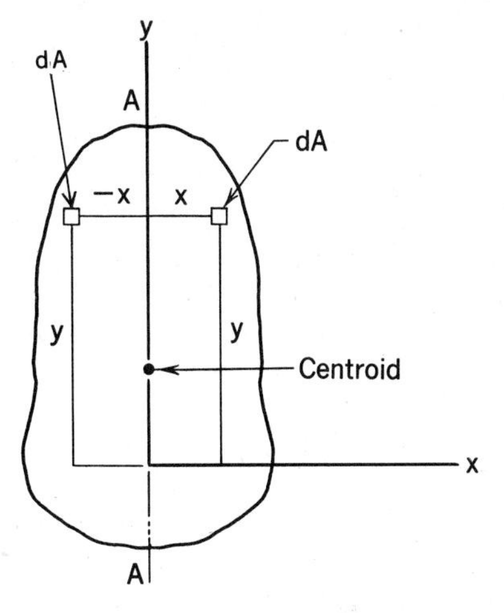


**Figure 8.20.** Area symmetric about $y$ axis.

## 8.6 Transfer Theorems

We shall now set forth a theorem that will be of great use in computing second moments and products of area for areas that can be decomposed into simple parts (composite areas). With this theorem, we can find second moments or products of area about any axis in terms of second moments or products of area about a parallel set of axes going through the centroid of the area in question.

An $x$ axis is shown in Fig. 8.21 parallel to and at a distance $d$ from an axis $x'$ going through the centroid of the area. The latter axis you will recall is a *centroidal axis*. The second moment of area about the $x$ axis is

$$I_{xx} = \int_A y^2\,dA = \int_A (y' + d)^2\,dA$$

where the distance $y$ has been replaced by $(y' + d)$. Integration leads to the result

$$I_{xx} = \int_A y'^2\,dA + 2d\int_A y'\,dA + Ad^2$$

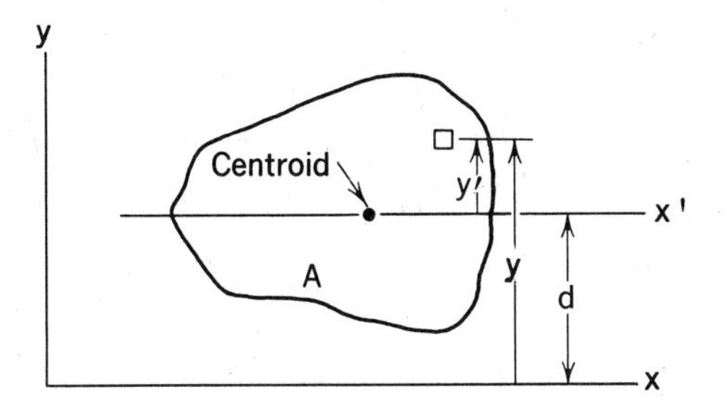


**Figure 8.21.** $x$ and $x'$ are parallel axes.

The first term on the right-hand side is clearly $I_{x'x'}$. The second term involves the first moment of area about the $x'$ axis. But the $x'$ axis here is a centroidal axis, and so the second term is zero. We can now state the transfer theorem (frequently called the parallel-axis theorem):

$$I_{\text{about any axis}} = I_{\substack{\text{about a parallel} \\ \text{axis at centroid}}} + Ad^2 \tag{8.21}$$

where $d$ is the perpendicular distance between the axis for which $I$ is being computed and the parallel centroidal axis.

In strength of materials, a course generally following statics, second moments of area about noncentroidal axes are commonly used. The areas involved are complicated and not subject to simple integration. Accordingly, in structural handbooks, the areas and second moments about various centroidal axes are listed for many of the practical configurations with the understanding that designers will use the parallel-axis theorem for axes not at the centroid.

Let us now examine the product of area in order to establish a parallel-axis theorem for this quantity. Accordingly, two references are shown in Fig. 8.22, one $(x', y')$

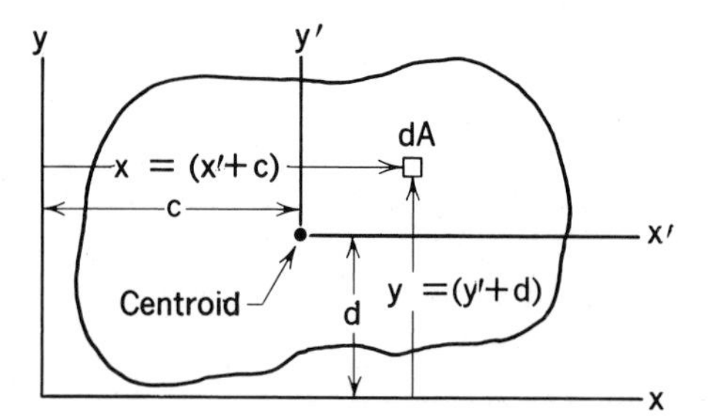


**Figure 8.22.** *c* and *d* measured from *xy*.

at the centroid and the other $(x, y)$ positioned arbitrarily but *parallel* relative to $xy$. Note that $c$ and $d$ are the $x$ and $y$ *coordinates* of the centroid of $A$ as measured from reference $xy$. These coordinates accordingly must have the proper signs, dependent on what quadrant the centroid of $A$ is in relative to $xy$. The product of area about the noncentroidal axes $xy$ can then be given as

$$I_{xy} = \int_A xy\, dA = \int_A (x' + c)(y' + d)dA$$

Carrying out the multiplication, we get

$$I_{xy} = \int_A x'y'\, dA + c\int_A y'\, dA + d\int_A x'\, dA + Adc$$

Clearly, the first term on the right side is $I_{x'y'}$, whereas the next two terms are zero, since $x'$ and $y'$ are centroidal axes. Thus, we arrive at a parallel-axis theorem for products of area of the form:

$$I_{xy \text{ for any set of axes}} = I_{x'y' \substack{\text{ for a parallel set of} \\ \text{axes at centroid}}} + Adc \tag{8.22}$$

It is important to remember that $c$ and $d$ are measured *from the xy axes to the centroid* and must have the appropriate sign. This will be carefully pointed out again in the examples of Section 8.7.

## 8.7 Computations Involving Second Moments and Products of Area

We shall now examine examples for the computation of second moments and products of area.

### EXAMPLE 8.6

A rectangle is shown in Fig. 8.23. Compute the second moment and product of area about the centroidal $x'y'$ axes as well as about the $xy$ axes.

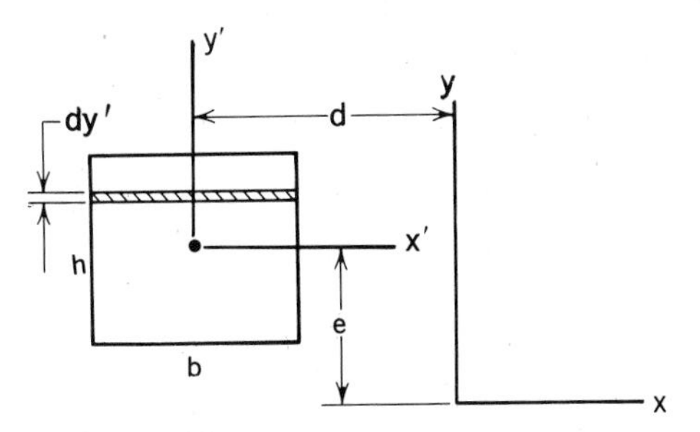


Figure 8.23. Rectangle: base $b$, height $h$.

$I_{x'x'}, I_{y'y'}, I_{x'y'}$. For computing $I_{x'x'}$, we can use a strip of width $dy'$ at a distance $y'$ from the $x'$ axis. The area $dA$ then becomes $b\,dy'$. Hence, we have

$$I_{x'x'} = \int_{-h/2}^{+h/2} y'^2 b\,dy = b\frac{y'^3}{3}\bigg|_{-h/2}^{+h/2} = \frac{b}{3}\left(\frac{h^3}{8} + \frac{h^3}{8}\right) = \frac{1}{12}\,bh^3 \tag{a}$$

This is a common result and should well be remembered since it occurs so often. Verbally, for such an axis, the second moment is equal to $\frac{1}{12}$ the base $b$ times the height $h$ cubed. The second moment of area for the $y'$ axis can immediately be written as

$$I_{y'y'} = \tfrac{1}{12}\,hb^3 \tag{b}$$

where the base and height have simply been interchanged.

As a result of the previous statements on symmetry, we immediately note that

$$I_{x'y'} = 0 \tag{c}$$

$I_{xx}, I_{yy}, I_{xy}$. Employing the transfer theorems, we get

$$I_{xx} = \tfrac{1}{12}\,bh^3 + bhe^2$$

$$I_{yy} = \tfrac{1}{12}\,hb^3 + bhd^2$$

In computing the product of area, we must be careful to employ the proper signs for the transfer distances. In checking the derivation of the transfer theorem, we see that these distances are measured from the noncentroidal axes to the centroid $C$. Therefore, in this problem the transfer distances are $(+e)$ and $(-d)$. Hence, the computation of $I_{xy}$ becomes

$$I_{xy} = 0 + (bh)(+e)(-d) = -bhed$$

and is thus a negative quantity.

**EXAMPLE 8.7**

What are $I_{xx}$, $I_{yy}$, and $I_{xy}$ for the area under the parabolic curve shown in Fig. 8.24?

To find $I_{xx}$, we may use horizontal strips of width $dy$ as shown in Fig. 8.25. We can then say for $I_{xx}$:

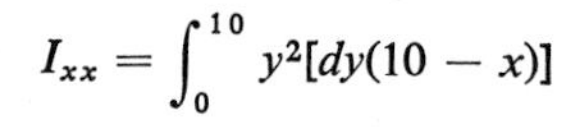

$$I_{xx} = \int_0^{10} y^2[dy(10 - x)]$$

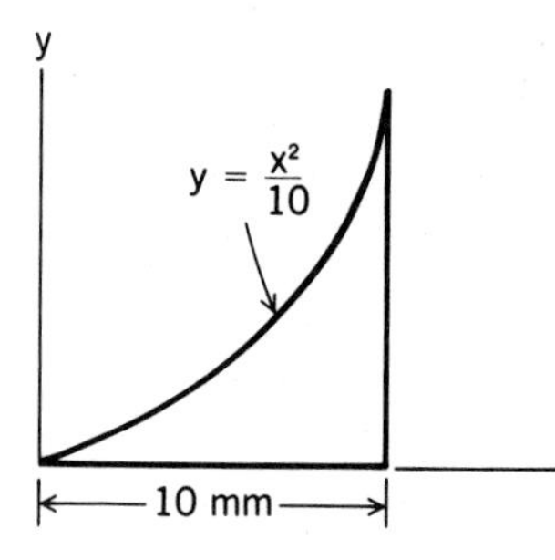


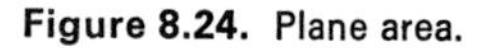
**Figure 8.24.** Plane area.

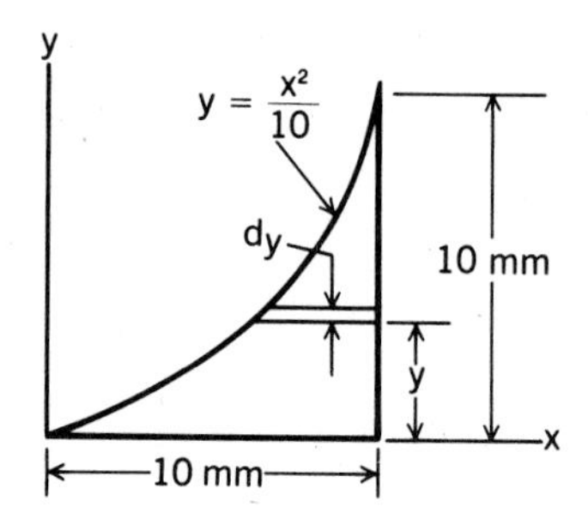


**Figure 8.25.** Horizontal strip.

But

$$x = \sqrt{10}\, y^{1/2}$$

Therefore,

$$\begin{aligned} I_{xx} &= \int_0^{10} y^2(10 - \sqrt{10}\, y^{1/2})\, dy \\ &= \left[10\frac{y^3}{3} - \sqrt{10}\, y^{7/2}\left(\frac{2}{7}\right)\right]\Bigg|_0^{10} \\ &= \frac{10(10^3)}{3} - \sqrt{10}(10^{7/2})\left(\frac{2}{7}\right) = 476.2 \text{ mm}^4 \end{aligned}$$

As for $I_{yy}$, we use vertical infinitesimal strips as shown in Fig. 8.26. We can, accordingly, say:

$$\begin{aligned} I_{yy} &= \int_0^{10} x^2(y\, dx) = \int_0^{10} \frac{x^4}{10}\, dx \\ &= \frac{x^5}{50}\Bigg|_0^{10} = 2000 \text{ mm}^4 \end{aligned}$$

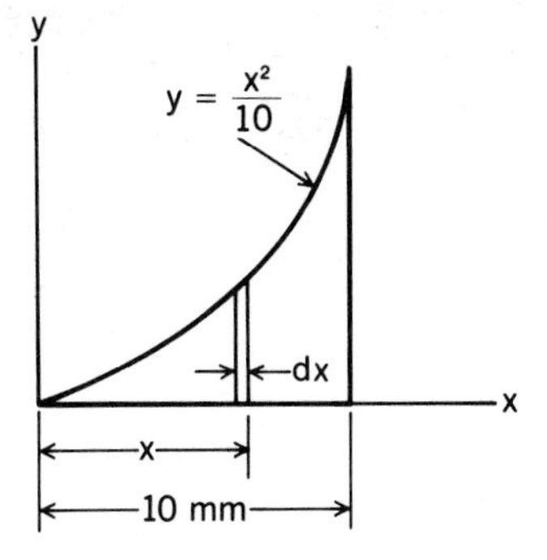


Figure 8.26. Vertical strip.

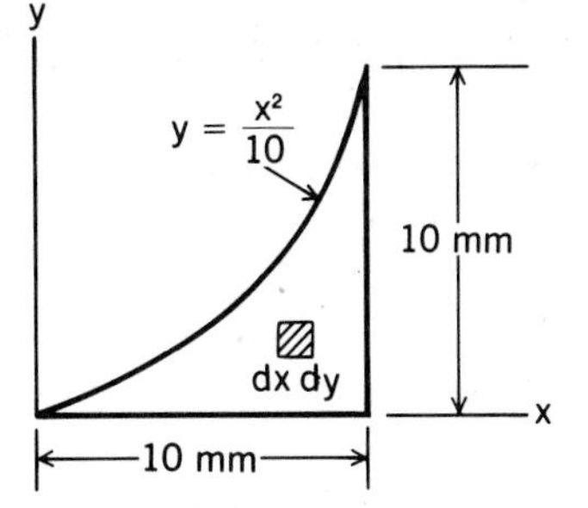


Figure 8.27. Element for multiple integration.

Finally, for $I_{xy}$ we use an infinitesimal area element $dx\,dy$ shown in Fig. 8.27. We must now perform multiple integration.[4] Thus, we have

$$I_{xy} = \int_0^{10} \int_{y=0}^{y=x^2/10} xy\,dy\,dx$$

Notice by holding $x$ constant and letting $y$ first run from $y = 0$ to the curve $y = x^2/10$ we cover the vertical strip of thickness $dx$ at position $x$ such as is shown in Fig. 8.26. Then by letting $x$ run from zero to 10, we cover the entire area. Accordingly, we first integrate with respect to $y$ holding $x$ constant. Thus,

$$I_{xy} = \int_0^{10} x\left(\frac{y^2}{2}\right)\Bigg|_0^{x^2/10} dx = \int_0^{10} \frac{x^5}{200}\,dx$$

Next, integrating with respect to $x$, we have

$$I_{xy} = \frac{x^6}{1200}\Bigg|_0^{10} = 833 \text{ mm}^4$$

**EXAMPLE 8.8**

Compute the second moment of area of a circular area about a diameter (Fig. 8.28).

Using polar coordinates, we have[5] for $I_{xx}$:

$$I_{xx} = \int_0^{D/2} \int_0^{2\pi} (r \sin\theta)^2 r\,d\theta\,dr = \int_0^{D/2} \pi r^3\,dr$$

---

[4]This multiple integration involves boundaries requiring some variable limits, in contrast to previous multiple integrations.

[5]The integral $\int_0^{2\pi} \sin^2\theta\,d\theta$ may be evaluated by methods of substitution or may readily be seen in the following manner. $\int_0^{2\pi} \sin^2\theta\,d\theta$ equals the area under the curve shown, which is half the area of the rectangle. Hence, this integral equals $\pi$.

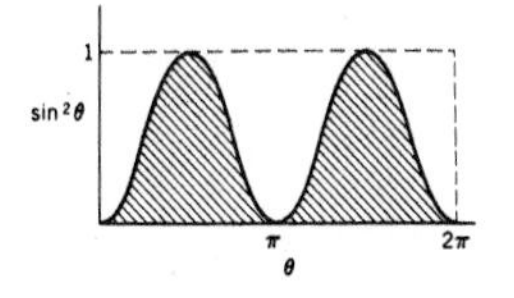

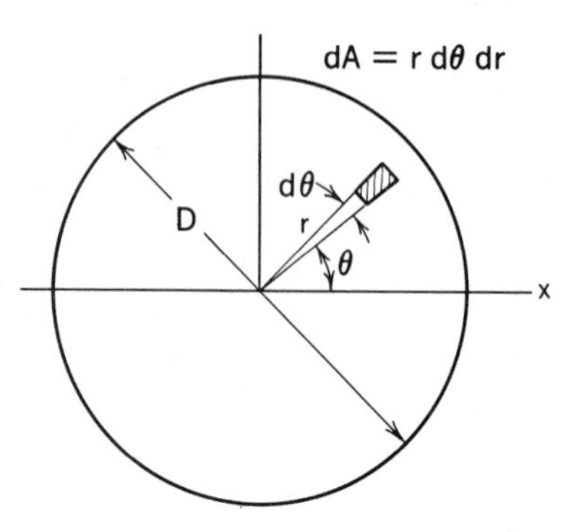


**Figure 8.28.** Circular area with polar coordinates.

Completing the integration, we have

$$I_{xx} = \frac{r^4}{4}\pi\Big|_0^{D/2} = \pi\frac{D^4}{64}$$

The product of area $I_{xy}$ must be zero, owing to symmetry of the area about the $xy$ axes.

In the previous examples, we computed second moments and products of area using the calculus. Many problems of interest involve an area that may be subdivided into simpler component areas. Such an area has been referred to in earlier discussions as a *composite* area. The second moments and products of area for certain centroidal axes of many simple areas may be found in engineering handbooks (also see the inside covers). Using these formulas plus the parallel-axis theorems, we can easily compute desired second moments and products of area for composite areas as we have done earlier for first moments of area. The following example illustrates this procedure.

**EXAMPLE 8.9**

Find the centroid of the area of the unequal-leg Z section shown in Fig. 8.29. Next, determine the second moment of area about the centroidal axes parallel to the sides of the Z section. Finally, determine the product of area for the aforementioned centroidal axes.

We shall subdivide the Z section into three rectangular areas, as shown in Fig. 8.30. Also, we shall insert a convenient reference $xy$, as shown in the diagram. To find the centroid, we proceed in the following manner:

| | $A_i$ (in.$^2$) | $\bar{x}_i$ (in.) | $\bar{y}_i$ (in.) | $A_i\bar{x}_i$ (in.$^3$) | $A_i\bar{y}_i$ (in.$^3$) |
|---|---|---|---|---|---|
| 1. | (2)(1) = 2 | 1 | 7.50 | 2 | 15 |
| 2. | (8)(1) = 8 | 2.50 | 4 | 20 | 32 |
| 3. | (4)(1) = 4 | 5 | .50 | 20 | 2 |
| | $\sum_i A_i = 14$ | | | $\sum_i A_i\bar{x}_i = 42$ | $\sum_i A_i\bar{y}_i = 49$ |

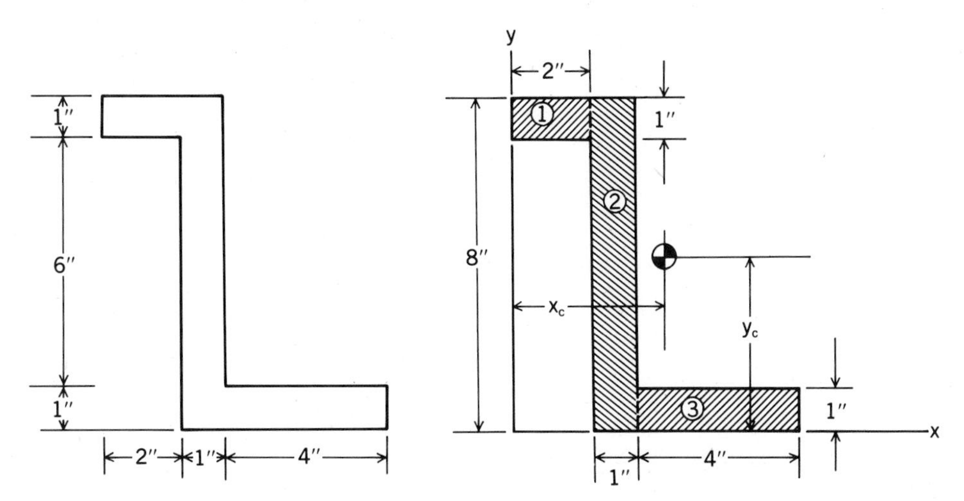

**Figure 8.29.** Unequal-leg Z section.

**Figure 8.30.** Composite area.

Therefore,

$$x_c = \frac{\sum_i A_i \bar{x}_i}{\sum_i A_i} = \frac{42}{14} = 3 \text{ in.}$$

$$y_c = \frac{\sum_i A_i \bar{y}_i}{\sum_i A_i} = \frac{49}{14} = 3.5 \text{ in.}$$

We have shown the centroidal axes $x_c y_c$ in Fig. 8.31. We now find $I_{x_c x_c}$ and $I_{y_c y_c}$ using the parallel-axis theorem and the formula $\frac{1}{12} bh^3$ for the second moment of area about a centroidal axis of symmetry of a rectangle.

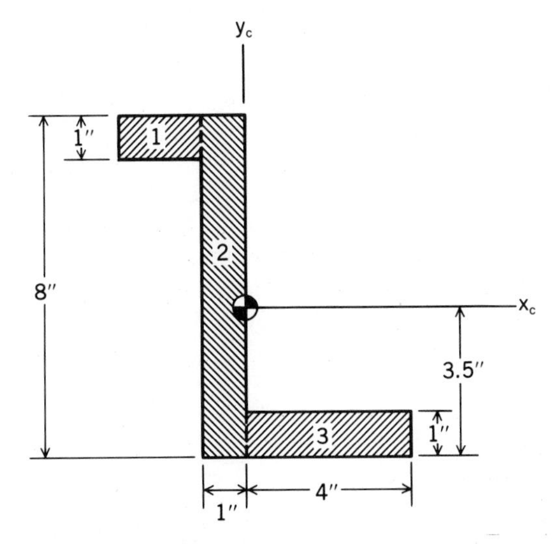


**Figure 8.31.** Centroidal axes $x_c y_c$.

$$I_{x_cx_c} = \underbrace{[(\tfrac{1}{12})(2)(1^3) + (2)(4^2)]}_{①} + \underbrace{[(\tfrac{1}{12})(1)(8^3) + (8)(\tfrac{1}{2})^2]}_{②}$$
$$+ \underbrace{[(\tfrac{1}{12})(4)(1^3) + (4)(3^2)]}_{③} = 113.2 \text{ in.}^4$$

$$I_{y_cy_c} = \underbrace{[(\tfrac{1}{12})(1)(2^3) + (2)(2^2)]}_{①} + \underbrace{[(\tfrac{1}{12})(8)(1^3) + (8)(\tfrac{1}{2})^2]}_{②}$$
$$+ \underbrace{[(\tfrac{1}{12})(1)(4^3) + (4)(2^2)]}_{③} = 32.67 \text{ in.}^4$$

Finally, we consider the product of area $I_{x_cy_c}$. Here we must be cautious in using the parallel-axis theorem. Remember that $x_cy_c$ are centroidal axes for the *entire* area of the $Z$ section. In using the parallel-axis theorem for a *subarea*, we must note that $x_cy_c$ are *not* centroidal axes for the subarea. The centroidal axes to be used in this problem for subareas are the axes of symmetry of each subarea. In short, $x_cy_c$ are simply axes about which we are computing the product of area of each subarea. Therefore, in the parallel axis theorem, the transfer distances $c$ and $d$ are measured *from the $x_cy_c$ axes to the centroid* in each subarea, as noted in the development of the parallel-axis theorem. The proper sign must be assigned each time to the transfer distances with this in mind. We have for $I_{x_cy_c}$:

$$I_{x_cy_c} = \underbrace{[0 + (2)(-2)(4)]}_{①} + \underbrace{[0 + (8)(-\tfrac{1}{2})(\tfrac{1}{2})]}_{②}$$
$$+ \underbrace{[0 + (4)(2)(-3)]}_{③} = -42 \text{ in.}^4$$

## Problems

**8.48.** Find $I_{xx}$, $I_{yy}$, and $I_{xy}$ for the triangle shown. Give the results in feet.

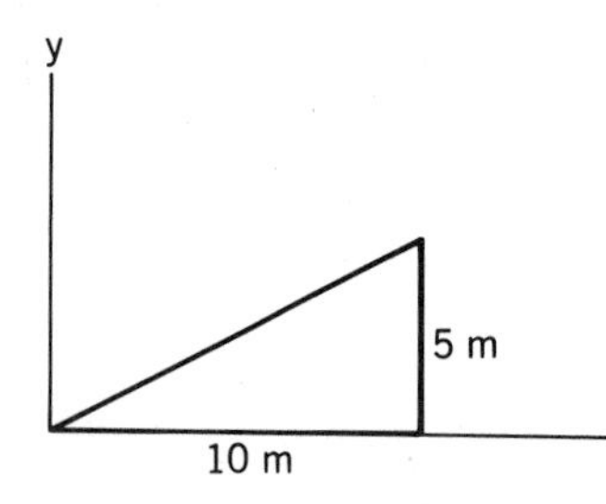


Figure P.8.48

**8.49.** What are the second moments and products of area of the ellipse for reference $xy$? (*Hint:* Can you work with one quadrant and then multiply by 4 for the second moments?)

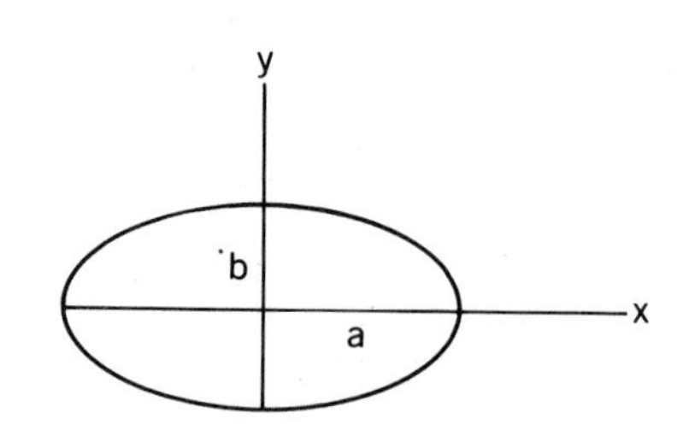


Figure P.8.49

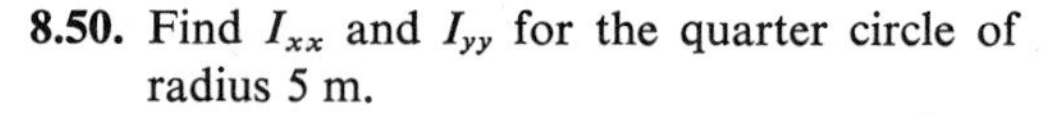

**8.50.** Find $I_{xx}$ and $I_{yy}$ for the quarter circle of radius 5 m.

Figure P.8.50

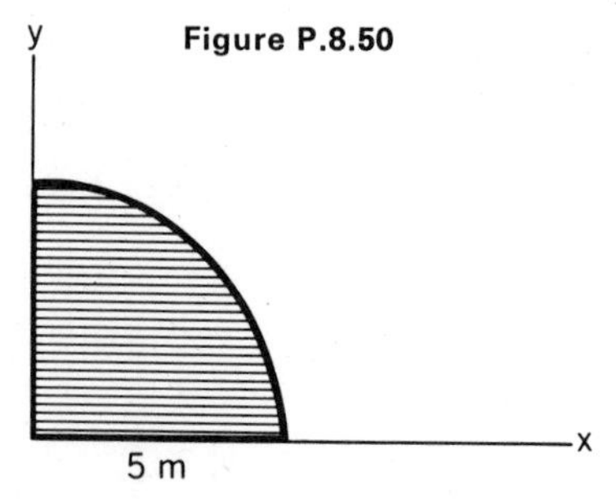

**8.51.** Find $I_{xx}$, $I_{yy}$, and $I_{xy}$ for the shaded area.

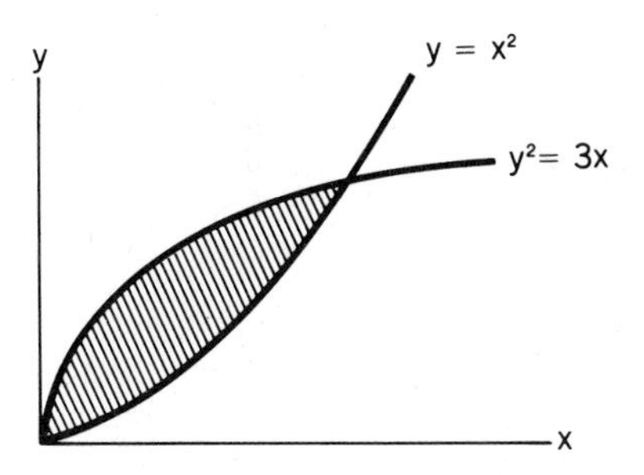


Figure P.8.51

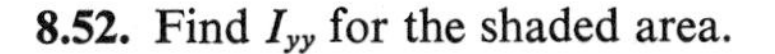

**8.52.** Find $I_{yy}$ for the shaded area.

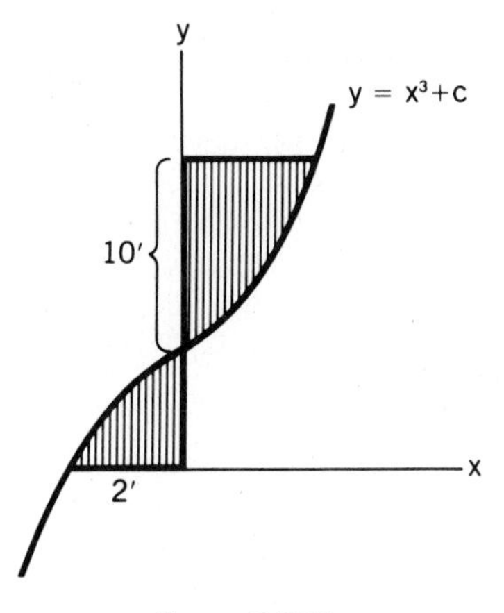

Figure P.8.52

**8.53.** Find $I_{yy}$ for the area between the curves

$$y = 2 \sin x \text{ ft}$$

$$y = \sin 2x \text{ ft}$$

from $x = 0$ to $x = \pi$ ft.

**8.54.** Find $I_{yy}$ for the areas enclosed between curves $y = \cos x$ and $y = \sin x$ and the lines $x = 0$ and $x = \pi/2$.

**8.55.** Show that $I_{xx} = bh^3/12$, $I_{yy} = b^3h/12$, and $I_{xy} = b^2h^2/24$ for the right triangle.

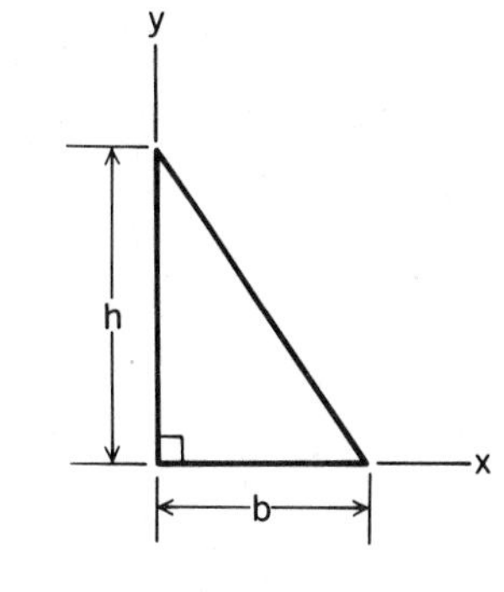


Figure P.8.55

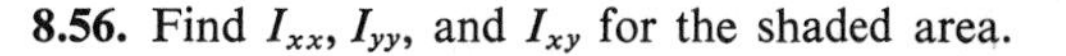

**8.56.** Find $I_{xx}$, $I_{yy}$, and $I_{xy}$ for the shaded area.

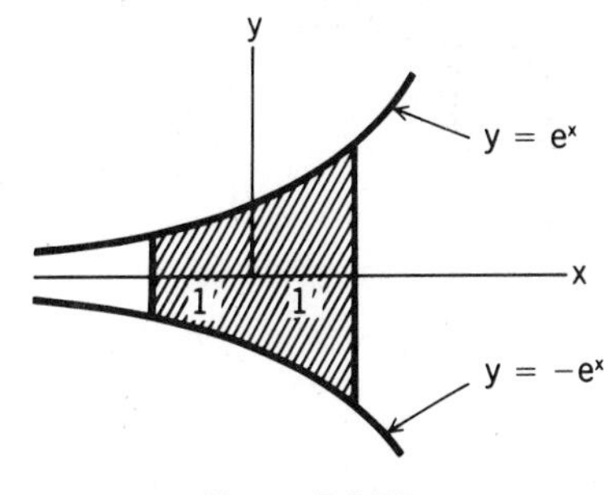


Figure P.8.56

**8.57.** Find $I_{xx}$, $I_{yy}$, and $I_{xy}$ for the area of Problem 8.4. The equation of the curve is $y^2 = 20x - 60$.

**8.58.** Find $I_{xx}$, $I_{yy}$, and $I_{xy}$ for the cross section shown.

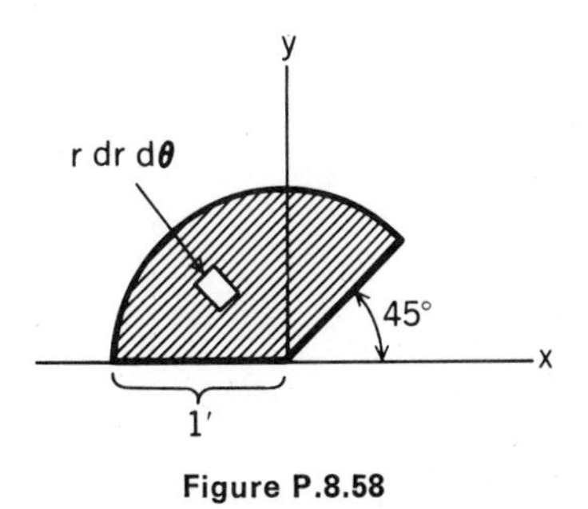


Figure P.8.58

**8.59.** Find $I_{xx}$ and $I_{yy}$ for the area of Problem 8.5. The equation of the parabola is $y = (x^2/9) - 5$. (*Hint:* The area of a vertical element in the region below the $x$ axis is $(0 - y)\,dx$.)

**8.60.** In Problem 8.59, determine $I_{xy}$ using multiple integration.

**8.61.** Find the second moments of area about axes $xy$ for the shaded area shown.

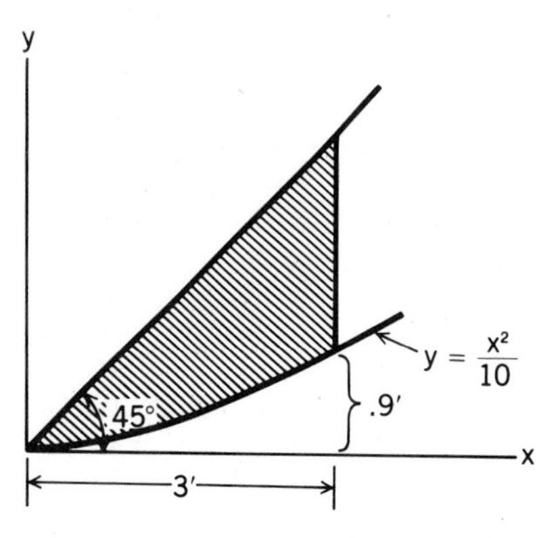


Figure P.8.61

**8.62.** If the second moment of area about axis $A$–$A$ is known to be 600 ft⁴, what is the second moment of area about a parallel axis $B$–$B$ a distance 3 ft from $A$–$A$, for an area of 10 ft²? The centroid of this area is 4 ft from $B$–$B$.

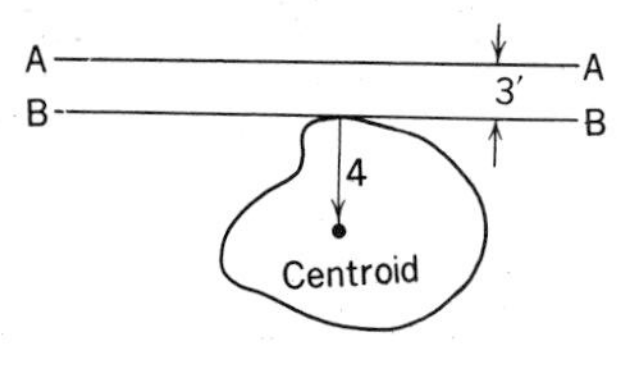


Figure P.8.62

**8.63.** Using the results of Problem 8.55, show that $I_{x_cx_c} = bh^3/36$, $I_{y_cy_c} = hb^3/36$, and $I_{x_cy_c} = -b^2h^2/72$ for the right triangle shown.

Figure P.8.63

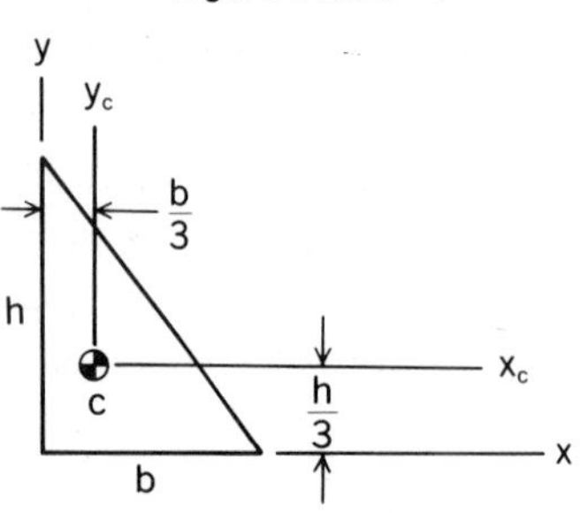


**8.64.** Show that $I_{xx} = bh^3/12$, $I_{yy} = (hb/12)(b^2 + ab + a^2)$, and $I_{xy} = (h^2b/24)(2a + b)$ for the triangle. (*Hint:* Break the triangle into two right triangles for which the various moments are known. (See Problem 8.63).)

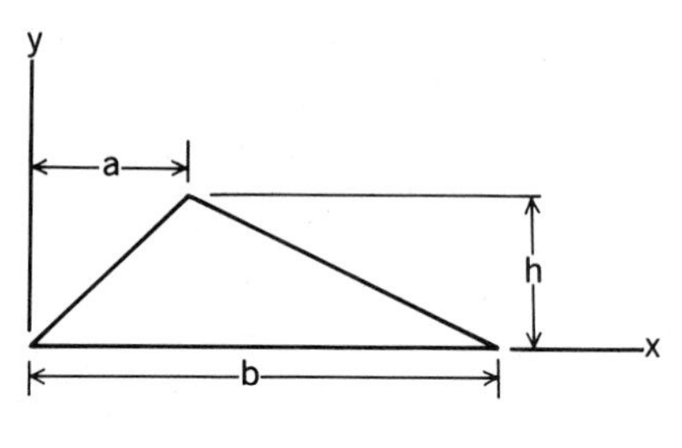


Figure P.8.64

**8.65.** In Problem 8.64, show that $I_{x_cx_c} = bh^3/36$, $I_{y_cy_c} = (bh/36)(b^2 - ab + a^2)$, and $I_{x_cy_c} = (h^2b/72)(2a - b)$ for the triangle. (*Hint:* Use the results of Problems 8.8 and 8.64 and the parallel-axis theorem.)

**8.66.** Find $I_{xx}$, $I_{yy}$, and $I_{xy}$ of the extruded section. Disregard all rounded edges.

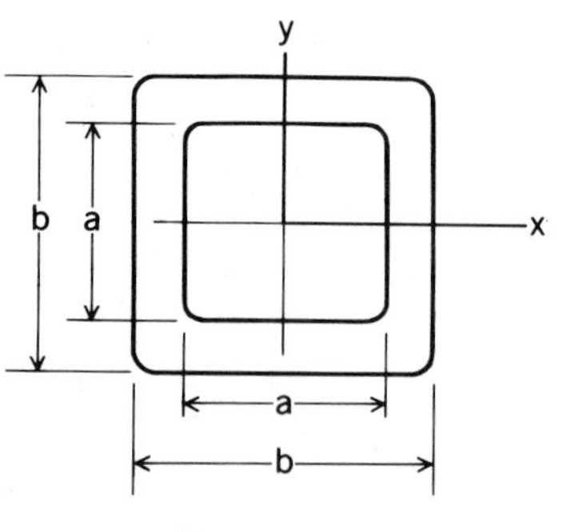


Figure P.8.66

**8.67.** Find the second moment of area of the rectangle (with a hole) about the base of the rectangle. Also, determine the product of area about the base and left side.

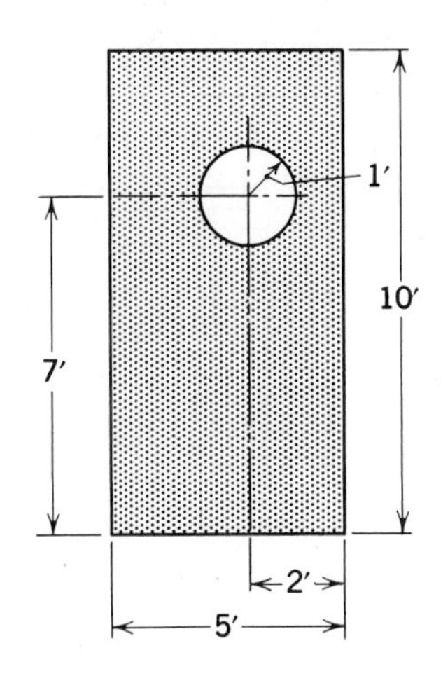


Figure P.8.67

**8.68.** Find $I_{xx}$, $I_{yy}$, $I_{x_cx_c}$, and $I_{y_cy_c}$ for the structural "hat" section. Disregard all rounded edges.

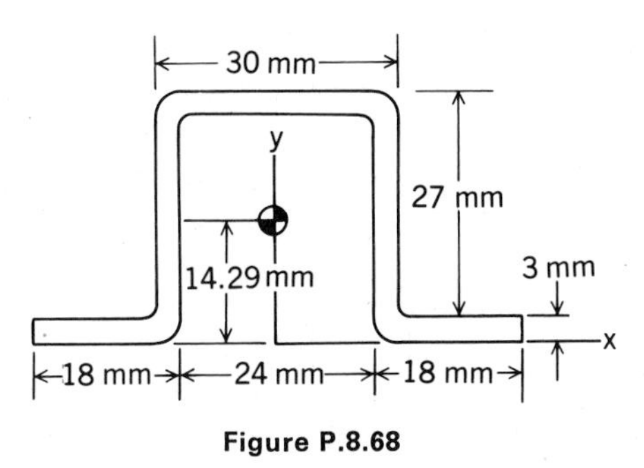


Figure P.8.68

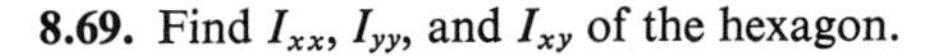

**8.69.** Find $I_{xx}$, $I_{yy}$, and $I_{xy}$ of the hexagon.

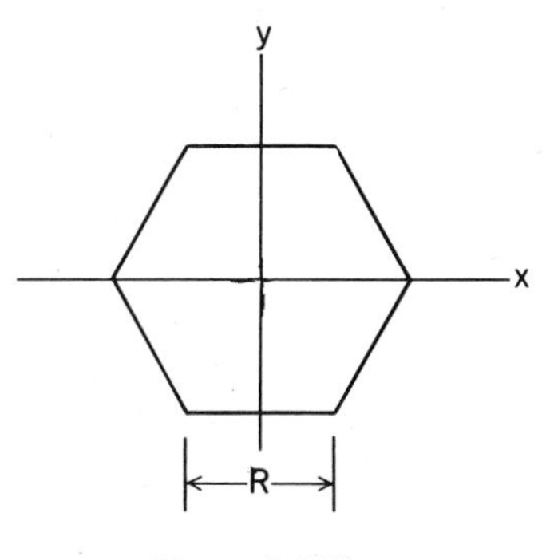


Figure P.8.69

**8.70.** A beam cross section is made up of an I-shaped section with an additional thick plate welded on. Find the second moments of area for the centroidal axes $x_cy_c$ of the beam cross section. What is $I_{x_cy_c}$? Give the results in millimeters.

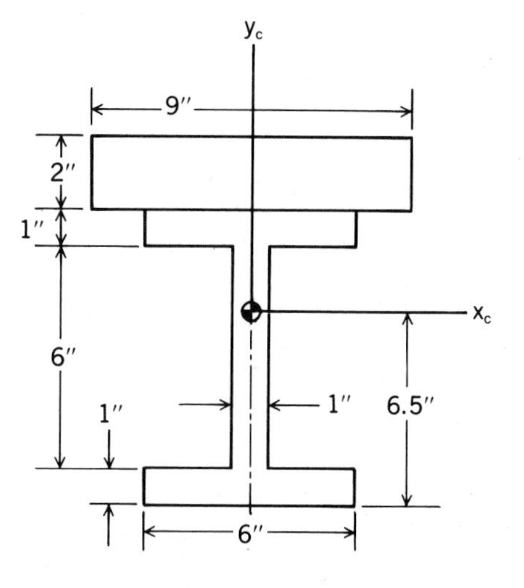

Figure P.8.70

**8.71.** Find the second moments of the area shown about centroidal axes parallel to the $x$ and $y$ axes. That is, find $I_{x_cx_c}$ and $I_{y_cy_c}$. Give the results in millimeters.

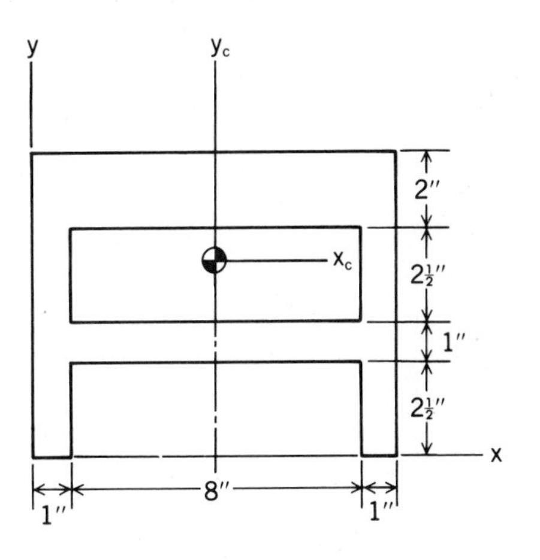


Figure P.8.71

**8.72.** Find $I_{xx}$, $I_{yy}$, $I_{xy}$, $I_{x_cx_c}$, $I_{y_cy_c}$, and $I_{x_cy_c}$ of the unequal-leg rolled channel section. Disregard all rounded edges.

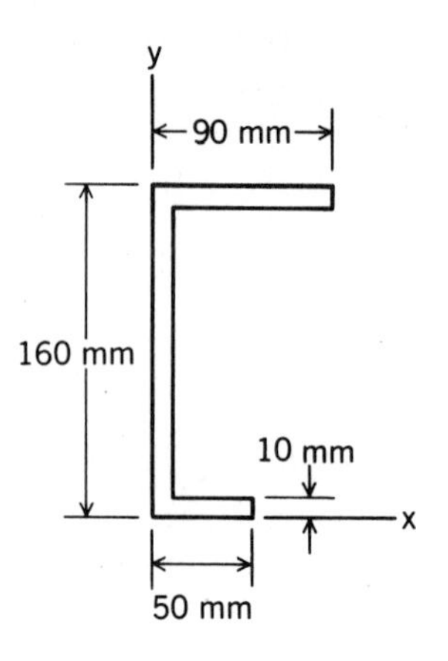


Figure P.8.72

## 8.8 Relation Between Second Moments and Products of Area

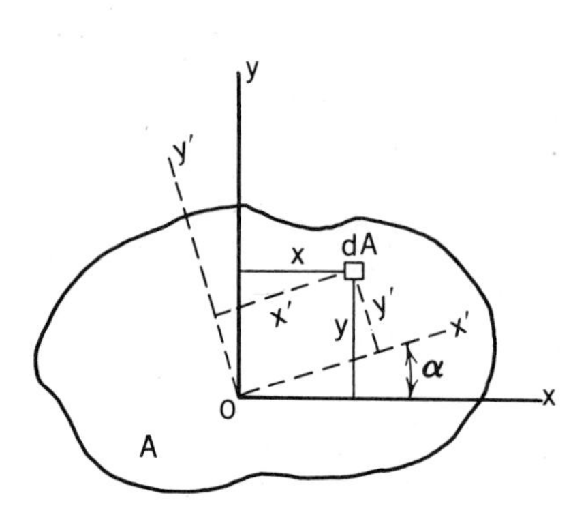


Figure 8.32. Rotation of axes.

We shall now show that we can ascertain second moments and product of area relative to a rotated reference $x'y'$ if we know these quantities for reference $xy$ that has the *same origin*. Such a reference $x'y'$ rotated an angle $\alpha$ from $xy$ is shown in Fig. 8.32. We shall assume that the second moments and product of area for the unprimed reference are known.

Before proceeding, we must know the relation between the coordinates of the area elements $dA$ for the two references. From Fig. 8.32, it is clear that

$$x' = x \cos \alpha + y \sin \alpha \tag{8.23a}$$

$$y' = -x \sin \alpha + y \cos \alpha \tag{8.23b}$$

With relation 8.23b, we can express $I_{x'x'}$ in the following manner:

$$I_{x'x'} = \int_A (y')^2 \, dA = \int_A (-x \sin \alpha + y \cos \alpha)^2 \, dA \tag{8.24}$$

Carrying out the square, we have

$$I_{x'x'} = \sin^2 \alpha \int_A x^2 \, dA - 2 \sin \alpha \cos \alpha \int_A xy \, dA + \cos^2 \alpha \int_A y^2 \, dA$$

Therefore,

$$I_{x'x'} = I_{yy} \sin^2 \alpha + I_{xx} \cos^2 \alpha - 2I_{xy} \sin \alpha \cos \alpha \tag{8.25}$$

A more common form of the desired relation can be formed by using the following trigonometric identities:

$$\cos^2 \alpha = \tfrac{1}{2}(1 + \cos 2\alpha) \quad \text{(a)}$$

$$\sin^2 \alpha = \tfrac{1}{2}(1 - \cos 2\alpha) \quad \text{(b)}$$

$$2 \sin \alpha \cos \alpha = \sin 2\alpha \quad \text{(c)}$$

We then have

$$I_{x'x'} = \frac{I_{xx} + I_{yy}}{2} + \frac{I_{xx} - I_{yy}}{2} \cos 2\alpha - I_{xy} \sin 2\alpha \quad (8.26)$$

To determine $I_{y'y'}$, we need only replace the $\alpha$ in the preceding result by $(\alpha + \pi/2)$. Thus,

$$I_{y'y'} = \frac{I_{xx} + I_{yy}}{2} + \frac{I_{xx} - I_{yy}}{2} \cos (2\alpha + \pi) - I_{xy} \sin (2\alpha + \pi)$$

Note that $\cos (2\alpha + \pi) = -\cos 2\alpha$ and $\sin (2\alpha + \pi) = -\sin 2\alpha$. Hence, the equation above becomes

$$I_{y'y'} = \frac{I_{xx} + I_{yy}}{2} - \frac{I_{xx} - I_{yy}}{2} \cos 2\alpha + I_{xy} \sin 2\alpha \quad (8.27)$$

Next, the product of area $I_{x'y'}$ can be computed in a similar manner:

$$I_{x'y'} = \int_A x'y' \, dA = \int_A (x \cos \alpha + y \sin \alpha)(-x \sin \alpha + y \cos \alpha) \, dA$$

This becomes

$$I_{x'y'} = \sin \alpha \cos \alpha \, (I_{xx} - I_{yy}) + (\cos^2 \alpha - \sin^2 \alpha) I_{xy}$$

Utilizing the previously defined trigonometric identities, we get

$$I_{x'y'} = \frac{I_{xx} - I_{yy}}{2} \sin 2\alpha + I_{xy} \cos 2\alpha \quad (8.28)$$

Thus, we see that, if we know the quantities $I_{xx}$, $I_{yy}$, and $I_{xy}$ for some reference $xy$ at point $O$, the second moments and products of area for *every* set of axes at point $O$ can be computed. And if, in addition, we employ the transfer theorems, we can compute second moments and products of area for *any* reference in the plane of the area.

## 8.9 Polar Moment of Area

In the previous section, we saw that the second moments and product of area for an orthogonal reference determined all such quantities for *any* orthogonal reference

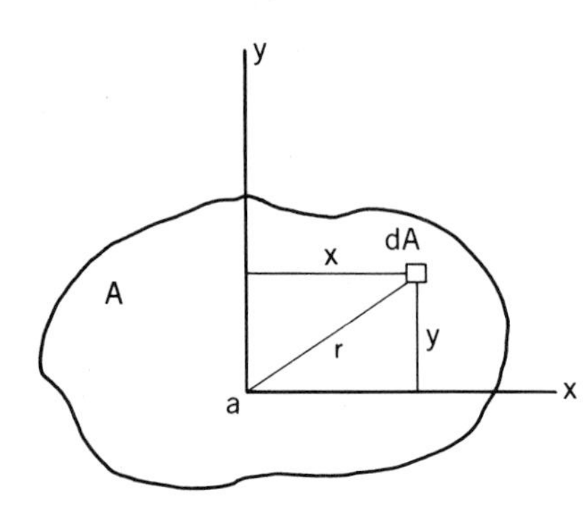

**Figure 8.33.** $J = I_{xx} + I_{yy}$.

having the same origin. We shall now show that the sum of the pairs of second moments of area is a constant for all such references at a point. Thus, in Fig. 8.33 we have a reference $xy$ associated with point $a$. Summing $I_{xx}$ and $I_{yy}$, we have

$$I_{xx} + I_{yy} = \int_A y^2\,dA + \int_A x^2\,dA$$
$$= \int_A (x^2 + y^2)\,dA = \int_A r^2\,dA$$

Since $r^2$ is independent of the inclination of the coordinate system, the sum $I_{xx} + I_{yy}$ is independent of the inclination of the reference. Therefore, the sum of second moments of area about orthogonal axes is a function only of the position of the origin $a$ for the axes. This sum is termed the *polar moment of area*, $J$.[6] We can then consider $J$ to be a scalar field. Mathematically, this statement is expressed as

$$J = J(x', y') \tag{8.29}$$

where $x'$ and $y'$ are the coordinates of the origin $a$ in the plane as measured from some convenient reference $x'y'$.

That $(I_{xx} + I_{yy})$ does not change on rotation of axes can also be deduced from the transformation equations 8.26 and 8.27. This group of terms is accordingly termed an *invariant*. We can similarly show that $(I_{xx}I_{yy} - I_{xy}^2)$ is also invariant under a rotation of axes.

## 8.10 Principal Axes

Still other conclusions may be drawn about second moments and products of area associated with a point in an area. In Fig. 8.34 is an area with a reference $xy$ having its origin at point $a$. We shall assume that $I_{xx}$, $I_{yy}$, and $I_{xy}$ are known for this reference, and shall ask at what angle $\alpha$ we shall find an axis having the *maximum* second moment of area. Since the sum of the second moments of area is constant for any reference with origin at $a$, the *minimum* second moment of area must then correspond to an axis at *right angles* to the axis having the maximum second moment. Since second moments of area have been expressed in Eqs. 8.26 and 8.27 as functions of the variable $\alpha$ at a point, these extremes may readily be determined by setting the partial derivative of $I_{x'x'}$ with respect to $\alpha$ equal to zero. Thus,

$$\frac{\partial I_{x'x'}}{\partial \alpha} = (I_{xx} - I_{yy})(-\sin 2\alpha) - 2I_{xy}\cos 2\alpha = 0$$

---

[6]Quite often $I_p$ is used for the polar moment of area.

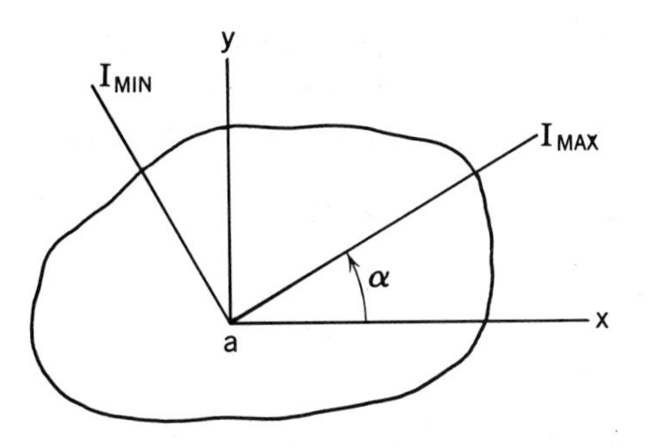


**Figure 8.34.** Principal axes.

If we denote the value of $\alpha$ that satisfies the equation above as $\tilde{\alpha}$, we have

$$(I_{yy} - I_{xx}) \sin 2\tilde{\alpha} - 2I_{xy} \cos 2\tilde{\alpha} = 0$$

Hence,

$$\boxed{\tan 2\tilde{\alpha} = \frac{2I_{xy}}{I_{yy} - I_{xx}}} \tag{8.30}$$

This formulation gives us the angle $\tilde{\alpha}$, which corresponds to an extreme value of $I_{x'x'}$ (i.e., to a maximum or minimum value). Actually, there are two possible values of $2\tilde{\alpha}$ which are $\pi$ radians apart that will satisfy the equation above. Thus,

$$2\tilde{\alpha} = \beta \qquad \text{where } \beta = \tan^{-1} \frac{2I_{xy}}{I_{yy} - I_{xx}}$$

or

$$2\tilde{\alpha} = \beta + \pi$$

This means that we have two values of $\tilde{\alpha}$, given as

$$\tilde{\alpha}_1 = \frac{\beta}{2}, \qquad \tilde{\alpha}_2 = \frac{\beta}{2} + \frac{\pi}{2}$$

Thus, there are two axes orthogonal to each other having extreme values for the second moment of area at $a$. On one of these axes is the maximum second moment of area and, as pointed out earlier, the minimum second moment of area must appear on the other axis. These axes are called the *principal axes*.

Let us now substitute the angle $\tilde{\alpha}$ into Eq. 8.28 for $I_{x'y'}$:

$$I_{x'y'} = \frac{I_{xx} - I_{yy}}{2} \sin\left(\tan^{-1} \frac{2I_{xy}}{I_{yy} - I_{xx}}\right) + I_{xy} \cos\left(\tan^{-1} \frac{2I_{xy}}{I_{yy} - I_{xx}}\right)$$

This becomes

$$I_{x'y'} = -(I_{yy} - I_{xx}) \frac{I_{xy}}{[(I_{yy} - I_{xx})^2 + 4I_{xy}^2]^{1/2}} + I_{xy} \frac{I_{yy} - I_{xx}}{[(I_{yy} - I_{xx})^2 + 4I_{xy}^2]^{1/2}}$$

Hence,

$$I_{x'y'} = 0 \tag{8.31}$$

Thus, we see that the *product of area corresponding to the principal axes is zero.* If we set $I_{x'y'}$ equal to zero in Eq. 8.28, you can demonstrate the converse of the preceding statement by solving for $\alpha$ and comparing the result with Eq. 8.30. That is, if the product of area is zero for a set of axes at a point, these axes must be principal axes. Consequently, if one axis of a set of axes is symmetrical for the area, the axes are principal axes.

The concept of principal axes will appear again in the following chapter in connection with the inertia tensor. Thus, the concept is not an isolated occurrence but is characteristic of a whole family of quantities. We shall, then, have further occasion to examine some of the topics introduced in this chapter from a more general viewpoint.

**EXAMPLE 8.10**

Find the principal second moments of area at the centroid of the Z section of Example 8.9.

We have from this example the following results that will be of use to us:

$$I_{x_cx_c} = 113.2 \text{ in.}^4$$
$$I_{y_cy_c} = 32.67 \text{ in.}^4$$
$$I_{x_cy_c} = -42.0 \text{ in.}^4$$

Hence, we have

$$\tan 2\tilde{\alpha} = \frac{2I_{x_cy_c}}{I_{y_cy_c} - I_{x_cx_c}} = \frac{(2)(-42.0)}{32.67 - 113.2} = 1.043$$
$$2\tilde{\alpha} = 46.21°; \quad 226.2°$$

For $2\tilde{\alpha} = 46.21°$:

$$I_1 = \frac{113.2 + 32.67}{2} + \frac{113.2 - 32.67}{2} \cos(46.21°) - (-42) \sin 46.21°$$
$$= 72.9 + 27.9 + 30.3 = 131.1 \text{ in.}^4$$

For $2\tilde{\alpha} = 226.2°$:

$$I_2 = 72.9 - 27.9 - 30.3 = 14.75 \text{ in.}^4$$

As a check on our work, we note that the sum of the second moments of area are invariant at a point for a rotation of axes. This means that

$$I_{x_cx_c} + I_{y_cy_c} = I_1 + I_2$$
$$113.2 + 32.7 = 131.1 + 14.75$$

Therefore,

$$145.9 = 145.9$$

We thus have a check on our work.

Before closing, we wish to point out that there is a graphical construction called *Mohr's circle* relating second moments and products of area for all possible axes at a point. However, in this text we shall use the analytical relations thus far presented rather than Mohr's circle. You will see Mohr circle construction in your strength of materials course where its use in conjunction with plane stress and plane strain is very helpful.[7]

## Problems

**8.73.** It is known that area $A$ is 10 ft$^2$ and has the following moments and products of area for the centroidal axes shown:
$I_{xx} = 40$ ft$^4$, $I_{yy} = 20$ ft$^4$, $I_{xy} = -4$ ft$^4$
Find the moments and products of area for the $x'y'$ reference at point (a).

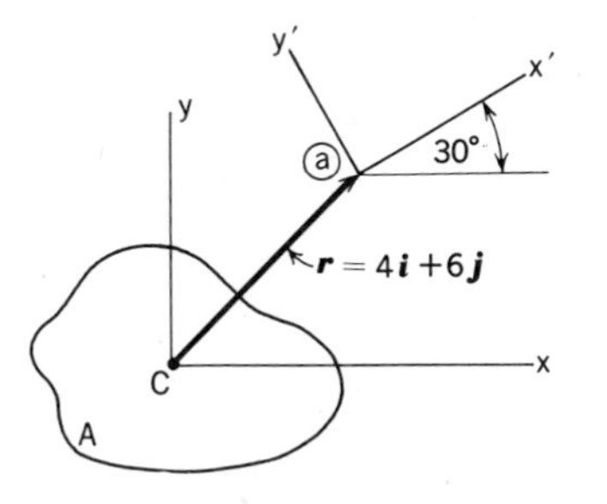


Figure P.8.73

**8.74.** The cross section of a beam is shown. Compute $I_{x'x'}$, $I_{y'y'}$, and $I_{x'y'}$ in the simplest way without using formulas for second moments and products of area for a triangle.

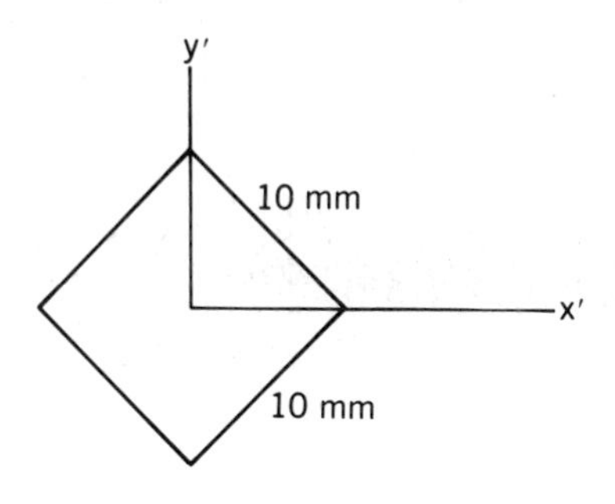


Figure P.8.74

**8.75.** Find $I_{x'x'}$, $I_{y'y'}$, and $I_{x'y'}$ for the cross section of the beam shown. The origin of $x'y'$ is at the centroid of the cross section.

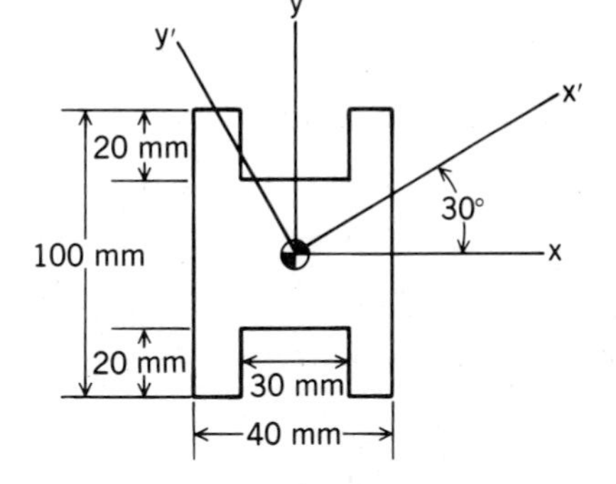


Figure P.8.75

**8.76.** Find $I_{xx}$, $I_{yy}$, and $I_{xy}$ for the rectangle. Also, compute the polar moment of area at points (a) and (b).

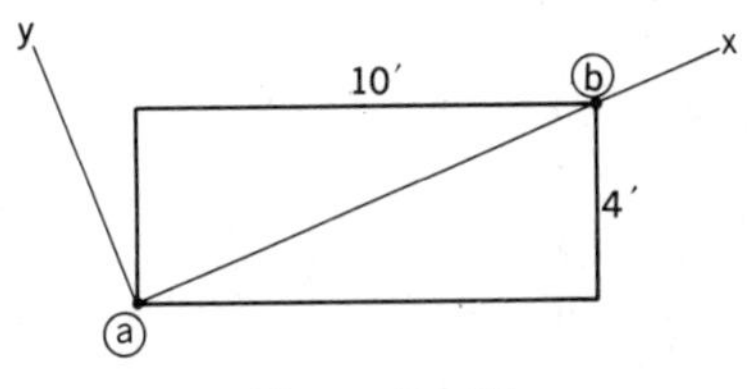

Figure P.8.76

**8.77.** Express the polar moment of area of the square as a function of $x$, $y$, the coordinates of points about which the polar moment is taken.

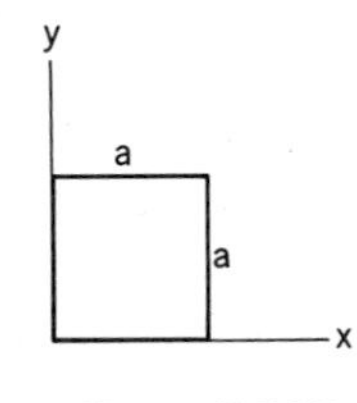

Figure P.8.77

[7]See I. H. Shames, *Introduction to Solid Mechanics*, Prentice-Hall, Inc., Englewood Cliffs, N.J., 1975.

**8.78.** Use the calculus to show that the polar moment of area of a circular area of radius $r$ is $\pi r^4/2$ at the center.

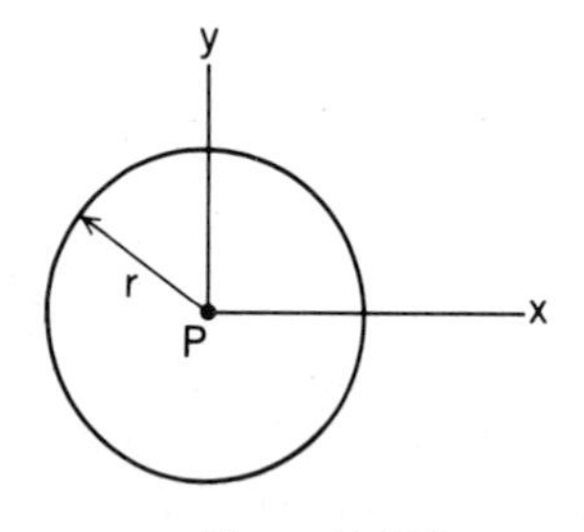


Figure P.8.78

**8.79.** Find the direction of the principal axes for the angle section about point $A$.

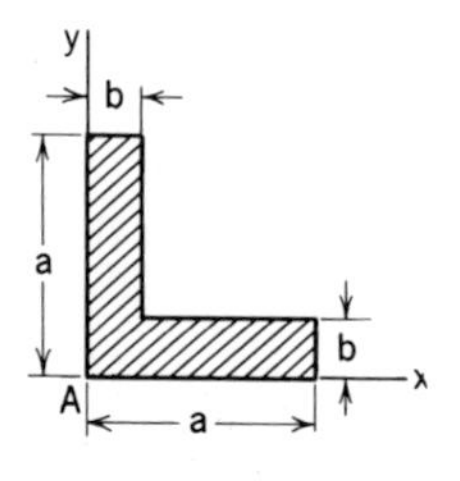


Figure P.8.79

**8.80.** What are the principal second moments of area for the area of Example 8.7?

**8.81.** Find the principal second moments of area at the centroid for the area shown.

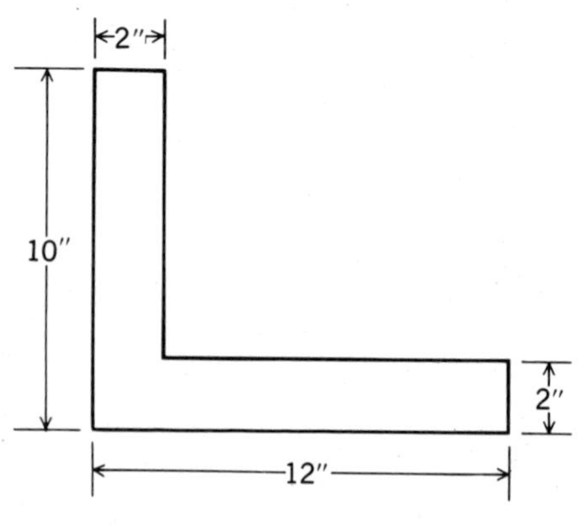


Figure P.8.81

**8.82.** Determine the principal second moments of area at point $A$.

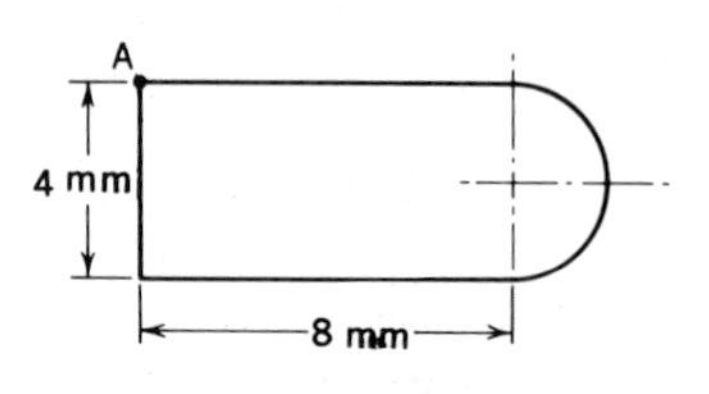


Figure P.8.82

**8.83.** A rectangular area has two holes cut out. What is the maximum second moment of area at $A$? What is it at $B$ at the center of the area?

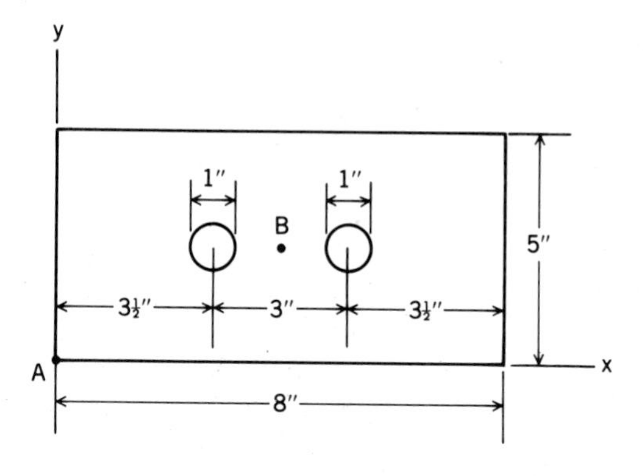

Figure P.8.83

**8.84.** Show that the axes for which the product of area is a maximum are rotated from $xy$ by an angle $\alpha$ so that

$$\tan 2\alpha = \frac{I_{xx} - I_{yy}}{2I_{xy}}$$

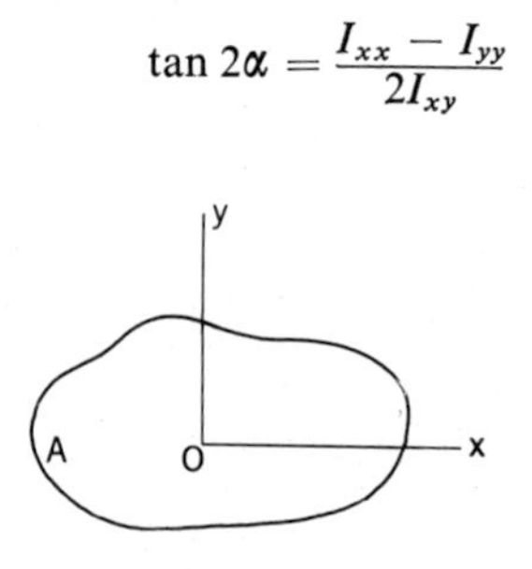


Figure P.8.84

## 8.11 Closure

In this chapter, we discussed primarily the first and second moments of plane areas as well as the product of plane areas. These formulations give certain kinds of evaluations of the distribution of area relative to a plane reference $xy$. You will most certainly make much use of these quantities in your later courses in strength of materials.

We also briefly discussed in this chapter the first moment of mass. Another set of quantities that will prove to be indispensible in rigid-body dynamics are the second moments of *mass* (or inertia) as well as products of *mass* (or inertia). They, like the first moment of mass, represent certain measures of mass distribution relative to a reference $xyz$. We shall consider such quantities in Chapter 9 and we shall see that the second moments and products of area represent a special case of second moments and products of mass (or inertia).

## Review Problems

**8.85.** Find the position of the centroid of the shaded area under the curve $y = \sin^2 x$ m. Find $M_{x'}$ and $M_{y'}$ of this area.

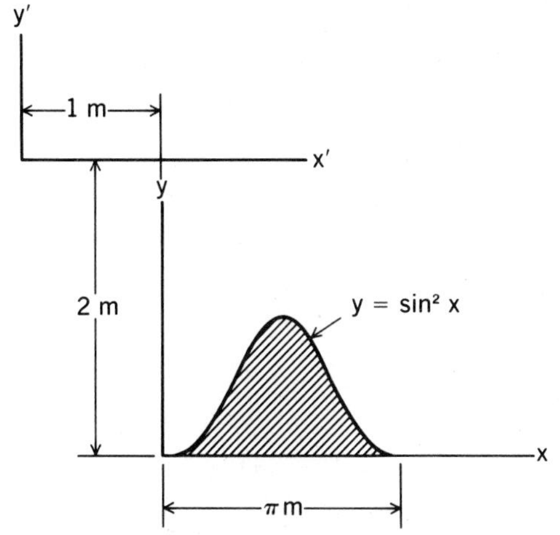

Figure P.8.85

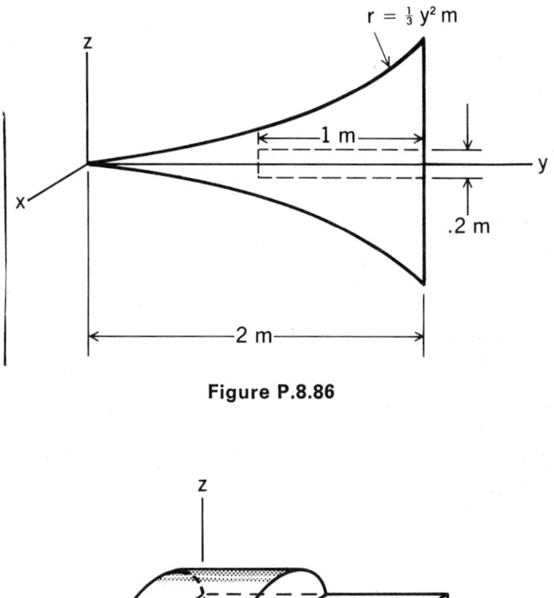

**8.86.** Find the center of volume of the body of revolution with a cylindrical cavity.



Figure P.8.86

**8.87.** Locate the center of volume, center of mass, and center of gravity of the wooden rectangular block and the plastic semicylinder. The wood weighs .0003 N/mm³ and the plastic weighs .0005 N/mm³.

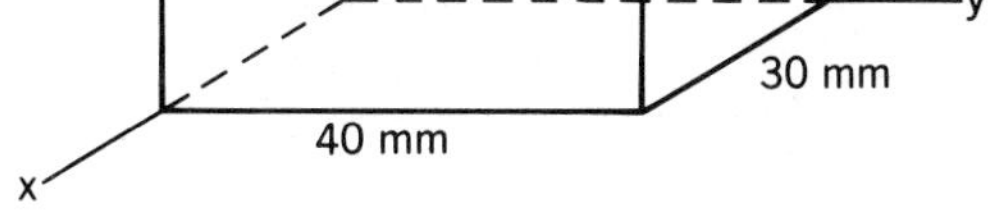

Figure P.8.87

**8.88.** Using the theorems of Pappus and Guldinus, find the centroid of the area of a quarter-circle.

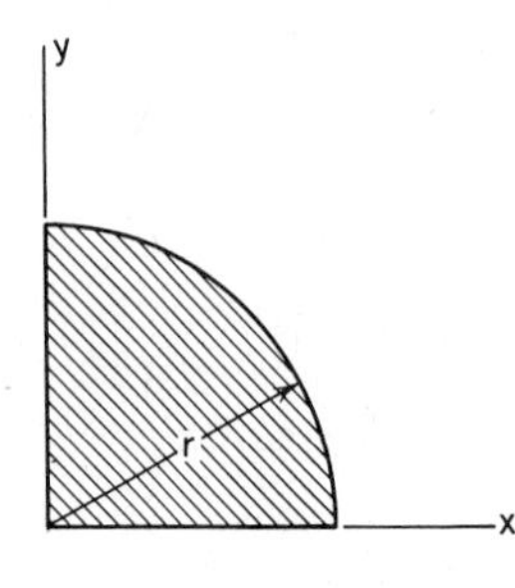


Figure P.8.88

**8.89.** A tank has a semispherical dome at the left end. Using the theorems of Pappus and Guldinus, compute the surface and volume of the tank. Give the results in meters.

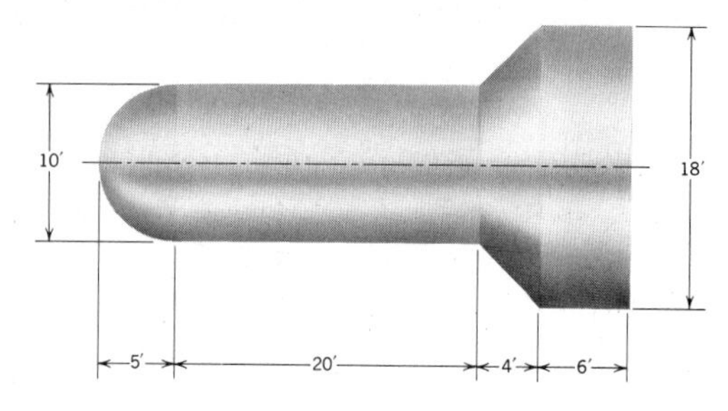

Figure P.8.89

**8.90.** Find $I_{x'x'}$, $I_{y'y'}$, and $I_{x'y'}$ at point $A$ for the rectangular area.

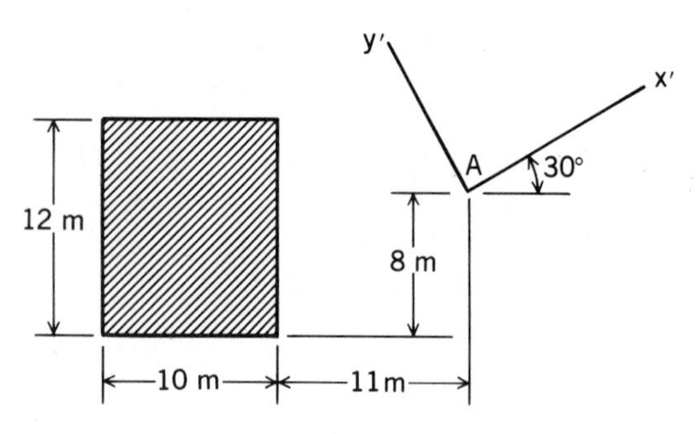


Figure P.8.90

**8.91.** Find the centroid of the area, and then find the second moments of the area about centroidal axes parallel to the sides of the area.

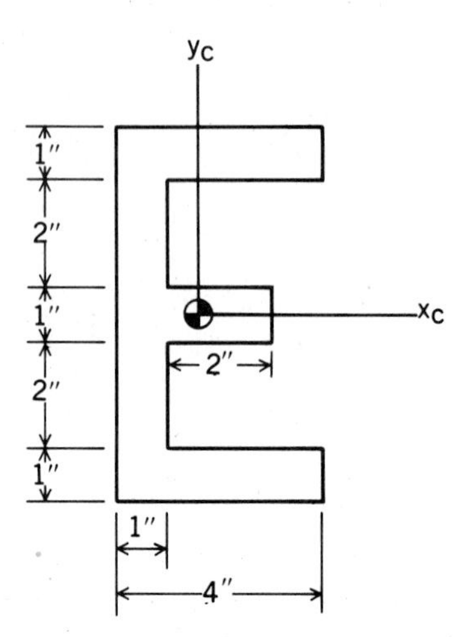


Figure P.8.91

**8.92.** Find the principal second moments of area at a point where $I_{xy} = 321$ in.$^4$, $I_{xx} = 118.4$ in.$^4$ and $I_{yy} = 1028$ in.$^4$.

**8.93.** Find the polar moment of area at $O$ for the shaded area.

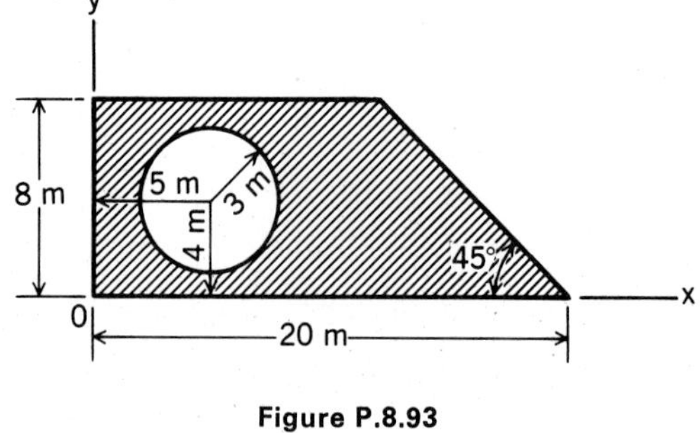


Figure P.8.93

## 9.1 Introduction

In this chapter, we shall consider certain measures of mass distribution relative to a reference. These quantities are vital for the study of the dynamics of rigid bodies. Because these quantities are so closely related to second moments and products of area, we shall consider them at this early stage rather than wait for dynamics. We shall also discuss the fact that these measures of mass distribution—the second moments of inertia of mass and the products of inertia of mass—are components of what we call a second-order tensor. Recognizing this fact early will make more simple and understandable your future studies of stress and strain, since these quantities also happen to be second-order tensors.

## 9.2 Formal Definition of Inertia Quantities

We shall now formally define a set of quantities that give information about the distribution of mass of a body relative to a Cartesian reference. For this purpose, a body of mass $M$ and a reference $xyz$ are presented in Fig. 9.1. This reference and the body may have any motion whatever relative to each other. The ensuing discussion then holds for the instantaneous orientation shown at time $t$. We shall consider that the body is composed of a continuum of particles each of which has a mass given by $\rho\, dv$, where

[1]This chapter may be covered at a later stage when studying dynamics. In that case, it should be covered directly after Chapter 15.

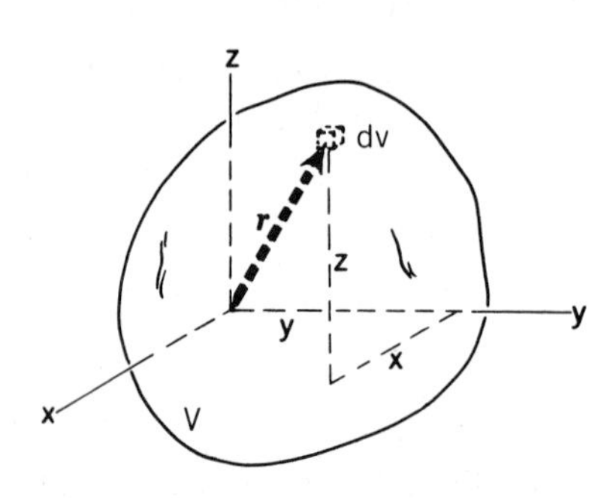


Figure 9.1. Body and reference at time $t$.

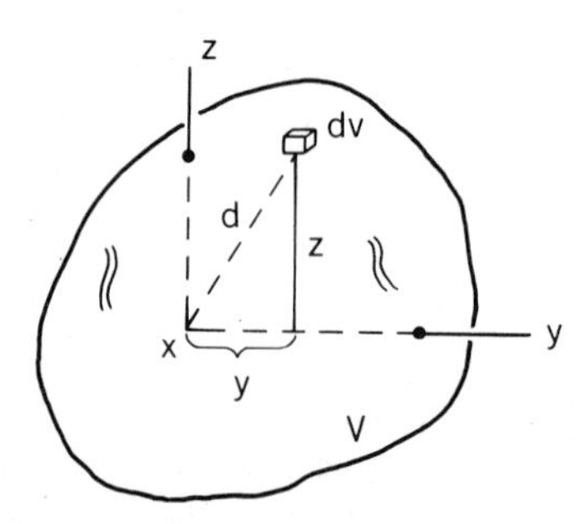


Figure 9.2. View of body along $x$ axis.

$\rho$ is the mass density and $dv$ is a volume element. We define the second moments and products of inertia of the body $M$ for the reference $xyz$ at time $t$ in the following manner[2]:

$$I_{xx} = \iiint\limits_V (y^2 + z^2)\rho \, dv \tag{9.1a}$$

$$I_{yy} = \iiint\limits_V (x^2 + z^2)\rho \, dv \tag{9.1b}$$

$$I_{zz} = \iiint\limits_V (x^2 + y^2)\rho \, dv \tag{9.1c}$$

$$I_{xy} = \iiint\limits_V xy\rho \, dv \tag{9.1d}$$

$$I_{xz} = \iiint\limits_V xz\rho \, dv \tag{9.1e}$$

$$I_{yz} = \iiint\limits_V yz\rho \, dv \tag{9.1f}$$

The terms $I_{xx}$, $I_{yy}$, and $I_{zz}$ in the set above are called the *mass moments of inertia* of the body about the $x$, $y$, and $z$ axes, respectively. Note that in each such case we are integrating the mass elements, $\rho \, dv$, times the *perpendicular distance squared* from the mass elements to the coordinate axis about which we are computing the moment of inertia. Thus, if we look along the $x$ axis toward the origin in Fig. 9.1, we would have the view shown in Fig. 9.2. The quantity $y^2 + z^2$ used in Eq. 9.1a for $I_{xx}$ is clearly $d^2$, the perpendicular distance squared from $dv$ to the $x$ axis (now seen as a dot). Each of the terms with mixed indices is called the *mass product of inertia* about the pair of axes given by the indices. Clearly, from the definition of the product of inertia, we

[2]We use the same notation as was used for second moments and products of area, which are also sometimes called moments and products of inertia. This is standard practice in mechanics. There need be no confusion in using these quantities if we keep the context of discussions clearly in mind.

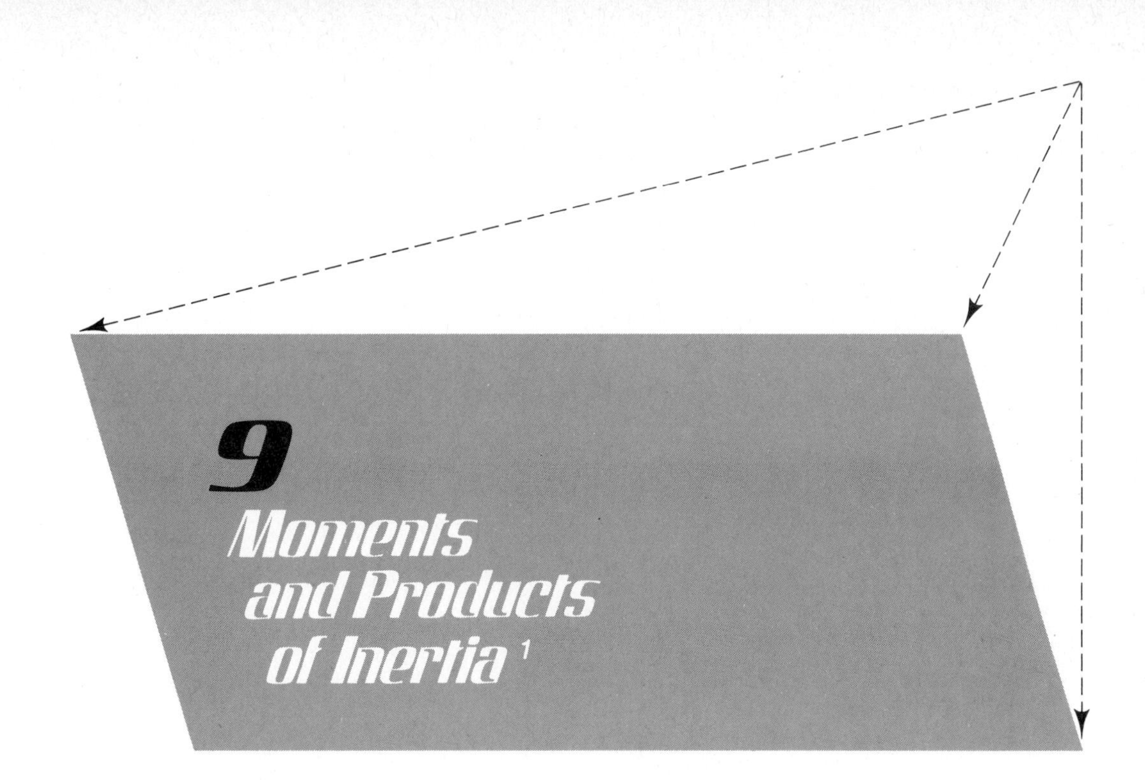


## 9.1 Introduction

In this chapter, we shall consider certain measures of mass distribution relative to a reference. These quantities are vital for the study of the dynamics of rigid bodies. Because these quantities are so closely related to second moments and products of area, we shall consider them at this early stage rather than wait for dynamics. We shall also discuss the fact that these measures of mass distribution—the second moments of inertia of mass and the products of inertia of mass—are components of what we call a second-order tensor. Recognizing this fact early will make more simple and understandable your future studies of stress and strain, since these quantities also happen to be second-order tensors.

## 9.2 Formal Definition of Inertia Quantities

We shall now formally define a set of quantities that give information about the distribution of mass of a body relative to a Cartesian reference. For this purpose, a body of mass $M$ and a reference $xyz$ are presented in Fig. 9.1. This reference and the body may have any motion whatever relative to each other. The ensuing discussion then holds for the instantaneous orientation shown at time $t$. We shall consider that the body is composed of a continuum of particles each of which has a mass given by $\rho\, dv$, where

[1]This chapter may be covered at a later stage when studying dynamics. In that case, it should be covered directly after Chapter 15.

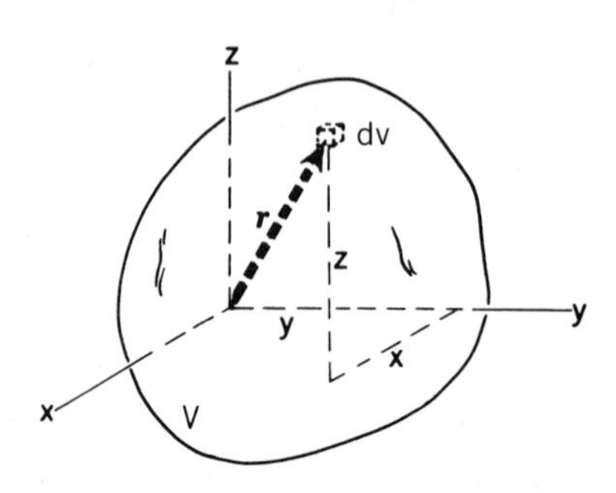


Figure 9.1. Body and reference at time $t$.

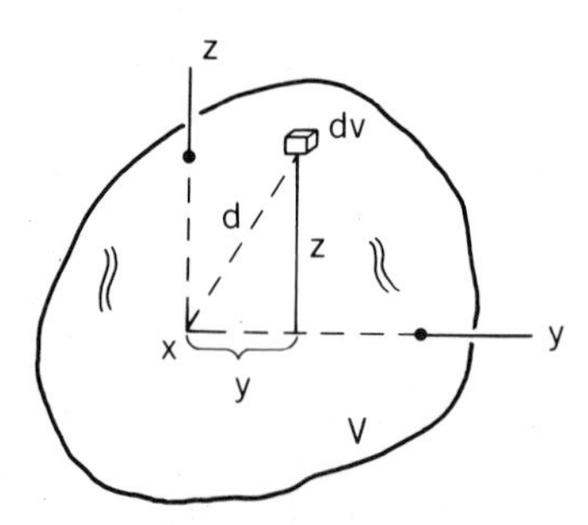


Figure 9.2. View of body along $x$ axis.

$\rho$ is the mass density and $dv$ is a volume element. We define the second moments and products of inertia of the body $M$ for the reference $xyz$ at time $t$ in the following manner[2]:

$$I_{xx} = \iiint_V (y^2 + z^2)\rho \, dv \tag{9.1a}$$

$$I_{yy} = \iiint_V (x^2 + z^2)\rho \, dv \tag{9.1b}$$

$$I_{zz} = \iiint_V (x^2 + y^2)\rho \, dv \tag{9.1c}$$

$$I_{xy} = \iiint_V xy\rho \, dv \tag{9.1d}$$

$$I_{xz} = \iiint_V xz\rho \, dv \tag{9.1e}$$

$$I_{yz} = \iiint_V yz\rho \, dv \tag{9.1f}$$

The terms $I_{xx}$, $I_{yy}$, and $I_{zz}$ in the set above are called the *mass moments of inertia* of the body about the $x$, $y$, and $z$ axes, respectively. Note that in each such case we are integrating the mass elements, $\rho \, dv$, times the *perpendicular distance squared* from the mass elements to the coordinate axis about which we are computing the moment of inertia. Thus, if we look along the $x$ axis toward the origin in Fig. 9.1, we would have the view shown in Fig. 9.2. The quantity $y^2 + z^2$ used in Eq. 9.1a for $I_{xx}$ is clearly $d^2$, the perpendicular distance squared from $dv$ to the $x$ axis (now seen as a dot). Each of the terms with mixed indices is called the *mass product of inertia* about the pair of axes given by the indices. Clearly, from the definition of the product of inertia, we

[2]We use the same notation as was used for second moments and products of area, which are also sometimes called moments and products of inertia. This is standard practice in mechanics. There need be no confusion in using these quantities if we keep the context of discussions clearly in mind.

could reverse indices and thereby form three additional products of inertia for a reference. The additional three quantities formed in this way, however, are equal to the corresponding quantities of the original set. That is,

$$I_{xy} = I_{yx}, \qquad I_{xz} = I_{zx}, \qquad I_{yz} = I_{zy}$$

We now have nine inertia terms at a point for a given reference at this point. The values of the set of six independent quantities will, for a given body, depend on the *position* and *inclination* of the reference relative to the body. You should also understand that the reference may be established anywhere in space and *need not* be situated in the rigid body of interest. Thus there will be nine inertia terms for reference $xyz$ at point $O$ outside the body (Fig. 9.3) computed using Eqs. 9.1, where the domain of integration is the volume $V$ of the body. As will be explained later, the nine moments and products of inertia are components of the inertia tensor.

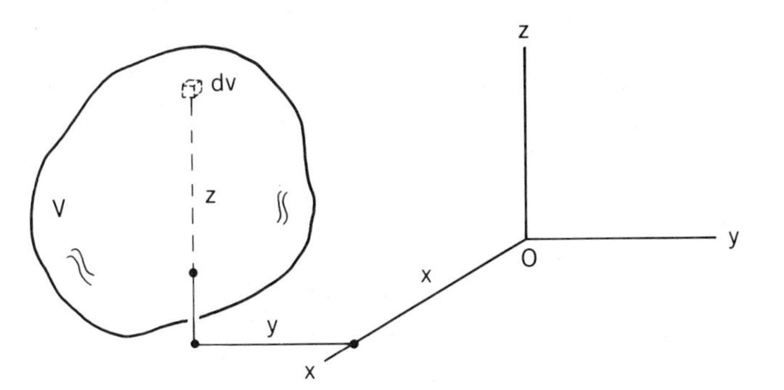


**Figure 9.3.** Origin of *xyz* outside body.

It will be convenient, when referring to the nine moments and products of inertia for reference $xyz$ at a point, to list them in a matrix array, as follows:

$$I_{ij} = \begin{pmatrix} I_{xx} & I_{xy} & I_{xz} \\ I_{yx} & I_{yy} & I_{yz} \\ I_{zx} & I_{zy} & I_{zz} \end{pmatrix}$$

Notice that the first subscript gives the row and the second subscript gives the column in the array. Furthermore, the left-to-right downward diagonal in the array is composed of mass moment of inertia terms while the products of inertia, oriented at mirror-image positions about this diagonal, are equal. For this reason we say that the array is *symmetric*.

We shall now show that the sum of the mass moments of inertia for a set of orthogonal axes is independent of the orientation of the axes and depends only on the position of the origin. Examine the sum of such a set of terms:

$$I_{xx} + I_{yy} + I_{zz} = \iiint_V (y^2 + z^2)\rho \, dv + \iiint_V (x^2 + z^2)\rho \, dv + \iiint_V (x^2 + y^2)\rho \, dv$$

Combining the integrals and rearranging, we get

$$I_{xx} + I_{yy} + I_{zz} = \iiint_V 2(x^2 + y^2 + z^2)\rho \, dv = \iiint_V 2|\mathbf{r}|^2 \rho \, dv \tag{9.2}$$

But the magnitude of the position vector from the origin to a particle is *independent* of the inclination of the reference at the origin. Thus, *the sum of the moments of inertia at a point in space for a given body clearly is an invariant with respect to rotation of axes.*

Clearly, on inspection of the definitions 9.1, the moments of inertia must always exceed zero, while the products of inertia may have any value. Of interest is the case where one of the coordinate planes is a *plane of symmetry* for the mass distribution of the body. Such a plane is the $zy$ plane shown in Fig. 9.4 cutting a body into two parts, which, by definition of symmetry, are mirror images of each other. For the computation of $I_{xz}$, each half will give a contribution of the same magnitude but of opposite sign. We can most readily see that this is so by looking along the $y$ axis toward the origin. The plane of symmetry then appears as a line coinciding with the $z$ axis (see Fig. 9.5). We can consider the body to be composed of pairs of mass elements $dm$ which are mirror images of each other with respect to position and shape about the plane of symmetry. The product of inertia $I_{xz}$ for such a pair is then

$$xz \, dm - xz \, dm = 0$$

Thus, we can conclude that

$$I_{xz} = \underbrace{\int xz \, dm}_{\substack{\text{right}\\ \text{domain}}} - \underbrace{\int xz \, dm}_{\substack{\text{left}\\ \text{domain}}} = 0$$

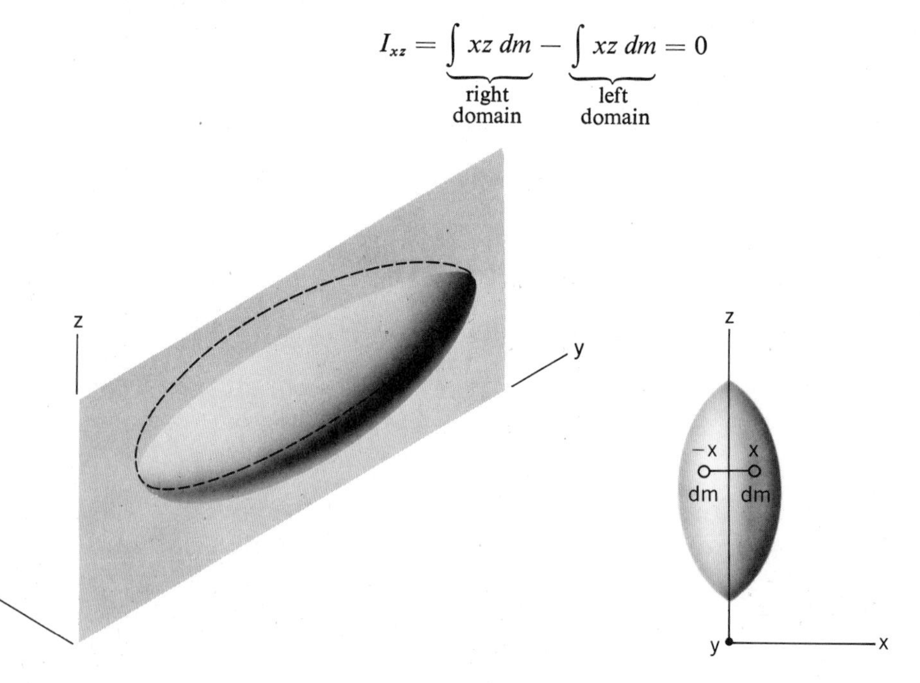


**Figure 9.4.** *zy* is plane of symmetry.

**Figure 9.5.** View along *y* axis.

This conclusion is also true for $I_{xy}$. We can say that $I_{xy} = I_{xz} = 0$. But on consulting Fig. 9.4, you should be able to readily decide that the term $I_{zy}$ will have a positive value. Note that those products of inertia having $x$ as an index are zero and that the $x$ coordinate axis is normal to the plane of symmetry. Thus, we can conclude that *if two axes form a plane of symmetry for the mass distribution of a body, the products of inertia having as an index the coordinate that is normal to the plane of symmetry will be zero.*

Consider next a body of *revolution*. Take the $z$ axis to coincide with the axis of symmetry. It is easy to conclude for the origin $O$ of $xyz$ anywhere along the axis of symmetry that

$$I_{xz} = I_{yz} = 0$$
$$I_{xx} = I_{yy} = \text{constant}$$

for all possible $xy$ axes formed by rotating about the $z$ axis at $O$. Can you justify these conclusions?

Finally, we define *radii of gyration* in a manner analogous to that used for second moments of area in Chapter 8. Thus:

$$I_{xx} = k_x^2 M$$
$$I_{yy} = k_y^2 M$$
$$I_{zz} = k_z^2 M$$

where $k_x$, $k_y$, and $k_z$ are the radii of gyration and $M$ is the total mass.

**EXAMPLE 9.1**

Find the nine components of the inertia tensor of a rectangular body of uniform density $\rho$ about point $O$ for a reference $xyz$ coincident with the edges of the block as shown in Fig. 9.6.

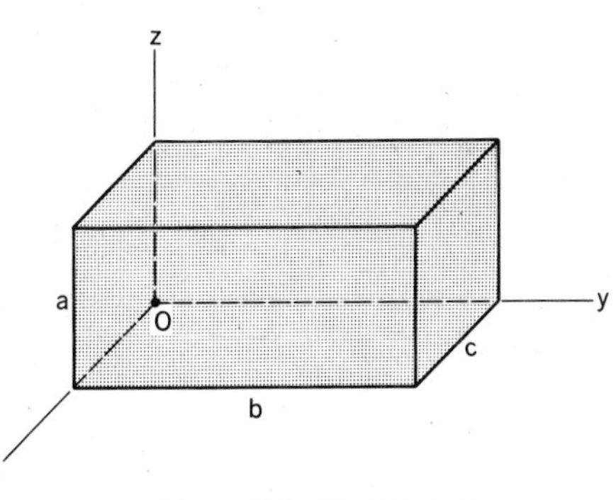


**Figure 9.6.** Find $I_{ij}$ at $O$.

We first compute $I_{xx}$. Using volume elements $dv = dx\,dy\,dz$, we get on using simple multiple integration:

$$
\begin{aligned}
I_{xx} &= \int_0^a \int_0^b \int_0^c (y^2 + z^2)\rho \, dx \, dy \, dz \\
&= \int_0^a \int_0^b (y^2 + z^2)c\rho \, dy \, dz = \int_0^a \left(\frac{b^3}{3} + z^2 b\right)c\rho \, dz \\
&= \left(\frac{ab^3c}{3} + \frac{a^3bc}{3}\right)\rho = \frac{\rho V}{3}(b^2 + a^2)
\end{aligned} \tag{a}
$$

where $V$ is the volume of the body. Permuting the terms, we can get $I_{yy}$ and $I_{zz}$ by inspection as follows:

$$I_{yy} = \frac{\rho V}{3}(c^2 + a^2) \tag{b}$$

$$I_{zz} = \frac{\rho V}{3}(b^2 + c^2) \tag{c}$$

We next compute $I_{xy}$.

$$
\begin{aligned}
I_{xy} &= \int_0^a \int_0^b \int_0^c xy\rho \, dx \, dy \, dz = \int_0^a \int_0^b \frac{c^2}{2} y\rho \, dy \, dz \\
&= \int_0^a \frac{c^2b^2}{4}\rho \, dz = \frac{ac^2b^2}{4}\rho = \frac{\rho V}{4}cb
\end{aligned} \tag{d}
$$

Permuting the terms, we get

$$I_{xz} = \frac{\rho V}{4}ac \tag{e}$$

$$I_{yz} = \frac{\rho V}{4}ab \tag{f}$$

We accordingly have, for the inertia tensor:

$$
I_{ij} = \begin{pmatrix}
\frac{\rho V}{3}(b^2 + a^2) & \frac{\rho V}{4}cb & \frac{\rho V}{4}ac \\
\frac{\rho V}{4}cb & \frac{\rho V}{3}(c^2 + a^2) & \frac{\rho V}{4}ab \\
\frac{\rho V}{4}ac & \frac{\rho V}{4}ab & \frac{\rho V}{3}(b^2 + c^2)
\end{pmatrix} \tag{g}
$$

**EXAMPLE 9.2**

Compute the components of the inertia tensor at the center of a solid sphere of uniform density $\rho$ as shown in Fig. 9.7.

We shall first compute $I_{yy}$. Using spherical coordinates, we have[3]

$$I_{yy} = \iiint_V (x^2 + z^2)\rho \, dv$$

---

[3]For those unfamiliar with spherical coordinates, we have shown in Fig.9.8 a more detailed study of the volume element used. The volume $dv$ is simply the product of the three edges of the element shown in the diagram.

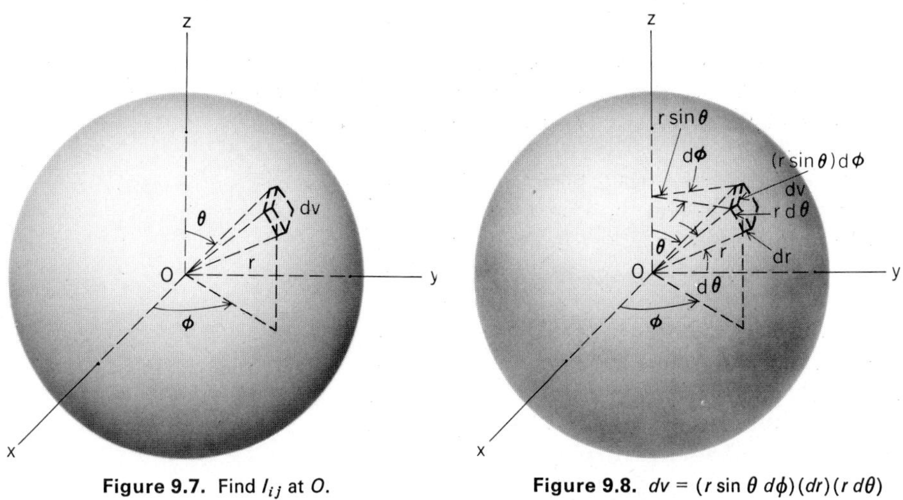


**Figure 9.7.** Find $I_{ij}$ at $O$.

**Figure 9.8.** $dv = (r \sin\theta\, d\phi)(dr)(r\, d\theta) = r^2 \sin\theta\, d\theta\, d\phi\, dr$.

$$= \int_0^R \int_0^{2\pi} \int_0^{\pi} [(r \sin\theta \cos\phi)^2 + (r\cos\theta)^2]\rho(r^2 \sin\theta\, d\theta\, d\phi\, dr)$$

$$= \int_0^R \int_0^{2\pi} \int_0^{\pi} (r^4 \sin^3\theta \cos^2\phi)\rho\, d\theta\, d\phi\, dr + \int_0^R \int_0^{2\pi} \int_0^{\pi} (r^4 \cos^2\theta \sin\theta)\rho\, d\theta\, d\phi\, dr$$

$$= \rho \int_0^R \int_0^{2\pi} (r^4 \cos^2\phi)\left(\int_0^{\pi} \sin^3\theta\, d\theta\right) d\phi\, dr$$

$$+ \rho \int_0^R \int_0^{2\pi} r^4 \left(\int_0^{\pi} \cos^2\theta \sin\theta\, d\theta\right) d\phi\, dr$$

With the aid of integration formulas from Appendix I, we have

$$I_{yy} = \rho \int_0^R \int_0^{2\pi} r^4 \cos^2\phi \left[-\tfrac{1}{3}\cos\theta\,(\sin^2\theta + 2)\right]\Big|_0^{\pi} d\phi\, dr$$

$$+ \rho \int_0^R \int_0^{2\pi} r^4 \left(-\frac{\cos^3\theta}{3}\right)\Big|_0^{\pi} d\phi\, dr$$

$$= \rho \int_0^R \int_0^{2\pi} r^4 \cos^2\phi\, \tfrac{4}{3}\, d\phi\, dr + \rho \int_0^R \int_0^{2\pi} (r^4)(\tfrac{2}{3})\, d\phi\, dr$$

Integrating next with respect to $\phi$, we get

$$I_{yy} = \rho \int_0^R (r^4)(\tfrac{4}{3})(\pi)\, dr + \rho \int_0^R r^4 (\tfrac{2}{3})(2\pi)\, dr$$

Finally, we get

$$I_{yy} = \rho \frac{R^5}{5}\frac{4}{3}\pi + \rho \frac{R^5}{5}\frac{4}{3}\pi$$

$$= \frac{8}{15}\rho\pi R^5$$

But

$$M = \rho\, \tfrac{4}{3}\pi R^3$$

Hence,

$$I_{yy} = \tfrac{2}{5} MR^2$$

Because of the point symmetry about point $O$, we can also say that

$$I_{xx} = I_{zz} = \tfrac{2}{5} MR^2$$

Because the coordinate planes are all planes of symmetry for the mass distribution, the products of inertia are zero. Thus, the inertia tensor can be given as

$$I_{ij} = \begin{pmatrix} \frac{2}{5}MR^2 & 0 & 0 \\ 0 & \frac{2}{5}MR^2 & 0 \\ 0 & 0 & \frac{2}{5}MR^2 \end{pmatrix}$$

## 9.3 Relation Between Mass-Inertia Terms and Area-Inertia Terms

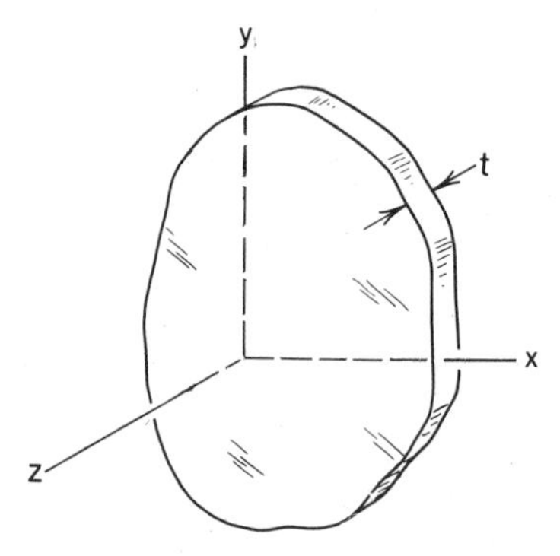


**Figure 9.9.** Plate of thickness $t$.

We now relate the second moment and product of area studied in Chapter 8 with the inertia tensor. To do this, consider a plate of constant thickness $t$ and uniform density $\rho$ (Fig. 9.9). A reference is picked so that the $xy$ plane is in the midplane of this plate. The components of the inertia tensor are rewritten for convenience as

$$\begin{aligned} I_{xx} &= \rho \iiint_V (y^2 + z^2)\, dv, & I_{xy} &= \rho \iiint_V xy\, dv \\ I_{yy} &= \rho \iiint_V (x^2 + z^2)\, dv, & I_{xz} &= \rho \iiint_V xz\, dv \\ I_{zz} &= \rho \iiint_V (x^2 + y^2)\, dv, & I_{yz} &= \rho \iiint_V yz\, dv \end{aligned} \tag{9.3}$$

Now consider that the thickness $t$ is *small* compared to the lateral dimensions of the plate. This means that $z$ is restricted to a range of values having a small magnitude. As a result, we can make two simplifications in the equations above. First, we shall set $z$ equal to zero whenever it appears on the right side of the equations above. Second, we shall express $dv$ as

$$dv = t\, dA$$

where $dA$ is an area element on the *surface* of the plate, as shown in Fig. 9.10. Equations 9.3 then become

$$I_{xx} = \rho t \iint_A y^2\, dA, \qquad I_{xy} = \rho t \iint_A xy\, dA$$

$$I_{yy} = \rho t \iint_A x^2\, dA, \qquad I_{xz} = 0$$

$$I_{zz} = \rho t \iint_A (x^2 + y^2)\, dA, \qquad I_{yz} = 0$$

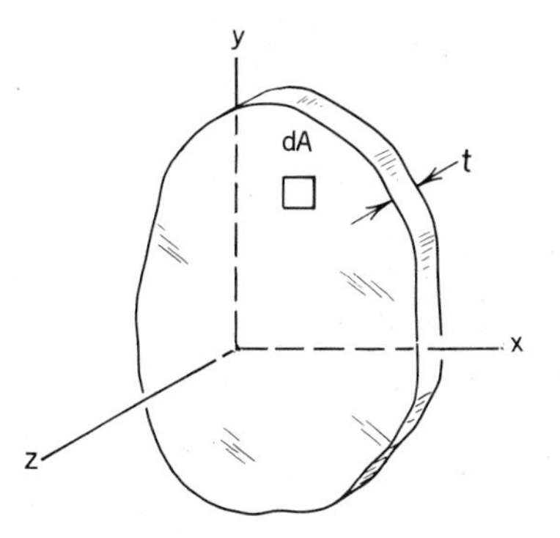

Figure 9.10. Use volume elements $t\, dA$.

Notice, now, that the integrals on the right sides of the equations above are moments and products of *area* presented in Chapter 8. Denoting mass-inertia terms with a subscript $M$ and area-inertia terms with a subscript $A$, we can then say for the nonzero expressions:

$$(I_{xx})_M = \rho t (I_{xx})_A$$
$$(I_{yy})_M = \rho t (I_{yy})_A$$
$$(I_{zz})_M = \rho t (J)_A$$
$$(I_{xy})_M = \rho t (I_{xy})_A$$

Thus, for a thin plate with a constant product $\rho t$ throughout, we can compute the inertia tensor components for reference $xyz$ (see Fig. 9.9) by using the moments and product of area of the surface of the plate relative to axes $xy$. It should be pointed out, in this regard, that $\rho t$ is the *mass per unit area* of the plate.[4] We now illustrate this procedure in the following example.

**EXAMPLE 9.3**

Determine the inertia tensor components for the thin plate (Fig. 9.11) relative to the indicated axes $xyz$. The weight of the plate is .002 N/mm². For the top edge, $y = 2\sqrt{x}$ with $x$ and $y$ in millimeters.

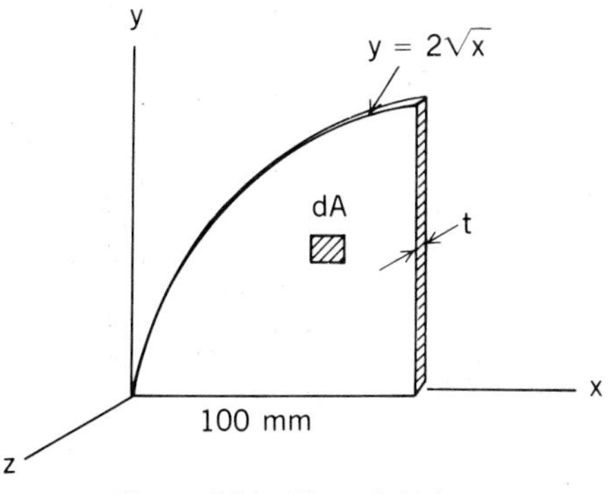

Figure 9.11. Plate of thickness $t$.

[4]If we let $t \longrightarrow 0$ and $\rho \longrightarrow \infty$ such that $\rho t \longrightarrow 1$, we see that the nonzero mass-inertia terms degenerate to the moments and product of area.

It is clear that for $pt$ we have

$$pt = \frac{.002}{9.81} = .000204 \text{ kg/mm}^2 \tag{a}$$

We now examine the moments and products of area for the surface of the plate about axes $xy$. Thus,[5]

$$\begin{aligned}
(I_{xx})_A &= \int_0^{100} \int_{y=0}^{y=2\sqrt{x}} y^2 \, dy \, dx \\
&= \int_0^{100} \frac{y^3}{3}\bigg|_0^{2\sqrt{x}} dx = \int_0^{100} \frac{8}{3} x^{3/2} \, dx \\
&= \frac{8}{3} \frac{x^{5/2}}{\frac{5}{2}}\bigg|_0^{100} = (\tfrac{8}{3})(\tfrac{2}{5})(100^{5/2}) \\
&= 1.067 \times 10^5 \text{ mm}^4
\end{aligned}$$

$$\begin{aligned}
(I_{yy})_A &= \int_0^{100} \int_{y=0}^{y=2\sqrt{x}} x^2 dy \, dx \\
&= \int_0^{100} x^2 y\bigg|_0^{2\sqrt{x}} dx = \int_0^{100} x^2(2\sqrt{x}) \, dx \\
&= 2\frac{x^{7/2}}{\frac{7}{2}}\bigg|_0^{100} = 2(\tfrac{2}{7})(100)^{7/2} \\
&= 5.71 \times 10^6 \text{ mm}^4
\end{aligned}$$

$$\begin{aligned}
(I_{xy})_A &= \int_0^{100} \int_{y=0}^{y=2\sqrt{x}} xy \, dy \, dx \\
&= \int_0^{100} x\frac{y^2}{2}\bigg|_0^{2\sqrt{x}} dx = \int_0^{100} 2x^2 \, dx \\
&= 2\left(\frac{100^3}{3}\right) = 6.67 \times 10^5 \text{ mm}^4
\end{aligned}$$

Using Eq. (a), we can then say for the nonzero inertia tensor components:

$$\begin{aligned}
(I_{xx})_M &= (.0204)(1.067 \times 10^5) = 2176 \text{ kg-mm}^2 \\
(I_{yy})_M &= (.0204)(5.71 \times 10^6) = 116{,}500 \text{ kg-mm}^2 \\
(I_{xy})_M &= (.0204)(6.67 \times 10^5) = 13{,}610 \text{ kg-mm}^2
\end{aligned}$$

Note that the nonzero inertia tensor components for a reference $xyz$ on a plate (see Fig. 9.9) are *proportional* through $pt$ to the corresponding area-inertia terms for the plate surface. This means that all the formulations of Chapter 8 apply to the aforementioned nonzero inertia tensor components. Thus, on rotating the axes about the $z$ axis we may use the transformation equations of Chapter 8. Consequently, the concept of *principal axes* in the midplane of the plate at a point applies. For such axes,

---

[5]Note we have multiple integration where one of the boundaries is variable. The procedure to follow should be evident from the example.

the product of inertia is zero. One such axis then gives the maximum moment of inertia for all axes in the midplane at the point, the other the minimum moment of inertia. We have presented such problems at the end of the section.

What about principal axes for the inertia tensor at a point in a general three-dimensional body? Those students who have time to study Section 9.6 will learn that there are *three principal axes* at a point in the general case. These axes are *mutually orthogonal* and the *products of inertia are all zero* for such a set of axes at a point.[6] Furthermore, one of the axes will have a maximum moment of inertia, another axis will have a minimum moment of inertia, while the third axis will have an intermediate value. The sum of these three inertia terms must have the value that is common for all sets of axes at the point.

If, perchance, a set of axes $xyz$ at a point is such that $xy$ and $xz$ form *two planes of symmetry* for the mass distribution of the body, then, as we learned earlier, since the $z$ axis and the $y$ axis are normal to planes of symmetry, $I_{xy} = I_{xz} = I_{yz} = 0$. Thus, all products of inertia are zero. This would also be true for *any* two sets of axes of $xyz$ forming two planes of symmetry. Clearly, axes forming two planes of symmetry must be *principal axes.* This information will suffice most instances when we have to identify principal axes. On the other hand, consider the case where there is only *one plane of symmetry* for the mass distribution of a body at some point $A$. Let the $xy$ plane at $A$ form this plane of symmetry. Then, clearly, the products of inertia between the $z$ axis that is normal to the plane of symmetry $xy$ and *any axis* in the $xy$ plane at $A$ must be zero, as pointed out earlier. Obviously, the $z$ axis must be a principal axis. The other two principal axes must be in the plane of symmetry, but generally cannot be located by inspection.

## Problems

**9.1.** A uniform homogeneous slender rod of mass $M$ is shown. Compute $I_{xx}$ and $I_{x'x'}$.

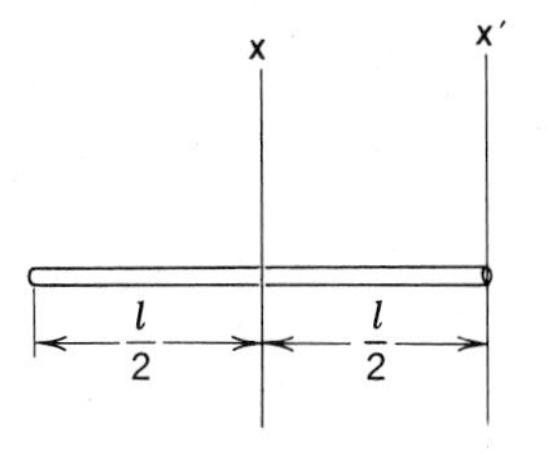


Figure P.9.1

**9.2.** Find $I_{xx}$ and $I_{x'x'}$ for the thin rod of Problem 9.1 for the case where the mass per unit length at the left end is 5 lbm/ft and increases linearly so that at the right end it is 8 lbm/ft. The rod is 20 ft in length.

**9.3.** Compute $I_{xy}$ for the thin homogeneous hoop of mass $M$.

---

[6]The third principal axis for the plate at a point in the midplane is the $z$ axis normal to the plate. Note that $(I_{zz})_M$ must always equal $(I_{xx})_M + (I_{yy})_M$. Why?

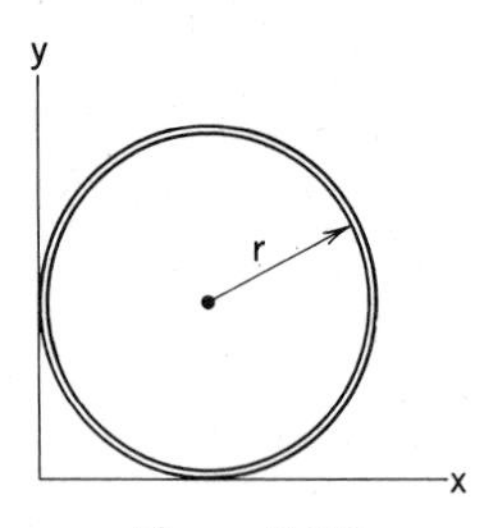


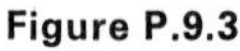
Figure P.9.3

**9.4.** Compute $I_{xx}$, $I_{yy}$, $I_{zz}$, and $I_{xy}$ for the homogeneous rectangular parallelepiped.

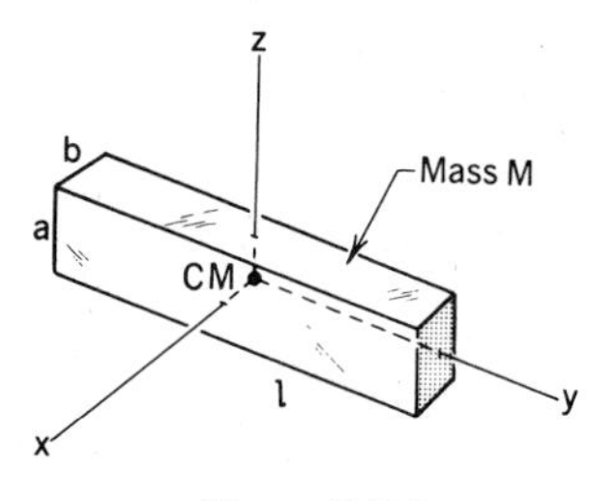


Figure P.9.4

**9.5.** A wire having the shape of a parabola is shown. The curve is in the $yz$ plane. If the mass of the wire is .3 N/m, what are $I_{yy}$ and $I_{xz}$? [*Hint:* Replace $ds$ along the wire by $\sqrt{(dy/dz)^2 + 1}\,dz$.]

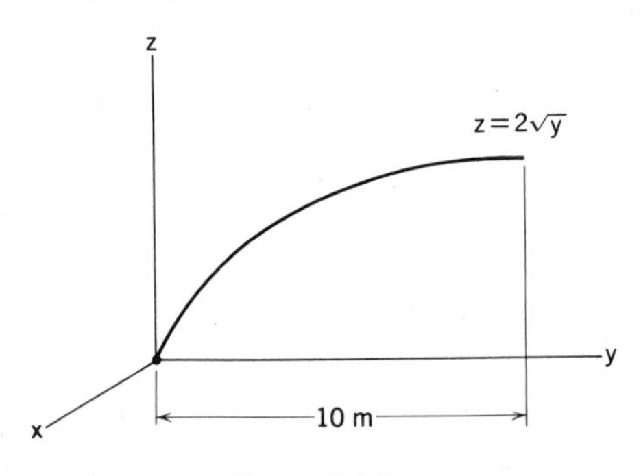


Figure P.9.5

**9.6.** Compute the moment of inertia, $I_{BB}$, for the half-cylinder shown. The body is homogeneous and has a mass $M$.

**9.7.** Find $I_{zz}$ and $I_{xx}$ for the homogeneous right circular cylinder of mass $M$.

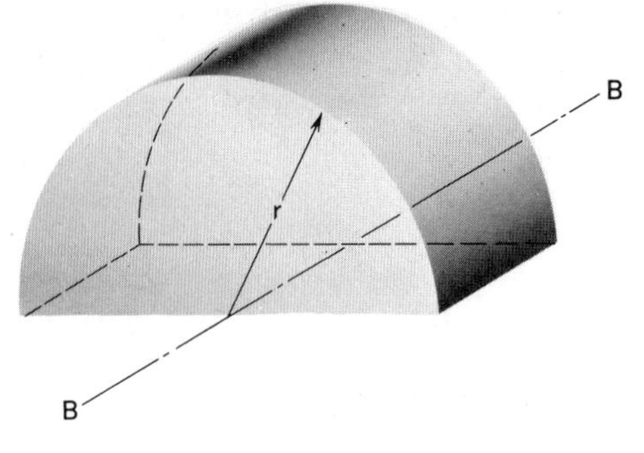

Figure P.9.6

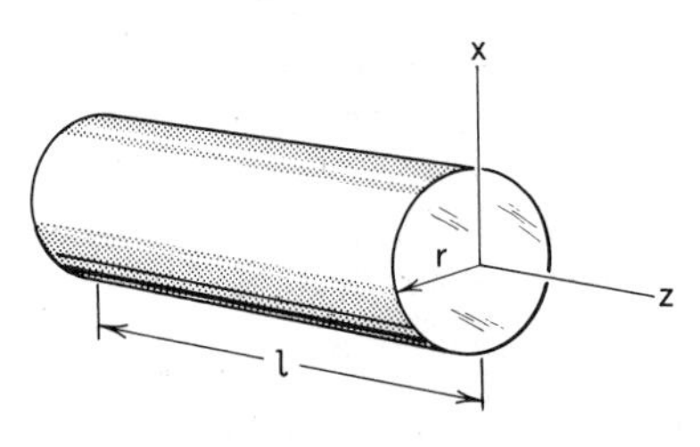


Figure P.9.7

**9.8.** For the cylinder in Problem 9.7, the density increases linearly in the $z$ direction from a value of .100 grams/mm$^3$ at the left end to a value of .180 grams/mm$^3$ at the right end. Take $r = 30$ mm and $l = 150$ mm. Find $I_{xx}$ and $I_{zz}$.

**9.9.** Show that $I_{zz}$ for the homogeneous right circular cone is $\frac{3}{10} MR^2$.

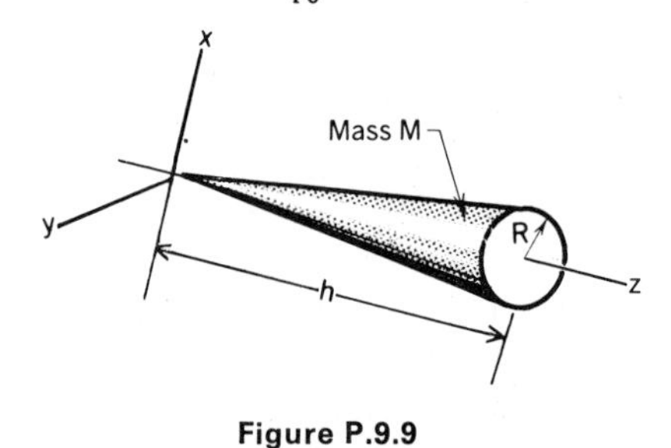


Figure P.9.9

**9.10.** In Problem 9.9, the density increases as the square of $z$ in the $z$ direction from a value of .200 grams/mm$^3$ at the left end to a value of .400 grams/mm$^3$ at the right end. If $r = 20$ mm and the cone is 100 mm in length, find $I_{zz}$.

**9.11.** A body of revolution is shown. The radial distance $r$ of the boundary from the $x$ axis is given as $r = .2x^2$ m. What is $I_{xx}$ for a uniform density of 1600 kg/m³?

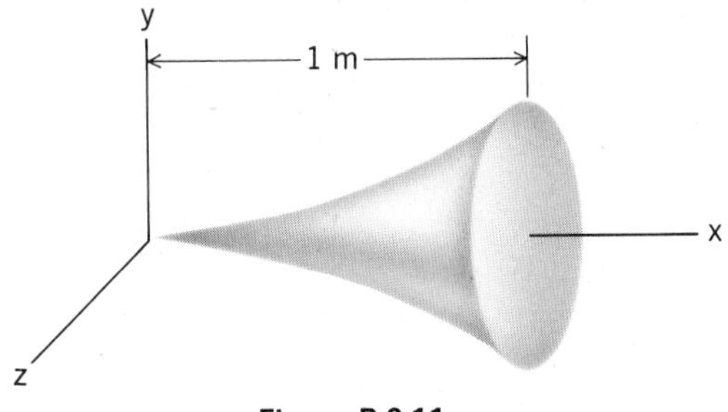


Figure P.9.11

**9.12.** A thick hemispherical shell is shown with an inside radius of 40 mm and an outside radius of 60 mm. If the density $\rho$ is 7000 kg/m³, what is $I_{yy}$?

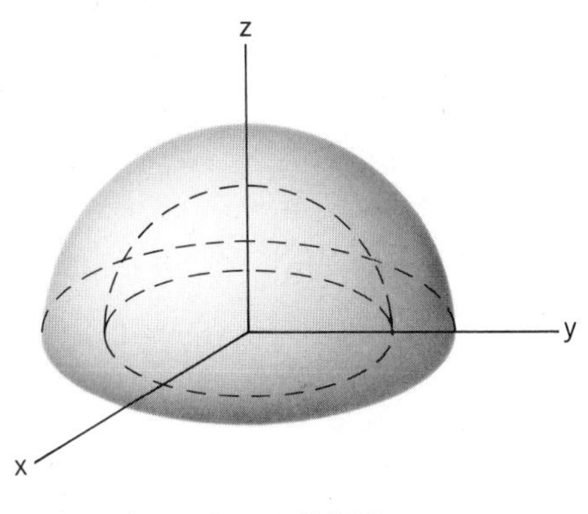

Figure P.9.12

**9.13.** Find the mass moment of inertia $I_{xx}$ for a very thin plate forming a quarter-sector of a circle. The plate weighs .4 N. What is the second moment of area about the $x$ axis? What is the product of inertia?

Figure P.9.13

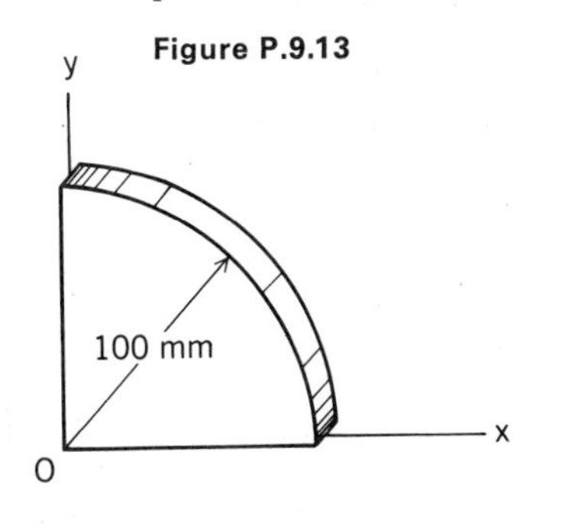


**9.14.** Find the second moment of area about the $x$ axis for the top surface of a very thin plate. If the weight of the plate is .02N/mm², find the second moments of mass about the $xy$ axes. What is the mass product of inertia $I_{xy}$?

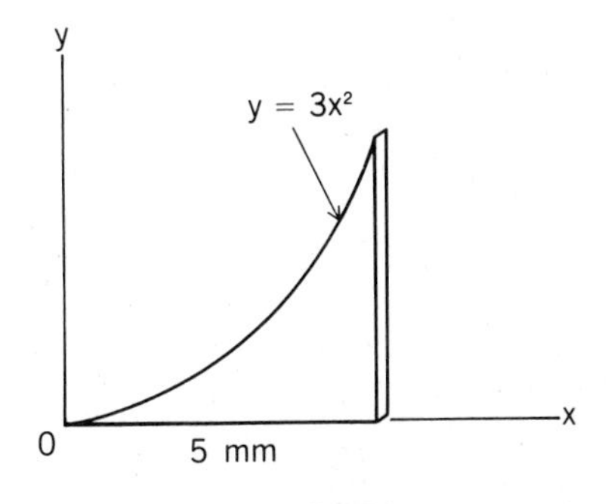


Figure P.9.14

***9.15.** A uniform tetrahedron is shown having sides of length $a$, $b$, and $c$, respectively, and a mass $M$. Show that $I_{yz} = \frac{1}{20} Mac$. (*Suggestion:* Let $z$ run from zero to surface $ABC$. Let $x$ run from zero to $AB$. Finally, let $y$ run from zero to $B$. Note that the equation of a plane surface is $z = \alpha x + \beta y + \gamma$, where $\alpha$, $\beta$, and $\gamma$ are constants. The mass of the tetrahedron is $\rho abc/6$. It will be simplest in expanding $(1 - x/b - y/c)^2$ to proceed in the form $[(1 - y/c) - (x/b)]^2$, keeping $(1 - y/c)$ intact. In the last integration replace $y$ by $[-c(1 - y/c) + c]$, etc.)

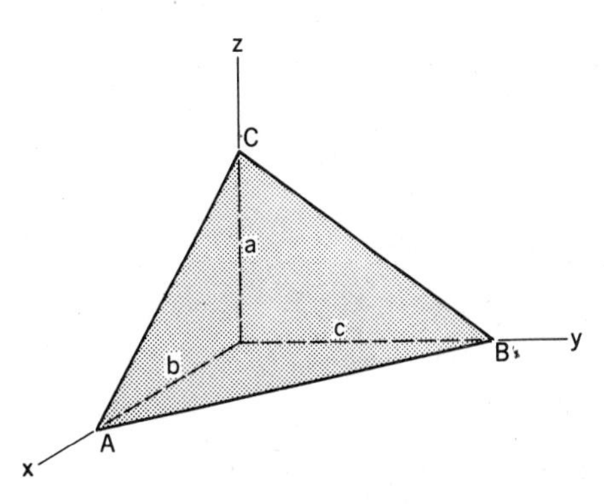


Figure P.9.15

**9.16.** In Problem 9.13, find the three principal mass moments of inertia at $O$. Use the results from this problem, that

$$(I_{xx})_M = 101.9 \text{ kg-mm}^2$$

$$(I_{xy})_M = 64.9 \text{ kg-mm}^2$$

**9.17.** In Problem 9.14, compute the values of the three principal mass moments of inertia at $O$. From Problem 9.14 we have the results

$$(I_{xx})_M = 20{,}500 \text{ kg-mm}^2$$

$$(I_{yy})_M = 382 \text{ kg-mm}^2$$

$$(I_{xy})_M = 2390 \text{ kg-mm}^2$$

**9.18.** Can you identify by inspection any of the principal axes of inertia at $A$? At $B$? Explain. The density of the material is uniform.

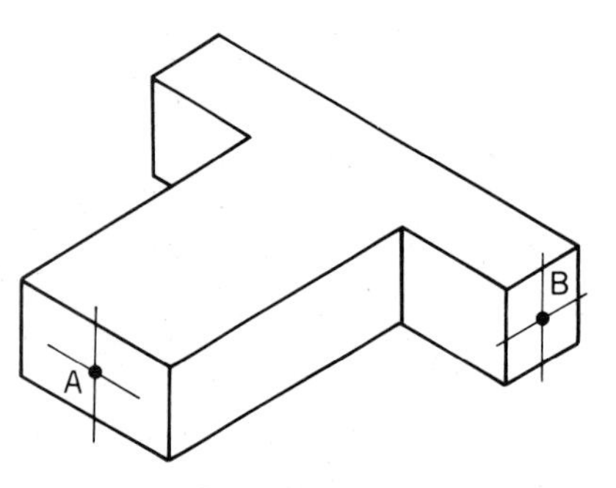


Figure P.9.18

**9.19.** By inspection, identify as many principal axes as you can for mass moments of inertia at positions $A$, $B$, and $C$. Explain your choices. The mass density of the material is uniform throughout.

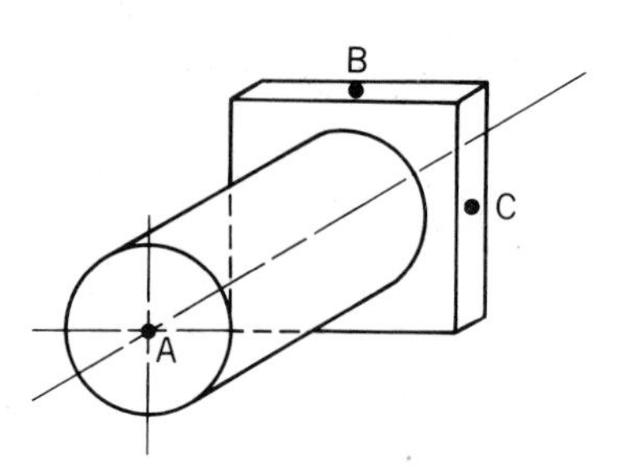

Figure P.9.19

## 9.4 Translation of Coordinate Axes

In this section, we will compute mass moment and product of inertia quantities for a reference $xyz$ that is displaced under a translation (no rotation) from a reference $x'y'z'$ at the center of mass (Fig. 9.12) for which the inertia terms are presumed known. Let us first compute the moment of inertia $I_{zz}$. Observing Fig. 9.12, we see that

$$\boldsymbol{r} = \boldsymbol{r}_c + \boldsymbol{r}'$$

Hence,

$$x = x_c + x'$$

$$y = y_c + y'$$

$$z = z_c + z'$$

We can now formulate $I_{zz}$ in the following way:

$$I_{zz} = \iiint_V (x^2 + y^2)\rho\, dv = \iiint_V [(x_c + x')^2 + (y_c + y')^2]\rho\, dv \tag{9.4}$$

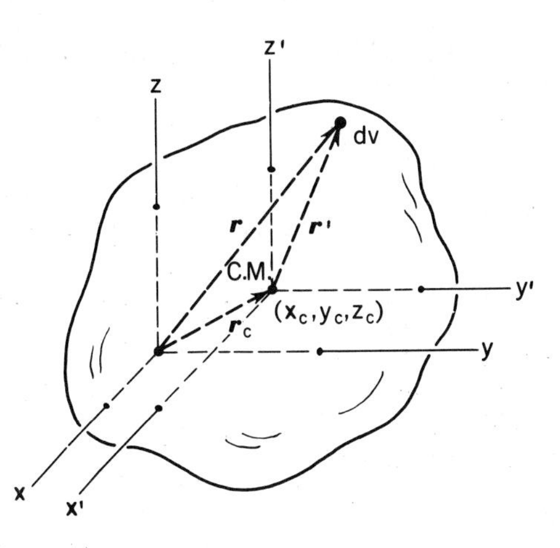


**Figure 9.12.** *xyz* translated from *x'y'z'* at C.M.

Carrying out the squares and rearranging, we have

$$I_{zz} = \iiint_V (x_c^2 + y_c^2)\rho\, dv + 2\iiint_V x_c x'\rho\, dv + 2\iiint_V y_c y'\rho\, dv + \iiint_V (x'^2 + y'^2)\rho\, dv \tag{9.5}$$

Note that the quantities bearing the subscript $c$ are constant for the integration and can be extracted from under the integral sign. Thus,

$$I_{zz} = M(x_c^2 + y_c^2) + 2x_c\iiint_V x'\, dm + 2y_c\iiint_V y'\, dm + \iiint_V (x'^2 + y'^2)\rho\, dv \tag{9.6}$$

where $\rho\, dv$ has been replaced in some terms by $dm$, and the integration $\iiint_V \rho\, dv$ in the first integration has been evaluated as $M$, the total mass of the body. The origin of the primed reference being at the center of mass requires of the first moments of mass that $\iiint x'\, dm = \iiint y'\, dm = \iiint z'\, dm = 0$. The middle two terms accordingly drop out of the expression above, and we recognize the last expression to be $I_{z'z'}$. Thus, the desired relation is

$$I_{zz} = I_{z'z'} + M(x_c^2 + y_c^2) \tag{9.7}$$

By observing the body in Fig. 9.12 along the $z$ and $z'$ axes (i.e., from directly above), we get a view as is shown in Fig. 9.13. From this diagram, we can see that $y_c^2 + x_c^2 = d^2$, where $d$ is the perpendicular distance between the $z'$ axis through the centroid and the $z$ axis about which we are taking moments of inertia. We may then give the result above as

$$\boxed{I_{zz} = I_{z'z'} + Md^2} \tag{9.8}$$

Let us generalize from the above statement.

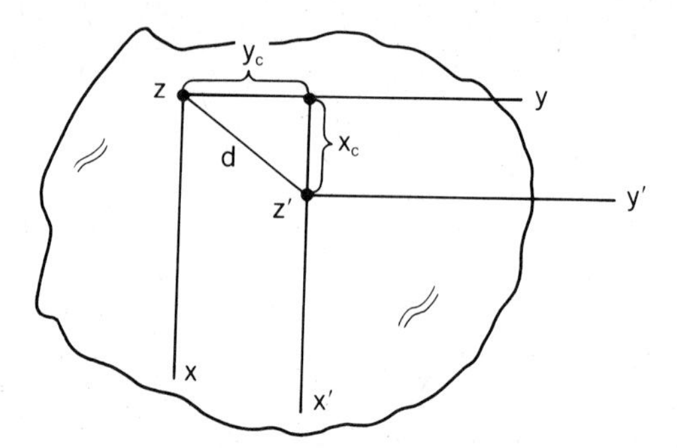


**Figure 9.13.** View along $z$ direction (from above).

*The moment of inertia of a body about any axis equals the moment of inertia of the body about a parallel axis that goes through the center of mass, plus the total mass times the perpendicular distance between the axes squared.*

We leave it to you to show that for products of inertia a similar relation can be reached. For $I_{xy}$, for example, we have

$$\boxed{I_{xy} = I_{x'y'} + Mx_cy_c} \tag{9.9}$$

Here, we must take care to put in the proper signs of $x_c$ and $y_c$ as measured *from* the $xyz$ reference. Equations 9.8 and 9.9 comprise the well-known *parallel-axis theorems* analogous to those formed in Chapter 8 for areas. You can use them to advantage for bodies composed of simple familiar shapes, as we now illustrate.

**EXAMPLE 9.4**

Find $I_{xx}$ and $I_{xy}$ for the body shown in Fig. 9.14. Take $\rho$ as constant for the body. Use the formulations for moments and products of inertia at the center of mass as given in the inside cover pages.

We shall consider first a solid rectangular prism having the outer dimensions given in Fig. 9.14, and we shall then subtract the contribution of the cylinder and the rectangular block that have been cut away. Thus, we have, for the overall rectangular block which we consider as body 1,

$$\begin{aligned}(I_{xx})_1 &= (I_{xx})_c + Md^2 = \tfrac{1}{12}M(a^2 + b^2) + Md^2 \\ &= \tfrac{1}{12}[(\rho)(20)(8)(15)](8^2 + 15^2) + [(\rho)(20)(8)(15)](4^2 + 7.5^2) \\ &= 231{,}200\rho\end{aligned} \tag{a}$$

From this, we shall take away the contribution of the cylinder, which we denote as body 2. Use formulas from inside cover pages.

$$\begin{aligned}(I_{xx})_2 &= \tfrac{1}{12}M(3r^2 + h^2) + Md^2 = \tfrac{1}{12}[\rho\pi(1)^2(15)][3(1^2) + 15^2] \\ &\qquad + [\rho\pi(1)^2(15)](6^2 + 7.5^2) \\ &= 5243\rho\end{aligned} \tag{b}$$

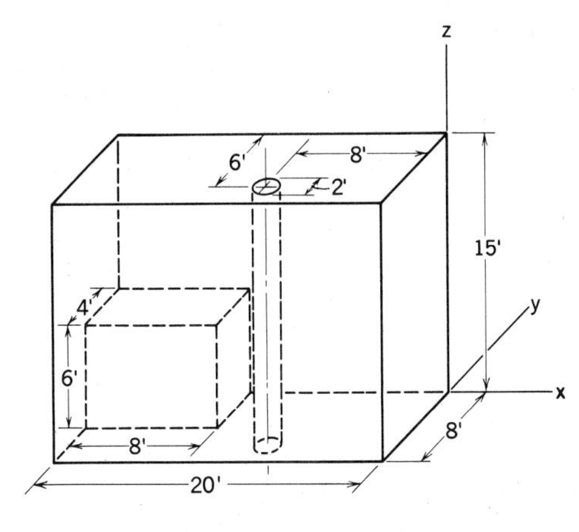


**Figure 9.14.** Find $I_{xx}$ and $I_{xy}$.

Also, we shall take away the contribution of the rectangular cutout (body 3):

$$(I_{xx})_3 = \tfrac{1}{12}M(a^2 + b^2) + Md^2 = \tfrac{1}{12}[(\rho)(8)(6)(4)](4^2 + 6^2) + [(\rho)(8)(6)(4)](2^2 + 3^2) \qquad \text{(c)}$$

$$= 3328\rho$$

We get, accordingly,

$$I_{xx} = (231{,}200 - 5243 - 3328)\rho = 223{,}000\rho \qquad \text{(d)}$$

We do the same for $I_{xy}$. Thus, for the block as a whole, we have

$$(I_{xy})_1 = (I_{xy})_c + Mx_cy_c$$

At the center of mass of the block, both the $(x')_1$ and $(y')_1$ axes are normal to planes of symmetry. Accordingly, $(I_{xy})_c = 0$. Hence,

$$(I_{xy})_1 = 0 + [\rho(20)(8)(15)](-4)(-10) = 96{,}000\rho \qquad \text{(e)}$$

For the cylinder, we note that both the $(x')_2$ and $(y')_2$ axes at the center of mass are normal to planes of symmetry. Hence, we can say that

$$(I_{xy})_2 = 0 + [\rho(\pi)(1^2)(15)](-8)(-6) = 2262\rho \qquad \text{(f)}$$

Finally, for the small cutout rectangular parallelepiped, we note that the $(x')_3$ and $(y')_3$ axes at the center of mass are perpendicular to planes of symmetry. Hence, we have

$$(I_{xy})_3 = 0 + [(\rho)(8)(6)(4)](-2)(-16) = 6144\rho \qquad \text{(g)}$$

The quantity $I_{xy}$ for the body with the rectangular and cylindrical cavities is then

$$I_{xy} = (96{,}000 - 2262 - 6144)\rho = 87{,}600\rho \qquad \text{(h)}$$

If $\rho$ is given in units of lbm/ft$^3$, the inertia terms have units lbm-ft$^2$.

## *9.5 Transformation Properties of the Inertia Terms

Let us assume that the six independent inertia terms are known for a given reference. What is the mass moment of inertia for an axis going through the origin of the reference and having the direction cosines $l$, $m$, and $n$ relative to the axes of this reference? The axis is designated as $kk$ in Fig. 9.15. From previous conclusions, we can say that

$$I_{kk} = \iiint_V [|\boldsymbol{r}|(\sin\phi)]^2 \rho\, dv \tag{9.10}$$

where $\phi$ is the angle between $kk$ and $\boldsymbol{r}$. We shall now put $\sin^2\phi$ into a more useful form by considering the right triangle formed by the position vector $\boldsymbol{r}$ and the axis $kk$. This triangle is shown enlarged in Fig. 9.16. The side $a$ of the triangle has a magnitude that can be given by the dot product of $\boldsymbol{r}$ and the unit vector $\boldsymbol{\epsilon}_k$ along $kk$. Thus,

$$a = \boldsymbol{r} \cdot \boldsymbol{\epsilon}_k = (x\boldsymbol{i} + y\boldsymbol{j} + z\boldsymbol{k}) \cdot (l\boldsymbol{i} + m\boldsymbol{j} + n\boldsymbol{k}) \tag{9.11}$$

Hence,

$$a = lx + my + nz$$

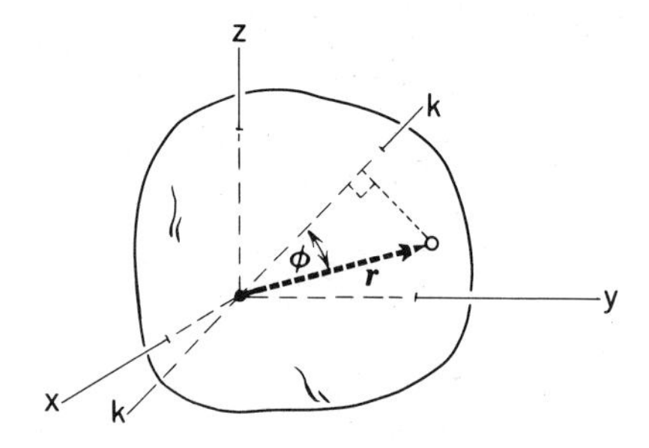


**Figure 9.15.** Find $I_{kk}$.

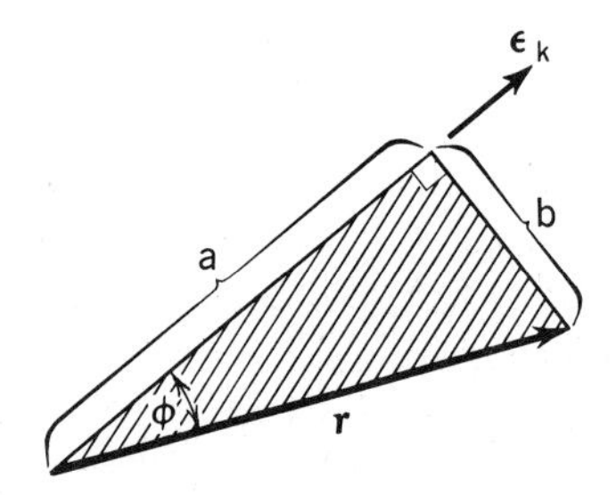


**Figure 9.16.** Right triangle formed by $r$ and $kk$.

Using the Pythagorean theorem, we can now give side $b$ as

$$b^2 = |\boldsymbol{r}|^2 - a^2 = (x^2 + y^2 + z^2) - (l^2x^2 + m^2y^2 + n^2z^2 + 2lmxy + 2lnxz + 2mnyz)$$

The term $\sin^2\phi$ may next be given as

$$\sin^2\phi = \frac{b^2}{r^2} = \frac{(x^2 + y^2 + z^2) - (l^2x^2 + m^2y^2 + n^2z^2 + 2lmxy + 2lnxz + 2mnyz)}{x^2 + y^2 + z^2} \tag{9.12}$$

Substituting back into Eq. 9.10, we get, on canceling terms,

$$I_{kk} = \iiint_V [(x^2 + y^2 + z^2) - (l^2x^2 + m^2y^2 + n^2z^2 + 2lmxy + 2lnxz + 2mnyz)]\rho\, dv$$

Since $l^2 + m^2 + n^2 = 1$, we can multiply the first bracketed expression in the integral by this sum:

$$I_{kk} = \iiint\limits_V [(x^2 + y^2 + z^2)(l^2 + m^2 + n^2)$$
$$- (l^2x^2 + m^2y^2 + n^2z^2 + 2lmxy + 2lnxz + 2mnyz)]\rho\, dv$$

Carrying out the multiplication and collecting terms, we get the relation

$$I_{kk} = l^2 \iiint\limits_V (y^2 + z^2)\rho\, av + m^2 \iiint\limits_V (x^2 + z^2)\rho\, dv + n^2 \iiint\limits_V (x^2 + y^2)\rho\, dv$$
$$- 2lm \iiint\limits_V (xy)\rho\, dv - 2ln \iiint\limits_V (xz)\rho\, dv - 2mn \iiint\limits_V (yz)\rho\, dv$$

Referring back to the definitions presented by relations 9.1, we reach the desired transformation equation:

$$I_{kk} = l^2 I_{xx} + m^2 I_{yy} + n^2 I_{zz} - 2lmI_{xy} - 2lnI_{xz} - 2mnI_{yz} \tag{9.13}$$

We next put this in a more useful form of the kind you will see in later courses in mechanics. Note first that $l$ is the direction cosine between the $k$ axis and the $x$ axis. It is common practice to identify this cosine as $a_{kx}$ instead of $l$. Note that the subscripts identify the axes involved. Similarly, $m = a_{ky}$ and $n = a_{kz}$. We can now express Eq. 9.13 in a form similar to a matrix array as follows on noting that $I_{xy} = I_{yx}$, etc.

$$\boxed{\begin{aligned} I_{kk} = \quad & I_{xx}a_{kx}^2 - I_{xy}a_{kx}a_{ky} - I_{xz}a_{kx}a_{kz} \\ & - I_{yx}a_{ky}a_{kx} + I_{yy}a_{ky}^2 - I_{yz}a_{ky}a_{kz} \\ & - I_{zx}a_{kz}a_{kx} - I_{zy}a_{kz}a_{ky} + I_{zz}a_{kz}^2 \end{aligned}} \tag{9.14}$$

This format is easily written by first writing the matrix array of $I$'s on the right side and then inserting the $a$'s remembering to insert minus signs for off-diagonal terms.

Let us next compute the product of inertia for a pair of mutually perpendicular axes, $Ok$ and $Oq$, as shown in Fig. 9.17. The direction cosines of $Ok$ we shall take as $l$, $m$, and $n$, whereas the direction cosines of $Oq$ we shall take as $l'$, $m'$, and $n'$. Since the axes are at right angles to each other, we know that

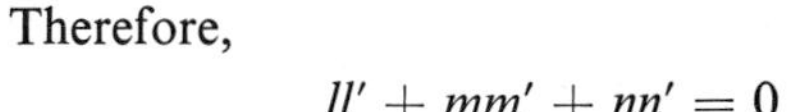

$$\boldsymbol{\epsilon}_k \cdot \boldsymbol{\epsilon}_q = 0$$

Therefore,

$$ll' + mm' + nn' = 0 \tag{9.15}$$

**Figure 9.17.** Find $I_{kq}$.

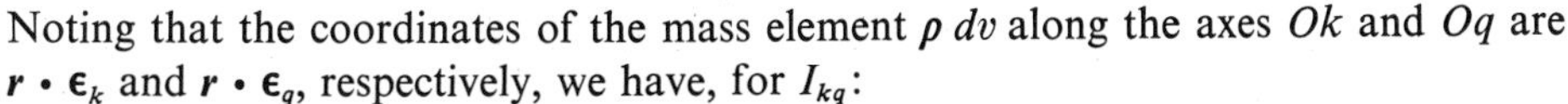

Noting that the coordinates of the mass element $\rho\, dv$ along the axes $Ok$ and $Oq$ are $\boldsymbol{r} \cdot \boldsymbol{\epsilon}_k$ and $\boldsymbol{r} \cdot \boldsymbol{\epsilon}_q$, respectively, we have, for $I_{kq}$:

$$I_{kq} = \iiint\limits_V (\boldsymbol{r} \cdot \boldsymbol{\epsilon}_k)(\boldsymbol{r} \cdot \boldsymbol{\epsilon}_q)\rho\, dv$$

Using $xyz$ components of $\mathbf{r}$ and the unit vectors, we have

$$I_{kq} = \iiint_V [(x\mathbf{i} + y\mathbf{j} + z\mathbf{k}) \cdot (l\mathbf{i} + m\mathbf{j} + n\mathbf{k})] \times [(x\mathbf{i} + y\mathbf{j} + z\mathbf{k}) \cdot (l'\mathbf{i} + m'\mathbf{j} + n'\mathbf{k})]\rho\, dv \quad (9.16)$$

Carrying out the dot products in the integrand above, we get the following result:

$$I_{kq} = \iiint_V (xl + ym + zn)(xl' + ym' + zn')\rho\, dv$$

Hence,

$$I_{kq} = \iiint_V (x^2ll' + y^2mm' + z^2nn' + xylm' + xzln' + yxml' + yzmn' + zxnl' + zynm')\rho\, dv \quad (9.17)$$

Noting from Eq. 9.15 that $(ll' + mm' + nn')$ is zero, we may for convenience add the term, $(-x^2 - y^2 - z^2)(ll' + mm' + nn')$, to the integrand in the equation above. After canceling some terms, we have

$$I_{kq} = \iiint_V (-x^2mm' - x^2nn' - y^2ll' - y^2nn' - z^2ll' - z^2mm' + xylm' + xzln' + yxml' + yzmn' + zxnl' + xynm')\rho\, dv$$

Collecting terms and bringing the direction cosines outside the integrations, we get

$$\begin{aligned} I_{kq} = &-ll' \iiint_V (y^2 + z^2)\rho\, dv - mm' \iiint_V (x^2 + z^2)\rho\, dv \\ &-nn' \iiint_V (y^2 + x^2)\rho\, dv + (lm' + ml') \iiint_V xy\rho\, dv \\ &+ (ln' + nl') \iiint_V xz\rho\, dv + (mn' + nm') \iiint_V yz\rho\, dv \end{aligned} \quad (9.18)$$

Noting the definitions in Eq. 9.1, we can state the desired transformation:

$$\begin{aligned} I_{kq} = &-ll'I_{xx} - mm'I_{yy} - nn'I_{zz} + (lm' + ml')I_{xy} \\ &+ (ln' + nl')I_{xz} + (mn' + nm')I_{yz} \end{aligned} \quad (9.19)$$

We can now rewrite the previous equation in a more useful and simple form using $a$'s as direction cosines. Thus, noting that $l' = a_{qx}$, etc.,

$$\boxed{\begin{aligned} -I_{kq} = &\quad I_{xx}a_{kx}a_{qx} - I_{xy}a_{kx}a_{qy} - I_{xz}a_{kx}a_{qz} \\ &-I_{yx}a_{ky}a_{qx} + I_{yy}a_{ky}a_{qy} - I_{yz}a_{ky}a_{qz} \\ &-I_{zx}a_{kz}a_{qx} - I_{zy}a_{kz}a_{qy} + I_{zz}a_{kz}a_{qz} \end{aligned}} \quad (9.20)$$

Again you will note that the right side can easily be set forth by first putting down the matrix array of $I_{ij}$ and then inserting the $a$'s with easily determined subscripts while remembering to insert minus signs for off-diagonal terms.

By making the axis $Ok$ in Fig. 9.17 an $x'$ axis at $O$ and using the direction cosines for this axis $(a_{x'x}, a_{x'y}, a_{x'z})$, we can formulate $I_{x'x'}$ from Eq. 9.14. By a similar procedure, we can formulate $I_{y'y'}$ and $I_{z'z'}$ for reference $x'y'z'$ at 0 rotated arbitrarily relative to $xyz$. Also, by considering the $Ok$ and $Oq$ axes to be $x'$ and $y'$ axes, respectively, with $a_{x'x}$, $a_{x'y}$, and $a_{x'z}$ as direction cosines for the $x'$ axis and $a_{y'x}$, $a_{y'y}$, and $a_{y'z}$ as direction cosines for the $y'$ axis, we can evaluate $I_{x'y'}$ at $O$ using Eq. 9.20. This approach can similarly be followed to find $I_{x'z'}$ and $I_{y'z'}$. Thus, employing Eqs. 9.14 and 9.20 as parent equations, we can develop equations for computing the nine inertia quantities for a reference $x'y'z'$ rotated arbitrarily relative to $xyz$ at $O$ in terms of the nine known inertia quantities for reference $xyz$ at $O$. Thus, once the nine inertia quantities are known for one reference at some point, they can be determined for *any* reference at that point. We say that the inertia terms *transform* from one set of components for $xyz$ at some point $O$ to another set of components for $x'y'z'$ at $O$ by means of certain transformations formed from Eqs. 9.14 and 9.20.

We now define a symmetric,[7] *second-order tensor as a set of nine components*

$$\begin{pmatrix} A_{xx} & A_{xy} & A_{xz} \\ A_{yx} & A_{yy} & A_{yz} \\ A_{zx} & A_{zy} & A_{zz} \end{pmatrix}$$

*which transforms with a rotation of axes according to the following parent equations.* For the diagonal terms,

$$\begin{aligned} A_{kk} = {} & A_{xx}a_{kx}^2 + A_{xy}a_{kx}a_{ky} + A_{xz}a_{kx}a_{kz} \\ & + A_{yx}a_{ky}a_{kx} + A_{yy}a_{ky}^2 + A_{yz}a_{ky}a_{kz} \\ & + A_{zx}a_{kz}a_{kx} + A_{zy}a_{kz}a_{ky} + A_{zz}a_{kz}^2 \end{aligned} \tag{9.21}$$

For the off-diagonal terms,

$$\begin{aligned} A_{kq} = {} & A_{xx}a_{kx}a_{qx} + A_{xy}a_{kx}a_{qy} + A_{xz}a_{kx}a_{kz} \\ & + A_{yx}a_{ky}a_{qx} + A_{yy}a_{ky}a_{qy} + A_{yz}a_{ky}a_{kz} \\ & + A_{zx}a_{kz}a_{qx} + A_{zy}a_{kz}a_{qy} + A_{zz}a_{kz}a_{qz} \end{aligned} \tag{9.22}$$

On comparing Eqs. 9.21 and 9.22, respectively, with Eqs. 9.14 and 9.20, we can conclude that the array of terms

$$I_{ij} = \begin{pmatrix} I_{xx} & -I_{xy} & -I_{xz} \\ -I_{yx} & I_{yy} & -I_{yz} \\ -I_{zx} & -I_{zy} & I_{zz} \end{pmatrix} \tag{9.23}$$

is a second-order tensor.

You will learn that because of the common transformation law identifying certain quantities as tensors, there will be extremely important common characteristics for these quantities which set them apart from other quantities. Thus, in order to learn these common characteristics in an efficient way and to understand them better, we do become involved with tensors as an entity in the engineering sciences, physics, and applied mathematics. You will soon be confronted with the stress and strain tensors in your courses in strength of materials

[7]The word "symmetric" refers to the condition $A_{12} = A_{21}$, etc., that is required if the transformation equation is to have the form given. We can have nonsymmetric second-order tensors, but since they are less common in engineering work, we shall not concern ourselves here with such possibilities.

and in solid mechanics. And in electromagnetic theory and nuclear physics, you will be introduced to the quadrupole tensor.[8]

**EXAMPLE 9.5**

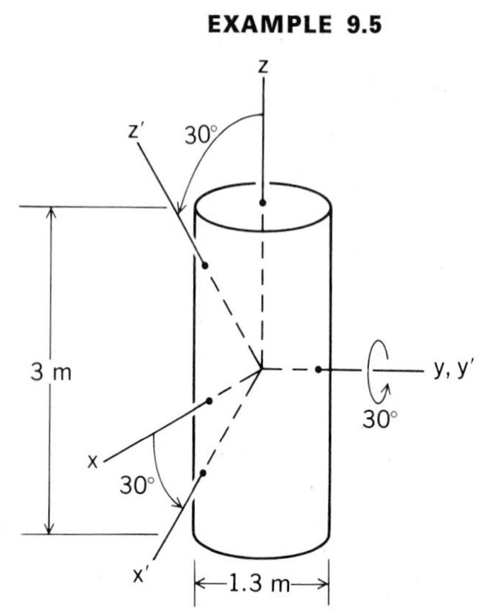


**Figure 9.18.** Find $I_{z'z'}$ and $I_{x'z'}$.

Find $I_{z'z'}$ and $I_{x'z'}$ for the cylinder shown in Fig. 9.18. The reference $x'y'z'$ is found by rotating about the $y$ axis an amount 30°, as shown in the diagram. The mass of the cylinder is 100 kg.

It is simplest to first get the inertia tensor components for reference $xyz$. Thus, using formulas from the inside cover pages we have

$$I_{zz} = \frac{1}{2}Mr^2 = \frac{1}{2}(100)\left(\frac{1.3}{2}\right)^2 = 21.13 \text{ kg-m}^2$$

$$I_{xx} = I_{yy} = \frac{1}{12}M(3r^2 + h^2)$$

$$= \frac{1}{12}(100)\left[(3)\left(\frac{1.3}{2}\right)^2 + 3^2\right]$$

$$= 85.6 \text{ kg-m}^2$$

Noting that the $xyz$ coordinate planes are planes of symmetry, we can conclude that

$$I_{xz} = I_{yx} = I_{yz} = 0$$

Next, evaluate the direction cosines of the $z'$ and the $x'$ axes relative to $xyz$. Thus,

For $z'$ axis:

$$a_{z'x} = \cos 60° = .500$$

$$a_{z'y} = \cos 90° = 0$$

$$a_{z'z} = \cos 30° = .866$$

For $x'$ axis:

$$a_{x'x} = \cos 30° = .866$$

$$a_{x'y} = \cos 90° = 0$$

---

[8]Vectors may be defined in terms of the way components of the vector for a new reference are related to the components of the old reference at a point $A$. Thus, for any direction $\boldsymbol{n}$, we have for component $A_n$:

$$A_n = A_x a_{nx} + A_y a_{ny} + A_z a_{nz} \tag{a}$$

Using Eq. (a), we can find components of vector $\boldsymbol{A}$ with respect to $x'y'z'$ rotated arbitrarily relative to $xyz$. Thus, all vectors must transform in accordance with Eq. (a) on rotation of the reference. Obviously, the vector, as seen from this point of view, is a special, simple case of the second-order tensor. We say, accordingly, that vectors are *first-order tensors.*

As for scalars, there is clearly no change in value when there is a rotation of axes at a point. Thus,

$$T(x', y', z') = T(x, y, z) \tag{b}$$

for $x'y'z'$ rotated relative to $xyz$. Scalars are a special form of tensor when considered from a transformation point of view. In fact, they are called *zero-order tensors.*

$$a_{x'z} = \cos 120° = -.500$$

First, we employ Eq. 9.14 to get $I_{z'z'}$.

$$I_{z'z'} = (85.6)(.500)^2 + (21.23)(.866)^2$$
$$= 37.32 \text{ kg-m}^2$$

Finally, we employ Eq. 9.20 to get $I_{x'z'}$.

$$-I_{x'z'} = (85.6)(.500)(.866) + (21.13)(.866)(-.500)$$

Therefore,

$$I_{x'z'} = -27.92 \text{ kg-m}^2$$

## Problems

*In the following problems, use the formulas for moments and products of inertia at the mass center to be found in the inside cover pages.*

**9.20.** What are the moments and products of inertia for the $xyz$ and $x'y'z'$ axes for the cylinder?

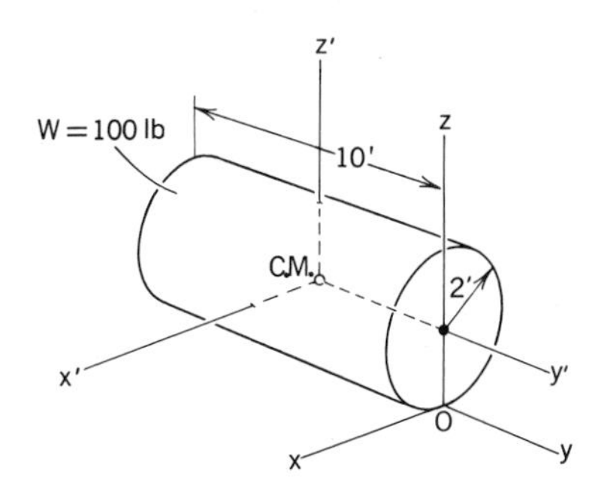

Figure P.9.20

**9.21.** For the uniform block, compute the inertia tensor at the center of mass, at point $a$, and at point $b$ for axes parallel to the $xyz$ reference. Take the mass of the body as $M$ kg.

**9.22.** Determine $I_{xx} + I_{yy} + I_{zz}$ as a function of $x$, $y$, and $z$ for all points in space for the uniform rectangular parallelepiped. Note that $xyz$ has its origin at the center of mass and is parallel to the sides.

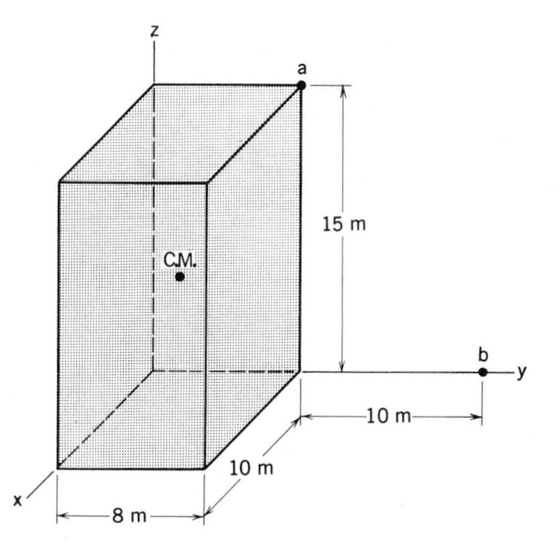

Figure P.9.21

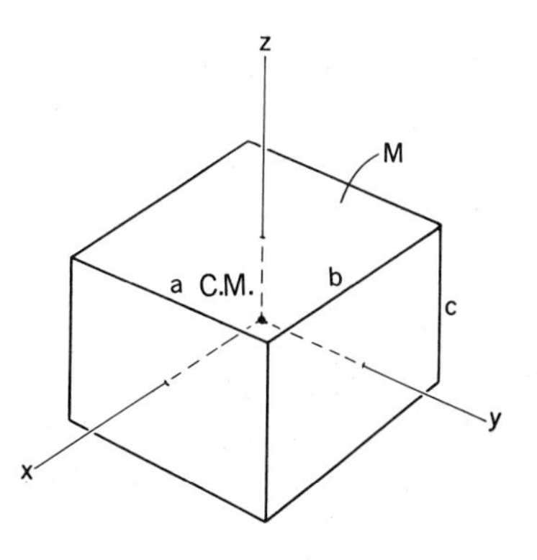


Figure P.9.22

**9.23.** A thin plate weighing 100 N has the following mass moments of inertia at mass center $O$:

$$I_{xx} = 15 \text{ kg-m}^2$$
$$I_{yy} = 13 \text{ kg-m}^2$$
$$I_{xy} = -10 \text{ kg-m}^2$$

What are the moments of inertia $I_{x'x'}$, $I_{y'y'}$, and $I_{z'z'}$ at point $P$ having the position vector:

$$\boldsymbol{r} = .5\boldsymbol{i} + .2\boldsymbol{j} + .6\boldsymbol{k} \text{ m}$$

Also determine $I_{x'z'}$ at $P$.

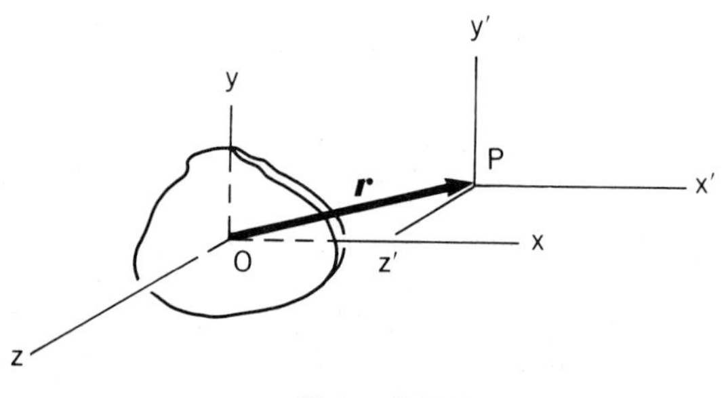


Figure P.9.23

**9.24.** A crate with its contents weighs 20 kN and has its center of mass at

$$\boldsymbol{r}_c = 1.3\boldsymbol{i} + 3\boldsymbol{j} + .8\boldsymbol{k} \text{ m}$$

It is known that at corner $A$,

$$I_{x'x'} = 5500 \text{ kg-m}^2$$
$$I_{x'y'} = -1500 \text{ kg-m}^2$$

for primed axes parallel to $xyz$. At point $B$, find $I_{x''x''}$ and $I_{x''y''}$ for double-primed axes parallel to $xyz$.

Figure P.9.24

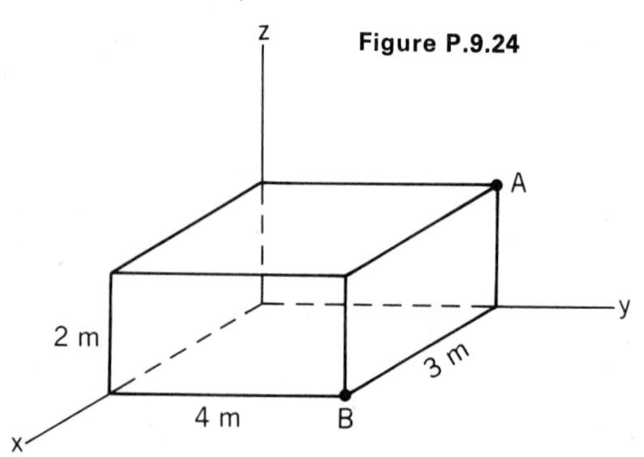


**9.25.** A cylindrical crate and its contents weigh 500 N. The center of mass is at

$$\boldsymbol{r}_c = .6\boldsymbol{i} + .7\boldsymbol{j} + 2\boldsymbol{k} \text{ m}$$

It is known that at $A$,

$$(I_{yy})_A = 85 \text{ kg-m}^2$$
$$(I_{yz})_A = -22 \text{ kg-m}^2$$

Find $I_{yy}$ and $I_{zy}$ at $B$.

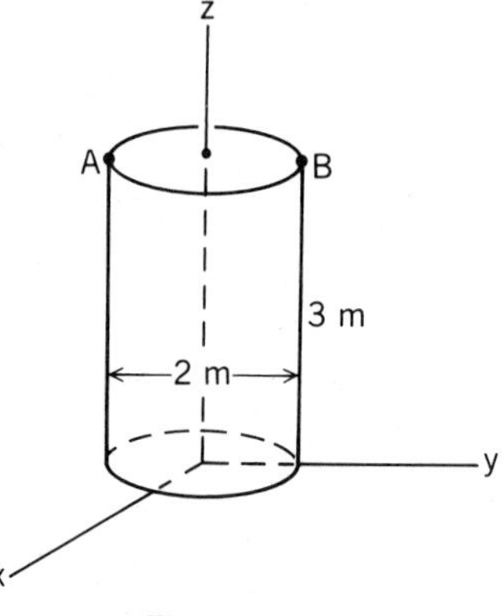

Figure P.9.25

**9.26.** A block having a uniform density of 5 grams/cm³ has a hole of diameter 40 mm cut out. What are the principal moments of inertia at point $A$ at the centroid of the right face of the block?

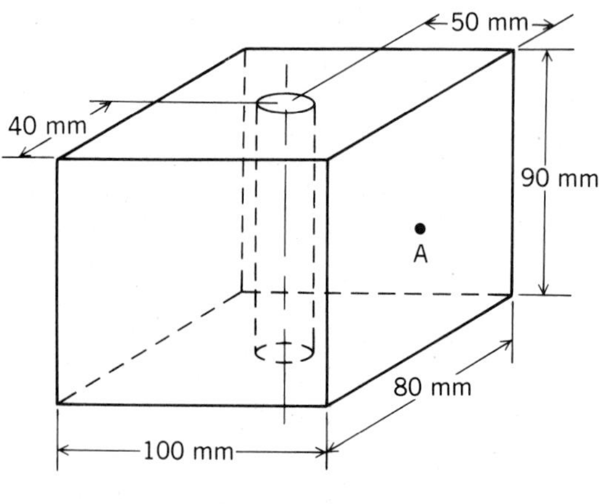


Figure P.9.26

**9.27.** Find maximum and minimum moment of inertia at point $A$. The block weighs 20 N and the cone weighs 14 N.

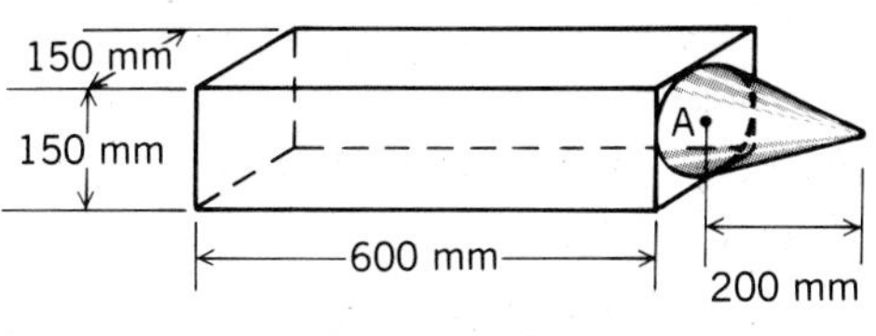


**Figure P.9.27**

**9.28.** Solid spheres $C$ and $D$ each weighing 25 N and having radius of 50 mm are attached to a thin solid rod weighing 30 N. Also, solid spheres $E$ and $G$ each weighing 20 N and having radii of 30 mm are attached to a thin rod weighing 20 N. The rods are attached to be orthogonal to each other. What are the principal moments of inertia at point $A$?

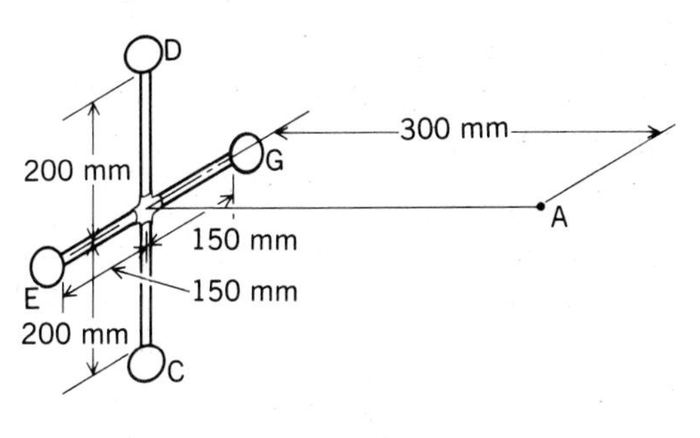


**Figure P.9.28**

**9.29.** A cylinder is shown having a conical cavity oriented parallel to the axis $A$–$A$ and a cylindrical cavity oriented normal to $A$–$A$. If the density of the material is 7200 kg/m³, what is $I_{AA}$?

**Figure P.9.29**

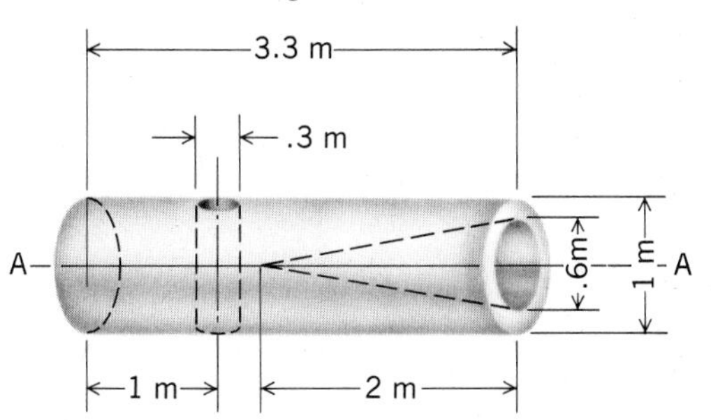


**9.30.** A flywheel is made of steel having a specific weight of 490 lb/ft³. What is the moment of inertia about its geometric axis? What is the radius of gyration?

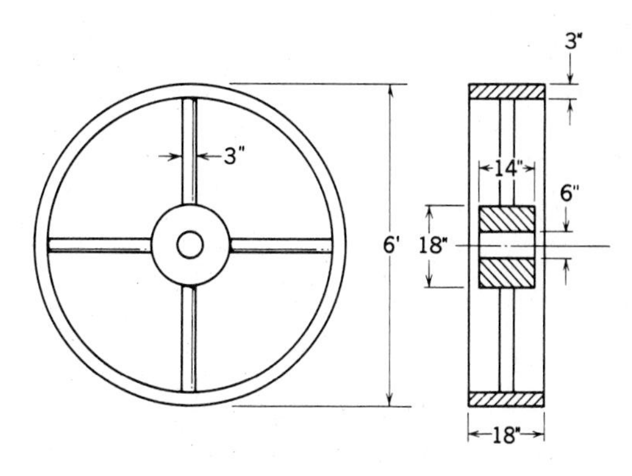


**Figure P.9.30**

**9.31.** Compute $I_{yy}$ and $I_{xy}$ for the right circular cylinder, which has a mass of 50 kg, and the square rod, which has a mass of 10 kg, when the two are joined together so that the rod is radial to the cylinder.

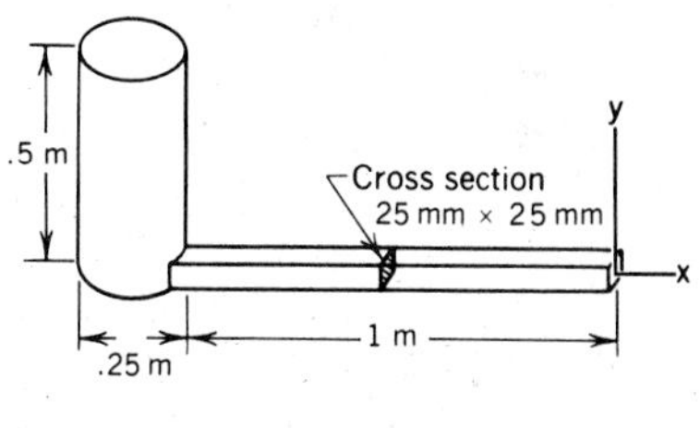


**Figure P.9.31**

**9.32.** Compute the moments and products of inertia for the $xy$ axes. The specific weight is 490 lb/ft³ throughout.

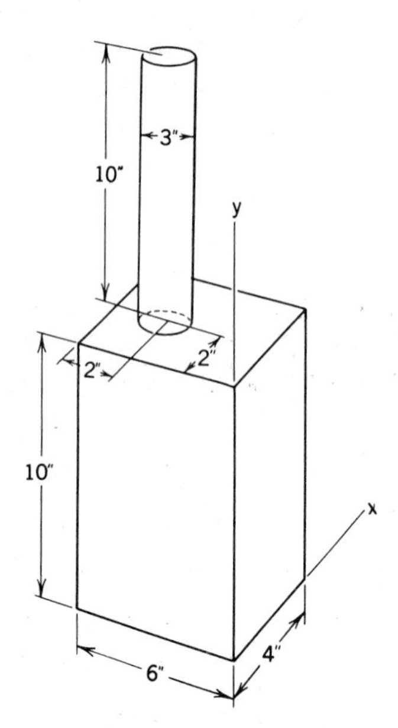


Figure P.9.32

**9.33.** A disc $A$ is mounted on a shaft such that its normal is oriented 10° from the centerline of the shaft. The disc has a diameter of 2 ft, is 1 in. in thickness, and weighs 100 lb. Compute the moment of inertia of the disc about the centerline of the shaft.

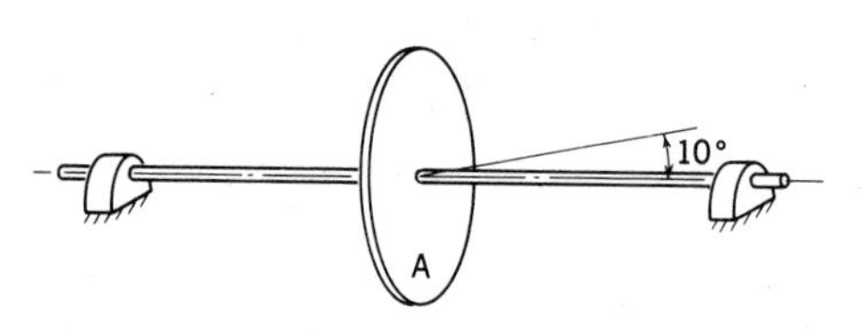


Figure P.9.33

**9.34.** A gear $B$ having a mass of 25 kg rotates about axis $C$–$C$. If the rod $A$ has a mass distribution of 7.5 kg/m, compute the moment of inertia of $A$ and $B$ about the axis $C$–$C$.

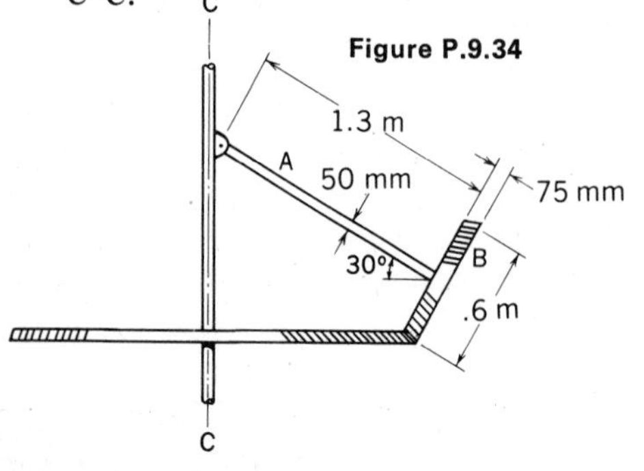


Figure P.9.34

**9.35.** A block weighing 100 N is shown. Compute the moment of inertia about the diagonal $D$–$D$.

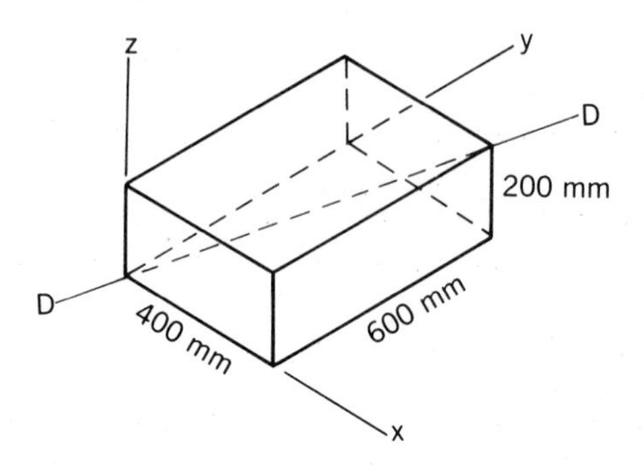


Figure P.9.35

**9.36.** A solid sphere $A$ of diameter 1 ft and weight 100 lb is connected to the shaft $B$–$B$ by a solid rod weighing 2 lb/ft and having a diameter of 1 in. Compute $I_{z'z'}$ for the rod and ball.

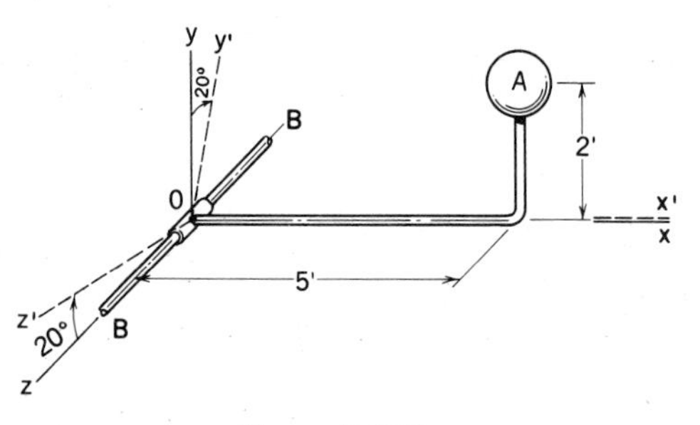

Figure P.9.36

**9.37.** In Problem 9.13, we found the following results for the thin plate:

$$I_{xx} = I_{yy} = .1019 \text{ grams-m}^2$$

$$I_{xy} = .0649 \text{ grams-m}^2$$

Find all components for the inertia tensor for reference $x'y'z'$. Axes $x'y'$ lie in mid-plane of the plate.

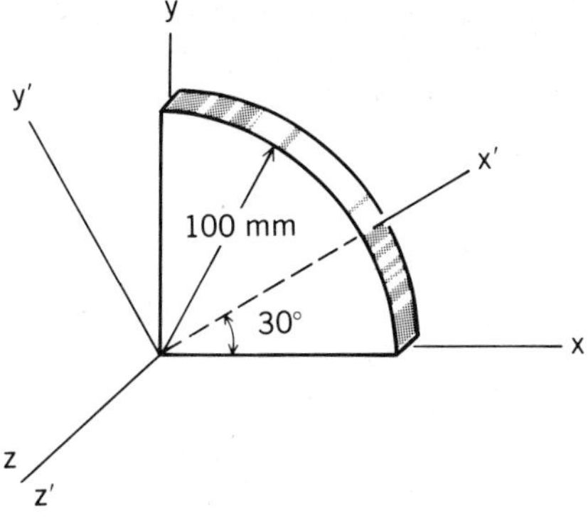

Figure P.9.37

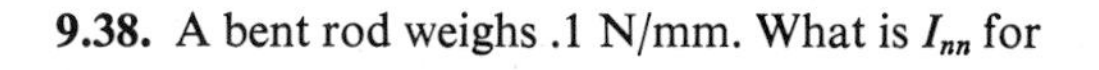
**9.38.** A bent rod weighs .1 N/mm. What is $I_{nn}$ for

$$\epsilon_n = .30\boldsymbol{i} + .45\boldsymbol{j} + .841\boldsymbol{k}?$$

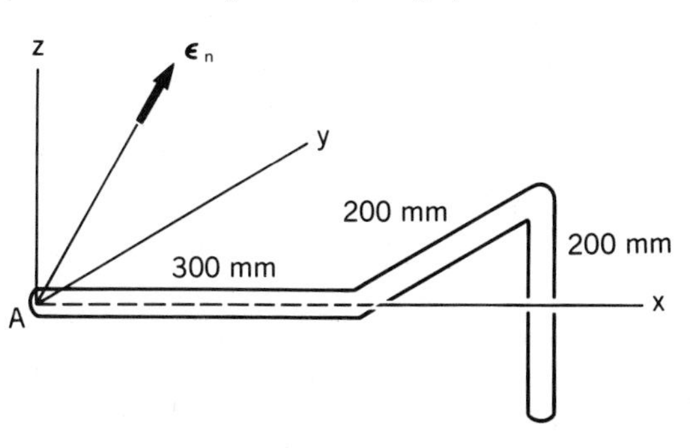

Figure P.9.38

**9.39.** Evaluate the matrix of direction cosines for the primed axes relative to the unprimed axes.

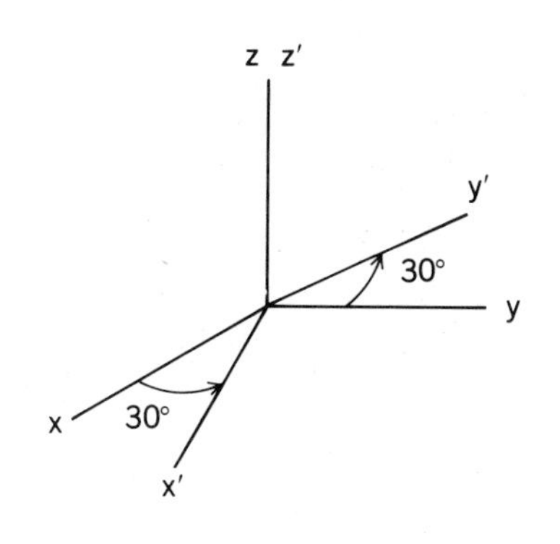


Figure P.9.39

$$a_{ij} = \begin{pmatrix} a_{x'x} & a_{x'y} & a_{x'z} \\ a_{y'x} & a_{y'y} & a_{y'z} \\ a_{z'x} & a_{z'y} & a_{z'z} \end{pmatrix}$$

**9.40.** The block is uniform in density and weighs 10 N. Find $I_{y'z'}$.

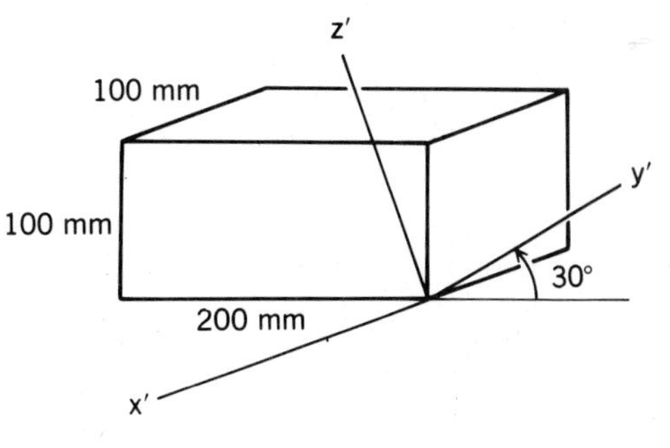


Figure P.9.40

***9.41.** A thin rod of length 300 mm and weight 12 N is oriented relative to $x'y'z'$ such that

$$\epsilon_n = .4\boldsymbol{i}' + .3\boldsymbol{j}' + .866\boldsymbol{k}'$$

What is $I_{x'y'}$?

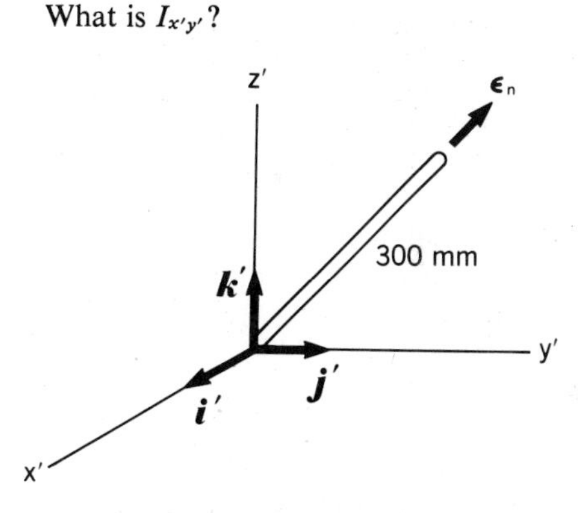


Figure P.9.41

***9.42.** Show that the transformation equation for the inertia tensor components at a point when there is a rotation of axes (i.e., Eqs. 9.14 and 9.20) can be given as follows:

$$I_{kq} = \sum_j \sum_i a_{ki} a_{qj} I_{ij}$$

where $k$ can be $x'$, $y'$, or $z'$ and $q$ can be $x'$, $y'$, or $z'$, and where $i$ and $j$ go from $x$ to $y$ to $z$. The equation above is a compact definition of *second-order tensors*. Remember that in the inertia tensor you must have a minus sign in front of each product of inertia term (i.e., $-I_{xy}$, $-I_{yz}$, etc.). (*Hint:* Let $i = x$; then sum over $j$; then let $i = y$ and sum again over $j$; etc.)

***9.43.** In Problem 9.42, express the transformation equation to get $I_{y'z'}$ in terms of the inertia tensor components for reference $xyz$ having the same origin as $x'y'z'$.

## *9.6 The Inertia Ellipsoid and Principal Moments of Inertia

Equation 9.14 gives the moment of inertia of a body about an axis $k$ in terms of the direction cosines of that axis measured from an orthogonal reference with an origin $O$ on the axis, and in terms of six independent inertia quantities for this reference. We wish to explore the nature of the variation of $I_{kk}$ at a point $O$ in space as the direction of $k$ is changed. (The $k$ axis and the body are shown in Fig. 9.19, which we shall call the physical diagram.) To do this, we will employ a geometric representation of moment of inertia at a point that is developed in the following manner. Along the axis $k$, we lay off as a distance the quantity $OA$ given by the relation

$$OA = \frac{d}{\sqrt{I_{kk}/M}} \tag{9.24}$$

where $d$ is any arbitrary constant that has a dimension of length that will render $OA$ dimensionless, as the reader can verify. The term $\sqrt{I_{kk}/M}$ is the *radius of gyration* and was presented earlier. To avoid confusion, this operation is shown in another diagram, called the inertia diagram (Fig. 9.20), where the new $\xi$, $\eta$, and $\zeta$ axes are *parallel* to $x$, $y$, and $z$ axes of the physical diagram. Considering all possible directions of $k$, we observe that some surface will be formed about the point $O'$, and this surface is related to the shape of the body through Eq. 9.14. We can express the equation of this surface quite readily. Suppose that we call $\xi$, $\eta$, and $\zeta$ the coordinates of point $A$. Since $O'A$ is

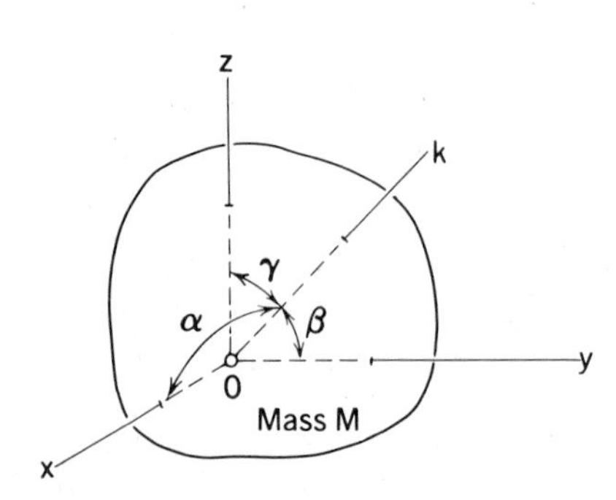


Figure 9.19. Physical diagram.

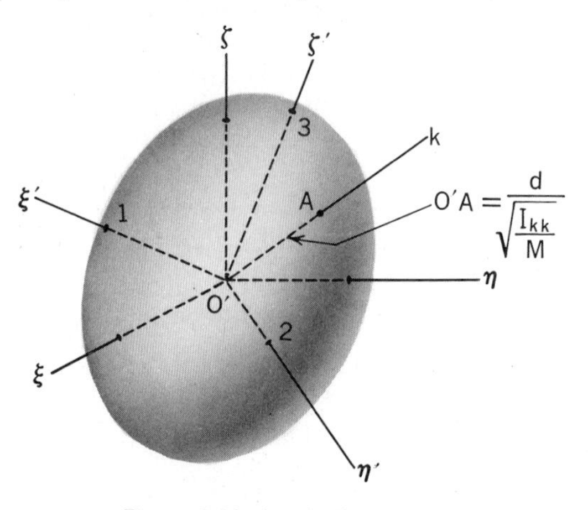


Figure 9.20. Inertia diagram.

parallel to the line $k$ and thus has the direction cosines $a_{kx}$, $a_{ky}$, and $a_{kz}$ that are associated with this line, we can say that

$$a_{kx} = \frac{\xi}{O'A} = \frac{\xi}{d\sqrt{M/I_{kk}}}$$
$$a_{ky} = \frac{\eta}{O'A} = \frac{\eta}{d\sqrt{M/I_{kk}}} \tag{9.25}$$
$$a_{kz} = \frac{\zeta}{O'A} = \frac{\zeta}{d\sqrt{M/I_{kk}}}$$

Now replace the direction cosines in Eq. 9.13, using the relations above:

$$\begin{aligned} I_{kk} &= \frac{\xi^2}{Md^2/I_{kk}} I_{xx} + \frac{\eta^2}{Md^2/I_{kk}} I_{yy} + \frac{\zeta^2}{Md^2/I_{kk}} I_{zz} \\ &+ 2\frac{\xi\eta}{Md^2/I_{kk}}(-I_{xy}) + 2\frac{\xi\zeta}{Md^2/I_{kk}}(-I_{xz}) + 2\frac{\eta\zeta}{Md^2/I_{kk}}(-I_{yz}) \end{aligned} \tag{9.26}$$

We can see that $I_{kk}$ cancels out of the preceding equation, leaving an equation involving the coordinates $\xi$, $\eta$, and $\zeta$ of the surface and the inertia terms of the body itself. Rearranging the terms, we then have

$$\frac{\xi^2}{Md^2/I_{xx}} + \frac{\eta^2}{Md^2/I_{yy}} + \frac{\zeta^2}{Md^2/I_{zz}} + \frac{2\xi\eta}{Md^2}(-I_{xy}) + \frac{2\xi\zeta}{Md^2}(-I_{xz}) + \frac{2\eta\zeta}{Md^2}(-I_{yz}) = 1 \tag{9.27}$$

Considering analytic geometry, we know that the surface is that of an ellipsoid (see Fig. 9.20), and is thus called the *ellipsoid of inertia*. The distance squared from $O'$ to any point $A$ on the ellipsoid is inversely proportional to the moment of inertia (see Eq. 9.24) about an axis in the body at $O$ having the same direction as $O'A$. We can conclude that the inertia tensor for any point of a body can be represented geometrically by such a second-order surface, and this surface may be thought of as analogous to the arrow used to represent a vector graphically. The size, shape, and inclination of the ellipsoid will vary for each point in space for a given body. (Since all second-order tensors may be represented by second-order surfaces, you will, if you study elasticity, also encounter the ellipsoids of stress and strain.[9])

An ellipsoid has three orthogonal axes of symmetry, which have a common point at the center, $O'$. In the diagram, these axes are shown as $O'1$, $O'2$, and $O'3$. We pointed out that the shape and inclination of the ellipsoid of inertia depend on the mass distribution of the body about the *origin* of the $xyz$ reference, and they have nothing to do with the choice of *the orientation of the xyz* (and hence the $\xi\eta\zeta$) reference at the point. We can therefore imagine that the $xyz$ reference (and hence the $\xi\eta\zeta$ reference) can be chosen to have directions that coincide with the aforementioned symmetric axes, $O'1$,

[9]See I. H. Shames, *Mechanics of Deformable Solids*, Prentice-Hall, Inc., Englewood Cliffs, N.J., 1964, Chap. 2. Also, Krieger Publishing Co., N.Y., 1979.

$O'2$, and $O'3$. If we call such references $x'y'z'$ and $\xi'\eta'\zeta'$, respectively, we know from analytic geometry that Eq. 9.27 becomes

$$\frac{(\xi')^2}{Md^2/I_{x'x'}} + \frac{(\eta')^2}{Md^2/I_{y'y'}} + \frac{(\zeta')^2}{Md^2/I_{z'z'}} = 1 \tag{9.28}$$

where $\xi'$, $\eta'$, and $\zeta'$ are coordinates of the ellipsoidal surface relative to the new reference, and $I_{x'x'}$, $I_{y'y'}$, and $I_{z'z'}$ are mass moments of inertia of the body about the new axes. We can now draw several important conclusions from this geometrical construction and the accompanying equations. One of the symmetrical axes of the ellipsoid above is the longest distance from the origin to the surface of the ellipsoid, and another axis is the smallest distance from the origin to the ellipsoidal surface. Examining the definition in Eq. 9.24, we must conclude that the minimum moment of inertia for the point $O$ must correspond to the axis having the maximum length, and the maximum moment of inertia must correspond to the axis having the minimum length. The third axis has an intermediate value that makes the sum of the moment of inertia terms equal to the sum of the moment of inertia terms for all orthogonal axes at point $O$, in accordance with Eq. 9.2. In addition, Eq. 9.28 leads us to conclude that $I_{x'y'} = I_{y'z'} = I_{x'z'} = 0$. That is, the products of inertia of the mass about these axes must be zero. Clearly, these axes are the *principal axes* of inertia at the point $O$.

Since the preceding operations could be carried out at any point in space for the body, we can conclude that:

> *At each point there is a set of principal axes having the extreme values of moments of inertia for that point and having zero products of inertia.*[10] *The orientation of these axes will vary continuously from point to point throughout space for the given body.*

All second-order tensor quantities have the properties discussed above for the inertia tensor. By transforming from the original reference to the principal reference, we change the inertia tensor representation from

$$\begin{pmatrix} I_{xx} & (-I_{xy}) & (-I_{xz}) \\ (-I_{yx}) & I_{yy} & (-I_{yz}) \\ (-I_{zx}) & (-I_{zy}) & I_{zz} \end{pmatrix} \quad \text{to} \quad \begin{pmatrix} I_{x'x'} & 0 & 0 \\ 0 & I_{y'y'} & 0 \\ 0 & 0 & I_{z'z'} \end{pmatrix} \tag{9.29}$$

In mathematical parlance, we have "diagonalized" the tensor by the preceding operations.

## 9.7 Closure

In this chapter, we first introduced the nine components comprising the inertia tensor. Next, we considered the case of the very thin flat plate in which the $xy$ axes form the midplane of the plate. We found that the mass-inertia terms $(I_{xx})_M$, $(I_{yy})_M$, and $(I_{xy})_M$ for the plate are proportional respectively to $(I_{xx})_A$, $(I_{yy})_A$, and $(I_{xy})_A$, the second moments and product of area of the plate surface. As a result, we could set forth the

[10] A general procedure for computing principal moments of inertia is set forth in Appendix III.

concept of principal axes for the inertia tensor as an extension of the work in Chapter 8. Thus, we pointed out that for these axes the products of inertia will be zero. Furthermore, one principal axis corresponds to the maximum moment of inertia at the point while another of the principal axes corresponds to the minimum moment of inertia at the point. We pointed out that for bodies with two orthogonal planes of symmetry, the principal axes at any point on the line of intersection of the planes of symmetry must be along this line of intersection and normal to this line in the planes of symmetry.

Those readers who studied the starred sections from Section 9.5 onward will have found proofs of the extensions set forth earlier about principal axes from Chapter 8. Even more important is the disclosure that the inertia tensor components change their values when the axes are rotated at a point in exactly the same way as many other physical quantities having nine components. Such quantities are called second-order tensors. Because of the common transformation equation for such quantities, they have many important identical properties, such as principal axes. In your course in strength of materials you should learn that stress and strain are second-order tensors and hence have principal axes.[11] Additionally, you will find that a two-dimensional stress distribution called *plane stress* is related to the stress tensor exactly as the moments and products of area are related to the inertia tensor. The same situation exists with strain. Consequently, there are Mohr's-circle constructions for plane stress and the corresponding case for strain (plane strain). Thus, by taking the extra time to consider the mathematical considerations of Sections 9.5 and 9.6, you will find unity between Chapter 9 and some very important aspects of strength of materials to be studied later in your program.

In Chapter 10, we shall introduce another approach to studying equilibrium beyond what we have used thus far. This approach is valuable for certain important classes of statics problems and at the same time forms the groundwork for a number of advanced techniques that many students will study later in their programs.

## Review Problems

**9.44.** Find $I_{zz}$ for the body of revolution having uniform density of .2 N/mm³. The radial distances out from the $z$ axis to the surface is given as

$$r^2 = -4\,z \text{ mm}^2$$

where $z$ is in millimeters. (*Hint:* Make use of the formula for the moment of inertia about the axis of a disc, $\frac{1}{2}Mr^2$.)

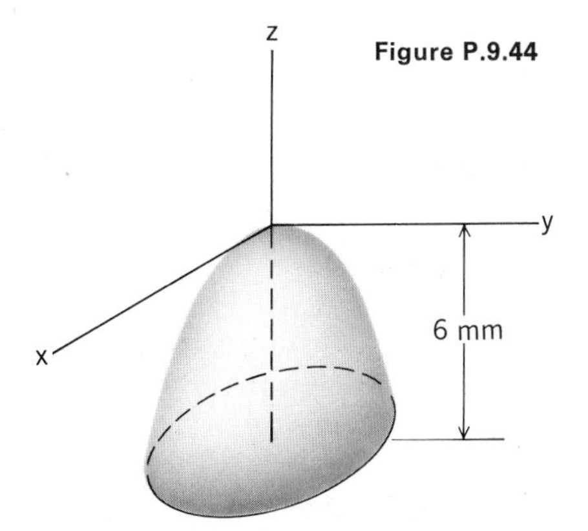


Figure P.9.44

[11]See I. H. Shames, *Introduction to Solid Mechanics*, Prentice-Hall, Inc., Englewood Cliffs, N.J., 1975.

**9.45.** In Problem 9.44, determine $I_{zz}$ without using the disc formula but using multiple integration instead.

**9.46.** What are the inertia tensor components for the thin plate about axes $xyz$? The plate weighs 2 N.

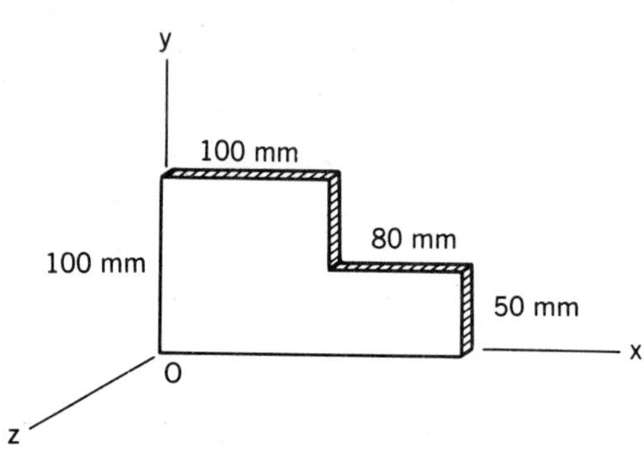

**Figure P.9.46**

**9.47.** In Problem 9.46, what are the principal axes and the principal moments of inertia for the intertia tensor at $O$?

**9.48.** What are the principal mass moments of inertia at point $O$? Block $A$ weighs 15 N. Rod $B$ weighs 6 N and solid sphere $C$ weighs 10 N. The density in each body is uniform. The diameter of the sphere is 50 mm.

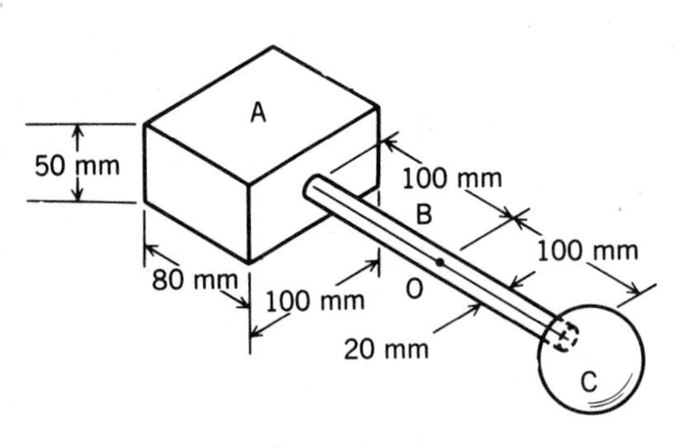


**Figure P.9.48**

**9.49.** The block has a density of 15 kg/m³. Find the moment of inertia about axis $AB$.

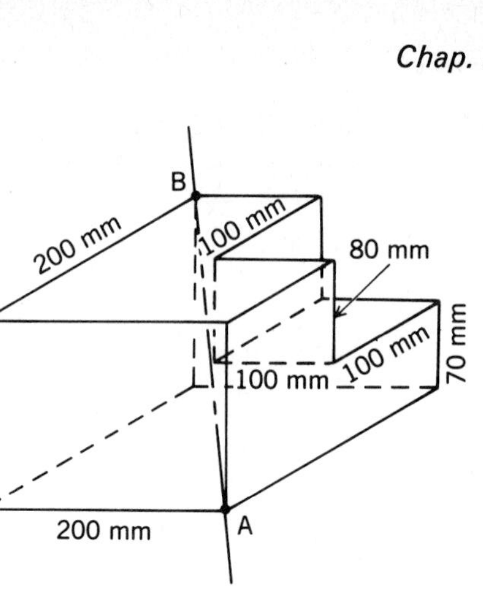

**Figure P.9.49**

**9.50.** A crate and its contents weighs 10 kN. The center of mass of the crate and its contents is at

$$\boldsymbol{r}_c = .40\boldsymbol{i} + .30\boldsymbol{j} + .60\boldsymbol{k} \text{ m}$$

If at $A$ we know that

$$I_{yy} = 800 \text{ kg-m}^2$$
$$I_{yz} = 500 \text{ kg-m}^2$$

find $I_{yy}$ and $I_{yz}$ at $B$.

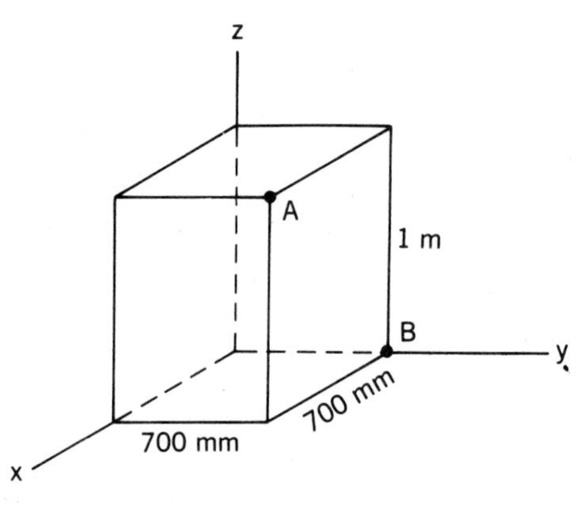


**Figure P.9.50**

**9.51.** A semicylinder weighs 50 N. What are the principal moments of inertia at $O$? What is the product of inertia $I_{y'x'}$? What conclusion can you draw about the direction of principal axes at $O$?

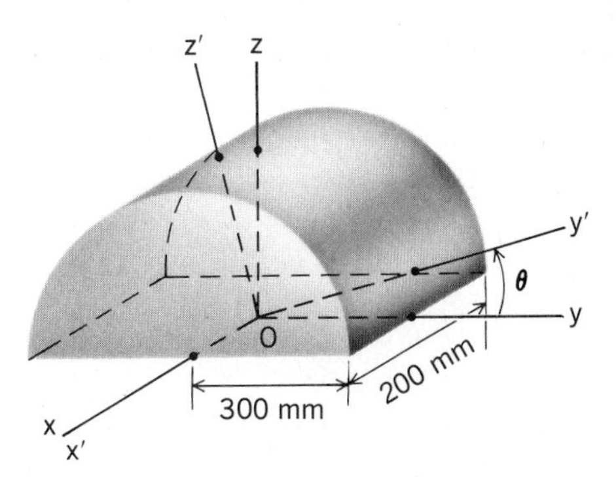


Figure P.9.51

# 10 *Methods of Virtual Work and Stationary Potential Energy

## 10.1 Introduction

In the study of statics thus far, we have followed the procedure of isolating a body to expose certain unknown forces and then formulating either scalar or vector equations of equilibrium that include *all* the forces acting on the body. At this time, alternative methods of expressing conditions of equilibrium, called the *method of virtual work* and, allied to it, the *method of stationary potential energy*, will be presented. These methods will yield equilibrium equations equivalent to those of preceding sections. Furthermore, these new equations include only certain forces on a body, and accordingly in some problems will provide a more simple means of solving for desired unknowns.

Actually, we are making a very modest beginning into a vast field of endeavor called *variational mechanics* or *energy methods* with important applications to both rigid-body and deformable-body solid mechanics. Indeed, more advanced studies in these fields will surely center around these methods.[1]

A central concept for energy methods is the work of a force. A differential amount of work $dW_k$ of a force $\boldsymbol{F}$ acting on a particle equals the component of this force in the direction of movement of the particle times the differential displacement of the particle:

$$dW_k = \boldsymbol{F} \cdot d\boldsymbol{r} \tag{10.1}$$

[1]For a treatment of energy methods for deformable solids, see C. Dym and I. H. Shames, *Solid Mechanics—A Variational Approach*, McGraw-Hill Book Company, New York, 1973.

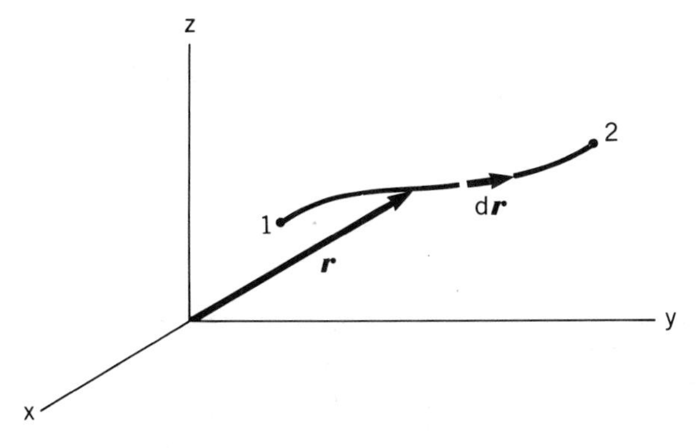


**Figure 10.1.** Path of particle on which $\boldsymbol{F}$ does work.

And the work $W_k$ on a particle by force $\boldsymbol{F}$ when the particle moves along some path (see Fig. 10.1) from point 1 to point 2 is then

$$W_k = \int_{r_1}^{r_2} \boldsymbol{F} \cdot d\boldsymbol{r} \tag{10.2}$$

Note that the value and direction of $\boldsymbol{F}$ can vary along the path. This fact must be taken into account during the integration. We shall have more to say about the concept of work in later sections.[2]

# Part A
# *Method of Virtual Work*

## 10.2 Principle of Virtual Work for a Particle

For our introduction to the principle of virtual work, we will first consider a particle acted on by external loads $\boldsymbol{K}_1, \boldsymbol{K}_2, \ldots, \boldsymbol{K}_n$, whose resultant force pushes the particle against a rigid constraining surface $S$ in space (Fig. 10.2). This surface $S$ is assumed to be frictionless and will thus exert a constraining force $\boldsymbol{N}$ on the particle which is normal to $S$. The forces $\boldsymbol{K}_i$ are called *active forces* in connection with the method of

---

[2]We could have defined work as

$$W_k = \int_{t_1}^{t_2} \boldsymbol{F} \cdot \boldsymbol{V}\, dt$$

where $\boldsymbol{V}$ is the velocity of the point of application of the force. When the force acts on a particular particle, the result above becomes $\int_{r_1}^{r_2} \boldsymbol{F} \cdot d\boldsymbol{r}$, where $\boldsymbol{r}$ is the position vector of the particle. There are times when the force acts on continually *changing* particles as time passes. The more general formulation above can then be used effectively.

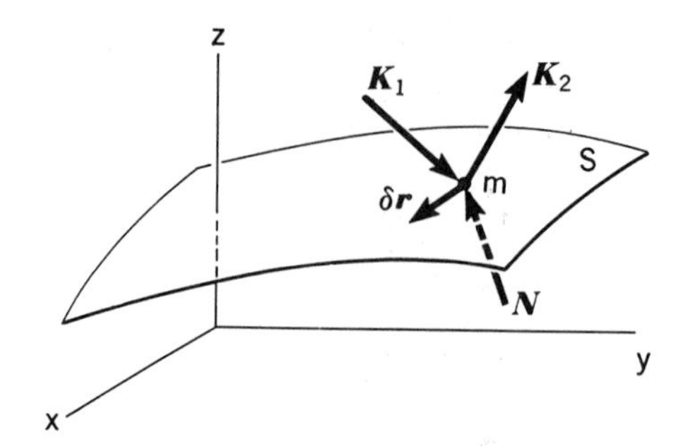


**Figure 10.2.** Particle on a frictionless surface.

virtual work, while $\boldsymbol{N}$ retains the identification of a *constraining force* as used previously. Employing the re ultant active force $\boldsymbol{K}_R$, we can give the necessary and sufficient[3] conditions for eq ıilibrium for the particle as

$$\boldsymbol{K}_R + \boldsymbol{N} = \boldsymbol{0} \tag{10.3}$$

We shall now prove that we can express the necessary and sufficient conditions of equilibrium in yet another way. Let us imagine that we give the particle an infinitesimal hypothetical displacement that is consistent with the constraints (i.e., along the surface), while keeping the forces $\boldsymbol{K}_R$ and $\boldsymbol{N}$ constant. Such a displacement is termed a *virtual displacement*, and will be denoted by $\delta\boldsymbol{r}$, in contrast to a real infinitesimal displacement, $d\boldsymbol{r}$, which might actually occur during a time interval $dt$. We can then take the dot product of the vector $\delta\boldsymbol{r}$ with the force vectors in the equation above:

$$\boldsymbol{K}_R \cdot \delta\boldsymbol{r} + \boldsymbol{N} \cdot \delta\boldsymbol{r} = 0 \tag{10.4}$$

Since $\boldsymbol{N}$ is normal to the surface and $\delta\boldsymbol{r}$ is tangential to the surface, the corresponding scalar product must be zero, leaving

$$\boldsymbol{K}_R \cdot \delta\boldsymbol{r} = 0 \tag{10.5}$$

The expression $\boldsymbol{K}_R \cdot \delta\boldsymbol{r}$ is called the *virtual work* of the system of forces and is denoted as $\delta W$. Thus, the virtual work by the active forces on a particle with frictionless constraints is *necessarily* zero for a particle in equilibrium for any virtual displacement consistent with the frictionless constraints.

We shall now show that this statement is also *sufficient* to ensure equilibrium for the case of a particle initially at rest (relative to an inertial reference) at the time of application of the active loads. To demonstrate this, *assume that Eq. 10.5 holds but that the particle is not in equilibrium*. If the particle is not in equilibrium, it must move in a direction that corresponds to the direction of the resultant of all forces acting on the particle. Consider that $d\boldsymbol{r}$ represents the initial displacement during the time interval $dt$. The work done by the forces must exceed zero for this movement. Since the normal force $\boldsymbol{N}$ cannot do work for this displacement,

$$\boldsymbol{K}_R \cdot d\boldsymbol{r} > 0 \tag{10.6}$$

[3]The sufficiency condition applies to an initially stationary particle.

However, we can choose a *virtual* displacement $\delta \boldsymbol{r}$ to be used in Eq. 10.5 that is *exactly* equal to the proposed $d\boldsymbol{r}$ stated above, and so we see that, by admitting nonequilibrium, we arrive at a result (10.6) that is in *contradiction to the starting known condition* (Eq. 10.5). We can then conclude that the conjecture that the particle is not in equilibrium is false. Thus, Eq. 10.3 is not only a necessary condition of equilibrium, but, for an initially stationary particle, is in itself sufficient for equilibrium. Thus, Eq. 10.5 is completely equivalent to the equation of equilibrium, 10.3.

We can now state the principle of virtual work for a particle.

> *The necessary and sufficient condition for equilibrium of an initially stationary particle with frictionless constraints requires that the virtual work for all virtual displacements consistent with the constraints be zero.*[4]

The case of a particle that is not constrained is a special case of the situation discussed above. Here $\boldsymbol{N} = \boldsymbol{0}$, so that Eq. 10.5 is applicable for *all* infinitesimal displacements as a criterion for equilibrium.

## 10.3 Principle of Virtual Work for Rigid Bodies

We now examine a rigid body in equilibrium acted on by active forces $\boldsymbol{K}_i$ and constrained without the aid of friction (Fig. 10.3). The constraining forces $\boldsymbol{N}_i$ arise from direct contact with other immovable bodies (in which case the constraining forces are oriented normal to the contact surface) or from contact with immovable bodies through pin and ball-joint connections. We shall consider the body to be made up of elementary particles for the purposes of discussion.

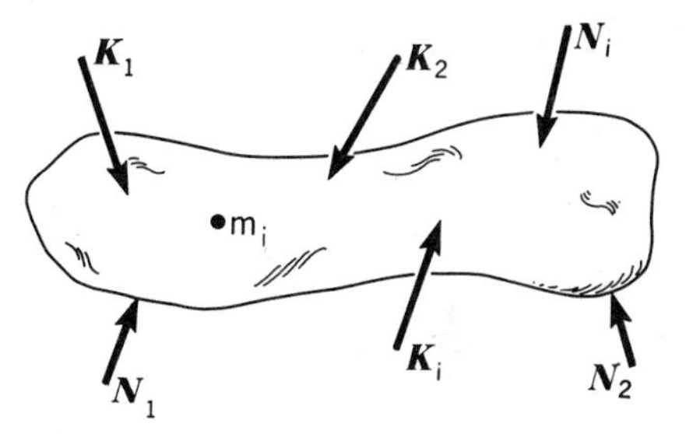


Figure 10.3. Rigid body with active forces and ideal constraining forces.

Now consider such a particle of mass $m_i$. Active loads, external constraining forces, and forces from other particles may possibly be acting on the particle. The forces from other particles are internal forces $\boldsymbol{S}_i$ which

[4]This test breaks down for a particle that is moving. Consider a particle constrained to move in a circular path in a horizontal plane, as shown in the diagram. The particle is moving with constant speed. There are no active forces, and we consider the constraints as frictionless. The work for a vir-

tual displacement consistent with the constraints at any time $t$ gives us a zero result. Nevertheless, the particle is not in equilibrium, since clearly there is at time $t$ an acceleration toward the center of curvature. Thus, we had to restrict the sufficiency condition to particles that are initially stationary.

maintain the rigidity of the body. Using the resultants of these various forces on the particle, we may state from Newton's law that the necessary and sufficient[5] condition for equilibrium of the $i$th particle is

$$(\boldsymbol{K}_R)_i + (\boldsymbol{N}_R)_i + (\boldsymbol{S}_R)_i = \boldsymbol{0} \tag{10.7}$$

Now, we give the particle a virtual displacement $\delta\boldsymbol{r}_i$ that is consistent with the exterior constraints and with the condition that the body is rigid. Taking the dot product of the vectors in the equation above with $\delta\boldsymbol{r}_i$, we get

$$(\boldsymbol{K}_R)_i \cdot \delta\boldsymbol{r}_i + (\boldsymbol{N}_R)_i \cdot \delta\boldsymbol{r}_i + (\boldsymbol{S}_R)_i \cdot \delta\boldsymbol{r}_i = 0 \tag{10.8}$$

Clearly, $(\boldsymbol{N}_R)_i \cdot \delta\boldsymbol{r}_i$ must be zero, because $\delta\boldsymbol{r}_i$ is normal to $\boldsymbol{N}_i$ for constraint stemming from direct contact with immovable bodies or because $\delta\boldsymbol{r}_i = \boldsymbol{0}$ for constraint stemming from pin and ball-joint connections with immovable bodies. Let us sum the equations of the form 10.8 for all the particles that are considered to make up the body. We have, for $n$ particles,

$$\sum_{i=1}^{n} (\boldsymbol{K}_R)_i \cdot \delta\boldsymbol{r}_i + \sum_{i=1}^{n} (\boldsymbol{S}_R)_i \cdot \delta\boldsymbol{r}_i = 0 \tag{10.9}$$

Let us now consider in more detail the internal forces in order to show that the second quantity on the left-hand side of the equation above is zero. The force on $m_i$ from particle $m_j$ will be equal and opposite to the force on particle $m_j$ from particle $m_i$, according to Newton's third law. The internal forces on these particles are shown as $\boldsymbol{S}_{ij}$ and $\boldsymbol{S}_{ji}$ in Fig. 10.4. The first subscript identifies the particle on which a force acts, while the second subscript identifies the particle exerting this force. We can then say that

$$\boldsymbol{S}_{ij} = -\boldsymbol{S}_{ji} \tag{10.10}$$

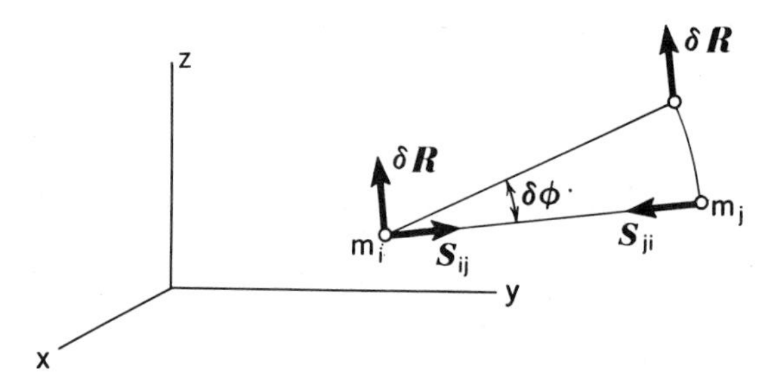


**Figure 10.4.** Two particles of a body undergoing displacement $\delta\boldsymbol{R}$ and rotation $\delta\phi$.

Any virtual motion we give to any pair of particles must maintain a constant distance between the particles. This requirement stems from the rigid body condition and will be true if:

1. Both particles are given the same displacement $\delta\boldsymbol{R}$.

---

[5]The sufficiency requirement again applies to an initially stationary particle.

2. The particles are rotated $\delta\phi$ relative to each other.[6]

We now consider the general case where both motions are present: that is, both $m_i$ and $m_j$ are given a virtual displacement $\delta\boldsymbol{R}$, and furthermore, $m_j$ is rotated through some angle $\delta\phi$ about $m_i$ (Fig. 10.4). The work done in the rotation must be zero, since $\boldsymbol{S}_{ji}$ is at right angles to the motion of the mass $m_j$. Also, the work done on each particle during the equal displacement of both masses must be equal and opposite since the forces move through equal displacements and are themselves equal and opposite. The mutual effect of all particles of the body is of the type described. Thus, we can conclude that the internal work done for a rigid body during a virtual displacement is zero. Hence, a *necessary* condition for equilibrium is

$$\sum_{i=1}^{n} (\boldsymbol{K}_R)_i \cdot \delta\boldsymbol{r}_i = \delta W = 0 \tag{10.11}$$

*Thus, the virtual work done by active forces on a rigid body having frictionless constraints during virtual displacements consistent with the constraints is zero if the body is in equilibrium.*

We can readily prove that Eq. 10.11 is a *sufficient* condition for equilibrium of an initially stationary body by reasoning in the same manner that we did in the case of the single particle. We shall *state first that Eq. 10.11 is valid for a body*. If the body is not in equilibrium, it must begin to move. Let us say that each particle $m_i$ moves a distance $d\boldsymbol{r}_i$ consistent with the constraints under the action of the forces. The work done on particle $m_i$ is

$$(\boldsymbol{K}_R)_i \cdot d\boldsymbol{r}_i + (\boldsymbol{N}_R)_i \cdot d\boldsymbol{r}_i + (\boldsymbol{S}_R)_i \cdot d\boldsymbol{r}_i > 0 \tag{10.12}$$

But $(\boldsymbol{N}_R)_i \cdot d\boldsymbol{r}_i$ is necessarily zero because of the nature of the constraints. When we sum the terms in the equation above for all particles, $\sum_i (\boldsymbol{S}_R)_i \cdot d\boldsymbol{r}$ must also be zero because of the condition of rigidity of the body. Therefore, we may state that the supposition of no equilibrium leads to the following inequality:

$$\sum_{i=1}^{n} (\boldsymbol{K}_R)_i \cdot d\boldsymbol{r}_i > 0 \tag{10.13}$$

But we can conceive a virtual displacement $\delta\boldsymbol{r}_i$ *equal* to $d\boldsymbol{r}_i$ for each particle to be used in Eq. 10.11, thus bringing us to a contradiction between this equation and Eq. 10.13. Since we have taken Eq. 10.11 to apply, we conclude that the supposition of non-equilibrium which led to Eq. 10.13 must be invalid, and so the body must be in equilibrium. This logic proves the sufficiency condition for the principle of virtual work in the case of a rigid body with ideal constraints that is initially stationary at the time of application of the active forces.

Consider now *several* movable rigid bodies that are interconnected by smooth pins and ball joints or that are in direct frictionless contact with each other (Fig. 10.5). Some of these bodies are also ideally constrained by immovable rigid bodies in the manner described above. Again, we may examine the system of particles $m_i$ making

[6]The virtual displacements $\delta\boldsymbol{r}_i$ of each of the two particles must then be the result of the superposition of $\delta\boldsymbol{R}$ and $\delta\phi$.

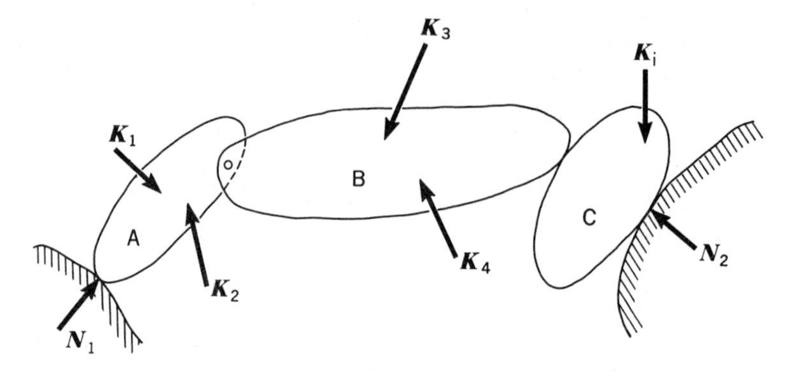


**Figure 10.5.** System of ideally constrained rigid bodies acted on by forces $\boldsymbol{K}_i$.

up the various rigid bodies. The only new kind of force to be considered is a force at the connecting point between bodies. The force on one such particle on body $A$ will be equal and opposite to the force on the corresponding particle in body $B$ at the contact point; and so on. Since such pairs of contiguous particles have the same virtual displacement, clearly the virtual work at all connecting points between bodies is zero for any virtual displacement of the system consistent with the constraints. Hence, we can say *for a system of initially stationary rigid bodies, the necessary and sufficient condition of equilibrium is that the virtual work of the active forces be zero for all possible virtual displacements consistent with the constraints.* We may then use the following equation instead of equilibrium:

$$\sum_i (\boldsymbol{K}_R)_i \cdot \delta \boldsymbol{r}_i = \delta W = 0 \tag{10.14}$$

where $(\boldsymbol{K}_R)_i$ are the active forces on the system of rigid bodies and $\delta \boldsymbol{r}_i$ are the movements of the application of these forces during a virtual displacement of the system consistent with the constraints.

## 10.4 Degrees of Freedom and the Solution of Problems

We have developed equations sufficient for equilibrium of initially stationary systems of bodies by using the concept of virtual work for virtual displacements consistent with the constraints. These equations do not involve reactions or connecting forces, and when these forces are not of interest, the method is quite useful. Thus, we may solve for as many unknown *active* forces as there are *independent* equations stemming from virtual displacements. Then our prime interest is to know how many independent equations can be written for a system stemming from virtual displacements.

For this purpose, we define *the number of degrees of freedom of a system as the number of generalized coordinates*[7] *which is required to fully specify the configuration of the system.* Thus, for the pendulum in Fig. 10.6, which is restricted to move in a plane, one *independent* coordinate $\theta$ locates the pendulum. Hence, this system has but one degree of freedom. We may ask: Can't we specify $x$ and $y$ of the bob, and thus aren't there two degrees of freedom? The answer is no, because when we specify $x$ or $y$, the other coordinate is *determined* since the pendulum support, being inextensible, must sweep out a known circle as shown in the diagram. In Fig. 10.7, the piston and crank arrangement, the four-bar linkage,[8] and the balance require only one coordinate and thus have but one degree of freedom. On the other hand, the double pendulum has two degrees of freedom and a particle in space has three degrees of freedom. The number of degrees of freedom may usually be readily determined by inspection.

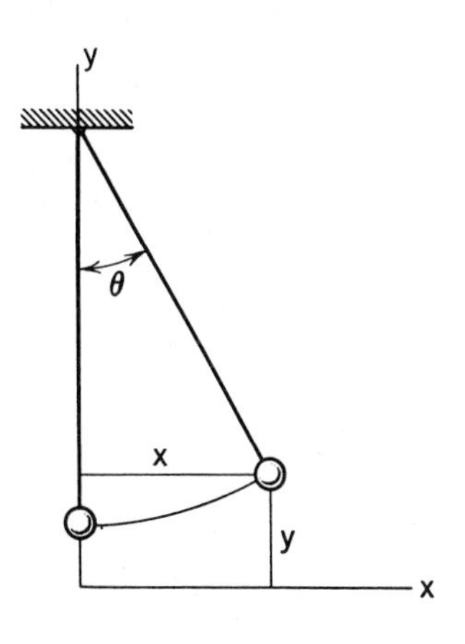


**Figure 10.6.** Plane pendulum.

Since each degree of freedom represents an independent coordinate, we can, for an $n$-degree-of-freedom system, institute $n$ unique virtual displacements by varying each coordinate separately. This procedure will then give $n$ independent equations of equilibrium from which $n$ unknowns related to the active forces can be solved. We shall examine several problems to illustrate the method of virtual work and its advantages.

Before considering the examples, we wish to point out that a torque $\boldsymbol{M}$ undergoing a virtual displacement $\delta\boldsymbol{\phi}$ does an amount of virtual work $\delta W$ equal

$$\delta W = \boldsymbol{M} \cdot \delta\boldsymbol{\phi} \tag{10.15}$$

The proof of this is asked for in Problem 10.30.

### EXAMPLE 10.1

In Fig. 10.8 is a device for compressing metal scrap, namely a compactor. A horizontal force $P$ is exerted on joint $B$. The piston at $C$ then compresses the scrap material. For a given force $P$ and a given angle $\theta$, what is the force $F$ developed on the scrap by the piston $C$? Neglect the friction between the piston and the cylinder wall, and consider the pin joints to be ideal.

We see by inspection that one coordinate $\theta$ describes the configuration of the system. The device therefore has one degree of freedom. We shall neglect the weight

---

[7]*Generalized coordinates* are any set of *independent* numbers that can fully specify the configuration of a system. Generalized coordinates can include any of usual coordinates, such as Cartesian coordinates or cylindrical coordinates, but need not. We shall only consider these cases where the usual coordinates serve as the generalized coordinates.

[8]The fourth bar is the base.

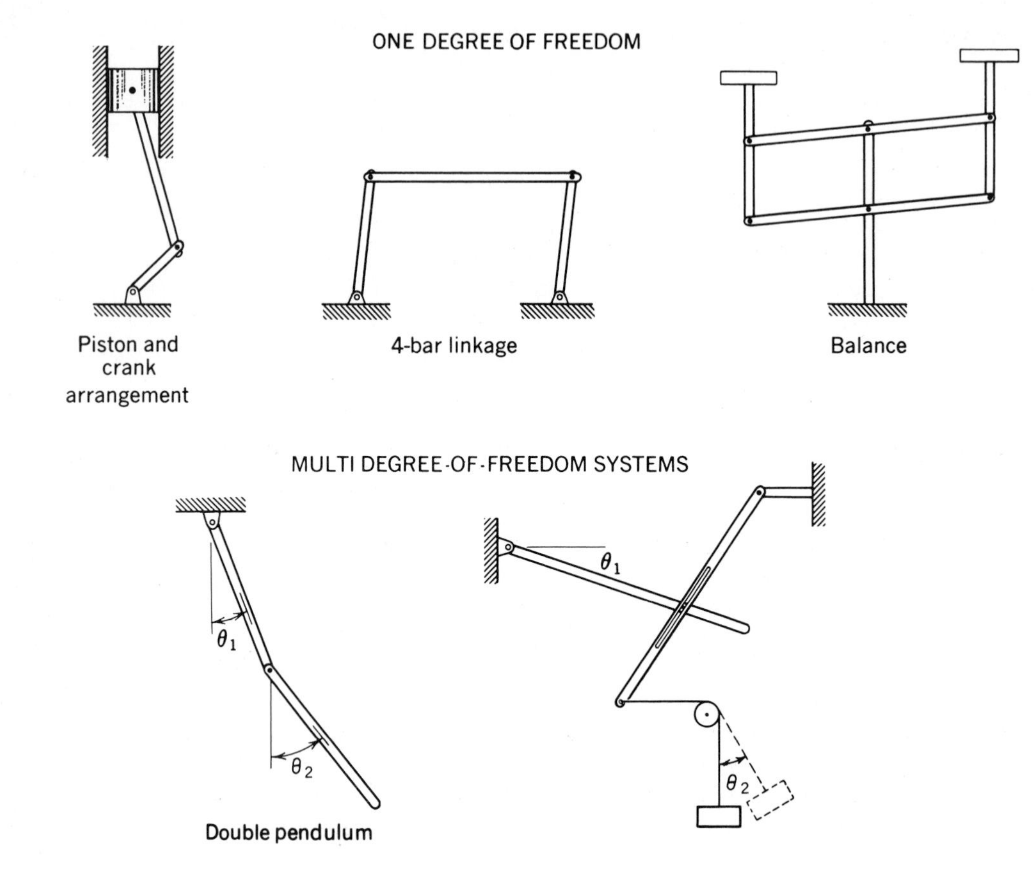


**Figure 10.7.** Various systems illustrating degrees of freedom.

of the members, and so only two active forces are present, $P$ and $F$. By assuming a virtual displacement $\delta\theta$, we will involve in the principle of virtual work only those quantities that are of interest to us, $P$, $F$, and $\theta$.[9] Let us then compute the virtual work of the active forces.

**Force *P*.** The virtual displacement $\delta\theta$ is such that force $P$ has a motion in the horizontal direction of ($l\,\delta\theta \cos\theta$) as can readily be deduced from Fig. 10.9 by elementary trigonometric considerations. There is yet another way of deducing this horizontal motion, which, sometimes, is more desirable. Using an $xy$ coordinate system at $A$ as shown in Figs. 10.8 and 10.9, we can say for joint $B$:

$$y_B = l \sin\theta \tag{a}$$

Now take the differential of both sides of the equation to get

$$dy_B = l \cos\theta \, d\theta \tag{b}$$

[9]If we had used a free-body approach, we would have had to bring in force components at $A$ and at $C$, and we would have had to dismember the system. To appreciate the method of virtual work even for this simple problem, we urge you to at least set up the problem by the use of free-body diagrams.

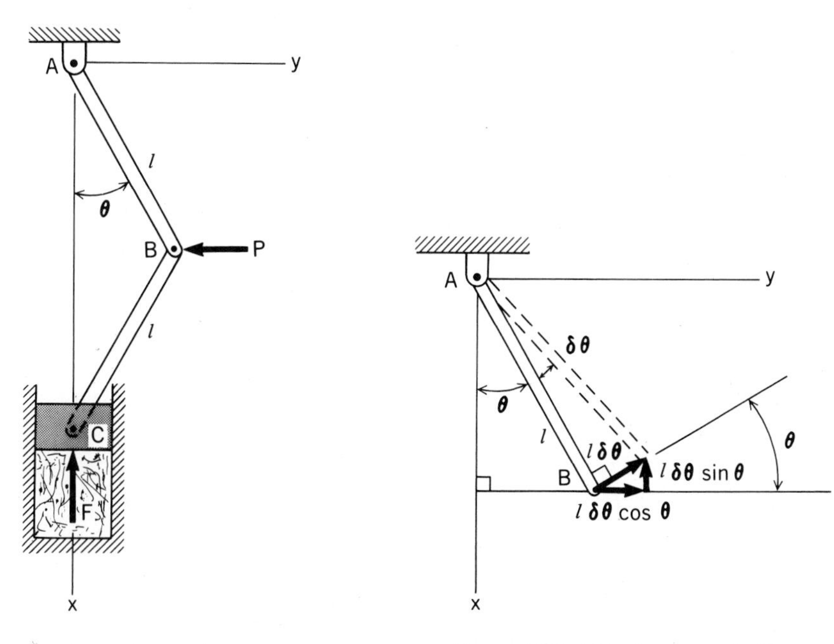


**Figure 10.8.** Compacting device.

**Figure 10.9.** Virtual movement of leg *AB*.

A differential of a quantity $A$, namely $dA$, is very similar to a variation of the quantity, $\delta A$. The former might actually take place in a process; the latter takes place in the mind of the engineer. Nevertheless, the relation between differential quantities should be the same as the relation between varied quantities. Accordingly, from Eq. (b), we can say:

$$\delta y_B = l \cos\theta \, \delta\theta \tag{c}$$

Note that the same horizontal movement of $B$ for $\delta\theta$ is thus computed as at the outset using trigonometry.

For the variation $\delta\theta$ chosen, the force $P$ has opposite direction as $\delta y_B$, and so this virtual-work contribution is negative. Thus, we have

$$\delta W_P = -Pl \cos\theta \, \delta\theta \tag{d}$$

**Force *F*.** We can use the differential approach to get the virtual displacement of piston $C$. That is,

$$x_c = l\cos\theta + l\cos\theta = 2l\cos\theta$$

$$dx_c = -2l\sin\theta \, d\theta$$

Therefore,

$$\delta x_c = -2l\sin\theta \, \delta\theta \tag{e}$$

Since the force $F$ is in the same direction as $\delta x_c$, we should have a positive result for work. Accordingly, we have

$$\delta W_c = F(2l\sin\theta)\,\delta\theta \tag{f}$$

We may now employ the principle of virtual work which recall is sufficient here for ensuring equilibrium. Thus, we can say that

$$-Pl\cos\theta\,\delta\theta + F(2l\sin\theta)\,\delta\theta = 0 \tag{g}$$

Canceling $l\,\delta\theta$ and solving for $F$, we get

$$F = \frac{P}{2\tan\theta} \tag{h}$$

For any given values of $P$ and $\theta$, we now know the amount of compressive force that the compactor can develop.

In the next example, we consider a case where we cannot conveniently use the differential approach in arriving at the virtual displacements.

**EXAMPLE 10.2**

A tractor with a bulldozer (Fig. 10.10) is used to push at constant speed an earthmover. If the tractor force on the earthmover is 150 kN, what is the hydraulic ram force $F$ at joint $B$?

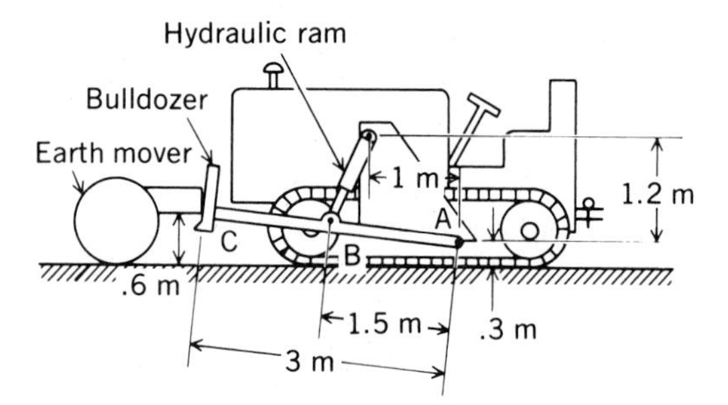


**Figure 10.10.** Tractor with bulldozer.

Member $ABC$ (the bulldozer) has a single degree of freedom relative to the tractor frame. We show this member diagrammatically in Fig. 10.11, with independent coordinate $\theta$ and active forces $F$ and 150 kN. The 1.492 m dimension in Fig.

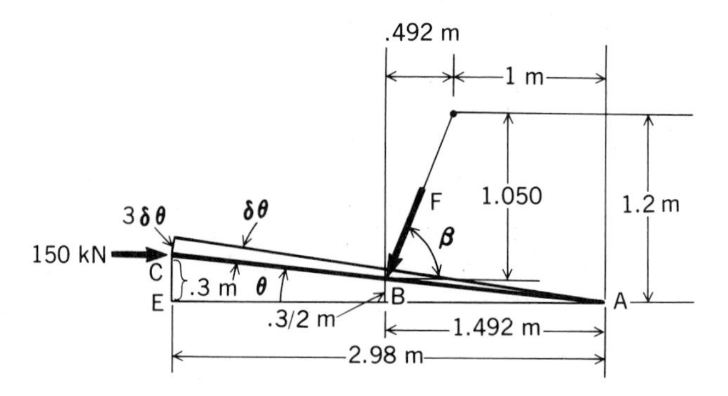


**Figure 10.11.** Bulldozer portion of device.

10.11 is determined using the Pythagorean theorem. That is, first noting in Fig. 10.10 that $\overline{AB} = 1.5$ m, we have on observing Fig. 10.11:

$$\left[\overline{AB}^2 - \left(\frac{.6 - .3}{2}\right)^2\right]^{\frac{1}{2}} = (1.5^2 - .15^2)^{\frac{1}{2}} = 1.492 \text{ m}$$

The .492 m dimension in the upper part of Fig. 10.11 follows directly. Furthermore, $\overline{AE} = 2(1.492) = 2.98$ m. The angle $\theta$ is next easily determined.

$$\theta = \tan^{-1}\frac{.3}{2.98} = 5.74°$$

The 1.050 m dimension can next be determined. Finally the angle $\beta$ is computed as

$$\beta = \tan^{-1}\frac{1.050}{.492} = 64.9°$$

In computing the virtual work of the ram force $F$, it is simplest not to use the differential approach that worked so nicely in the preceding example, but instead to use simple geometry and trigonometry. The reason for the geometrical approach here is due to the fact that $F$ (see Fig. 10.12) is not in a convenient coordinate direction in relation to $\theta$ as was the case for the active forces in the earlier cases. We can now use the method of virtual work as follows.

$$\delta W = (150)[(3)(\delta\theta)\sin\theta] - F[1.5\delta\theta\cos(90 - \theta - \beta)] = 0$$
$$(150)(3)(\delta\theta)\sin(5.74°) - F[1.5\delta\theta\cos(90 - 5.74 - 64.9)] = 0$$

Therefore,

$$F = 31.8 \text{ kN}$$

The decision as to using differentials or as to using simple geometry and trigonometry depends on the problem.

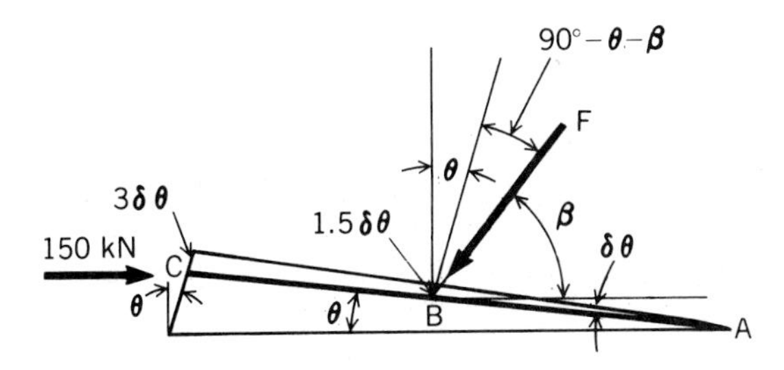


**Figure 10.12.** Give $AC$ virtual displacement $\delta\theta$.

***EXAMPLE 10.3**

A hydraulic-lift platform for loading trucks in shown in Fig. 10.13(a). Only one side of the system is shown; the other side is identical. If the diameter of the piston in the cylinder is 4 in., what pressure $p$ is needed to support a load $W$ of 5000 lb when $\theta = 60°$? The following additional data apply:

$$l = 24 \text{ in.}$$
$$d = 60 \text{ in.}$$
$$e = 10 \text{ in.}$$

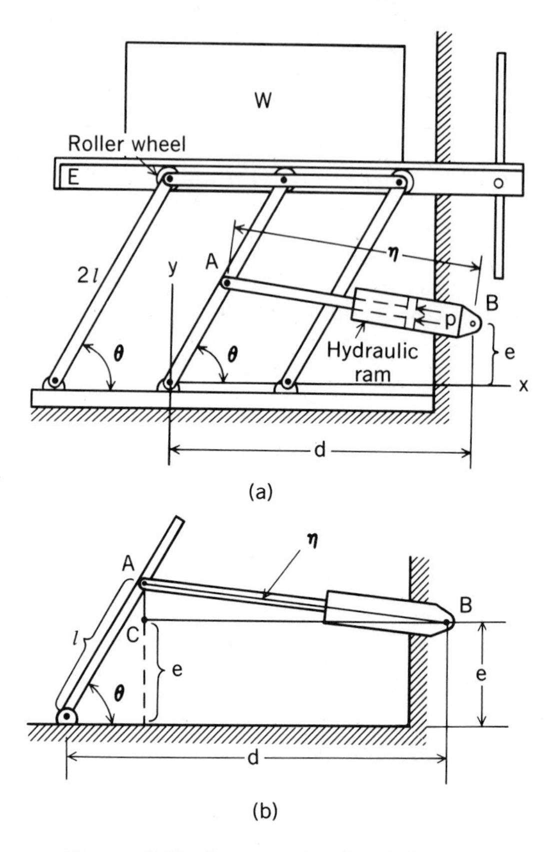


**Figure 10.13.** Pneumatic loading platform.

Pin $A$ is at the center of the rod.

We have here a system with one degree of freedom characterized by the angle $\theta$. The active forces that do work during a virtual displacement $\delta\theta$ are the weight $W$ and the force from the hydraulic ram. Accordingly, the virtual movements of both the platform and joint $A$ of the pump must be found. Note first using reference $xy$:

$$y_E = 2l \sin \theta$$

Therefore,

$$\delta y_E = 2l \cos \theta \; \delta\theta \tag{a}$$

For the ram force, we want the movement of pin $A$ in the direction of the axis of the pump, namely $\delta\eta$, where $\eta$ is shown in Fig. 10.13(a). Observing Fig. 10.13(b) we can say for $\eta$:

$$\begin{aligned} \eta^2 &= \overline{AC}^2 + \overline{CB}^2 \\ &= [(l \sin \theta) - e]^2 + (d - l \cos \theta)^2 \end{aligned} \tag{b}$$

Hence, we have

$$2\eta \; \delta\eta = 2(l \sin \theta - e)(l \cos \theta) \; \delta\theta + 2(d - l \cos \theta)(l \sin \theta) \; \delta\theta \tag{c}$$

Solving $\delta\eta$, we get

$$\begin{aligned}\delta\eta &= \frac{l}{\eta}[(l\sin\theta - e)\cos\theta + (d - l\cos\theta)\sin\theta]\,\delta\theta \\ &= \frac{l}{\eta}(l\sin\theta\cos\theta - e\cos\theta + d\sin\theta - l\sin\theta\cos\theta)\,\delta\theta \qquad \text{(d)} \\ &= \frac{l}{\eta}(d\sin\theta - e\cos\theta)\,\delta\theta\end{aligned}$$

The principle of virtual work is now applied to ensure equilibrium. Thus, considering one side of the system and using half the load, we have

$$-\frac{W}{2}(\delta y_B) + \left[p\frac{\pi(4^2)}{4}\right]\delta\eta = 0$$

Hence,

$$-(2500)(2l\cos\theta\,\delta\theta) + p(4\pi)\left[\frac{l}{\eta}(d\sin\theta - e\cos\theta)\right]\delta\theta = 0 \qquad \text{(e)}$$

The value of $\eta$ at the configuration of interest may be determined from Eq. (b). Thus,

$$\eta^2 = [(24)(.866) - 10]^2 + [60 - (24)(.5)]^2$$

Therefore,

$$\eta = 49.2 \text{ in.}$$

Now canceling $\delta\theta$ and substituting known data into Eq. (e), we may then determine $p$ for equilibrium:

$$-(2500)(2)(24)(.5) + p(4\pi)\left\{\frac{24}{49.2}[(60)(.866) - (10)(.5)]\right\} = 0$$

Therefore,

$$p = 208 \text{ psi} \qquad \text{(f)}$$

In a few of the homework problems, you have to use simple kinematics of a cylinder rolling without slipping (see Fig. 10.14). You will recall from physics that the cylinder is actually rotating about the point of contact $A$. If the cylinder rotates an angle $\delta\theta$, then $\delta C = -r\,\delta\theta$. We shall consider kinematics of rigid bodies in detail later in the text.

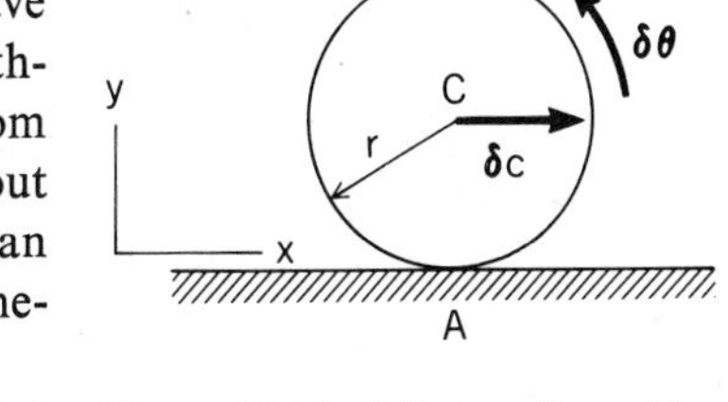


**Figure 10.14.** Cylinder rolling without slipping.

In concluding this section, we wish to point out that the method of virtual work is actually *not* restricted to ideal systems. Furthermore, it is permissible to give virtual displacements that *violate* one or more constraints. We then proceed by considering those friction forces and torques that perform virtual work as active forces. And, where a constraint is violated, we consider the corresponding constraining force or torque to be active. We point out that the method of virtual work offers generally no advantage in situations where there is friction and where constraints are violated. Furthermore, the extensions of virtual work to other useful theories are primarily restricted to ideal systems. Accordingly we shall consider only ideal systems and shall take virtual displacements that do not violate constraints.

## Problems

**10.1.** How many degrees of freedom do the following systems possess? What coordinates can be used to locate the system?

(a) A rigid body not constrained in space.

(b) A rigid body constrained to move along a plane surface.

(c) The board *AB* in the diagram (a).

(d) The spherical bodies shown in diagram (b) may slide along shaft *C–C*, which in turn rotates about axis *E–E*. Shaft *C–C* may also slide along *E–E*. The spindle *E–E* is on a rotating platform. Give the number of degrees of freedom and coordinates for a sphere, shaft *C–C*, and spindle *E–E*.

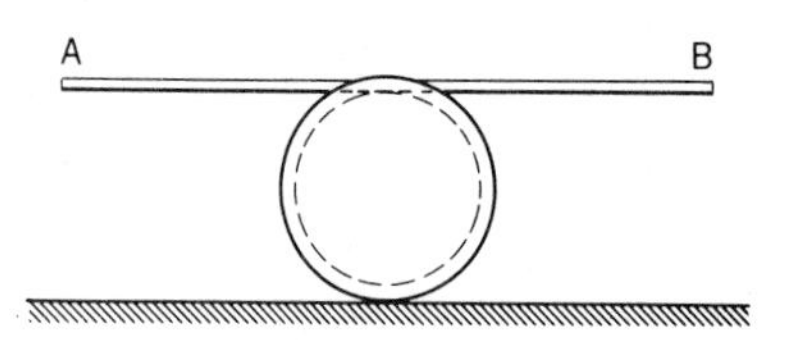

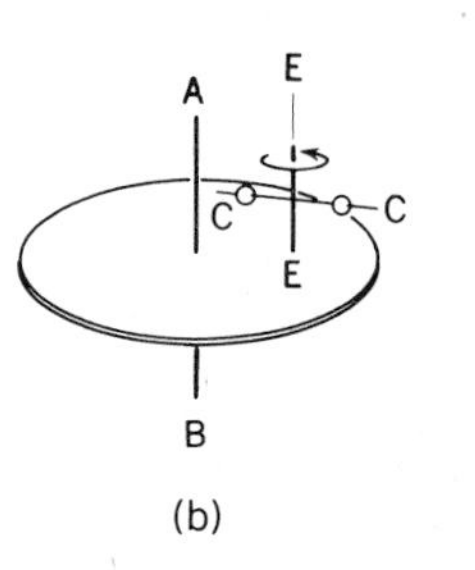


Figure P.10.1

**10.2.** A parking-lot gate arm weighs 150 N. Because of the taper, the weight can be regarded as concentrated at a point 1.25 m from the pivot point. What is the solenoid force to lift the gate? What is the solenoid force if a 300-N counterweight is placed .25 m to left of the pivot point?

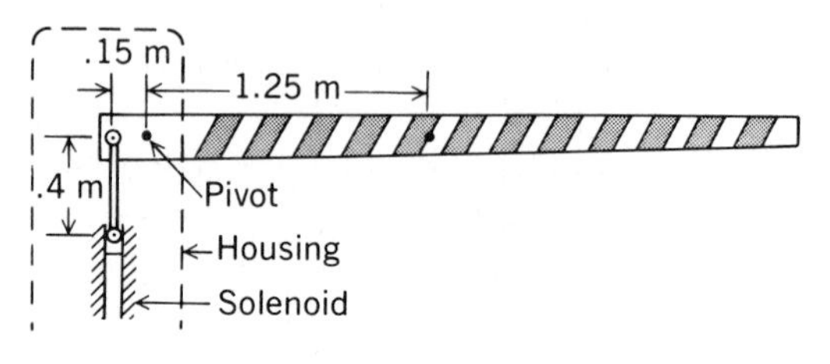

Figure P.10.2

**10.3.** What is the longest portion of pipe weighing 400 lb/ft that can be lifted without tipping the 12,000-lb tractor?

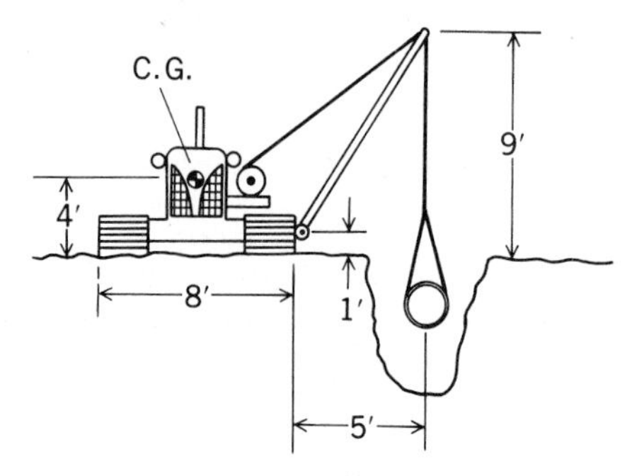


Figure P.10.3

**10.4.** If $W_1 = 100$ N and $W_2 = 150$ N, find the angle $\theta$ for equilibrium.

Figure P.10.4

**10.5.** The triple pulley sheave and the double pulley sheave weigh 150 N and 100 N, respectively. What rope force is necessary to lift a 3500-N engine?

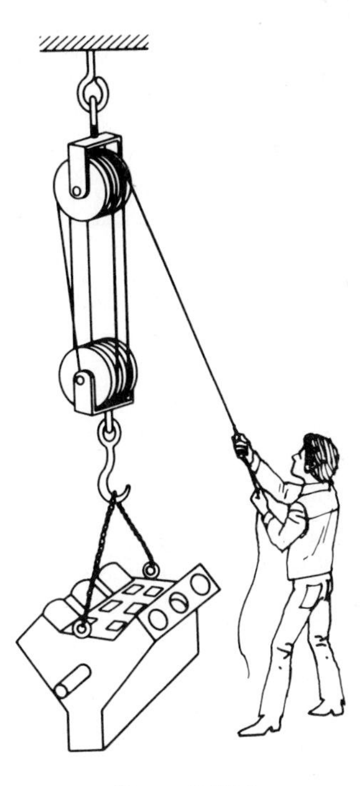

Figure P.10.5

**10.6.** What weight $W$ can be lifted with the A-frame hoist in the position shown if the cable tension is $T$?

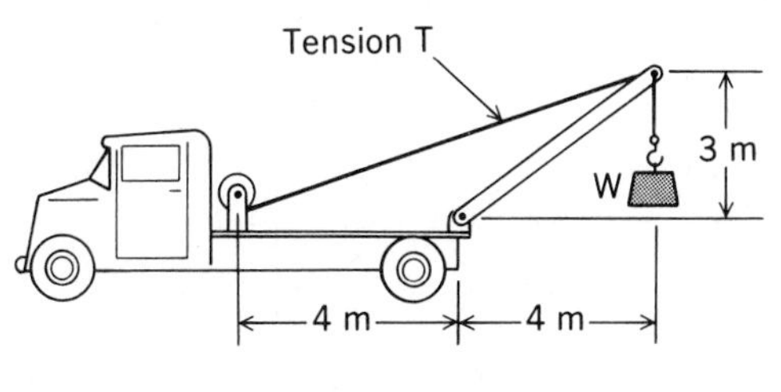


Figure P.10.6

**10.7.** A small hoist has a lifting capacity of 20 kN. What is the maximum cable tension?

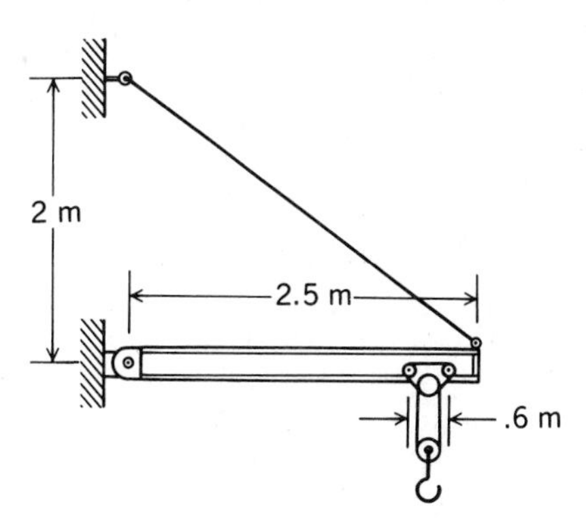

Figure P.10.7

**10.8.** If $W = 1000$ N and $P = 300$ N, find the angle $\theta$ for equilibrium.

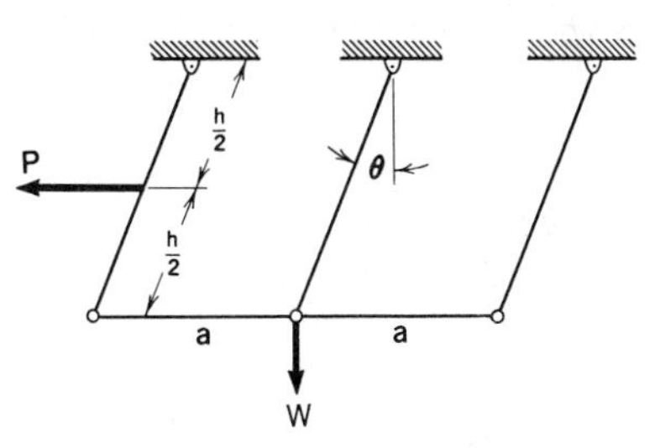


Figure P.10.8

**10.9.** What is the tension in the cables of a 10-ft-wide 12-ft-long 6000-lb drawbridge when the bridge is first raised? When the bridge is at 45°?

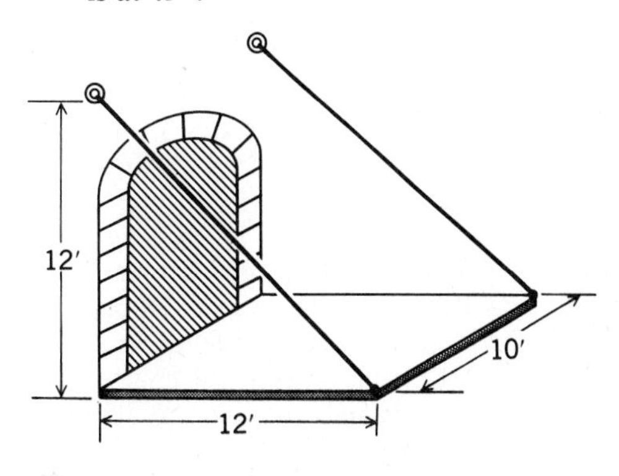


Figure P.10.9

**10.10.** Assuming frictionless contacts, determine the magnitude of $P$ for equilibrium.

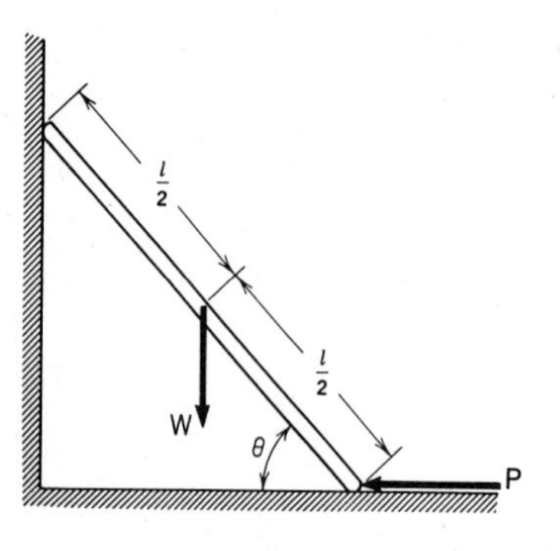


Figure P.10.10

**10.11.** A rock crusher is shown in action. If $p_1 = 50$ psig and $p_2 = 100$ psig, what is the force on the rock at the configuration shown? The diameter of the pistons is 4 in.

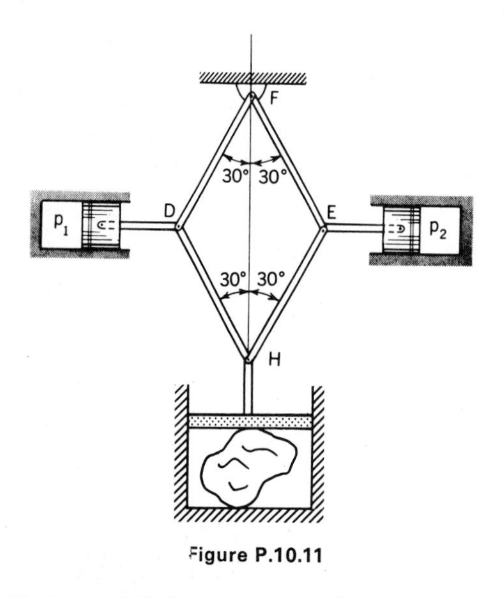


Figure P.10.11

**10.12.** A 20-lb-ft torque is applied to a scissor jack. If friction is disregarded throughout, what weight can be maintained in equilibrium? Take the pitch of the screw threads to be .3 in. in opposite senses. All links are of equal length, 1 ft.

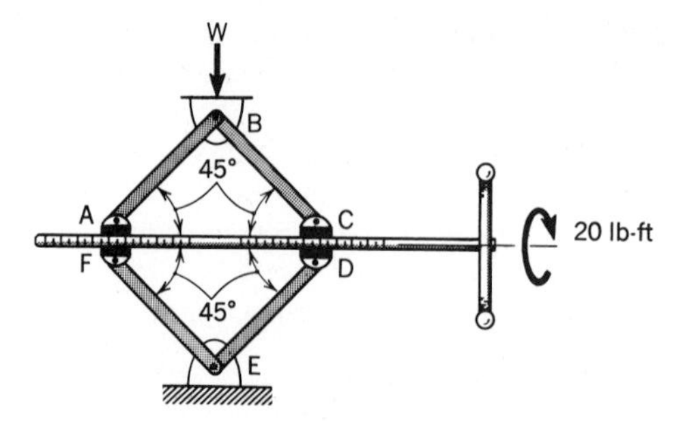

Figure P.10.12

**10.13.** The 5000-lb van of an airline food catering truck rises straight up until its floor is level with the airplane floor. What is the ram force in that position?

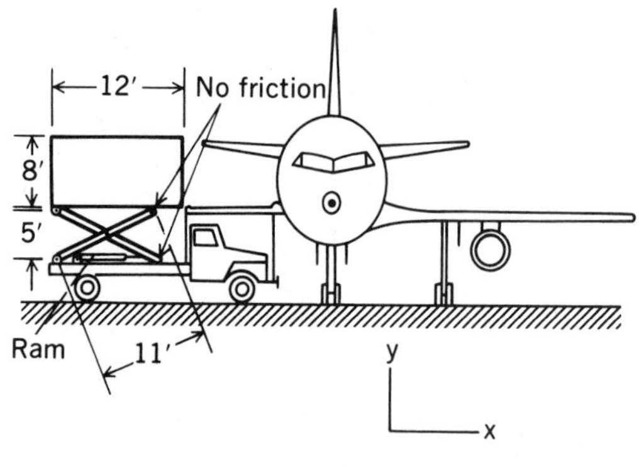

Figure P.10.13

**10.14.** What are the cable tensions when the arms of the power shovel are in the position shown? Arm $AC$ weighs 13 kN, arm $DF$ weighs 11 kN, and the shovel and payload weigh 9 kN.

Figure P.10.14

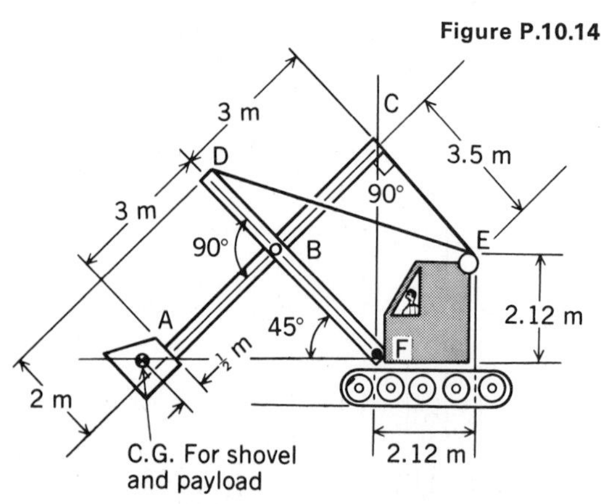

**10.15.** A hydraulically actuated gate in a 2-m-square water-carrying tunnel under a dam is held in place with a vertical beam $AC$. What is the force in the hydraulic ram if the specific weight of water is 9818 $N/m^3$? (*Hint:* See the statement just before Problem 4.59 for assistance in computing the resultant force from the water on the gate.)

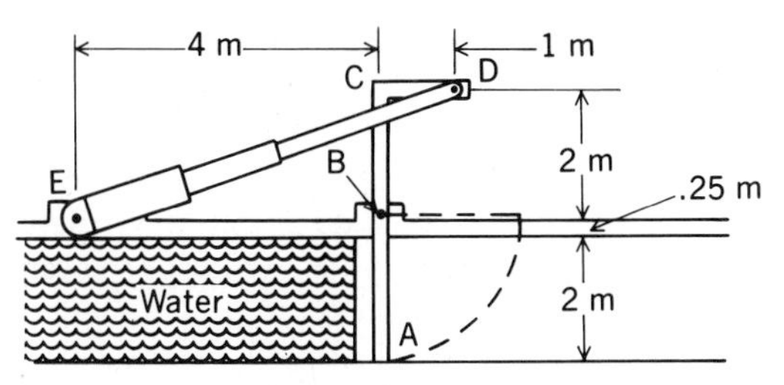


Figure P.10.15

**10.16.** Find the angle $\beta$ for equilibrium in terms of the parameters given in the diagram. Neglect friction and the weight of the beam.

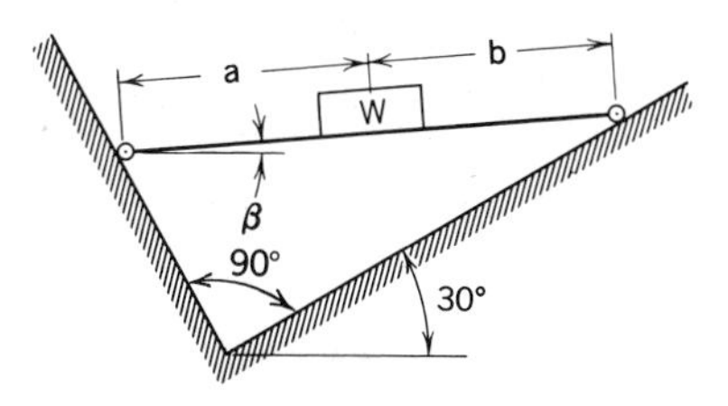


Figure P.10.16

**10.17.** Do Problem 5.55 by the method of virtual work.

**10.18.** Do Problem 5.56 by the method of virtual work.

**10.19.** What is the relation among $P$, $Q$, and $\theta$ for equilibrium?

Figure P.10.19

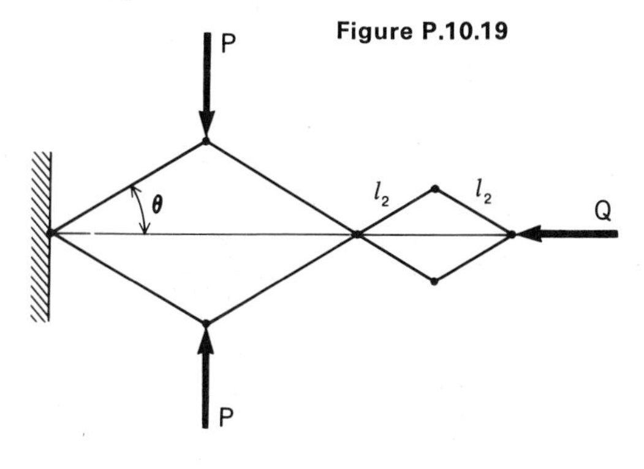


**10.20.** A paper collater is shown with the weight $Q$ of the collated papers equal to .2 N. The collater rests on a smooth surface and, accordingly, can slide on this surface with no resistance. What force $P$ is needed to keep the system in equilibrium for the position shown?

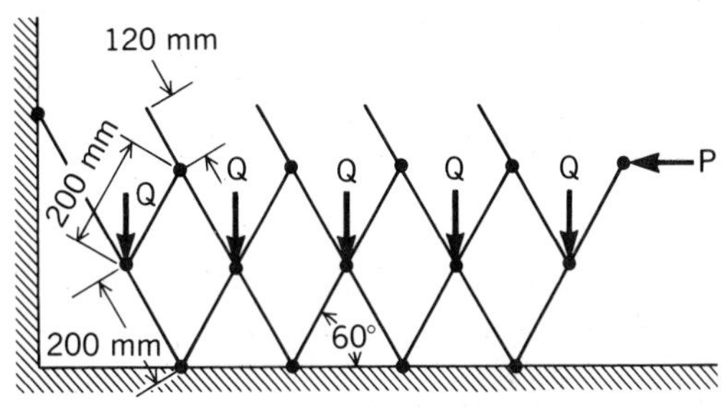


Figure P.10.20

**10.21.** A stepped cylinder of weight 500 lb is connected to vehicle $A$ weighing 300 lb and to sheave $B$ weighing 50 lb. Sheave $B$ supports a weight $C$. What is the value of the weight of $C$ for equilibrium? Neglect friction.

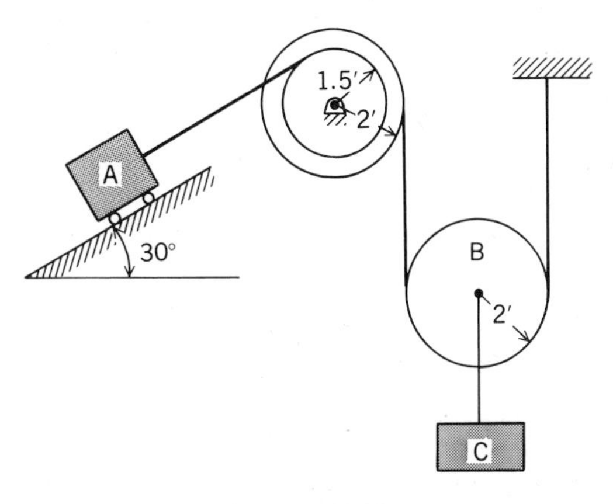


Figure P.10.21

**10.22.** Do the first part of Problem 5.71 by the method of virtual work.

**10.23.** Compute the weight $W$ that can be lifted by the *differential pulley* system for an applied force $F$. Neglect the weight of the lower pulley.

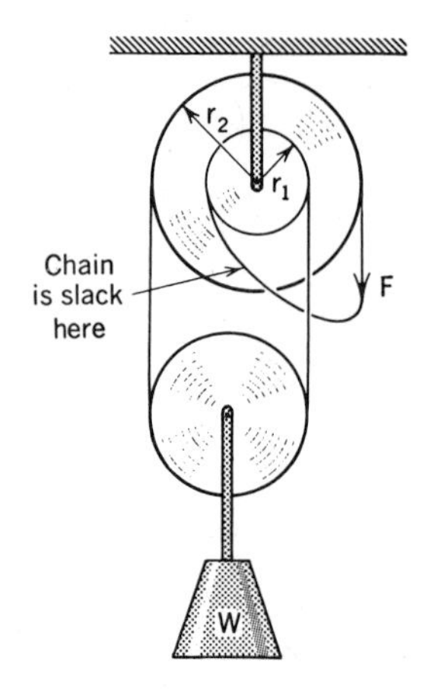

Figure P.10.23

**10.24.** The pressure $p$ driving a piston of diameter 100 mm is 1 N/mm². At the configuration shown, what weight $W$ will the system hold if we neglect friction?

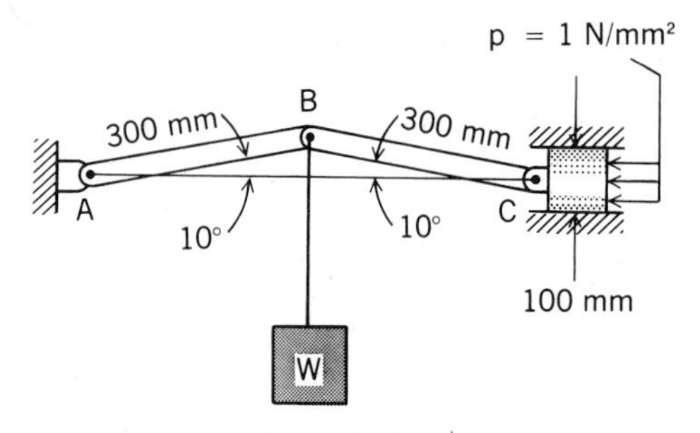


Figure P.10.24

**10.25.** Blocks $A$ and $B$ weigh 200 N and 150 N, respectively. They are connected at their base by a light cord. At what position $\theta$ is there equilibrium if we disregard friction?

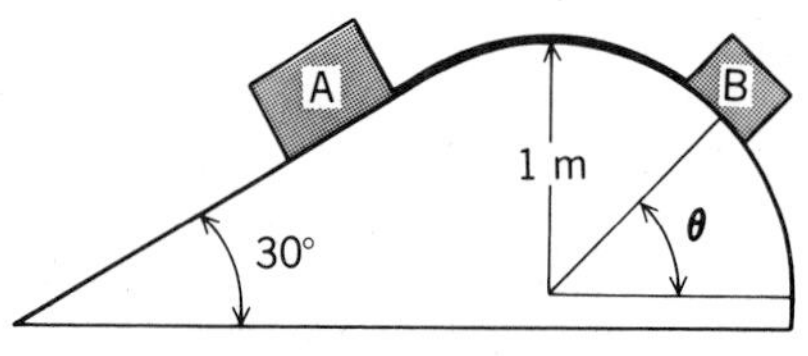


Figure P.10.25

**10.26.** If $A$ weighs 500 N, and if $B$ weighs 100 N, determine the proper weight of $C$ for equilibrium.

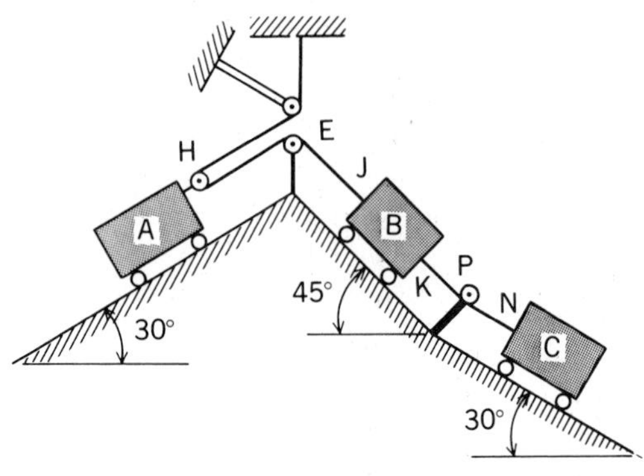


Figure P.10.26

**10.27.** An embossing device imprints an image at $D$ on metal stock. If a force $F$ of 200 N is exerted by the operator, what is the force at $D$ on the stock? The lengths of $AB$ and $BC$ are each 150 mm.

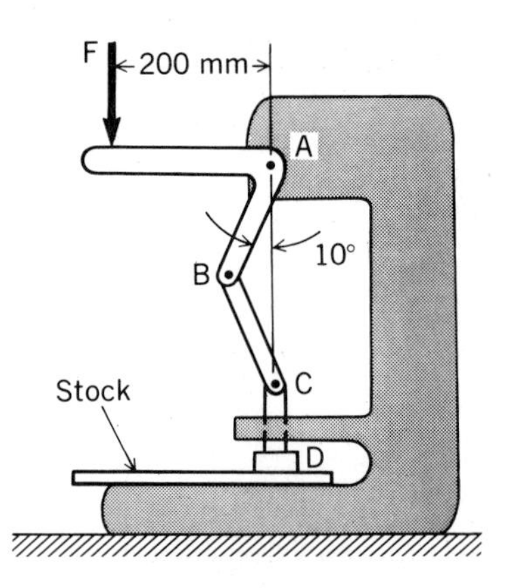

Figure P.10.27

**10.28.** A support system holds a 500-N load. Without the load, $\theta = 45°$ and the spring is not compressed. If $K$ for the spring is 10,000 N/m, how far down $d$ will the 500-N load depress the upper platform if the load is applied slowly and carefully? Neglect all other weights. $DB = BE = AB = CB =$ 400 mm. (*Note:* The force from the spring is $K$ times its contraction.)

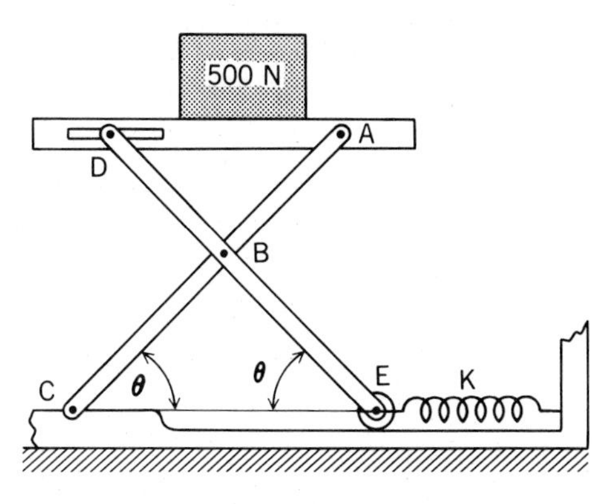


Figure P.10.28

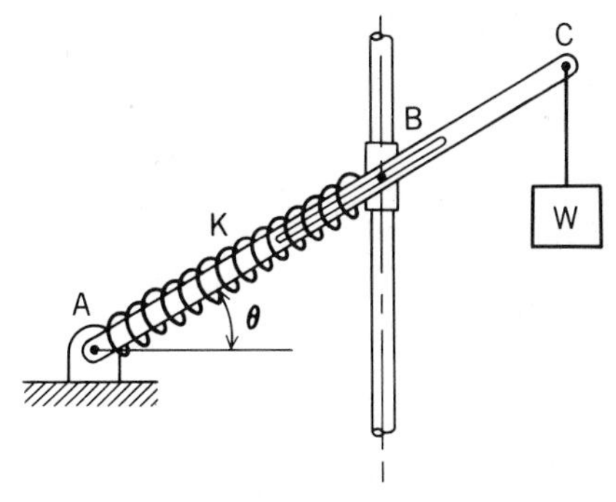


Figure P.10.29

**10.29.** Rod *ABC* is connected through a pin and slot to a sleeve which slides on a vertical rod. Before the weight *W* of 100 N is applied at *C*, the rod is inclined at an angle of 45°. If *K* of the spring is 8000 N/m, what is the angle $\theta$ for equilibrium? The length of *AB* is 300 mm and the length of *BC* is 200 mm when $\theta = 45°$. Neglect friction and all weights other than *W*. (*Note:* The force from the spring is *K* times its contraction.)

**10.30.** Show that the virtual work of a couple moment $\boldsymbol{M}$ for a rotation $\delta\boldsymbol{\phi}$ is given as:

$$\delta W = \boldsymbol{M} \cdot \delta\boldsymbol{\phi}$$

(*Hint:* Decompose $\boldsymbol{M}$ into components normal to and collinear with $\delta\boldsymbol{\phi}$.)

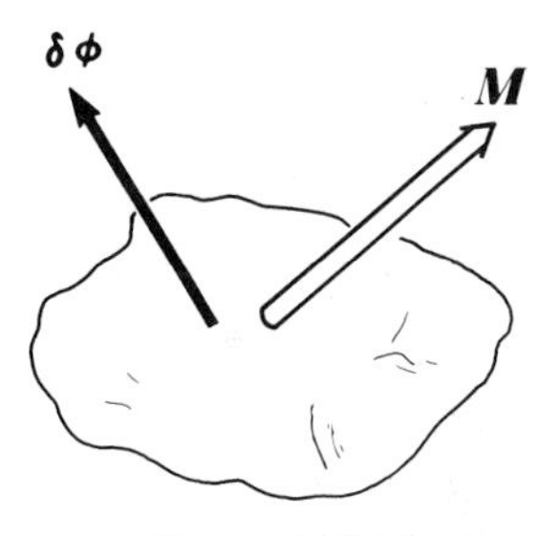


Figure P.10.30

# Part B
# Method of Total Potential Energy

## 10.5 Conservative Systems

We shall restrict ourselves in this section to certain types of active forces. This restriction will permit us to arrive at some additional very useful relations.

Consider first a body acted on only by gravity force *W* as an active force and moving along a frictionless path from position 1 to position 2, as shown in Fig. 10.15. The work done by gravity, $W_{1-2}$, is then

$$W_{1-2} = \int_1^2 \boldsymbol{F} \cdot d\boldsymbol{r} = \int_1^2 (-W\boldsymbol{j}) \cdot d\boldsymbol{r} = -W\int_1^2 dy = -W(y_2 - y_1) \\ = W(y_1 - y_2) \qquad (10.16)$$

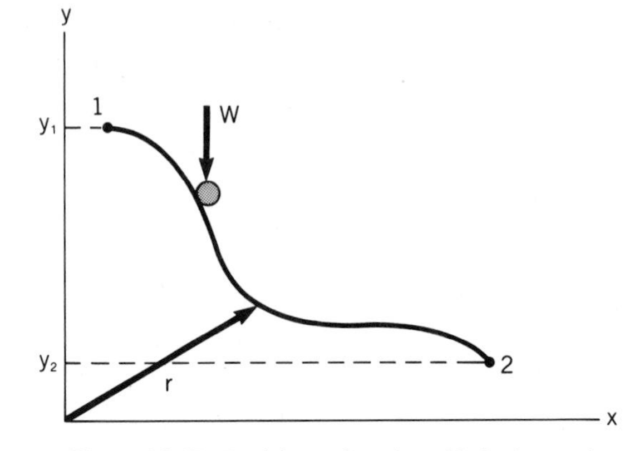


**Figure 10.15.** Particle moving along frictionless path.

Note that the work done *does not depend* on the path, but depends only on the positions of the *end points* of the path. *Force fields which are functions of position and whose work like gravity is independent of the path are called conservative force fields.* In general, we can say for a conservative force field $\boldsymbol{F}(x, y, z)$ that, along a path between positions 1 and 2, the work is analogous to that in Eq. 10.16

$$W_{1-2} = \int_1^2 \boldsymbol{F} \cdot d\boldsymbol{r} = V_1(x, y, z) - V_2(x, y, z) \tag{10.17}$$

where $V$, a scalar function evaluated at the end points, is called the *potential energy function.*[10] We may rewrite the equation above as follows:

$$-\int_1^2 \boldsymbol{F} \cdot d\boldsymbol{r} = V_2(x, y, z) - V_1(x, y, z) = \Delta V \tag{10.18}$$

From the result above we can say that *the change in potential energy*, $\Delta V$, associated with a force field is *the negative of the work done by this force field in going from position 1 to position 2 along any path.* For any *closed* path, the work done by a conservative force field $\boldsymbol{F}$ is

$$\oint \boldsymbol{F} \cdot d\boldsymbol{r} = 0 \tag{10.19}$$

How is the potential energy function $V$ related to $\boldsymbol{F}$? To answer this query, consider that an arbitrary infinitesimal path $d\boldsymbol{r}$ starts from point 1. We can then give Eq. 10.18 as

$$\boldsymbol{F} \cdot d\boldsymbol{r} = -dV \tag{10.20}$$

Expressing the dot product on the left side in terms of components, and expressing $dV$ as a total differential, we get

$$F_x\,dx + F_y\,dy + F_z\,dz = -\left(\frac{\partial V}{\partial x}\,dx + \frac{\partial V}{\partial y}\,dy + \frac{\partial V}{\partial z}\,dz\right) \tag{10.21}$$

---

[10]The context of any discussion should make clear whether $V$ refers to potential energy, or to speed, or to volume.

We can conclude from the equation above that since $\boldsymbol{dr}$ is arbitrary

$$F_x = -\frac{\partial V}{\partial x}$$

$$F_y = -\frac{\partial V}{\partial y} \tag{10.22}$$

$$F_z = -\frac{\partial V}{\partial z}$$

Or, in other words,

$$\begin{aligned} \boldsymbol{F} &= -\left(\frac{\partial V}{\partial x}\boldsymbol{i} + \frac{\partial V}{\partial y}\boldsymbol{j} + \frac{\partial V}{\partial z}\boldsymbol{k}\right) \\ &= -\left(\frac{\partial}{\partial x}\boldsymbol{i} + \frac{\partial}{\partial y}\boldsymbol{j} + \frac{\partial}{\partial z}\boldsymbol{k}\right)V \\ &= -\mathbf{grad}\, V = -\boldsymbol{\nabla} V \end{aligned} \tag{10.23}$$

The operator we have introduced is called the *gradient* operator and is given as follows for rectangular coordinates:

$$\mathbf{grad} \equiv \boldsymbol{\nabla} \equiv \frac{\partial}{\partial x}\boldsymbol{i} + \frac{\partial}{\partial y}\boldsymbol{j} + \frac{\partial}{\partial z}\boldsymbol{k} \tag{10.24}$$

We can now say, as an alternative definition, that *a conservative force field must be a function of position and expressible as the gradient of a scalar field function.* The inverse to this statement is also valid. *That is, if a force field is a function of position and the gradient of a scalar field, it must then be a conservative force field.*

Two examples of conservative force fields will now be presented and discussed.

**Constant Force Field.** If the force field is constant at all positions, it can always be expressed as the gradient of a scalar function of the form $V = -(ax + by + cz)$, where $a$, $b$, and $c$ are constants. The constant force field, then, is $\boldsymbol{F} = a\boldsymbol{i} + b\boldsymbol{j} + c\boldsymbol{k}$.

In limited changes of position near the earth's surface (a common situation), we can consider the gravitational force on a particle of mass, $m$, as a constant force field given by $-mg\boldsymbol{k}$. Thus, the constants for the general force field given above are $a = b = 0$ and $c = -mg$. Clearly, $V = mgz$ for this case.

**Force Proportional to Linear Displacements.** Consider a body limited by constraints to move along a straight line. Along this line a force is developed directly proportional to the displacement of the body from some point on the line. If this line is the $x$ axis, we give this force as

$$\boldsymbol{F} = -Kx\boldsymbol{i} \tag{10.25}$$

where $x$ is the displacement from the point. The constant $K$ is a positive number, so that, with the minus sign in this equation, a positive displacement $x$ from the origin means that the force is negative and is then directed back to the origin. A displacement in the negative direction from the origin (negative $x$) means that the force is positive and is directed again toward the origin. Thus, the force given above is a

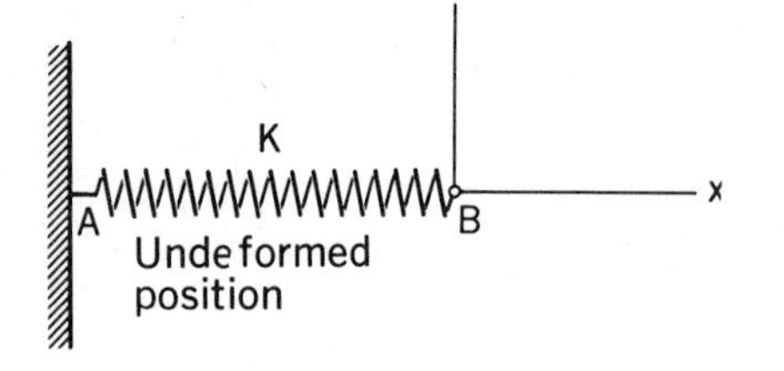


Figure 10.16. Linear spring.

*restoring* force about the origin. An example of this force is that of a linear spring (Fig. 10.16). The force that the spring exerts will be directly proportional to the amount of elongation or compression in the $x$ direction beyond the unextended position which is taken at the origin of the $x$ axis. Furthermore, the force is a restoring force. The constant $K$ in this situation is called the *spring constant.*

The change in potential energy due to the displacements from the origin to some position $x$, therefore, is

$$V = \frac{Kx^2}{2} \tag{10.26}$$

The *change* in potential energy has been defined as the *negative* of the work done by a conservative force as we go from one position to another. Clearly, the potential energy change is then *directly equal* to the work done by the *reaction* to the conservative force during this displacement. In the case of the spring, the reaction force would be the force *from* the surroundings acting *on* the spring at point $B$ (Fig. 10.16). During extension or compression of the spring from the undeformed position, this force (from the surroundings) clearly must do a positive amount of work. This work must as noted above equal the potential energy change. We now note that we can consider this work (or in other words the change in potential energy) to be a measure of the energy *stored* in the spring. That is, when allowed to return to its original position, the spring will do this amount of positive work *on* the surroundings at $B$, provided that the return motion is slow enough to prevent oscillations, etc. The reason for employing the name "potential energy" for $V$ may now be more apparent.

## 10.6 Condition of Equilibrium for a Conservative System

Let us now consider a system of rigid bodies that is ideally constrained and acted on by conservative active forces. For a virtual displacement from a configuration of equilibrium, the virtual work done by the active forces, which are maintained constant during the virtual displacement, must be zero. We shall now show that the condition of equilibrium can be stated in yet another way for this system.

Specifically, suppose that we have $n$ conservative forces acting on the system of bodies. The increment of work for a real infinitesimal movement of the system can be given as follows:

$$\begin{aligned} dW &= \sum_{i=1}^{n} \boldsymbol{F}_i \cdot d\boldsymbol{r}_i \\ &= \sum_{i=1}^{n} \left[ -\left( \frac{\partial V_i}{\partial x_i}\boldsymbol{i} + \frac{\partial V_i}{\partial y_i}\boldsymbol{j} + \frac{\partial V_i}{\partial z_i}\boldsymbol{k} \right) \right] \cdot (dx_i\boldsymbol{i} + dy_i\boldsymbol{j} + dz_i\boldsymbol{k}) \\ &= \sum_{i=1}^{n} \left[ -\left( \frac{\partial V_i}{\partial x_i} dx_i + \frac{\partial V_i}{\partial y_i} dy_i + \frac{\partial V_i}{\partial z_i} dz_i \right) \right] \end{aligned}$$

We can conclude from the equation above that since $\boldsymbol{dr}$ is arbitrary

$$F_x = -\frac{\partial V}{\partial x}$$
$$F_y = -\frac{\partial V}{\partial y} \tag{10.22}$$
$$F_z = -\frac{\partial V}{\partial z}$$

Or, in other words,

$$\begin{aligned} \boldsymbol{F} &= -\left(\frac{\partial V}{\partial x}\boldsymbol{i} + \frac{\partial V}{\partial y}\boldsymbol{j} + \frac{\partial V}{\partial z}\boldsymbol{k}\right) \\ &= -\left(\frac{\partial}{\partial x}\boldsymbol{i} + \frac{\partial}{\partial y}\boldsymbol{j} + \frac{\partial}{\partial z}\boldsymbol{k}\right)V \\ &= -\textbf{grad}\ V = -\boldsymbol{\nabla} V \end{aligned} \tag{10.23}$$

The operator we have introduced is called the *gradient* operator and is given as follows for rectangular coordinates:

$$\textbf{grad} \equiv \boldsymbol{\nabla} \equiv \frac{\partial}{\partial x}\boldsymbol{i} + \frac{\partial}{\partial y}\boldsymbol{j} + \frac{\partial}{\partial z}\boldsymbol{k} \tag{10.24}$$

We can now say, as an alternative definition, that *a conservative force field must be a function of position and expressible as the gradient of a scalar field function.* The inverse to this statement is also valid. *That is, if a force field is a function of position and the gradient of a scalar field, it must then be a conservative force field.*

Two examples of conservative force fields will now be presented and discussed.

**Constant Force Field.** If the force field is constant at all positions, it can always be expressed as the gradient of a scalar function of the form $V = -(ax + by + cz)$, where $a$, $b$, and $c$ are constants. The constant force field, then, is $\boldsymbol{F} = a\boldsymbol{i} + b\boldsymbol{j} + c\boldsymbol{k}$.

In limited changes of position near the earth's surface (a common situation), we can consider the gravitational force on a particle of mass, $m$, as a constant force field given by $-mg\boldsymbol{k}$. Thus, the constants for the general force field given above are $a = b = 0$ and $c = -mg$. Clearly, $V = mgz$ for this case.

**Force Proportional to Linear Displacements.** Consider a body limited by constraints to move along a straight line. Along this line a force is developed directly proportional to the displacement of the body from some point on the line. If this line is the $x$ axis, we give this force as

$$\boldsymbol{F} = -Kx\boldsymbol{i} \tag{10.25}$$

where $x$ is the displacement from the point. The constant $K$ is a positive number, so that, with the minus sign in this equation, a positive displacement $x$ from the origin means that the force is negative and is then directed back to the origin. A displacement in the negative direction from the origin (negative $x$) means that the force is positive and is directed again toward the origin. Thus, the force given above is a

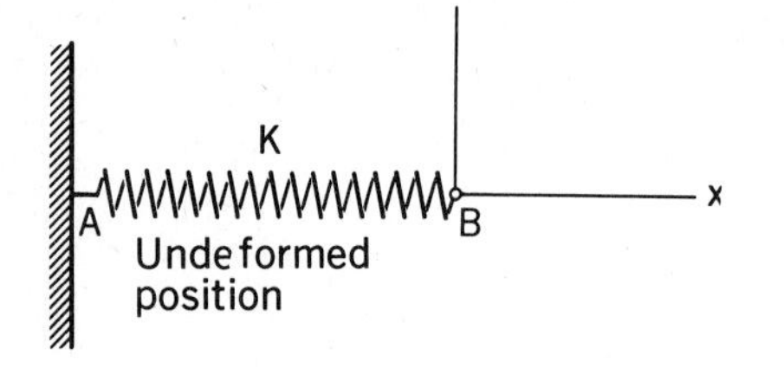

**Figure 10.16.** Linear spring.

*restoring* force about the origin. An example of this force is that of a linear spring (Fig. 10.16). The force that the spring exerts will be directly proportional to the amount of elongation or compression in the $x$ direction beyond the unextended position which is taken at the origin of the $x$ axis. Furthermore, the force is a restoring force. The constant $K$ in this situation is called the *spring constant.*

The change in potential energy due to the displacements from the origin to some position $x$, therefore, is

$$V = \frac{Kx^2}{2} \tag{10.26}$$

The *change* in potential energy has been defined as the *negative* of the work done by a conservative force as we go from one position to another. Clearly, the potential energy change is then *directly equal* to the work done by the *reaction* to the conservative force during this displacement. In the case of the spring, the reaction force would be the force *from* the surroundings acting *on* the spring at point $B$ (Fig. 10.16). During extension or compression of the spring from the undeformed position, this force (from the surroundings) clearly must do a positive amount of work. This work must as noted above equal the potential energy change. We now note that we can consider this work (or in other words the change in potential energy) to be a measure of the energy *stored* in the spring. That is, when allowed to return to its original position, the spring will do this amount of positive work *on* the surroundings at $B$, provided that the return motion is slow enough to prevent oscillations, etc. The reason for employing the name "potential energy" for $V$ may now be more apparent.

## 10.6 Condition of Equilibrium for a Conservative System

Let us now consider a system of rigid bodies that is ideally constrained and acted on by conservative active forces. For a virtual displacement from a configuration of equilibrium, the virtual work done by the active forces, which are maintained constant during the virtual displacement, must be zero. We shall now show that the condition of equilibrium can be stated in yet another way for this system.

Specifically, suppose that we have $n$ conservative forces acting on the system of bodies. The increment of work for a real infinitesimal movement of the system can be given as follows:

$$\begin{aligned} dW &= \sum_{i=1}^{n} \mathbf{F}_i \cdot d\mathbf{r}_i \\ &= \sum_{i=1}^{n} \left[ -\left( \frac{\partial V_i}{\partial x_i}\mathbf{i} + \frac{\partial V_i}{\partial y_i}\mathbf{j} + \frac{\partial V_i}{\partial z_i}\mathbf{k} \right) \right] \cdot (dx_i\mathbf{i} + dy_i\mathbf{j} + dz_i\mathbf{k}) \\ &= \sum_{i=1}^{n} \left[ -\left( \frac{\partial V_i}{\partial x_i} dx_i + \frac{\partial V_i}{\partial y_i} dy_i + \frac{\partial V_i}{\partial z_i} dz_i \right) \right] \end{aligned}$$

$$= -\sum_{i=1}^{n} dV_i = -d\left(\sum_{i=1}^{n} V_i\right) = -dV$$

where $V$ without subscripts refers to *total* potential energy. By treating $\delta \boldsymbol{r}_i$ like $d\boldsymbol{r}_i$ in the equations above, we can express the virtual work $\delta W$ as

$$\begin{aligned}
\delta W &= \sum_{i=1}^{n} \boldsymbol{F}_i \cdot \delta \boldsymbol{r}_i \\
&= \sum_{i=1}^{n} \left[-\left(\frac{\partial V_i}{\partial x}\boldsymbol{i} + \frac{\partial V_i}{\partial y}\boldsymbol{j} + \frac{\partial V_i}{\partial z}\boldsymbol{k}\right)\right] \cdot (\delta x_i \boldsymbol{i} + \delta y_i \boldsymbol{j} + \delta z_i \boldsymbol{k}) \\
&= \sum_{i=1}^{n} \left[-\left(\frac{\partial V_i}{\partial x_i}\delta x_i + \frac{\partial V_i}{\partial y_i}\delta y_i + \frac{\partial V_i}{\partial z_i}\delta z_i\right)\right] \\
&= -\sum_{i=1}^{n} \delta V_i = -\delta\left(\sum_{i=1}^{n} V_i\right) = -\delta V
\end{aligned}$$

But we know that for equilibrium $\delta W = 0$, and so we can similarly say for equilibrium:

$$\boxed{\delta V = 0} \tag{10.27}$$

Mathematically, this means that *the potential energy has a stationary or an extremum value at a configuration of equilibrium*, or, putting it another way, *the variation of V is zero at a configuration of equilibrium.*[11] Thus, we have another criterion which we may use to solve problems of equilibrium for conservative force systems with ideal constraints.

To use this formulation for solving problems, we proceed in the following manner. First, determine the potential energy of the system using a convenient set of

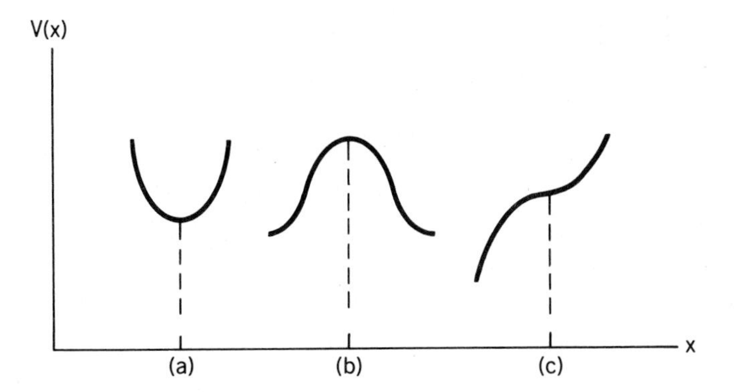


**Figure 10.17.** Stationary or extremum points.

[11] To further understand this, consider $V$ as a function of only one variable, $x$. A *stationary* value (or, as we may say, an extremum) might be a local minimum ($a$ in Fig. 10.17), a local maximum ($b$ in the figure), or an inflection point ($c$ in the figure). Note for these points that for a differential movement, $\delta x$, there is zero first-order change in $V$ (i.e., $\delta V = 0$).

independent coordinates to locate the system. Then, take the variation, $\delta$, of the potential energy. This operation is, for our purposes, the same as taking the differential. Thus, suppose that $V$ is a function of independent variables $q_1, q_2, \ldots, q_n$, thereby having $n$ degrees of freedom. The variation of $V$ becomes

$$\delta V = \frac{\partial V}{\partial q_1}\delta q_1 + \frac{\partial V}{\partial q_2}\delta q_2 + \ldots + \frac{\partial V}{\partial q_n}\delta q_n \tag{10.28}$$

For equilibrium, we set this variation equal to zero according to Eq. 10.27. For the right side of the equation above to be zero, the coefficient of each $\delta q_i$ must be zero, since the $\delta q_i$ are independent of each other. Thus,

$$\boxed{\begin{aligned} \frac{\partial V}{\partial q_1} &= 0 \\ \frac{\partial V}{\partial q_2} &= 0 \\ &\cdot \\ &\cdot \\ &\cdot \\ \frac{\partial V}{\partial q_n} &= 0 \end{aligned}} \tag{10.29}$$

We now have $n$ independent equations, which we can now solve for $n$ unknowns. This method of approach is illustrated in the following examples.

**EXAMPLE 10.4**

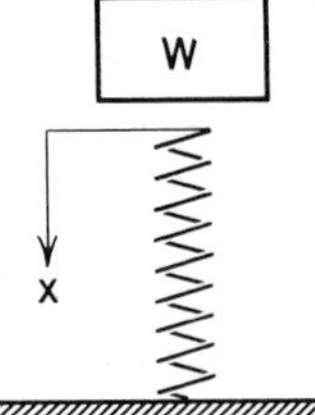


**Figure 10.18.** Mass placed on a linear spring.

A block weighing $W$ lb is placed slowly on a spring having a spring contant of $K$ lb/ft (see Fig. 10.18). Calculate how much the spring is compressed at the equilibrium configuration.

This is a simple problem and could be solved by using the definition of the spring constant, but we shall take advantage of the simplicity to illustrate the preceding comments. Notice only conservative forces act on $W$—gravity and the spring force. Using the unextended position of the spring as the datum for gravitational potential energy and measuring $x$ from this position we have, for the potential energy of the system:

$$V = -Wx + \tfrac{1}{2}Kx^2$$

Consequently, for equilibrium, we have since there is only one degree of freedom

$$\frac{dV}{dx} = -W + Kx = 0$$

Solving for $x$, we have

$$x = \frac{W}{K}$$

a circular surface. Use the top surface as a datum.)

r $AB$ is 10 ft and member $BC$ is w that the angle $\theta$ corresponding rium is 34.5° if the spring con- s 10 lb/in. Neglect the weight of bers and friction everywhere. $= 30°$ for the configuration where g is unstretched.

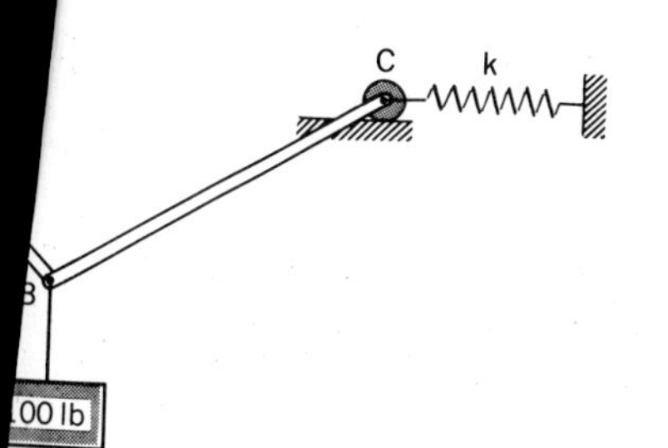


Figure P.10.42

mbination of spring and torsion-bar nsion is shown. The spring has a constant of 150 N/mm. The torsion s shown on end at $A$ and has a tor- l resistance to rotation of rod $AB$ of N-m/rad. If the load is zero, the cal spring is of length 450 mm, and $AB$ is horizontal. What is the angle $\alpha$ n the suspension supports a weight of N? Rod $AB$ is 400 mm in length.

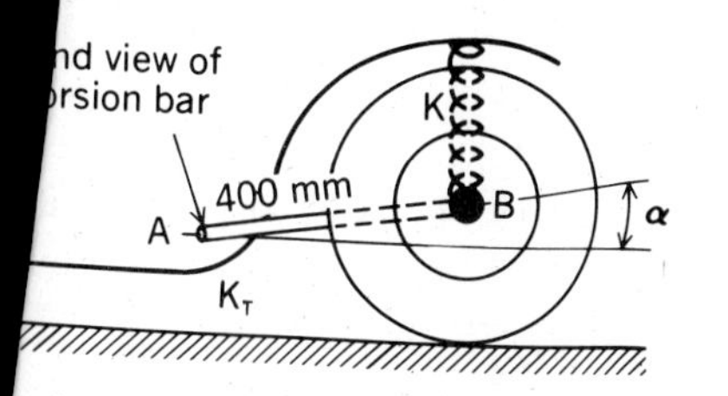


Figure P.10.43

ight rods $AB$ and $CB$ are pinned together t $B$ and pass through frictionless bearings D and $E$. These bearings are connected to he ground by ball-and-socket connections and are free to rotate about these joints. Springs, each having a spring constant $K$ $= 800$ N/m, restrain the rods as shown. The springs are unstretched when $\theta = 45°$. Show that the deflection of $B$ is .440 m when a 500-N load is attached slowly to pin $B$. The rods are each 1 m in length, and each unstretched spring is .250 m in length. Neglect the weight of the rods.

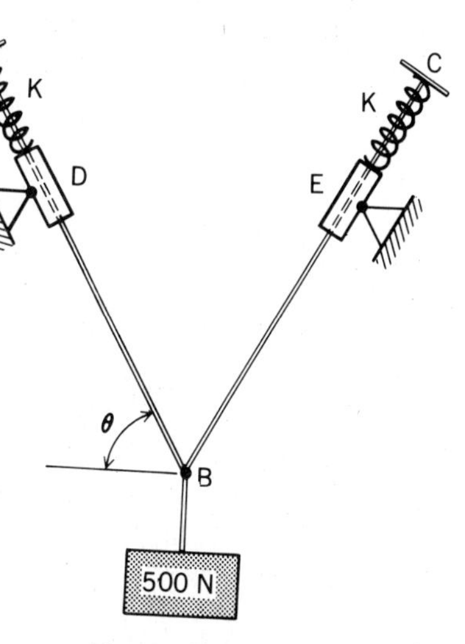


Figure P.10.44

**10.45.** Do Problem 10.26 by the method of total potential energy. (*Hint:* Use $E$ as a datum and get lengths $EJ$, $KP$, and $PN$ in terms of length $HE$, including unknown constants.)

**10.46.** An elastic band is originally 1 m long. Applying a tension force of 30 N, the band will stretch .8 m in length. What deflection $a$ does a 10-N load induce on the band when the load is applied slowly at the center of the band? Consider the force vs. elongation of the band to be linear like a spring. (*Hint:* If you consider half of the band, you double the "spring constant.")

Figure P.10.46

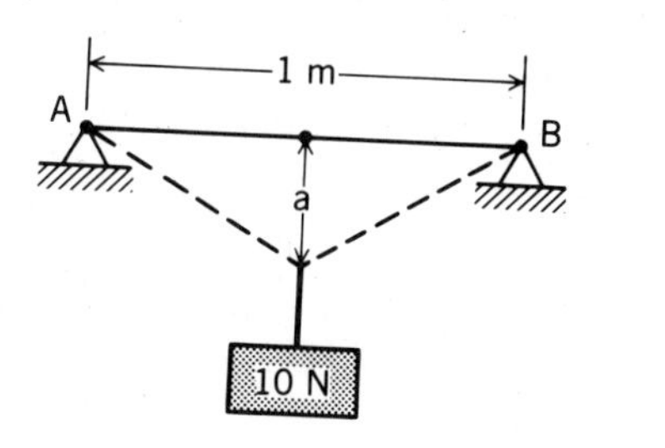




**EXAMPLE 10.5**

A mechanism shown in Fig. 10.19 consists of two weights $W$, four pinned linkage rods of length $a$, and a spring $K$ connecting the linkage rods. The spring is unextended when $\theta = 45°$. If friction and the weights of the linkage rods are negligible, what are the equilibrium configurations for the system of linkage rods and weights?

Only conservative forces can perform work on the system, and so we may use the stationary potential-energy criterion for equilibrium. We shall compute the potential energy as a function of $\theta$ (clearly, there is but one degree of freedom) using the configuration $\theta = 45°$ as the source of datum levels for the various energies. Observing Fig. 10.20, we can say that

$$V = -2Wd + \tfrac{1}{2}K(2d)^2 \tag{a}$$

As for the distance $d$, we can say (see Fig. 10.20)

$$d = a\cos 45° - a\cos\theta \tag{b}$$

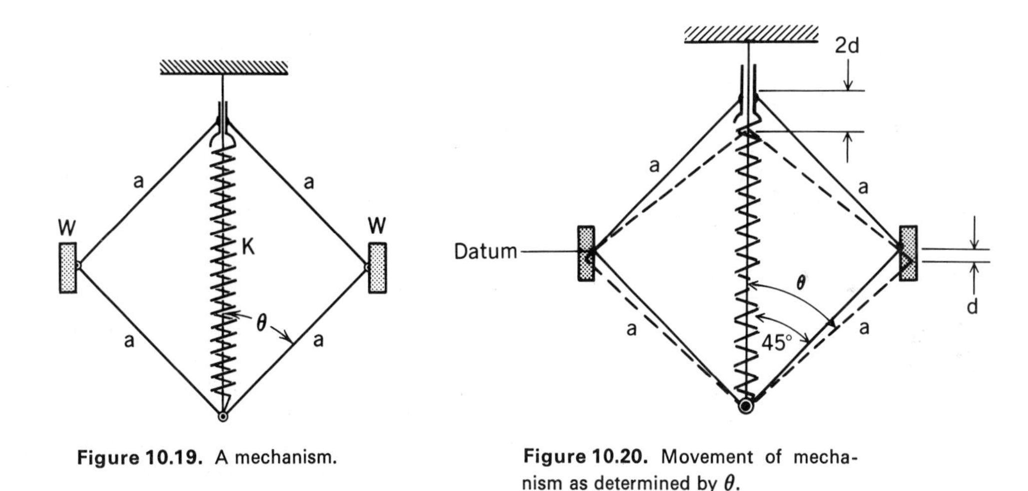


**Figure 10.19.** A mechanism.

**Figure 10.20.** Movement of mechanism as determined by $\theta$.

Hence, we have, for Eq. (a),

$$V = -2Wa(\cos 45° - \cos\theta) + \tfrac{1}{2}K4a^2(\cos 45° - \cos\theta)^2$$

For equilibrium, we require that

$$\frac{dV}{d\theta} = 0 = -2Wa\sin\theta + 4Ka^2(\cos 45° - \cos\theta)(\sin\theta) \tag{c}$$

We can then say

$$\sin\theta\left[-W - 2Ka\left(\cos\theta - \frac{1}{\sqrt{2}}\right)\right] = 0 \tag{d}$$

We have here two possibilities for satisfying the equation. First, $\sin\theta = 0$ is a solution, so we may say that $\theta_1 = 0$ (this may not be mechanically possible) is a configuration of equilibrium. Clearly, another solution can be reached by setting the bracketed terms equal to zero:

$$-W - 2Ka\left(\cos\theta - \frac{1}{\sqrt{2}}\right) = 0$$

Therefore,

$$\cos\theta = \frac{1}{\sqrt{2}} - \frac{W}{2Ka} \tag{e}$$

The solution for $\theta$ then is

$$\theta_2 = \cos^{-1}\left(\frac{1}{\sqrt{2}} - \frac{W}{2Ka}\right) \tag{f}$$

We have here two possible equilibrium configurations.

## Problems

**10.31.** A 50-kg block is placed carefully on a spring. The spring is nonlinear. The force to deflect the spring a distance $x$ mm is proportional to the square of $x$. Also, we know that 5 N deflects the spring 1 mm. By method of minimum potential energy, what will be the compression of the spring? Check the result using simple calculation based on the behavior of the spring.

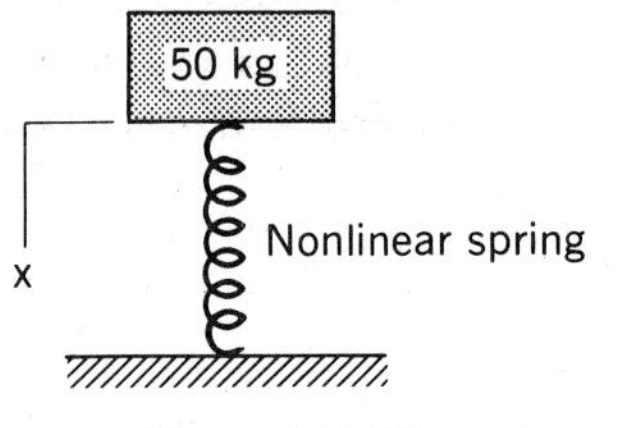

Figure P.10.31

**10.32.** A cylinder of radius 2 ft has wrapped around it a light, inextensible cord which is tied to a 100-lb block $B$ on a 30° inclined surface. The cylinder $A$ is connected to a *torsional spring*. This spring requires a torque of 1000 ft-lb/rad of rotation and it is linear and, of course, restoring. If $B$ is connected to $A$ when the torsional spring is unstrained, and if $B$ is allowed to move slowly down the incline, what distance $d$ do you allow it to move to reach an equilibrium configuration? Use the method of stationary potential energy and then check the result by more elementary reasoning.

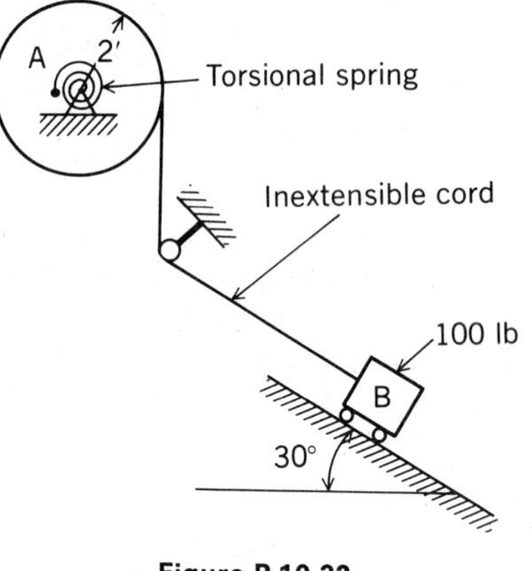


Figure P.10.32

**10.33.** Find the equilibrium configurations for the system of equal bars $W$ of length 3 m and mass 25 kg. The spring is unstretched when the bars are horizontal and has a spring constant of 1500 N/m.

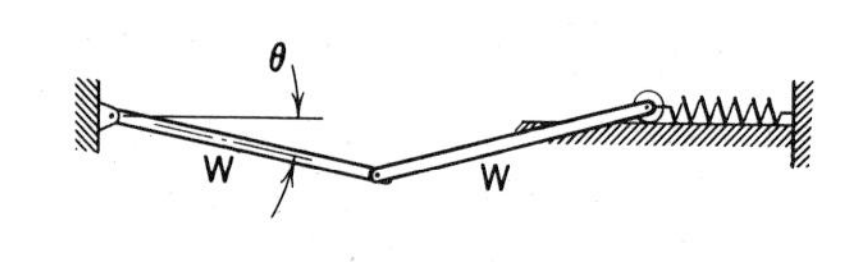


Figure P.10.33



**10.34.** The springs of the mechanism are unstretched when $\theta = \theta_0$. Show that $\theta = 17.10°$ when the weight $W$ is added. Take $W = 500$ N, $a = .3$ m, $K_1 = 1$ N/mm, $K_2 = 2$ N/mm, and $\theta_0 = 45°$. Neglect the weight of the members.

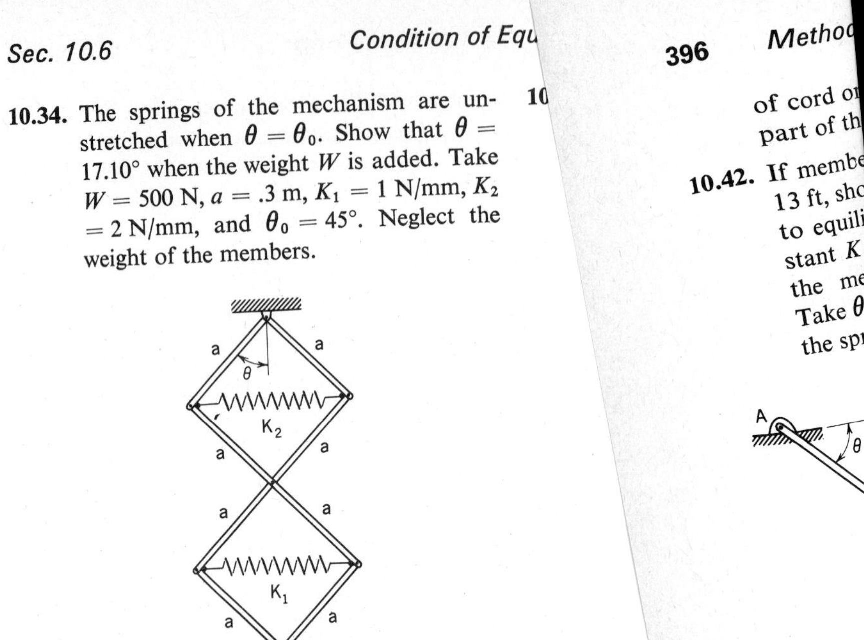


Figure P.10.34

**10.35.** At what elevation $h$ must body $A$ be for equilibrium? Neglect friction. (*Hint:* What is the differential relation between $\theta$ and $l$ to the positions of the blocks along the surface? Integrate to get the relations themselves.)

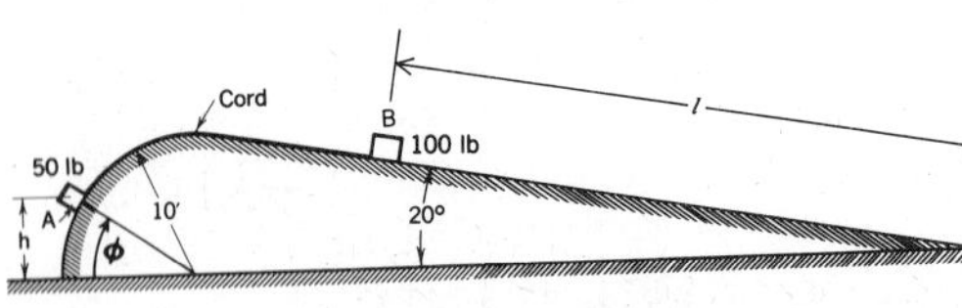


Figure P.10.35

**10.36.** Show that the position of equilibrium is $\theta = 77.3°$ for the 20-kg rod $AB$. Neglect friction.

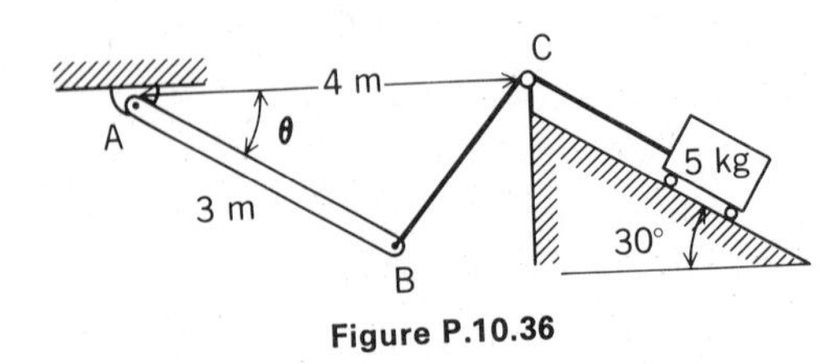


Figure P.10.36

**10.38.** Light ... load. ... end $C$ ... spring ... to $A$ an... $\theta = 45$... is 1066 ... supporte...

2 m A θ Fi...

**10.39.** Work Proble... total potentia...

**10.40.** Work Problem... total potential...

**10.41.** Do Problem 10... potential energy...



of cord o... part of th...

**10.42.** If membe... 13 ft, sho... to equili... stant $K$... the mem... Take $\theta$... the spri...

**10.43.** A co... suspe... sprin... bar ... sion... 5000... vert... rod ... whe... 5 k...

**10.44.** ...

**10.47.** In Problem 10.46, the band is first stretched and then tied while stretched to supports $A$ and $B$ so that there is an initial tension in the band of 15 N. What is then the deflection $a$ caused by the 10-N load?

**10.48.** A rubber band of length .7 m is stretched to connect to points $A$ and $B$. A tension force of 40 N is thereby developed in the band. A 20-N weight is then attached to the band at $C$. Find the distance $a$ that point $C$ moves downward if the 20-N weight is constrained to move vertically downward along a frictionless rod. (*Hint:* If you consider part of the band, the "spring constant" for it will be greater than that of the whole band.)

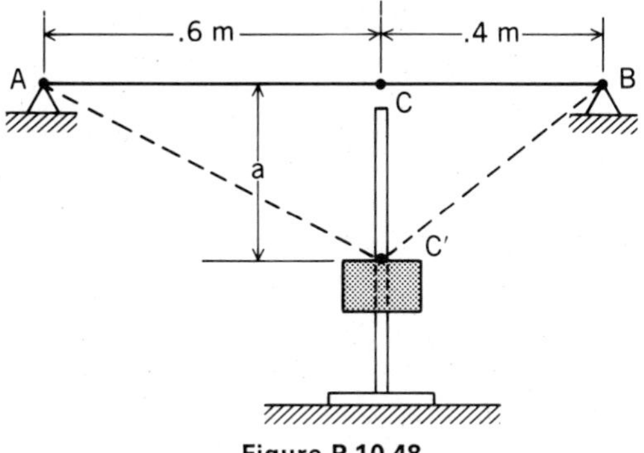


Figure P.10.48

**10.49.** The spring connecting bodies $A$ and $B$ has a spring constant $K$ of 3 N/mm. The unstretched length of the spring is 450 mm. If body $A$ weighs 60 N and body $B$ weighs 90 N, what is the stretched length of the spring for equilibrium? (*Hint:* $V$ will be a function of two variables.)

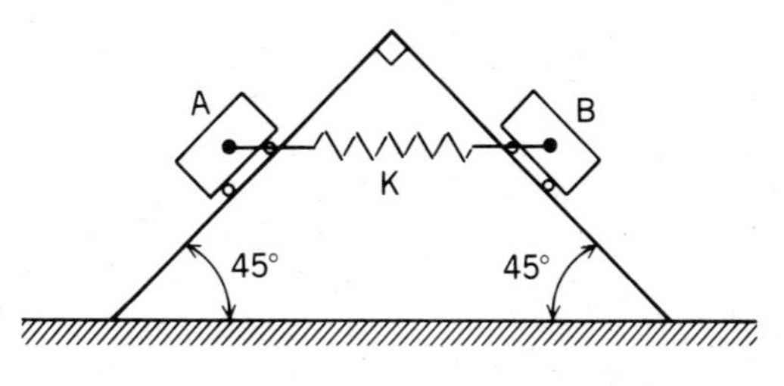


Figure P.10.49

## 10.7 Stability

Consider a cylinder resting on various surfaces (Fig. 10.21). If we neglect friction, the only active force is that of gravity. Thus, we have here conservative systems for which Eq. 10.27 is valid. The only virtual displacement for which contact with the surfaces is maintained is along the path. In each case, $dy/dx$ is zero. Thus, for an infinitesimal virtual displacement, the first-order change in elevation is zero. Hence, the change in potential energy is zero for the first-order considerations. The bodies, therefore, are in *equilibrium,* according to the previous section. However, distinct physical differences exist between the states of equilibrium of the four cases.

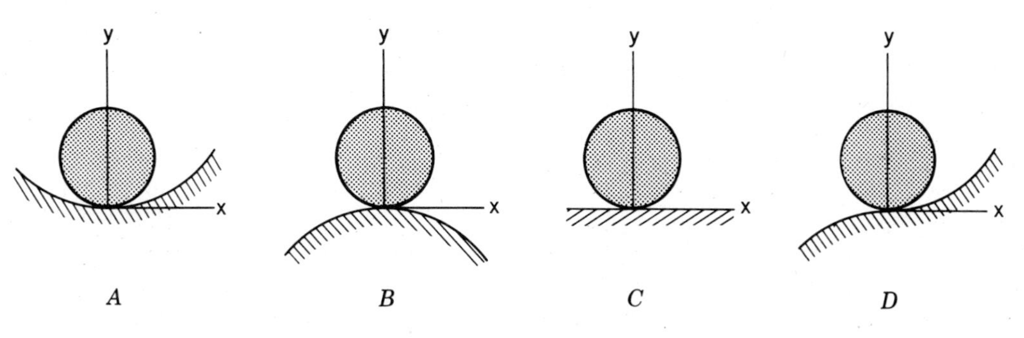


**Figure 10.21.** Different equilibrium configurations.

**Case A.** The equilibrium here is said to be *stable* in that an actual displacement from this configuration is such that the forces tend to return the body to its equilibrium configuration. Notice that the potential energy is at a *minimum* for this condition.

**Case B.** The equilibrium here is said to be *unstable* in that an actual displacement from the configuration is such that the forces aid in increasing the departure from the equilibrium configuration. The potential energy is at a *maximum* for this condition.

**Case C.** The equilibrium here is said to be *neutral.* Any displacement means that another equilibrium configuration is established. The potential energy is a constant for all possible positions of the body.

**Case D.** This equilibrium state is considered *unstable* since any displacement to the left of the equilibrium configuration will result in an increasing departure from this position.

How can we tell whether a system is stable or unstable at its equilibrium configuration other than by physical inspection, as was done above? Consider again a simple situation where the potential energy is a function of only one space coordinate $x$. That is, $V = V(x)$. We can expand the potential energy in the form of a Maclaurin series about the position of equilibrium.[12] Thus,

$$V = V_{\text{eq}} + \left(\frac{dV}{dx}\right)_{\text{eq}} x + \frac{1}{2!}\left(\frac{d^2V}{dx^2}\right)_{\text{eq}} x^2 + \ldots \tag{10.30}$$

We know from Eq. 10.29 applied to one variable that at the equilibrium configuration $(dV/dx)_{\text{eq}} = 0$. Hence, we can restate the equation above:

$$V - V_{\text{eq}} = \Delta V = \frac{1}{2!}\left(\frac{d^2V}{dx^2}\right)_{\text{eq}} x^2 + \frac{1}{3!}\left(\frac{d^3V}{dx^3}\right)_{\text{eq}} x^3 + \ldots \tag{10.31}$$

For small enough $x$, say $x_0$, the sign of $\Delta V$ will be determined by the sign of the first term in the series, $(1/2!)(d^2V/dx^2)_{\text{eq}}x^2$.[13] For this reason this term is called the *dominant* term in the series. Hence, the sign of $(d^2V/dx^2)_{\text{eq}}$ is vital in determining the sign of $\Delta V$ for small enough $x$. If $(d^2V/dx^2)_{\text{eq}}$ is positive, then $\Delta V$ is positive for any value of $x$ smaller than $x_0$. This means that $V$ is a *local minimum* at the equilibrium configuration, and we have *stable equilibrium.*[14] If $(d^2V/dx^2)_{\text{eq}}$ is negative, then $V$ is a *local maximum* at the equilibrium configuration and we have *unstable equilibrium.* Finally, if $(d^2V/dx^2)_{\text{eq}}$ is zero, we must investigate the next higher-order derivative in the expansion, and so forth.

For cases where the potential energy is known in terms of several variables, the determination of the kind of equilibrium for the system is correspondingly more complex. For example, if the function $V$ is known in terms of $x$ and $y$, we have from the calculus of several variables:

For minimum potential energy and therefore for stability:

---

[12]Note that in a Maclaurin series the coefficients of the independent variable $x$ are evaluated at $x = 0$, which for us is the equilibrium position. We denote this position with the subscript eq.

[13]As $x$ gets smaller than unity, $x^2$ will become increasingly larger than $x^3$ and powers of $x$ higher than 3. Hence, depending on the values of derivatives of $V$ at equilibrium, there will be a value of $x$—say $x_0$—for which the first term in the series will be larger than the sum of all other terms for values of $x < x_0$.

[14]That is, if the body is displaced a distance $x < x_0$, the body will return to equilibrium on release.

$$\frac{\partial V}{\partial x} = \frac{\partial V}{\partial y} = 0 \tag{10.32a}$$

$$\left(\frac{\partial^2 V}{\partial x\,\partial y}\right)^2 - \frac{\partial^2 V}{\partial x^2}\frac{\partial^2 V}{\partial y^2} < 0 \tag{10.32b}$$

$$\frac{\partial^2 V}{\partial x^2} + \frac{\partial^2 V}{\partial y^2} > 0 \tag{10.32c}$$

For maximum potential energy and therefore for instability:

$$\frac{\partial V}{\partial x} = \frac{\partial V}{\partial y} = 0 \tag{10.33a}$$

$$\left(\frac{\partial^2 V}{\partial x\,\partial y}\right)^2 - \frac{\partial^2 V}{\partial x^2}\frac{\partial^2 V}{\partial y^2} < 0 \tag{10.33b}$$

$$\frac{\partial^2 V}{\partial x^2} + \frac{\partial^2 V}{\partial y^2} < 0 \tag{10.33c}$$

The criteria become increasingly more complex for three and more independent variables.

**EXAMPLE 10.6**

A thick plate whose bottom edge is that of a circular arc of radius $R$ is shown in Fig. 10.22. The center of gravity of the plate is a distance $h$ above the ground when the plate is in the vertical position as shown in the diagram. What relation must be satisfied by $h$ and $R$ for stable equilibrium?

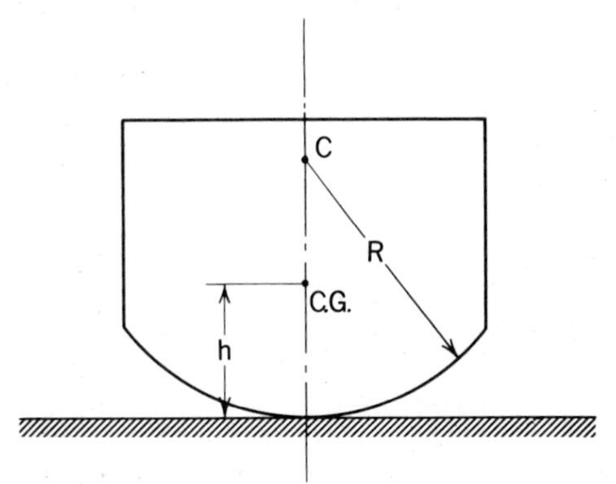


**Figure 10.22.** Plate with circular bottom edge.

The plate has one degree of freedom under the action of gravity and we can use the angle $\theta$ (Fig. 10.23) as the independent coordinate. We can express the potential energy $V$ of the system relative to the ground as a function of $\theta$ in the following manner (see Fig. 10.24):

$$V = W[R - (R - h)\cos\theta] \tag{a}$$

where $W$ is the weight of the plate. Clearly, $\theta = 0$ is a position of equilibrium since

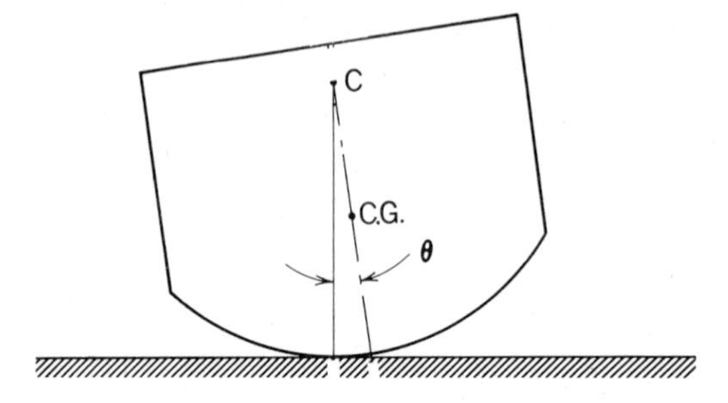


**Figure 10.23.** One degree of freedom.

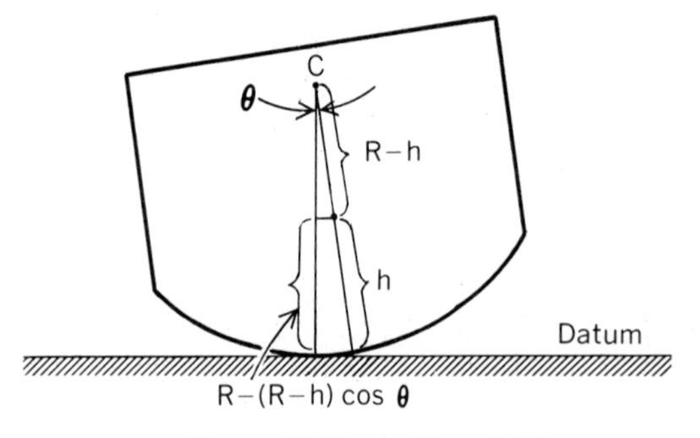


**Figure 10.24.** Position of C.G.

$$\left(\frac{dV}{d\theta}\right)_{\theta=0} = [W(R-h)\sin\theta]_{\theta=0} = 0 \tag{b}$$

Now consider $d^2V/d\theta^2$ at $\theta = 0$. We have

$$\left(\frac{d^2V}{d\theta^2}\right)_{\theta=0} = W(R-h) \tag{c}$$

Clearly, when $R > h$, $(d^2V/d\theta^2)_{\theta=0}$ is positive, and so this is the desired requirement for stable equilibrium.

## Problems

**10.50.** A rod *AB* is connected to the ground by a frictionless ball-and-socket connection at *A*. The rod is free to rest on the inside edge of a plate as shown in the diagram. The square *abcd* has its center directly over *A*. The curve *efg* is a semicircle. Without resorting to mathematical calculations, identify positions on this inside edge where equilibrium is possible for the rod *AB*. Describe the nature of the equilibrium and supply supporting arguments. Assume the edge of plate is frictionless.

**Figure P.10.50**

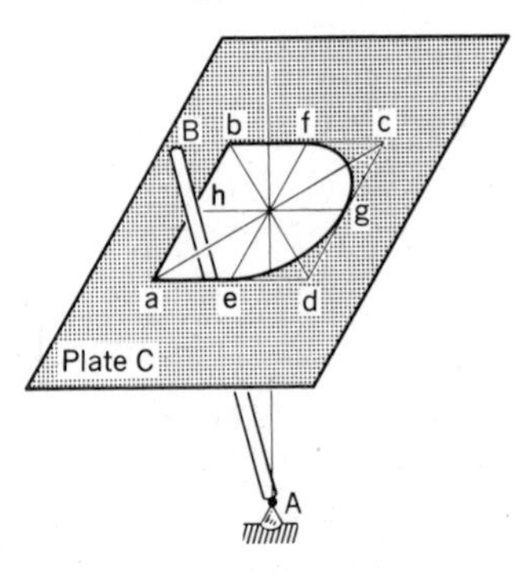

**EXAMPLE 10.5**

A mechanism shown in Fig. 10.19 consists of two weights $W$, four pinned linkage rods of length $a$, and a spring $K$ connecting the linkage rods. The spring is unextended when $\theta = 45°$. If friction and the weights of the linkage rods are negligible, what are the equilibrium configurations for the system of linkage rods and weights?

Only conservative forces can perform work on the system, and so we may use the stationary potential-energy criterion for equilibrium. We shall compute the potential energy as a function of $\theta$ (clearly, there is but one degree of freedom) using the configuration $\theta = 45°$ as the source of datum levels for the various energies. Observing Fig. 10.20, we can say that

$$V = -2Wd + \tfrac{1}{2}K(2d)^2 \tag{a}$$

As for the distance $d$, we can say (see Fig. 10.20)

$$d = a\cos 45° - a\cos\theta \tag{b}$$

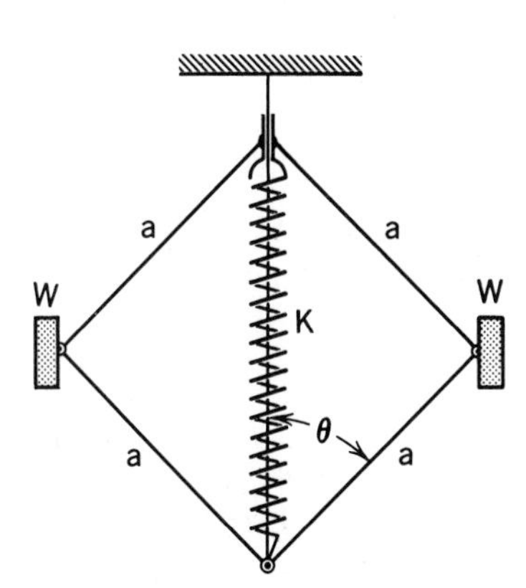


Figure 10.19. A mechanism.

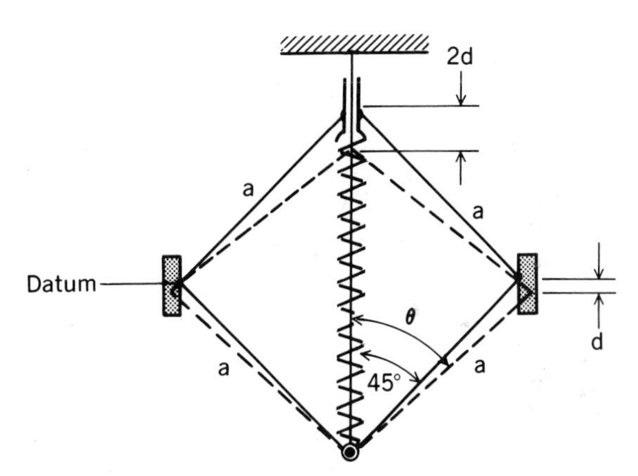


Figure 10.20. Movement of mechanism as determined by $\theta$.

Hence, we have, for Eq. (a),

$$V = -2Wa(\cos 45° - \cos\theta) + \tfrac{1}{2}K4a^2(\cos 45° - \cos\theta)^2$$

For equilibrium, we require that

$$\frac{dV}{d\theta} = 0 = -2Wa\sin\theta + 4Ka^2(\cos 45° - \cos\theta)(\sin\theta) \tag{c}$$

We can then say

$$\sin\theta\left[-W - 2Ka\left(\cos\theta - \frac{1}{\sqrt{2}}\right)\right] = 0 \tag{d}$$

We have here two possibilities for satisfying the equation. First, $\sin\theta = 0$ is a solution, so we may say that $\theta_1 = 0$ (this may not be mechanically possible) is a configuration of equilibrium. Clearly, another solution can be reached by setting the bracketed terms equal to zero:

$$-W - 2Ka\left(\cos\theta - \frac{1}{\sqrt{2}}\right) = 0$$

Therefore,

$$\cos\theta = \frac{1}{\sqrt{2}} - \frac{W}{2Ka} \tag{e}$$

The solution for $\theta$ then is

$$\theta_2 = \cos^{-1}\left(\frac{1}{\sqrt{2}} - \frac{W}{2Ka}\right) \tag{f}$$

We have here two possible equilibrium configurations.

## Problems

**10.31.** A 50-kg block is placed carefully on a spring. The spring is nonlinear. The force to deflect the spring a distance $x$ mm is proportional to the square of $x$. Also, we know that 5 N deflects the spring 1 mm. By method of minimum potential energy, what will be the compression of the spring? Check the result using simple calculation based on the behavior of the spring.

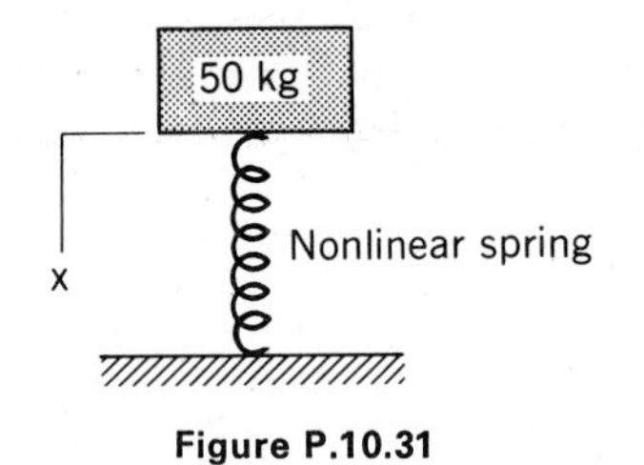


Figure P.10.31

**10.32.** A cylinder of radius 2 ft has wrapped around it a light, inextensible cord which is tied to a 100-lb block $B$ on a 30° inclined surface. The cylinder $A$ is connected to a *torsional spring*. This spring requires a torque of 1000 ft-lb/rad of rotation and it is linear and, of course, restoring. If $B$ is connected to $A$ when the torsional spring is unstrained, and if $B$ is allowed to move slowly down the incline, what distance $d$ do you allow it to move to reach an equilibrium configuration? Use the method of stationary potential energy and then check the result by more elementary reasoning.

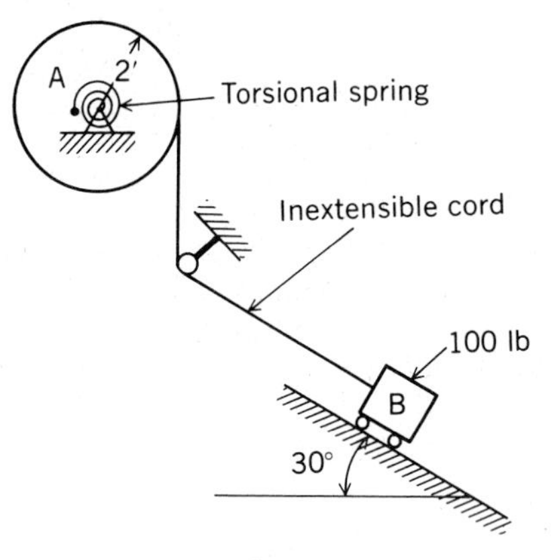

Figure P.10.32

**10.33.** Find the equilibrium configurations for the system of equal bars $W$ of length 3 m and mass 25 kg. The spring is unstretched when the bars are horizontal and has a spring constant of 1500 N/m.

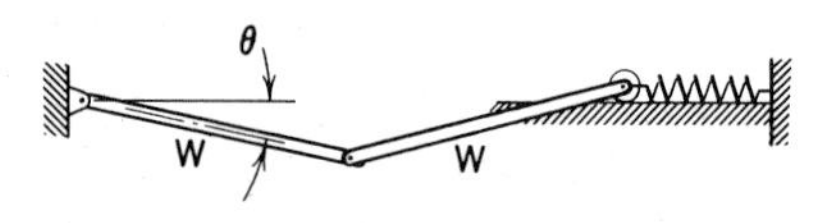


Figure P.10.33

**10.34.** The springs of the mechanism are unstretched when $\theta = \theta_0$. Show that $\theta =$ 17.10° when the weight $W$ is added. Take $W = 500$ N, $a = .3$ m, $K_1 = 1$ N/mm, $K_2 = 2$ N/mm, and $\theta_0 = 45°$. Neglect the weight of the members.

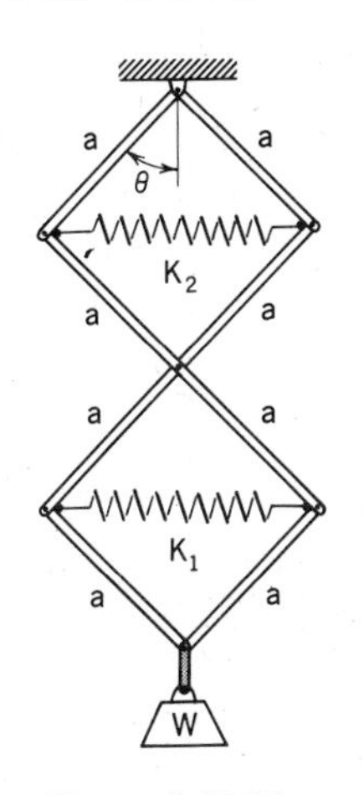


Figure P.10.34

**10.35.** At what elevation $h$ must body $A$ be for equilibrium? Neglect friction. (*Hint:* What is the differential relation between $\theta$ and $l$ to the positions of the blocks along the surface? Integrate to get the relations themselves.)

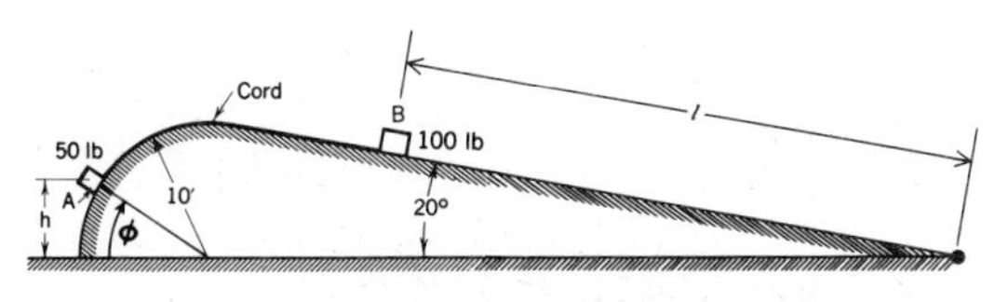


Figure P.10.35

**10.36.** Show that the position of equilibrium is $\theta = 77.3°$ for the 20-kg rod $AB$. Neglect friction.

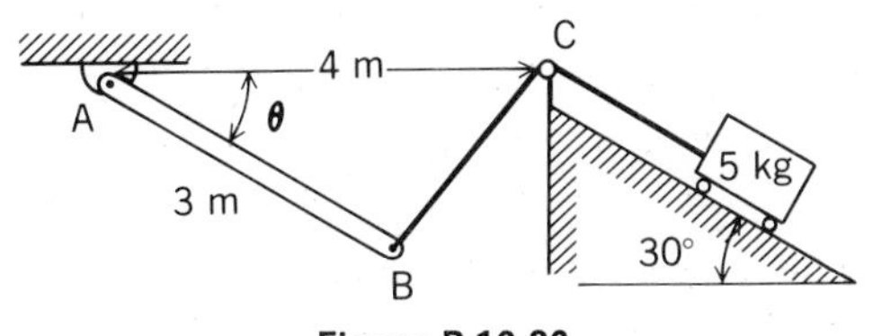


Figure P.10.36

**10.37.** A beam $BC$ of length 15 ft and weight 500 lb is placed against a spring (which has a spring constant of 10 lb/in.) and smooth walls and allowed to come to rest. If the end of the spring is 5 ft away from the vertical wall when it is not compressed, show by energy methods that the amount that the spring will be compressed is .889 ft.

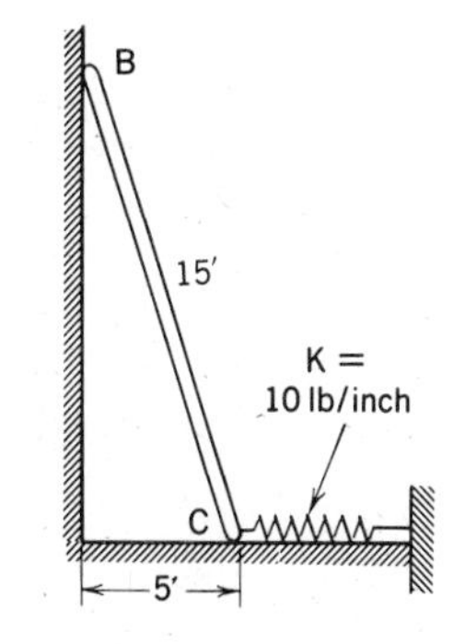


Figure P.10.37

**10.38.** Light rods $AB$ and $BC$ support a 500-N load. End $A$ of rod $AB$ is pinned, whereas end $C$ is on a roller. A spring having a spring constant of 1000 N/m is connected to $A$ and $C$. The spring is unstretched when $\theta = 45°$. Show that the force in the spring is 1066 N when the 500-N load is being supported.

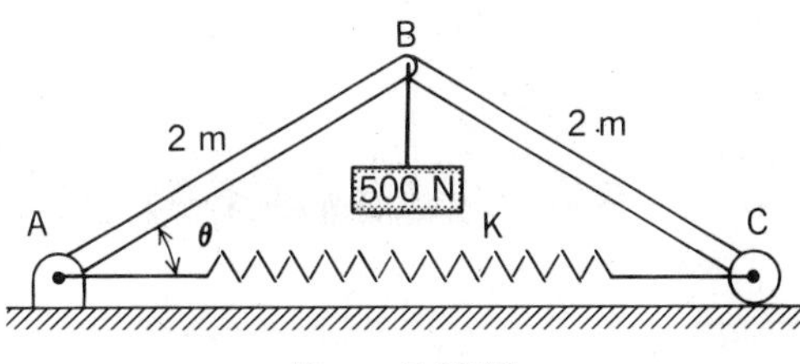


Figure P.10.38

**10.39.** Work Problem 10.28 using the method of total potential energy.

**10.40.** Work Problem 10.29 using the method of total potential energy.

**10.41.** Do Problem 10.25 by the method of total potential energy. (*Hint:* Consider a length of cord on a circular surface. Use the top part of the surface as a datum.)

**10.42.** If member $AB$ is 10 ft and member $BC$ is 13 ft, show that the angle $\theta$ corresponding to equilibrium is 34.5° if the spring constant $K$ is 10 lb/in. Neglect the weight of the members and friction everywhere. Take $\theta = 30°$ for the configuration where the spring is unstretched.

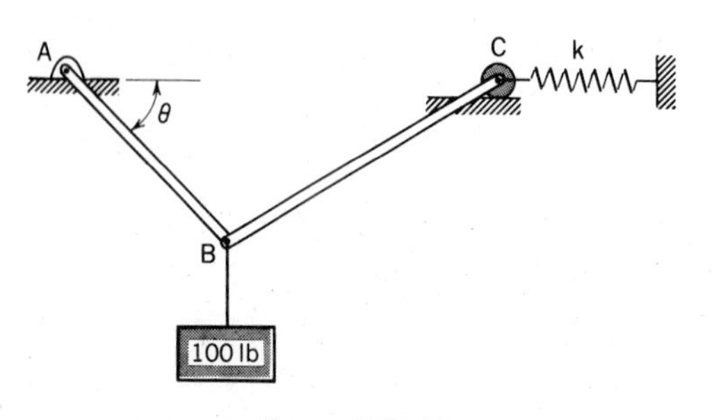


**Figure P.10.42**

**10.43.** A combination of spring and torsion-bar suspension is shown. The spring has a spring constant of 150 N/mm. The torsion bar is shown on end at $A$ and has a torsional resistance to rotation of rod $AB$ of 5000 N-m/rad. If the load is zero, the vertical spring is of length 450 mm, and rod $AB$ is horizontal. What is the angle $\alpha$ when the suspension supports a weight of 5 kN? Rod $AB$ is 400 mm in length.

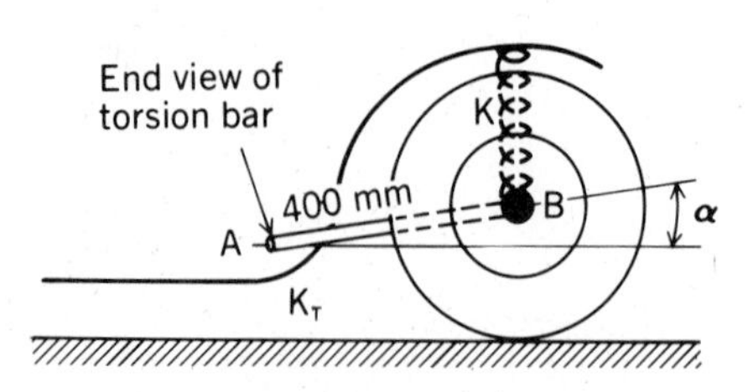


**Figure P.10.43**

**10.44.** Light rods $AB$ and $CB$ are pinned together at $B$ and pass through frictionless bearings $D$ and $E$. These bearings are connected to the ground by ball-and-socket connections and are free to rotate about these joints. Springs, each having a spring constant $K = 800$ N/m, restrain the rods as shown. The springs are unstretched when $\theta = 45°$. Show that the deflection of $B$ is .440 m when a 500-N load is attached slowly to pin $B$. The rods are each 1 m in length, and each unstretched spring is .250 m in length. Neglect the weight of the rods.

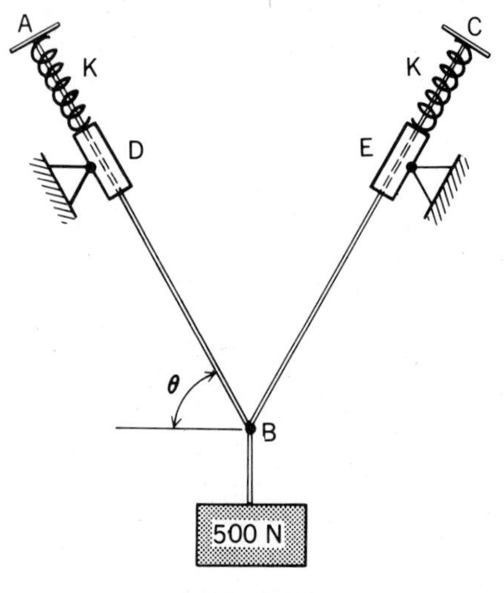


**Figure P.10.44**

**10.45.** Do Problem 10.26 by the method of total potential energy. (*Hint:* Use $E$ as a datum and get lengths $EJ$, $KP$, and $PN$ in terms of length $HE$, including unknown constants.)

**10.46.** An elastic band is originally 1 m long. Applying a tension force of 30 N, the band will stretch .8 m in length. What deflection $a$ does a 10-N load induce on the band when the load is applied slowly at the center of the band? Consider the force vs. elongation of the band to be linear like a spring. (*Hint:* If you consider half of the band, you double the "spring constant.")

**Figure P.10.46**

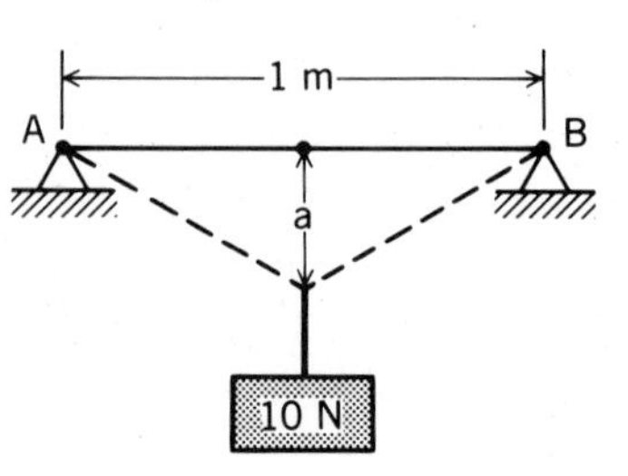

**10.51.** In Problem 10.50, show mathematically that position $h$ is a position of unstable equilibrium for the rod.

**10.52.** Rod $AB$ is supported by a frictionless ball-and-socket joint at $A$ and leans against the inside edge of a plate. What is the nature of the equilibrium position $a$ for the rod? Assume that the edge of the plate is frictionless.

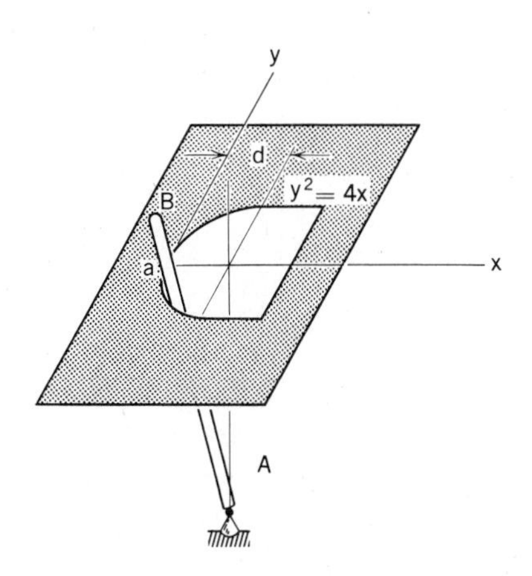


**Figure P.10.52**

**10.53.** Consider that the potential energy of a system is given by the formulation: $V = 8x^3 + 6x^2 - 7x$. What are the equilibrium positions? Indicate whether these positions are stable or not.

**10.54.** A section of a cylinder is free to roll on a horizontal surface. If $\gamma$ of a triangular portion of the cylinder is 180 lb/ft³ and that of the semicircular portion of the cylinder is 100 lb/ft³, is the configuration shown in the diagram in stable equilibrium?

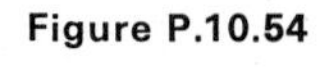

**Figure P.10.54**

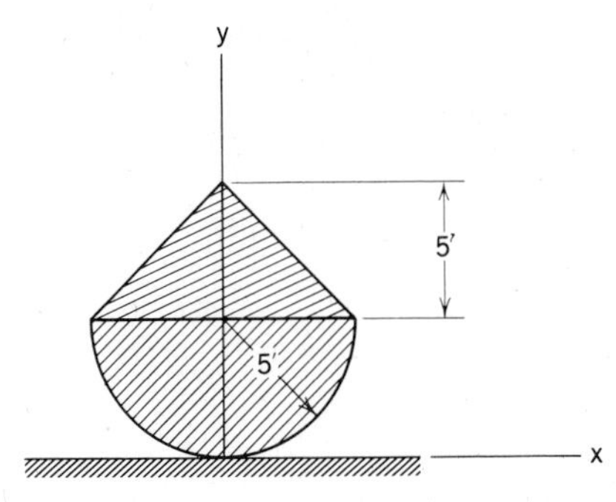


**10.55.** A system of springs and rigid bodies $AB$ and $BC$ is acted on by a weight $W$ through a pin connection at $A$. If $K$ is 50 N/mm, what is the range of the value of $W$ so that the system has an unstable equilibrium configuration when the rods $AB$ and $BC$ are collinear? Neglect the weight of the rods.

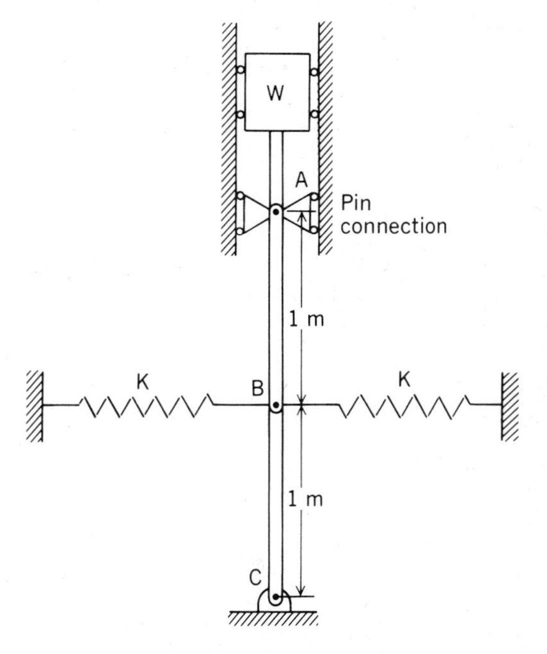


**Figure P.10.55**

**10.56.** A weight $W$ is welded to a light rod $AB$. At $B$ there is a torsional spring for which it takes 500 ft-lb to rotate 1 rad. The torsional spring is linear and restoring and is, for rotation, the analog of the ordinary linear spring for extension or contraction. If the torsional spring is unstrained when the rod is vertical, what is the largest value of $W$ for which we have stable equilibrium in the vertical direction?

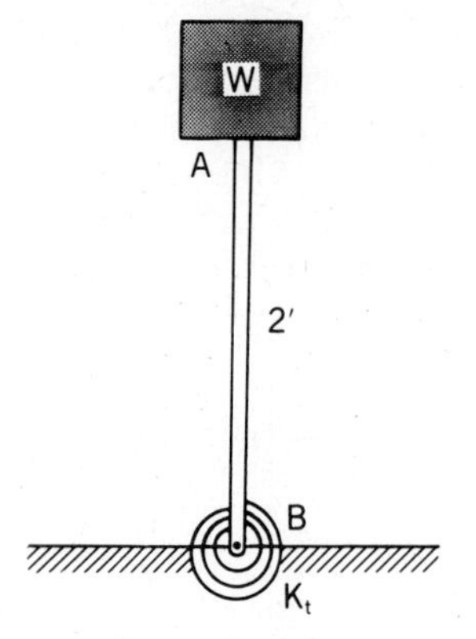


Figure P.10.56

**10.57.** A light rod $AB$ is pinned to a block of weight $W$ at $A$. Also at $A$ are two identical springs $K$. Show that, for $W$ less than $2Kl$, we have stable equilibrium in the vertical position and, for $W > 2Kl$, we have unstable equilibrium. The value $W = 2Kl$ is called a *critical load* for reasons that will be seen in Problem 10.58.

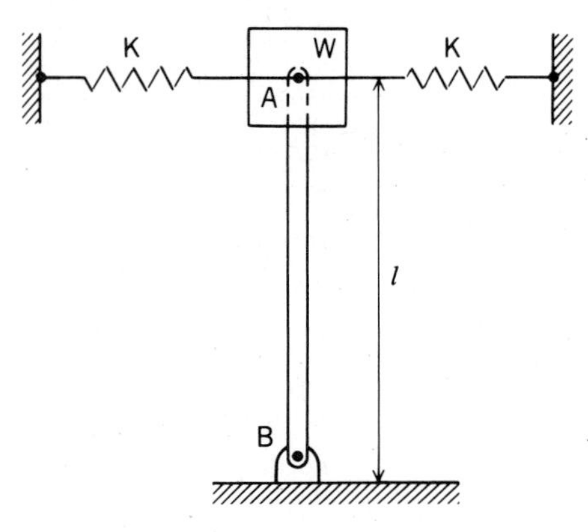


Figure P.10.57

**10.58.** In Problem 10.57, apply a small transverse force $F$ to body $A$ as shown. Compute the horizontal deflection $\delta$ of point $A$ for a position of equilibrium by using ordinary statics as developed in earlier chapters. Now show that when $W = 2Kl$ (i.e., the critical weight), the deflection $\delta$ mathematically blows up to infinity. This shows that, even if $W < 2Kl$ and we have stable equilibrium with $F = 0$, we get increasingly very large deflections as the weight $W$ approaches its critical value and a side load $F$, however small, is introduced. The study of stability of equilibrium configuration therefore is an important area of study in mechanics. Most of you will encounter this in the strength of materials course.

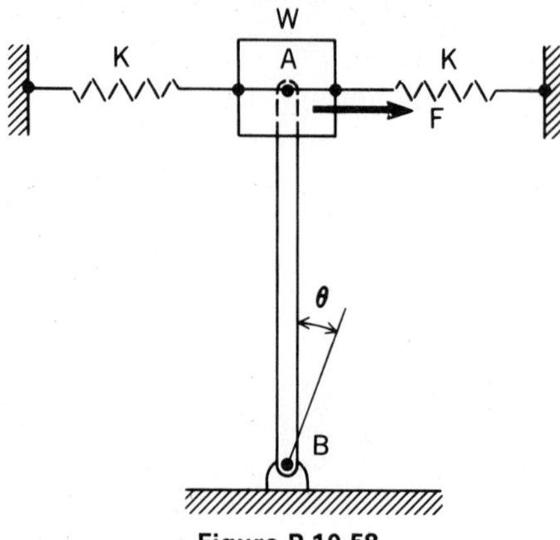


Figure P.10.58

**10.59.** Cylinders $A$ and $B$ have semicircular cross sections. Cylinder $A$ supports a rectangular solid shown as $C$. If $\rho_A = 1600 \text{ kg/m}^3$ and $\rho_C = 800 \text{ kg/m}^3$, ascertain whether arrangement shown is in stable equilibrium. (*Hint:* Make use of point $O$ in computing $V$.)

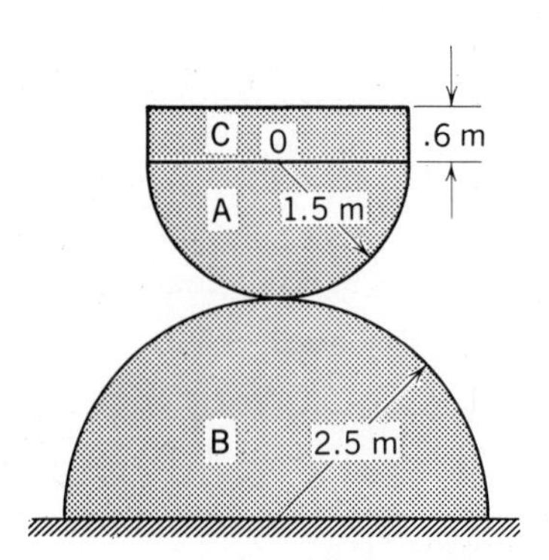


Figure P.10.59

## 10.8 Closure

In this chapter, we have taken an approach that differs radically from the approach used earlier in the text. In earlier chapters, we isolated a body for the purpose of writing equilibrium equations using all the forces acting on the body. This is the approach we often call *vectorial mechanics*. In this chapter, we have compared mathematically the equilibrium configuration with admissible neighboring configurations. We concluded that the equilibrium configuration was one from which there is zero virtual work under a virtual displacement. Or, equivalently for conservative active forces, the equilibrium configuration was the configuration having the minimum potential energy when compared to admissible configurations in the neighborhood. We call such an approach *variational mechanics*. The variational mechanics point of view is no doubt strange to you at this stage of study and far more subtle and mathematical than the vectorial mechanics approach.

Shifts like the one from the more physically acceptable vectorial mechanics to the more abstract variational mechanics take place in other engineering sciences. Variational methods and techniques are used in the study of plates and shells, elasticity, quantum mechanics, orbital mechanics, statistical thermodynamics, and electromagnetic theory. The variational methods and viewpoints thus are important and even vital in more advanced studies in the engineering sciences, physics, and applied mathematics.

## Review Problems

**10.60.** At what position must the operator of the counterweight crane locate the 50-kN counterweight when he lifts the 10-kN load of steel?

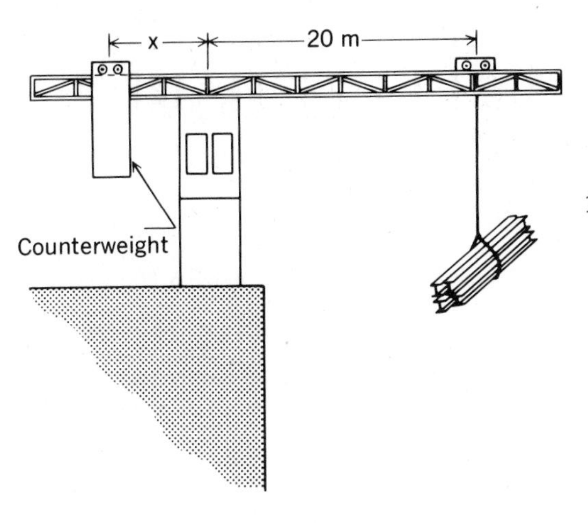


Figure P.10.60

**10.61.** What is the relation between $P$ and $Q$ for equilibrium?

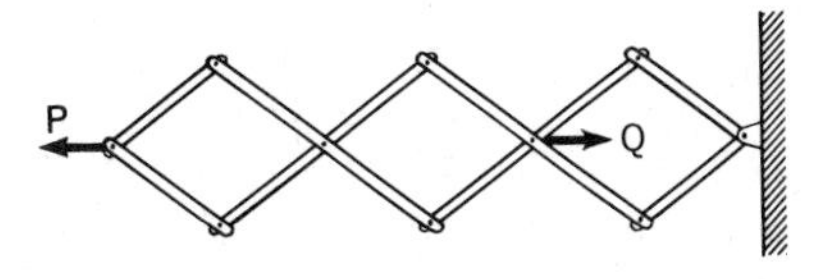


Figure P.10.61

**10.62.** A 50 lb-ft torque is applied to a press. The pitch of the screw is .5 in. If there is no friction on the screw, and if the base of the screw can rotate frictionlessly in a base plate $A$, what is the force $P$ imposed by the base plate on body $B$?

Figure P.10.62

**10.63.** The spring is unstretched when $\theta = 30°$. At any position of the pendulum, the spring remains horizontal. If the spring constant is 50 lb/in., at what position will the system be in equilibrium?

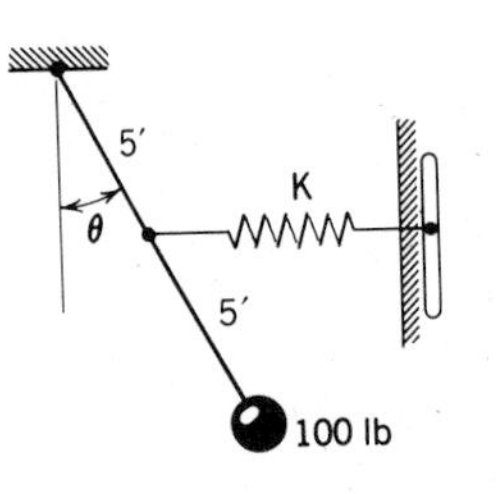


Figure P.10.63

**10.64.** If the springs are unstretched when $\theta = \theta_0$, find the angle $\theta$ when the weight $W$ is placed on the system. Use the method of minimizing potential energy.

Figure P.10.64

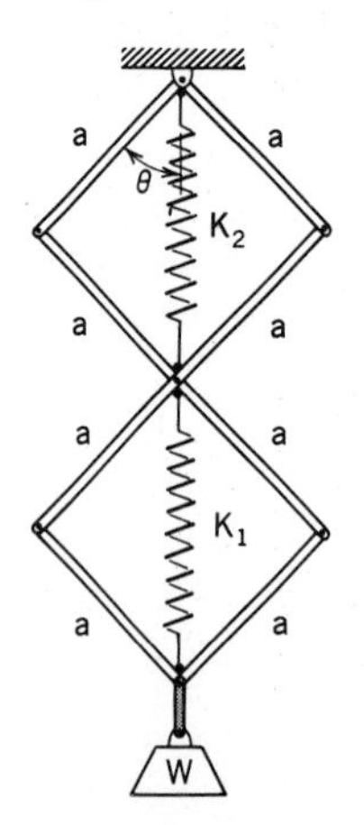


**10.65.** A mass $M$ of 20 kg slides with no friction along a vertical rod. Two internal springs $K_1$ of spring constant 2 N/mm and an external spring $K_2$ of spring constant 3 N/mm restrain the weight $W$. If all springs are unstrained at $\theta = 30°$, show that the equilibrium configuration $\theta$ is 27.8°.

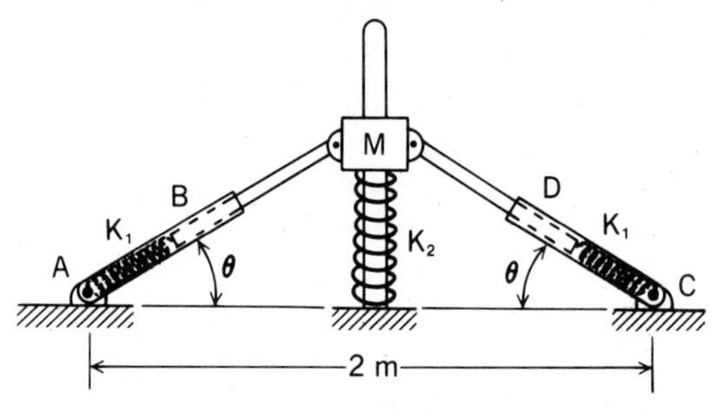


Figure P.10.65

**10.66.** When rod $AB$ is in the vertical position, the spring attached to the wheel by a flexible cord is unstretched. Determine all the possible angles $\theta$ for equilibrium. Show which are stable and which are not stable. The spring has a spring constant of 8 lb/in.

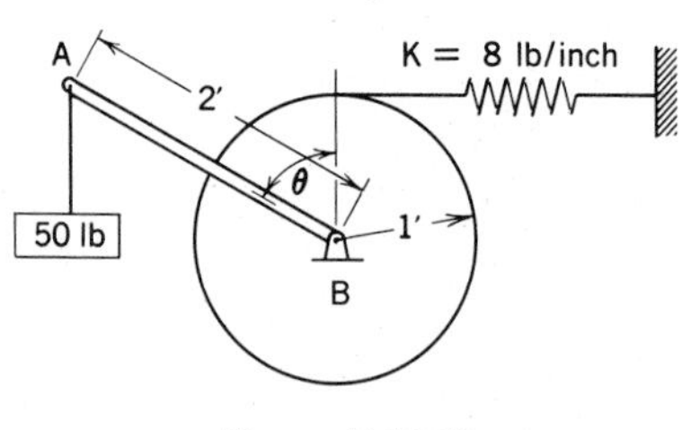


Figure P.10.66

**10.67.** Two identical rods are pinned together at $B$ and are pinned at $A$ and $C$. At $B$ there is a torsional spring requiring 500 N-m/rad of rotation. What is the maximum weight $W$ that each rod can have for a case of stable equilibrium when the rods are collinear?

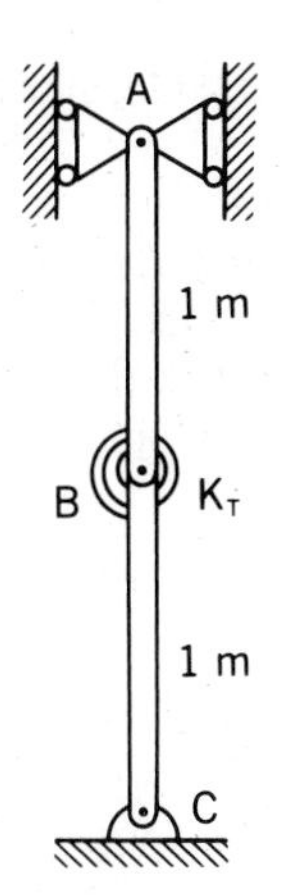

**Figure P.10.67**

**10.68.** A rectangular solid body of height $h$ rests on a cylinder with a semicircular section. Set up criteria for stable and unstable equilibrium in terms of $h$ and $R$ for the position shown.

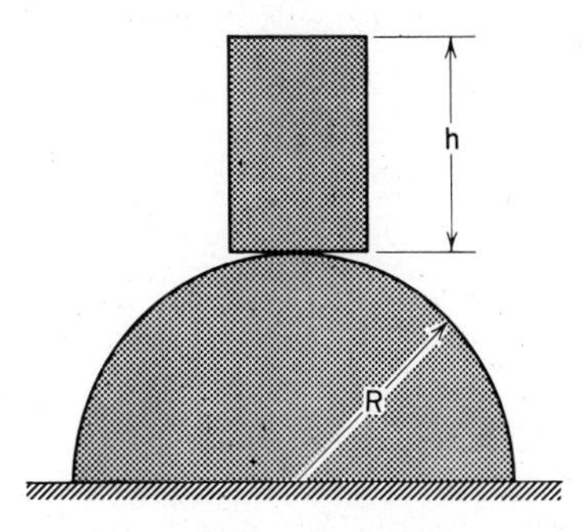


**Figure P.10.68**

# Appendices

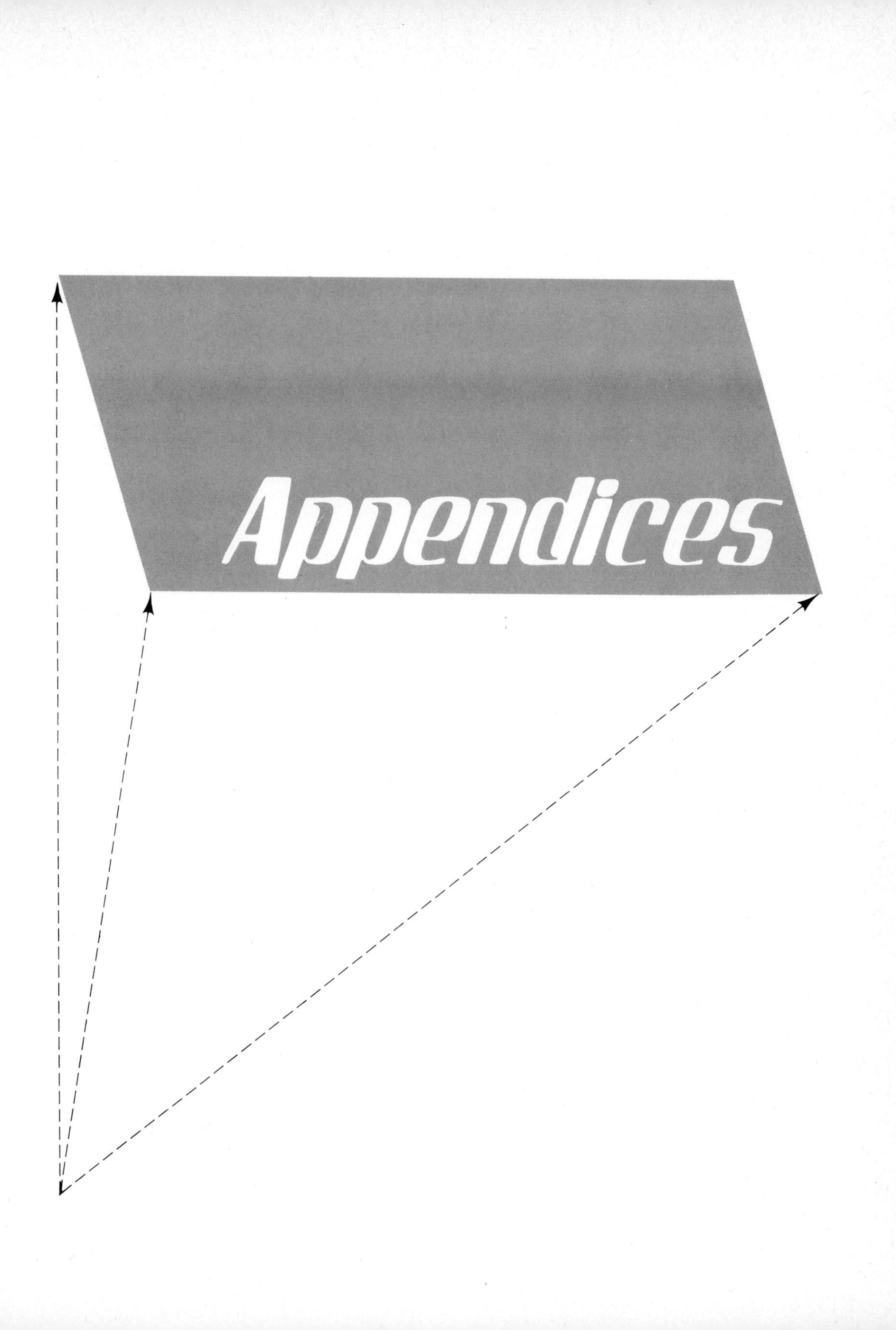

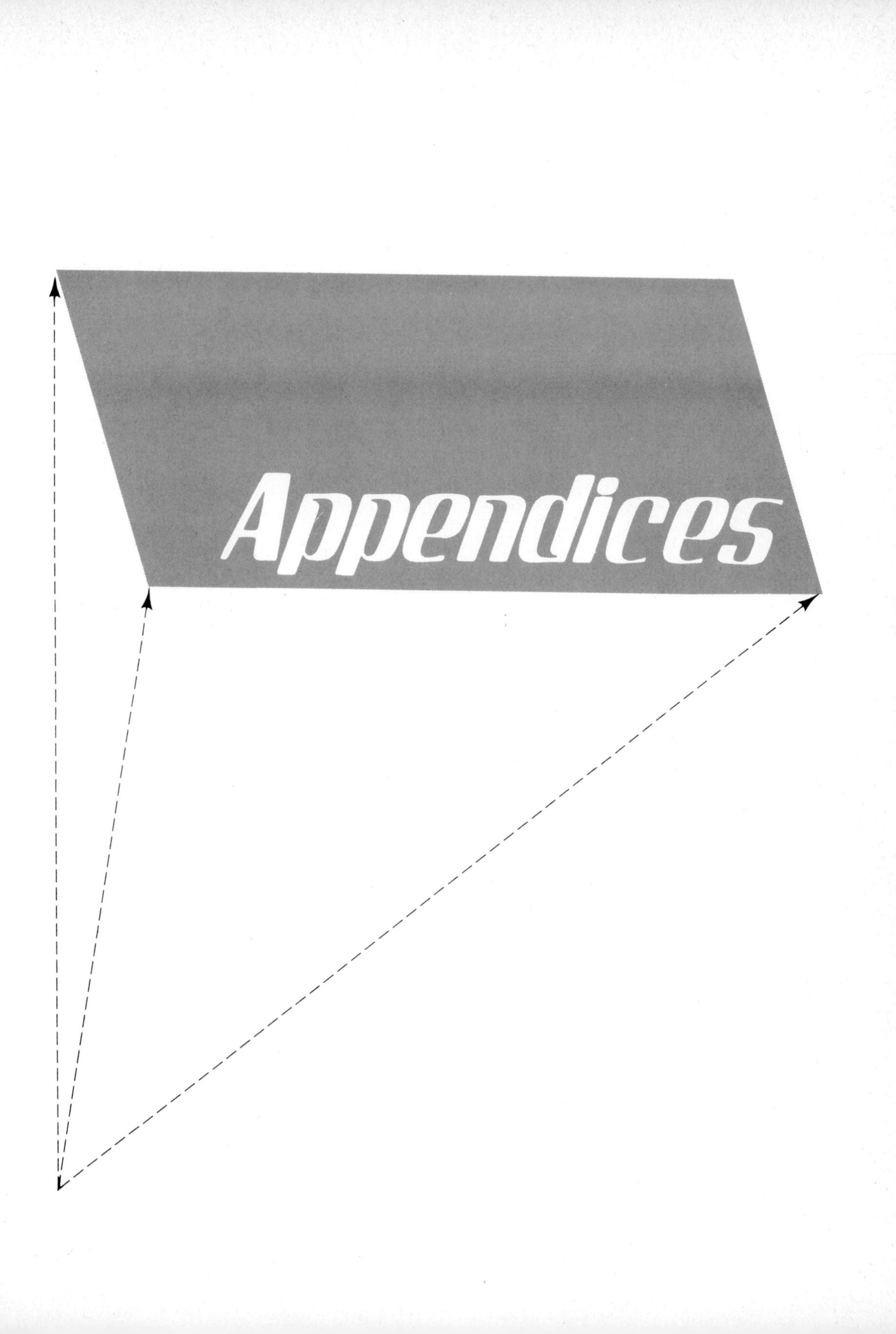
Appendices

# Appendix I
# Integration Formulas

**1.** $\displaystyle\int \frac{x\,dx}{a+bx} = \frac{1}{b^2}[a + bx - a\ln(a+bx)]$

**2.** $\displaystyle\int \frac{dx}{a^2 - x^2} = \frac{1}{2a}\ln\left(\frac{a+x}{a-x}\right)$

**3.** $\displaystyle\int \sqrt{x^2 \pm a^2}\,dx = \tfrac{1}{2}[x\sqrt{x^2 \pm a^2} \pm a^2 \ln(x + \sqrt{x^2 \pm a^2})]$

**4.** $\displaystyle\int \sqrt{a^2 - x^2}\,dx = \tfrac{1}{2}\left(x\sqrt{a^2 - x^2} + a^2 \sin^{-1}\frac{x}{a}\right)$

**5.** $\displaystyle\int x\sqrt{a^2 - x^2}\,dx = -\tfrac{1}{3}\sqrt{(a^2 - x^2)^3}$

**6.** $\displaystyle\int x\sqrt{a + bx}\,dx = -\frac{2(2a - 3bx)\sqrt{(a+bx)^3}}{15b^2}$

**7.** $\displaystyle\int x^2\sqrt{a^2 - x^2}\,dx = -\frac{x}{4}\sqrt{(a^2 - x^2)^3} + \frac{a^2}{8}\left(x\sqrt{a^2 - x^2} + a^2\sin^{-1}\frac{x}{a}\right)$

**8.** $\displaystyle\int x^2\sqrt{a^2 \pm x^2}\,dx = \frac{x}{4}\sqrt{(x^2 \pm a^2)^3} \mp \frac{a^2}{8}x\sqrt{x^2 \pm a^2} - \frac{a^4}{8}\ln(x + \sqrt{x^2 \pm a^2})$

**9.** $\displaystyle\int \frac{dx}{\sqrt{a^2 - x^2}} = \sin^{-1}\frac{x}{a}$

**10.** $\displaystyle\int \frac{dx}{\sqrt{x^2 + a^2}} = \ln(x + \sqrt{x^2 + a^2}) = \sinh^{-1}\frac{x}{a}$

**11.** $\displaystyle\int x^m e^{ax}\,dx = \frac{x^m e^{ax}}{a} - \frac{m}{a}\int x^{m-1}e^{ax}\,dx$

**12.** $\displaystyle\int x^m \ln x\,dx = x^{m+1}\left[\frac{\ln x}{m+1} - \frac{1}{(m+1)^2}\right]$

**13.** $\displaystyle\int \sin^2\theta\,d\theta = \tfrac{1}{2}\theta - \tfrac{1}{4}\sin 2\theta$

**14.** $\displaystyle\int \cos^2\theta\,d\theta = \tfrac{1}{2}\theta + \tfrac{1}{4}\sin 2\theta$

**15.** $$\int \sin^3\theta\, d\theta = -\tfrac{1}{3}\cos\theta(\sin^2\theta + 2)$$

**16.** $$\int \cos^m\theta \sin\theta\, d\theta = -\frac{\cos^{m+1}\theta}{m+1}$$

**17.** $$\int \sin^m\theta \cos\theta\, d\theta = \frac{\sin^{m+1}\theta}{m+1}$$

**18.** $$\int \sin^m\theta\, d\theta = -\frac{\sin^{m-1}\theta\cos\theta}{m} + \frac{m-1}{m}\int \sin^{m-2}\theta\, d\theta$$

**19.** $$\int \theta^2 \sin\theta\, d\theta = 2\theta\sin\theta - (\theta^2 - 2)\cos\theta$$

**20.** $$\int \theta^2 \cos\theta\, d\theta = 2\theta\cos\theta + (\theta^2 - 2)\sin\theta$$

**21.** $$\int \theta \sin^2\theta\, d\theta = \tfrac{1}{4}[\sin\theta\,(\sin\theta - 2\theta\cos\theta) + \theta^2]$$

**22.** $$\int \sin m\theta \cos m\theta\, d\theta = -\frac{1}{4m}\cos 2m\theta$$

**23.** $$\int \frac{d\theta}{(a + b\cos\theta)^2} = \frac{1}{(a^2 - b^2)}\left(\frac{-b\sin\theta}{a + b\cos\theta} + \frac{2a}{\sqrt{a^2 - b^2}}\tan^{-1}\frac{\sqrt{a^2 - b^2}\tan\dfrac{\theta}{2}}{a + b}\right)$$

# Appendix II Computation of Principal Moments of Inertia

We now turn to the problem of computing the principal moments of inertia and the directions of the principal axes for the case where we do not have planes of symmetry. It is unfortunate that a careful study of this important calculation is beyond the level of this text. However, we shall present enough material to permit the computation of the principal moments of inertia and the directions of their respective axes.

The procedure that we shall outline is that of extremizing the mass moment of inertia at a point where the inertia-tensor components are known for a reference $xyz$. This will be done by varying the direction cosines $l$, $m$, and $n$ of an axis $k$ so as to extremize $I_{kk}$ as given by Eq. 9.13. We accordinaly set the differential of $I_{kk}$ equal to zero as follows:

$$\begin{aligned} dI_{kk} = \quad & 2lI_{xx}\,dl + 2mI_{yy}\,dm + 2nI_{zz}\,dn \\ & - 2lI_{xy}\,dm - 2mI_{xy}\,dl - 2lI_{xz}\,dn \\ & - 2nI_{xz}\,dl - 2mI_{yz}\,dn - 2nI_{yz}\,dm = 0 \end{aligned} \tag{II.1}$$

Collecting terms and canceling the factor 2, we get

$$(lI_{xx} - mI_{xy} - nI_{zz})\,dl + (-lI_{xy} + mI_{yy} - nI_{yz})\,dm + (-lI_{xz} - mI_{yz} + nI_{zz})\,dn = 0 \tag{II.2}$$

If the differentials $dl$, $dm$, and $dn$ were *independent* we could set their respective coefficients equal to zero to satisfy the equation. However, they are not independent because the equation

$$l^2 + m^2 + n^2 = 1 \tag{II.3}$$

must at all times be satisfied. Accordingly, the differentials of the direction cosines must be related as follows[1]:

$$l\,dl + m\,dm + n\,dn = 0 \tag{II.4}$$

We can of course consider any two differentials as independent. The third is then established in accordance with the equation above.

We shall now introduce the *Lagrange multiplier* $\lambda$ to facilitate the extremizing process. This constant is an arbitrary constant at this stage of the calculation. Multiplying Eq. II.4 by $\lambda$ and subtracting Eq. II.4 from Eq. II.2 we get when collecting terms:

$$\begin{aligned} [(I_{xx} - \lambda)l - I_{xy}m - I_{xz}n]\,dl + [-I_{xy}l + (I_{yy} - \lambda)m - I_{yz}n]\,dm \\ + [-I_{xz}l - I_{yz}m + (I_{zz} - \lambda)n]\,dn = 0 \end{aligned} \tag{II.5}$$

Let us next consider that $m$ and $n$ are independent variables and consider the value of $\lambda$ so chosen that the coefficient of $dl$ is zero. That is,

$$(I_{xx} - \lambda)l - I_{xy}m - I_{xz}n = 0 \tag{II.6}$$

---

[1]We are thus extremizing $I_{kk}$ in the presence of a constraining equation.

With the first term Eq. II.5 disposed of in this way, we are left with differentials $dm$ and $dn$, which are independent. Accordingly, we can set their respective coefficients equal to zero in order to satisfy the equation. Hence, we have in addition to Eq. II.6 the following equations:

$$\begin{aligned} -I_{xy}l + (I_{yy} - \lambda)m - I_{yz}n &= 0 \\ -I_{xz}l - I_{yz}m + (I_{zz} - \lambda)n &= 0 \end{aligned} \tag{II.7}$$

A necessary condition for the solution of a set of direction cosines $l$, $m$, and $n$, from Eqs. II.6 and II.7, which does not violate Eq. II.3[2] is that the determinant of these variables be zero. Thus:

$$\begin{vmatrix} I_{xx} - \lambda & -I_{xy} & -I_{xz} \\ -I_{xy} & I_{yy} - \lambda & -I_{yz} \\ -I_{xz} & -I_{yz} & I_{zz} - \lambda \end{vmatrix} = 0 \tag{II.8}$$

This results in a cubic equation for which we can show there are three real roots for $\lambda$. Substituting these roots into any two of Eqs. II.6 and II.7 plus Eq. II.3, we can determine three direction cosines for each root. These are the direction cosines for the principal axes measured relative to $xyz$. We could get the principal moments of inertia next by substituting a set of these direction cosines into Eq. 9.13 and solving for $I_{kk}$. However, that is not necessary, since it can be shown that the three Lagrange multipliers *are* the principal moments of inertia.

[2]This precludes the possibility of a trivial solution $l = m = n = 0$.

# Answers to Problems

# Answers to Problems

The answers to even numbered problems are presented here as well as the answers to *all* review problems. The author has used the same coordinate axes that were used in the text in various sections. If perchance you use a different set of axes from the author's (for example, if your $x$ axis corresponds to the author's $y$ axis, etc.), you should still have no difficulty in deciding, after comparison of the numbers and a moment's thought, whether there is agreement or not.

**CHAPTER 2**

**2.2** 38.5 N at 66.6°

**2.6** $B = 22.9$ N; $D$ at 66.4°

**2.8** 221 N; 685 N

**2.10** 17.66 kg

**2.12** 66.5 lb

**2.14** 36.4 lb; 81.51 lb

**2.16** $.590\ F$; $.630\ F$

**2.18** 134.5 lb; 109.8 lb

**2.20** 37.4 lb; $l = .267$, $m = .535$, $n = -.802$

**2.22** $\gamma = 68.6°$
$OA = 17.10$ m
$OB = 43.3$ m
$EC = 18.24$ m

**2.24** $F_A = 56.6$ N comp.
$F_B = 70.7$ N comp.
$F_C = 42.4$ N tens.

**2.26** $25.7i + 24.7j + 16k$ lb

**2.28** $A = \pm 5\sqrt{2}\,i \mp 5\sqrt{2}\,j$

**2.30** $\hat{f} = .465i + .814j + .349k$
$F = 46.5i + 81.4j + 34.9k$ N

**2.32** $-.467$; $-10.5$

**2.34** $D = 10i - .769j - 3.77k$

**2.38** $16q$ joules

**2.40** $A = 2.5$ N; $\alpha = 45.7°$

**2.42** $-75\ \text{ft}^2$; 95.9°

**2.44** 52.2°

**2.46** $18i + 20j - 42k$; 47

**2.50** $n = .804i - .465j + .372k$

**2.52** $A_1 = 500k - 800j\ \text{m}^2$
$A_\xi = 640\ \text{m}^2$

**2.54** Results should be the same.

**2.56** $2600\ \text{ft}^3$

**2.58**
$$\begin{cases} .303n - .256m = 0 \\ .256l - 1.610n = 0 \\ 1.610m - .303l = 0 \end{cases}$$
$l = .971$; $m = .1828$; $n = .1545$

**2.59** 35.7 N

**2.60** 55.95°; 45.6°

**2.61** 130.7 km

**2.62** 57.1 N; 342 N; 971 N

**2.63** 209 lb; $l = .479$, $m = .866$, $n = -.1437$

**2.64** 32.4 ft-lb

**2.65** $F_z = -80.1 \times 10^{-13}$ N
$F_x = F_y = 0$

**2.66** $-156i + 94j + 28k\ \text{m}^2/\text{sec}$

**2.67** $251\ \text{ft}^2$

## CHAPTER 3

**3.2** $\varrho = 4i - 16j - 3k$ ft

**3.4** $r = 6i + 10.16j - 2.4k$ m

**3.6** $r = \pm 2\sqrt{2z}\, j + zk$

**3.8** $-35{,}000i + 10{,}000j - 1333k$ yd

**3.10** 149,000 ft-lb

**3.12** $M_A = 2820$ lb-ft
$M_B = 1879$ lb-ft

**3.14** $M = 300$ kN-m

**3.16** $M = -84i + 94j - 46k$ N-m

**3.18** $M_A = M_B = \dfrac{10}{\sqrt{3}} ak - \dfrac{10}{\sqrt{3}} aj$ lb-ft

$M_E = \dfrac{-10}{\sqrt{3}} ai + \dfrac{10}{\sqrt{3}} ak$ lb-ft

**3.20** $M = -6971j + 4024i$ kN-m

**3.22** $M_A = .60i$ kN-m
$T_A = .60$ kN-m

**3.24** $-18.19$ N-m

**3.26** $F = 146.2$ lb
$M_x = 77.0$ lb-ft

**3.28** 5769 lb-ft

**3.30** $-56.3$ kN-m; 26.25 kN-m

**3.32** 43.5 lb; 128.3 ft-lb

**3.34** $d_{max} = 2$ ft

**3.36** 60 lb; 39 lb

**3.38** $-50i$; 0

**3.40** 60 lb; 127.3 lb-in.

**3.42** $M_A = M_P = -261i - 261j$ N-m

**3.44** $M_P = 48i - 36j - 226k$ N-m
$M_\varepsilon = -146.9$ N-m

**3.46** $-27.2$ N-m

**3.48** 3635 lb-in. at 68.2° to horizontal

**3.50** $35.35i + 22.4j + 80.1k$ N-m

**3.52** $10{,}600i + 4500j - 5000k$ m
$|r| = 12{,}550$ m

**3.53** $M_A = -2165i + 850j - 6500k$ lb-ft
$M_B = -11{,}230i - 4330j - 4330k$ lb-ft

**3.54** 3725 N-m

**3.55** 120 ft-lb; 160 ft-lb; 60 lb

**3.56** 1750 N-m

**3.57** 83.7 lb-ft

**3.58** $C = 600i - 600j - 800k$

**3.59** 1079.5 N-m

## CHAPTER 4

**4.2** $F = 15$ kN
$C = 30$ kN-m

**4.4** $F = 150$ N
$C = 187.5$ N-m

**4.6** $F = -10j$ kN
$M_A = 8k$ kN-m
$M_B = 22.7k$ kN-m

**4.8** Move 200-lb force 5 ft to left.

**4.10** $F = -150k - 200j$ N
$C = 34i + 26k$ N-m

**4.12** $F = 20i - 60j + 30k$ N
$C_A = 900i + 680j + 760k$ N-m

**4.14** $F_A = -440k$ lb
$C_A = -1950i - 2080j$ ft-lb

**4.16** $F = -16.77i + 9.68j + 134.2k$ kN
$C = 1936i - 3354j$ N-m

**4.18** $F = -25i - 43.3k$ lb
$C_A = -1210i - 1106j + 700k$ ft-lb
$M_y = -1106$ ft-lb

**4.20** $F = -100k$ kN
$(C_R)_1 = -120i - 600j$ kN-m
$(C_R)_2 = -120i + 1400j$ kN-m

**4.22** $F_R = -31.8i - 238j$ lb
$C_R = -690i + 775j - 1297k$ lb-ft

**4.24** $F_R = -250i - 2300j$ lb
$\bar{x} = 15.66$ ft

**4.26** $F_R = 37.5i - 495j$ lb
$\bar{x} = 18.66$ ft

**4.28** $F_R = 39.4i + 74.7j$ N
$\bar{x} = .669$ m

**4.30** (a) $F_R = -100k$ m
$\bar{x} = 2.5$ m; $\bar{y} = 2.2$ m
(b) $F_R = \mathbf{0}$
$C_R = 280i + 450j$ N-m

**4.32** $F_R = \mathbf{0}$
$C_R = 860i + 900j$ N-mm

**4.34** $F = -43000j$ lb
$\bar{x} = 23.2$ ft

**4.38** $F = 50{,}625k$ oz.

**4.40** $\bar{x} = 6.67$ m

**4.42** $\bar{z} = h/4$

**4.46** $2.37 \times 10^8$ kN

**4.48** $\bar{x} = 2.615$ m; $\bar{y} = 1.598$ m

**4.50** $F = -37.5k$ kN
$\bar{x} = .844$ m, $\bar{y} = 1.067$ m } from left front corner
5.25 kN or 3.5 m$^3$

**4.52** 61.1 lb-ft
$\bar{x} = 12.33$ ft
$F = 4950$ lb

**4.54** $F_R = -P_0 \dfrac{ab}{2} k$

$\bar{y} = \dfrac{2}{3}a$

**4.56** 315 K
$\bar{x} = 48.0$ ft; $\bar{y} = 21.7$ ft

**4.58** $F_R = 45{,}000$ N
$\bar{x} = 13.33$ m; $\bar{y} = 5.185$ m

**4.60** 13,230 lb

**4.62** $F_R = 1.699 \times 10^{10}$ N
24.04 m from base

**4.64** $\bar{x} = 19.4$ ft
$\bar{y} = 8.12$ ft

**4.66** $F = 2200$ lb
$\bar{x} = 4.136$ ft from origin

**4.68** $F = 720$ lb
$\bar{x} = 14$ ft

**4.70** $F = 86.6j + 50k$ lb
9.42 ft above $B$

**4.71** $F = -400j$ kN
$M_1 = -900k$ kN-m
$M_2 = -300k$ kN-m

**4.72** $F = 10i + j - 2k$ lb
$C = 13i - 22j + 32k$ ft-lb

**4.73** $F_R = 4.70i - 10.09j$ kN
$M_R = -.3i + 3.23k$ kN-m

**4.74** $F_R = -90j$ lb
$\bar{x} = 21.2$ ft

**4.75** $F_R = 1033$ kN
$\bar{x} = 2.62$ m

**4.76** $F_R = 58{,}900$ lb
$\bar{x} = 3.92$ ft from left of post

**4.77** $F_R = -120j$ N
$\bar{x} = 6.97$ m; $\bar{z} = .471$ m

**4.78** $\bar{x} = 1.20$ ft; $\bar{y} = 1.125$ ft;

$\bar{z} = \dfrac{t}{2}$ ft

**4.79** $F = 91.5$ kN
Distance above $A$ is .49 m.

**4.80** $\bar{y} = 1.299$ ft

**4.81** $\bar{x} = 5.32$ m

**4.82** $F = -1550$ lb
$\bar{x} = 20.04$ ft

## CHAPTER 5

**5.18** $T_{BA} = 37.9$ lb
$T_{BC} = 26.8$ lb

**5.20** 9000 N
$T_1 = 13{,}135$ N
$T_2 = 13{,}126$ N

**5.24** $BN = 4905$ N
$F = 1380$ N

**5.26** $2T\left[\left(\dfrac{T}{5000} + \dfrac{1}{2}\right) - \left(\dfrac{1}{2}\right)^2\right]^{1/2} =$

$\left(\dfrac{T}{5000} + \dfrac{1}{2}\right)(1000)$

**5.28** 7.46 mm

**5.30** (a) $T = 2121$ lb
$H = 3000$ lb; $V = 3000$ lb
(b) $T = 1148$ lb
$H = 2121$ lb; $V = 5121$ lb

**5.32** $DB = 353$ lb
$AC = 289$ lb
$EC = 394$ lb

**5.34** 63.75 kN-m

**5.36** 260 N

**5.38** $A_y = B_y = 68{,}170$ N
$A_x = 6366$ N

**5.40** $A = 260$ tons
$B = 276$ tons
$C = 259$ tons
$D = 0$

**5.42** $B_x = 288$ N
$B_y = 433$ N
$M_B = 252$ N-m

**5.44** $A_x = 650$ N
$A_y = -138.9$ N
$M_A = -1500$ N-m

**5.46** $A_x = 388$ N
$A_y = -31$ N
$B_x = 388$ N
$B_y = 431$ N

**5.48** $A_x = -377$ lb
$A_y = 244$ lb
$C_y = 206$ lb

**5.50** $T = 55.3$ ft-lb

**5.52** $T = 11.36$ ft-lb

**5.54** $T = 443$ N-m

**5.56** $T_E = -40$ N-m

**5.58** $B_x = 5620$ N
$B_y = -27{,}620$ N
$T_D = 25{,}900$ N

**5.62** (a) $R = 4.69$ kN
$F = 5.11$ kN
(b) $R = 3.75$ kN
$F = 8.05$ kN

**5.64** 24 ft

**5.66** $D_x = 0$; $D_y = 4.1\ K$;
$C_y = 2.9\ K$

**5.68** 320 N; 530 N; 2025 N-m

**5.70** 73.5 kN; 106.5 kN

**5.72** 1.521 m

**5.74** $F_1 = 7.33$ kN
$F_2 = .667$ kN
$F_3 = 12.00$ kN

**5.76** $F_A = -625$ lb
$F_B = 6250$ lb
$F = 5625$ lb

**5.78** $F_1 = W$
$C_2 = .6W$
$F_3 = .1316C_1$

**5.80** $A_x = 0$ $B_y = -100$ kN
$A_y = -600$ kN $B_z = 0$
$A_z = 200$ kN $C_y = 600$ kN

**5.82** $A = -10\boldsymbol{i} + 3\boldsymbol{j} + 500\boldsymbol{k}$ N
$M_A = -560\boldsymbol{i} - 4300\boldsymbol{j} + 10\boldsymbol{k}$ N-m

**5.84** $F = 16.67$ lb $B_x = 0$
$A_x = 0$ $B_y = 63.3$ lb
$A_y = 53.3$ lb

**5.86** $T_G = 7490$ N $A_y = 21{,}020$ N
$T_K = 15{,}060$ N $A_z = -1521$ N
$A_x = -587$ N

**5.88** $P = 20{,}800$ N $B_z = 150$ N
$B_x = 25{,}300$ N $A_y = -8913$ N
$B_y = 1670$ N $A_x = 8705$ N

**5.90** $T_E = 1688$ N
$T_D = 2840$ N

**5.92** $T_R = 1928$ N $l = .0828$
$T_P = 709$ N $m = .782$
$T_A = 2129$ N $n = .618$

**5.94** $A = 36.7\boldsymbol{i} + 8.73\boldsymbol{j} + 15.91\boldsymbol{k}$ N

**5.96** 750 N

**5.98** 69.3 kN

**5.100** $T_1 = 649$ N
$T_2 = 0$
$A = 591$ N

**5.102** $CE = 4.36$ kN
$A_x = .800$ kN
$A_y = 196.7$ kN

**5.104** 2440 N

**5.106** $F_A = 2617$ N
$F_B = -617$ N
$C = 2801$ N

**5.108** $CD = 600$ lb
$F = 1620$ lb

**5.110** $T = .975$ compression
$F = 1.378$ N upward

**5.112** $C_y = 0$ $A = 312.5$ lb
$C_x = 600$ lb $D = 1033$ lb
$C_z = -320$ lb

**5.114** 208 psi

**5.116** $A = 433\text{i} - 533\boldsymbol{j} - 19.00\boldsymbol{k}$ lb
$E = 240\boldsymbol{k}$ lb
$H = -433\boldsymbol{i} + 433\boldsymbol{j} - 520\boldsymbol{k}$
$K = \mathbf{0}$

**5.118** $F_{BC} = 4963$ N
$F_{EB} = 6541$ N
$F_{GE} = 4920$ N

**5.119** $G_x = 750$ lb
$G_y = 650$ lb

**5.120** $T = 30$ ft-tons
$B_y = 6.8$ tons

**5.121** $F_x = A_x = 806$ lb
$F_y = A_y = 500$ lb

**5.122** 335 N-m

**5.123** $R_1 = 41$ kN
$R_2 = 88$ kN
$R_3 = 141$ kN

**5.124** $B = 12.50\boldsymbol{i} + 100\boldsymbol{j} - 18.75\boldsymbol{k}$ lb
$A = 12.50\boldsymbol{i} - 18.75\boldsymbol{k}$ lb

**5.125** $F_{CD} = 1694$ lb
$F_{EF} = 4931$ lb

**5.126** $R = 7930$ lb
$C = 9640$ ft-lb

**5.127** $\theta = 43.73°$

**5.128** $AB = 800$ lb
$E_x = 5000$ lb
$E_y = 6928$ lb
$M_E = 5000$ ft-lb

**5.129** $E_x = -23.2$ kN
$E_y = -30.18$ kN
$D = 46.4$ kN

## CHAPTER 6

**6.2** $CD = 3606$ C. $AB = 2670$ C.
$ED = 2000$ T. $BF = 0$
$EF = 2000$ T. $BC = 2670$ C.
$CE = 2000$ T. $FC = 1202$ T.
$AF = 4808$ T.

**6.4** $CH = 12{,}480$ lb C.
$AB = DE = 18{,}720$ lb C.
$EF = AJ = 24{,}960$ lb C.
$AI = EG = 26{,}474$ lb T.
$DG = BI = 18{,}720$ lb C.
$DH = HB = 8{,}825$ lb T.
$HI = HG = 18{,}720$ lb T.

**6.6** $AB = DE = 26.8$ kN C.
$AG = FE = 19$ kN T.
$BG = DF = 3.14$ kN T.
$GC = CF = 3.39$ kN C.
$BC = CD = 20$ kN C.
$GF = 21.36$ kN C.

**6.8** $AB = 353$ kN C.
$AC = 250$ kN T.
$LK = 250$ kN T.
$LB = 100$ kN T.
$BC = 316$ kN C.
$BK = 70.8$ kN T.
$KJ = 300$ kN T.
$CK = 49.9$ kN T.
$CD = 304$ kN C.
$CJ = .251$ kN C.
$DE = 304$ kN C.
$DJ = 100.4$ kN T.

**6.10** $JI = 25$ kN T.
$AI = 44.2$ kN T.
$IH = 56.25$ kN T.
$BH = 8.84$ kN C.
$CD = 50$ kN C.
$HD = 8.84$ kN T.
$GD = 6.25$ kN C.
$AJ = 31.25$ kN C.
$AB = 56.25$ kN C.
$BI = 6.25$ kN T.
$BC = 50$ kN C.
$HC = 0$
$HG = 43.75$ kN T.
$DE = 43.75$ kN C.
$GE = 61.9$ kN T.
$EF = 43.75$ kN C.
$GF = 0$

**6.12** $CD = 583$ lb C.
$AC = BC = AD = 0$
$BD = 1158$ lb T.

**6.14** $BA = BC = BE = AF = EF =$
$CE = AE = 0$; $AC = 14.14$ kN T.
$CD = 14$ kN T.
$DE = 2$ kN
$AD = 24.5$ kN C.

**6.16** $CB = 250$ N C.
$BE = 500$ N C.

**6.18** $KU = 15$ kN T.

**6.20** $CD = 7.07$ kN C.
$DG = 0$
$HG = 25$ kN T.

**6.22** $LK = 0$
$KJ = 2000$ lb T.

**6.24** $HE = 4.72$ K T.
$FH = 4.72$ K T.
$FE = 2.5$ K T.
$FC = 0$

**6.26** $FI = .558$ K C.
$EH = 1.118$ K C.
$DH = 1.118$ K T.

**6.28** $EG = 916{,}000$ lb C.
$FH = 836{,}000$ lb T.
$IJ = 144{,}400$ lb T.

**6.30** $\underline{0 < x < 5}$
$V = M = 0$

$\underline{5 < x \leqslant 10}$
$V = 500$ lb
$M = -500x + 2500$ ft-lb

$\underline{10 \leqslant x < 25}$
$V = 200 + 30x$ lb
$M = -15x^2 - 200x + 1000$ ft-lb

**6.32** (a) $V = -737.5$ lb
$M = 3690$ ft-lb
(b) $V = 262.5$ lb
$M = 6850$ ft-lb
(c) $V = 262.5$ lb
$M = 1312$ ft-lb

**6.34** $\underline{0 < s < 14.14}$
$V = -37.1$ lb
$H = -37.1$ lb
$M = 37.1s$ ft-lb

$\underline{14.14 < s \leqslant 29.14}$
$V = -52.5$ lb
$H = 0$
$M = 52.5s - 217$ ft-lb

$\underline{29.14 \leqslant s < 44.14}$
$V = 10s - 343.9$ lb
$H = 0$
$M = -5s^2 + 344s - 4462$ ft-lb

$\underline{44.14 < s < 58.28}$
$V = 68.9$ lb
$H = -68.9$ lb
$M = -68.9s + 4018$ ft-lb

**6.36** Section $CD$
$\underline{0 \leqslant s < 10}$
$V_x = 0$
$H = 0$
$V_y = -20s$ N
$M_x = -10s^2$ N-m
$M_y = M_z = 0$

Section $BC$
$\underline{10 < s < 20}$
$V_y = 200$ N
$H = V_z = 0$
$M_x = 1000$ N-m
$M_y = 0$
$M_z = 200s - 2000$ N-m

Section $AB$
$\underline{20 < s < 30}$
$V_x = 0$
$V_z = -1000$ N
$H = 200$ N
$M_x = -1000s + 21{,}000$ N-m
$M_y = 0$
$M_z = 2000$ N-m

**6.38** $\underline{0 < \theta < \pi / 4}$
$V = -70.7 \sin\theta$ N
$H = -70.7 \cos\theta$ N
$M = 424 - 424 \cos\theta$ N-m

$\underline{(\pi / 4) < \theta < (\pi / 2)}$
$V = 29.3 \sin\theta$ N
$H = 29.3 \cos\theta$ N
$M = 176 \cos\theta$ N-m

**6.40** Section $AB$
$H = 0$
$V_x = 0$
$V_z = 11.36s - 262$ lb
$M_x = -5.68s^2 + 262s - 2194$ ft-lb
$M_y = 410$ ft-lb
$M_z = 0$

Section $BC$
$V = -130.0 + 5.64s$ lb
$H = 227.5 - 9.87s$ lb
$M_{z'} = -123.3$ ft-lb
$M_{x'} = -70.47$ ft-lb

Section $CD$
$V_z = -262 + 11.36s$
$M_y = 0$
$M_x = -3021 + 262s - 5.68s^2$ ft-lb

**6.42** $\underline{0 < x < 5}$
$V = M = 0$

$\underline{5 < x \leqslant 10}$
$V = 500$ lb
$M = -500x + 2500$ ft-lb

$\underline{10 \leqslant x < 25}$
$V = 30x + 200$ lb
$M = -15x^2 - 200x + 1000$ ft-lb

**6.44** $\underline{0 < x < 10}$
$V = -737.5$ lb
$M = 737.5x$ ft-lb

$\underline{10 < x < 25}$
$V = 262.5$ lb
$M = -262.5x + 10{,}000$ ft-lb

$\underline{25 < x < 40}$
$V = 262.5$ lb
$M = -262.5x + 10{,}500$ ft-lb

**6.46** $\underline{0 < x \leqslant 5}$
$V = -1333$ N
$M = 1333x$ N

$\underline{5 \leqslant x \leqslant 15}$
$V = 20x^2 + 200x - 2833$ N
$M = 1333x - 200\,(x-5)^2 + \dfrac{20}{3}(x-5)^3$ N-m

$\underline{15 \leqslant x < 25}$
$V = 667$ N
$M = 667x + 23{,}340$ N-m

**6.54** $M_{\max} = 400 / \pi^2$ ft-lb

**6.56** (a) 2 ft from left end
(b) Just to left of 8000-lb force

**6.58** $\underline{0 < x < 30}$
$V = 1000$ lb
$M = -1000x$ ft-lb

$\underline{30 < x \leqslant 60}$
$V = -500 + \dfrac{5}{3}(x-30)^2$ lb

$M = 500x - 45{,}000 - \dfrac{5}{9}(x-30)^3$ ft-lb

$\underline{60 \leqslant x < 90}$
$V = 1000$ lb
$M = -1000x + 30{,}000$ ft-lb

$90 < x < 120$
$V = 0$
$M = -60{,}000$ ft-lb

**6.60** $y = .000702x^2 + .08727x$
$T_{max} = 2.02 \times 10^6$ N

**6.62** $T_{max} = \dfrac{5l^2\sqrt{l^1 + (3h)^2}}{6h}$ N

**6.64** 300 ft; 89.7 ft

**6.66** 11.34 m; $T_{max} = 840$ N

**6.68** 53 ft; 432.6 ft

**6.70** *AB:* $\alpha_1 = 56.3°$; 18 ft
*BC:* $\alpha_2 = 45°$; 28.3 ft
*DE:* $\alpha_3 = 45°$; 28.3 ft

**6.71** $DB = BE = AE = 0$
$BC = 45.3$ kN C.
$DC = 32$ kN C.
$DE = 32$ kN C.
$BA = 45.3$ kN T.

**6.72** $AB = 85$ kN C.
$AH = 73.0$ kN T.
$HG = 78.6$ kN T.
$HB = 29.2$ kN C.
$BC = 85.0$ kN C.
$BG = 0$
$CD = BC = 85.0$ kN C.
$CG = 58.3$ kN T.
Others via symmetry.

**6.73** $CD = 1500$ lb T.
$CA = CE = 1060$ lb C.
$AE = 1250$ lb T.
$ED = 1030$ lb C.
$AD = AE = 1250$ lb T.

**6.74** $BG = 5.59$ K C.
$BF = 7.50$ K T.
$CE = 15$ K T.

**6.75** $0 \leqslant x < 3$
$V = 10x$ N
$M = -5x^2$ N-m
$3 < x < 9$
$V = 10x - 19.44$ N
$M = -5x^2 + 19.44\,(x - 3)$ N-m
$9 < x < 12$
$V = 70.6$ N
$M = 19.44\,(x - 3) - 90(x - 4.5)$
$+ 500$ N-m

**6.76** $0 \leqslant x < 2$
$V = 0$
$M = 0$
$2 < x \leqslant 5$
$V = 0$
$M = 300$ N-m
$5 \leqslant x < 12$
$V = \dfrac{500}{14}(x - 5)^2$ N
$M = 300 - \dfrac{500}{42}(x - 5)^3$ N-m

**6.77** $0 \leqslant x < 10$
$V = M = 0$
$10 < x < 20$
$V = -650$ lb
$M = 650\,(x - 10)$ ft-lb
$20 < x < 40$
$V = 350$ lb
$M = 650(x - 10) - 1000(x - 20)$ ft-lb
$40 < x < 65$
$V = 0$
$M = -500$ ft-lb
$M = 0$ occurs at 28.6 ft to right of left support.

**6.78** $R_1 = 518$ N
$R_2 = 432$ N
$V_{max} = -518$ N
$M_{max} = 1354$ N-m

**6.79** $y = h\left(1 - \cos\dfrac{\pi x}{l}\right)$

**6.80** $\alpha = 30.7°$
$W = 122.6$ lb

## CHAPTER 7

**7.4** 600 lb; yes

**7.6** 151.0 lb; 1390 lb

**7.8** $F = 1176$ lb; $T = 15.59$ ton-ft

**7.10** 1447 N

**7.12** $\beta = 58.8°$

**7.14** 8.33 lb

**7.16** $T = 229.8$ lb; no slipping

**7.18** $f = 115.5$ N; $\mu = .578$

**7.22** $C$ moves 1.5 m to left

**7.24** 3520 lb

**7.28** $c = \dfrac{.483a + .1294d}{.978\mu_s + .259} - d$

**7.30** $T = 38.5$ kN-m
$P = 44.5$ kN

**7.32** 15.8 in.

**7.34** $\mu_s = .620$
**7.36** 1.753 m
**7.38** 6.81 ft
**7.40** .991 in.
**7.42** 50 mm
**7.44** $\alpha = 7.26°$
**7.48** 43.3 N-mm
**7.50** 2.73$P$ N-mm
**7.52** 2144 N
**7.54** 3 turns
**7.56** $\mu = .292$
**7.58** $F_1 = .652W$
$F_2 = .348W$
**7.60** 197.1 ft-lb
**7.62** Large wheel: 313 ft-lb
Small wheel: 156.6 ft-lb
**7.64** 20 lb
$T$ at right wheel: 53.2 lb
$T$ at left wheel: 33.2 lb
**7.68** 217 N
**7.70** 3687 N
**7.72** $T_1 = 513$ N
$T_1' = 324$ N
37% error
**7.74** 315 N-mm
**7.76** 1906 in.-lb
**7.78** (a) $M_z = 6790$ N-mm
(b) $M_z = 7599$ N-mm
**7.82** 381 N-mm
**7.83** 128.2 N
**7.84** 50 lb
**7.85** $\theta = 30.96°$
$T = 292$ lb
**7.86** $M_2 = 840$ N-m
$M_1 = 16.80$ N-m
105 N
**7.87** 3333 N
**7.88** 267 N
**7.89** 94.7 mm
**7.90** 1065 in.-lb
**7.91** (a) $T_1 = 272$ lb
$P = 13.22$ lb
(b) $T_1 = 272$ lb
$P = 163.2$ lb
**7.92** 20 in.-lb
$T_1 = 18.65$ lb
$T_2 = 8.65$ lb
$T_3 = 32.8$ lb
$T_4 = 12.77$ lb
**7.93** 40.7 N-m
**7.94** 74.7 N

## CHAPTER 8

**8.4** $M_x = 125$ ft³
$M_y = 200$ ft³
**8.10** $x_c = 1.712$ mm
$y_c = 3.75$ mm
**8.12** $x_c = 3.79$ mm
$y_c = 7.58$ mm
$M_{AA} = 234$ mm³
**8.14** $x_c = y_c = 424$ mm
**8.16** $x_c = 1.717$ m
$y_c = .1722$ m
**8.18** $x_c = 2.89$ m
$y_c = 2.67$ m
**8.20** $y_c = 315.7$ mm
**8.22** $x_c = 40.1$ mm
$y_c = 19.51$ mm
**8.24** $\bar{x}_c = 12.92$ ft
**8.26** $x_c = a/2$
$y_c = \frac{2}{3}b$
$z_c = c/3$
**8.28** $z_c = \frac{3}{8}a$
**8.30** $r_c = 1.179\boldsymbol{i} + .955\boldsymbol{j} + .284\boldsymbol{k}$ m
**8.32** $x_c = 2.57$ m
**8.34** $x_c = 8.22$ m
$y_c = z_c = 0$
**8.36** Center of volume
$x_c = -9.75$ mm
$y_c = 122.6$ mm
$z_c = 0$
Center of mass and gravity
$\bar{x} = -32.8$ mm
$\bar{y} = 447$ mm
$\bar{z} = 0$
**8.38** $x_c = 67.65$ mm
$y_c = 608$ mm
$z_c = 493$ mm
**8.40** $r_c = .742\boldsymbol{j} - .1720\boldsymbol{k}$ m
**8.42** $y_c = \frac{2}{3}\frac{b}{\pi}$

**8.44** $A = 1751 \text{ in.}^2$
$V = 5242 \text{ in.}^3$

**8.46** $A = .8624 \text{ m}^2$
$V = .0633 \text{ m}^3$

**8.48** $I_{xx} = 11{,}426 \text{ ft}^4$
$I_{yy} = 137{,}100 \text{ ft}^4$
$I_{xy} = 34{,}280 \text{ ft}^4$

**8.50** $I_{xx} = 122.7 \text{ m}^4$
$I_{yy} = 122.7 \text{ m}^4$

**8.52** $27.3 \text{ ft}^4$

**8.54** .763

**8.56** $I_{xx} = 1.758$
$I_{yy} = 4.45$
$I_{xy} = 0$

**8.58** $I_{xx} = .357 \text{ ft}^4$
$I_{yy} = .232 \text{ ft}^4$
$I_{xy} = -.0625 \text{ ft}^4$

**8.60** $I_{xy} = 5906 \text{ ft}^4$

**8.62** $270 \text{ ft}^4$

**8.66** $I_{xx} = I_{yy} = \dfrac{b-a}{12}(a^3 + ab^2 + a^2b + b^3)$
$I_{xy} = 0$

**8.68** $I_{xx} = 11.28 \times 10^4 \text{ mm}^4$
$I_{yy} = 8.36 \times 10^4 \text{ mm}^4$
$I_{x_cx_c} = 4.298 \text{ mm}^4$
$I_{y_cy_c} = 8.36 \times 10^4 \text{ mm}^4$

**8.70** $I_{x_cx_c} = 1.652 \times 10^7 \text{ mm}^4$
$I_{y_cy_c} = 6.58 \times 10^7 \text{ mm}^4$
$I_{x_cy_c} = 0$

**8.72** $I_{xx} = 3.29 \times 10^7 \text{ mm}^4$
$I_{yy} = 2.89 \times 10^6 \text{ mm}^4$
$I_{xy} = 6.90 \times 10^6 \text{ mm}^4$
$I_{x_cx_c} = 9.85 \times 10^6 \text{ mm}^4$
$I_{y_cy_c} = 1.607 \times 10^6 \text{ mm}^4$
$I_{x_cy_c} = 1.457 \times 10^6 \text{ mm}^4$

**8.74** $I_{x'x'} = 833 \text{ mm}^4$
$I_{y'y'} = 833 \text{ mm}^4$
$I_{x'y'} = 0$

**8.76** $I_{xx} = 91.9 \text{ ft}^4$
$I_{yy} = 1455 \text{ ft}^4$
$I_{xy} = -96.5 \text{ ft}^4$
$(I_p)_a = 1547 \text{ ft}^4$
$(I_p)_b = 1547 \text{ ft}^4$

**8.80** $I_1 = 109.2 \text{ mm}^4$
$I_2 = 2367 \text{ mm}^4$

**8.82** $I_1 = 1299 \text{ mm}^4$
$I_2 = 79.1 \text{ mm}^4$

**8.85** $y_c = \dfrac{3}{8} \text{ m}$
$x_c = \pi / 2 \text{ m}$
$M_{x'} = -2.55 \text{ m}^3$
$M_{y'} = 4.04 \text{ m}^3$

**8.86** $y_c = 1.671 \text{ m}$

**8.87** (a) Center of volume:
$r_C = 15\boldsymbol{i} + 17.72\boldsymbol{j} + 13.72\boldsymbol{k}$ mm
(b) Center of mass and center of gravity:
$\boldsymbol{r} = 15\boldsymbol{i} + 16.71\boldsymbol{j} + 15.39\boldsymbol{k}$ mm

**8.88** $x_c = y_c = \dfrac{4}{3}\dfrac{r}{\pi}$

**8.89** $A = 155.4 \text{ m}^2$
$V = 117.8 \text{ m}^3$

**8.90** $I_{x'x'} = 6044 \text{ m}^4$
$I_{y'y'} = 27{,}590 \text{ m}^4$
$I_{x'y'} = 14{,}820 \text{ m}^4$

**8.91** $I_{x_cx_c} = 83.25 \text{ in.}^4$
$I_{y_cy_c} = 19.25 \text{ in.}^4$
$I_{x_cy_c} = 0$

**8.92** $I_1 = 16.53 \text{ ft}^4$
$I_2 = 1130 \text{ ft}^4$

**8.93** $6294 \text{ m}^4$

## CHAPTER 9

**9.2** $I_{xx} = 4333 \text{ lbm-ft}^2$
$I_{x'x'} = 15{,}333 \text{ lbm-ft}^2$

**9.4** $I_{xx} = \dfrac{1}{12}M(a^2 + l^2)$
$I_{yy} = \dfrac{1}{12}M(a^2 + b^2)$
$I_{zz} = \dfrac{1}{12}M(b^2 + l^2)$
$I_{xy} = 0$

**9.6** $I_{BB} = \dfrac{1}{2}Mr^2$

**9.8** $I_{xx} = 3.95 \times 10^8 \text{ gram-mm}^2$
$I_{zz} = 2.67 \times 10^7 \text{ gram-mm}^2$

**9.10** $I_{zz} = 1.723 \times 10^6 \text{ gramm-mm}^2$

**9.12** $I_{yy} = 3959 \text{ kg-mm}^2$

**9.14** $1.004 \times 10^9 \text{ mm}^4$
$20{,}500 \text{ kg-mm}^2$
$3820 \text{ kg-mm}^2$
$2390 \text{ kg-mm}^2$

**9.16** $I_1 = 37.0 \text{ kg-mm}^2$
$I_2 = 166.8 \text{ kg-mm}^2$
$I_3 = 204 \text{ kg-mm}^2$

**9.20** $I_{x'x'} = I_{z'z'} = 29.0$ slug-ft²
$I_{y'y'} = 6.21$ slug-ft²
$I_{x'y'} = I_{x'z'} = I_{y'z'} = 0$
$I_{xx} = 119.1$ slug-ft²
$I_{yy} = 18.63$ slug-ft²
$I_{zz} = 106.6$ slug-ft²
$I_{xy} = I_{zx} = 0$
$I_{yz} = -31.1$ slug-ft²

**9.22** $(M/6)(a^2+b^2+c^2)+2M(x^2+y^2+c^2)$

**9.24** $I_{x''x''} = 3870$ kg-m²
$I_{x''y''} = 4615$ kg-m²

**9.26** $1.258\times10^{-6}$ kg-m²
$1.239\times10^{-6}$ kg-m²
$3.91\times10^{-5}$ kg-m²

**9.28** 1.650 kg-m²
1.417 kg-m²
.432 kg-m²

**9.30** 28,400 lbm-ft²
2.54 ft

**9.32** $I_{xx} = 8093$ lbm-in.²
$I_{yy} = 1603$ lbm-in.²
$I_{xy} = 1282$ lbm-in.²

**9.34** 50.8 kg-m²

**9.36** $I_{z'z'} = 3026$ lbm-ft²

**9.38** .499 kg-m²

**9.40** −.01206 kg-m²

**9.44** 362 kg-mm²

**9.45** 362 kg-mm²

**9.46** $I_{xx} = 534$ kg-mm²
$I_{yy} = 1659$ kg-mm²
$I_{zz} = 2190$ kg-mm²
$I_{xy} = 568$ kg-mm²
$I_{xy} = I_{yz} = \approx 0$

**9.47** 2190 kg-mm²
1896 kg-mm²
297 kg-mm²
$\alpha_1 = 22.65°$
$\alpha_2 = 112.65°$
$z$ axis

**9.48** .0493 kg-m²
.00187 kg-m²
.0503 kg-m²

**9.49** .00295 kg-m²

**9.50** $I_{yy} = 1075$ kg-m²
$I_{yz} = 92.35$ kg-m²

**9.51** $I_{y'x'} = 0$ for all $\theta$

**CHAPTER 10**

**10.2** $S = 1250$ N
$S = 750$ N

**10.4** $\theta = 19.48°$

**10.6** $W = .351\ T$

**10.8** $\theta = 8.53°$

**10.10** $P = (W/2)\cot\beta$

**10.12** 2513 lb

**10.14** $T_{CE} = 7.42$ kN
$T_{DE} = 29.3$ kN

**10.16** $\tan\beta = \dfrac{1}{\sqrt{3}}[(a-3b)/(b+a)]$

**10.18** $T' = 40$ N-m

**10.20** $P = .0545$ N

**10.22** $W = 270$ lb

**10.24** $W = 2770$ N

**10.26** $C = 108.6$ N

**10.28** $d = 72.0$ mm

**10.32** $d = .2$ ft

**10.40** $\theta = 19.22°$

**10.46** $a = .358$ m

**10.48** $a = .1126$ m

**10.52** When $d > 2$, stable equilibrium.
When $d < 2$, unstable equilibrium.

**10.56** $W_{max} = 250$ lb

**10.60** $x = 4$ m

**10.61** $Q = 3P$

**10.62** $P = 7540$ lb

**10.63** $\theta = 27.7°$

**10.64** $\cos\theta = \dfrac{\cos\theta_0(aK_1+aK_2)+W}{aK_1+aK_2}$

**10.66** $\theta = 0$ (unstable)
$\theta = 28.1°$ (stable)

**10.67** $W_{max} = 1000$ N

**10.68** $R > (h/2)$ for stable equilibrium.
$R < (h/2)$ for unstable equilibrium.

**8.44** $A = 1751$ in.$^2$
$V = 5242$ in.$^3$

**8.46** $A = .8624$ m$^2$
$V = .0633$ m$^3$

**8.48** $I_{xx} = 11{,}426$ ft$^4$
$I_{yy} = 137{,}100$ ft$^4$
$I_{xy} = 34{,}280$ ft$^4$

**8.50** $I_{xx} = 122.7$ m$^4$
$I_{yy} = 122.7$ m$^4$

**8.52** 27.3 ft$^4$

**8.54** .763

**8.56** $I_{xx} = 1.758$
$I_{yy} = 4.45$
$I_{xy} = 0$

**8.58** $I_{xx} = .357$ ft$^4$
$I_{yy} = .232$ ft$^4$
$I_{xy} = -.0625$ ft$^4$

**8.60** $I_{xy} = 5906$ ft$^4$

**8.62** 270 ft$^4$

**8.66** $I_{xx} = I_{yy} = \dfrac{b-a}{12}(a^3 + ab^2 + a^2b + b^3)$
$I_{xy} = 0$

**8.68** $I_{xx} = 11.28 \times 10^4$ mm$^4$
$I_{yy} = 8.36 \times 10^4$ mm$^4$
$I_{x_cx_c} = 4.298$ mm$^4$
$I_{y_cy_c} = 8.36 \times 10^4$ mm$^4$

**8.70** $I_{x_cx_c} = 1.652 \times 10^7$ mm$^4$
$I_{y_cy_c} = 6.58 \times 10^7$ mm$^4$
$I_{x_cy_c} = 0$

**8.72** $I_{xx} = 3.29 \times 10^7$ mm$^4$
$I_{yy} = 2.89 \times 10^6$ mm$^4$
$I_{xy} = 6.90 \times 10^6$ mm$^4$
$I_{x_cx_c} = 9.85 \times 10^6$ mm$^4$
$I_{y_cy_c} = 1.607 \times 10^6$ mm$^4$
$I_{x_cy_c} = 1.457 \times 10^6$ mm$^4$

**8.74** $I_{x'x'} = 833$ mm$^4$
$I_{y'y'} = 833$ mm$^4$
$I_{x'y'} = 0$

**8.76** $I_{xx} = 91.9$ ft$^4$
$I_{yy} = 1455$ ft$^4$
$I_{xy} = -96.5$ ft$^4$
$(I_p)_a = 1547$ ft$^4$
$(I_p)_b = 1547$ ft$^4$

**8.80** $I_1 = 109.2$ mm$^4$
$I_2 = 2367$ mm$^4$

**8.82** $I_1 = 1299$ mm$^4$
$I_2 = 79.1$ mm$^4$

**8.85** $y_c = \dfrac{3}{8}$ m
$x_c = \pi / 2$ m
$M_{x'} = -2.55$ m$^3$
$M_{y'} = 4.04$ m$^3$

**8.86** $y_c = 1.671$ m

**8.87** (a) Center of volume:
$r_C = 15i + 17.72j + 13.72k$ mm
(b) Center of mass and center of gravity:
$r = 15i + 16.71j + 15.39k$ mm

**8.88** $x_c = y_c = \dfrac{4}{3}\dfrac{r}{\pi}$

**8.89** $A = 155.4$ m$^2$
$V = 117.8$ m$^3$

**8.90** $I_{x'x'} = 6044$ m$^4$
$I_{y'y'} = 27{,}590$ m$^4$
$I_{x'y'} = 14{,}820$ m$^4$

**8.91** $I_{x_cx_c} = 83.25$ in.$^4$
$I_{y_cy_c} = 19.25$ in.$^4$
$I_{x_cy_c} = 0$

**8.92** $I_1 = 16.53$ ft$^4$
$I_2 = 1130$ ft$^4$

**8.93** 6294 m$^4$

## CHAPTER 9

**9.2** $I_{xx} = 4333$ lbm-ft$^2$
$I_{x'x'} = 15{,}333$ lbm-ft$^2$

**9.4** $I_{xx} = \dfrac{1}{12}M(a^2 + l^2)$
$I_{yy} = \dfrac{1}{12}M(a^2 + b^2)$
$I_{zz} = \dfrac{1}{12}M(b^2 + l^2)$
$I_{xy} = 0$

**9.6** $I_{BB} = \dfrac{1}{2}Mr^2$

**9.8** $I_{xx} = 3.95 \times 10^8$ gram-mm$^2$
$I_{zz} = 2.67 \times 10^7$ gram-mm$^2$

**9.10** $I_{zz} = 1.723 \times 10^6$ gramm-mm$^2$

**9.12** $I_{yy} = 3959$ kg-mm$^2$

**9.14** $1.004 \times 10^9$ mm$^4$
20,500 kg-mm$^2$
3820 kg-mm$^2$
2390 kg-mm$^2$

**9.16** $I_1 = 37.0$ kg-mm$^2$
$I_2 = 166.8$ kg-mm$^2$
$I_3 = 204$ kg-mm$^2$

**9.20** $I_{x'x'} = I_{z'z'} = 29.0$ slug-ft²
$I_{y'y'} = 6.21$ slug-ft²
$I_{x'y'} = I_{x'z'} = I_{y'z'} = 0$
$I_{xx} = 119.1$ slug-ft²
$I_{yy} = 18.63$ slug-ft²
$I_{zz} = 106.6$ slug-ft²
$I_{xy} = I_{zx} = 0$
$I_{yz} = -31.1$ slug-ft²

**9.22** $(M/6)(a^2+b^2+c^2)+2M(x^2+y^2+c^2)$

**9.24** $I_{x''x''} = 3870$ kg-m²
$I_{x''y''} = 4615$ kg-m²

**9.26** $1.258 \times 10^{-6}$ kg-m²
$1.239 \times 10^{-6}$ kg-m²
$3.91 \times 10^{-5}$ kg-m²

**9.28** 1.650 kg-m²
1.417 kg-m²
.432 kg-m²

**9.30** 28,400 lbm-ft²
2.54 ft

**9.32** $I_{xx} = 8093$ lbm-in.²
$I_{yy} = 1603$ lbm-in.²
$I_{xy} = 1282$ lbm-in.²

**9.34** 50.8 kg-m²

**9.36** $I_{z'z'} = 3026$ lbm-ft²

**9.38** .499 kg-m²

**9.40** −.01206 kg-m²

**9.44** 362 kg-mm²

**9.45** 362 kg-mm²

**9.46** $I_{xx} = 534$ kg-mm²
$I_{yy} = 1659$ kg-mm²
$I_{zz} = 2190$ kg-mm²
$I_{xy} = 568$ kg-mm²
$I_{xy} = I_{yz} = \approx 0$

**9.47** 2190 kg-mm²
1896 kg-mm²
297 kg-mm²
$\alpha_1 = 22.65°$
$\alpha_2 = 112.65°$
$z$ axis

**9.48** .0493 kg-m²
.00187 kg-m²
.0503 kg-m²

**9.49** .00295 kg-m²

**9.50** $I_{yy} = 1075$ kg-m²
$I_{yz} = 92.35$ kg-m²

**9.51** $I_{y'x'} = 0$ for all $\theta$

## CHAPTER 10

**10.2** $S = 1250$ N
$S = 750$ N

**10.4** $\theta = 19.48°$

**10.6** $W = .351\ T$

**10.8** $\theta = 8.53°$

**10.10** $P = (W/2) \cot \beta$

**10.12** 2513 lb

**10.14** $T_{CE} = 7.42$ kN
$T_{DE} = 29.3$ kN

**10.16** $\tan \beta = \frac{1}{\sqrt{3}} [(a-3b)/(b+a)]$

**10.18** $T' = 40$ N-m

**10.20** $P = .0545$ N

**10.22** $W = 270$ lb

**10.24** $W = 2770$ N

**10.26** $C = 108.6$ N

**10.28** $d = 72.0$ mm

**10.32** $d = .2$ ft

**10.40** $\theta = 19.22°$

**10.46** $a = .358$ m

**10.48** $a = .1126$ m

**10.52** When $d > 2$, stable equilibrium.
When $d < 2$, unstable equilibrium.

**10.56** $W_{max} = 250$ lb

**10.60** $x = 4$ m

**10.61** $Q = 3P$

**10.62** $P = 7540$ lb

**10.63** $\theta = 27.7°$

**10.64** $\cos \theta = \dfrac{\cos \theta_0 (aK_1 + aK_2) + W}{aK_1 + aK_2}$

**10.66** $\theta = 0$ (unstable)
$\theta = 28.1°$ (stable)

**10.67** $W_{max} = 1000$ N

**10.68** $R > (h/2)$ for stable equilibrium.
$R < (h/2)$ for unstable equilibrium.

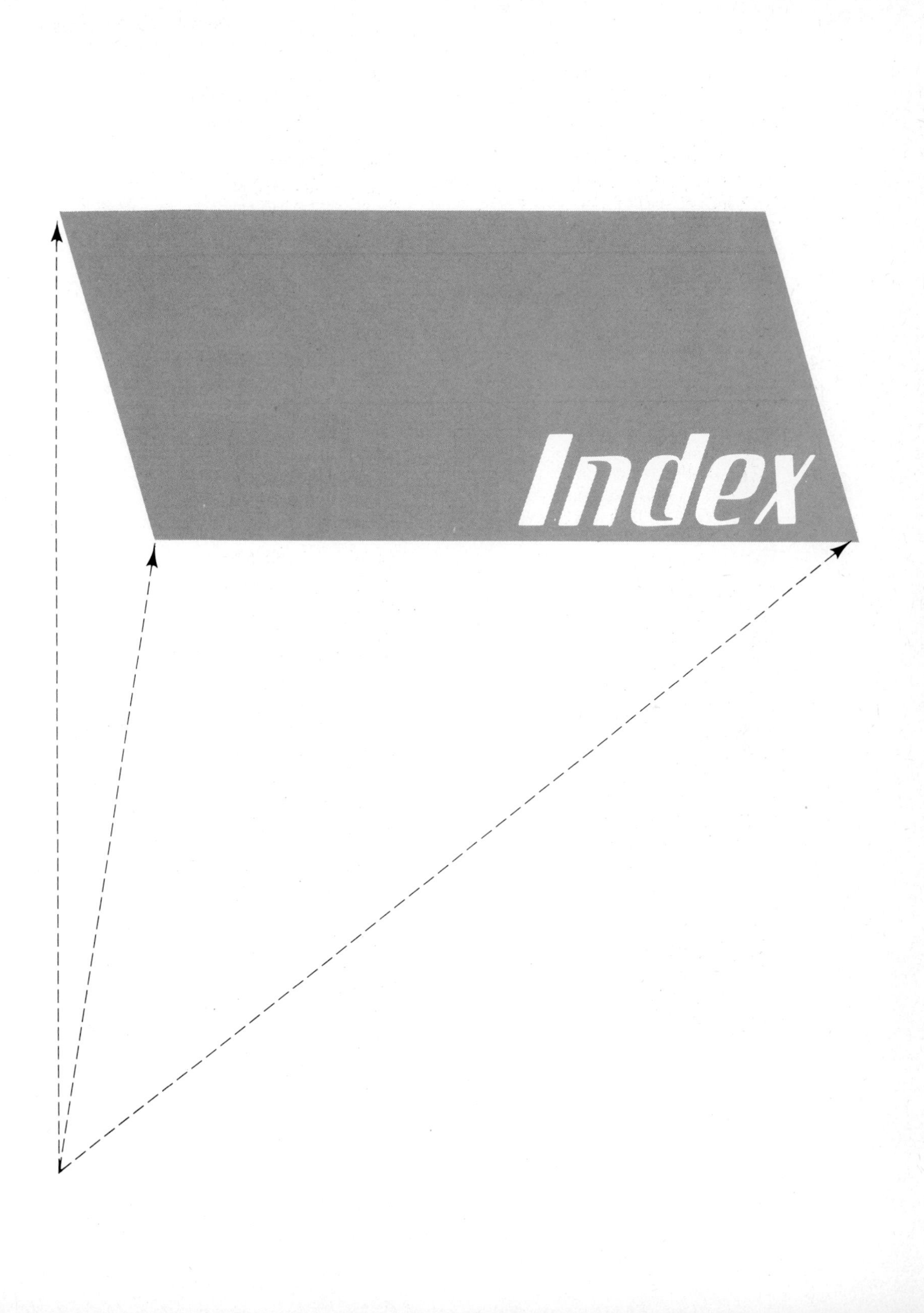
Index

# Index

## SI UNIT PREFIXES

| Multiplication Factor | Prefix | Symbol | Pronunciation | Term |
|---|---|---|---|---|
| 1 000 000 000 000 = $10^{12}$ | tera | T | as in *terra*ce | one trillion |
| 1 000 000 000 = $10^{9}$ | giga | G | jig ′a (*a* as in *a*bout) | one billion |
| 1 000 000 = $10^{6}$ | mega | M | as in *mega*phone | one million |
| 1 000 = $10^{3}$ | kilo | k | as in *kilo*watt | one thousand |
| 100 = $10^{2}$ | hecto | h | heck ′toe | one hundred |
| 10 = 10 | deka | da | deck ′a (*a* as in *a*bout) | ten |
| 0.1 = $10^{-1}$ | deci | d | as in *deci*mal | one tenth |
| 0.01 = $10^{-2}$ | centi | c | as in *senti*ment | one hundredth |
| 0.001 = $10^{-3}$ | milli | m | as in *mili*tary | one thousandth |
| 0.000 001 = $10^{-6}$ | micro | $\mu$ | as in *micro*phone | one millionth |
| 0.000 000 001 = $10^{-9}$ | nano | n | nan ′oh (*an* as in *an*t) | one billionth |
| 0.000 000 000 001 = $10^{-12}$ | pico | p | peek ′oh | one trillionth |

# PROPERTIES OF VARIOUS AREAS

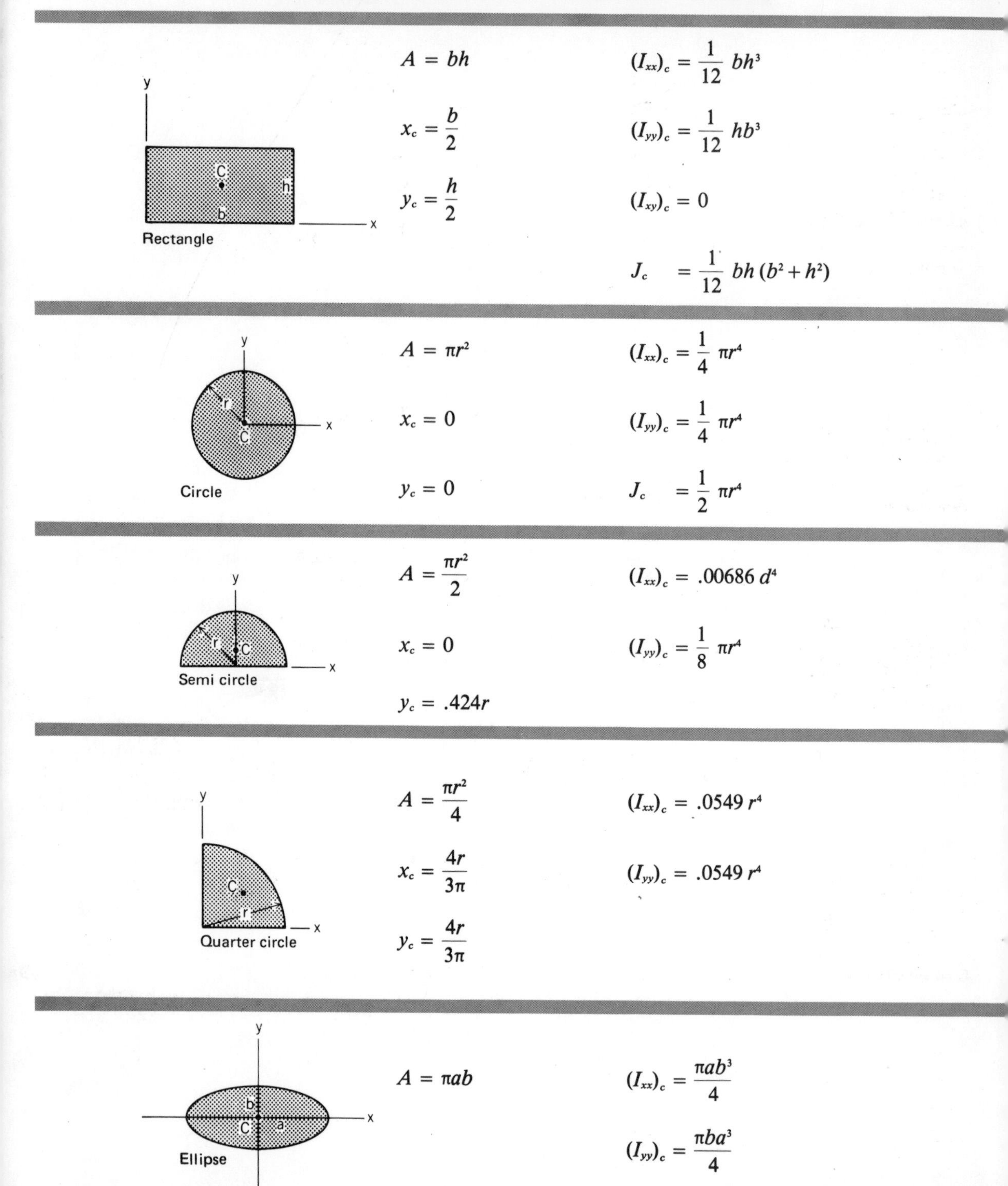